Studies in Advanced Mathematics

Wavelets: Mathematics and Applications

Studies in Advanced Mathematics

Titles Included in the Series

Real Analysis and Foundations, *Steven G. Krantz*

CR Manifolds and the Tangential Cauchy–Riemann Complex, *Albert Boggess*

Elementary Introduction to the Theory of Pseudodifferential Operators, *Xavier Saint Raymond*

Fast Fourier Transforms, *James S. Walker*

Measure Theory and Fine Properties of Functions, *L. Craig Evans and Ronald Gariepy*

Partial Differential Equations and Complex Analysis, *Steven G. Krantz*

The Cauchy Transform, Potential Theory, and Conformal Mapping, *Steven R. Bell*

Several Complex Variables and the Geometry of Real Hypersurfaces, *John P. D'Angelo*

Modern Differential Geometry of Curves and Surfaces, *Alfred Gray*

An Introduction to Operator Algebras, *Kehe Zhu*

Wavelets: Mathematics and Applications, *John J. Benedetto* and *Michael W. Frazier* (editors)

Vibration and Damping in Distributed Systems, Vol. 1 Analysis, Estimation, Attenuation, and Design; Vol. 2 WKB and Wave Methods, Visualization, and Experimentation, *Goong Chen* and *Jianxin Zhou*

JOHN J. BENEDETTO
University of Maryland

MICHAEL W. FRAZIER
Michigan State University

Wavelets: Mathematics and Applications

CRC PRESS

Boca Raton   *London*   *Tokyo*

Library of Congress Cataloging-in-Publication Data

Wavelets: mathematics and applications / [edited by] John J. Benedetto, Michael W. Frazier.
      p. cm. — (Studies in advanced mathematics)
      Includes bibliographical references and index.
      ISBN 0-8493-8271-8
      1. Wavelets (Mathematics) I. Benedetto, John
  II. Frazier, Michael, 1956– III. Series
  QA403.3.W4 1993
  515′.2433–dc20   93-12835
                                                         CIP

Direct all inquiries to CRC Press, Inc., 2000 Corporate Blvd., N.W., Boca Raton, Florida 33431.

© 1994 by CRC Press, Inc.

No claim to original U.S. Government works

International Standard Book Number 0-8493-8271-8

Library of Congress Card Number 93-12835

Printed in the United States of America   3 4 5 6 7 8 9 0

Printed on acid-free paper

Contents

Preface

Notwithstanding the possible plethora of wavelet books, we decided to edit this volume because of the rapid development of basic wavelet theory, because of the vital relationship between wavelet theory and ideas from signal processing, and because of the importance of wavelet theory in the study of partial differential operators. This book is divided into three parts, which reflect these three issues. Our Introduction sets the stage for the rest of the book by outlining the background for wavelet theory and by motivating its emergence onto the engineering, mathematical, and scientific scene. Further, the Introduction provides perspective for the material in the various chapters.

We deeply appreciate the excellent contributions by the authors, and it has been a pleasure working with them. There was also the generous and friendly assistance of Ingrid Daubechies, Björn Jawerth, Yves Meyer, and Guido Weiss in critiquing our Introduction in a constructive way. Further, we are grateful to Dr. Wayne Yuhasz and Carol Whitehead of CRC Press for all of their help at every phase of this venture. Finally, we thank Steven Krantz, who is the editor of the Studies in Advanced Mathematics series, for encouraging us to produce this book. We hope you enjoy it.

> — John J. Benedetto
>   College Park, Maryland
> — Michael W. Frazier
>   East Lansing, Michigan

Contributors

Lars Andersson Department of Mathematics, Royal Institute of Technology, Stockholm, Sweden

Pascal Auscher Institut de Recherche Mathématique de Rennes, Université de Rennes I, Campus de Beaulieu, Rennes, France

John J. Benedetto Department of Mathematics, University of Maryland at College Park, College Park, Maryland *and* The MITRE Corporation, McLean, Virginia *and* Prometheus, Inc., Newport, Rhode Island

G. Beylkin Program in Applied Mathematics, University of Colorado, Boulder, Colorado

Ronald R. Coifman Department of Mathematics, Yale University, New Haven, Connecticut

David Colella The MITRE Corporation, McLean, Virginia

Hans G. Feichtinger Mathematisches Institut, Universität Wien, Wien, Austria

Michael W. Frazier Department of Mathematics, Michigan State University, East Lansing, Michigan

Karlheinz Gröchenig Department of Mathematics, The University of Connecticut, Storrs, Connecticut

Christopher Heil The MITRE Corporation, McLean, Virginia *and* Department of Mathematics, The Massachusetts Institute of Technology, Cambridge, Massachusetts

Frédéric Heurtaux Department of Mathematics, Washington University in St. Louis

Christian Houdré Department of Statistics, Stanford University, Stanford, California

C.-c. Hsiao Department of Mathematics, National University of Taiwan, Taiwan

Stéphane Jaffard Ecole Nationale des Points et Chausees, Noisy-le-Grand, France

Björn Jawerth Department of Mathematics, University of South Carolina, Columbia, South Carolina

Contributors

Arun Kumar Southwestern Bell Technology Resources, Inc., St. Louis, Missouri

B.J. Lucier Department of Mathematics, Purdue University, West Lafayette, Indiana

Marius Mitrea Department of Mathematics, University of South Carolina, Columbia, South Carolina

Fabrice Planchon Department of Mathematics, Washington University in St. Louis

Richard Rochberg Department of Mathematics, Washington University in St. Louis

Robert S. Strichartz Mathematics Department, Cornell University, Ithaca, New York

Rodolfo Torres Department of Mathematics, University of Michigan, Ann Arbor, Michigan

David F. Walnut The MITRE Corporation, McLean, Virginia

Mladen Victor Wickerhauser Department of Mathematics, Washington University in St. Louis

X. M. Yu Department of Mathematics, University of South Carolina, Columbia, South Carolina

Introduction

John J. Benedetto and Michael W. Frazier[*]

CONTENTS

1 General remarks

A great deal of mathematical and engineering analysis depends on methods for representing a complex phenomenon in terms of elementary, well-understood phenomena. Examples include the use of Fourier expansions in studying partial differential equations, Sturm–Liouville expansions in ordinary differential equations, and Riemann's decomposition of the number theoretic psi function in terms of elementary waveforms [Bo], used to investigate zeros of the Riemann zeta function. Recently wavelet theory has provided a new method for decomposing a function or signal. In this introduction we hope to motivate the study of wavelets by indicating the circumstances in which a wavelet representation may be useful, and by describing some of the connections between wavelet theory and various topics in mathematics and engineering.

We shall do this primarily by considering the prehistory of wavelets, as seen from the perspective of several different disciplines, each of which has been dramatically enriched by wavelet theory. First, however, we would like to trace a few of the high points of the proper history of (orthonormal) wavelets, which

[*]The first author was partially supported by NSF Grant DMS-9002420. The second author was partially supported by NSF Grant DMS-9204323.

we regard as starting in the mid-1980s with the sensational first construction of a smooth orthonormal wavelet basis on $\mathbb{R}$ by Meyer [Me1] and on $\mathbb{R}^n$ by Lemarié and Meyer [LeMe]. Soon afterward, Mallat and Meyer ([M], [Me2]) placed this singular calculation within a general framework by formulating the notion of multiresolution analysis. Daubechies [D1] used this approach to construct families of compactly supported wavelets with certain degrees of smoothness. Moreover, her work greatly clarified the relation between wavelets in the continuous context of $\mathbb{R}$ (familiar to mathematical analysts) and wavelets in the discrete context of $\mathbb{Z}$ or $\mathbb{Z}_N$, as needed in digital signal analysis. This helped to spur a tremendous interdisciplinary effort to apply wavelet methods to speech and image coding; interactions of this type have in our view, unfortunately, been rare in modern times. Beylkin, Coifman, and Rokhlin [Beyetal] pioneered the use of wavelets in numerical analysis problems related to partial differential equations. Motivated by acoustic signal compression, Coifman, Meyer, Quake, and Wickerhauser [Coietal] developed wavelet packets, and Coifman and Meyer introduced the local sine and cosine transforms ([CoiMe]). These works illustrate the transition from individual *tours de force* to the realization that many of the "new" wavelet ideas had been developing in various disciplines under various disguises for many years.

There is already an extensive and diverse literature on the subject of wavelets. Two of the preeminent treatises are those of Meyer [Me3] and Daubechies [D2]. Meyer's book focuses on mathematical applications of wavelet theory in harmonic analysis. (Guy David's excellent book [Da] has a similar focus.) Daubechies' book gives a thorough presentation of techniques for constructing wavelet bases with desired properties, along with a variety of methods for mathematical signal analysis. There are also several collections of articles on wavelets ([Chu2], [Ruetal], [ComGrTc], [Le]) which exhibit the wide variety of applications of wavelet analysis and present many of the key developments of the theory.

We believe that the future of wavelet theory will continue to be marked by the synthesis of traditionally distinct disciplines in mathematics, engineering, and the sciences. This will require adjustments by all parties, including an increased willingness to address applications on the part of mathematicians, and an increased attention to mathematical rigor on the part of the engineering community. We hope this book will contribute toward this future. We have been fortunate to obtain articles which are at once mathematically precise and, for the most part, accessible to a general scientific and engineering audience. We have tried to stress exposition with broad appeal. The first part of this book, titled "Core Material," presents basic aspects of wavelet theory that we consider fundamental for future developments. The articles in this part are of an expository nature, although some contain significant new results. The second part deals with signal processing, while the third deals with analysis of operators, especially in reference to partial differential equations. We regard these as two of the most promising areas for further important applications of wavelets. We hope that, taken altogether, these articles provide a coherent overview of wavelet theory and its applications.

A remarkable feature of the subject of wavelets is that it is an area in which engineers, mathematicians, and physicists are communicating in a productive way. We are reminded of Dyson's sobering Gibb's Lecture, "Missed Opportunities" [Dy], about the "unwillingness to listen to one another." We regard wavelet theory as one of Dyson's "open opportunities." We hope this book will contribute to this interdisciplinary communication and collaboration.

To lay the groundwork for the subject, we make some further remarks about the history and prehistory of wavelets. It is beyond the scope of this introduction, and its authors, to give a definitive or comprehensive historical account. We refer to other sources, such as [Me3], [D2], or [RiVe], for more historical data. Our purpose is only to provide a coherent, albeit selective, motivation for the study and application of wavelet theory. In particular, we do not hesitate to interpret past results with the clarity of hindsight, and with a focus that may have been only partially present in the original work.

One of the great strengths of wavelet theory is its role in bringing a number of apparently disparate streams of research in engineering, mathematics, and physics more in touch with each other. In fact, wavelet theory can be regarded as a natural next step in many of these areas. We shall make some more detailed remarks about a few of these streams. First, however, we would like to take note of some of the other directions leading to wavelets that we shall not be able to discuss in detail.

Orthonormal wavelets provide explicit unconditional bases for many of the important function spaces in harmonic analysis, partial differential equations (PDE), and Banach space theory. There has been a long history of the explicit construction of such bases. Some of this work, including results by Haar (1910), Walsh (1923), Franklin (1928), and Pelcynski and Ciesielski in the modern period, has had a direct influence on wavelet theory. Carleson's construction of an unconditional basis for the Hardy space H^1 in 1980 led Strömberg (1981) ([St]) to develop an orthonormal spline wavelet basis several years before Meyer's decisive contributions to wavelet theory.

Spline theory, of course, also has a long and fruitful history. The connection between splines and wavelets continues to the present day, for example in further constructions of spline wavelet bases, see, e.g., [Chu1]. Both of these topics are naturally connected to the huge area of approximation theory. Recent work relating approximation theory and wavelets has been done by De Vore, Jawerth, Popov, de Boor, and Ron ([DeVJaP], [deetal]).

The "atomic decomposition" theory of Coifman and others ([Coi], [L], [Coi-Wei]) for the Hardy spaces H^p, $0 < p \leq 1$, had a very strong influence on the development of wavelet theory. For example, it is another motivation for the work of Carleson and Strömberg noted above. It and its generalization to the "molecular theory" (see [CoiRo], [TWei]) were also a motivation for the smooth atomic and smooth molecular decompositions to be discussed below.

Another direct forerunner of wavelet theory was work in mathematical physics related to coherent states by Weyl, von Neumann, and Klauder. This approach leads to Gabor theory, which will play a prominent role in our discussion. Work related to group representations in mathematical physics by Grossmann, Morlet, Paul, and others, e.g., [GrMoP], was a direct motivation for Meyer's work. Another very important early contribution to wavelet theory from mathematical physics was work of Guy Battle [Ba], who constructed orthonormal wavelets using renormalization group ideas. In addition, the theory of nonharmonic Fourier Series [DuSc] can be viewed as having a significant connection to wavelets.

We forego the temptation to continue this list, turning instead to a more detailed discussion of two particular streams of development leading up to wavelets. The first is a stream in harmonic analysis starting with Fourier, continuing through the Haar functions to the Calderón formula and related decompositions. The second is the topic of time-frequency and time-scale methods for signal analysis.

2 From Fourier to wavelets: A short course

Motivated by PDE, in particular the heat equation, Fourier (1807) believed that a "general" function f on $[0, 1)$ could be represented by its Fourier series $\sum_{n \in \mathbb{Z}} \hat{f}(n) e^{2\pi i n x}$ (in modern notation), where

$$\hat{f}(n) = \langle f(x), e^{2\pi i n x} \rangle \equiv \int_0^1 f(x) e^{-2\pi i n x} dx.$$

(This can already be regarded as "data compression," replacing the uncountable data $\{f(x) : x \in [0, 1)\}$ by the countable data $\{\hat{f}(n) : n \in \mathbb{Z}\}$.) A great deal of beautiful mathematics has gone into trying to make Fourier's belief precise ([Z]). The advantage of the Fourier development for PDE is that $(e^{2\pi i n x})' = 2\pi i n e^{2\pi i n x}$, i.e., the complex exponentials are eigenfunctions of the derivative operator. This advantage is exactly the advantage of doing matrix computations in a basis that diagonalizes the matrix.

In fact, more is true. The trigonometric system $\{e^{2\pi i n x} : n \in \mathbb{Z}\}$ diagonalizes all translation invariant operators on $L^2([0, 1))$. Since the corresponding type of analysis using wavelets is the focus of the third section of this book, we shall take the time to make this statement precise. Suppose T is a linear operator, bounded on $L^2([0, 1))$. We assume boundedness for simplicity–to avoid discussing the domain of T, but the result holds for unbounded operators such as d/dx as well. Let $f \in L^2([0, 1))$. Then

$$Tf(x) = T\left(\sum_{m \in \mathbb{Z}} \hat{f}(m) e^{2\pi i m \cdot}\right)(x) = \sum_{m \in \mathbb{Z}} \hat{f}(m) T(e^{2\pi i m \cdot})(x)$$

$$= \sum_{m \in \mathbb{Z}} \hat{f}(m) \sum_{n \in \mathbb{Z}} \langle T(e^{2\pi i m \cdot})(y), e^{2\pi i n y} \rangle e^{2\pi i n x} = \sum_{n \in \mathbb{Z}} (A\hat{f})(n) e^{2\pi i n x},$$

where $(A\hat{f})(n) \equiv \sum_{m \in \mathbb{Z}} a_{n,m}\hat{f}(m)$ for $a_{n,m} \equiv \langle T(e^{2\pi im\cdot})(y), e^{2\pi iny} \rangle$. Thus the action of T on $L^2([0,1))$ is equivalent to the action of the infinite matrix $A \equiv \{a_{n,m} : m, n \in \mathbb{Z}\}$ on $\ell^2(\mathbb{Z})$.

Now suppose T is translation invariant. To define this notion, let g be any function on $[0, 1)$. Extend g periodically to $\mathbb{R}$ with period 1. For any fixed $t \in \mathbb{R}$, define the translate by t of g as $\tau_t g(x) = g(x - t)$. T is *translation invariant* (or *shift invariant*) if $T(\tau_t g) = \tau_t(Tg)$, i.e., $T(g(\cdot - t))(x) = (Tg)(x - t)$ for all t and all g in the domain of T. For a fixed $m \in \mathbb{Z}$, we can write

$$T(e^{2\pi im\cdot})(x) = \sum_{n \in \mathbb{Z}} c_n e^{2\pi inx}.$$

Then, by assumption,

$$T(e^{2\pi im(\cdot - t)})(x) = T(e^{2\pi im\cdot})(x - t)$$
$$= \sum_{n \in \mathbb{Z}} c_n e^{2\pi in(x-t)} = \sum_{n \in \mathbb{Z}} c_n e^{-2\pi int} e^{2\pi inx}.$$

On the other hand, by linearity,

$$T(e^{2\pi im(\cdot - t)})(x) = e^{-2\pi imt} T(e^{2\pi im\cdot})(x) = \sum_{n \in \mathbb{Z}} c_n e^{-2\pi imt} e^{2\pi inx}.$$

Equating Fourier coefficients gives $c_n = 0$ if $n \neq m$, i.e., $T(e^{2\pi im\cdot})(x) = c_m e^{2\pi imx}$. Thus, T acts as a multiplier on the Fourier side, and the matrix A corresponding to T is diagonal since $a_{n,m} = \langle T(e^{2\pi im\cdot})(y), e^{2\pi iny} \rangle = 0$ if $n \neq m$. This is a great advantage both theoretically and computationally.

This simple result is the starting point in harmonic analysis for a large body of research leading up to the modern theory of Calderón–Zygmund operators. In signal analysis it is restated as the basic fact that the action of a linear, shift-invariant system (e.g., an ideal amplifier) is given by a convolution with its impulse response K, where $\hat{K}(m) = a_{m,m}, m \in \mathbb{Z}$. Thus $Tf = K * f$, so that $K = K * \delta = T\delta$, justifying the term "impulse response". This is an example of the not-so-rare phenomenon that harmonic analysts and signal analysts know the same results but via different terminology.

Despite the advantages of the Fourier system, it is not well adapted to the local analysis of a function. More precisely, since $\hat{f}(n) = \int_0^1 f(x)e^{-2\pi inx}dx$ and $|e^{2\pi inx}| = 1$ for each x and n, a local perturbation of f may significantly affect all coefficients $\hat{f}(n)$. Therefore no terms of the expansion can be safely ignored. One would prefer to have an effective local analysis in any number of circumstances, for example in studying singularities or shock formation in PDE, or in image compression to represent a local change in an image using only a small percentage of the data bits needed to represent the full image.

These remarks about Fourier series also apply to the Fourier transform on $\mathbb{R}^n$, once we replace sums with integrals to reflect the nonperiodic structure. The

Fourier transform of $f \in L^1(\mathbb{R}^n)$ is the function $\hat{f}$ defined by

$$\hat{f}(\gamma) = \int_{\mathbb{R}^n} f(x) e^{-2\pi i x \cdot \gamma} dx.$$

In 1910, Haar [H] constructed an orthonormal basis for $L^2([0,1))$, now known as the Haar system, which provides a local analysis in the sense we have just described. Schauder (1928) proved that the Haar system is a basis for $L^p([0,1))$ for $p \geq 1$. We define the Haar system in $L^2(\mathbb{R})$, where it is also an orthonormal basis. For $m, n \in \mathbb{Z}$, let I_{mn} be the closed interval

$$I_{mn} = [2^{-m}n, 2^{-m}(n+1)] \subseteq \mathbb{R}.$$

Such intervals are called *dyadic intervals*. The collection $\{I_{mn} : m, n \in \mathbb{Z}\}$ of all dyadic intervals has the *nesting property*: if the interiors of I_{mn} and I_{pq} have nonempty intersection, then either $I_{mn} \subseteq I_{pq}$ or $I_{pq} \subseteq I_{mn}$. The *Haar functions* $\{h_{m,n} : m, n \in \mathbb{Z}\}$ on $\mathbb{R}$ are defined as

$$h_{m,n}(x) = \begin{cases} 2^{m/2} & \text{if } 2^{-m}n \leq x < 2^{-m}(n+1/2), \\ -2^{m/2} & \text{if } 2^{-m}(n+1/2) \leq x < 2^{-m}(n+1), \\ 0 & \text{otherwise.} \end{cases}$$

Each $h_{m,n}$ is supported on $I_{m,n}$, and $\|h_{m,n}\|_{L^2(\mathbb{R})} = 1$. From the nesting property of $\{I_{m,n}\}$, it is easy to see that $\{h_{m,n}\}$ is an orthogonal set, i.e.,

$$\langle h_{m,n}, h_{p,q} \rangle \equiv \int_{\mathbb{R}} h_{m,n}(x) \overline{h_{p,q}(x)} dx = 0,$$

unless $m = p$ and $n = q$. The Haar system $\{h_{m,n} : m, n \in \mathbb{R}\}$ is complete in $L^2(\mathbb{R})$, so we have the identity

$$f = \sum_{m,n \in \mathbb{Z}} \langle f, h_{m,n} \rangle h_{m,n} \text{ in } L^2(\mathbb{R}).$$

This expansion is local in the sense that if f is 0 on $I_{m,n}$, then $\langle f, h_{m,n} \rangle = 0$. Moreover, to analyze f at a point x_0, we can restrict attention to the $m, n \in \mathbb{Z}$ for which $x_0 \in I_{m,n}$, since $h_{m,n}(x) = 0$ for $x \notin I_{m,n}$. The disadvantage of the Haar functions is that they are not smooth, in fact, not even continuous. Consequently they are not readily suited to the study of PDE.

Notice that if we let $h \equiv h_{0,0}$, then for each $m, n \in \mathbb{Z}$, $h_{m,n}(x) = 2^{m/2} h(2^m x - n)$. This is the characteristic structure of wavelet bases, viz., all basis elements are obtained by certain translations and dilations of one element, called a *wavelet*. Thus the Haar system is regarded as the simplest example of a wavelet basis; h is known as the *Haar wavelet*.

Another elementary example of a wavelet basis can be obtained by methods closely related to the Classical (or Shannon) Sampling Theorem, which is also known in complex analysis as an interpolation result (the cardinal series) for functions of exponential type. Let $\varphi \in L^2(\mathbb{R})$ be such that

$$\hat{\varphi}(\gamma) = \begin{cases} 1 & \text{if } 1/2 \le |\gamma| \le 1, \\ 0 & \text{otherwise.} \end{cases}$$

For $m, n \in \mathbb{Z}$, let $\varphi_{m,n}(x) = 2^{m/2}\varphi(2^m x - n)$. It is fairly easy to see that $\{\varphi_{m,n} : m, n \in \mathbb{Z}\}$ is an orthonormal basis for $L^2(\mathbb{R})$; see, e.g., Chapter 6. Since $\hat{\varphi}_{m,n}(\gamma) = 0$ unless $2^{m-1} \le |\gamma| \le 2^m$, we regard $\{\varphi_{m,n}\}$ as well localized in frequency. However, $\{\varphi_{m,n}\}$ is not so well localized in space since

$$\varphi(x) = \frac{\sin \pi x}{\pi x}(2\cos \pi x - 1).$$

The Haar functions have the opposite property of being well localized spatially but poorly localized in frequency: $|\hat{h}_{m,n}(\gamma)| \approx |\gamma|^{-1}$ as $|\gamma| \to \infty$.

The last two examples illustrate a basic dilemma. Since a pure Fourier basis diagonalizes translation invariant linear operators, we expect a basis that is well localized in frequency to nearly diagonalize such operators, in the sense that the matrix entries will decay rapidly away from the diagonal. On the other hand, we would like our basis to be well localized in space for an effective local analysis. We might try to achieve both by smoothing out the Haar functions or the functions $(\varphi_{m,n})^\wedge$, but this destroys the disjoint support properties that lead to orthogonality (however, see Chapter 5 for an approach along similar lines that does work). More fundamentally, however, the uncertainty principle is an obstruction to obtaining the best of both worlds. The Classical Uncertainty Principle Inequality (see, e.g., Chapter 7) states that for any $x_0, \gamma_0 \in \mathbb{R}$ and any $\varphi \in L^2(\mathbb{R})$ with $\|\varphi\|_{L^2(\mathbb{R})} = 1$,

$$\int_{\mathbb{R}} (x - x_0)^2 |\varphi(x)|^2 dx \int_{\mathbb{R}} (\gamma - \gamma_0)^2 |\hat{\varphi}(\gamma)|^2 d\gamma \ge \frac{1}{16\pi^2}.$$

Clearly the integrals in this inequality measure the variance, or dispersion, of φ about x_0 and $\hat{\varphi}$ about γ_0, respectively. Thus this inequality shows that φ and $\hat{\varphi}$ cannot be arbitrarily well localized simultaneously. Our examples $\{h_{m,n}\}$ and $\{\varphi_{m,n}\}$ above illustrate this fact. However, if $\varphi \in \mathcal{S}(\mathbb{R})$ (i.e., φ is infinitely differentiable and φ and all its derivatives decay faster than the reciprocal of any polynomial at $\pm\infty$), then $\hat{\varphi} \in \mathcal{S}(\mathbb{R})$ also. Thus it is possible to obtain much greater simultaneous localization than our examples so far indicate.

An identity known as the Calderón formula ([Ca1]) can be regarded as a step in the direction we are considering. After Calderón's work (1964), the formula has been refined by many authors, e.g., [CTo], [ChF], [U]. In one form, the Calderón formula states that for a function $\varphi \in \mathcal{S}(\mathbb{R})$, with $\varphi(x) = 0$ for $|x| > 1$, which is even, real valued, and satisfies $\int_0^\infty (\hat{\varphi}(t))^2 dt/t = 1$, then, for "general" f,

$$f(x) = \int_0^\infty \int_{\mathbb{R}} \varphi_t(x - y)\varphi_t * f(y) dy \frac{dt}{t}, \tag{1}$$

where $\varphi_t(x) = (1/t)\varphi(x/t)$ is the dilation of φ by t. Although we have an integral rather than a sum, we can think of (1) as expressing f as a "linear combination" of the translates and dilates $\varphi_t(x - y)$ of a fixed $\varphi \in \mathcal{S}(\mathbb{R})$, with "coefficients" $\varphi_t * f(y)$. Since $\varphi, \hat{\varphi} \in \mathcal{S}(\mathbb{R})$, we have a space-frequency localized representation formula for f.

For many circumstances, especially involving computation, we would prefer to have a discrete identity. By breaking up the integral over $(0, \infty) \times \mathbb{R}$ in (1) appropriately, we can obtain a discrete "smooth atomic" decomposition of f of the form $f(x) = \sum_{m,n \in \mathbb{Z}} b_{m,n}(f) a_{m,n}(x)$, where each $b_{m,n}(f)$ is a scalar, each $a_{m,n} \in \mathcal{S}(\mathbb{R})$, and $a_{m,n}(x) = 0$ if x does not belong to the triple of the dyadic interval $I_{m,n}$ (see, e.g., the discussion in [FrJaWei]). We call $a_{m,n}$ a *smooth atom*. One advantage of this and related decompositions, including wavelet decompositions, is that the function space characteristics of f are well reflected in the size properties of the coefficients $b_{m,n}(f)$. This is a consequence of a mathematical development known as Littlewood-Paley-Stein theory. A basic problem with this decomposition is that the family $\{a_{m,n}\}$ is not fixed; it varies with the function f being represented. Thus we cannot even talk about a matrix associated to a linear operator by this identity.

The φ-transform identity avoids this difficulty, while maintaining good time-frequency localization ([FrJa1], [FrJa2], [FrJa3]). Using a discrete version of the Calderón formula and the Classical Sampling Theorem, one can obtain the identity $f = \sum_{m,n \in \mathbb{Z}} \langle f, \varphi_{m,n} \rangle \psi_{m,n}$, for certain $\varphi, \psi \in \mathcal{S}(\mathbb{R})$, where $\varphi_{m,n}(x) = 2^{m/2} \varphi(2^m x - n)$ and $\psi_{m,n}(x) = 2^{m/2} \psi(2^m x - n)$. Thus $\{\varphi_{m,n}\}$ and $\{\psi_{m,n}\}$ are fixed families obtained by translating and dilating functions φ and ψ, which we regard as simultaneously space and frequency localized since $\varphi, \psi \in \mathcal{S}(\mathbb{R})$. The identity can be used to study operators via their matrices determined by this identity. However, the sequence $\{\psi_{m,n}\}$ above is not orthonormal; in fact, a nontrivial representation $0 = \sum_{m,n \in \mathbb{Z}} c_{m,n} \psi_{m,n}$ is possible. This redundancy in $\{\psi_{m,n}\}$ can cause some computational difficulties; for example, the matrix associated to the identity operator by the φ-transform representation is not the identity matrix. On the other hand, there has been a long and fruitful development of biorthogonal decompositions, and more generally frame decompositions, which can be compared with the φ-transform. Moreover, there are many physical systems with overlapping filter banks, such as the cochlear process, which are most effectively analyzed by non-orthonormal "harmonics", see, e.g., [BeTe].

This sets the stage for Meyer's discovery [Me1] of an orthonormal basis for $L^2(\mathbb{R})$ of the form $\{\psi_{m,n}\}$, where $\psi_{m,n}(x) = 2^{m/2} \psi(2^m x - n)$ for some $\psi \in \mathcal{S}(\mathbb{R})$. These $\{\psi_{m,n}\}$ are known as the Littlewood–Paley wavelets. This yields the identity $f = \sum_{m,n \in \mathbb{R}} \langle f, \psi_{m,n} \rangle \psi_{m,n}$ with good space-frequency localization and without redundancy. (A more detailed discussion of these ideas, including the construction of the Littlewood–Paley wavelets, can be found in [FrJaWei].) At this point we have reached the beginning of what we called above the proper history of wavelets. Now we shall backtrack and consider the prehistory of wavelets from another point of view.

3   Time-frequency methods in signal analysis

A different perspective which exhibits the importance of wavelet theory comes from signal analysis. Both Fourier analysts and signal analysts think of the Fourier transform variable γ as representing frequency; but whereas Fourier analysts think of the original variable x as representing position, people in signal analysis prefer to regard x as the time variable for an evolving signal $f(x)$. Hence in this section the term "space-frequency" gives way to "time-frequency," with the same meaning mathematically. There has been a long history of simultaneous time and frequency methods in signal analysis prior to the advent of wavelet theory. As examples, we mention Walsh functions, short time Fourier transforms (STFTs), Gabor decompositions, and quadrature mirror filters (QMFs).

In 1922 Rademacher [R] defined an orthonormal sequence $\{r_m : m = 0, 1, 2, \ldots\}$ of functions on $[0, 1)$ as follows. Let $r_0 = h$, the Haar wavelet. For $m > 0$, let $r_m(x) = h(2^m x)$ for $x \in [0, 2^{-m})$, and extend r_m periodically with period 2^{-m} to $[0, 1)$. The *Walsh system* $\{w_m : m = 0, 1, \ldots\}$ on $[0, 1)$ is defined by setting $w_0 = 1$ and, for $m \geq 1$,

$$w_m = r_{m_1} r_{m_2} \ldots r_{m_k},$$

where $0 \leq m_1 < m_2 < \cdots < m_k$ and $m = 2^{m_1} + 2^{m_2} + \cdots + 2^{m_k}$. Walsh (1923) ([W]) proved that $\{w_m\}$ is an orthonormal basis of $L^2([0, 1))$; it is then easy to see that $\{\tau_n w_m : n \in \mathbb{Z}, m \in \{0\} \cup \mathbb{N}\}$ is an orthonormal basis of $L^2(\mathbb{R})$, where τ_n denotes translation by n.

About 1900, engineers designed transposition schemes in open-wired lines based on the Walsh system, and used these schemes to minimize channel "crosstalk". The Walsh system has continued to be a relevant tool in communications theory for most of this century; see, e.g., [SchS].

The Walsh system can be conveniently ordered according to the number of sign changes of its elements. For example, w_1, w_3, w_2, and w_6 have, respectively, 1, 2, 3, and 4 sign changes. The importance of zero crossings in applications is well known, and so the *sequency ordering* of the Walsh system is a useful device. The frequency localization of the Fourier system and the time localization of the Haar system can be compared with the corresponding "sequency localization" of the Walsh system. For wavelet theory, the Walsh system is a prototype of wavelet packets ([Coietal], [CoiMeWi]), just as the Haar system is a prototype of orthonormal wavelet bases.

The short time Fourier transform (STFT) was developed to obtain a frequency analysis which is also local in time for a given signal. Let g be a localized function such as $g = \mathbf{1}_{(T)}$, where $\mathbf{1}_{(T)}(x) = 1$ if $x \in [-T, T]$ and 0 otherwise. Such a g is called a *window* function. The STFT is defined by

$$S_g f(\gamma) = (fg)^{\wedge}(\gamma) = \int_{\mathbb{R}} f(x) g(x) e^{-2\pi i x \gamma} dx.$$

Thus the spectral data $S_g f$ provides frequency information about the restriction of f to $[-T, T]$. In the case $g = \tau_t \mathbf{1}_{(T)}$, we write

$$Sf(t, \gamma) \equiv \int_{\mathbb{R}} f(x) \tau_t \mathbf{1}_{(T)}(x) e^{-2\pi i x \gamma} dx.$$

The *spectrogram* $|Sf(t, \gamma)|^2$ has proved to be an effective tool in signal analysis, despite the loss of information incurred by dealing only with amplitudes. There are brilliant "readers" of spectograms who, using context, experience, and talent, can reconstruct f from $|Sf(t, \gamma)|^2$ with remarkable accuracy. In the modern age, spectrograms must give way to more sophisticated, adaptive time-frequency methods; see, e.g., [Co].

The periodogram is a closely related notion. Along with high resolution methods such as maximum entropy, it has been a critical tool in spectral estimation; see, e.g., [Pr]. Given a suitably constrained stationary stochastic process, consisting of *sample functions* $f(\cdot, \alpha)$, then

$$S_g f(\gamma, \alpha) = \left| \int_{\mathbb{R}} f(x, \alpha) g(x) e^{-2\pi i x \gamma} dx \right|^2$$

is the *periodogram* associated with the process f and the *data window* g. The *power spectrum* of f is the positive measure S defined as the Fourier transform of the expectation $P(t) = E\{f(t+x)\overline{f(x)}\}$. Bartlett and Tukey were responsible for the early adaptation of expectations of periodograms as an effective tool in estimating S. The deterministic theory goes back to Wiener's Generalized Harmonic Analysis (1930), cf. [Be1, Chapter 2], [Wie].

Motivated by the role of the Gaussian in minimizing the product of the variances in the Classical Uncertainty Principle Inequality, Gabor [G] sought to represent signals as discrete decompositions of modulates and translates of the Gaussian. It turns out that, for a given signal f, the Gaussian can be replaced by other "harmonics," better suited to providing an understanding of f in terms of its Gabor decomposition. Gabor decompositions and their relations to various spaces and uncertainty principle inequalities are a significant part of Chapters 3 and 7, so we do not go into any detail now.

Quadrature mirror filters (QMFs) were introduced in the 1970s to deal with speech coding problems ([CroEGa], [CrWeFl], [EGa], [SmB]). One version of the definition is that a QMF $H \in L^2([0, 1))$ is a Fourier series $\sum_{n \in \mathbb{Z}} \hat{H}(n) e^{2\pi i n \gamma}$ for which

$$|H(\gamma)|^2 + |H(\gamma + 1/2)|^2 = 2 \text{ a.e.}$$

Such filters give rise to subband coding schemes which provide unaliased perfect reconstruction of transmitted signals ([Be2, Definition 78-Theorem 83], [SmB], [RiVe], [V], [VeHe]). An image-compression technique, analogous to a QMF, is the Laplace-pyramid code formulated by Burt and Adelson [BuA] (1983) to remove redundant image correlation of neighboring pixels. There are related ideas in image processing dealing with multiscale analysis, see, e.g., [Ma] (1983).

There is a close relationship between these ideas from speech and image processing, the idea of multiresolution analysis (MRA), and orthonormal wavelets. As discussed in detail in Chapter 1, an MRA consists of a sequence of subspaces of $L^2(\mathbb{R})$ and a *scaling function* φ which satisfy certain conditions. *Meyer's algorithm* is a procedure to construct an orthonormal wavelet basis of $L^2(\mathbb{R})$ from any given MRA (see, e.g., [Me2], [M]). In this construction a certain Fourier series H arises. For example (see Chapter 1), it is easy to obtain an MRA such that $\varphi(x) = 1$ if $x \in [0, 1)$ and 0 otherwise. In this case the wavelet basis is the Haar system and $H(\gamma) = (1/\sqrt{2})(1 + e^{2\pi i \gamma})$, which is easily seen to be a QMF.

It turns out that every MRA defines a QMF in this way. Therefore, in a formal sense, the notion of a QMF is more general than an MRA. The reality is more complicated, in that the MRAs and wavelet bases constructed by Daubechies, Lemarié, Mallat, Meyer, et al. were designed to be smooth and to solve different problems than the QMFs previously studied for speech analysis.

Thus Walsh functions, STFTs, Gabor decompositions, and QMFs are mathematical tools arising in signal processing which can be thought of as precursors of various aspects of wavelet theory, broadly interpreted (e.g., to include general Gabor representations).

Given the multitude of notions that we have discussed, it is worthwhile to emphasize that this book is devoted to wavelet and Gabor theory and their applications. With this in mind, we would like to give some perspective on the relation between these two topics. Our view on this is consistent with Feichtinger and Gröchenig's work from 1986 ([FeGrö]). Recall that a wavelet decomposition has the form

$$f(x) = \sum_{m,n \in \mathbb{Z}} c_{m,n}(f) \psi_{m,n}(x) \tag{2}$$

where $\psi_{m,n}(x) = 2^{m/2} \psi(2^m x - n)$. A *Gabor* (or *coherent state* or *Weyl–Heisenberg*) *decomposition* has the form

$$f(x) = \sum_{m,n \in \mathbb{Z}} d_{m,n}(f) g_{m,n}(x), \tag{3}$$

where $g_{m,n}(x) = e^{2\pi i x m b} g(x - na)$. The dilations and translations in the wavelet case are associated with the locally compact nonabelian affine group. The modulations and translations in the Gabor case are associated with the locally compact Heisenberg group. Locally compact groups play a fundamental role in understanding the symmetries and invariants in harmonic analysis; see, e.g., [HewRos].

In wavelets and coherent states, there is an alternative approach based on decompositions rather than symmetries. For example, the continuous "oversampled" versions of (2) and (3) both have the form

$$f(t) = \int_G \theta_{x,w}(t) \Theta f(x, w) dx dw, \tag{4}$$

where G is the underlying group with Haar measure "$dxdw$", $\theta_{x,w}$ corresponds to either $\psi_{m,n}$ or $g_{m,n}$, and Θf is generally an inner product corresponding to the coefficients in the decomposition, similarly to (1). If G is replaced by a less symmetric domain, (4) allows us to consider other transforms and decompositions depending on the choice of $\theta_{x,w}$. An effective understanding of a signal may require tilings of the time-frequency plane other than those for the wavelet and Gabor decompositions. Thus the group-theoretic perspective not only unifies the wavelet and Gabor theory mathematically, it inspires a departure from the group context for the sake of applications.

4   Description of the chapters

The first of the three sections of this book is titled "Core Material." The five chapters making up this section lay out what we regard as the fundamentals of wavelet theory (in the wider sense). The material in these articles will be drawn upon as background in subsequent chapters.

The first chapter, by Robert Strichartz, presents the basic theory of multiresolution analysis and the construction of orthonormal wavelets in $\mathbb{R}^n$, $n \geq 1$. As an application, Daubechies' construction [D1] of orthonormal wavelets of compact support is given. Strichartz also includes a discussion of subband coding, the fast method for computing the orthonormal wavelet coefficients. One advantage of the development of multiresolution analysis in this article is that it is given in a general form that allows lattices other than the usual dyadic grid.

The next chapter, by Michael W. Frazier and Arun Kumar, is a tutorial in orthonormal wavelet theory on the discrete sets $\mathbb{Z}$, the integers, and $\mathbb{Z}_N$, the integers modulo N $\{0, 1, \ldots, N-1\}$. This theory has the same relation to the continuous theory discussed in Strichartz's article that the discrete Fourier transform (DFT) has to the continuous Fourier transform on $\mathbb{R}$ or $\mathbb{R}^n$. This article was written with an engineering audience in mind, for whom the discrete case may be of paramount interest. The z-transform is used to develop the theory because of its familiarity to engineers and because it allows a simultaneous treatment of the finite and infinite cases, $\mathbb{Z}_N$ and $\mathbb{Z}$, respectively. Theoretical aspects underlying the results are not stressed; instead a direct path leading to an algorithm for constructing discrete orthonormal wavelets is presented. The fast computation of the discrete wavelet coefficients (and similarly of the signal reconstruction), via a sequence of filter banks, is emphasized from the start. The multidimensional case is treated at the end.

Chapter 3, by John J. Benedetto and David Walnut, presents the general theory of frames, and then specializes to Gabor frame decompositions. The Gabor or coherent state theory is based on modulations and translations, whereas wavelet decompositions are based on dilations and translations. Besides expositing the classical Gabor theory for L^2, Chapter 3 develops the authors' theory of Gabor decompositions for L^1 and for Sobolev spaces and their generalizations. Another feature of this chapter is a study of frames consisting only of translations.

Each multiresolution analysis determines an L^2 solution of a "dilation equation" $f(t) = \sum_n \hat{H}(n)f(2t - n)$, whose coefficients $\hat{H}(n)$ are the Fourier coefficients of a QMF. The theory of dilation equations is an essential component in the analysis and construction of orthonormal wavelet bases satisfying prescribed properties such as degree of smoothness. In Chapter 4, David Colella and Christopher Heil present a thorough treatment of this theory, synthesizing their own results with those of other contributors to the field. They present the algebraic and analytic tools required for further work in this area.

Recently, two refinements of the windowed Fourier transform have been developed which give orthonormal bases on $\mathbb{R}$. Wilson bases are a variation on Gabor frames. Their construction has been simplified by Daubechies, Jaffard, and Journé [DJJou]. The local sine and cosine bases, or Malvar bases, were developed independently by Malvar and by Coifman and Meyer. Chapter 5, by Pascal Auscher, begins with a new characterization of Wilson bases, which also yields the main results of [DJJou]. As a consequence, an example of a Wilson basis with compact support is obtained. After a quick review of Malvar bases, the general notion of local Fourier bases is presented. Auscher shows that this gives a unified framework, which yields the Wilson and Malvar bases as special cases depending on a choice of polarity. For comparison, the Littlewood–Paley wavelets of Meyer can also be obtained from local Fourier bases.

Following the core material, the second section of this book deals with topics relevant to signal processing. The first article in this section, Chapter 6, by Michael W. Frazier and Rodolfo Torres, is an introduction to sampling theory. It begins with a general version of the Shannon sampling theorem for "B-transversal" sets. This leads to the φ-transform identity and to the "Shannon basis," the simple orthonormal wavelet basis noted earlier with wavelet $(\sin \pi x)(2 \cos \pi x - 1)/(\pi x)$. Versions of these results are given in the continuous settings of $\mathbb{R}$ and the unit circle $\mathbb{T}$, and in the discrete settings $\mathbb{Z}$ and $\mathbb{Z}_N$.

Chapter 7, by John J. Benedetto, explores the natural relationships between sampling formulas and uncertainty principle inequalities. The sampling formulas are in terms of frame decompositions. The uncertainty principle arises in providing the most efficient tiling of the information plane for the problem at hand. The sampling results deal with both regularly and irregularly spaced data, as well as local sampling. The uncertainty principle inequalties deal with a variety of cases generalizing the Classical (Heisenberg) Uncertainty Principle Inequality. Results about L^2 frames are required that go beyond the general Hilbert space theory of frames given in Chapter 3.

Hans Feichtinger and Karl-Heinz Gröchenig deal with their theory of irregular sampling and its implementation in Chapter 8. Irregular sampling is an active research area, and there have been a number of parallel developments. The authors' methods are based on analytical ideas they have developed, which include weighted frames and the so-called adaptive weights method. Their reconstruction formula from local averages circumvents to some extent the impact of the uncertainty principle and the related technical limitations of making pointwise sampling measurements. A significant part of Chapter 8 is devoted to the numerical implementation of the algorithm derived from the authors' theory.

In Chapter 9, Christian Houdré's study of the role of wavelets in probability and statistics focuses on three problems. These problems are not only important in theoretical signal processing, but also provide new results for both probability and wavelet theory. First, weak stationarity and similar properties of certain processes are characterized in terms of their wavelet transforms. Wiener's Tauberian theorem is used in an essential way in the proof. Second, Houdré presents an irregular sampling theory for stationary band-limited processes. Third, he presents basic ideas necessary for applying wavelet methods to nonparametric statistics. There are a number of other related results, including new characterizations of Bessel sequences and exact frames.

The subject of Chapter 10, by Ronald Coifman and Victor Wickerhauser, is the fundamental relationship between computational efficiency and meaningful signal representations. The authors establish this relationship in the context of data compression and pattern extraction, and in terms of their Adapted Waveform Transform (AWT) method. The AWT method provides collections of orthonormal bases, such as wavelets, wavelet packets, and windowed trigonometric waveforms, as well as associated fast numerical algorithms. The process is initiated by comparing segments of given data with stored waveforms, and choosing a basis associated with the data in a prescribed way. Various *libraries* of such bases are computed, including collections of windowed trigonometric bases and wavelet packets. The authors show that wavelet analysis corresponds to windowing frequency space. Using this format and these libraries, there is an associated phase space analysis for dealing with best bases, denoising, and compression problems. Finally, multidimensional questions are addressed.

The second part of this book concludes with Chapter 11, by Chia-chang Hsiao, Björn Jawerth, Bradley Lucier, and Xiangming Yu. It deals with mathematical aspects of wavelet-based compression. This is a good example of engineering problems leading to interesting new directions in theoretical mathematics, which in turn have importance for applications. The basic problem in compression is to approximate a signal or image as well as possible using a small amount of data. Compression has enormous importance in signal processing. Wavelets yield a natural, experimentally promising approach. The theory of function spaces gives a mathematical framework allowing precise measurement of the quality of approximation. Fortunately, most of the natural function spaces are explicitly characterized in terms of their wavelet expansions by Littlewood–Paley theory. This allows the authors to show that a simple, easily implemented selection of coefficients based on magnitudes gives a near-optimal compression rate. Characterizations of functions having certain rates of approximation are also obtained, as well as methods for compressing linear operators using their matrices in a wavelet basis. These issues are related to the classical problem of comparing L^p and ℓ^p as Banach spaces.

The third section of this book deals with the natural link between wavelet theory and the study of operators arising in PDE. Historically, this link reflects the strong connections between Littlewood–Paley theory and Calderón–Zygmund operators. Wavelet analysis can be seen as a particularly explicit and effective formulation of Littlewood–Paley theory. As such, it is an incisive theoretical tool for operator analysis in the Calderón–Zygmund tradition. Moreover, wavelets are so simple and explicit that they allow for the transformation of deep theoretical insights into useful computational procedures. The prototype for this is the groundbreaking work of Beylkin, Coifman, and Rokhlin [Beyetal], who showed that integral operators used in PDE are represented by sparse matrices in wavelet coordinates, leading to fast computation.

In Chapter 12, Gregory Beylkin considers a second-order, two-point Dirichlet boundary condition differential equation on the interval $[0, 1]$. He applies wavelet analysis directly to the discretization of this equation, rather than to a corresponding integral equation as in [Beyetal]. This converts the differential equation to a linear algebraic system with a sparse matrix A. However, the condition number of A grows rapidly with the number of points N in the discretization, which leads to computational instability. Applying a standard preconditioning matrix can reduce to a problem with a condition number which is $O(N)$. Beylkin provides an explicit computational solution method using wavelets which, after preconditioning with a diagonal matrix, gives a condition number which is bounded independent of N. This method maintains the sparsity of the matrices involved and allows for computation of the Green's function (the solution operator) in $O(N)$ steps.

Stephane Jaffard shows, in Chapter 13, that properly chosen orthonormal wavelet bases can be used to obtain sharp local analysis of nonlinear phenomena. For example, suppose that a function f satisfies a Hölder condition locally. Then there are sufficiently precise estimates for its wavelet coefficients that, viewed as averages and accompanied by a Tauberian condition, they essentially characterize the Hölder property. Applications and extensions of such techniques provide results on pointwise regularity of elliptic operators, as well as results characterizing singularities of elements in various Sobolev spaces. Jaffard's analysis leads to the construction of prescribed multifractal functions, which can arise in turbulence, and to the wavelet decomposition of nonlinear quantities used to resolve certain nonlinear PDEs.

Chapter 14 is due to Frédéric Heurtaux, Fabrice Planchon, and Victor Wickerhauser. It is an incisive glimpse into some of the intricacies associated with turbulence. A fundamental perception in characterizing turbulent behavior is that energy cascades from large wavelengths to small ones. This transition, from large-scale velocity fields to activity in small-scale eddies, poses basic problems dealing with a meaningful definition of energy at a given scale. The authors replace a classical Fourier wave number interpretation of energy by a highly motivated wavelet alternative. Using this new idea, they address the issue of energy transfer between scales in evolving turbulent flows. Burgers' one-dimensional evolution equation is studied numerically in this context. For a wavelet decomposition of the solution of Burgers' equation, it is seen how the small-scale wavelets collect energy as the evolution proceeds, approaching a singularity. The analysis is extended to other important cases where exact analytic solutions are not available. Besides particular examples and algorithms to deal with them, the general theme of the chapter is to improve numerical solutions of fluid dynamics problems using such adaptive algorithms.

In Chapter 15 by Lars Andersson, Björn Jawerth, and Marius Mitrea, wavelet methods are applied to prove the L^2 boundedness of the Cauchy singular integral operator (CSIO). This result can be applied to solve the Dirichlet problem for the Laplacian on a Lipschitz domain in $\mathbb{R}^n$ using the double layer potential. The L^2 boundedness in one dimension, on a Lipschitz curve, is the famous result of Calderón [Ca2] and Coifman, McIntosh, and Meyer [CoiMcMe]. Recently a very simple proof of this result, using the Haar wavelet, was given by Coifman, Jones, and Semmes [CoiJoSe]. Anderson, Jawerth, and Mitrea prove L^2 boundedness in n dimensions by obtaining boundedness for the corresponding operator taking values in the 2^n-dimensional Clifford algebra $\mathbb{R}_{(n)}$, which in some sense plays the role in higher dimensions of the complex numbers. Clifford analysis, for example, the Clifford version of the Cauchy integral formula, enables the authors to apply methods similar to those in [CoiJoSe]. In particular, they create a set of Clifford-valued wavelets similar to the Haar system on $\mathbb{R}$. The proof then follows the traditions of Littlewood–Paley theory. This paper can be regarded as another step in the continuing program of using more refined Littlewood–Paley techiques, in this case wavelets, to give simple approaches to the study of operators which are related to, but do not exactly fall within, the classical Calderón–Zygmund framework.

The final chapter (Chapter 16), by Richard Rochberg, presents a general discussion of methods for applying decomposition theorems in the analysis of operators. This includes continuous or nonorthogonal decompositions like (1) as well as orthonormal wavelet representations. One way to use these results to study operators on functions is to decompose a given function f and consider the action of the operator on the pieces comprising f. Rochberg applies this method to a result of Cwikel on the singular values of certain generalized differential operators. A second approach is to decompose the operator itself. This method is applied to obtain eigenvalue estimates for the Schrödinger operator. Further, these types of approaches lead to an extension of the theory to a general class of operators called Calderón–Toeplitz operators. An appealing aspect of Rochberg's development is that he shows how complicated operators can be understood in terms of relatively simple model operators obtained using wavelet decompositions.

These chapters demonstrate the intensive interplay between wavelet theory and topics such as operator analysis, PDE, sampling theory, signal analysis, probability and statistics, and numerical analysis. The depth and diversity of these connections illustrate the richness and vitality of wavelet theory. We hope this book will provide an accessible introduction to the growing area of wavelet analysis and applications.

Bibliography

[Ba]   G. Battle. 1987. "A block spin construction of ondelettes. Part I: Lemarié functions," *Commun. Math. Phys.* **110**: 601–615.

[Be1]   J. Benedetto. 1975. *Spectral Synthesis*, New York: Academic Press.

[Be2]   J. Benedetto. 1992. "Irregular sampling and frames," in *Wavelets: A Tutorial in Theory and Applications*, C.K. Chui, ed., Boston: Academic Press, pp. 445–507.

[BeTe]   J. Benedetto and A. Teolis. 1993. "A wavelet auditory model and data compression," *Appl. Comp. Harmonic Analysis*, to appear.

[Beyetal] G. Beylkin, R. Coifman, and V. Rokhlin. 1991. "Fast wavelet transforms and numerical algorithms I," *Commun. Pure and Appl. Math.* **44**: 141–183.

[Bo]   E. Bombieri. 1992. "Prime territory,'" *The Sciences*, Sept./Oct., 30–36.

[BuA]   P.J. Burt and E.H. Adelson. 1983. "The Laplacian pyramid as a compact image code," *IEEE Trans. Commun.* **31**: 532–540.

[C1]   A.P. Calderón. 1964. "Intermediate spaces and interpolation, the complex method," *Studia Math.* **24**: 113–190.

[C2]   A.P. Calderón. 1977. "Cauchy integrals on Lipschitz curves and related operators," *Proc. Natl. Acad. Sci. USA* **74**: 1324–1327.

[CTo]   A.P. Calderón and A. Torchinsky. 1975 & 1977. "Parabolic maximal functions associated with a distribution I, II," *Adv. Math.* **16**: 1–64 and **24**: 101–171.

[ChF]   S.-Y.A. Chang and R. Fefferman. 1980. "A continuous version of duality of H^1 and BMO on the bidisc," *Ann. Math.* **112**: 179–201.

[Chu1]   C.K. Chui. 1992. *An Introduction to Wavelets*, New York: Academic Press.

[Chu2]   C.K. Chui (ed.). 1992. *Wavelets: A Tutorial in Theory and Applications*, New York: Academic Press.

[Co]   L. Cohen. 1989. "Time-frequency distributions—a review," *Proc. IEEE* **77**: 941–981.

[Coi]   R. Coifman. 1974. "A real variable characterization of H^p," *Studia Math.* **51**: 269–274.

[CoiJoSe]   R. Coifman, P. Jones, and S. Semmes. 1989. "Two elementary proofs of the L^2 boundedness of the Cauchy integral on Lipschitz curves," *J. A.M.S.* **2**: 553–564.

[CoiMcMe]   R. Coifman, A. McIntosh, and Y. Meyer. 1982. "L'intégrale de Cauchy définit un opérateur borné sur L^2 pour les courbes lipschitziennes," *Ann. Math.* **116**: 361–387.

[CoiMe]   R. Coifman and Y. Meyer. 1991. "Remarques sur l'analyse de Fourier à fenêtre," *C. R. Acad. Sci.*, sér. 1 **312** 259–261.

[Coietal]   R. Coifman, Y. Meyer, S. Quake, and M. Wickerhauser. 1990. "Signal processing and compression with wavelet packets," Numerical Algorithms Research Group, Yale University.

[CoiMeWi]   R. Coifman, Y. Meyer, and M. Wickerhauser. 1992. "Wavelet analysis and signal processing," in *Wavelets and Their Applications*, Ruskai et al., eds., Boston: Jones and Bartlett, pp. 153–178.

[CoiRo]   R. Coifman and R. Rochberg. 1980. "Representation theorems for holomorphic and harmonic functions in L^p," *Astérisque* **77**: 11–66.

[CoiWei]   R. Coifman and G. Weiss. 1977. "Extensions of Hardy spaces and their use in analysis," *Bull. Amer. Math. Soc.* **83**: 569–645.

[ComGrTc]   J.M. Combes, A. Grossmann, and P. Tchamitchian (eds.). 1989. *Wavelets, Time-Frequency Methods and Phase Space*, Proc. Int. Conf., Marseille, Dec. 1987, Berlin: Springer-Verlag,

[CrWeFl]   R. Crochiere, S. Weber, and J. Flanagan. 1976. "Digital coding of speech in subbands," *Bell System Tech. J.* **55**(Oct.): 1069–1085.

[CroEGa]   A. Croisier, D. Esteban, and C. Galand. 1976. "Perfect channel splitting by use of interpolation, decimation, tree decomposition techniques," Int. Conf. Information Sciences/Systems, Patras, August, pp. 443–446.

[D1]   I. Daubechies. 1988. "Orthonormal bases of compactly supported wavelets," *Commun. Pure Appl. Math.* **41**: 909–996.

[D2]   I. Daubechies. 1992. *Ten Lectures on Wavelets*, CBMS-NSF Reg. Conf. Ser. Appl. Math. **61**, Philadelphia: Soc. Ind. Appl. Math,

[DJJou]   I. Daubechies, S. Jaffard, and J.-L. Journé. 1991. "A simple Wilson orthonormal basis with exponential decay," *SIAM J. Math. Anal.* **22**: 554–572.

[Da]   G. David. 1991. *Wavelets and Singular Integrals on Curves and Surfaces*, Lect. Notes Math. **1465**, Berlin: Springer-Verlag.

[deetal]   C. de Boor, R. DeVore, and A. Ron. 1991. "Approximation from shift-invariant subspaces of $L_2(\mathbb{R}^d)$," CM-TSR University of Wisconsin-Madison 92-2, pp. 1–20.

[DeVJaP]   R. DeVore, B. Jawerth, and V. Popov. 1992. "Compression of wavelet decompositions," *Amer. J. Math.* **114**: 737–785.

[DuSc]   R. Duffin and A. Schaeffer. 1952. "A class of nonharmonic Fourier series," *Trans. Amer. Math. Soc.* **72**: 341–366.

[Dy]   F. Dyson. 1972. "Missed opportunities," *Bull. Amer. Math. Soc.* **78**: 635–652.

[EGa]   D. Esteban and C. Galand. 1977. "Applications of quadrature mirror filters to split-band voice coding schemes," *Int. Conf. Acoust. Speech Signal Process.*, Hartford, CT (May 1977), pp. 191–195.

[FeGrö]   H. Feichtinger and K. Gröchenig. 1989. "Banach spaces related to integrable group representations and their atomic decompositions I," *J. Funct. Anal.* **86**: 307–340.

[FrJa1]   M. Frazier and B. Jawerth. 1985. "Decomposition of Besov spaces," *Indiana Univ. Math. J.* **34**: 777–799.

[FrJa2]   M. Frazier and B. Jawerth. 1988. "The φ-transform and applications to distribution spaces," in *Function Spaces and Applications*, M. Cwikel et al., eds., Lect. Notes Math. **1302** Berlin: Springer-Verlag, pp. 223–246.

[FrJa3]   M. Frazier and B. Jawerth. 1990. "A discrete transform and decompositions of distribution spaces," *J. Funct. Anal.* **93**: 34–170.

[FrJaWei]   M. Frazier, B. Jawerth, and G. Weiss. 1991. *Littlewood-Paley Theory and the Study of Function Spaces*, CBMS Conf. Lect. Notes **79**, Providence: Amer. Math. Soc.

[G]   D. Gabor. 1946. "Theory of communication," *J. IEE (London)* **93**: 429–457.

[GrMoP]   A. Grossmann, J. Morlet, and T. Paul. 1985 & 1986. "Transforms associated to square integrable group representations I. General results," *J. Math. Phys.* **26**: 2473–2479; "II. Examples," *Ann. Inst. Henri Poincaré* **45**: 293–309.

[H]   A. Haar. 1910. "Zur theorie der orthogonalen Funktionensysteme," *Math. Ann.* **69**: 331–371.

[HewRos]   E. Hewitt and K. Ross. 1963. *Abstract Harmonic Analysis*, Berlin: Springer-Verlag.

[L]   R.H. Latter. 1978. "A decomposition of $H^p(\mathbb{R}^n)$ in terms of atoms," *Studia Math.* **62**: 92–101.

[Le]   P.G. Lemarié (ed.). 1990. *Les ondelettes en 1989*, Lect. Notes Math. **1438**, Berlin: Springer-Verlag.

[LeMe]   P.G. Lemarié and Y. Meyer. 1986. "Ondelettes et bases hilbertiennes," *Rev. Mat. Iberoamericana* **2**: 1–18.

[M]   S. Mallat. 1989. "Multiresolution approximations and wavelet orthonormal bases of $L^2(\mathbb{R})$," *Trans. Amer. Math. Soc.* **315**: 69–87.

[Ma]   D. Marr. 1982. *Vision*, New York: Freeman.

[Me1]   Y. Meyer. 1985–1986. "Principe d'incertitude, bases Hilbertiennes et albgèbres d'opérateurs," *Séminaire Bourbaki* **662**: 1–15.

[Me2]   Y. Meyer. 1987. "Ondelettes, fonctions splines et analyses graduées," *Rapport Ceremade* **8703**.

[Me3]   Y. Meyer. 1990. *Ondelettes et Opérateurs*, Paris: Hermann.

[Pr]   M. Priestley. 1981. *Spectral Analysis and Time Series*, New York: Academic Press.

[R]   H. Rademacher. 1922. "Einige Sätze über Reihen von allgemeinen Orthogonal-funktionen," *Math. Ann.* **87**: 112–138.

[RiVe]   O. Rioul and M. Vetterli. 1991. "Wavelets and signal processing," *IEEE Signal Process. Mag.* October: 14–38.

[Ruetal] M. Ruskai, G. Beylkin, R. Coifman, I. Daubechies, S. Mallat, Y. Meyer, L. Raphael (eds.). 1992. *Wavelets and Their Applications*, Boston: Jones and Bartlett.

[SchS]   H. Schreiber and G. Sandy (eds.). 1974. *Applications of Walsh Functions and Sequency Theory*, New York: IEEE.

[SmB]   M. Smith and T. Barnwell. 1986. "Exact reconstruction for tree-structured subband codes," *IEEE Trans. Acoust. Speech Signal Process.* **34**: 434–441.

[St]   J.-O. Strömberg. 1983. "A modified Franklin system and higher order spline systems on $\mathbb{R}^n$ as unconditional bases for Hardy spaces," in *Conference on Harmonic Analysis in Honor of Antoni Zygmund II*, Beckner et al., eds., Belmont, CA: Wadsworth, pp. 475–494.

[TWei]   M. Taibleson and G. Weiss. 1980. "The molecular characterization of certain Hardy spaces," *Astérisque* **77**: 67–149.

[U]   A. Uchiyama. 1982. "A constructive proof of the Fefferman-Stein decomposition of BMO ($\mathbb{R}^n$)," *Acta Math.* **148**: 215–241.

[V]   P. Vaidyanathan. 1990. "Multirate digital filters, filter banks, polyphase networks, and applications: a tutorial," *Proc. IEEE* **78**: 56–93.

[VeHe]   M. Vetterli and C. Herley. 1992. "Wavelets and filter banks: theory and design," *IEEE Trans. Signal Processing* **40**: 2207–2232.

[W]   J. Walsh. 1923. "A closed set of normal orthogonal functions," *Amer. J. Math.* **45**: 5–24.

[Wie]   N. Wiener. 1981. *Collected Works*, P. Masani, ed., Cambridge, MA: MIT Press.

[Z]   A. Zygmund. 1959. *Trigonometric Series*, 2nd ed., Cambridge: Cambridge Univ. Press.

Part I

Core Material

1

Construction of orthonormal wavelets

Robert S. Strichartz

ABSTRACT   This paper gives a self-contained presentation of Meyer and Mallat's basic theory of multiresolution analysis and the construction of orthonormal wavelet bases in $\mathbb{R}^n$. In fact, this is done in the more general setting of an arbitrary lattice in $\mathbb{R}^n$ with a strictly expanding matrix which preserves this lattice. The most familiar case is the integer lattice $\mathbb{Z}^n$ with matrix equal to twice the identity. This general theory is applied to obtain a construction of Daubechies' compactly supported wavelets. Also included is a discussion of subband coding, which gives an efficient algorithm for computing wavelet coefficients and for reconstructing a function from these coefficients.

CONTENTS

1.1 Introduction

Our goal is to describe the basic theory of orthonormal wavelets and multiresolution analyses, and the contruction of compactly supported wavelets. We try to give a coherent account that explains the motivation, and provides all the definitions and most of the proofs. This is by no means a complete account, but it should serve the reader who wants a rapid introduction to the "why," "what," and "how" of the subject. We have chosen to present the theory of multiresolution analyses in a somewhat more general form than is actually needed for the wavelet construction we give, because this seems to be a direction that is being rapidly developed. We have not attempted to give the correct attribution to the many ideas presented here (the "who")—except for the general disclaimer that none of them should be attributed to the author. For a more detailed account of the subject see [Retal], [C1], [C2], [D2], [FJaW], or [M]. A shorter version of this paper, emphasizing the one-dimensional theory, will appear in the American Mathematical Monthly under the title "How to make wavelets" [St].

Orthonormal wavelet expansions are an attempt to improve on Fourier series and other classical expansions. To explain the kind of improvement we are after, let's examine a simple case. Consider a real-valued function f(x) on the interval [0, 1]. You can expand it in a Fourier series

$$f(x) = b_0 + \sum_1^\infty (b_k \cos 2\pi kx + a_k \sin 2\pi kx), \qquad (1.1)$$

or you can expand it in a Haar function series

$$f(x) = c_0 + \sum_{j=0}^\infty \sum_{k=0}^{2^j-1} c_{jk}\psi(2^j x - k), \qquad (1.2)$$

where $\psi(x)$ is the function defined by

$$\psi(x) = \begin{cases} 1 & \text{if } 0 \le x < \dfrac{1}{2}, \\ -1 \\ 0 & \text{if } \dfrac{1}{2} \le x < 1, \\ & \text{otherwise} \end{cases} \qquad (1.3)$$

(see Figure 1.1).

Both series are examples of expansions in terms of orthogonal functions in $L^2(0,1)$. Thus there are simple formulas for the coefficients. (Exercise: Show that $\{\psi(2^j x - k)\}$ are orthogonal, but not normalized.) But the Fourier series is not well localized in space; if you are interested in the behavior of $f(x)$ on a subinterval [a, b] you need to involve all the Fourier coefficients. On the other hand, the Haar series is very well localized in that to restrict attention to the subinterval [a, b] you need only take the sum in (1.2) over those indices for which the interval

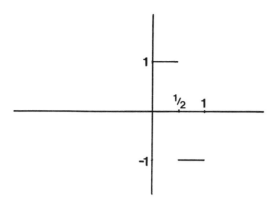

FIGURE 1.1
The graph of the generator of the Haar functions.

$I_{jk} = [2^{-j}k, 2^{-j}(k+1)]$ (the support of $\psi(2^j x - k)$) intersects $[a, b]$. Furthermore, the partial sums of the Haar series (summing $0 \leq j \leq N$) clearly represent an approximation to f taking into account details on the order of magnitude 2^{-N} or greater. These two properties, *localization in space*, and *scaling*, are the hallmarks of wavelet expansions. In addition, the Haar functions are created out of a single function ψ by dyadic dilations and integer translations. Essentially the same property is shared by all the wavelet bases we will discuss, and may in fact be taken as an approximate definition of a wavelet expansion.

The wavelet expansions we are going to construct can be thought of as generalizations of the Haar series, in which the function ψ is replaced by smoother cousins. Before we can say exactly what properties we want these functions to have, and how we can go about constructing them, it is useful to backtrack and see exactly how the Haar functions arise. It will turn out to be easier if we consider the whole line as the domain of our functions.

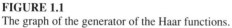

1.2   The rough-and-ready Haar wavelets

We begin with the function φ, the characteristic function of the unit interval $[0, 1]$. Surely this is one of the simplest functions one can imagine, but it is chosen because it has two important properties:

i. the translates of φ by integers, $\varphi(x - k), k \in \mathbb{Z}$ form an orthonormal set of functions for $L^2(\mathbb{R})$;

ii. φ is *self-similar*. If you cut the graph in half then each half can be expanded to recover the whole graph. This property can be expressed algebraically by the *scaling identity*

$$\varphi(x) = \varphi(2x) + \varphi(2x - 1). \tag{1.4}$$

We will call φ the *scaling* function. (In the French literature it is sometimes called "le pére" and ψ is called "la mére", but this shows a scandalous misunderstanding of human reproduction; in fact, the generation of wavelets more closely resembles the reproductive life style of amoebas.) The scaling identity essentially determines φ up to a constant multiple (exercise). The significance of the scaling identity is the following: Let V_0 denote the linear span of the functions $\varphi(x - k), k \in \mathbb{Z}$ (or by abuse of notation the closure in $L^2(\mathbb{R})$ of this span, $\sum_{k=-\infty}^{\infty} a_k \varphi(x - k)$ with $\sum |a_k|^2 < \infty$). This is a natural space to consider in view of (i), since the functions $\varphi(x - k)$ form an orthonormal basis for V_0. Of course V_0 is not all of L^2, it is the subspace of piecewise constant functions with jump discontinuities at $\mathbb{Z}$. We can get a larger space by rescaling. Let $1/2\mathbb{Z}$ denote the lattice of half-integers $k/2, k \in \mathbb{Z}$, and let V_1 denote the subspace of L^2 of piecewise constant functions with jumps at $1/2\mathbb{Z}$. It is clear that $f(x) \in V_0$ if and only if $f(2x) \in V_1$, and the functions $2^{1/2}\varphi(2x - k)$ form an orthonormal basis for V_1 (the factor $2^{1/2}$ is thrown in to make the normalization $\|2^{1/2}\varphi(2x - k)\|_2 = 1$ hold). The scaling identity (1.4), or more propertly its translated version

$$\varphi(x - k) = \varphi(2x - 2k) + \varphi(2x - 2k - 1), \tag{1.4'}$$

says exactly $V_0 \subseteq V_1$, since a basis for V_0 is explicitly represented as linear combinations of basis elements of V_1. (Of course the containment $V_0 \subseteq V_1$ is clear from the description of the spaces V_0 and V_1 in terms of locations of jump discontinuities, but in the generalizations to come there will be no such simple description. However, there will be a scaling identity.)

The whole story can now be iterated, both up and down the dyadic scale. The result is an increasing sequence of subspaces V_j for $j \in \mathbb{Z}$, where V_j consists of the piecewise-constant L^2 functions with jumps at $2^{-j}\mathbb{Z}$. The functions $2^{j/2}\varphi(2^j x - k)$ for $k \in \mathbb{Z}$ form an orthonormal basis for V_j. We can pass back and forth among the space V_j by rescaling: $f(x) \in V_j$ if and only if $f(2^{k-j}x) \in V_k$, and the scaling identity (1.4), suitably rescaled, says $V_j \subseteq V_k$ if $j \leq k$. The sequence $\{V_j\}$ is an example of what is called a *multiresolution analysis*. There are two other properties of $\{V_j\}$ that are significant, namely

$$\bigcap_{j \in \mathbb{Z}} V_j = \{0\} \tag{1.5}$$

and

$$\bigcup_{j \in \mathbb{Z}} V_j \text{ is dense in } L^2 \tag{1.6}$$

(exercise). In view of (1.6) it would seem tempting to try to combine all the orthonormal bases $\{2^{j/2}\varphi(2^j x - k)\}$ of V_j into one orthonormal basis for $L^2(\mathbb{R})$. But although $V_j \subseteq V_{j+1}$, the orthonormal basis $\{2^{j/2}\varphi(2^j x - k)\}$ for V_j is not contained in the orthonormal basis $\{2^{(j+1)/2}\varphi(2^{j+1}x - k)\}$ for V_{j+1}. (Indeed, there are distinct elements in the two orthonormal bases that are not orthogonal to each other.) So our first naive attempt to obtain an orthonormal basis for $L^2(\mathbb{R})$ is flawed. Can we fix it up?

Back to the drawing boards! Since $V_0 \subseteq V_1$ and we have an orthonormal basis for V_0 of the form $\{\varphi(x - k)\}$, why don't we try to complete an orthonormal basis of V_1 by adjoining functions of the form $\{\psi(x - k)\}$ for some function ψ? This is the same thing as asking for an orthonormal basis of the desired form for the orthogonal complement of V_0 in V_1, which we denote W_0, so $V_1 = V_0 \oplus W_0$ (Hilbert space direct sum).

The answer is easy: we want to take ψ exactly to be the Haar function generator defined in Section 1.1. Note that ψ can be expressed in terms of φ by

$$\psi(x) = \varphi(2x) - \varphi(2x - 1), \tag{1.7}$$

which is very reminiscent of the scaling identity. (Exercise: show that $\{\psi(x - k)\}$ forms an orthonormal basis for W_0.) But now we can rescale the space W_0, so

$$V_{j+1} = V_j \oplus W_j \tag{1.8}$$

and $\{2^{j/2}\psi(2^jx - k)\}_{k \in \mathbb{Z}}$ is an orthonormal basis for W_j. If we combine conditions (1.5), (1.6), and (1.8), we obtain

$$L^2(\mathbb{R}) = \bigoplus_{j=-\infty}^{\infty} W_j. \tag{1.9}$$

Since the spaces W_j are all mutually orthogonal, we can now refine our naive attempt and combine all the orthonormal bases for W_j into one grand orthonormal basis $\{2^{j/2}\psi(2^jx - k)\}_{j \in \mathbb{Z}, k \in \mathbb{Z}}$ for $L^2(\mathbb{R})$. (The only change is that we have replaced the scaling function φ by the wavelet ψ.) This gives the Haar series basis for the whole line. There is a minor variation on this theme that is perhaps more closely related to the Haar expansion on the unit interval: instead of (1.9), we can also write

$$L^2(\mathbb{R}) = V_0 \oplus \left\{ \bigoplus_{j=0}^{\infty} W_j \right\} \tag{1.9'}$$

and then combine the basis $\{\varphi(x - k)\}_{k \in \mathbb{Z}}$ for V_0 with the bases $\{2^{j/2}\psi(2^{1/2}x - k)\}_{k \in \mathbb{Z}}$ for W_j, with $j \geq 0$, to obtain an orthonormal basis for $L^2(\mathbb{R})$.

1.3 Multiresolution analysis

The moral of the story so far is that we want to first build a scaling function φ and associated multiresolution analysis $\cdots \subseteq V_{-1} \subseteq V_0 \subseteq V_1 \subseteq \cdots$ before constructing the wavelets. In the last section these objects were based on the integer lattice

$\mathbb{Z} \subseteq \mathbb{R}$ and the dyadic dilation $x \rightarrow 2x$, but we can gain a great deal of generality at very little cost if we consider instead a lattice $\Gamma \subseteq \mathbb{R}^n$ (a discrete subgroup given by integral linear combinations of a vector space basis $v_1, \ldots, v_n$ of $\mathbb{R}^n$) and an $n \times n$ matrix M satisfying

$$M\Gamma \subset \Gamma \ (M \text{ preserves the lattice}) \tag{1.10}$$

$$\text{all eigenvalues } \lambda \text{ of } M \text{ satisfy } |\lambda| > 1 \tag{1.11}$$

$$(M \text{ is strictly expanding}).$$

It is an easy exercise to show that (1.10) and (1.11) imply that $|\det M| = m$ is an integer, $m \geq 2$, and $\Gamma/M\Gamma$ is a group of order m. The simplest example is $\Gamma = \mathbb{Z}^n$ and $M = 2I$, with $m = 2^n$. Here are some others:

(a) $n = 2, \Gamma = \mathbb{Z}^2, M = \begin{pmatrix} 1 & -1 \\ 1 & 1 \end{pmatrix}, m = 2;$

(b) $n = 2, \ \Gamma = \left\{ \left(j + \frac{1}{2}k, \frac{k\sqrt{3}}{2} \right) : (j, k) \in \mathbb{Z}^2 \right\}, \ M = \begin{pmatrix} \frac{5}{2} & \frac{-\sqrt{3}}{2} \\ \frac{\sqrt{3}}{2} & \frac{5}{2} \end{pmatrix},$

$m = 7.$

DEFINITION 1.1 *A multiresolution analysis $\cdots \subseteq V_{-1} \subseteq V_0 \subseteq V_1 \subseteq \cdots$ with scaling function φ is an increasing sequence of subspaces of $L^2(\mathbb{R}^n)$ satisfying the following four conditions:*

i. *(density) $\bigcup_j V_j$ is dense in $L^2(\mathbb{R}^n)$,*

ii. *(separation) $\bigcap_j V_j = \{0\}$.*

iii. *(scaling) $f(x) \in V_j \leftrightarrow f(M^{-j}x) \in V_0$.*

iv. *(orthonormality) $\{\varphi(x - \gamma)\}_{\gamma \in \Gamma}$ is an orthonormal basis for V_0.*

It follows easily from the definition that $\{m^{j/2}\varphi(M^j x - \gamma)\}_{\gamma \in \Gamma}$ forms an orthonormal basis for V_j. Since $\varphi \in V_0 \subseteq V_1$ we must have

$$\varphi(x) = \sum_{\gamma \in \Gamma} a(\gamma)\varphi(Mx - \gamma) \tag{1.12}$$

for some coefficients $a(\gamma)$ satisfying

$$\sum_{\gamma \in \Gamma} |a(\gamma)|^2 = m, \tag{1.13}$$

and, in fact,

$$a(\gamma) = m \int \varphi(x)\overline{\varphi(Mx - \gamma)}dx. \tag{1.14}$$

Equation (1.12) is the analogue of (1.4), and we will refer to it as the *scaling identity*.

It is clear from the definition that the scaling function determines the multiresoution analysis, but not conversely. A more difficult question is how to characterize those functions φ which are scaling functions for a multiresolution analysis. Here we expect the scaling identity to play a crucial role, but before we can say more we need to examine certain algebraic conditions on the coefficients $a(\gamma)$ that follow from the definition.

First, there is a consistency condition that arises from (iv) and (1.12). We know from (iv) that

$$\int \varphi(x - \gamma)\overline{\varphi(x)}dx = \delta(\gamma, 0) \tag{1.15}$$

(Kronecker δ). If we use (1.12) to substitute for $\varphi(x - \gamma)$ and $\overline{\varphi(x)}$ in (1.15) we obtain

$$\sum_{\gamma' \in \Gamma} \sum_{\gamma'' \in \Gamma} a(\gamma')\overline{a(\gamma'')} \int \varphi(Mx - M\gamma - \gamma')\overline{\varphi(Mx - \gamma'')}dx$$

$$= m^{-1} \sum_{\gamma'' = M\gamma + \gamma'} \sum a(\gamma')\overline{a(\gamma'')} = \delta(\gamma, 0)$$

after the change of variable $x \to M^{-1}x$ and use of (1.15). We rewrite this as

$$\sum_{\gamma' \in \Gamma} a(\gamma')\overline{a(M\gamma + \gamma')} = m\delta(\gamma, 0). \tag{1.16}$$

Note that (1.16) contains (1.13) as a special case.

Another algebraic condition arises if we assume φ is integrable and $\int \varphi(x)dx \neq 0$ (this condition holds for all the scaling functions that arise in practice). Then we integrate (1.12) and make a change of variable to obtain

$$\int \varphi(x)dx = \sum_{\gamma \in \Gamma} a(\gamma) \int \varphi(Mx - \gamma)dx$$

$$= \sum_{\gamma \in \Gamma} a(\gamma)m^{-1} \int \varphi(x)dx.$$

Hence

$$\sum_{\gamma \in \Gamma} a(\gamma) = m. \tag{1.17}$$

Now we would like to reverse the procedure. *Step 1* will be to produce solutions $a(\gamma)$ to the algebraic identities (1.16) and (1.17). *Step 2* will be to define the scaling function via the scaling identity (1.12). Notice that (1.12) says that φ is a fixed point of the linear transformation

$$Sf(x) = \sum_{\gamma \in \Gamma} a(\gamma)f(Mx - \gamma) \tag{1.18}$$

so it is reasonable to try to construct φ by iterating S,

$$\varphi = \lim_{n \to \infty} S^n f \tag{1.19}$$

for some reasonable initial function f. In a later section we will discuss another method for solving (1.12). *Step 3* will be to prove that the function φ that solves (1.12) (normalized so $\|\varphi\|_2 = 1$) generates a multiresolution analysis. This is the trickiest step, because there are simple counterexamples to show that it is not always true (try $n=1$; $\Gamma = \mathbb{Z}$; $M = 2$; $a(\gamma)$ equal to 1 for $\gamma = 0, 3$, and otherwise $a(\gamma) = 0$; and $\varphi = \chi_{[0,3)}$, which violates (iv)). Nevertheless, many choices of $a(\gamma)$ do yield a multiresolution analysis. The difficult condition to verify is the orthonormality (iv), and we will have to postpone the discussion of when and why this holds to a later section. In the remainder of this section we will show how to establish the density (i) and separation (ii), given orthonormality and the additional normalization condition

$$\int \varphi(x)dx = \sqrt{\mathrm{vol}(\mathbb{R}^n/\Gamma)} \tag{1.20}$$

(here $\mathrm{vol}(\mathbb{R}^n/\Gamma)$ denotes the volume of the torus $\mathbb{R}^n/\Gamma$, or equivalently the volume of a fundamental domain for Γ). The significance of this condition will emerge in the proof of density. The separation condition will follow from

LEMMA 1.2
Let V_0 be any subspace of $L^2(\mathbb{R}^n)$ which is contained in $L^\infty(\mathbb{R}^n)$ and which has the property that

$$\|f\|_\infty \leq c\|f\|_2 \quad \text{for all } f \in V_0. \tag{1.21}$$

Define V_j by the scaling condition (iii) (no assumption of the sort $V_j \subseteq V_{j+1}$ is necessary). Then (ii) holds.

PROOF   The scaling condition and a simple change of variable transforms (1.21) into

$$\|f\|_\infty \leq cm^{j/2}\|f\|_2 \quad \text{for all } f \in V_j. \tag{1.22}$$

If $f \in \cap V_j$ then (1.22) holds for all j, and letting $j \to -\infty$ we obtain $\|f\|_\infty = 0$, hence $f = 0$.   ∎

The estimate (1.21) is easy to obtain in our case. For simplicity assume φ is bounded and has compact support, which will be the case in all our examples. Then by the orthonormality (iv) we have

$$f(x) = \sum_{\gamma \in \Gamma} \varphi(x - \gamma) \int f(y)\overline{\varphi(y - \gamma)}dy = \int K(x, y)f(y)dy$$

where $K(x,y) = \sum_{\gamma \in \Gamma} \varphi(x-\gamma)\overline{\varphi(y-\gamma)}$, so

$$|f(x)| \leq \left(\int |K(x,y)|^2 dy \right)^{1/2} \|f\|_2 = \left(\sum_{\gamma \in \Gamma} |\varphi(x-\gamma)|^2 \right)^{1/2} \|f\|_2$$

and $\sum_{\gamma \in \Gamma} |\varphi(x-\gamma)|^2$ is uniformly bounded (of course much weaker conditions on φ, such as rapid decrease, will also imply this).

LEMMA 1.3
Assume φ has compact support and satisfies (1.12), (1.20), and the orthonormality condition (iv). Then the density condition (i) holds.

SKETCH OF PROOF (IN THE CASE THAT M IS A SIMILARITY)   Let

$$P_j f(x) = m^j \sum_{\gamma \in \Gamma} \varphi(M^j x - \gamma) \int f(y)\overline{\varphi(M^j y - \gamma)}dy$$

denote the orthogonal projection onto V_j. We need to show $\lim_{j \to \infty} P_j f = f$ in L^2 for all $f \in L^2$, which is equivalent to $\lim_{j \to \infty} \|P_j f\|_2^2 = \|f\|_2^2$ by the Pythagorean theorem. It suffices to prove this for $f = \chi_A, A$ any rectangle, by a density argument. But

$$\|P_j \chi_A\|_2^2 = m^j \sum_{\gamma \in \Gamma} \left| \int_A \varphi(M^j y - \gamma)dy \right|^2 = m^{-j} \sum_{\gamma \in \Gamma} \left| \int_{M^j A} \varphi(y-\gamma)dy \right|^2.$$

For large $j, M^j A$ will be a large parallelopiped, so essentially either $\int_{M^j A} \varphi(y - \gamma)dy = 0$ if $\gamma \notin M^j A$ or $\int_{M^j A} \varphi(y-\gamma)dy = \sqrt{\text{vol}(\mathbb{R}^n/\Gamma)}$ if $\gamma \in M^j A$ by (1.20) (for γ in a small neighborhood of the boundary of $M^j A$ this is not quite correct, but in the limit we can ignore this detail since we are assuming that M is a similarity). Thus

$$\|P_j \chi_A\|_2^2 \approx m^{-j} \text{vol}(\mathbb{R}^n/\Gamma)\#\{\gamma \in M^j A\} \approx \text{vol}(A) = \|\chi_A\|_2^2$$

and in the limit this becomes equality. ∎

Notice that we could essentially reverse the argument to deduce the necessity of the normalization condition (1.20).

Now we are ready to move on to *Step 4*, which is the construction of the wavelets themselves.

1.4   The wavelets

We will need m functions altogether, denoted $\psi_0, \psi_1, \ldots, \psi_{m-1}$, with $\psi_0 = \varphi$ the scaling function and $\psi_1, \ldots, \psi_{m-1}$ being the wavelet generators. We would like the functions $\{\psi_k(x-\gamma)\}_{\gamma \in \Gamma, k=0,\ldots,m-1}$ to be an orthonormal basis for V_1.

Since the functions $\{\varphi(Mx - \gamma)\}_{\gamma \in \Gamma}$ already form an orthogonal basis for V_1, the functions ψ_k must be linear combinations of $\varphi(Mx - \gamma)$, and hence must satisfy an identity

$$\psi_k(x) = \sum_{\gamma \in \Gamma} a_k(\gamma)\varphi(Mx - \gamma) \tag{1.23}$$

which generalizes (1.12) (of course $a_0(\gamma) = a(\gamma)$). Notice that for $k = 1, \dots,$ $m - 1$ (1.23) is an explicit formula, so there is nothing to solve. But what kind of conditions should we put on the coefficients $a_k(\gamma)$? The same reasoning that led to (1.16) leads to

$$\sum_{\gamma' \in \Gamma} a_j(\gamma')\overline{a_k(M\gamma + \gamma')} = m\delta(j,k)\delta(\gamma,0). \tag{1.24}$$

On the other hand, the condition $\int \varphi(x)dx \neq 0$ is not something we can expect to hold for ψ_k (think of the example of Haar functions), so condition (1.17) can only be recopied in our new notation

$$\sum_{\gamma \in \Gamma} a_0(\gamma) = m. \tag{1.25}$$

LEMMA 1.4
If $\{\varphi(x - \gamma)\}_{\gamma \in \Gamma}$ is an orthonormal set and if $a_j(\gamma)$ satisfy (1.24) and (1.25), then $\{\psi_k(x - \gamma)\}_{\gamma \in \Gamma, k=0,\dots,m-1}$ is an orthonormal set.

PROOF It suffices to show

$$\int \psi_j(x)\overline{\psi_k(x - \gamma)}dx = \delta(j,k)\delta(\gamma,0). \tag{1.26}$$

But

$$\int \psi_j(x)\overline{\psi_k(x - \gamma)}dx = \sum_{\gamma' \in \Gamma}\sum_{\gamma'' \in \Gamma} a_j(\gamma')\overline{a_k(\gamma'')}\int \varphi(Mx - \gamma')\overline{\varphi(Mx - M\gamma - \gamma'')}dx.$$

But the integral is $(1/m)\delta(\gamma', M\gamma + \gamma'')$ by the orthonormality of $\varphi(x - \gamma)$, so (1.26) reduces to (1.24). ∎

REMARK 1.5 We have omitted the justification of the interchange of series and integrals, but in most of our examples the series are actually finite sums.

Thus $\{\psi_k(x - \gamma)\}_{\gamma \in \Gamma, k=0,\dots,m-1}$ is an orthonormal set of functions in V_1. Is it a basis? (A kind of pseudo-dimension-counting argument makes this very plausible.) To show that it is a basis it suffices to represent the function $\varphi(Mx - \tilde{\gamma})$, for each $\tilde{\gamma} \in \Gamma$, as a linear combination. We know the coefficients will have to be

$$\int \varphi(Mx - \tilde{\gamma})\overline{\psi_k(x - \gamma)}dx = \sum_{\gamma' \in \Gamma}\overline{a_k(\gamma')}\int \varphi(Mx - \tilde{\gamma})\overline{\varphi(Mx - M\gamma - \gamma')}dx$$

$$= \frac{1}{m}\overline{a_k(\tilde{\gamma} - M\gamma)}.$$

Thus we need to show that

$$\frac{1}{m} \sum_{k=0}^{m-1} \sum_{\gamma \in \Gamma} \overline{a_k(\tilde{\gamma} - M\gamma)} \psi_k(x - \gamma) \qquad (1.27)$$

is equal to $\varphi(Mx - \tilde{\gamma})$. But if we substitute (1.23) into (1.27) we obtain

$$\sum_{\gamma \in \Gamma} \left(\frac{1}{m} \sum_{k=0}^{m-1} \sum_{\gamma' \in \Gamma} a_k \overline{(M\gamma' + \tilde{\gamma})} a_k(M\gamma' + \gamma) \right) \varphi(Mx - \gamma),$$

so it suffices to show

$$\sum_{k=0}^{m-1} \sum_{\gamma' \in \Gamma} \overline{a_k(M\gamma' + \tilde{\gamma})} a_k(M\gamma' + \gamma) = m\delta(\gamma, \tilde{\gamma}). \qquad (1.28)$$

LEMMA 1.6
The equation (1.28) always holds, hence $\{\psi_k(x - \gamma)\}_{\gamma \in \Gamma, k=0,\dots,m-1}$ is an orthonormal basis for V_1.

Although this is a purely algebraic statement, we postpone the proof until the next section.

THEOREM 1.7
Suppose φ generates a multiresolution analysis and $a_k(\gamma)$ satisfy (1.24) and (1.25) with ψ_k defined by (1.23) and $\psi_0 = \varphi$. Then the functions $\{m^{j/2}\psi_k(M^j x - \gamma)\}$ for $j \in \mathbb{Z}$, $k = 1, \dots, m-1$, and $\gamma \in \Gamma$ form an orthonormal basis of $L^2(\mathbb{R}^n)$.

PROOF As before, let W_0 denote the orthogonal complement of V_0 in V_1, $V_1 = V_0 \oplus W_0$. We claim $\{\psi_k(x - \gamma)\}_{\gamma \in \Gamma, k=1,\dots,m-1}$ is an orthonormal basis for W_0. This is obvious because we have merely taken the basis for V_1 given by Lemma 1.6 and removed $\{\psi_0(x - \gamma)\}_{\gamma \in \Gamma}$ which is a basis for V_0. By scaling we obtain

$$V_{j+1} = V_j \oplus W_j$$

and

$$\{m^{j/2}\psi_k(M^j x - \gamma)\}_{\gamma \in \Gamma, k=1,\dots,m-1}$$

is an orthonormal basis for W_j. But

$$L^2(\mathbb{R}^n) = \bigoplus_{j \in \mathbb{Z}} W_j$$

by the density condition. ∎

As a simple variation on the theme, which we leave as an exercise to the reader, the set of functions $\{\varphi(x - \gamma)\}$ for $\gamma \in \Gamma$ together with $\{m^{j/2}\psi_k(M^j x - \gamma)\}$ for $j \geq 0$, $k = 1, \dots, m-1$, $\gamma \in \Gamma$ form an orthonormal basis of $L^2(\mathbb{R}^n)$. The advantage of this variant is that we scale only to finer and finer resolutons ($j \to +\infty$ and take care of all the coarser resolutions ($j < 0$) by the single family $\{\varphi(x - \gamma)\}_{\gamma \in \Gamma}$.

In summary, we have reduced the construction of wavelets to the solution of the algebraic identities (1.24) and (1.25), modulo some technical conditions to ensure the orthonormality condition (iv). *Step 5* will be to actually produce the solutions to (1.24) and (1.25), and *step 6* will be to establish various properties of the wavelet functions: regularity, decay at infinity, and moment conditions.

The reason we have postponed some of the details in the construction so far is that they require a new technique, or rather an old technique: the Fourier transform.

1.5   The view from the Fourier transform side

Suppose we take the Fourier transform of everything in sight. Because most of our identities have a convolutional structure, we expect a simplification, with multiplicative identities arising in their place. Before doing so, let us return to the orthonormality question, because here the Fourier transform viewpoint gives us an entirely new handle on the problem. Given $\varphi \in L^2$, how can we tell from $\hat{\varphi}$ whether or not $\{\varphi(x - \gamma)\}_{\gamma \in \Gamma}$ is orthonormal?

It will simplify matters if we adopt the convention (as in [SW]) that

$$\hat{\varphi}(x) = \int e^{2\pi i x \cdot y} \varphi(y) dy \tag{1.29}$$

so that the Fourier inversion formula is just

$$\hat{\hat{\varphi}}(x) = \varphi(-x) \tag{1.20}$$

and the Plancherel formula is

$$\|\varphi\|_2 = \|\hat{\varphi}\|_2 \tag{1.31}$$

(warning: not all the references use this convention!). We define the *dual lattice* Γ^* to be the set of points $\gamma^* \in \mathbb{R}^n$ such that $\gamma \cdot \gamma^* \in \mathbb{Z}$ for all $\gamma \in \Gamma$ (so $e^{2\pi i \gamma \cdot \gamma^*} = 1$ for all $\gamma \in \Gamma$). Note that if $\Gamma = \mathbb{Z}^n$ then so does Γ^*. Also, we have $\text{vol}(\mathbb{R}^n/\Gamma) = (\text{vol}(\mathbb{R}^n/\Gamma^*))^{-1}$.

LEMMA 1.8
$\{\varphi(x - \gamma)\}_{\gamma \in \Gamma}$ *is an orthonormal set if and only if*

$$\sum_{\gamma^* \in \Gamma^*} |\hat{\varphi}(\xi + \gamma^*)|^2 = \text{vol}(\mathbb{R}^n/\Gamma) \quad \text{for all } \xi. \tag{1.32}$$

PROOF   By the Plancherel formula, $\{\varphi(x - \gamma)\}_{\gamma \in \Gamma}$ is orthonormal if and only if

$$\int e^{2\pi i \xi \cdot \gamma} |\hat{\varphi}(\xi)|^2 d\xi = \delta(\gamma, 0). \tag{1.33}$$

But the integral over $\mathbb{R}^n$ can be broken up into an integral over $\mathbb{R}^n/\Gamma^*$ (strictly speaking over a fundamental domain) and a sum over Γ^*. Since $e^{2\pi i \xi \cdot \gamma}$ is periodic for Γ^* we obtain

$$\int_{\mathbb{R}^n/\Gamma^*} e^{2\pi i \xi \cdot \gamma} \sum_{\gamma^* \in \Gamma^*} |\hat{\varphi}(\xi + \gamma^*)|^2 d\xi = \delta(\gamma, 0)$$

which means that the function $\sum_{\gamma^* \in \Gamma^*} |\hat{\varphi}(\xi + \gamma^*)|^2$ on $\mathbb{R}^n/\Gamma^*$ has as Fourier coefficients $\delta(\gamma, 0)$, hence must be constant. To evaluate the constant set $\gamma = 0$ in the above integral. ∎

Now the scaling identity (1.23) transcribes easily into the condition

$$\hat{\psi}_k(\xi) = A_k(M^{*-1}\xi)\hat{\varphi}(M^{*-1}\xi), \tag{1.34}$$

where

$$A_k(\xi) = \frac{1}{m} \sum_{\gamma \in \Gamma} a_k(\gamma) e^{2\pi i \gamma \cdot \xi} \tag{1.35}$$

(exercise, using the definition of the Fourier transform and a change of variable). Notice that $A_k(\xi)$ is Γ^* periodic, and is smooth if we assume the sum is finite (or more generally, if $a_k(\gamma)$ is rapidly decreasing). Then (1.25) says

$$A_0(0) = 1 \tag{1.36}$$

and (1.20) says

$$\hat{\varphi}(0) = \sqrt{\text{vol}(\mathbb{R}^n/\Gamma)}. \tag{1.37}$$

By iterating (1.34) for $k = 0$ (remember $\psi_0 = \varphi$) we obtain the infinite product representation

$$\hat{\varphi}(\xi) = \sqrt{\text{vol}(\mathbb{R}^n/\Gamma)} \prod_{j=1}^{\infty} A_0((M^*)^{-j}\xi), \tag{1.38}$$

where (1.36) and a simple argument justify the local uniform convergence of the infinite product. Substituting (1.38) back into (1.34) we obtain

$$\hat{\psi}_k(\xi) = \sqrt{\text{vol}(\mathbb{R}^n/\Gamma)} A_k(M^{*-1}\xi) \prod_{j=2}^{\infty} A_0((M^*)^{-j}\xi). \tag{1.39}$$

Thus the functions A_k completely and explicitly determine the wavelets.

The most intricate part of the transcription process is the identity (1.24) that the coefficients $a_j(\gamma)$ must satisfy. What does this tell us about the functions A_k? Rather than deal with this question directly (try it as an exercise, after the fact) we repeat the process which led to (1.24)—namely, the consistency of (1.23), alias (1.34), with the orthonormality, alias (1.32). In other words, if $\{\varphi(x - \gamma)\}_{\gamma \in \Gamma}$

is orthonormal then (1.32) must hold, and if (1.34) defines $\hat{\psi}_k$ then we want the analogue of (1.32):

$$\sum_{\gamma^* \in \Gamma^*} \hat{\psi}_k(\xi + \gamma^*) \overline{\hat{\psi}_j(\xi + \gamma^*)} = \delta_{jk} \operatorname{vol}(\mathbb{R}^n/\Gamma). \tag{1.40}$$

Before doing the computation we introduce the necessary notation. $M^{*-1}\Gamma^*$ is a lattice which contains Γ^* and $M^{*-1}\Gamma^*/\Gamma^*$ is a group of order m. Choose $\eta_1, \ldots, \eta_m$ in $M^{*-1}\Gamma^*$ to be representatives of the m distinct cosets of Γ^*. Then points of the lattice Γ^* can be represented uniquely as $M^*(\gamma^* + \eta_p)$ as γ^* varies in Γ^* and $p = 1, \ldots, m$. Then

$$\sum_{\gamma^* \in \Gamma^*} \hat{\psi}_k(\xi + \gamma^*) \overline{\hat{\psi}_j(\xi + \gamma^*)} = \sum_{p=1}^{m} \sum_{\gamma^* \in \Gamma^*} \hat{\psi}_k(\xi + M^*(\gamma^* + \eta_p)) \overline{\hat{\psi}_j(\xi + M^*(\gamma^* + \eta_p))}$$

by the above parametrization of Γ^*, and if we substitute (1.34) and use the Γ^* periodicity of A_k, we obtain

$$\sum_{p=1}^{m} A_k(M^{*-1}\xi + \eta_p) \overline{A_j(M^{*-1}\xi + \eta_p)} \sum_{\gamma^* \in \Gamma^*} |\hat{\varphi}(M^{*-1}\xi + \eta_p + \gamma^*)|^2.$$

The inner sum over Γ^* yields the constant $\operatorname{vol}(\mathbb{R}^n/\Gamma)$, and so (1.40) yields the consistency condition

$$\sum_{p=1}^{m} A_k(\xi + \eta_p) \overline{A_j(\xi + \eta_p)} = \delta_{jk} \tag{1.41}$$

This is the Fourier transform equivalent of (1.24). Note that (1.41) implies

$$|A_k(\xi)| \leq 1 \tag{1.42}$$

which implies the boundedness of the Fourier transforms $\hat{\psi}_k$.

We can now easily supply the missing proof of Lemma 1.6. Notice that (1.41) says that for every ξ, the $m \times m$ matrix $\{A_k(\xi + \eta_p)\}_{k,p}$ is unitary by rows. But this is equivalent to being unitary by columns,

$$\sum_{k=0}^{m-1} A_k(\xi + \eta_p) \overline{A_k(\xi + \eta_q)} = \delta_{pq} \tag{1.43}$$

Now substituting (1.35) into (1.43) we obtain

$$\sum_{\gamma \in \Gamma} \left(\frac{1}{m^2} \sum_{k=0}^{m-1} \sum_{\gamma' \in \Gamma} a_k(\gamma' + \gamma) \overline{a_k(\gamma')} e^{2\pi i \gamma \cdot \eta_p} e^{2\pi i \gamma' \cdot (\eta_p - \eta_q)} \right) e^{2\pi i \gamma \cdot \xi} = \delta_{pq}.$$

Regarding this as an identity between Fourier series expansions on $\mathbb{R}^n/\Gamma$ we can equate coefficients to conclude

$$\frac{1}{m^2} \sum_{k=0}^{m-1} \sum_{\gamma' \in \Gamma} a_k(\gamma' + \gamma) \overline{a_k(\gamma')} e^{2\pi i \gamma \cdot \eta_p} e^{2\pi i \gamma' \cdot (\eta_p - \eta_q)} = \delta_{pq} \delta(\gamma, 0).$$

which becomes

$$\frac{1}{m^2} \sum_{k=0}^{m-1} \sum_{\gamma' \in \Gamma} a_k(\gamma' + \gamma)\overline{a_k(\gamma' + \tilde{\gamma})} e^{2\pi i(\gamma+\gamma')\cdot\eta_p} e^{-2\pi i(\tilde{\gamma}+\gamma')\cdot\eta_q} = \delta_{pq}\delta(\gamma, \tilde{\gamma})$$

after the change of variables $\gamma \to \gamma - \tilde{\gamma}, \gamma' \to \gamma' + \tilde{\gamma}$. Choosing η_p to be the representative of the identity coset Γ^* (i.e., $\eta_p \in \Gamma^*$) so that $e^{2\pi i(\gamma+\gamma')\cdot\eta_p} = 1$, multiplying by $e^{2\pi i\tilde{\gamma}\cdot\eta_q}$, and summing over q, we obtain (1.28) since

$$\sum_{q=1}^{m} e^{-2\pi i\gamma'\cdot\eta_q} = \begin{cases} m & \text{if } \gamma' \in M\Gamma, \\ 0 & \text{otherwise} \end{cases}.$$

The time has come to grasp the bull by the horns and prove the orthonormality of $\{\varphi(x - \gamma)\}_{\gamma \in \Gamma}$ directly. For this we will need an additional hypothesis.

THEOREM 1.9

Suppose there exists a compact neighborhood of the origin K satisfying

(a) $M^{*-1}K \subseteq K$

(b) $K + M^{*-1}\Gamma^* = \mathbb{R}^n$

(c) $A_0(\xi) \neq 0$ for $\xi \in K$.

Then $\{\varphi(x - \gamma)\}_{\gamma \in \Gamma}$ is orthonormal.

PROOF Choose a subset $K_0 \subseteq K$ such that M^*K_0 is a fundamental domain for Γ^*. Define φ_0 by

$$\hat{\varphi}_0(\xi) = \sqrt{\text{vol}(\mathbb{R}^n/\Gamma)}\chi_{M^*K_0}(\xi). \tag{1.44}$$

We claim $\{\varphi_0(\xi - \gamma)\}_{\gamma \in \Gamma}$ is orthonormal. This follows immediately from Lemma 1.8 because (1.32) has exactly one non-zero term since M^*K_0 is a fundamental domain.

Inductively define functions φ_j by

$$\hat{\varphi}_j(\xi) = A_0(M^{*-1}\xi)\hat{\varphi}_{j-1}(M^{*-1}\xi). \tag{1.45}$$

We claim that $\{\varphi_j(x - \gamma)\}_{\gamma \in \Gamma}$ is again orthonormal. This follows immediately from (1.41) with $j = k = 0$ and Lemma 1.8. It can also be deduced from

$$\varphi_j(x) = \sum_{\gamma \in \Gamma} a_0(\gamma)\varphi_{j-1}(Mx - \gamma) \tag{1.46}$$

which is the non-Fourier transform version of (1.45), and (1.24). Note that

$$\hat{\varphi}_j(\xi) = \sqrt{\text{vol}(\mathbb{R}^n/\Gamma)} \left(\prod_{k=1}^{j} A_0((M^*)^{-k}\xi) \right) \chi_{(M^*)^{j+1}K_0}(\xi) \tag{1.47}$$

so that $\hat{\varphi}_j \to \hat{\varphi}$ pointwise, by (1.38).

We would like to show $\varphi_j \to \varphi$ in L^2 norm. This will suffice to complete the proof, because the norm limit of orthonormal sets is an orthonormal set. This is the key point of the proof, where hypothesis (c) must be used. (As an interesting exercise, see how the argument breaks down for the counterexample given in Section 1.3.)

By the Plancherel formula it suffices to show $\hat{\varphi}_j \to \hat{\varphi}$ in L^2 norm, and since we have pointwise convergence we would like to use the dominated convergence theorem. Note first that $\hat{\varphi} \in L^2$ by Fatou's theorem, since it is the pointwise limit of $\hat{\varphi}_j$ and $\|\hat{\varphi}_j\|_2 = 1$. Thus we can use a multiple of $\hat{\varphi}$ as a dominator. By comparing (1.47) and (1.38) we see

$$
\hat{\varphi}_j(\xi) = \begin{cases} \dfrac{\hat{\varphi}(\xi)\sqrt{\mathrm{vol}(\mathbb{R}^n/\Gamma}}{\hat{\varphi}((M^*)^{-j}\xi)} & \text{if } (M^*)^{-j}(\xi) \in M^* K_0, \\ 0 & \text{otherwise.} \end{cases} \tag{1.48}
$$

We claim that $\hat{\varphi}$ is bounded from below on $M^* K_0$. The point is that $\hat{\varphi}$ is continuous, and by (a) and (c) $A_0((M^*)^{-j}\xi) \neq 0$ for $\xi \in M^* K$. Thus $\hat{\varphi}$ doesn't vanish on $M^* K$, so $|\hat{\varphi}_j(\xi)| \leq c|\hat{\varphi}(\xi)|$ for $c = \sqrt{\mathrm{vol}(\mathbb{R}^n/\Gamma)}(\inf_{M^* K} |\hat{\varphi}|)^{-1}$. ∎

1.6   The recipe

We have now indicated all the major steps in the construction, but we have left the first until last. We need to find actual solutions to the algebraic identities (1.24) and (1.25), or their Fourier transform equivalents (1.41) and (1.36), such that the hypotheses of Theorem 1.9 are satisfied. There are several different approaches to this problem. We describe one that is due to Ingrid Daubechies [D1].

Now we return to the special case with which we started: $n = 1$, $\Gamma = \mathbb{Z}$, and $M = 2I$. We look for solutions with only a finite number of $a_k(\gamma)$ different from zero, which means $A_k(\xi)$ are trigonometric polynomials. This implies that the scaling function φ and wavelet ψ_1 have compact support. (This can be seen most easily from the iteration procedure (1.18) and (1.19), because if $a(\gamma) = 0$ unless $\gamma \in [-M, M]$, then if f has support in $[-M, M]$ and so does Sf.) We concentrate first on finding the function A_0, which must satisfy three conditions:

$$
A_0(0) = 1 \tag{1.49}
$$

$$
|A_0(\xi)|^2 + \left|A_0\left(\xi + \frac{1}{2}\right)\right|^2 = 1, \tag{1.50}
$$

$$
A_0(\xi) \neq 0 \quad \text{for } |\xi| \leq \frac{1}{4} \tag{1.51}
$$

Here (1.49) is (1.36), (1.50) is (1.41) for $j = k = 0$ and $\eta_p = 0$ and $1/2$, and (1.51) is hypothesis c) of Theorem 1.9 for $K = [-1/4, 1/4]$. And, of course, A_0 must be of the form

$$A_0(\xi) = \frac{1}{2} \sum_{\gamma \in \mathbb{Z}} a_0(\gamma) e^{2\pi i \gamma \xi} \text{ (finite sum).} \tag{1.52}$$

Note that $|A_0(\xi)|^2$ is then of the same form.

Now we already know one solution, namely

$$A_0(\xi) = \frac{1}{2}(1 + e^{2\pi i \xi}) = e^{\pi i \xi} \cos \pi \xi,$$

which yields the Haar wavelets. This was deemed unsatisfactory because the wavelets are not continuous. One way to create continuity and even differentiability is to take convolution powers, or on the Fourier transform side to take ordinary powers. Thus we are tempted to try $A_0(\xi) = (e^{\pi i \xi} \cos \pi \xi)^N$ for some large N. Unfortunately (1.50) no longer holds, but we can fix this up. Note that $\cos \pi(\xi + 1/2) = -\sin \pi \xi$, so that is why $|\cos \pi \xi|^2 + |\cos \pi(\xi + 1/2)|^2 = 1$.

Now take the identity $\cos^2 \pi \xi + \sin^2 \pi \xi = 1$ and raise it to an odd power, say

$$\begin{aligned}
1 = (\cos^2 \pi \xi + \sin^2 \pi \xi)^5 &= \cos^{10} \pi \xi + 5 \cos^8 \pi \xi \sin^2 \pi \xi \\
&\quad + 10 \cos^6 \pi \xi \sin^4 \pi \xi + 10 \cos^4 \pi \xi \sin^6 \pi \xi + 5 \cos^2 \pi \xi \sin^8 \pi \xi \\
&\quad + \sin^{10} \pi \xi.
\end{aligned}$$

Take the first half of the terms for $|A_0|^2$,

$$|A_0(\xi)|^2 = \cos^{10} \pi \xi + 5 \cos^8 \pi \xi \sin^2 \pi \xi + 10 \cos^6 \pi \xi \sin^4 \pi \xi. \tag{1.53}$$

Replacing ξ by $\xi + 1/2$ turns these into the second half of the terms, so (1.50) is automatic, and (1.49) and (1.51) are easy. This gives a recipe for producing $|A_0|^2$, and it remains to take a square root of the form (1.52). We would also like to take the coefficients $a_0(\gamma)$ in (1.52) to be real, for that will yield a real-valued scaling function (and in the end real-valued wavelets as well). There is a general theorem of F. Riesz that asserts that this is possible, but in this case it is easy enough to accomplish by trial and error. Since

$$\begin{aligned}
|A_0(\xi)|^2 &= (\cos^4 \pi \xi + 5 \cos^2 \pi \xi \sin^2 \pi \xi + 10 \sin^4 \pi \xi) \cos^6 \pi \xi \\
&= ((\cos^2 \pi \xi - \sqrt{10} \sin^2 \pi \xi)^2 + (5 + 2\sqrt{10}) \cos^2 \pi \xi \sin^2 \pi \xi) \cos^6 \pi \xi
\end{aligned}$$

we can take

$$\begin{aligned}
A_0(\xi) &= (e^{\pi i \xi} \cos \pi \xi)^3 (\cos^2 \pi \xi - \sqrt{10} \sin^2 \pi \xi + i\sqrt{5 + 2\sqrt{10}} \cos \pi \xi \sin \pi \xi) \\
&= \frac{1}{8}(e^{2\pi i \xi} + 1)^3 \left(\frac{1 - \sqrt{10}}{2} + \frac{1 + \sqrt{10}}{4}(e^{2\pi i \xi} + e^{-2\pi i \xi}) \right. \\
&\quad \left. + \frac{1}{4}\sqrt{5 + 2\sqrt{10}}(e^{2\pi i \xi} - e^{-2\pi i \xi}) \right), \tag{1.54}
\end{aligned}$$

which is clearly of the form (1.52) with $a_0(\gamma)$ real and $a_0(\gamma) \neq 0$ only if $-1 \leq \gamma \leq 4$.

To complete the story we need to find $A_1(\xi)$, also of the form (1.52), which satisfies

$$|A_1(\xi)|^2 + \left|A_1\left(\xi + \frac{1}{2}\right)\right|^2 = 1 \tag{1.55}$$

and

$$A_0(\xi)\overline{A_1(\xi)} + A_0\left(\xi + \frac{1}{2}\right)\overline{A_1\left(\xi + \frac{1}{2}\right)} = 0 \tag{1.56}$$

(these are the remaining conditions of (1.41)). Fortunately this is easy, just take

$$A_1(\xi) = e^{2\pi i \xi}\overline{A_0\left(\xi + \frac{1}{2}\right)}. \tag{1.57}$$

which amounts to setting

$$a_1(\gamma) = (-1)^{\gamma+1}\overline{a_0(1 - \gamma)}. \tag{1.58}$$

Then (1.55) and (1.56) are simple exercises in algebra using the periodicity of A_0 and (1.50). Note also that $a_1(\gamma)$ are real valued if $a_0(\gamma)$ are.

In the special case we are considering there is just one wavelet generator ψ_1, and its Fourier transform is given by (1.39), which now reads

$$\hat{\psi}_1(\xi) = A_1\left(\frac{1}{2}\xi\right)\prod_{j=2}^{\infty} A_0(2^{-j}\xi) \tag{1.59}$$

with A_0 given by (1.54) and A_1 by (1.57). If we want to obtain the wavelet ψ_1 itself rather than its Fourier transform we first find $\psi_0 = \varphi$ by iterating the mapping

$$Sf(x) = \sum a_0(j)f(2x - j). \tag{1.60}$$

(In practice, it is only necessary to use the iteration method to obtain the values $\varphi(x)$ for x an integer. Then the scaling equation can be used recursively to find the values of $\varphi(x)$ for $x = 2^{-k}$ times an integer.) Then we find ψ_1 via

$$\psi_1(x) = \sum a_1(j)\varphi(2x - j). \tag{1.61}$$

See Figures 1.2 and 1.3.

There is an alternative approach to constructing the scaling function that yields a different wavelet basis. It has the advantage of requiring less algebra, but the disadvantage of producing wavelets that are not compactly supported. We start with the Haar basis scaling function $\chi_{[0,1]}$, whose Fourier transform is $e^{\pi i \xi}(\sin \pi\xi)/(\pi\xi)$, and take the N-fold convolution product

$$g = \chi_{[0,1]} * \chi_{[0,1]} * \cdots * \chi_{[0,1]} \ (N \text{ factors}) \tag{1.62}$$

so that

$$\hat{g}(\xi) = \left(e^{\pi i \xi}\frac{\sin \pi\xi}{\pi\xi}\right)^N. \tag{1.63}$$

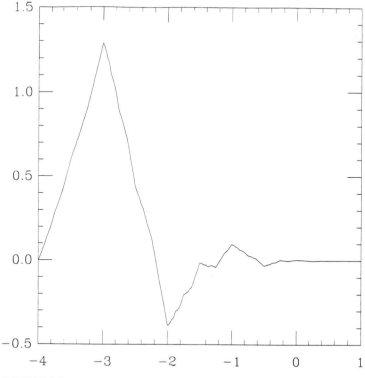

FIGURE 1.2
The graph of the scaling function φ, courtesy of David Aronstein.

It is easy to see that $g \in C^{N-1}$, but of course we have destroyed the orthonormality of translates by $\mathbb{Z}$ that $\chi_{[0,1]}$ had. Too bad, but this is easily fixed. Write

$$h(\xi) = \left(\sum_{k \in \mathbb{Z}} |\hat{g}(\xi + k)|^2 \right)^{1/2}, \tag{1.64}$$

and observe that h is periodic and

$$0 < c_1 \leq h(\xi) \leq c_2 < \infty \tag{1.65}$$

Then we have only to take

$$\hat{\varphi}(\xi) = \hat{g}(\xi)/h(\xi) \tag{1.66}$$

and (1.32) is automatic, so we have the orthonormality of $\{\varphi(x - k)\}_{k \in \mathbb{Z}}$. Notice that $\hat{g}(0) = 1$ and $\hat{g}(k) = 0$ for $k \neq 0$ so $\hat{\varphi}(0) = 1$ as required. And it is not difficult to show that $\varphi \in C^{N-1}$.

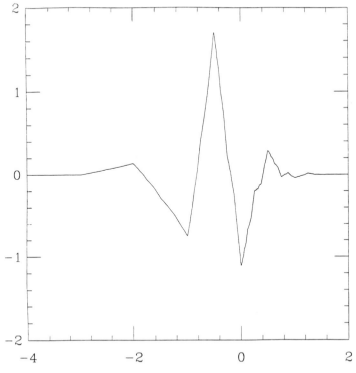

FIGURE 1.3
The graph of the wavelet generator ψ_1, courtesy of David Aronstein.

What about the scaling identity? Well, it certainly holds for g, namely

$$\hat{g}(\xi) = B(\xi/2)\hat{g}(\xi/2), \tag{1.67}$$

where

$$B(\xi) = (e^{\pi i \xi} \cos \pi \xi)^N \tag{1.68}$$

has the required form (1.52). It then follows that

$$\hat{\varphi}(\xi) = A_0(\xi/2)\hat{\varphi}(\xi/2), \tag{1.69}$$

where

$$A_0(\xi) = B(\xi)h(\xi)/h(2\xi). \tag{1.70}$$

Now A_0 is periodic of period one, so it must have the form (1.52), but the sum is no longer finite. This is where we lose the compact support of φ. On the other hand A_0 is clearly smooth, so the Fourier coefficients in (1.52) must be rapidly decreasing, which implies that φ is rapidly decreasing.

The construction of $A_1(\xi)$ and the wavelet Fourier transform $\hat{\psi}_1(\xi)$ then proceeds via (1.57) and (1.59) as before.

1.7 Smoothness of wavelets

How smooth are our wavelets? Since we understand them best on the Fourier transform side, we will use the principle that decay at infinity of $\hat{\varphi}$ implies smoothness of φ (we will establish smoothness of the scaling function and pass it on to the wavelets via (1.61)). For example, it is easy to show

$$|\hat{\varphi}(\xi)| \leq c(1+|\xi|)^{-N-1-\epsilon} \tag{1.71}$$

implies $\varphi \in C^N$. So how do we establish (1.71)?

We have the infinite product representation (1.38), which says

$$\hat{\varphi}(\xi) = \prod_{k=1}^{\infty} A_0(2^{-k}\xi) \tag{1.72}$$

and A_0 is periodic. Since each factor does not decay at infinity, why should the product? This is a mystery, which is best solved by looking at the simplest case, $A_0(\xi) = \cos \pi \xi$. Then

$$\prod_{k=1}^{\infty} \cos 2^{-k}\pi\xi = \frac{\sin \pi \xi}{\pi \xi} \tag{1.73}$$

does decay at the rate $0(|\xi|^{-1})$. (Formula (1.73) was proved by Euler, but special cases were known by Francois Viète in the late 1500's. A proof in the spirit of wavelets can be obtained by considering the Fourier transform of $\chi_{[-1/2,1/2]}$ and its scaling properties.)

Clearly, for most choices of ξ, the values of $\cos 2^{-k}\pi\xi$ will occasionally become small, and that makes the product (1.73) small. You might try to get around this by taking $\xi = 2^N$ for large N. Thus $\cos 2^{-k}\pi\xi = \pm 1$ for $k = 1, \ldots, N$, so there is no decay, but then $\cos 2^{-N-1}\pi\xi = 0$ wipes you out. You can try to quantify this line of reasoning, but there is no great payoff in showing, for example, that $\sin \pi\xi/\pi\xi = 0(|\xi|^{-2/3})$. So we will take (1.73) as our starting point.

The expression (1.54) for A_0, or any of its more complicated cousins, contains $\cos \pi \xi$ as a factor many times. Thus $\hat{\varphi}(\xi)$ contains $\sin \pi\xi/\xi$ as a factor many times, hence we expect decay. Unfortunately, the other factor grows. It is easier to work with $|A_0|^2$ given by (1.53), if we remember to take the square root at the end. We have, for the special case considered,

$$|A_0(\xi)|^2 = (\cos \pi\xi)^6(\cos^4 \pi\xi + 5\cos^2 \pi\xi \sin^2 \pi\xi + 10\sin^4 \pi\xi).$$

The first factor produces decay $0(|\xi|^{-6})$. The second factor can be written $1 + 3\sin^2 \pi\xi + 6\sin^4 \pi\xi$ so it clearly has a maximum value 10 at $\xi = 1/2$. We can obtain a crude estimate for the growth rate produced by the second factor by the following reasoning: if $|\xi| \approx 2^N$ then there will be about N factors where $2^{-k}|\xi|$ is large, so an upper bound for the product is a constant times 10^N. But $10^N \approx \xi|^\alpha$ for $\alpha = \log 10/\log 2 \approx 3.32$. So the growth rate is at most $0(|\xi|^{3.32})$, and the combination gives $0(|\xi|^{-2.68})$ for $|\hat{\varphi}(\xi)|^2$, hence $0(|\xi|^{-1.34})$ for $\hat{\varphi}(\xi)$.

This is a disappointing estimate. According to (1.71) it suffices only to show that φ is continuous. It can be improved, but not by a lot. To see why, consider $\xi = 2^N/3$. Then for each of the N factors $2^{-k}\xi = 2^{N-k}/3, 1 \leq k \leq N$, we have $1 + 3\sin^2(2^{N-k}\pi/3) + 6\sin^4(2^{N-k}\pi/3) = 1 + 3 \cdot (\sqrt{3}/2)^2 + 6(\sqrt{3}/2)^4 = 6.625$. So a lower bound for α is $\log 6.625/\log 2$, which yields $0(|\xi|^{-1.636})$ as the optimal improvement.

If we consider the family of wavelets constructed as outlined in Section 1.6, we will have $|A_0(\xi)|^2$ written as the product of higher and higher powers of $\cos \pi\xi$ by more and more complicated second factors. Thus we have faster decay times faster growth in $\hat{\varphi}(\xi)$. Which wins? Well, it is a close race! It turns out that the decay wins, but the crude method of estimating the growth used above is not good enough to show this. The final result ([D1], [Co2]) is that to create wavelets of class C^N we need to carry out the construction starting with $(\cos^2 \pi\xi + \sin^2 \pi\xi)^M = 1$ for M on the order of $5(N + 1)$. This means that there is a rather high price to pay in terms of complexity (the algebra required to pass from $|A_0|^2$ to A_0, for example) in order to gain a moderate amount of smoothness. (More recently, better techniques have been found to estimate the smoothness directly, without involving the Fourier transform [DL].) Figure 1.4 shows the graph of $\hat{\varphi}(\xi)$. See [JRSt] for a discussion of the surprising self-similarity properties of this function.

In addition to smoothness, another important property of wavelets is the vanishing moment conditions

$$\int_{-\infty}^{\infty} x^k \psi_1(x)dx = 0, k = 0, 1, \ldots, N, \tag{1.74}$$

which are equivalent to the vanishing of the Fourier transform to high order at the origin:

$$\left(\frac{d}{d\xi}\right)^k \hat{\psi}_1(0) = 0, k = 0, 1, \ldots, N. \tag{1.75}$$

In contrast to smoothness, however, it is only the wavelet, not the scaling function, which enjoys this property. The significance of this condition is that it implies a weak form of localization in the frequency (Fourier transform) variable, since the Fourier transform of $\psi_1(2^j x - k)$ is mainly concentrated around values of $|\xi|$ on the order of 2^j. (There is yet another family of wavelets in which the Fourier transform is actually supported in an annular region $c_1 2^j \leq |\xi| \leq c_2 2^j$. See [M] for a description of these "Littlewood–Paley" type wavelets.) For our wavelets the verification of (1.75) is easy. From (1.59) we see that $\hat{\psi}_1$ has a factor $A_1(\xi/2)$, and from (1.57) we see that A_1 at $\xi = 0$ has the same order zero as A_0 at $\xi = 1/2$. But A_0 has a factor of $\cos \pi\xi$ to a power, hence vanishes at $\xi = 1/2$ to order

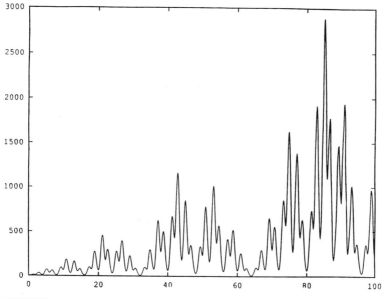

FIGURE 1.4
The graph of $\hat{\varphi}$, after factoring out a power of $\sin \pi x/\pi x$, courtesy of Prem Janardhan and David Rosenblum.

3 in our particular example, and to order M if we start our construction with $(\cos^2 \pi x + \sin^2 \pi x)^{2M-1} = 1$. Note that in general conditions (1.49) and (1.50) imply that $A_0(1/2) = 0$, and the flatter we make A_0 near $\xi = 0$, the more it vanishes near $\xi = 1/2$.

1.8 Higher dimensions via tensor products

Suppose we want wavelets in $\mathbb{R}^n$. We can obtain them easily from our $\mathbb{R}^1$ wavelets by a simple tensor product construction. We will use the lattice $\Gamma = \mathbb{Z}^n$ in $\mathbb{R}^n$ and the matrix $M = 2I$ for scaling. Since $\det M = m = 2^n$ we expect $2^n - 1$ wavelet generators.

Let $\psi_0(x)$ and $\psi_1(x)$ be an $\mathbb{R}^1$-scaling function and wavelet generator. Consider all products

$$\psi_{j_1}(x_1)\psi_{j_2}(x_2)\cdots\psi_{j_n}(x_n), \tag{1.76}$$

where $j_1, j_2, \ldots, j_n$ are 0 or 1. There are 2^n functions altogether, and it should come as no surprise that we want to take $j_1 = j_2 = \cdots = j_n = 0$ for the scaling function and the remaining $2^n - 1$ for wavelet generators. The scaling function corresponds to the multiresolution analysis where the new V_j is the tensor product of the old V_j's. Everything that needs to be verified is easy. Try it!

1.9 Subband coding

Given a function f, how do we actually compute the coefficients of its wavelet expansion? Do we have to perform all those numerical integrations? No! It turns out that there is a simple recursive algorithm for doing the computations algebraically, which goes under the name *subband coding*, and which works just as well in the reverse order, giving the function back from its wavelet expansion. The algorithm works entirely with the scaling coefficients $a_k(\gamma)$, so that strictly speaking it is not even necessary to compute the wavelets or scaling functions.

For simplicity we explain the algorithm in the one-dimensional setting. Suppose the function f is supported in the unit interval $[0, 1]$, and is given numerically by the values $f(x)$ at $x = k/2^J$ for some fixed integer J, with $0 \le k \le 2^J - 1$.

Now consider the space V_J in the multiresolution analysis. It has, roughly speaking, the same degree of resolution as our data for f, so it is reasonable to expect a good approximation to f of the form

$$\sum_k c_k \varphi(2^J x - k). \tag{1.77}$$

Since $2^J \int \varphi(2^J x - k)dx = 1$ by (1.20) and $\varphi(2^J x - k)$ is supported in a small interval about $x = k/2^J$, we will simply take

$$\sum_{k=0}^{2^J - 1} f(k/2^J) 2^J \varphi(2^J x - k) \tag{1.78}$$

as the approximation to f in V_J. There is really no point worrying about how good an approximation it is, because we can clearly recover all our data, $f(k/2^J)$, from the coefficients of the expansion. In other words, our initial data can be regarded as the coefficients c_k in the expansion (1.77) of a general function in V_J, subject to the support constraint that $c_k = 0$ unless $0 \le k \le 2^J - 1$.

Now our wavelet construction tells us that $V_J = V_{J-1} \oplus W_{J-1}$ where V_{J-1} is spanned by $\psi_0(2^{J-1}x - k)$ while W_{J-1} is spanned by $\psi_1(2^{J-1}x - k)$. Thus, a function of the form (1.77) can also be written

$$\sum_k a_k \psi_0(2^{J-1}x - k) + \sum_k b_k \psi_1(2^{J-1}x - k). \tag{1.79}$$

Subband coding tells us how to go from the $\{c_k\}$ coefficients in (1.77) to the $\{a_k\}$ and $\{b_k\}$ coefficients in (1.79) (encoding), and conversely how to go back from the $\{a_k\}$ and $\{b_k\}$ coefficients in (1.79) to the $\{c_k\}$ coefficients in (1.77) (decoding).

If we denote by f the function represented by (1.77) and (1.79), then the orthogonality of the bases yields

$$a_k = 2^{J-1} \int f(x)\overline{\psi_0(2^{J-1}x - k)}dx, \tag{1.80}$$

$$b_k = 2^{J-1} \int f(x)\overline{\psi_1(2^{J-1}x - k)}dx, \tag{1.81}$$

and

$$c_k = 2^J \int f(x)\overline{\psi(2^J x - k)}dx \tag{1.82}$$

But the scaling identity (1.23) tells us how to express $\psi_0(2^{J-1}x-k)$ and $\psi_1(2^{J-1}x-k)$ in terms of $\varphi(2^J x - k)$, namely

$$\psi_0(2^{J-1}x - k) = \sum_\gamma a_0(\gamma)\varphi(2^J x - 2k - \gamma) \tag{1.83}$$

$$\psi_1(2^{J-1}x - k) = \sum_\gamma a_1(\gamma)\varphi(2^J x - 2k - \gamma). \tag{1.84}$$

If we substitute (1.83) in (1.80), (1.84) in (1.81), and use (1.82) we obtain

$$a_k = \frac{1}{2} \sum_\gamma \overline{a_0(\gamma)}c_{2k+\gamma}, \tag{1.85}$$

and

$$b_k = \frac{1}{2} \sum_\gamma \overline{a_1(\gamma)}c_{2k+\gamma}, \tag{1.86}$$

which is our encoding algorithm.

The decoding algorithm is based on the identity

$$\begin{aligned}\varphi(2^J x - k) = \frac{1}{2} \sum_\gamma \overline{a_0(k - 2\gamma)}\psi_0(2^{J-1}x - \gamma), \\ + \frac{1}{2} \sum_\gamma \overline{a_1(k - 2\gamma)}\psi_1(2^{J-1}x - \gamma),\end{aligned} \tag{1.87}$$

which expresses elements of the basis $\{\varphi(2^J x - k)\}$ for V_J as linear combinations of elements of the bases $\{\psi_0(2^{J-1}x - \gamma)\}$ and $\{\psi_1(2^{J-1}x - \gamma)\}$ of V_{J-1} and W_{J-1}, respectively. The proof of (1.87) is based on inverting (1.83) and (1.84) using the orthogonality properties of the functions. If we multiply the complex conjugate of (1.83) by $\varphi(2^J x - k')$ and integrate, we obtain (after relabeling indices)

$$\int \varphi(2^J x - k)\overline{\psi_0(2^{J-1}x - \gamma)}dx = 2^{-J}\overline{a_0(k - 2\gamma)},$$

and similarly

$$\int \varphi(2^J x - k)\overline{\psi_1(2^{J-1}x - \gamma)}dx = 2^{-J}\overline{a_1(k - 2\gamma)}.$$

Together with the orthonormality of $\{2^{(J-1)/2}\psi_0(2^{J-1}x - \gamma)\}$ and $\{2^{(J-1)/2} \times \psi_1(2^{J-1}x - \gamma)\}$ this establishes (1.87). Then by substituting (1.87) in (1.82) and using (1.80) and (1.81) we obtain the decoding algorithm

$$c_k = \sum_{\gamma} a_0(k - 2\gamma)a_\gamma + \sum_{\gamma} a_1(k - 2\gamma)b_\gamma. \tag{1.88}$$

An equivalent formulation is

$$c_k = {\sum_{\gamma}}' a_0(\gamma)a_{(k-\gamma)/2} + {\sum_{\gamma}}' a_1(\gamma)b_{(k-\gamma)/2} \tag{1.88'}$$

where $\sum'$ means we sum over those γ with the same parity as k, so $(k - \gamma)/2$ is an integer.

Notice that since $\{c_k\}$ is supported in $0 \le k \le 2^J - 1$, if we say assume $a_0(\gamma)$ and $a_1(\gamma)$ are supported in $-N \le \gamma \le N$, then $\{a_k\}$ and $\{b_k\}$ are supported in $-N/2 \le k \le 2^{J-1} + (N - 1)/2$ by (1.85) and (1.86). Conversely this is the range in which we need to know $\{a_k\}$ and $\{b_k\}$ in order to apply (1.88) or (1.88') to recover all the initial data $\{c_k\}$. Thus there is a slight increase, by $2N$, in the number of encoded data as compared with the initial data. Presumably $2N$ is small in comparison with 2^J.

Now we iterate. The first sum in (1.79) is a function in V_{J-1}, which we know is equal to $V_{J-2} \oplus W_{J-2}$, and the same procedure can be applied to it, with J replaced by $J - 1$. By iterating this procedure and stopping at level $J = 0$, we obtain schematically

$$V_J = V_0 \oplus W_0 \oplus W_1 \oplus \cdots \oplus W_{J-1} \tag{1.89}$$

with an exact expression for the coefficients in each summand. It is perhaps better to change notation as follows:

$$f(x) = \sum_k c_{0,k}\psi_0(x - k) + \sum_{j=0}^{J-1}\sum_k b_{j,k}\psi_1(2^j x - k). \tag{1.90}$$

Our encoding algorithm reads

$$c_{j-1,k} = \frac{1}{2}\sum_{\gamma} \overline{a_0(\gamma)}c_{j,2k+\gamma} \tag{1.91}$$

$$b_{j-1,k} = \frac{1}{2}\sum_{\gamma} \overline{a_1(\gamma)}c_{j,2k+\gamma} \tag{1.92}$$

for obtaining the coefficients $\{c_{j-1,k}\}_k$ and $\{b_{j-1,k}\}_k$ at level $j - 1$ from the coefficients $\{c_{j,k}\}_k$ at level j. The initial data $\{c_{J,k}\}_k$ is what we called $\{c_k\}$ in (1.77). The intermediate levels $\{c_{j,k}\}_k$ for $j \ne 0$ are discarded in the end, but are needed in the encoding algorithm.

The encoded data is the collection of coefficients $\{c_{0,k}\}_k$ and $\{b_{j,k}\}_k$ for $0 \le j \le J - 1$ in (1.90), which is our approximation to the expansion given in the variant of Theorem 1.7. The range of k's for which we have nonzero data is on the order of 2^j at each level j, with an overlap of at most $4N$. Thus the set of encoded data is only slightly (about $4JN$) larger than the initial set of data. We decode iteratively via the analogue of (1.88′), namely

$$c_{j,k} = \sum_{\gamma}{}' a_0(\gamma)c_{j-1,(k-\gamma)/2} + \sum_{\gamma}{}' a_1(\gamma)b_{j-1,(k-\gamma)/2} \qquad (1.93)$$

($\sum_{\gamma}'$ denotes summing over γ with $\gamma - k$ even), which eventually recovers the initial data $\{c_{J,k}\}_k$ from the encoded data. Also note that the number of operations involved in either algorithm is of the same order as the number of data points.

Why bother? Well, if we believe the philosophy of wavelets, the encoded data should be relatively sparse. Most of the wavelet coefficients should be close enough to zero that we can set them equal to zero without appreciably changing the function. Thus we can achieve data *compression*; we store much less in the way of information, but retain a usual approximation. Further compression can be obtained by storing only approximations to the coefficients, which requires fewer bits of storage. For a fuller discussion of data compression see [DeJaLu].

Bibliography

[C1]   C.K. Chui. 1992. *An Introduction to Wavelets*, New York: Academic Press.

[C2]   C.K. Chui (ed.). 1992. *Wavelets: A Tutorial in Theory and Applications*, New York: Academic Press.

[Co1]   A. Cohen. 1990. "Ondelettes, analyses multi-résolutions et filtres miroir en quadrature," Ann. Inst. Henri Poincaré, Analyse non linéaires **7**: 439–459.

[Co2]   A. Cohen. 1990. "Construction de bases d'ondelettes α-Hölderiennes," *Rev. Mat. Iberoamericana* **6**: 91–108.

[D1]   I. Daubechies. 1988. "Orthonormal bases of compactly supported wavelets," *Commun. Pure Appl. Math.* **41**: 909–996.

[D2]   I. Daubechies. 1992. *Ten Lectures on Wavelets*, CBMS-NSF Reg. Conf. Ser. Appl. Math. **61**, Philadelphia: Soc. Ind. Appl. Math,

[DL]   I. Daubechies and J. Lagarias. 1991. "Two-scale difference equations: I and II," *SIAM J. Math. Anal.* **22**: 1388–1410; 1992. **23**: 1031–1079.

[DeJaLu] R. DeVore, B. Jawerth, and B. Lucier. 1992. "Image compression through wavelet transform coding," *IEEE J. Inform. Theory* **38**: 719–747.

[FJaW]   M. Frazier, B. Jawerth, and G. Weiss. 1991. *Littlewood–Paley Theory and the Study of Function Spaces*, CBMS Reg. Conf. Lect. Notes **79**, Providence: Amer. Math. Soc.

[JRSt]   P. Janardhan, D. Roseblum, and R.S. Strichartz. 1993. "Numerical experiments in Fourier asymptotics of Cantor measures and wavelets," *Exp. Math.*, to appear.

[M]   Y. Meyer. 1990. *Ondelettes et Operateurs*, 3 vols., Paris: Herman.

[Retal] M. Ruskai, G. Beylkin, R. Coifman, I. Daubechies, S. Mallat, Y. Meyer, L. Raphael (eds.). 1992. *Wavelets and Their Applications*, Boston: Jones and Bartlett.

[SW]   E. Stein and G. Weiss. 1971. *An Introduction to Fourier Analysis on Euclidean Spaces*, Princeton: Princeton Univ. Press.

[St]   R. Strichartz. 1993. "How to make wavelets," *Amer. Math. Monthly*, to appear.

2

An introduction to the orthonormal wavelet transform on discrete sets

Michael W. Frazier* and **Arun Kumar**

ABSTRACT   In this paper z-transform theory is used to develop the discrete orthonormal wavelet transform for multidimensional signals. The tone is tutorial and expository. Some rudimentary knowledge of z-transforms and vector spaces is assumed. The wavelet transform of a signal consists of a sequence of inner products of a signal computed against the elements of a complete orthonormal set of basis vectors. The signal is recovered as a weighted sum of the basis vectors. This paper addresses the necessary and sufficient conditions that such a basis must respect. An algorithm for the design of a proper basis is derived from the orthonormality and perfect reconstruction conditions. In the interest of simplicity the case of multidimensional signals is treated separately. The exposition lays bare the structure of hardware and software implementations.

CONTENTS

*Michael W. Frazier was supported in part by NSF postdoctoral fellowship DMS 8705935.

51

2.1   Introduction

In this paper we discuss the wavelet transform on discrete sets in a simple but
rigorous manner. Two underlying themes are developed simultaneously. The first
is that of signal recovery from a sequence of inner products. The second theme
is operational, and concerns the design of filters needed to implement the wavelet
transform in hardware or software. Necessary and sufficient conditions for orthonormal decomposition and perfect reconstruction are established. Algorithms
for the design of wavelet filters are presented together with illustrative examples.
A rudimentary knowledge of z-transforms, vector spaces, and inner-products is
assumed. The exposition is otherwise complete and self-contained. This paper is
a slightly updated version of an earlier technical report [FK], and of some material
in [K].

We do not attempt here to describe the involved history of the subject. Nearly
all the results discussed below (except possibly Section 2.3.2 and the second
example in Section 2.4.6) are known from the subband coding, filter bank, and
wavelet literature, and are not due to the authors. We refer to such excellent
expositions as [D2] and [RVe] for the history and for appropriate attributions. We
wish to present a simple and practical, but mathematically rigorous, guide to the
design of orthonormal wavelet bases on discrete sets, and to the computation of
wavelet transforms via recursive filter banks. To set a direct course for this goal,
we avoid discussing associated theoretical notions as much as possible. We hope
to be accessible to the engineering, scientific, and mathematical audiences, and to
provide sufficient reference information for the nonspecialists who would like to
use wavelet analysis in their work.

What we present here is only a brief introduction to a vast and still-growing area.
For recent articles specifically stressing the connections between wavelets and
filter banks we suggest [RVe], [VeH], and [V], which provide further information
and many references. The subject of wavelets in general is so vast that we only
refer to the introduction to this volume for further references.

In Section 2.2 we discuss the construction of "first generation" wavelet bases
for discrete finite and infinite-dimensional vector spaces. The first generation
basis consists of even translates of two functions f_L and f_H, usually thought
of as low- and high-pass filters, respectively. The fundamental result here is
Theorem 2.5 which gives a necessary and sufficient condition for the construction
of such an orthonormal basis. The decomposition and reconstruction of an input
signal in terms of this wavelet basis is best implemented through a one-stage
filter-bank structure. For such a structure there is a simple condition relating
the "synthesizing filters" g_L and g_H to the "analyzing filters" f_L and f_H, which is
necessary and sufficient for the perfect reconstruction of every input signal. This
result is also derived in Section 2.2.

We discuss the practical design of first generation wavelet bases in Section 2.3.
An explicit algorithm is given for obtaining all possible f_L and f_H satisfying
the condition of Theorem 2.5. We give some examples of such wavelet bases, in

particular the "Shannon" and the "real Shannon" bases, which are highly localized in the frequency domain. We also discuss the problem of finding the nearest (in the l^2 sense) first generation wavelet to a given vector. This problem has a simple solution based on our wavelet-construction algorithm.

In Section 2.4 we discuss the construction of higher-generation wavelet bases. The basic result is Theorem 2.10 which describes how to pass from a wavelet basis at one generation to the next. In the finite case this requires the construction of new filters; but these can be obtained from the first-generation filters by a simple process described in Lemma 2.11.

In Section 2.5 we state and prove the results necessary to extend the development to multidimensional signals lying in discrete finite and infinite-dimensional vector spaces. Since the main ideas are laid out in Sections 2.2 to 2.4, Section 2.5 is brief. The purpose is merely to provide a complete reference. The new difficulties are mostly notational.

We use the z-transform as a basic tool in our discussion. Among other things, it provides a convenient notation, allowing us to deal with the finite and infinite-dimensional cases simultaneously.

2.1.1 Notation

By $\mathbb{C}, \mathbb{R}, \mathbb{Z}$, and $\mathbb{Z}^+$, are meant the set of complex numbers, the set of real numbers, the set of integers (positive, negative, and zero), and the set of positive integers. If $a, b \in \mathbb{Z}$ are integers, then $a \mid b$ means that a divides b. $\mathbb{Z}_N$ denotes the ring of integers $\{0, \ldots, N-1\}$. For any discrete set U, $l^1(U)$ denotes the vector space of all absolutely summable sequences defined from U to $\mathbb{C}$, and $l^2(U)$ the vector space of all absolutely square summable sequences defined from U to $\mathbb{C}$. If f is a function from a discrete set U to $\mathbb{C}$, then the *support* of f is defined as that subset of U where f does not vanish: $\text{supp}(f) = \{u \in U : f(u) \neq 0\}$.

If A is a set of vectors in some vector space V, then $\text{span}(A)$ is the subspace of V consisting of all finite linear combinations of the elements of A. For any set $A \subseteq V$ the topological closure of A in V will be written $\text{cl}(A)$.

If α is a string over an alphabet A we will write $\alpha \in A^*$ where "$*$," in this context, denotes the Kleene operator. If α and β are strings, then $\alpha\beta$ is their concatenation. We write $\alpha = \alpha(1)\alpha(2)\ldots\alpha(n)$ where $\alpha(i)$ is the ith character in α. By $|\alpha| \in \{0\} \cup \mathbb{Z}^+$ will be meant the length of the string α. The string α is the *null string* if $|\alpha| = 0$. If $\alpha, \beta, \gamma \in A^*$, and $\alpha = \beta\gamma$, then β is called a *prefix* of α. By $\beta = \text{pfx}(\alpha, i), 0 \leq i \leq |\alpha|$, we will mean that β is a prefix of α and $|\beta| = i$.

2.2 Two-band orthonormal decomposition and perfect reconstruction of discrete one-dimensional signals

In this section we describe how an arbitrary signal in the discrete infinite-dimensional vector space $l^2(\mathbb{Z})$ or the discrete N-dimensional vector space $\mathbb{C}^N$

(where we assume $2 \mid N$) can be written as a weighted sum of certain "elementary" synthesizing functions. The expression of a signal x as a weighted sum of certain synthesizing functions is called a *decomposition* of x. In order to set up the problem consider the arrangement of filters, downsamplers, and upsamplers in Figure 2.1. By "$\downarrow 2$" is meant *downsampling*, or the deletion of every other number from the input sequence. By "$\uparrow 2$" is meant *upsampling*, or the insertion of a zero between every pair of numbers in the input sequence. The symbols $\mathcal{D}$ and $\mathcal{U}$ will also be used to denote the down and upsampling operators. The symbols f_L and f_H denote *analyzing filters*, while g_L and g_H are *synthesizing filters*. The subscripts "H" and "L" could be thought of as abbreviations for "high-pass" and "low-pass," respectively; because it is usual, though not necessary, that f_H and g_H are high-pass filters, and f_L and g_L low-pass filters.

If the signal x lies in infinite-dimensional vector space $V = l^2(\mathbb{Z})$, then we will assume that $f_L, f_H, g_L, g_H \in l^1(\mathbb{Z})$. It follows that for $x \in l^2(\mathbb{Z})$, the sequences u_L, $v_L, w_L, y_L, u_H, v_H, w_H$, and y_H all belong to $l^2(\mathbb{Z})$. For x in the finite-dimensional signal (vector) space $V = \mathbb{C}^N = l^2(\mathbb{Z}_N)$, we will assume $f_L, f_H, g_L, g_H \in \mathbb{C}^N$. Then for $x \in \mathbb{C}^N$ the sequences u_L, w_L, y_L, u_H, w_H, and y_H belong to $\mathbb{C}^N$; while the sequences v_L and v_H belong to $\mathbb{C}^{N/2}$.

The sequences v_H and v_L are said to define a discrete orthonormal wavelet transform of the signal x, if the analyzing-filter sequences f_H and f_L are chosen according to rules stated in Subsection 2.2.2. The synthesizing-filter sequences g_H and g_L will be chosen so as to attain a perfect reconstruction of x. Perfect reconstruction will be seen to define a decomposition of x.

The main results of this section are stated in two theorems that place constraints on the design of filter sequences f_H, f_L, g_H, and g_L. Another goal of this section is the unification of the engineering and the mathematical points of view. While the problem is so far stated as a filter-design problem, it could also be studied as a pure vector-space problem. We will see that the process of filtering is equivalent

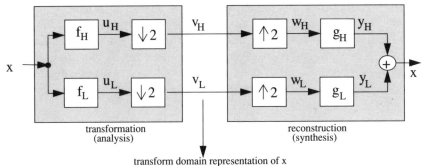

FIGURE 2.1
An arrangement of filters, downsamplers, and upsamplers.

to the process of the computation of a series of inner products, and there exists a close relation between the engineering and the mathematical points of view.

2.2.1 z-Transform notation

For an infinite sequence $x = (\ldots, x(-1), x(0), x(1), \ldots) \in l^2(\mathbb{Z})$, $x(i) \in \mathbb{C}$ for all i, the z-transform $\hat{x}(z)$ of x is written:

$$\hat{x}(z) = \sum_{n=-\infty}^{\infty} x(n)z^{-n}, \qquad (2.1)$$

where the indeterminate z ranges over the unit circle $\mathbf{T}$ in the complex plane. In the standard definition of the z-transform, the indeterminate z ranges over the entire complex plane. Here we will use restrictions of the standard definition. Note that $x(n)$ is the nth Fourier coefficient of the function that maps $\omega \in [-\pi, \pi]$ to $\hat{x}(e^{-j\omega})$; in particular then,

$$x(n) = \frac{1}{2\pi} \int_{-\pi}^{\pi} \hat{x}(e^{-j\omega}) e^{-jn\omega} \, d\omega. \qquad (2.2)$$

From (2.2), if $\hat{x}(z) = \hat{y}(z)$ for all $z \in \mathbf{T}$, then $x = y$; hence the z-transform is one-to-one.

If x is a finite sequence $x = (x(0), \ldots, x(N-1)) \in \mathbb{C}^N$ for some $N \in \mathbb{Z}^+$, then the z-transform $\hat{x}(z)$ of x is written:

$$\hat{x}(z) = \sum_{n=0}^{N-1} x(n)z^{-n}. \qquad (2.3)$$

The function $\hat{x}$ is defined for $z \in \mathbf{W}_N = \{e^{-j2\pi m/N} : m \in \mathbb{Z}_N\}$, the set of all the Nth roots of unity. The complex number $\hat{x}(e^{-j2\pi m/N})$, $m \in \mathbb{Z}_N$, is the mth Discrete Fourier Transform (DFT) coefficient of x. By DFT inversion,

$$x(n) = \frac{1}{N} \sum_{m=0}^{N-1} \hat{x}(e^{-j2\pi m/N}) e^{-j2\pi mn/N}. \qquad (2.4)$$

From (2.4), if $\hat{x}(z) = \hat{y}(z)$ for all $z \in \mathbf{W}_N$, then $x = y$.

If x lies in $l^2(\mathbb{Z})$, then by $(\mathcal{F}x)(\omega)$ we will mean the Fourier transform of x evaluated at $\omega \in [-\pi, \pi]$, $(\mathcal{F}x)(\omega) = \hat{x}(e^{-j\omega})$. If x lies in $\mathbb{C}^N$, then by $(\mathcal{F}x)(m)$ we will mean the DFT of x evaluated at $m \in \mathbb{Z}_N$: $(\mathcal{F}x)(m) = \hat{x}(e^{-j2\pi m/N})$.

When the range of the index of summation is not specified in some particular equation, it will mean that the equation holds equally for sums over $\mathbb{Z}$ and $\mathbb{Z}_N$. Similarly, when we make a statement about some property of $\hat{x}(z)$ that holds "for all z" or "$\forall z$" we will mean "$\forall z \in \mathbf{T}$" or "$\forall z \in \mathbf{W}_N$" depending on whether x belongs to the signal space $V = l^2(\mathbb{Z})$ or to $V = \mathbb{C}^N$.

In case of finite sequences x, since $z \in \mathbf{W}_N$, all arithmetic on the powers of the indeterminate z^{-1} in $\hat{x}(z)$ will be done in the ring $\mathbb{Z}_N$. Thus, if $N = 4$, then

$az^{-2} + bz^{-5} + cz^9 = az^{-2} + bz^{-1} + cz^{-3}$. When the z-transform of a finite sequence is written only in terms of z^{-1} raised to some number in $\mathbb{Z}_N$, then we will say that the z-transform is written in the *canonical form*. For infinite sequences the z-transform is always in the canonical form. If we assume that all z-transforms are always written in the canonical form, then the mathematical development here is valid, simultaneously, for the finite and infinite-dimensional signal spaces $\mathbb{C}^N$ and $l^2(\mathbb{Z})$.

For $x, y \in l^2(\mathbb{Z})$ or $x, y \in \mathbb{C}^N$, their convolution $(x * y)$ is defined by

$$(x * y)(n) = \sum_m x(m)y(n - m). \tag{2.5}$$

If $x, y \in l^2(\mathbb{Z})$, then n and m range over $\mathbb{Z}$. If $x, y \in \mathbb{C}^N$, then n and m range over $\mathbb{Z}_N$. In this latter case, the index $(n - m)$ in (2.5) may *appear* to go out of the domain $\mathbb{Z}_N$ of y. However, when dealing with finite convolutions, we will do all arithmetic on sequence indices in the ring $\mathbb{Z}_N$. This is as in the case of the powers of z^{-1} in z-transforms. Then there is no problem with indices going out of range, and all finite convolutions are "circular" convolutions in the standard terminology of digital signal processing.

The z-transform of the convolution $(x * y)$ of x and y is the product of the z-transforms of x and y: $(x * y)^\wedge(z) = \hat{x}(z)\hat{y}(z)$. By $\mathrm{coeff}_k(\hat{x}(z))$ we will mean the coefficient of z^{-k} in the canonical form $\hat{x}(z)$. By (2.2) or (2.4), this coefficient is well defined. Define the *inner product* of $f, h \in l^2(\mathbb{Z})$ or $f, h \in \mathbb{C}^N$ by

$$\langle f, h \rangle = \sum_n f(n)\bar{h}(n), \tag{2.6}$$

where the index n runs over $\mathbb{Z}$ or $\mathbb{Z}_N$ depending upon the signal space under consideration. This inner product can be written in the z-transform notation.

LEMMA 2.1

If f and h are complex sequences and $z \in \mathbf{T}$ or $z \in \mathbf{W}_N$, then $\langle f, h \rangle = \mathrm{coeff}_0\left(\hat{f}(z)\overline{\hat{h}(z)}\right)$.

PROOF OF LEMMA 2.1

$$\hat{f}(z)\overline{\hat{h}(z)} = \left(\sum_n f(n)z^{-n}\right)\overline{\left(\sum_m h(m)z^{-m}\right)} \tag{2.7}$$

$$= \sum_n \sum_m f(n)\bar{h}(m)z^{m-n}. \tag{2.8}$$

In going from (2.7) to (2.8) we have used the fact that for $z \in \mathbf{T}$ or $z \in \mathbf{W}_N$, $\overline{z^{-1}} = z$. Equation 2.8 implies the lemma. ∎

2.2.2 Orthonormal decomposition

Let R denote a *right-shift operator* that acts upon a sequence $x \in l^2(\mathbb{Z})$ or $x \in \mathbb{C}^N$ such that $(Rx)(n) = x(n-1)$, or, equivalently,

$$(Rx)^{\wedge}(z) = z^{-1}\hat{x}(z). \tag{2.9}$$

In the finite-dimensional signal space $\mathbb{C}^N$, the right shift operator defined in (2.9) wraps sequences around; i.e., when x is shifted right once, then the number $x(N-1)$ moves to occupy the place where $x(0)$ was. This is a consequence of using $\mathbb{Z}_N$ arithmetic upon indices for x. The z-transform notation reflects this wrap-around accurately when the canonical form is employed.

Let f be the impulse response of a filter with input x and output y. Define $\tilde{f}(n) = \bar{f}(-n)$. The sequence $\tilde{f}$ is the complex conjugate of the time reversal of the sequence f. Then,

$$y(k) = \sum_n x(n)f(k-n) = \sum_n x(n)\bar{\tilde{f}}(n-k) = \sum_n x(n)\left(R^k\bar{\tilde{f}}\right)(n) \tag{2.10}$$

$$= \sum_n x(n)\overline{\left(R^k\tilde{f}\right)}(n) = \langle x, R^k\tilde{f}\rangle. \tag{2.11}$$

Equation (2.11) follows from (2.10) because the right-shifted conjugate of the sequence $\tilde{f}$ is the same as the conjugate of the right-shifted sequence $\tilde{f}$. From (2.10) and (2.11) it is evident that the process of filtering a signal x with a filter f is *equivalent* to the computation of a sequence of inner products. One inner product is computed for each number $y(k)$ produced by the filter.

A *multiset* (or a *bag*) is a "set" which may contain multiple copies of one or more of its elements. In Figure 2.1, the sequence u_L is made up of the inner products of x against elements in the *multiset* $\{R^k\tilde{f}_L\}_{k\in\mathbb{Z} \text{ or } k\in\mathbb{Z}_N}$. The reason why $\{R^k\tilde{f}_L\}_k$ is declared to be a multiset, and not a set, is that it is possible that for some $k_1 \neq k_2$, $R^{k_1}\tilde{f}_L = R^{k_2}\tilde{f}_L$. The sequence v_L is obtained from u_L by discarding every other number, and therefore consists of the inner products of x against elements in the multiset

$$\tilde{B}_L = \{R^{2k}\tilde{f}_L\}_k, \tag{2.12}$$

where $k \in \mathbb{Z}$ or $k \in \mathbb{Z}_{N/2}$ depending upon the signal space.

Define also the multisets

$$\tilde{B}_H = \{R^{2k}\tilde{f}_H\}_k \tag{2.13}$$

$$\tilde{B} = \tilde{B}_L \cup \tilde{B}_H. \tag{2.14}$$

The union in (2.14) is a multiset union that preserves the multiplicity of multiset elements.

In orthonormal wavelet analysis, $\tilde{B}$ is required to define an orthonormal basis of the signal space $V = l^2(\mathbb{Z})$ or $V = \mathbb{C}^N$. The multisets $\tilde{B}_L$ and $\tilde{B}_H$ are required to define orthonormal bases of the mutually orthogonal subspaces V_L and V_H of V. Then $\tilde{B}_L, \tilde{B}_H$, and $\tilde{B}$ must each be a *set*. We will see that the orthonormality of $\tilde{B}$ will lead to particularly simple conditions for the decomposition and perfect reconstruction of signals.

We next deduce the consequences of the orthonormality of $\tilde{B}$. Before that, however, we state a lemma that will result in somewhat simpler notation.

LEMMA 2.2
Define the multisets

$$B_L = \{R^{2k}f_L\}_k \tag{2.15}$$

$$B_H = \{R^{2k}f_H\}_k \tag{2.16}$$

$$B = B_L \cup B_H. \tag{2.17}$$

Then $\tilde{B}_L$ is orthonormal if and only if B_L is; $\tilde{B}_H$ is orthonormal if and only if B_H is; and $\tilde{B}$ is orthonormal if and only if B is.

PROOF OF LEMMA 2.2   For $f \in l^2(\mathbb{Z})$ or $f \in \mathbb{C}^N$,

$$(\tilde{f})^{\wedge}(z) = \sum_n \tilde{f}(n)z^{-n} = \sum_n \bar{f}(-n)z^{-n} = \sum_{m=-n} \bar{f}(m)z^m = \overline{\left(\sum_m f(m)z^{-m}\right)} = \bar{\hat{f}}(z). \tag{2.18}$$

By Lemma 2.1 and (2.18),

$$\langle R^{2k}\tilde{f}, R^{2l}\tilde{h}\rangle = \mathrm{coeff}_0\left(z^{-2k}(\tilde{f})^{\wedge}(z)z^{2l}\overline{(\tilde{h})^{\wedge}(z)}\right) = \mathrm{coeff}_0\left(z^{-2k}\bar{\hat{f}}(z)z^{2l}\hat{h}(z)\right)$$

$$= \langle R^{-2l}h, R^{-2k}f\rangle. \tag{2.19}$$

If $\tilde{B}_L$ is orthonormal, then

$$\langle R^{2k}\tilde{f}_L, R^{2l}\tilde{f}_L\rangle = \delta(k - l); \tag{2.20}$$

where δ is the Kronecker delta:

$$\delta(i) \overset{\triangle}{=} \begin{cases} 1, & i = 0 \\ 0, & i \neq 0. \end{cases} \tag{2.21}$$

Then from (2.19) and (2.20), $\langle R^{-2l}f_L, R^{-2k}f_L\rangle = \delta(k - l)$, and B_L is orthonormal. The lemma follows by similar arguments. ∎

It is easy to see that the multiset $B = \{R^{2k}f_L\}_k \cup \{R^{2k}f_H\}_k$ (and therefore $\tilde{B}$) is orthonormal if and only if the following equations hold $\forall k \in \mathbb{Z}$ or $\forall k \in \mathbb{Z}_{N/2}$:

$$\langle f_L, R^{2k}f_L\rangle = \delta(k) \tag{2.22}$$

$$\langle f_H, R^{2k}f_H\rangle = \delta(k) \tag{2.23}$$

$$\langle f_L, R^{2k}f_H\rangle \equiv 0. \tag{2.24}$$

Equations (2.22)–(2.24) will be called the *orthonormality conditions*.

The two lemmas that follow will be used in the proof of the main theorem governing orthonormal decompositions. A definition is necessary before the statement of the lemmas. A function $\hat{h}(z)$ is said to be *odd* in z if and only if $\hat{h}(z) = -\hat{h}(-z)$. We will call a function $\hat{g}(z)$ an *almost-odd* function of z if and only if $\hat{g}(z)$ is the sum of the constant function "1" with an odd function.

LEMMA 2.3
Let $f \in l^2(\mathbb{Z})$ or $\mathbb{C}^N$. The following are equivalent:

P1 For all $k \in \mathbb{Z}$ or $k \in \mathbb{Z}_{N/2}$, $\langle f, R^{2k}f \rangle = \delta(k)$.
P2 $\forall z$, $|\hat{f}(z)|^2$ is an almost-odd function of z.
P3 $\forall z$, $|\hat{f}(z)|^2 + |\hat{f}(-z)|^2 = 2$.

PROOF OF LEMMA 2.3   By Lemma 2.1,

$$\langle f, R^{2k}f \rangle = \text{coeff}_0 \left(\hat{f}(z)\overline{(R^{2k}f)^{\wedge}(z)} \right) = \text{coeff}_0 \left(\hat{f}(z)\overline{z^{-2k}\hat{f}(z)} \right) \qquad (2.25)$$

$$= \text{coeff}_0 \left(z^{2k}|\hat{f}(z)|^2 \right) = \text{coeff}_{2k} \left(|\hat{f}(z)|^2 \right). \qquad (2.26)$$

Define

$$\hat{\alpha}(z) = \sum_n \alpha(n)z^{-n} = |\hat{f}(z)|^2. \qquad (2.27)$$

By (2.26), **P1** is true if and only if

$$\hat{\alpha}(z) = 1 + \sum_n \alpha(2n+1)z^{-(2n+1)}. \qquad (2.28)$$

Equation (2.28) implies that $\hat{\alpha}$ is an almost-odd function of z. Hence **P1** $\Rightarrow$ **P2**. Conversely, if **P2** holds, then (2.2) and (2.4) and a symmetry argument show that $\alpha(2n) = 0$ for $n \neq 0$. Hence (2.26) shows that **P2** $\Rightarrow$ **P1**.

By **P2**, $|\hat{f}(z)|^2 = 1 + \hat{\beta}(z)$, where $\hat{\beta}(z)$ is some function that is odd in z. By direct substitution, **P3** is equivalent to $\hat{\beta}(z) + \hat{\beta}(-z) = 0$. Hence **P2** $\Leftrightarrow$ **P3**.   ∎

LEMMA 2.4
Let $f, h \in l^2(\mathbb{Z})$ or $\mathbb{C}^N$. The following are equivalent:

P1 For all $k \in \mathbb{Z}$ or $k \in \mathbb{Z}_{N/2}$, $\langle f, R^{2k}h \rangle = 0$.

P2 $\forall z$, $(\hat{f}\overline{\hat{h}})$ is an odd function of z.

P3 $\forall z$, $\hat{f}(z)\overline{\hat{h}(z)} + \hat{f}(-z)\overline{\hat{h}(-z)} = 0$.

PROOF OF LEMMA 2.4   From Lemma 2.1,

$$\langle f, R^{2k}h \rangle = \text{coeff}_{2k} \left(\hat{f}(z)\overline{\hat{h}(z)} \right). \qquad (2.29)$$

Arguing as in the proof of Lemma 2.3, **P1** is true if, and only if, $(\hat{f}\overline{\hat{h}})$ is an odd function of z. Therefore **P1** $\Leftrightarrow$ **P2**. Also **P2** $\Leftrightarrow$ **P3** by definition.   ∎

In other words, Lemma 2.3 says that the sets $B_L = \{R^{2k}f_L\}_k$ and $B_H = \{R^{2k}f_H\}_k$ (and therefore $\tilde{B}_L$ and $\tilde{B}_H$) are each orthonormal if, and only if, the polynomials

$|\hat{f}_L(z)|^2$ and $|\hat{f}_H(z)|^2$ are almost-odd as functions of $z \in \mathbf{T}$ or $z \in \mathbf{W}_N$. Lemma 2.4 says that the sets B_L and B_H (and therefore $\tilde{B}_L$ and $\tilde{B}_H$) are mutually orthogonal if, and only if, the polynomial $\hat{f}(z)\overline{\hat{g}}(z)$ is an odd function of $z \in \mathbf{T}$ or $z \in \mathbf{W}_N$. These lemmas give us the following interesting result which is an analog of condition (1.41) in Strichartz [S].

THEOREM 2.5
The set $\tilde{B} = \{R^{2k}\tilde{f}_L\}_k \cup \{R^{2k}\tilde{f}_H\}_k$ is orthonormal if and only if the matrix

$$A(z) = \frac{1}{\sqrt{2}} \begin{pmatrix} \hat{f}_L(z) & \hat{f}_H(z) \\ \hat{f}_L(-z) & \hat{f}_H(-z) \end{pmatrix}. \tag{2.30}$$

is unitary for all $z \in \mathbf{T}$ or $\mathbf{W}_N$.

PROOF OF THEOREM 2.5   From the orthonormality conditions (2.22)–(2.24) and Lemmas 2.2–2.4, we see that the set $\tilde{B}$ is orthonormal if and only if the following equations hold for all $z \in \mathbf{T}$ or $\mathbf{W}_N$.

$$|\hat{f}_L(z)|^2 + |\hat{f}_L(-z)|^2 = 2 \tag{2.31}$$

$$|\hat{f}_H(z)|^2 + |\hat{f}_H(-z)|^2 = 2 \tag{2.32}$$

$$\hat{f}_L(z)\overline{\hat{f}_H}(z) + \hat{f}_L(-z)\overline{\hat{f}_H}(-z) = 0. \tag{2.33}$$

Equations (2.31)–(2.33) are equivalent to the matrix (2.30) being unitary.   ∎

Equations (2.31)–(2.33) may in fact be said to define the unitarity of $A(z)$ in (2.30). Equations (2.31) and (2.32) assert that the two columns of $A(z)$ must each have *norm* (or "size," or "length") of unity in some appropriately defined inner product space W. Equation (2.33) says that the columns of $A(z)$ regarded as vectors in W must be orthogonal, in that their inner product in W is required to vanish. The matrix $A(z)$ in (2.30) will be encountered over and over again, and will be called the *system matrix*.

In the finite case (2.30) is required to be unitary only for $z \in \mathbf{W}_N$. Since $\hat{f}(e^{-j2\pi m/N}) = (\mathcal{F}f)(m)$, Theorem 2.5 shows that the multiset $\tilde{B}$ is orthonormal if and only if

$$A(e^{-j2\pi m/N}) = \frac{1}{\sqrt{2}} \begin{pmatrix} (\mathcal{F}f_L)(m) & (\mathcal{F}f_H)(m) \\ (\mathcal{F}f_L)\left(m \pm \dfrac{N}{2}\right) & (\mathcal{F}f_H)\left(m \pm \dfrac{N}{2}\right) \end{pmatrix} \tag{2.34}$$

is unitary for $m = 0, \ldots, (N-1)$, where the sign in (2.34) is chosen such that $m \pm \frac{N}{2} \in \{0, \ldots, (N-1)\}$. In order to determine the validity of a filter pair (f_L, f_H) for the orthonormal decomposition of an N-dimensional signal space we only need check the unitarity of $N/2$ (2×2)–matrices. The remaining $N/2$ (2×2)–matrices are row-reversed copies of the first $N/2$, and their unitarity is automatic.

2.2.3 Perfect reconstruction

We have so far established the necessary and sufficient conditions for orthonormal decomposition. We now establish conditions such that the arrangement in Figure 2.1 will yield a perfect reconstruction of the input signal. The downsampling operation can result in the violation of the Nyquist sampling criterion in each branch of Figure 2.1. It is remarkable that these violations do not prohibit the faithful recovery of a signal. From the low-pass (i.e., lower) branch of Figure 2.1 we have:

$$\hat{u}_L(z) = \hat{f}_L(z)\hat{x}(z) \tag{2.35}$$

$$\hat{v}_L(z) = (\mathcal{D}u_L)^{\wedge}(z) = \frac{1}{2}\left(\hat{u}_L(z^{1/2}) + \hat{u}_L(-z^{1/2})\right) \tag{2.36}$$

$$\hat{w}_L(z) = (\mathcal{U}v_L)^{\wedge}(z) = \hat{v}_L(z^2) \tag{2.37}$$

$$\hat{y}_L(z) = \hat{g}_L(z)\hat{w}_L(z). \tag{2.38}$$

Equations (2.36) and (2.37) describe the down and upsampling operations, respectively, and may in fact be considered to define those operations. Alternatively, we may define the downsampling operator $\mathcal{D} : l^2(\mathbb{Z}) \to l^2(\mathbb{Z})$ or $\mathcal{D} : \mathbb{C}^N \to \mathbb{C}^{N/2}$ by

$$(\mathcal{D}u)(n) = u(2n), n \in \mathbb{Z} \text{ or } n \in \mathbb{Z}_{N/2}. \tag{2.39}$$

The definitions of $\mathcal{D}$ in (2.36) and (2.39) are equivalent because

$$\hat{u}(z^{1/2}) + \hat{u}(-z^{1/2}) = \sum_n u(n)z^{-n/2} + \sum_n u(n)(-1)^n z^{-n/2} \tag{2.40}$$

$$= 2\sum_{n \text{ even}} u(n)z^{-n/2} = 2\sum_m u(2m)z^{-m}. \tag{2.41}$$

Similarly, the upsampling operator $\mathcal{U} : l^2(\mathbb{Z}) \to l^2(\mathbb{Z})$ or $\mathcal{U} : \mathbb{C}^{N/2} \to \mathbb{C}^N$ may be defined as:

$$(\mathcal{U}v)(n) = \begin{cases} v(n/2), & n \text{ even} \\ 0, & \text{otherwise.} \end{cases} \tag{2.42}$$

The definitions of $\mathcal{U}$ in (2.37) and (2.42) are equivalent.
From (2.35)–(2.38),

$$\hat{y}_L(z) = \frac{1}{2}\hat{g}_L(z)\hat{f}_L(z)\hat{x}(z) + \frac{1}{2}\hat{g}_L(z)\hat{f}_L(-z)\hat{x}(-z). \tag{2.43}$$

Similarly, from the high-pass branch in Figure 2.1,

$$\hat{y}_H(z) = \frac{1}{2}\hat{g}_H(z)\hat{f}_H(z)\hat{x}(z) + \frac{1}{2}\hat{g}_H(z)\hat{f}_H(-z)\hat{x}(-z). \tag{2.44}$$

Perfect reconstruction holds in Figure 2.1 if and only if $\hat{x}(z) = \hat{y}_L(z) + \hat{y}_H(z)$; i.e., if and only if

$$\hat{x}(z) = \frac{1}{2}\left(\hat{g}_L(z)\hat{f}_L(z) + \hat{g}_H(z)\hat{f}_H(z)\right)\hat{x}(z) + \frac{1}{2}\left(\hat{g}_L(z)\hat{f}_L(-z) + \hat{g}_H(z)\hat{f}_H(-z)\right)\hat{x}(-z). \tag{2.45}$$

We also have the following:

LEMMA 2.6
If $P(z), Q(z)$ are fixed polynomials in z and for all $\hat{x}(z)$

$$\hat{x}(z) = P(z)\hat{x}(z) + Q(z)\hat{x}(-z), \tag{2.46}$$

then $P(z) = 1$ and $Q(z) = 0$.

PROOF OF LEMMA 2.6   Substituting $\hat{x}(z) = 1$ and $\hat{x}(z) = z$ into (2.46) we have

$$z = zP(z) + zQ(z) \tag{2.47}$$

$$z = zP(z) - zQ(z). \tag{2.48}$$

By adding and subtracting (2.47) and (2.48) we have the lemma.   ∎

From (2.45) and Lemma 2.6 we deduce that perfect reconstruction holds if and only if the following equations are true:

$$\hat{g}_L(z)\hat{f}_L(z) + \hat{g}_H(z)\hat{f}_H(z) = 2 \tag{2.49}$$

$$\hat{g}_L(z)\hat{f}_L(-z) + \hat{g}_H(z)\hat{f}_H(-z) = 0. \tag{2.50}$$

The system of equations (2.49) and (2.50) can be written as the single matrix equation below.

$$\frac{1}{\sqrt{2}}\begin{pmatrix} \hat{f}_L(z) & \hat{f}_H(z) \\ \hat{f}_L(-z) & \hat{f}_H(-z) \end{pmatrix}\begin{pmatrix} \hat{g}_L(z) \\ \hat{g}_H(z) \end{pmatrix} = A(z)\begin{pmatrix} \hat{g}_L(z) \\ \hat{g}_H(z) \end{pmatrix} = \begin{pmatrix} \sqrt{2} \\ 0 \end{pmatrix}, \tag{2.51}$$

where $A(z)$ is the system matrix in (2.30).

The condition (2.51) is necessary and sufficient for perfect reconstruction, *whether or not $\tilde{B}$ is orthonormal*. If $\tilde{B}$ *is* orthonormal then by Theorem 2.5 the system matrix $A(z)$ is unitary and is, therefore, particularly easy to invert: $A^{-1}(z) = A^*(z) = \bar{A}^T(z)$. Hence (2.51) is easily solved:

$$\begin{pmatrix} \hat{g}_L(z) \\ \hat{g}_H(z) \end{pmatrix} = A^*(z)\begin{pmatrix} \sqrt{2} \\ 0 \end{pmatrix} = \begin{pmatrix} \overline{\hat{f}_L(z)} \\ \overline{\hat{f}_H(z)} \end{pmatrix} \tag{2.52}$$

From (2.52) and (2.18), when $\tilde{B}$ is orthonormal, perfect reconstruction requires that $g_L = \tilde{f}_L$ and $g_H = \tilde{f}_H$.

This is hardly a surprising result in view of the fact that given an orthonormal basis $\tilde{B}$ of an inner product space V, $x \in V$ can be decomposed as a weighted sum of vectors in $\tilde{B}$, where the weights are given by the inner products of x computed against the basis vectors. Recall from (2.10)–(2.11) that $(x * f)(2k) = \langle x, R^{2k}\tilde{f} \rangle$. By perfect reconstruction in Figure 2.1,

$$x(n) = (g_L * (\mathcal{U}\mathcal{D}(x * f_L)))(n) + (g_H * (\mathcal{U}\mathcal{D}(x * f_H)))(n) \qquad (2.53)$$

$$= \sum_k g_L(n - 2k)\langle x, R^{2k}\tilde{f}_L \rangle + \sum_k g_H(n - 2k)\langle x, R^{2k}\tilde{f}_H \rangle \qquad (2.54)$$

$$= \sum_k \langle x, R^{2k}\tilde{f}_L \rangle (R^{2k}g_L)(n) + \sum_k \langle x, R^{2k}\tilde{f}_H \rangle (R^{2k}g_H)(n). \qquad (2.55)$$

For $x \in \mathbb{C}^N$, the index k in (2.54) and (2.55) runs over $\mathbb{Z}_{N/2}$. In view of the decomposition (2.55) of x and the orthonormality of $\tilde{B}$, it is not surprising that $g_L = \tilde{f}_L$ and $g_H = \tilde{f}_H$. Equation (2.55), along with the equations $g_L = \tilde{f}_L$ and $g_H = \tilde{f}_H$, demonstrates the *completeness* of $\tilde{B}$. These three equations show that every vector $x \in l^2(\mathbb{Z})$ or $\mathbb{C}^N$ can be written as a sum of the elements of $\tilde{B}$. The sequences $\tilde{f}_L$ and $\tilde{f}_H$ are *first-generation wavelets*.

It is worth remarking that if we do not assume the orthonormality of $\tilde{B}$, it is possible for (2.53)–(2.55) to hold with $g_L \neq \tilde{h}_L$ and $g_H \neq \tilde{h}_H$. This is the first step in the development of "biorthogonal wavelets." We refer to [C] and [CDFe] for further information on the subject.

In summary, we have proved:

THEOREM 2.7
The following are equivalent:

P1 *The set $\tilde{B} = \{R^{2k}\tilde{f}_L\}_k \cup \{R^{2k}\tilde{f}_H\}_k$ is orthonormal, and we have perfect reconstruction in Figure 2.1.*

P2 *The system matrix $A(z)$ in (2.30) is unitary for all $z \in \mathbf{T}$ or $z \in \mathbf{W}_N$, $g_L = \tilde{f}_L$, and $g_H = \tilde{f}_H$.*

2.3 The design of wavelet filters

In this section we discuss the design of the analysis filters f_L and f_H, and of the synthesis filters g_L and g_H, in the signal spaces $l^2(\mathbb{Z})$ and $\mathbb{C}^N$. The requirements of orthonormality determine the unitarity of the system matrix $A(z)$ by Theorem 2.5. By the unitarity condition the norm of the first column of the system matrix $A(z)$ must be unity. Therefore,

$$|\hat{f}_L(z)|^2 + |\hat{f}_L(-z)|^2 = 2. \qquad (2.56)$$

From (2.56) and Lemma 2.3, $|\hat{f}_L(z)|^2$ is almost-odd. Then $|\hat{f}_L(z)|^2 = 1 + \hat{h}(z)$, where $\hat{h}$ is some odd function of z. Moreover $\hat{h}(z)$ is real valued because $|\hat{f}_L(z)|^2 - 1$ is. As the squared modulus of a complex number, $|\hat{f}_L(z)|^2$ is bounded below by zero for all $z \in \mathbf{T}$. From (2.56), $|\hat{f}_L(z)|^2$ is also bounded above: $|\hat{f}_L(z)|^2 \le 2$, $\forall z \in \mathbf{T}$ or $\mathbf{W}_N$. Then $-1 \le \hat{h}(z) \le 1$, $\forall z \in \mathbf{T}$ or $\mathbf{W}_N$. These observations yield a recipe for the construction of $\hat{f}_L(z)$.

Let $\hat{h}(z)$ be any real-valued function defined upon the complex unit circle $\mathbf{T}$ or the roots $\mathbf{W}_N$ of unity, such that $\hat{h}(z) = -\hat{h}(-z)$, and $-1 \le \hat{h}(z) \le 1$. Let $\rho(z)$ be another arbitrary real-valued function defined on $\mathbf{T}$ or $\mathbf{W}_N$. Define

$$\hat{f}_L(z) = \sqrt{1 + \hat{h}(z)}e^{j\rho(z)}. \tag{2.57}$$

The form of $\hat{f}_L(z)$ in (2.56) is the most general possible. Since $\rho(z)$ need not be a polynomial in z, we do not write $\hat{\rho}(z)$.

Because $A(z)$ is unitary, so is $A^T(z)$. Because the norm of the first column of $A^T(z)$ must be unity, $|\hat{f}_L(z)|^2 + |\hat{f}_H(z)|^2 = 2$. Then $\hat{f}_H(z)$ must have the form:

$$\hat{f}_H(z) = \sqrt{1 - \hat{h}(z)}e^{j\sigma(z)}, \tag{2.58}$$

where $\sigma(z)$ is a real-valued function of z.

Substituting (2.57) and (2.58) into the system matrix (2.30), and using the oddness of $\hat{h}(z)$, we have:

$$A(z) = \frac{1}{\sqrt{2}} \begin{pmatrix} \sqrt{1 + \hat{h}(z)}e^{j\rho(z)} & \sqrt{1 - \hat{h}(z)}e^{j\sigma(z)} \\ \sqrt{1 - \hat{h}(z)}e^{j\rho(-z)} & \sqrt{1 + \hat{h}(z)}e^{j\sigma(-z)} \end{pmatrix}. \tag{2.59}$$

Theorem 2.5 also requires the orthogonality of the two column vectors in $A^T(z)$. In other words, the inner product of the column vectors in $A^T(z)$ must vanish, then

$$\sqrt{1 - \hat{h}^2(z)} \left(e^{j(\rho(z) - \rho(-z))} + e^{j(\sigma(z) - \sigma(-z))} \right) = 0. \tag{2.60}$$

If $\hat{h}(z) \ne \pm 1$, then (2.60) requires that:

$$\sigma(z) - \sigma(-z) = (2k + 1)\pi + \rho(z) - \rho(-z), \tag{2.61}$$

for some $k \in \mathbb{Z}$, k depending possibly on z. If $\hat{h}(z) = \pm 1$, then $\sigma(z)$ and $\rho(z)$ are unconstrained.

We have shown that if $A(z)$ is unitary then $A(z)$ has the form (2.59) for $\hat{h}(z)$, $\sigma(z)$, and $\rho(z)$, as described. Conversely, any such matrix is easily seen to be unitary. Finally, the synthesis filter sequences g_L and g_H must be chosen so as to satisfy Theorem 2.7. The three steps **S1**, **S2**, and **S3**, in the construction of the filter sequences f_L, f_H, g_L, and g_H, are summarized below.

S1 Construct an arbitrary real-valued function $\hat{h}(z)$, $z \in \mathbf{T}$ or $\mathbf{W}_N$, such that $-1 \le \hat{h}(z) \le 1$ and $\hat{h}(z) = -\hat{h}(-z)$. Construct an arbitrary real-valued function $\rho(z)$, $z \in \mathbf{T}$ or $\mathbf{W}_N$. Define $\hat{f}_L(z) = \sqrt{1 + \hat{h}(z)}e^{j\rho(z)}$, $z \in \mathbf{T}$ or $\mathbf{W}_N$.

S2 Construct a real-valued function $\sigma(z)$, $z \in \mathbf{T}$ or $\mathbf{W}_N$, such that $\sigma(z) - \sigma(-z) = (2k + 1)\pi + \rho(z) - \rho(-z)$, if $\hat{h}(z) \neq \pm 1$. Here $k \in \mathbb{Z}$ is an arbitrary integer that possibly depends on z. Define $\hat{f}_H(z) = \sqrt{1 - \hat{h}(z)} e^{j\sigma(z)}$, $z \in \mathbf{T}$ or $\mathbf{W}_N$.

S3 Define $\hat{g}_L(z) = \overline{\hat{f}_L(z)}$; $\hat{g}_H(z) = \overline{\hat{f}_H(z)}$.

Theorem 2.7 shows that any and all wavelet filters at the first stage can be constructed with this algorithm.

2.3.1 Examples of wavelet filter construction

Let $x \in \mathbb{C}^4$. Let $\omega = e^{-j2\pi/4} = -j$; $\mathbf{W}_4 = \{\omega^0, \omega^1, \omega^2, \omega^3\} = \{1, -j, -1, j\}$. Choose

$$\hat{h}(1) = 1, \quad \hat{h}(-j) = -1, \quad \hat{h}(-1) = -1, \quad \hat{h}(j) = 1. \tag{2.62}$$

Choose $\rho(z) \equiv 0$. Then from $\hat{f}_L(z) = \sqrt{1 + \hat{h}(z)}$ we have

$$\hat{f}_L(1) = \sqrt{2}, \quad \hat{f}_L(-j) = 0, \quad \hat{f}_L(-1) = 0, \quad \hat{f}_L(j) = \sqrt{2}. \tag{2.63}$$

The filter sequence f_L itself can be computed as the inverse DFT of the vector $\mathcal{F}f_L = (\sqrt{2}, 0, 0, \sqrt{2})$: $f_L = 2^{-3/2}(2, (1 + j), 0, (1 - j))$. Knowing f_L, we can write the polynomial $\hat{f}_L(z)$:

$$\hat{f}_L(z) = \frac{1}{\sqrt{2}} + \frac{1}{2\sqrt{2}}(1 + j)z^{-1} + \frac{1}{2\sqrt{2}}(1 - j)z^{-3}. \tag{2.64}$$

In order to compute $\hat{f}_H(z)$ we need to choose the function $\sigma(z)$. Since $\hat{h}(z) = \pm 1$ for $z = 1, j, -1, -j$, the choice of $\sigma(z)$ is unconstrained by $\rho(z)$. We choose $\sigma(z) \equiv 0$. Then the DFT of f_H, computed from $\hat{f}_H(z) = \sqrt{1 - \hat{h}(z)}$, is:

$$\mathcal{F}f_H = (0, \sqrt{2}, \sqrt{2}, 0), \tag{2.65}$$

and

$$\hat{f}_H(z) = \frac{1}{\sqrt{2}} - \frac{1}{2\sqrt{2}}(1 + j)z^{-1} - \frac{1}{2\sqrt{2}}(1 - j)z^{-3}. \tag{2.66}$$

By Step 3 of the filter construction algorithm we have $\hat{g}_L(z) = \overline{\hat{f}_L(z)} = \hat{f}_L(z)$ and $\hat{g}_H(z) = \overline{\hat{f}_H(z)} = \hat{f}_H(z)$.

2.3.1.1 A second example: The Shannon wavelet basis We can generalize the last example for signal spaces $\mathbb{C}^N$, where $4 \mid N$. Let A and B be the sets

$$A = \{0, 1, \ldots, N/4 - 1\} \cup \{3N/4, \ldots, N - 1\} \tag{2.67}$$

$$B = \{N/4, \ldots, 3N/4 - 1\}. \tag{2.68}$$

Define

$$(\mathcal{F}h)(m) = \begin{cases} 1, & m \in A \\ -1, & m \in B \end{cases} \tag{2.69}$$

$$\sigma(z) = \rho(z) \equiv 0. \tag{2.70}$$

This gives:

$$(\mathcal{F}f_L)(m) = \hat{f}_L(e^{-j2\pi m/N}) = \begin{cases} \sqrt{2}, & m \in A \\ 0, & m \in B \end{cases} \tag{2.71}$$

$$(\mathcal{F}f_H)(m) = \hat{f}_H(e^{-j2\pi m/N}) = \begin{cases} 0, & m \in A \\ \sqrt{2}, & m \in B. \end{cases} \tag{2.72}$$

The filters f_H and f_L have disjoint frequency supports. We call the basis defined by this pair of filters the *Shannon wavelet basis* because the filters are similar to the sinc functions that appear in the Shannon sampling theorem.

2.3.1.2 A third example: The real Shannon wavelet basis If the signals being dealt with are real valued, then it is sometimes advantageous to have real-valued filters so that all inner products are real too. This saves computation time and storage. We can modify the last example to yield a real-valued basis. A function $f \in \mathbb{C}^N$ is real valued if and only if $(\mathcal{F}f)(m) = \overline{(\mathcal{F}f)}(N - m), \forall m \in \mathbb{Z}_N$. The filter sequences $\mathcal{F}f_L$ and $\mathcal{F}f_H$ in the last example each fail this criterion at $m = N/4$. However, Theorem 2.5 tells us how to alter f_L and f_H in order to obtain real filters.

If f_L is to be real, we require $(\mathcal{F}f_L)(N/4) = \overline{(\mathcal{F}f_L)}(3N/4) \Rightarrow \hat{f}_L(e^{-j\pi/2}) = \overline{\hat{f}_L(e^{-j3\pi/2})} \Rightarrow \hat{f}_L(-j) = \overline{\hat{f}_L(j)}$. Moreover, by Theorem 2.5 the system matrix is required to be unitary for all z in $\mathbf{W}_N$. For $z = -j$,

$$A(-j) = \frac{1}{\sqrt{2}} \begin{pmatrix} \hat{f}_L(-j) & \hat{f}_H(-j) \\ \hat{f}_L(j) & \hat{f}_H(j) \end{pmatrix}. \tag{2.73}$$

If we choose, $\hat{f}_L(-j) = j, \hat{f}_L(j) = -j$, and $\hat{f}_H(-j) = \hat{f}_H(j) = 1$, then $A(-j)$ is unitary, and $A(j)$ is too. For $\pm j \neq z \in \mathbf{W}_N$, $A(z)$ is unitary for $\hat{f}_L(z)$ and $\hat{f}_H(z)$ as defined in the last example. Therefore, the following pair of filters defines a valid wavelet basis:

$$(\mathcal{F}f_L)(m) = \hat{f}_L(e^{-j2\pi m/N})$$

$$= \begin{cases} \sqrt{2}, & 0 \le m \le N/4 - 1 \text{ or } 3N/4 + 1 \le m \le N - 1 \\ j, & m = N/4 \\ -j, & m = 3N/4 \\ 0, & N/4 + 1 \le m \le 3N/4 - 1 \end{cases} \tag{2.74}$$

$$(\mathcal{F}f_H)(m) = \hat{f}_H(e^{-j2\pi m/N})$$

$$= \begin{cases} 0, & 0 \leq m \leq N/4 - 1 \text{ or } 3N/4 + 1 \leq m \leq N - 1 \\ 1, & m = N/4 \text{ or } m = 3N/4 \\ \sqrt{2}, & N/4 + 1 \leq m \leq 3N/4 - 1. \end{cases} \quad (2.75)$$

We have $\text{supp}(\mathcal{F}f_L) \cap \text{supp}(\mathcal{F}f_H) = \{N/4, 3N/4\}$, a minimal overlap in support. We call the basis defined by this filter pair the *real Shannon wavelet basis*.

2.3.2 Time-frequency localization

Wavelets and other methods of time-frequency analysis, like the φ-transform (see, e.g., [KFuFJ]), have many practical applications. Some of these applications require that the filter sequences or "analyzing functions" be well localized simultaneously in the time (or space) and the frequency domains (see, e.g., [K]). We have seen that it is easy to construct wavelet bases that are well localized in the frequency domain: viz., the Shannon and the real Shannon bases. Now we discuss the problem of *simultaneous* time-frequency localization.

In Figure 2.2 are drawn the subspaces V_T and V_B of $V = l^2(\mathbb{Z})$ or $V = \mathbb{C}^N$. The subspace V_T consists of all sequences $x \in V$, with a fixed $\text{supp}(x)$, which is some proper subset of V. The subspace V_B consists of all sequences $x \in V$ with a fixed support for $\mathcal{F}x$. $P_T : V \to V_T$ and $P_B : V \to V_B$ are projection operators.

Let $V_S \subseteq V$ be the set of all valid wavelets f_L and f_H. It is easy to check that V_S is neither a subspace nor an affine space of V, hence any mapping $P_S : V \to V_S$ cannot be a projection. V_S is in fact a manifold that lies embedded in the surface of the unit sphere in $l^2(\mathbb{Z})$ or $\mathbb{C}^N$. Define $P_S : V \to V_S$ to be an operator that maps

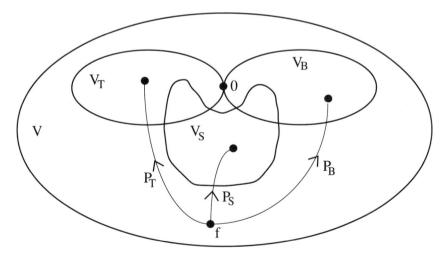

FIGURE 2.2
The signal space V, the subspaces V_T and V_B, the manifold V_S, and mappings from V into V_T, V_B, and V_S.

any given $f \in V$ to $\psi = P_S f \in V_S$, such that distance $\|f - \psi\|_V = \langle f - \psi, f - \psi \rangle^{1/2}$ between f and ψ is a minimum. We now construct this operator for $V = \mathbb{C}^N$.

Given any $f \in \mathbb{C}^N$ we would like to find a wavelet ψ that is closest to f. If ψ is a wavelet that is closest to f, then $\|f - \psi\|^2_{\mathbb{C}^N}$ is a minimum.

$$\|f - \psi\|^2_{\mathbb{C}^N} = \frac{1}{N} \|(\mathcal{F}f) - (\mathcal{F}\psi)\|^2_{\mathbb{C}^N} = \frac{1}{N} \sum_{m=0}^{N-1} |(\mathcal{F}f)(m) - (\mathcal{F}\psi)(m)|^2 \quad (2.76)$$

$$= \frac{1}{N} \sum_{m=0}^{N/2-1} \left(|(\mathcal{F}f)(m) - (\mathcal{F}\psi)(m)|^2 \right.$$

$$\left. + |(\mathcal{F}f)(m + N/2) - (\mathcal{F}\psi)(m + N/2)|^2 \right). \quad (2.77)$$

For $m \in \mathbb{Z}_{N/2}$; define

$$A(m) = |(\mathcal{F}f)(m) - (\mathcal{F}\psi)(m)|^2 + |(\mathcal{F}f)(m + N/2) - (\mathcal{F}\psi)(m + N/2)|^2. \quad (2.78)$$

From (2.76)–(2.78),

$$\|f - \psi\|^2_{\mathbb{C}^N} = \frac{1}{N} \sum_{m=0}^{N/2-1} A(m). \quad (2.79)$$

In order to minimize $\|f - \psi\|^2_{\mathbb{C}^N}$, we minimize $A(m)$ for each $m \in \mathbb{Z}_{N/2}$.

In order for ψ to be a valid wavelet filter f_L or f_H, the following equation must hold for $m \in \mathbb{Z}_N$:

$$(\mathcal{F}\psi)(m) = \sqrt{1 + (\mathcal{F}h)(m)} e^{j\rho(e^{-j2\pi m/N})}, \quad (2.80)$$

by the conditions set forth in step **S1** of the wavelet construction algorithm. Also it must be that $(\mathcal{F}h)(m + N/2) = -(\mathcal{F}h)(m) \in [-1, 1]$, and $\rho(e^{-j2\pi m/N})$ real, for $m \in \mathbb{Z}_N$.

Define $a(m)$, $b(m)$, $\theta(m)$, and $\gamma(m)$ by

$$(\mathcal{F}f)(m) = a(m)e^{j\theta(m)}; \quad m = 0, \ldots, (N/2 - 1) \quad (2.81)$$

$$(\mathcal{F}f)(m + N/2) = b(m)e^{j\gamma(m)}; \quad m = 0, \ldots, (N/2 - 1), \quad (2.82)$$

where $a(m), b(m) \in \mathbb{R}$; $a(m), b(m) \geq 0$; and $\theta(m), \gamma(m) \in [-\pi, \pi]$ for $m \in \mathbb{Z}_{N/2}$. From (2.78) and (2.80)–(2.82), for $m = 0, \ldots, (N/2 - 1)$,

$$A(m) = \left| a(m)e^{j\theta(m)} - \sqrt{1 + (\mathcal{F}h)(m)} e^{j\rho(e^{-j2\pi m/N})} \right|^2 +$$

$$\left| b(m)e^{j\gamma(m)} - \sqrt{1 - (\mathcal{F}h)(m)} e^{j\rho(-e^{-j2\pi m/N})} \right|^2. \quad (2.83)$$

If $a(m) = b(m) = 0$ for some $m = m_0$ then (2.83) tells us that $A(m_0) = 2$, and we have all the freedom we want in the choice of $(\mathcal{F}h)(m_0) \in [-1, 1]$, $\rho(e^{-j2\pi m_0/N}) \in [-\pi, \pi]$, and $\rho(-e^{-j2\pi m_0/N}) \in [-\pi, \pi]$. It follows that there may be infinitely many wavelets ψ that are closest to a given $f \in V$.

Now assume that $a(m)$ and $b(m)$ are not both zero. From (2.83) we note that, in order to minimize $A(m)$, we want $\rho(e^{-j2\pi m/N}) = \theta(m)$ and $\rho(-e^{-j2\pi m/N}) = \gamma(m)$. Then (2.83) reduces to:

$$A(m) = 2 + a^2(m) + b^2(m) - 2\left(a(m)\sqrt{1 + (\mathcal{F}h)(m)} + b(m)\sqrt{1 - (\mathcal{F}h)(m)}\right). \quad (2.84)$$

In (2.84) $a(m)$ and $b(m)$ are fixed, and we wish to choose $(\mathcal{F}h)(m) \in [-1, 1]$ so as to minimize $A(m)$. This will be done if we choose $(\mathcal{F}h)(m)$ so as to maximize $a(m)\sqrt{1 + (\mathcal{F}h)(m)} + b(m)\sqrt{1 - (\mathcal{F}h)(m)}$.

By calculus, if $a, b \geq 0$ are fixed and are not both zero, then the function $f(x) = a\sqrt{1 + x} + b\sqrt{1 - x}$ attains a maximum on $[-1, 1]$ at $x = (a^2 - b^2)/(a^2 + b^2)$. Hence, if not both $a(m)$ and $b(m)$ are zero, then $a(m)\sqrt{1 + (\mathcal{F}h)(m)} + b(m)\sqrt{1 - (\mathcal{F}h)(m)}$ is maximized (and $A(m)$ minimized) by the following choice of $(\mathcal{F}h)(m) \in [-1, 1]$, $m \in \mathbb{Z}_{N/2}$:

$$(\mathcal{F}h)(m) = \frac{a^2(m) - b^2(m)}{a^2(m) + b^2(m)} = \frac{|(\mathcal{F}f)(m)|^2 - |(\mathcal{F}f)(m + N/2)|^2}{|(\mathcal{F}f)(m)|^2 + |(\mathcal{F}f)(m + N/2)|^2}; \quad m \in \mathbb{Z}_{N/2}. \quad (2.85)$$

Hence the DFT coefficient $(\mathcal{F}\psi)(m)$, $m \in \mathbb{Z}_{N/2}$, of the wavelet ψ that is closest to $f \in V$ is

$$(\mathcal{F}\psi)(m) = \sqrt{1 + (\mathcal{F}h)(m)}\, e^{j\theta(m)} = \frac{\sqrt{2}\, a(m) e^{j\theta(m)}}{\sqrt{a^2(m) + b^2(m)}} \quad (2.86)$$

$$= \frac{\sqrt{2}(\mathcal{F}f)(m)}{\sqrt{|(\mathcal{F}f)(m)|^2 + |(\mathcal{F}f)(m + N/2)|^2}}, \quad m \in \mathbb{Z}_{N/2}. \quad (2.87)$$

For $N/2 \leq m < N$, step **S1** in the wavelet construction algorithm tell us that $(\mathcal{F}h)(m) = -(\mathcal{F}h)(m - N/2)$. Then from (2.85), for $N/2 \leq m < N$,

$$(\mathcal{F}h)(m) = -(\mathcal{F}h)(m - N/2) = -\frac{|(\mathcal{F}f)(m - N/2)|^2 - |(\mathcal{F}f)(m)|^2}{|(\mathcal{F}f)(m - N/2)|^2 + |(\mathcal{F}f)(m)|^2}. \quad (2.88)$$

By $\mathbb{Z}_N$ arithmetic upon indices into $(\mathcal{F}f)$, $(\mathcal{F}f)(m - N/2) = \mathcal{F}(m + N/2)$, and we note from (2.85) and (2.88) that (2.85) and (2.87) hold not only for $m \in \mathbb{Z}_{N/2}$, but for the full range $m \in \mathbb{Z}_N$.

In summary, we present the following simple three-step algorithm for the construction of a wavelet ψ that is closest to a given vector $f \in \mathbb{C}^N, f \neq 0$:

S1 If $(\mathcal{F}f)(m) = (\mathcal{F}f)(m + N/2) = 0$, then assign any value $C \in [0, \sqrt{2}]$ to $|(\mathcal{F}\psi)(m)|$. To $|(\mathcal{F}\psi)(m + N/2)|$ assign $\sqrt{2 - C^2}$. To each of the two phase terms $\rho(e^{-j2\pi m/N})$ and $\rho(-e^{-j2\pi m/N})$ of $(\mathcal{F}\psi)(m)$ and $(\mathcal{F}\psi)(m + N/2)$ assign any value in $[-\pi, \pi]$.

S2 If not both $(\mathcal{F}f)(m)$ and $(\mathcal{F}f)(m + N/2)$ vanish, then assign

$$(\mathcal{F}\psi)(m) = \frac{\sqrt{2}(\mathcal{F}f)(m)}{\sqrt{|(\mathcal{F}f)(m)|^2 + |(\mathcal{F}f)(m + N/2)|^2}}. \quad (2.89)$$

S3 From $\mathcal{F}\psi$ compute ψ by DFT inversion.

This algorithm is an operational description of the operator $P_S : V \to V_S$, $V = \mathbb{C}^N$. A similar description of P_S is possible for the signal space $l^2(\mathbb{Z})$.

We note from the above algorithm that if $\mathcal{F}f$ is real valued, then we can always find a wavelet ψ closest to f such that $\mathcal{F}\psi$ is also real-valued.

In case of φ-transform analyzing functions [KFuFJ], time-frequency localization involves the computation of an eigenvector of the double projection operator $(P_B P_T)$. This eigenvector may not lie in V_S and may not be a valid wavelet. One approach to the construction of a localized wavelet may be to map the eigenvector ϕ of $(P_B P_T)$ into V_S using P_S. Let ψ be the image of ϕ under P_S. If the support of ϕ is not too severely restricted in the frequency domain, i.e., if not both $(\mathcal{F}\phi)(m)$ and $(\mathcal{F}\phi)(m+N/2)$ vanish for any m, then (2.89) tells us that ψ will preserve, precisely, the frequency localization of f. If the support of $\mathcal{F}\phi$ *is* severely restricted then we can still, to some extent, control the support of $\mathcal{F}\psi$ through a judicious choice of the constants C in **S1**. It is not, however, possible to say anything about the time-localization of ψ on the basis of the analysis in this section.

It is possible to choose ϕ such that both ϕ and $\mathcal{F}\phi$ are real valued. Then we can produce a wavelet ψ closest to ϕ that is also real valued in both the time and the frequency domains.

Define $(\mathcal{F}g)(m) = \sqrt{2} \left(|(\mathcal{F}\phi)(m)|^2 + |(\mathcal{F}\phi)(m+N/2)|^2 \right)^{-1/2}$. Then, from (2.89), $\psi = \phi * g$. It follows that if ϕ and g are well localized in time, then ψ will be also. The function ϕ is well localized by design, and that leaves us with questions concerning the time-localization of g. This question remains open.

2.4 Wavelet recursion for one-dimensional signals

If in Figure 2.1 we subject the sequence v_L to the same treatment to which x was subjected, then we obtain the situation depicted in Figure 2.3. If we decompose and reconstruct both v_L and v_H, we have the situation in Figure 2.4. Like the level–1 filters f_L, f_H, g_L, and g_H; the level–2 filters $f_{LL}, f_{LH}, f_{HL}, f_{HH}, g_{LL}, g_{LH}, g_{HL}$,

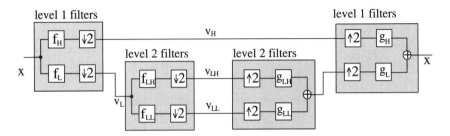

FIGURE 2.3
Recursion in the lower branch.

and g_{HH} must also obey the requirements set forth in Theorem 2.7 if we require orthogonal decomposition and perfect reconstruction. The matrices

$$\frac{1}{\sqrt{2}} \begin{pmatrix} \hat{f}_{LL}(z) & \hat{f}_{LH}(z) \\ \hat{f}_{LL}(-z) & \hat{f}_{LH}(-z) \end{pmatrix}, \text{ and } \frac{1}{\sqrt{2}} \begin{pmatrix} \hat{f}_{HL}(z) & \hat{f}_{HH}(z) \\ \hat{f}_{HL}(-z) & \hat{f}_{HH}(-z) \end{pmatrix},$$

must be unitary; and the reconstruction filters g_{LL}, g_{LH}, g_{HL}, and g_{HH} must equal $\tilde{f}_{LL}, \tilde{f}_{LH}, \tilde{f}_{HL}$, and $\tilde{f}_{HH}$, respectively. The process can be repeated for v_{LL}, v_{LH}, v_{HL}, or v_{HH}.

2.4.1 Wavelets, filter banks, and wavelet packets

In classical wavelet analysis, recursion is performed only in the lower-most branch. Then wavelet decomposition generates the sequence of graphs in Figure 2.5, where each node represents a single filter pair with their associated up-sample-by-2 or down-sample-by-2 operators. The number of filters at each level of analysis and synthesis is constant.

In case of full recursion at every level, we have the sequence of graphs in Figure 2.6. The number of filters doubles at each level of analysis and synthesis. This is the approach adopted by the filter bank school. Yet a third approach to recursion is that of "best-adapted wavelet-packets" pioneered by Coifman, Meyer, Quake, and Wickerhauser (see, e.g., [CoMW]), and characterized by an arrangement of the sort in Figure 2.7. In the best-adapted wavelet-packet method recursion is or is not performed at a certain level in the transformation tree depending upon some criterion of the optimality of representation.

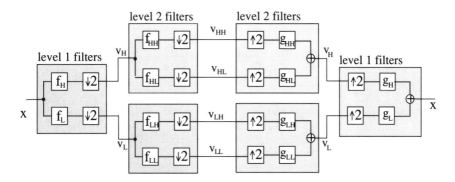

FIGURE 2.4
Recursion in both branches.

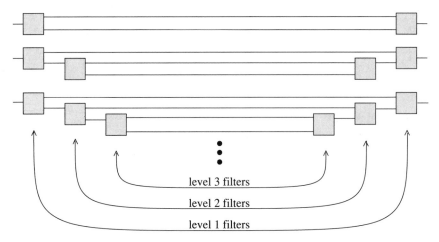

FIGURE 2.5
The sequence of graphs generated in classical wavelet analysis.

2.4.2   Equivalent non-recursive structures

The recursive structures in Figures 2.3 and 2.4 have nonrecursive equivalents that can be constructed with the help of the theorem stated and proved in this subsection. A study of the equivalent nonrecursive structures deepens our understanding of the nature of wavelet-based signal decomposition.

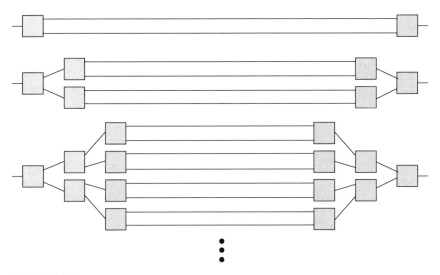

FIGURE 2.6
The sequence of graphs generated in classical filter-bank analysis.

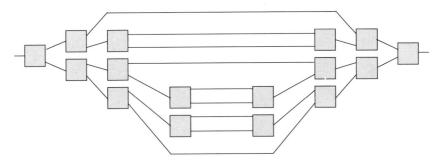

FIGURE 2.7
Irregular recursion is characteristic of wavelet-packet representations.

Define the "bigstar" notation $\bigstar$ to do for convolution what the bigcup notation does for union:

$$\bigstar_{i=n_1}^{n_2} h_i = h_{n_1} * h_{n_1+1} * \ldots * h_{n_2}. \tag{2.90}$$

Let $\alpha \in \{L, H\}^*$ be a string. For $|\alpha| > 0$, define

$$F_\alpha = \bigstar_{i=1}^{|\alpha|} \mathcal{U}^{i-1} f_{\text{pfx}(\alpha,i)} \tag{2.91}$$

$$G_\alpha = \bigstar_{i=1}^{|\alpha|} \mathcal{U}^{i-1} g_{\text{pfx}(\alpha,i)}. \tag{2.92}$$

Here is an example to illustrate the notation: $F_{HLL} = f_H * (\mathcal{U} f_{HL}) * (\mathcal{U}^2 f_{HLL})$. The next theorem shows that the filters F_α and G_α can be used in place of a series of filters; so that, for example, the nonrecursive structure in Figure 2.8 is equivalent to the recursive structure in Figure 2.3. The analyzing filters in the nonrecursive structure are F_{LL}, F_{LH} and F_H, while the synthesizing filters are G_{LL}, G_{LH} and G_H.

For the infinite-dimensional case define $X_i = \mathbb{Z}$, $i \in \mathbb{Z}^+$. For the finite-dimensional case define $X_i = \mathbb{Z}_{N2^{1-i}}$, $i \in \mathbb{Z}^+$. With X_i so defined, the following

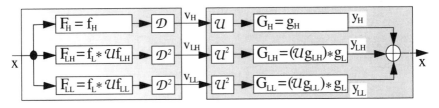

FIGURE 2.8
Direct decomposition into three orthogonal subspaces.

lemma and theorem hold simultaneously for both finite and infinite-dimensional signal spaces.

LEMMA 2.8
For the finite-dimensional case assume $2^n \mid N$. *Let* $i \in \{1, \ldots, n\}$. *Then,*

(a) *For* $f \in l^2(X_i)$ *and* $h \in l^1(X_{i+1})$, $(\mathcal{D}f) * h = \mathcal{D}(f * \mathcal{U}h) \in l^2(X_{i+1})$.

(b) *For* $f \in l^2(X_{i+1})$ *and* $h \in l^1(X_{i+1})$, $\mathcal{U}(f * h) = (\mathcal{U}f) * (\mathcal{U}h) \in l^2(X_i)$.

PROOF OF LEMMA 2.8

(a) For $n \in X_{i+1}$,

$$(\mathcal{D}(f * \mathcal{U}h))(n) = (f * \mathcal{U}h)(2n) = \sum_{m \in X_i} f(2n - m)(\mathcal{U}h)(m) = \sum_{k \in X_{i+1}} f(2n - 2k)h(k) \tag{2.93}$$

$$= \sum_{k \in X_{i+1}} (\mathcal{D}f)(n - k)h(k) = ((\mathcal{D}f) * h)(n). \tag{2.94}$$

(b) Suppose $n \in X_i$. If $n = 2m$ for some $m \in X_{i+1}$, then

$$(\mathcal{U}(f * h))(n) = (f * h)(m) = \sum_{k \in X_{i+1}} f(m - k)h(k) = \sum_{k \in X_{i+1}} (\mathcal{U}f)(2m - 2k)(\mathcal{U}h)(2k) \tag{2.95}$$

$$= \sum_{l \in X_i} (\mathcal{U}f)(n - l)(\mathcal{U}h)(l) = ((\mathcal{U}f) * (\mathcal{U}h))(n). \tag{2.96}$$

Equation (2.96) follows from (2.95) because $(\mathcal{U}h)(l) = 0$ unless $l = 2k$ for some $k \in X_{i+1}$.

If n is not of the form $2m$, $m \in X_{i+1}$, then $(\mathcal{U}(f * h))(n) = 0$. However, in this case, $((\mathcal{U}f) * (\mathcal{U}h))(n) = \sum_{l \in X_i}(\mathcal{U}f)(n - l)(\mathcal{U}h)(l)$ is zero since $(\mathcal{U}h)(l) \neq 0$ only when $l = 2k$, some $k \in X_{i+1}$, in which case $(\mathcal{U}f)(n - l) = 0$. ∎

THEOREM 2.9
Let $\alpha \in \{L, H\}^*$, $|\alpha| = n > 0$, *be fixed. Let* $2^n \mid N$. *Define* $\beta_i = \mathrm{pfx}(\alpha, i)$, $i = 1, \ldots, n$.

(a) *Suppose* $x \in l^2(X_1)$ *and* $f_{\beta_i} \in l^1(X_i)$. *Then*

$$\mathcal{D}(f_{\beta_n} * \mathcal{D}(f_{\beta_{n-1}} * \ldots * \mathcal{D}(f_{\beta_2} * \mathcal{D}(f_{\beta_1} * x)) \ldots))$$

$$= \mathcal{D}^n \left(x * \left(\underset{i=1}{\overset{n}{\ast}} \left(\mathcal{U}^{i-1} f_{\beta_i} \right) \right) \right) \tag{2.97}$$

$$= \mathcal{D}^n (x * F_\alpha). \tag{2.98}$$

(b) *Suppose* $v_\alpha \in l^2(X_{n+1})$ *and* $g_{\beta_i} \in l^1(X_i)$. *Then*

$$g_{\beta_1} * (\mathcal{U}(g_{\beta_2} * \ldots * \mathcal{U}(g_{\beta_{n-1}} * (\mathcal{U}(g_{\beta_n} * \mathcal{U}v_\alpha)) \ldots))$$

$$= (\mathcal{U}^n v_\alpha) * \left(\underset{i=1}{\overset{n}{\ast}} \left(\mathcal{U}^{i-1} g_{\beta_i} \right) \right) \tag{2.99}$$

$$= (\mathcal{U}^n v_\alpha) * G_\alpha. \tag{2.100}$$

Each part of Theorem 2.9 follows easily by the repeated use of the corresponding part of Lemma 2.8. This theorem explains the method for determining the nonrecursive equivalent of a recursive structure. In particular, it establishes the equivalence of Figure 2.3 and Figure 2.8.

2.4.3 The wavelet basis

The knowledge of equivalent nonrecursive filters makes it easy to determine the basis vectors whose inner products with the signal yield the transform-domain representation of the signal.

Let $\alpha \in \{L, H\}^*$ be a string. Define

$$\tilde{B}_\alpha = \left\{ R^{2^{|\alpha|}k} \tilde{F}_\alpha \right\}_{k \in X_{|\alpha|+1}} \tag{2.101}$$

$$V_\alpha = \mathrm{cl} \left(\mathrm{span} \left(\tilde{B}_\alpha \right) \right). \tag{2.102}$$

The definition of $\tilde{B}_\alpha$ in (2.101) is consistent with the definitions of $\tilde{B}_L$ and $\tilde{B}_H$ in (2.12) and (2.13). $\tilde{B}_\alpha$ consists of time-shifted copies of the conjugate time-reversed nonrecursive equivalent filter $\tilde{F}_\alpha$. V_α is defined to be the subspace of $V = l^2(\mathbb{Z})$ or $\mathbb{C}^N$ spanned by $\tilde{B}_\alpha$. We have seen in Section 2.2 that $\tilde{B}_L, \tilde{B}_H$ form the bases for $V_L, V_H \subseteq V$. We now show that recursion splits a subspace V_α into an orthogonal direct sum of subspaces $V_{\alpha L}$ and $V_{\alpha H}$.

THEOREM 2.10 The Iteration Theorem
Let $\alpha \in \{L, H\}^*$, $|\alpha| > 0$, be fixed. In the finite-dimensional case let $2^{(|\alpha|+1)} \mid N$. Let $\beta \in \{L, H\}$. Suppose that $\tilde{B}_\alpha$ is an orthonormal basis for V_α, $V_\alpha \subseteq l^2(X_1)$. Suppose that $f_{\alpha L}, f_{\alpha H} \in l^1(X_{|\alpha|+1})$ are chosen such that the matrix

$$\frac{1}{\sqrt{2}} \begin{pmatrix} \hat{f}_{\alpha L}(z) & \hat{f}_{\alpha H}(z) \\ \hat{f}_{\alpha L}(-z) & \hat{f}_{\alpha H}(-z) \end{pmatrix} \tag{2.103}$$

is unitary for $\forall z \in \mathbf{T}$ or $\mathbf{W}_{N2^{-|\alpha|}}$ (as in the statement of Theorem 2.5). Let

$$\tilde{F}_{\alpha\beta} = \tilde{F}_\alpha * \mathcal{U}^{|\alpha|} \tilde{f}_{\alpha\beta}. \tag{2.104}$$

Let $\tilde{B}_{\alpha\beta}$ and $V_{\alpha\beta}$ be as defined in (2.101) and (2.102). Then V_α is the orthogonal direct sum of $\{V_{\alpha\beta}\}_\beta$, $V_\alpha = V_{\alpha L} \oplus V_{\alpha H}$. Also $\tilde{B}_{\alpha\beta}$ is an orthonormal basis for $V_{\alpha\beta}$.

In Theorem 2.10 we may regard $\tilde{F}_\alpha$ as being determined by the filters $\{f_\gamma\}_\gamma$ where γ is some prefix of α; but this is not essential. This theorem will be proved in its multidimensional form as Theorem 2.23.

In any recursive structure that is characterized by a set of analysis or synthesis branches labeled with distinct strings α, the nonrecursive equivalent filters $\{F_\alpha\}_\alpha$, taken together, are said to generate the *wavelet basis*

$$\tilde{B} = \bigcup_\alpha \tilde{B}_\alpha = \bigcup_\alpha \left\{ R^{2^{|\alpha|}k} \tilde{F}_\alpha \right\}_{k \in X_{|\alpha|+1}} \tag{2.105}$$

of the signal space $V = l^2(\mathbb{Z})$ or $\mathbb{C}^N$. For example, in Figures 2.3 and 2.8 the wavelet basis is $\tilde{B} = \tilde{B}_{LL} \cup \tilde{B}_{LH} \cup \tilde{B}_H$. We show below that the sequences v_α consists of numbers called *wavelet transform coefficients* that are the inner products of the signal x computed against the vectors in $\tilde{B}_\alpha$. The numbers in $\{v_\alpha\}_\alpha$ completely determine the signal x. We show also that y_α's are linear combinations of the vectors in $\tilde{B}_\alpha$, and the weighting coefficients are the elements of v_α. Then, by (2.102), $y_\alpha \in V_\alpha$.

If x is an input signal and if the filters along a recursion path are $f_{\beta_1}, \ldots, f_{\beta_{|\alpha|}}$ in the terminology of Theorem 2.9, then the analysis or transformation part consists of the computation of the following sequences: first $x * f_{\beta_1}$, then $\mathcal{D}(x * f_{\beta_1})$, then $f_{\beta_2} * (\mathcal{D}(x * f_{\beta_1}))$, then $\mathcal{D}(f_{\beta_2} * (\mathcal{D}(x * f_{\beta_1})))$, and so on. The result after $|\alpha|$ filtering and downsampling steps is $\mathcal{D}^{|\alpha|}(x * F_\alpha)$ by (2.98). Note that for $k \in X_{|\alpha|+1}$,

$$v_\alpha(k) = \left(\mathcal{D}^{|\alpha|}(x * F_\alpha) \right)(k) = (x * F_\alpha)\left(2^{|\alpha|}k\right)$$

$$= \sum_{l \in X_1} x(l) F_\alpha \left(2^{|\alpha|}k - l \right) \tag{2.106}$$

$$= \sum_{l \in X_1} x(l) \overline{\tilde{F}_\alpha} \left(l - 2^{|\alpha|}k \right) = \sum_{l \in X_1} x(l) \overline{\left(R^{2^{|\alpha|}k} \tilde{F}_\alpha \right)}(l) \tag{2.107}$$

$$= \left\langle x, R^{2^{|\alpha|}k} \tilde{F}_\alpha \right\rangle. \tag{2.108}$$

The output v_α of the analysis part is exactly the set of coefficients in the wavelet expansion of x corresponding to the wavelet basis functions $\tilde{B}_\alpha = \{R^{2^{|\alpha|}k} \tilde{F}_\alpha\}_{k \in X_{|\alpha|+1}}$.

Let $v_\alpha = (\mathcal{D}^{|\alpha|}(x * F_\alpha))$ be the output of the analyzing step. Consider the output of the corresponding synthesizing or reconstruction step. This output is the result of the following sequence of computations:

$$v_\alpha \to \mathcal{U}v_\alpha \to g_{\beta_{|\alpha|}} * \mathcal{U}v_\alpha \to \mathcal{U}\left(g_{\beta_{|\alpha|}} * \mathcal{U}v_\alpha\right) \to g_{\beta_{|\alpha|-1}} * \mathcal{U}\left(g_{\beta_{|\alpha|}} * \mathcal{U}v_\alpha\right) \to \cdots. \tag{2.109}$$

By part B of Theorem 2.9, the final output after $|\alpha|$ steps is $(\mathcal{U}^{|\alpha|}v_\alpha) * G_\alpha$. Note that for $n \in X_1$,

$$\left((\mathcal{U}^{|\alpha|}v_\alpha) * G_\alpha \right)(n) = \sum_{l \in X_1} (\mathcal{U}^{|\alpha|}v_\alpha)(l) G_\alpha(n - l)$$

$$= \sum_{k \in X_{|\alpha|+1}} v_\alpha(k) G_\alpha\left(n - 2^{|\alpha|}k\right) \tag{2.110}$$

$$= \sum_{k \in X_{|\alpha|+1}} v_\alpha(k) \left(R^{2^{|\alpha|}k} G_\alpha \right)(n), \tag{2.111}$$

that is,

$$\left(\left(\mathcal{U}^{|\alpha|} v_\alpha \right) * G_\alpha \right) = \sum_{k \in X_{|\alpha|+1}} v_\alpha(k) \left(R^{2^{|\alpha|}k} G_\alpha \right). \qquad (2.112)$$

If we had selected $f_{\beta_1}, \ldots, f_{\beta_\alpha}$ so as to obtain orthonormal wavelets $\{R^{2^{|\alpha|}k}\tilde{F}_\alpha\}_{k \in X_{|\alpha|+1}}$, and if according to the perfect reconstruction condition we choose $g_{\beta_i} = \tilde{f}_{\beta_i}$ so that $G_\alpha = \tilde{F}_\alpha$, then from (2.108) and (2.112) the output of the α branch can be written:

$$y_\alpha = \sum_{k \in X_{|\alpha|+1}} \left\langle x, R^{2^{|\alpha|}k}\tilde{F}_\alpha \right\rangle R^{2^{|\alpha|}k}\tilde{F}_\alpha. \qquad (2.113)$$

2.4.4 Recursion with repeated filters

Consider the choice of the level–2 filters $f_{LL}, f_{LH}, f_{HL}, f_{HH}$; of the level–3 filters $f_{LLL}, f_{LLH}, f_{LHL}$, etc.; and so on. It is possible to construct the filters independently and differently at each level according to the prescription in Theorem 2.5. It is possible also to derive the filters at levels higher than the first from the level–1 filters f_L and f_H.

In case of filters $f \in l^1(\mathbb{Z})$, we can clearly choose $f_{\alpha L} = f_L$ and $f_{\alpha H} = f_H$, where α is some string over the alphabet $\{L, H\}$. This does not quite make sense in the case of finite signals since, for example, if f_L lies in $\mathbb{C}^N$ then $f_{\alpha L}$ lies in $\mathbb{C}^{N2^{-|\alpha|}}$. If f_L and f_H are such that $A(z)$ in (2.30) is unitary $\forall z \in \mathbf{W}_N$, then (2.103) will be unitary $\forall z \in \mathbf{W}_{N2^{-|\alpha|}}$ if we can determine $f_{\alpha L}$ and $f_{\alpha H}$ so that $\hat{f}_{\alpha L}(z) = \hat{f}_L(z)$ and $\hat{f}_{\alpha H}(z) = \hat{f}_H(z)$ for all $z \in \mathbf{W}_{N2^{-|\alpha|}}$. The next two lemmas tells us how to do this.

LEMMA 2.11 The Folding Lemma
Let $f \in \mathbb{C}^N$, $h \in \mathbb{C}^{N/2}$. Then the following are equivalent:

P1 $\hat{h}(z) = \hat{f}(z)$, $\forall z \in \mathbf{W}_{N/2}$.

P2 $(\mathcal{F}h)(m) = (\mathcal{F}f)(2m)$, $\forall m \in \mathbb{Z}_{N/2}$.

P3 $h(n) = f(n) + f(n + N/2)$, $\forall n \in \mathbb{Z}_{N/2}$.

The folding lemma will be proved in a general form as Lemma 2.24 when we discuss the wavelet decomposition of multidimensional signals. This lemma indicates that we can obtain $f_{LL} \in \mathbb{C}^{N/2}$ from $f_L \in \mathbb{C}^N$, for example, by "folding" f_L; i.e., by breaking f_L into two halves, placing one half on top of the other, and summing pairwise: $f_{LL}(n) = f_L(n) + f_L(n + N/2)$. Similarly, to obtain f_{LH} we compute $f_H(n) + f_H(n + N/2)$. If $4 \mid N$, then we can continue by defining, for example, $f_{HHL} \in \mathbb{C}^{N/4}$ by $f_{HHL}(n) = f_{HL}(n) + f_{HL}(n + N/4) = [f_L(n) + f_L(n + N/2)] + [f_L(n + N/4) + f_L(n + 3N/4)]$, for $n \in \mathbb{Z}_{N/4}$. The following lemma is immediate from Lemma 2.11.

LEMMA 2.12
Let $2^{|\alpha|} \mid N$, $f \in \mathbb{C}^N$, $h \in \mathbb{C}^{N2^{-|\alpha|}}$. Then the following are equivalent:

P1 $\hat{h}(z) = \hat{f}(z), \forall z \in \mathbf{W}_{N2^{-|\alpha|}}$.

P2 $(\mathcal{F}h)(m) = (\mathcal{F}f)(2^{|\alpha|}m), \forall m \in \mathbb{Z}_{N2^{-|\alpha|}}$.

P3 $h(n) = \sum_{i=0}^{[2^{|\alpha|}-1]} f(n + iN2^{-|\alpha|}), \forall n \in \mathbb{Z}_{N2^{-|\alpha|}}$.

2.4.5 A summary theorem for classical wavelet analysis

The following theorem summarizes the results of this section for the case of *classical* wavelet analysis, where recursion is performed only in the low-pass branch using repeated filters. In this special case the sequence of nonrecursive analyzing filters is $F_H, F_{LH}, F_{LLH}, \ldots, F_{\alpha H}, F_{\alpha L}$, where $\alpha \in \{L\}^*$.

THEOREM 2.13

Suppose $N \in \mathbb{Z}$, $\alpha \in \{L\}^*$, $2^{(|\alpha|+1)} \mid N$, and β is some prefix of α, $0 \le |\beta| \le |\alpha|$. Suppose $f_L, f_H \in l^1(\mathbb{Z})$ or $\mathbb{C}^N$ are such that the system matrix in (2.30) is unitary for all $z \in \mathbf{T}$ or $\mathbf{W}_N$. Suppose that $F_{\alpha L}$ and $F_{\beta H}$ satisfy

$$\hat{F}_{\alpha L}(z) = \prod_{i=1}^{|\alpha|+1} \hat{f}_L\left(z^{2^{(i-1)}}\right) \tag{2.114}$$

$$\hat{F}_{\beta H}(z) = \hat{f}_H\left(z^{2^{|\beta|}}\right) \prod_{i=1}^{|\beta|} \hat{f}_L\left(z^{2^{(i-1)}}\right). \tag{2.115}$$

Then

$$\tilde{B} = \tilde{B}_{\alpha L} \cup \left(\bigcup_{\substack{\beta \\ 0 \le |\beta| \le |\alpha|}} \tilde{B}_{\beta H} \right) \tag{2.116}$$

is an orthonormal wavelet basis for $V = l^2(\mathbb{Z})$ or $\mathbb{C}^N$.

PROOF OF THEOREM 2.13   Define $f_{\alpha L} \in l^1(X_{|\alpha|+1})$ and $f_{\beta H} \in l^1(X_{|\beta|+1})$ such that $\hat{f}_{\alpha L}(z) = \hat{f}_L(z)$ and $\hat{f}_{\beta H}(z) = \hat{f}_H(z)$. Then (by Lemma 2.12 and (2.37)) (2.114) and (2.115) are consistent with (2.91). The result follows by an induction argument based on the iteration theorem, where we consistently factor the subspace $V_{\beta L}$ through recursion. ∎

In the finite-dimensional case, the following expressions for the Fourier transforms of $F_{\alpha L}$ and $F_{\beta H}$ are immediate from the equations (2.114) and (2.115):

$$(\mathcal{F}F_{\alpha L})(m) = \prod_{i=1}^{|\alpha|+1} (\mathcal{F}f_L)(2^{(i-1)}m) \tag{2.117}$$

$$(\mathcal{F}F_{\beta H})(m) = (\mathcal{F}F_H)(2^{|\beta|}m) \prod_{i=1}^{|\beta|} (\mathcal{F}F_L)(2^{(i-1)}m). \tag{2.118}$$

Note the similarity of (2.117) and (2.118) with (1.38) and (1.39) in Strichartz [S].

2.4.6 Shannon and real Shannon wavelet bases in classical wavelet analysis

Let β be any prefix of $\alpha \in \{L\}^*$. For $F_{\beta H}, F_{\alpha L} \in \mathbb{C}^N$, $2^{(|\alpha|+2)} \mid N$, an induction argument using (2.117) and (2.118) gives the following values for the nonrecursive Shannon filters:

$$(\mathcal{F}F_{\alpha L})(m) = \begin{cases} 2^{(|\alpha|+1)/2}, & 0 \leq m \leq \left(N2^{-(|\alpha|+2)} - 1\right) \text{ or} \\ & \left(1 - 2^{-(|\alpha|+2)}\right)N \leq m \leq (N-1) \quad (2.119) \\ 0, & \text{otherwise} \end{cases}$$

$$(\mathcal{F}F_{\beta H})(m) = \begin{cases} 2^{(|\beta|+1)/2}, & N2^{-(|\beta|+2)} \leq m \leq \left(N2^{-(|\beta|+1)} - 1\right) \text{ or} \\ & \left(1 - 2^{-(|\beta|+1)}\right)N \leq m \\ & \qquad \leq \left(\left(1 - 2^{-(|\beta|+2)}\right)N - 1\right) \quad (2.120) \\ 0, & \text{otherwise.} \end{cases}$$

Notice that the supports of $\{F_{\alpha L}, F_{\beta H}\}_\beta$ are disjoint in the frequency domain. Similarly, for the real Shannon filters we have:

$$(\mathcal{F}F_{\alpha L})(m) = \begin{cases} 2^{(|\alpha|+1)/2}, & 0 \leq m \leq \left(N2^{-(|\alpha|+2)} - 1\right) \text{ or} \\ & \left(\left(1 - 2^{-(|\alpha|+2)}\right)N + 1\right) \leq m \leq (N-1) \\ j2^{|\alpha|/2}, & m = N2^{-(|\alpha|+2)} \quad (2.121) \\ -j2^{|\alpha|/2}, & m = \left(1 - 2^{-(|\alpha|+2)}\right)N \\ 0, & \text{otherwise} \end{cases}$$

$$(\mathcal{F}F_{\beta H})(m) = \begin{cases} 2^{|\beta|/2}, & m = N2^{-(|\beta|+2)} \text{ or } m = \left(1 - 2^{-(|\beta|+2)}\right)N \\ 2^{(|\beta|+1)/2}, & \left(N2^{-(|\beta|+2)} + 1\right) \leq m \leq \left(N2^{-(|\beta|+1)} - 1\right) \text{ or} \\ & \left(\left(1 - 2^{-(|\beta|+1)}\right)N + 1\right) \leq m \\ & \qquad \leq \left(\left(1 - 2^{-(|\beta|+2)}\right)N - 1\right) \quad (2.122) \\ j2^{|\beta|/2}, & m = N2^{-(|\beta|+1)} \\ -j2^{|\beta|/2}, & m = \left(1 - 2^{-(|\beta|+1)}\right)N \\ 0, & \text{otherwise,} \end{cases}$$

if $|\beta| \geq 1$, and define $\mathcal{F}F_H = \mathcal{F}f_H$ by (2.75). The real Shannon filters $\{F_{\alpha L}, F_{\beta H}\}_\beta$ are real valued; and if any two overlap in the frequency domain they do so only at their end points.

2.4.7 Computational complexity

In this subsection we discuss signal spaces $\mathbb{C}^N$. The case of classical wavelet analysis is considered, where recursion is performed only in the low-pass branch.

Assume that forward and inverse N-point DFTs can be performed using a Fast Fourier Transform (FFT) algorithm that requires $(N/2)\log N$ complex multiplications, where the logarithm is computed to the base 2. Let M be the number of levels of analysis, e.g., $M = 2$ in Figures 2.3 and 2.8.

Suppose that a nonrecursive structure like that in Figure 2.8 is employed, and convolution is performed through the computation of dot products in the frequency domain. Suppose further that the sequences $\mathcal{F}F_{\alpha H}$ and $\mathcal{F}F_{\alpha L}$ are precomputed and stored. Then the computation of the DFTs of all wavelet transform sequences v_β requires

$$\frac{N}{2}\log N + (M+1)N = \mathbf{O}(MN + N\log N) = \mathbf{O}(N\log N) \qquad (2.123)$$

complex multiplications. The last equality in (2.123) holds because $M \leq \log N$.

The downsampling can be done on the DFT side using the identity $(\mathcal{F}(\mathcal{D}u))(m) = 1/2((\mathcal{F}u)(m) + (\mathcal{F}u)(m + N/2))$ for $u \in \mathbb{C}^N$, which is proved similarly to Lemma 2.11. To get the actual wavelet coefficients we need to compute the inverse DFTs of the downsampled sequences. This requires

$$\frac{1}{2}\sum_{i=1}^{M}\left(\frac{N}{2^i}\log\frac{N}{2^i}\right) + \frac{N}{2^{M+1}}\log\frac{N}{2^M} = \mathbf{O}(N\log N) \qquad (2.124)$$

complex multiplications. Thus the computation of the wavelet coefficients using a nonrecursive structure like that in Figure 2.8 requires $\mathbf{O}(N\log N)$ complex multiplications in all.

Computation using recursive structures like that in Figure 2.3 require

$$\frac{N}{2}\log N + 2\sum_{i=0}^{M-1}\frac{N}{2^i} = \mathbf{O}(N\log N) \qquad (2.125)$$

complex multiplications in order to obtain the DFTs of the sequences v_β. Then DFT inversion requires the same number of multiplications as in (2.124). Hence the computational complexity for the recursive computation of the wavelet coefficients is also $\mathbf{O}(N\log N)$.

In both cases signal reconstruction from the transform coefficients in v_β takes $\mathbf{O}(N\log N)$ complex multiplications.

If the filters selected are short in the time domain, i.e., if they have a small support in the time domain, then the operation count can be made linear in N. An example of such a filter pair is the four-point *Daubechies filter pair*. The Daubechies filters are defined by equations (2.129) and (2.130) below:

$$a = -\sqrt{3}; \quad c = \frac{1-a}{1+a} \qquad (2.126)$$

$$h_L(n) = \begin{cases} 1, & n = 0 \\ a, & n = 1 \\ -ac, & n = 2 \\ c, & n = 3 \\ 0, & 4 \leq n < N \end{cases} \tag{2.127}$$

$$h_H(n) = \begin{cases} 1, & n = 0 \\ a, & n = 1 \\ a/c, & n = 2 \\ -1/c, & n = 3 \\ 0, & 4 \leq n < N \end{cases} \tag{2.128}$$

$$f_L = \frac{h_L}{\|h_L\|_{\mathbb{C}^N}} \tag{2.129}$$

$$f_H = \frac{h_H}{\|h_H\|_{\mathbb{C}^N}}. \tag{2.130}$$

With short filters recursive computation is much more efficient than nonrecursive computation if convolutions are performed directly in the time domain. In this case, the number of complex multiplications needed to compute the sequences v_β is $TN \sum_{i=0}^{M-1} 2^{-i} \leq 2TN$, where T is the filter length. Recursive signal reconstruction also takes a like number of multiplications.

2.5 Orthonormal wavelets for multidimensional signals

Let P be a fixed positive integer. Let $x(n)$ be a P-dimensional signal in $l^2(\mathbb{Z}^P)$ or in $\mathbb{C}^{N^P} = l^2(\mathbb{Z}_N^P) = l^2((\mathbb{Z}_N)^P)$, N even. A signal in $\mathbb{C}^{N^2}$, for example, could be thought of as representing a square picture of size N pixels by N pixels. While it is trivial to extend the treatment here to cover rectangular "multidimensional pictures," we will rest content with square pictures in order to keep the notation simple. The dimensionality of a signal and the dimensionality of the vector space in which the signal lies are two distinct concepts that must be kept apart. For example, a two-dimensional picture in $\mathbb{C}^{N^2}$ lies in an N^2-dimensional space. We will use multiindex notation when convenient, writing

$$\hat{x}(z) = \sum_n x(n) z^{-n} \tag{2.131}$$

for

$$\hat{x}(z_1, z_2, \ldots, z_P) = \sum_{n_1} \sum_{n_2} \cdots \sum_{n_P} x(n_1, \ldots, n_P) z_1^{-n_1} \cdots z_P^{-n_P}. \tag{2.132}$$

In (2.131) z ranges over $\mathbf{T}^P$ or $\mathbf{W}_N^P$, and n ranges over $\mathbb{Z}^P$ or $\mathbb{Z}_N^P$, depending upon whether x belongs to $l^2(\mathbb{Z}^P)$ or to $\mathbb{C}^{N^P}$.

The first lemma below states a result of general utility.

LEMMA 2.14
For any $n \in \mathbb{Z}^P$ or $n \in \mathbb{Z}_N^P$,

$$\sum_{k \in \{0,1\}^P} (-1)^{(k_1 n_1 + \ldots + k_P n_P)} = \sum_{k \in \mathbb{Z}_2^P} (-1)^{k \cdot n} = \begin{cases} 0, & \text{if for some } i, \ n_i \text{ is odd} \\ 2^P, & \text{otherwise,} \end{cases}$$

(2.133)

where $k \cdot n$ denotes the dot product $k \cdot n = k_1 n_1 + \ldots + k_P n_P$.

PROOF OF LEMMA 2.14   If $n = 2m = (2m_1, \ldots, 2m_P)$ for $m \in \mathbb{Z}^P$ or $m \in \mathbb{Z}_{N/2}^P$, then each term in the sum is 1, and the sum is 2^P. Otherwise, if n_l is odd for some l, write

$$\sum_{k \in \mathbb{Z}_2^P} (-1)^{k \cdot n} = \sum_{k_1, \ldots, k_{l-1}, k_{l+1}, \ldots, k_P \in \mathbb{Z}_2} (-1)^{(k_1 n_1 + \ldots + k_{l-1} n_{l-1} + k_{l+1} n_{l+1} + \cdots + k_P n_P)} \left(\sum_{k_l = 0}^{1} (-1)^{k_l n_l} \right).$$

(2.134)

Because n_l is odd, the last sum and, hence, the entire sum vanishes. ∎

Define the downsampling operator $\mathcal{D}$ by $(\mathcal{D}f)(n) = f(2n)$, $n \in \mathbb{Z}^P$ or $n \in \mathbb{Z}_{N/2}^P$. $\mathcal{D} : l^2(\mathbb{Z}^P) \to l^2(\mathbb{Z}^P)$ or $\mathcal{D} : \mathbb{C}^{N^P} \to \mathbb{C}^{(N/2)^P}$. By way of example we draw a signal in $\mathbb{C}^{4^2}$ in Figure 2.9. The points retained by the downsampling operation are dark circled. The next lemma states a property of the downsampling operator concerning its action upon a filter sequence in $l^2(\mathbb{Z}^P)$ or $\mathbb{C}^{N^P}$.

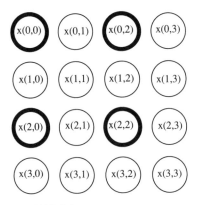

FIGURE 2.9
The downsampling operator retains the elements with dark circles.

LEMMA 2.15
For $z \in \mathbf{T}^P$ or $\mathbf{W}_N^P$,

$$(\mathcal{D}f)^\wedge(z) = 2^{-P} \sum_{k \in \mathbb{Z}_2^P} \hat{f}\left((-1)^{k_1} z_1^{1/2}, \ldots, (-1)^{k_P} z_P^{1/2}\right) = 2^{-P} \sum_{k \in \mathbb{Z}_2^P} \hat{f}((-1)^k z^{1/2}).$$
(2.135)

PROOF OF LEMMA 2.15 For $n \in \mathbb{Z}^P$ or $n \in \mathbb{Z}_N^P$,

$$\hat{f}((-1)^k z^{1/2}) = \sum_n f(n)((-1)^k z^{1/2})^{-n} = \sum_n f(n)(-1)^{k \cdot n} z^{-n/2} \quad (2.136)$$

$$= \sum_n f(n)(-1)^{(k_1 n_1 + \ldots + k_P n_P)} z_1^{-n_1/2} \cdots z_P^{-n_P/2}. \quad (2.137)$$

Therefore,

$$2^{-P} \sum_{k \in \mathbb{Z}_2^P} \hat{f}((-1)^k z^{1/2}) = 2^{-P} \sum_{k \in \mathbb{Z}_2^P} \sum_n f(n)(-1)^{k \cdot n} z_1^{-n_1/2} \cdots z_P^{-n_P/2}. \quad (2.138)$$

Interchanging the order of summation in (2.138) and using Lemma 2.14, we have

$$2^{-P} \sum_{k \in \mathbb{Z}_2^P} \hat{f}((-1)^k z^{1/2}) = \sum_{n=2m} f(n_1, \ldots, n_P) z_1^{-n_1/2} \cdots z_P^{-n_P/2} \quad (2.139)$$

$$= \sum_m f(2m_1, \ldots, 2m_P) z_1^{-m_1} \cdots z_P^{-m_P} \quad (2.140)$$

$$= \sum_m (\mathcal{D}f)(m_1, \ldots, m_P) z_1^{-m_1} \cdots z_P^{-m_P}$$

$$= (\mathcal{D}f)^\wedge(z). \quad \blacksquare \quad (2.141)$$

Define the upsampling operator $\mathcal{U} : l^2(\mathbb{Z}^P) \to l^2(\mathbb{Z}^P)$ or $\mathcal{U} : \mathbb{C}^{N^P} \to \mathbb{C}^{(2N)^P}$ by

$$(\mathcal{U}f)(n) = \begin{cases} f(n/2), & \text{if } n = 2m, \ m \in \mathbb{Z}^P \text{ or } m \in \mathbb{Z}_N^P \\ 0, & \text{otherwise.} \end{cases} \quad (2.142)$$

It is easy to see that $(\mathcal{U}f)^\wedge(z) = \hat{f}(z^2)$, i.e., $(\mathcal{U}f)^\wedge(z_1, \ldots, z_P) = \hat{f}(z_1^2, \ldots, z_P^2)$.
Define the shift operator R so that $(R^k f)(n) = f(n - k)$, or

$$(R^{(k_1, \ldots, k_P)} f)(n_1, \ldots, n_P) = f(n_1 - k_1, \ldots, n_P - k_P). \quad (2.143)$$

It is easy to see that $(R^k f)^\wedge(z) = z^{-k} \hat{f}(z) = z_1^{-k_1} \ldots z_P^{-k_P} \hat{f}(z_1, \ldots, z_P)$. By way of example, Figure 2.10 shows the effect of the operator $R^{(2,3)}$ upon a sequence in $\mathbb{C}^{4^2}$. $R^{(2,3)}$ shifts the input sequence by two in the "direction" of z_1 (down) and by three in the "direction" of z_2 (right). The choice of the "directions" was dictated by the fact that we would like $x(i, k)$ to be the coefficient of $z_1^{-i} z_2^{-k}$.

By $\text{coeff}_k \hat{f}(z)$, $k \in \mathbb{Z}^P$ or $k \in \mathbb{Z}_N^P$, will be meant the coefficient of $z^{-k} = z_1^{-k_1} \cdots z_P^{-k_P}$ in the canonical form of the polynomial $\hat{f}(z)$. For $f = \{f(n)\}_n$ and

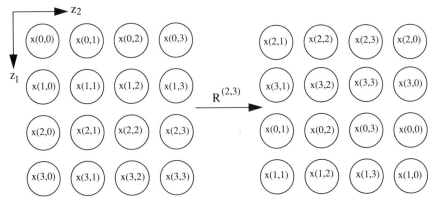

FIGURE 2.10
An example of the shift operation.

$g = \{g(n)\}_n$, $\langle f, g \rangle = \sum_n f(n)\bar{g}(n)$. It is easy to see that $\langle f, R^k g \rangle = \text{coeff}_k(\hat{f}(z)\bar{\hat{g}}(z))$, $z \in \mathbf{T}^P$ or $\mathbf{W}_N^P$.

2.5.1 The orthonormality condition

In this section we present three lemmas that lead to a multidimensional version of Theorem 2.5.

LEMMA 2.16
Let $f = f(n)$, $n \in \mathbb{Z}^P$ or $n \in \mathbb{Z}_N^P$. Then the following are equivalent:

P1 $\langle f, R^{2k}f \rangle = \delta(k) = \delta(k_1) \ldots \delta(k_P)$.

P2 $\sum_{l \in \mathbb{Z}_2^P} |\hat{f}((-1)^l z)|^2 = 2^P$, $\forall z \in \mathbf{T}^P$ or $\mathbf{W}_N^P$.

Here $\hat{f}((-1)^l z) = \hat{f}((-1)^{l_1} z_1, \ldots, (-1)^{l_P} z_P)$ for $l = (l_1, \ldots, l_P) \in \mathbb{Z}_2^P$.

PROOF OF LEMMA 2.16   **P1** is equivalent to:

$$\text{coeff}_{2k}\left(|\hat{f}(z)|^2\right) = \delta(k). \tag{2.144}$$

Define

$$\hat{h}(z) = |\hat{f}(z)|^2. \tag{2.145}$$

For $m \in \mathbb{Z}^P$ or $m \in \mathbb{Z}_{N/2}^P$, (2.144) is equivalent to:

$$\sum_{n=2m} h(n)z^{-n} = 1. \tag{2.146}$$

But $\sum_{n=2m} h(n)z^{-n} = (\mathcal{U}\mathcal{D}h)^\wedge(z) = (\mathcal{D}h)^\wedge(z^2)$. Therefore, by Lemma 2.15,

$$\sum_{n=2m} h(n)z^{-n} = (\mathcal{D}h)^\wedge(z^2) = 2^{-P}\sum_{l\in\mathbb{Z}_2^P}\hat{h}((-1)^l z). \qquad (2.147)$$

From (2.147), (2.146) is equivalent to:

$$\sum_{l\in\mathbb{Z}_2^P}\hat{h}((-1)^l z) = 2^P. \qquad (2.148)$$

From (2.145), $\hat{h}((-1)^l z) = |\hat{f}((-1)^l z)|^2$. Substituting for $\hat{h}((-1)^l z)$ in (2.148) we have the lemma. ∎

LEMMA 2.17
Let $f = f(n), g = g(n), n \in \mathbb{Z}^P$ or $n \in \mathbb{Z}_N^P$. Then the following are equivalent:

P1 $\langle f, R^{2k}g \rangle = 0, \forall k.$

P2 $\sum_{l\in\mathbb{Z}_2^P}\hat{f}((-1)^l z)\overline{\hat{g}}((-1)^l z) = 0, \forall z \in \mathbf{T}^P$ or $\mathbf{W}_N^P.$

PROOF OF LEMMA 2.17 The proof is similar to that of Lemma 2.16. **P1** is equivalent to:

$$\text{coeff}_{2k}\left(\hat{f}(z)\overline{\hat{g}}(z)\right) = 0. \qquad (2.149)$$

Define

$$\hat{h}(z) = \hat{f}(z)\overline{\hat{g}}(z). \qquad (2.150)$$

From (2.150), for $m \in \mathbb{Z}^P$ or $m \in \mathbb{Z}_{N/2}^P$, (2.149) is equivalent to:

$$\sum_{n=2m} h(n)z^{-n} = 0. \qquad (2.151)$$

By Lemma 2.15,

$$\sum_{n=2m} h(n)z^{-n} = (\mathcal{U}\mathcal{D}h)^\wedge(z) = (\mathcal{D}h)^\wedge(z^2) = 2^{-P}\sum_{l\in\mathbb{Z}_2^P}\hat{h}((-1)^l z). \qquad (2.152)$$

Substituting (2.150) in (2.152) we have the lemma. ∎

The proof of the following lemma is similar to the proof of Lemma 2.2.

LEMMA 2.18
For $i \in \mathbb{Z}_{2^P}$ define

$$\tilde{f}_i(n) = \overline{f}_i(-n) \qquad (2.153)$$

$$\tilde{B} = \bigcup_{i=0}^{2^P-1}\{R^{2k}\tilde{f}_i : k \in \mathbb{Z}^P \text{ or } k \in \mathbb{Z}_{N/2}^P\} \qquad (2.154)$$

$$B = \bigcup_{i=0}^{2^P-1}\{R^{2k}f_i : k \in \mathbb{Z}^P \text{ or } k \in \mathbb{Z}_{N/2}^P\}. \qquad (2.155)$$

The multiset $\tilde{B}$ is orthonormal if and only if the multiset B is.

Consider now the orthonormal decomposition of the signal x in Figure 2.11.

We need some new notation for a statement of the main theorem in this subsection. Let $j = (j_1, \ldots, j_P) \in \{0,1\}^P = \mathbb{Z}_2^P$. Since $\mathrm{card}(\{0,1\}^P) = 2^P$, we can enumerate all such j's as $j^{(0)}, \ldots, j^{(2^P-1)}$. Select this enumeration such that $j^{(0)} = (0,0,\ldots,0)$. Then, we write $j^{(i)} = (j_1^{(i)}, \ldots, j_P^{(i)})$, with $j_k^{(i)} \in \mathbb{Z}_2$ for all k. We also write $\hat{f}((-1)^{j^{(i)}}z) = \hat{f}((-1)^{j_1^{(i)}}z_1, \ldots, (-1)^{j_P^{(i)}}z_P)$.

THEOREM 2.19

Let $f_0(n) = f_0(n_1, \ldots, n_P), \ldots, f_{2^P-1}(n) = f_{2^P-1}(n_1, \ldots, n_P)$ be 2^P sequences, with $n \in \mathbb{Z}^P$ or $n \in \mathbb{Z}_N^P$. Define a $2^P \times 2^P$ matrix called the system matrix *$A(z) = (A_{k,i}(z))_{k,i \in \mathbb{Z}_{2^P}}, z \in \mathbf{T}^P$ or $\mathbf{W}_N^P$, such that $A_{k,i}(z) = 2^{-P/2}\hat{f}_i((-1)^{j^{(k)}}z)$, i.e.,*

$$A(z) = 2^{-P/2} \begin{pmatrix} \hat{f}_0((-1)^{j^{(0)}}z) & \hat{f}_1((-1)^{j^{(0)}}z) & \cdots & \hat{f}_{2^P-1}((-1)^{j^{(0)}}z) \\ \hat{f}_0((-1)^{j^{(1)}}z) & \hat{f}_1((-1)^{j^{(1)}}z) & \cdots & \hat{f}_{2^P-1}((-1)^{j^{(1)}}z) \\ \vdots & \vdots & \ddots & \vdots \\ \hat{f}_0((-1)^{j^{(2^P-1)}}z) & \hat{f}_1((-1)^{j^{(2^P-1)}}z) & \cdots & \hat{f}_{2^P-1}((-1)^{j^{(2^P-1)}}z) \end{pmatrix}.$$

(2.156)

Then $\tilde{B}$ is orthonormal if, and only if, $A(z)$ is unitary for all $z \in \mathbf{T}^P$ or $\mathbf{W}_N^P$.

PROOF OF THEOREM 2.19 By Lemma 2.18, the orthonormality of $\tilde{B}$ is equivalent to the orthonormality of B. By Lemma 2.16 the orthonormality of $\{R^{2k}f_i : k \in \mathbb{Z}^P \text{ or } k \in \mathbb{Z}_{N/2}^P\}$ for every fixed i is equivalent to the ith column of A having length "1." By Lemma 2.17 the cross-orthogonality of

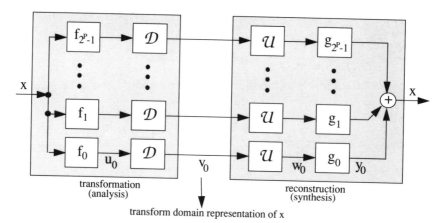

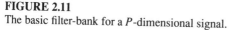

FIGURE 2.11
The basic filter-bank for a P-dimensional signal.

$\{R^{2k}f_{i_1} : k \in \mathbb{Z}^P \text{ or } k \in \mathbb{Z}^P_{N/2}\}$ and $\{R^{2k}f_{i_2} : k \in \mathbb{Z}^P \text{ or } k \in \mathbb{Z}^P_{N/2}\}$ is equivalent to the orthogonality of the i_1th and the i_2th columns of A. ∎

By way of example, the system matrix for the signal space C^{4^2} is as follows:

$$A(z_1, z_2) = \frac{1}{2} \begin{pmatrix} \hat{f}_0(z_1, z_2) & \hat{f}_1(z_1, z_2) & \hat{f}_2(z_1, z_2) & \hat{f}_3(z_1, z_2) \\ \hat{f}_0(-z_1, z_2) & \hat{f}_1(-z_1, z_2) & \hat{f}_2(-z_1, z_2) & \hat{f}_3(-z_1, z_2) \\ \hat{f}_0(z_1, -z_2) & \hat{f}_1(z_1, -z_2) & \hat{f}_2(z_1, -z_2) & \hat{f}_3(z_1, -z_2) \\ \hat{f}_0(-z_1, -z_2) & \hat{f}_1(-z_1, -z_2) & \hat{f}_2(-z_1, -z_2) & \hat{f}_3(-z_1, -z_2) \end{pmatrix}$$

(2.157)

2.5.2 Perfect reconstruction

From Figure 2.11 we can see that perfect reconstruction occurs if and only if the following equation holds for all input signals x:

$$\sum_{i=0}^{2^P-1} (g_i * \mathcal{U}\mathcal{D}(f_i * x))^\wedge (z) = \hat{x}(z). \tag{2.158}$$

Equation (2.158) will be called the *perfect reconstruction condition*.

LEMMA 2.20
The perfect reconstruction condition is satisfied if, and only if,

$$\sum_{i=0}^{2^P-1} \hat{f}_i((-1)^k z)\hat{g}_i(z) = 2^P \delta(k). \tag{2.159}$$

PROOF OF LEMMA 2.20   From Lemma 2.15,

$$(\mathcal{D}(f_i * x))^\wedge (z) = 2^{-P} \sum_{k \in \mathbb{Z}_2^P} \hat{f}_i((-1)^k z^{1/2})\hat{x}((-1)^k z^{1/2}). \tag{2.160}$$

From (2.160) it follows that

$$(\mathcal{U}\mathcal{D}(f_i * x))^\wedge (z) = 2^{-P} \sum_{k \in \mathbb{Z}_2^P} \hat{f}_i((-1)^k z)\hat{x}((-1)^k z). \tag{2.161}$$

Using (2.161), the perfect reconstruction condition (2.158) can be written as:

$$2^{-P} \sum_{i=0}^{2^P-1} \sum_{k \in \mathbb{Z}_2^P} \hat{f}_i((-1)^k z)\hat{x}((-1)^k z)\hat{g}_i(z) = \hat{x}(z). \tag{2.162}$$

Define

$$\hat{C}_k(z) = \sum_{i=0}^{2^P-1} \hat{f}_i((-1)^k z)\hat{g}_i(z). \tag{2.163}$$

The perfect reconstruction condition in (2.162) now reads:

$$2^{-P} \sum_{k \in \mathbb{Z}_2^P} \hat{C}_k(z)\hat{x}((-1)^k z) = \hat{x}(z). \tag{2.164}$$

Define $\hat{x}_l(z) = z^l$, for $l \in \mathbb{Z}_2^P$. Since the perfect reconstruction condition must hold for all $\hat{x}(z)$, it must *in particular* hold for all the $\hat{x}_l(z)$, $l \in \mathbb{Z}_2^P$. Then for every l, from (2.164),

$$\hat{x}_l(z) = z^l = 2^{-P} \sum_{k \in \mathbb{Z}_2^P} ((-1)^k z)^l \hat{C}_k(z) = 2^{-P} \sum_{k \in \mathbb{Z}_2^P} (-1)^{k \cdot l} z^l \hat{C}_k(z). \tag{2.165}$$

From (2.165) it follows that, for all $l \in \mathbb{Z}_2^P$,

$$2^{-P} \sum_{k \in \mathbb{Z}_2^P} (-1)^{k \cdot l} \hat{C}_k(z) = 1. \tag{2.166}$$

Let $K \in \mathbb{Z}_2^P$ be any fixed binary P-tuple. Multiplying both sides in (2.166) by $(-1)^{K \cdot l}$, and summing over all l we have:

$$\sum_{l \in \mathbb{Z}_2^P} (-1)^{K \cdot l} = 2^{-P} \sum_{l \in \mathbb{Z}_2^P} \sum_{k \in \mathbb{Z}_2^P} (-1)^{k \cdot l} (-1)^{K \cdot l} \hat{C}_k(z) \tag{2.167}$$

$$= 2^{-P} \sum_{l \in \mathbb{Z}_2^P} \sum_{k \in \mathbb{Z}_2^P} (-1)^{k \cdot l} (-1)^{-K \cdot l} \hat{C}_k(z) \tag{2.168}$$

$$= 2^{-P} \sum_{k \in \mathbb{Z}_2^P} \hat{C}_k(z) \sum_{l \in \mathbb{Z}_2^P} (-1)^{(k-K) \cdot l}. \tag{2.169}$$

By Lemma 2.14, because $K \in \mathbb{Z}_2^P$,

$$\sum_{l \in \mathbb{Z}_2^P} (-1)^{K \cdot l} = 2^P \delta(K) = 2^P \delta(K_1) \cdots \delta(K_P). \tag{2.170}$$

Also by Lemma 2.14,

$$\sum_{l \in \mathbb{Z}_2^P} (-1)^{(k-K) \cdot l} = 2^P \delta(k - K). \tag{2.171}$$

From (2.167)–(2.171),

$$\sum_{k \in \mathbb{Z}_2^P} \hat{C}_k(z)\delta(k - K) = 2^P \delta(K), \tag{2.172}$$

or

$$\hat{C}_K(z) = 2^P \delta(K). \tag{2.173}$$

From (2.163) and (2.173) we have the lemma.   ∎

The following theorem is immediate from Lemma 2.20.

THEOREM 2.21

Let $A(z)$ be the system matrix (2.156). The system in Figure 2.11 gives perfect reconstruction if and only if, $\forall z \in \mathbf{T}^P$ or $\mathbb{C}_N^P$,

$$
A(z) \begin{pmatrix} \hat{g}_0(z) \\ \hat{g}_1(z) \\ \vdots \\ \hat{g}_{2^P-1}(z) \end{pmatrix} = \begin{pmatrix} 2^{P/2} \\ 0 \\ \vdots \\ 0 \end{pmatrix}. \tag{2.174}
$$

In the following corollary we summarize the results of the orthonormal decomposition and perfect reconstruction of signals in $l^2(\mathbb{Z}^P)$ and $\mathbb{C}^{N^P}$.

COROLLARY 2.22

The following are equivalent:

P1 *The set $\tilde{B} = \bigcup_{i=0}^{2^P-1} \{R^{2k}\tilde{f}_i : k \in \mathbb{Z}^P$ or $k \in \mathbb{Z}_{N/2}^P\}$ is orthonormal, and we have perfect reconstruction in Figure 2.11.*

P2 *The system matrix $A(z)$ in (2.156) is unitary, and $\forall i \in \mathbb{Z}_{2^P}$, $g_i = \tilde{f}_i$.*

PROOF OF COROLLARY 2.22 (**P1** $\Rightarrow$ **P2**). The unitarity of $A(z)$ follows from Theorem 2.19. Hence $A^{-1}(z) = \bar{A}^T(z)$. From this observation and Theorem 2.21 follows the fact that, $\forall i \in \mathbb{Z}_{2^P}$, $\hat{g}_i(z) = \hat{\overline{f}}_i(z)$; or equivalently $g_i = \tilde{f}_i$.

(**P1** $\Leftarrow$ **P2**). By Theorem 2.19, B is orthonormal. Since $\hat{g}_i(z) = \hat{\overline{f}}_i(z)$, the equation

$$
A(z) \begin{pmatrix} 2^{-P/2}\hat{g}_0(z) \\ 2^{-P/2}\hat{g}_1(z) \\ \vdots \\ 2^{-P/2}\hat{g}_{2^P-1}(z) \end{pmatrix} = A(z) \begin{pmatrix} 2^{-P/2}\overline{\hat{f}_0}(z) \\ 2^{-P/2}\overline{\hat{f}_1}(z) \\ \vdots \\ 2^{-P/2}\overline{\hat{f}_{2^P-1}}(z) \end{pmatrix} = \begin{pmatrix} 1 \\ 0 \\ \vdots \\ 0 \end{pmatrix} \tag{2.175}
$$

is true, because the rows of $A(z)$ are orthonormal by the unitarity of $A(z)$. Then perfect reconstruction follows by Theorem 2.21. $\blacksquare$

2.5.3 Multidimensional product filters

Suppose $\hat{f}_L$ and $\hat{f}_H$ are one-dimensional filters such that

$$
\frac{1}{\sqrt{2}} \begin{pmatrix} \hat{f}_L(z) & \hat{f}_H(z) \\ \hat{f}_L(-z) & \hat{f}_H(-z) \end{pmatrix} \tag{2.176}
$$

is unitary for all $z \in \mathbf{T}$ or $\mathbf{W}_N$. Define

$$
f_0(n_1, n_2) = f_L(n_1)f_L(n_2) \tag{2.177}
$$

$$f_1(n_1, n_2) = f_H(n_1)f_L(n_2) \tag{2.178}$$

$$f_2(n_1, n_2) = f_L(n_1)f_H(n_2) \tag{2.179}$$

$$f_3(n_1, n_2) = f_H(n_1)f_H(n_2). \tag{2.180}$$

With this definition, the matrix (2.157) is a Kronecker product of two matrices of form (2.176):

$$A(z_1, z_2)$$

$$= \frac{1}{2} \begin{pmatrix} \hat{f}_0(z_1, z_2) & \hat{f}_1(z_1, z_2) & \hat{f}_2(z_1, z_2) & \hat{f}_3(z_1, z_2) \\ \hat{f}_0(-z_1, z_2) & \hat{f}_1(-z_1, z_2) & \hat{f}_2(-z_1, z_2) & \hat{f}_3(-z_1, z_2) \\ \hat{f}_0(z_1, -z_2) & \hat{f}_1(z_1, -z_2) & \hat{f}_2(z_1, -z_2) & \hat{f}_3(z_1, -z_2) \\ \hat{f}_0(-z_1, -z_2) & \hat{f}_1(-z_1, -z_2) & \hat{f}_2(-z_1, -z_2) & \hat{f}_3(-z_1, -z_2) \end{pmatrix} \tag{2.181}$$

$$= \frac{1}{\sqrt{2}} \begin{pmatrix} \hat{f}_L(z_2) & \hat{f}_H(z_2) \\ \hat{f}_L(-z_2) & \hat{f}_H(-z_2) \end{pmatrix} \otimes \frac{1}{\sqrt{2}} \begin{pmatrix} \hat{f}_L(z_1) & \hat{f}_H(z_1) \\ \hat{f}_L(-z_1) & \hat{f}_H(-z_1) \end{pmatrix}. \tag{2.182}$$

It is easy to check directly that the orthonormality of the set $\{R^{2k}f_L\}_k \cup \{R^{2k}f_H\}_k$ implies the orthonormality of the set $\tilde{B}$ in (2.154), for $f_0, \ldots, f_3$ given by (2.177)–(2.180). This is the usual way of constructing two-dimensional wavelets from one-dimensional wavelets. We call $f_0, \ldots, f_3$ *product wavelets*. This is consistent with Theorem 2.19, since the unitarity of the matrix (2.181) follows from the unitarity of the matrices in the Kronecker product formula in (2.182).

Thus, two-dimensional wavelet filters can be constructed as products of one-dimensional filters. However, Theorem 2.19 shows that there exist two-dimensional wavelet filters that are not product filters. This follows from the existence of the following unitary matrix, which cannot be expressed as a Kronecker product:

$$\frac{1}{4} \begin{pmatrix} \sqrt{2} & 0 & 0 & \sqrt{2} \\ \sqrt{2} & 0 & 0 & -\sqrt{2} \\ 0 & \sqrt{2} & \sqrt{2} & 0 \\ 0 & \sqrt{2} & -\sqrt{2} & 0 \end{pmatrix}. \tag{2.183}$$

These arguments can be extended to the construction of general multidimensional filters.

2.5.4 Recursion

Let $V = l^2(\mathbb{Z}^P)$ or C_N^P. Let $\alpha \in \mathbb{Z}_{2^P}^*$, $|\alpha| = n > 0$, be fixed. Define $\beta_i = \text{pfx}(\alpha, i)$, $i = 1, \ldots, n$. Let $F_\alpha, G_\alpha \in V$ be defined as in (2.91) and (2.92). For $k = 1, \ldots, n+1$, let $X_k = \mathbb{Z}^P$ in the infinite-dimensional case. In the finite-dimensional case assume $2^n \mid N$, and let $X_k = \mathbb{Z}_{N2^{1-k}}^P$ for $k = 1, \ldots, n+1$. Then Theorem 2.9 carries over to the multidimensional case.

Let $\tilde{B}_\alpha$ and V_α be defined exactly as in (2.101) and (2.102). We now prove the following analog to Theorem 2.10:

THEOREM 2.23   The Iteration Theorem

Let $\alpha \in \mathbb{Z}_{2^p}^*$, $|\alpha| > 0$, be fixed. In the finite-dimensional case let $2^{(|\alpha|+1)} \mid N$. Let $\beta \in \mathbb{Z}_{2^p}$. Suppose that $\tilde{B}_\alpha$ is an orthonormal basis for V_α, $V_\alpha \subseteq l^2(X_1)$. Suppose that $\{f_{\alpha\beta} \in l^1(X_{|\alpha|+1})\}_\beta = \{f_{\alpha 0}, f_{\alpha 1}, \ldots, f_{\alpha(2^p-1)}\}$ satisfy Theorem 2.19. Let

$$\tilde{F}_{\alpha\beta} = \tilde{F}_\alpha * \mathcal{U}^{|\alpha|} \tilde{f}_{\alpha\beta}. \tag{2.184}$$

Let $\tilde{B}_{\alpha\beta}$ and $V_{\alpha\beta}$ be as defined in (2.101) and (2.102). Then V_α is the orthogonal direct sum of $\{V_{\alpha\beta}\}_\beta$,

$$V_\alpha = \bigoplus_\beta V_{\alpha\beta}. \tag{2.185}$$

Also $\tilde{B}_{\alpha\beta}$ is an orthonormal basis for $V_{\alpha\beta}$.

PROOF OF THEOREM 2.23   We first demonstrate the orthonormality of $\bigcup_\beta \tilde{B}_{\alpha\beta}$. In order to do that it suffices to prove that

$$\left\langle \tilde{F}_{\alpha\beta_1}, R^{2^{(|\alpha|+1)}k} \tilde{F}_{\alpha\beta_2} \right\rangle = \begin{cases} 1, & k = 0 \text{ and } \beta_1 = \beta_2 \\ 0, & \text{otherwise.} \end{cases} \tag{2.186}$$

We have:

$$\left\langle \tilde{F}_{\alpha\beta_1}, R^{2^{(|\alpha|+1)}k} \tilde{F}_{\alpha\beta_2} \right\rangle = \mathrm{coeff}_0 \left(\left(\tilde{F}_{\alpha\beta_1} \right)^\wedge (z) \overline{\left(R^{2^{(|\alpha|+1)}k} \tilde{F}_{\alpha\beta_2} \right)^\wedge (z)} \right) \tag{2.187}$$

$$= \mathrm{coeff}_0 \left(\overline{\hat{F}_{\alpha\beta_1}}(z) z^{2^{(|\alpha|+1)}k} \hat{F}_{\alpha\beta_2}(z) \right) \tag{2.188}$$

$$= \mathrm{coeff}_{2^{(|\alpha|+1)}k} \left(\overline{\hat{F}_{\alpha\beta_1}}(z) \hat{F}_{\alpha\beta_2}(z) \right). \tag{2.189}$$

From (2.184),

$$\hat{F}_{\alpha\beta}(z) = \hat{F}_\alpha(z) \left(\mathcal{U}^{|\alpha|} f_{\alpha\beta} \right)^\wedge (z) = \hat{F}_\alpha(z) \hat{f}_{\alpha\beta} \left(z^{2^{|\alpha|}} \right). \tag{2.190}$$

From (2.189) and (2.190),

$$\left\langle \tilde{F}_{\alpha\beta_1}, R^{2^{(|\alpha|+1)}k} \tilde{F}_{\alpha\beta_2} \right\rangle = \mathrm{coeff}_{2^{(|\alpha|+1)}k} \left(\left| \hat{F}_\alpha \right|^2 (z) \overline{\hat{f}_{\alpha\beta_1}} \left(z^{2^{|\alpha|}} \right) \hat{f}_{\alpha\beta_2} \left(z^{2^{|\alpha|}} \right) \right). \tag{2.191}$$

By the orthonormality of $\tilde{B}_\alpha = \left\{ R^{2^{|\alpha|}k} \tilde{F}_\alpha \right\}_{k \in X_{|\alpha|+1}}$,

$$\left\langle \tilde{F}_\alpha, R^{2^{|\alpha|}k} \tilde{F}_\alpha \right\rangle = \mathrm{coeff}_{2^{|\alpha|}k} \left(\left| \hat{F}_\alpha \right|^2 (z) \right) = \delta(k). \tag{2.192}$$

From (2.192) it follows that

$$\left| \hat{F}_\alpha \right|^2 (z) = 1 + \sum_{\substack{l \in X_1 \\ 2^{|\alpha|} \nmid l}} a(l) z^{-l} \tag{2.193}$$

for some $a(l)$. Here by "$2^{|\alpha|} \nmid l$" we mean that $\exists i, 2^{|\alpha|} \nmid l_i$, where l_i is the ith component of the vector $l \in X_1 = \mathbb{Z}^P$ or $\mathbb{Z}_N^P$.

Similarly, by the orthonormality of $\{f_{\alpha\beta} \in l^2(X_{|\alpha|+1})\}_{\beta \in \mathbb{Z}_{2^P}}$,

$$\langle f_{\alpha\beta_1}, R^{2k} f_{\alpha\beta_2} \rangle = \text{coeff}_{2k} \left(\hat{f}_{\alpha\beta_1}(z)\overline{\hat{f}_{\alpha\beta_2}(z)} \right) = \delta(k)\delta(\beta_1 - \beta_2) \tag{2.194}$$

From (2.194) it follows that

$$\hat{f}_{\alpha\beta_1}(z)\overline{\hat{f}_{\alpha\beta_2}(z)} = \delta(\beta_1 - \beta_2) + \sum_{\substack{l \in X_{|\alpha|+1} \\ 2 \nmid l}} b(l)z^{-l} \tag{2.195}$$

for some $b(l)$. Again, by "$2 \nmid l$" we mean that $\exists i, 2 \nmid l_i$. From (2.191), (2.193), and (2.195) we have

$$\langle \tilde{F}_{\alpha\beta_1}, R^{2^{(|\alpha|+1)}k} \tilde{F}_{\alpha\beta_2} \rangle$$

$$= \text{coeff}_{2^{|\alpha|+1}k} \left(\delta(\beta_1 - \beta_2) \left(1 + \sum_{\substack{l \in X_1 \\ 2^{|\alpha|} \nmid l}} a(l)z^{-l} \right) \right.$$

$$\left. + \sum_{\substack{l \in X_{|\alpha|+1} \\ 2 \nmid l}} b(l)z^{-l2^{|\alpha|}} + \sum_{\substack{l \in X_1 \\ 2^{|\alpha|} \nmid l}} \sum_{\substack{m \in X_{|\alpha|+1} \\ 2 \nmid m}} a(l)b(m)z^{-(l+m2^{|\alpha|})} \right). \tag{2.196}$$

Equation (2.186) follows from (2.196), and $\bigcup_\beta \tilde{B}_{\alpha\beta}$ is orthonormal.

We now show that $\left(\bigcup_\beta \tilde{B}_{\alpha\beta} \right) \subseteq \text{cl}\left(\text{span}\left(\tilde{B}_\alpha \right) \right) = V_\alpha$. For $n \in X_1$,

$$\tilde{F}_{\alpha\beta}(n) = \left(\tilde{F}_\alpha * \mathcal{U}^{|\alpha|} \tilde{f}_{\alpha\beta} \right)(n) = \sum_{m \in X_1} \tilde{F}_\alpha(n - m) \left(\mathcal{U}^{|\alpha|} \tilde{f}_{\alpha\beta} \right)(m) \tag{2.197}$$

$$= \sum_{k \in X_{|\alpha|+1}} \tilde{F}_\alpha(n - 2^{|\alpha|}k)\tilde{f}_{\alpha\beta}(k) = \sum_{k \in X_{|\alpha|+1}} \tilde{f}_{\alpha\beta}(k) \left(R^{2^{|\alpha|}k} \tilde{F}_\alpha \right)(n). \tag{2.198}$$

In (2.198) we have $\tilde{F}_{\alpha\beta}$ expressed as a linear combination of the elements of $\tilde{B}_\alpha$. Hence $\tilde{F}_{\alpha\beta} \in V_\alpha$. Since $\tilde{B}_\alpha$ is invariant under the operator $R^{2^{|\alpha|}k}$, for all $k \in X_{|\alpha|+1}$, we obtain $\left(\bigcup_\beta \tilde{B}_{\alpha\beta} \right) \subseteq V_\alpha$. Hence $\text{cl}\left(\text{span}\left(\bigcup_\beta \tilde{B}_{\alpha\beta} \right) \right) = \bigoplus_\beta V_{\alpha\beta} \subseteq \text{cl}\left(\text{span}\left(\tilde{B}_\alpha \right) \right)$.

In the finite-dimensional case the proof is complete, since the cardinalities of the orthonormal sets $\tilde{B}_\alpha$ and $\bigcup_\beta \tilde{B}_{\alpha\beta}$ are equal, so that $\text{span}\left(\tilde{B}_\alpha \right) = \text{span}\left(\bigcup_\beta \tilde{B}_{\alpha\beta} \right)$. For the infinite-dimensional case, however, the inclusion $\tilde{B}_\alpha \subseteq \text{cl}\left(\text{span}\left(\bigcup_\beta \tilde{B}_{\alpha\beta} \right) \right) = \bigoplus_\beta V_{\alpha\beta}$ remains to be established. Using (2.198) note that for all $n, k \in \mathbb{Z}^P$ and $\beta \in \mathbb{Z}_{2^P}$,

$$\left(R^{2^{|\alpha|+1}k} \tilde{F}_{\alpha\beta} \right)(n) = \sum_{m \in \mathbb{Z}^P} \tilde{f}_{\alpha\beta}(m)\tilde{F}_\alpha \left(n - 2^{|\alpha|}(m + 2k) \right) \tag{2.199}$$

$$= \sum_{m \in \mathbb{Z}^P} \tilde{F}_\alpha \left(n - 2^{|\alpha|}m \right) \tilde{f}_{\alpha\beta}(m - 2k) \tag{2.200}$$

$$= \sum_{m \in \mathbb{Z}^P} \left(R^{2k} \tilde{f}_{\alpha\beta} \right)(m) \left(R^{2^{|\alpha|}m} \tilde{F}_\alpha \right)(n). \tag{2.201}$$

For $p \in \mathbb{Z}^P$ fixed, and $n \in \mathbb{Z}^P$, define δ_p by $\delta_p(n) = 1$ if $p = n$, $\delta_p(n) = 0$ otherwise. Expanding δ_p with respect to the basis $\{R^{2k}\tilde{f}_{\alpha\beta} : k \in \mathbb{Z}^P, \beta \in \mathbb{Z}_{2^P}\}$ gives

$$\delta_p = \sum_{\beta \in \mathbb{Z}_{2^P}} \sum_{k \in \mathbb{Z}^P} \overline{\tilde{f}_{\alpha\beta}}(p - 2k) R^{2k} \tilde{f}_{\alpha\beta}. \tag{2.202}$$

Hence, using (2.201), for each $p \in \mathbb{Z}^P$ we have

$$\sum_{\beta \in \mathbb{Z}_{2^P}} \sum_{k \in \mathbb{Z}^P} \overline{\tilde{f}_{\alpha\beta}}(p - 2k) R^{2^{|\alpha|+1}k} \left(\tilde{F}_\alpha * \mathcal{U}^{|\alpha|} \tilde{f}_{\alpha\beta} \right) \tag{2.203}$$

$$= \sum_{n \in \mathbb{Z}^P} \sum_{\beta \in \mathbb{Z}_{2^P}} \sum_{k \in \mathbb{Z}^P} \overline{\tilde{f}_{\alpha\beta}}(p - 2k) \left(R^{2k} \tilde{f}_{\alpha\beta} \right)(n) \left(R^{2^{|\alpha|}n} \tilde{F}_\alpha \right) \tag{2.204}$$

$$= \sum_{n \in \mathbb{Z}^P} \delta_p(n) \left(R^{2^{|\alpha|}n} \tilde{F}_\alpha \right) = R^{2^{|\alpha|}p} \tilde{F}_\alpha. \tag{2.205}$$

This expresses each element of $\tilde{B}_\alpha$ as a linear combination (converging in $l^2(\mathbb{Z}^P)$) since $\tilde{f}_{\alpha\beta} \in l^1(\mathbb{Z}^P)$) of the elements of $\bigcup_\beta \tilde{B}_{\alpha\beta}$. This completes the proof. ∎

In like manner to (2.108), (2.113) it is easy to see that for any signal $x \in X_1$ and for $k \in X_{|\alpha|+1}$,

$$v_\alpha(k) = \left\langle x, R^{2^{|\alpha|}k} \tilde{F}_\alpha \right\rangle \tag{2.206}$$

$$y_\alpha = \sum_{k \in X_{|\alpha|+1}} \left\langle x, R^{2^{|\alpha|}k} \tilde{F}_\alpha \right\rangle R^{2^{|\alpha|}k} \tilde{F}_\alpha. \tag{2.207}$$

The sum of y_α for different α yields a decomposition of x in terms of the wavelet bases.

Recursion can be performed with "repeated" filters, as discussed in the single-dimensional case in Subsection 2.4.4. We now prove the following folding lemma that allows us to construct filters at higher levels by "folding" the filters at the previous level. This lemma is analogous to Lemma 2.11.

LEMMA 2.24 *The Folding Lemma*
Suppose $M \in \mathbb{Z}^+, f = \{f(n)\}_{n \in \mathbb{Z}_{2M}^P}$, and $h = \{h(n)\}_{n \in \mathbb{Z}_M^P}$. Then the following are equivalent:

P1 $\hat{h}(z) = \hat{f}(z)$, $\forall z \in \mathbf{W}_M^P$.
P2 $(\mathcal{F}h)(m) = (\mathcal{F}f)(2m)$, $\forall m \in \mathbb{Z}_M^P$.
P3 $h(n) = \sum_{k \in \mathbb{Z}_2^P} f(n + Mk)$, $\forall n \in \mathbb{Z}_M^P$.

PROOF OF LEMMA 2.24 We have:

$$(\mathcal{F}h)(n) = \hat{h}(e^{-j2\pi n/M}), \qquad n \in \mathbb{Z}_M^P \tag{2.208}$$

$$(\mathcal{F}f)(2n) = \hat{f}\left(e^{-j2\pi 2n/2M}\right) = \hat{f}\left(e^{-j2\pi n/M}\right), \qquad n \in \mathbb{Z}_M^P. \tag{2.209}$$

Hence **P1** $\Leftrightarrow$ **P2**. Also,

$$\hat{f}(z) = \sum_{n \in \mathbb{Z}_{2M}^P} f(n)z^{-n} = \sum_{n \in \mathbb{Z}_M^P} \sum_{k \in \mathbb{Z}_2^P} f(n+Mk)z^{-n-Mk}. \qquad (2.210)$$

Therefore, for $z \in \mathbf{W}_M^P$,

$$\hat{f}(z) = \sum_{n \in \mathbb{Z}_M^P} \sum_{k \in \mathbb{Z}_2^P} f(n+Mk)z^{-n}. \qquad (2.211)$$

Thus **P3** $\Rightarrow$ **P1**, and **P1** $\Rightarrow$ **P3** since the z-transform is one-to-one on $\mathbb{Z}_M^P$. ∎

2.6   Conclusions and discussion

We can summarize the results presented here in the following way. Theorems 2.5 and 2.19 give the necessary and sufficient conditions for the construction of first generation wavelets. Lemmas 2.11 and 2.24 show how to obtain higher level auxiliary filters needed for the iteration step. Theorems 2.10 and 2.23 describe this iteration, showing how to obtain higher-generation wavelet bases. From a strictly mathematical point of view these results are all that is necessary to create discrete wavelet bases. However, the decomposition and reconstruction of a signal by a direct implementation of such a basis is not always efficient from a computational point of view, as discussed in Subsection 2.4.7. Faster computation is possible when short filters are employed in a recursive filter-bank structure. Theorem 2.9 and (2.206)–(2.207) establish that the recursive filter-bank structure does in fact implement the decomposition and reconstruction of a signal in terms of the orthonormal wavelet bases we have developed.

An algorithm is given for the construction of any and all wavelet filters for the decomposition of one-dimensional signals. Filters for higher-dimensional signal spaces can be realized as tensor products of one-dimensional filters, but the direct design of multidimensional filters with desirable properties remains a problem for further research.

The problems of frequency localization and simultaneous time-frequency localization of one-dimensional wavelet filters are discussed because of their importance to data compression and coding. While the mapping of a filter in the signal space to a closest wavelet is seen to preserve its frequency localization, the mapping destroys time localization. The problem of simultaneous time-frequency localization is an important problem so far as wavelets are concerned (see, e.g., [D1]).

Acknowledgments.   The authors would like to acknowledge helpful discussions with Professors Pascal Auscher, Jerry Cox, Dan Fuhrmann, and Guido Weiss at Washington University, St. Louis. The first author would also like to acknowledge helpful conversations with Jay Epperson, which took place while consulting at Daniel H. Wagner Associates.

Bibliography

[C]   A. Cohen. 1992. "Biorthogonal wavelets," in *Wavelets: A Tutorial in Theory and Applications*, C.K. Chui, ed., Boston: Academic Press, pp. 123–152.

[CDFe] A. Cohen, I. Daubechies, and J.-C. Feauveau. 1992. "Biorthogonal bases of compactly supported wavelets," *Commun. Pure Appl. Math.* **45**: 485–560.

[CoMW] R.R. Coifman, Y. Meyer, and V. Wickerhauser. 1992. "Wavelet analysis and signal processing," in *Wavelets and Their Applications*, M. Ruskai, G. Beylkin, R. Coifman, I. Daubechies, S. Mallat, Y. Meyer, and L. Raphael, eds., Boston: Jones and Bartlett, pp. 153–178.

[D1]   I. Daubechies. 1990. "The wavelet transform, time-frequency localization, and signal analysis," *IEEE Trans. Inform. Theory* **36**: 961–1005.

[D2]   I. Daubechies. 1992. *Ten Lectures on Wavelets*, Philadelphia: Soc. Ind. Appl. Math.

[FK]   M. Frazier and A. Kumar. 1990. "The discrete orthonormal wavelet transform: An introduction," Tech. Report No. WUCS-90-30, Dept. of Computer Sci., Washington Univ., St. Louis.

[K]   A. Kumar. 1990. "Time-frequency analysis and transform coding," Ph.D. Thesis, Dept. of Computer Sci., Washington Univ., St. Louis.

[KFuFJ] A. Kumar, D.R. Fuhrmann, M. Frazier, and B. Jawerth. 1992. "A new transform for time-frequency analysis," *IEEE Trans. Signal Process.* **40**: 1697–1707.

[RVe]   O. Raoul and M. Vetterli. 1992. "Wavelets and signal processing," *IEEE Signal Process. Mag.* **40**: 14–38.

[V]   P.P. Vaidyanathan. 1990. "Multirate digital filters, filter banks, polyphase networks, and applications: A tutorial," *Proc. IEEE* **78**: 56–93.

[VeH]   M. Vetterli and C. Herley. 1992. "Wavelets and filter banks: Theory and design," *IEEE Trans. Signal Process.* **40**: 2207–2232.

[S]   R. Strichartz. 1993. "Construction of orthonormal wavelets," this volume, Chap. 1.

3

Gabor frames for L^2 and related spaces

John J. Benedetto and David F. Walnut

ABSTRACT   The basic theory of frames is reviewed, and special topics dealing with Gabor frames and decompositions are developed.   These topics include Gabor decompositions of L^1 and of Bessel potential spaces. (Sobolev spaces are Bessel potential spaces.)   Frames of translates in L^2 are characterized; and the Balian–Low theorem for L^2 is proved. The former result is not only useful for the Gabor theory, but is the basis of multiresolution analysis frames; the latter result is related to the classical uncertainty principle inequality.

CONTENTS

0-8493-8271-8/94/$0.00 + $.50
© 1994 by CRC Press, Inc.

97

3.1 Introduction

The theory of frames was formulated by Duffin and Schaeffer [DuSc] to address difficult problems from the theory of nonharmonic Fourier series. This latter topic is an area of classical analysis whose central results are contained in the profound contributions of Paley and Wiener (1934) [PWi], Levinson (1940) [Le], and Beurling and Malliavin (1960s) [BeuMa1], [BeuMa2]. It is a subject which explicitly goes back to G.D. Birkhoff's work on Sturm–Liouville functions (1917), and subsequent trigonometric results by J.L. Walsh (1921). The field of nonharmonic Fourier series also has components anchored in the study of bases, e.g., the contributions of Köthe [Kö] and Bari [Ba]. A remarkable treatment of the whole subject, including the role of frames, is due to Young [Y].

The theory of frames re-emerged in 1986 at the dawn of the "wavelet era" in a compelling paper by Daubechies, Grossmann, and Meyer [DGroMe]. Since then, there have been further contributions by Daubechies emphasizing applications in signal processing, e.g., [D1], [D2, Chapter 3], and work by Feichtinger and Gröchenig [FGr2] firmly establishing the historical relationships from physics between frame-like decompositions and group representations. There is also interesting work by others, including a research tutorial by Heil and the second named author [HW].

The purpose of this chapter is to deal with topics in the theory of frames which do not play a major role in the aforementioned research, but which seem to us to be of fundamental interest for further developments. We have concentrated on Gabor frames. Generally, frames provide a mathematical tool for studying signal representation in cases where oversampling is inherent, such as some biological systems, and where stability is essential. Our emphasis on the Gabor case, as opposed to the wavelet case, is based on the fact that wavelets have received the lion's share of recent attention, e.g., [D2], [FrJaw], [FrJawWe], [Me], and on our opinion that decompositions based on translations and modulations, viz., Gabor decompositions, are a natural means of dealing with some of the most important spaces and operators in analysis.

Our point of view is decidedly mathematical, with the hope that this will ultimately make our presentation more useful and with the knowledge that there are more "applied" expositions of some of the topics. The table of contents spells out the material presented, and there is a brief explanatory discussion and/or motivation at the beginning of each section. For example, in Section 3.5 the relationship between the Balian–Low phenomenon and the classical uncertainty principle inequality is pointed out and referenced; and, early in Section 3.6, it is noted that the more popular Sobolev spaces are special cases of the lesser known Bessel potential spaces.

The prerequisites for this chapter are elementary functional analysis as found in [Ru, Part I] and elementary harmonic analysis as found in [StWe, Chapter I]. Besides the usual notation from analysis in these books and the books by Hörmander

[Hö] and Schwartz [S], we shall use the conventions and notation described in Section 3.2.

3.2 Notation

The integral over $\mathbb{R}^d$ is designated by "$\int$," and $\hat{\mathbb{R}}^d$ $(= \mathbb{R}^d)$ is the dual group of $\mathbb{R}^d$. In this chapter, the sum $\sum a_n$ over $\mathbb{Z}^d$ designates $\lim_{N \to \infty} \sum_{|n| \le N} a_n$, where $n = (n_1, \ldots, n_d)$ and $|n|^2 = \sum n_i^2$. The d-dimensional torus $\mathbb{T}^d$ is the compact group $\mathbb{R}^d/\mathbb{Z}^d$. The Fourier transform $\hat{f}$ of f on $\mathbb{R}^d$ is

$$\hat{f}(\gamma) = \int f(t) e^{-2\pi i t \cdot \gamma} \, dt, \quad \gamma \in \hat{\mathbb{R}}^d;$$

and the inverse Fourier transform of F on $\hat{\mathbb{R}}^d$ is designated $\check{F}$. The norm of $f \in L^p(\mathbb{R}^d)$ is $\|f\|_p$, and if $f \in X \ne L^p(\mathbb{R}^d)$ the norm of f is $\|f\|_X$. $C_c^\infty(\mathbb{R}^d)$ is the space of infinitely differentiable functions having compact support. We also use the following notation:

$\tau_a f(t) \equiv f(t - a)$,

$e_b(t) \equiv e^{2\pi i t \cdot b}$,

$|S|$ is the Lebesgue measure of the set $S \subseteq \mathbb{R}^d$,

$\#F$ is the cardinality (counting measure) of the set $F \subseteq \mathbb{Z}^d$,

$\mathbf{1}_S$ is the characteristic function of the set $S \subseteq \mathbb{R}^d$,

$S^\sim$ is the complement of the set $S \subseteq \mathbb{R}^d$,

$B(T)$ is the ball of radius T about the origin in $\mathbb{R}^d$,

$\delta_{m,n} \equiv \begin{cases} 1, & \text{if } m = n, \\ 0, & \text{if } m \ne n, \end{cases}$

span X is the vector space generated by the elements of the set X,

$\overline{\text{Span}}\, X$ is the closure of span X,

supp(f) is the support of f,

$Q \equiv [0, 1)^d \times [0, 1)^d$.

3.3 Frames for Hilbert spaces

The basic results in this section are due to Duffin and Schaeffer [DuSc] as part of their profound contribution to the theory of nonharmonic Fourier series.

Let H be a separable complex Hilbert space. A linear map $L: H_1 \to H_2$ between two such Hilbert spaces is a *topological isomorphism* if it is bijective and continuous, in which case its inverse L^{-1} is also continuous.

Definition/Remark 3.1

(a) A sequence $\{x_n : n \in \mathbb{Z}^d\}$ in a Hilbert space H is a *frame* for H if there exist $A, B > 0$ such that

$$\forall y \in H, \quad A\|y\|^2 \leq \sum |\langle y, x_n \rangle|^2 \leq B\|y\|^2,$$

where the norm of $y \in H$ is $\|y\| = \langle y, y \rangle^{1/2}$. A and B are the *frame bounds*, and a frame is *tight* if $A = B$. A frame is *exact* if it is no longer a frame whenever any one of its elements is removed.

(b) A sequence $\{x_n : n \in \mathbb{Z}^d\}$ in a Hilbert space H is a *Bessel sequence* if the *Bessel map*,

$$L : H \rightarrow \ell^2(\mathbb{Z}^d)$$

$$y \mapsto \{\langle y, x_n \rangle\},$$

is a well-defined linear map.

(c) If $\{x_n : n \in \mathbb{Z}^d\}$ is a Bessel sequence with Bessel map L, then it is elementary to prove that

$$\exists B > 0, \quad \text{such that } \forall y \in H, \; \sum |\langle y, x_n \rangle|^2 \leq B\|y\|^2, \tag{3.1}$$

and

$$\forall c = \{c_n\} \in \ell^2(\mathbb{Z}^d), \; \exists y \equiv L^* c = \lim_{N \to \infty} \sum_{|n| \leq N} c_n x_n \text{ in } H, \tag{3.2}$$

where $L^* : \ell^2(\mathbb{Z}^d) \rightarrow H$ is the adjoint of L and $|n|^2 = \sum n_j^2$, $n = (n_1, n_2, \ldots, n_d)$, e.g., [Be5, Section 3.5]. Bessel sequences play a role in Section 3.6 and in [Be5, Section 3.11].

(d) For a given sequence $\{x_n : n \in \mathbb{Z}^d\} \subseteq H$, S denotes the map,

$$S : H \rightarrow H$$

$$y \mapsto \sum \langle y, x_n \rangle x_n.$$

If $\{x_n : n \in \mathbb{Z}^d\} \subseteq H$ is a Bessel sequence with Bessel map L, then $L^* L$ is continuous and

$$S = L^* L, \tag{3.3}$$

cf. [Be5, Theorem 6.7]. Clearly, S is a positive operator, written $S \geq 0$, i.e.,

$$\forall y \in H, \; \langle Sy, y \rangle = \sum |\langle y, x_n \rangle|^2 \geq 0.$$

In particular, $S = S^*$.   ▯

THEOREM 3.2
Let $\{x_n : n \in \mathbb{Z}^d\} \subseteq H$.

(a) *The following are equivalent.*

    i. $\{x_n\}$ *is a frame for H with frame bounds A and B.*

    ii. $S: H \rightarrow H$ *is a topological isomorphism with norm bounds* $\|S\| \leq B$ *and* $\|S^{-1}\| \leq A^{-1}$.

(b) *In the case of either condition in part a, we have that*

$$AI \leq S \leq BI, \qquad B^{-1}I \leq S^{-1} \leq A^{-1}I,$$

$\{S^{-1}x_n\}$ *is a frame for H with frame bounds* B^{-1} *and* A^{-1}, *and, for all* $y \in H$,

$$y = \sum \langle y, S^{-1}x_n \rangle x_n \tag{3.4}$$

and

$$y = \sum \langle y, x_n \rangle S^{-1}x_n. \tag{3.5}$$

$\{S^{-1}x_n\}$ *is the dual frame of* $\{x_n\}$, *(3.4) is the frame decomposition of y, and (3.5) is the dual frame decomposition of y. (I is the identity map, and* $S \leq BI$ *means that* $\langle (BI - S)y, y \rangle \geq 0$ *for each* $y \in H.)$

PROOF

(a) Assume statement ii. Since $\|S\| \leq B$, we have

$$\langle Sy, y \rangle \leq \|Sy\| \|y\| \leq B\|y\|^2 = \langle By, y \rangle.$$

As a consequence the second inequality in the definition of frame is valid. Next, note that $S^{-1} \geq 0$ since

$$\langle S^{-1}y, y \rangle = \langle x, Sx \rangle = \langle Sx, x \rangle \geq 0.$$

Then, observe that $AI \leq S$ if and only if $S^{-1} \leq A^{-1}I$. (This is a consequence of a basic fact from functional analysis which allows us to preserve operator inequalities upon composition with a positive map as long as the map commutes with both sides of a given inequality, e.g., [He, page 269], [Ru, page 324].) $S^{-1} \leq A^{-1}I$ holds since

$$\langle S^{-1}y, y \rangle \leq \|S^{-1}y\| \|y\| \leq A^{-1}\|y\|^2 = A^{-1}\langle y, y \rangle.$$

Thus, $AI \leq S$, and, as a consequence, the first inequality in the definition of frame is valid. Statement i is proved.

Now assume statement i. We begin by setting

$$S_N y = \sum_{|n| \leq N} \langle y, x_n \rangle x_n,$$

and proving that $S: H \rightarrow H$ is a well-defined continuous linear map. If $N \geq M$ and $y \in H$, then, by the definition of norm, Hölder's inequality, and the upper frame bound, we compute

$$\|S_N y - S_M y\|^2 = \sup_{\|x\|=1} \left| \sum_{M < |n| \leq N} \langle y, x_n \rangle \langle x_n, x \rangle \right|^2$$

$$\leq \sup_{\|x\|=1} B\|x\|^2 \sum_{M<|n|\leq N} |\langle y,x_n\rangle|^2 = B \sum_{M<|n|\leq N} |\langle y,x_n\rangle|^2;$$

and the right side goes to 0 as $M,N \to \infty$ by another application of the second norm inequality. Thus, since H is complete,

$$\lim_{N\to\infty} S_N y \equiv Sy$$

exists in H. To check that S is continuous, we compute as above, and obtain

$$\|Sy\|^2 = \sup_{\|x\|=1} \left| \sum \langle y,x_n\rangle\langle x_n,x\rangle \right|^2$$

$$\leq \sup_{\|x\|=1} B\|x\|^2 B\|y\|^2 = B^2\|y\|^2.$$

S is injective because of the lower frame bound. In fact, if $Sy = 0$ then $\|y\| = 0$, since

$$\langle Sy,y\rangle = \sum |\langle y,x_n\rangle|^2 = 0$$

and by the frame condition. Thus, $y = 0$.

Finally, we shall verify that S is surjective and S^{-1} is continuous. There are several ways to prove this; we choose to use the Neumann expansion, e.g., [GoGol, page 70] or [Ru, page 231] (which emphasizes the Banach algebra property of the set of bounded linear operators on H). To this end, and using the consequence $AI \leq S \leq BI$ of our frame hypothesis, we have

$$\frac{1}{B}S \leq I \quad \text{and} \quad \frac{1}{B}S \geq \frac{A}{B}I,$$

so that

$$0 \leq I - \frac{1}{B}S \equiv P_1 \leq I - \frac{A}{B}I = \frac{B-A}{B}I \equiv P_2.$$

Since $P_1,P_2 \geq 0$ (and hence P_1,P_2 are self-adjoint) and $P_2 \geq P_1$, we have $\|P_2\| \geq \|P_1\|$; this is a consequence of the fact that

$$\|P\| = \sup\{|\langle Px,x\rangle|: \|x\| = 1\}$$

for all self-adjoint bounded operators P, e.g., [GoGol, page 110]. Therefore

$$\left\| I - \frac{1}{B}S \right\| \leq \left\| \frac{B-A}{B}I \right\| = \frac{B-A}{B} < 1;$$

and so, by the Neumann expansion (which goes back to Liouville),

$$\left(\frac{1}{B}S \right)^{-1} : H \to H$$

is a bounded linear operator. Consequently,

$$S^{-1} = \frac{1}{B}\left(\frac{1}{B}S \right)^{-1} = \frac{1}{B}\sum_{k=0}^{\infty}\left(I - \frac{1}{B}S \right)^k \qquad (3.6)$$

is a well-defined bounded linear operator on H, and our claims about S and S^{-1} are true. The convergence of the sum in (3.6) is uniform convergence over closed balls of H.

(b) The first three claims of part b are consequences of calculations in part a. For the frame decomposition, we use the definition of S and the self-adjointness of S^{-1} (since $S^{-1} \geq 0$) to obtain $y = SS^{-1}y$ as the right side of (3.4). For the dual frame decomposition, we use the definition of S and the continuity of S^{-1} to obtain $y = S^{-1}Sy$ as the right side of (3.5). ∎

REMARK 3.3 Clearly, an orthonormal basis $\{x_n : n \in \mathbb{Z}^d\}$ of H is a tight frame with $A = B = 1$. Conversely, if $\{x_n\}$ is a tight frame with $A = B = 1$ and if $\|x_n\| = 1$ for each n, then $\{x_n\}$ is an orthonormal basis of H. To verify this latter claim, we first use tightness and $A = 1$ to write,

$$\forall m, \ \|x_m\|^2 = \|x_m\|^4 + \sum_{n \neq m} |\langle x_m, x_n \rangle|^2;$$

and obtain that $\{x_n\}$ is orthonormal since each $\|x_n\| = 1$. To conclude the proof, we invoke the well-known fact that if $\|x_n\|$ is orthonormal then it is an orthonormal basis of H if and only if

$$\forall y \in H, \ \|y\|^2 = \sum |\langle y, x_n \rangle|^2,$$

e.g., [Be5, Theorems 3.1–3.4]. ∎

This result, characterizing tight frames which are orthonormal bases, is actually a special case of the following beautiful little result proved by Vitali [V] in 1921. He proved that an orthonormal sequence $\{x_n\} \subseteq L^2[a, b]$ is complete, and so $\{x_n\}$ is an orthonormal basis, if and only if

$$\forall t \in [a, b], \ \sum \left| \int_a^t x_n(u) \, du \right|^2 = t - a. \tag{3.7}$$

This is a stronger result than the aforementioned since (3.7) is tightness with $A = 1$ for functions $y = \mathbf{1}_{[a,t]}$.

Definition/Remark 3.4

(a) A sequence $\{x_n : n \in \mathbb{Z}^d\}$ in a Hilbert space H is *minimal* if

$$\forall n \in \mathbb{Z}^d, \ x_n \notin \overline{\text{Span}} \{x_k : k \neq n\}.$$

The sequences $\{x_n : n \in \mathbb{Z}^d\}$, $\{y_n : n \in \mathbb{Z}^d\} \subseteq H$ are *biorthonormal* if

$$\forall m, n \in \mathbb{Z}^d, \ \langle x_m, y_n \rangle = \delta_{m,n}.$$

(b) For a given sequence $\{x_n\} \subseteq H$, an elementary Hahn–Banach argument shows that there is a sequence $\{y_n\} \subseteq H$ so that $\{x_n\}, \{y_n\}$ are biorthonormal

if and only if $\{x_n\}$ is minimal. Furthermore, $\{y_n\}$ is uniquely determined if and only if $\{x_n\}$ is not only minimal in H but also $\overline{\text{Span}}\,\{x_n\} = H$.

(c) The notion of *stationary frames* was introduced in [Be4]. We mention this now since stationarity is used in conjunction with minimal and biorthonormal sequences, e.g., [Be5, Section 3.5g], and because there is a new characterization of frames using probabilistic ideas associated with stationarity [Hou].   ☐

THEOREM 3.5
Let $\{x_n : n \in \mathbb{Z}^d\} \subseteq H$ be a frame for H with frame bounds A and B.

(a) *For each $c \equiv \{c_n\} \in \ell^2(\mathbb{Z}^d)$, $y_c \equiv \sum c_n x_n$ converges in H and $\|y_c\|^2 \leq B\|c\|^2_{\ell^2(\mathbb{Z}^d)}$.*

(b) *For each $y \in H$, there is $c_y \equiv \{c_n\} \in \ell^2(\mathbb{Z}^d)$ such that $y = \sum c_n x_n \in H$. In fact, we can take $c_n = \langle y, S^{-1}x_n \rangle$ for each n.*

(c) *If $y \in H$ and $b_y \equiv \{b_n\} \subseteq \mathbb{C}$ have the property that $y = \sum b_n x_n \in H$ and $b_n \neq \langle y, S^{-1}x_n \rangle$ for some n, then $\{b_n\}$ is not of the form $\{\langle x, S^{-1}x_n \rangle\}$ for any $x \in H$, and*

$$\sum |b_n|^2 = \sum |\langle y, S^{-1}x_n \rangle|^2 + \sum |\langle y, S^{-1}x_n \rangle - b_n|^2. \qquad (3.8)$$

(d) *Let $X_m = \{x_n : x_n \neq x_m\}$, where m is fixed. Then X_m is a frame if $\langle x_m, S^{-1}x_m \rangle \neq 1$, and X_m is incomplete if $\langle x_m, S^{-1}x_m \rangle = 1$. Further, if $\langle x_m, S^{-1}x_m \rangle = 1$ then $\langle x_m, S^{-1}x_n \rangle = 0$ for all $n \neq m$.*

(e) *If $\{x_n\}$ is an exact frame then $\{x_n\}$, $\{S^{-1}x_n\}$ are biorthonormal, $\{S^{-1}x_n\}$ is the unique sequence in H which is biorthonormal to $\{x_n\}$, and X_m (defined in part d) is incomplete.*

(f) *$\{x_n\}$ is an exact frame if and only if it is minimal.*

PROOF   Letting $y_N = \sum_{|n| \leq N} c_n x_n$, part a is a consequence of the upper frame condition, the fact that H is complete, and the estimate,

$$\|y_N - y_M\|^2 = \sum_{M < |n| \leq N} c_n \langle x_n, y_N - y_M \rangle$$

$$\leq \left(\sum_{M < |n| \leq N} |c_n|^2 \right)^{1/2} \left(\sum |\langle y_N - y_M, x_n \rangle|^2 \right)^{1/2}$$

$$\leq B\|y_N - y_M\| \left(\sum_{M < |n| \leq N} |c_n|^2 \right)^{1/2}.$$

Part b is a corollary of Theorem 3.2.

For part c, it is obvious from (3.8) that there is no loss of generality in assuming that $b_y \in \ell^2(\mathbb{Z}^d)$. If $b_y \in \ell^2(\mathbb{Z}^d)$ and $c_n \equiv \langle y, S^{-1}x_n \rangle$ for a given $y = Sx$ and all n, then $c_n = \langle x, x_n \rangle$ and $0 = \sum (b_n - c_n) x_n$. Thus, if we take the inner product of this last equation with x, we obtain

$$\sum (b_n - c_n) \overline{c_n} = 0. \tag{3.9}$$

Equation (3.8) is a consequence of (3.9) since $\sum b_n \overline{c_n}$ is the right side of (3.8), which is seen by expanding this right side and using (3.9) (again). Finally (for part c), if $\{b_n\}$ is of the form $\{\langle x, S^{-1} x_n \rangle\}$ for some x, then $0 = \sum \langle x - y, S^{-1} x_n \rangle S^{-1} x_n$; and so, if S_{-1} is the "S-operator" for the frame $\{S^{-1} x_n\}$, then $S_{-1}(x - y) = 0$, which implies $x = y$. This contradicts our hypothesis on b_y.

Assume part d for the moment. Hence, since $\{x_n\}$ is exact, then $\langle x_m, S^{-1} x_m \rangle = 1$ for all m. Therefore, X_m is incomplete from the second statement of part d, and the biorthonormality is obtained from the last statement of part d. The uniqueness of $\{S^{-1} x_n\}$ utilizes part d again, as well as Definition/Remark 3.4b. Part e is proved. For part f, if $\{x_n\}$ is exact then part e and Definition/Remark 3.4b yield minimality. Conversely, if $\{x_n\}$ is minimal, then $x_n \notin \overline{\text{Span}} X_m$ for each m. To prove that $\{x_n\}$ is an exact frame, we must show that each X_m is not a frame. If any X_m were a frame then by Theorem 3.2b, $x_m = \sum_{n \neq m} c_n x_n$, and this fails by minimality. Part f is proved.

It remains to prove part d. For a fixed m, we define $c_n \equiv \langle x_m, S^{-1} x_n \rangle$, and have the decompositions $x_m = \sum c_n x_n = \sum \delta_{m,n} x_n$. Applying (3.8), for $b_n = \delta_{m,n}$, we have

$$\sum_{n \neq m} |c_n|^2 = \frac{1 - |c_m|^2 - |c_m - 1|^2}{2} < \infty.$$

If $c_m = 1$ then $\sum_{n \neq m} |c_n|^2 = 0$, i.e., $\langle x_n, S^{-1} x_m \rangle = \langle x_m, S^{-1} x_n \rangle = 0$ for each $n \neq m$. This is the last statement of part d. However, $S^{-1} x_m \neq 0$ since $c_m = 1$. Thus, X_m is incomplete since it annihilates a nonzero element.

If $c_m \neq 1$, then the frame decomposition of x_m yields

$$x_m = \frac{1}{1 - c_m} \sum_{n \neq m} c_n x_n.$$

Writing $\langle y, x_m \rangle$ in terms of this formula, setting $C \equiv 1 + |1 - c_m|^{-2} \sum_{n \neq m} |c_n|^2$, and using our frame hypothesis, we have

$$\forall y \in H, \quad \frac{A}{C} \|y\|^2 \leq \frac{1}{C} \sum |\langle y, x_n \rangle|^2 \leq \sum_{n \neq m} |\langle y, x_n \rangle|^2 \leq B \|y\|^2.$$

Thus, X_m is a frame with frame bounds A/C and B. ∎

DEFINITION 3.6

(a) *A sequence $\{x_n : n \in \mathbb{Z}^d\} \subseteq H$ is a Schauder basis or basis of H if each $y \in H$ has a unique decomposition $y = \sum c_n(y) x_n$. A basis $\{x_n\}$ is an unconditional basis if there exists $C > 0$ such that for all $F \subseteq \mathbb{Z}^d$, where*

$\#F < \infty$, *and for all* $b_n, c_n \in \mathbb{C}$, *where* $n \in F$ *and* $|b_n| \leq |c_n|$,

$$\left\| \sum_{n \in F} b_n x_n \right\| \leq C \left\| \sum_{n \in F} c_n x_n \right\|.$$

An unconditional basis $\{x_n\}$ *is bounded if there exist* $A, B > 0$ *such that for all* n,

$$A \leq \|x_n\| \leq B.$$

(b) *A sequence* $\{x_n : n \in \mathbb{Z}^d\} \subseteq H$ *is a Riesz basis if there is an orthonormal basis* $\{u_n\}$ *and a topological isomorphism* $L : H \to H$ *such that* $Lx_n = u_n$ *for each* n.

(c) *Note that a frame which is not an exact frame can not be a basis. To see this, choose* m *so that* X_m *(defined in Theorem 3.5d) is a frame for* H. *In particular,* X_m *is complete in* H, *which implies that* $\langle x_m, S^{-1}x_m \rangle \neq 1$ *by Theorem 3.5d. Thus, the frame decomposition and trivial representation* $(x_m = x_m)$ *in* H *are different; and this nonuniqueness violates the definition of basis.*

Köthe (1936) [Kö] proved the implication, b implies c, of the following theorem, cf. [Y, pages 32–36]. The implication, c implies b, is straightforward; and the equivalence of a and c is found in [Y, pages 188–189]. Nina Bari's contributions (1951) [Ba] not only influence all aspects of the theorem, but have a broader impact on subtle issues ultimately dealing with Riemann's sets of uniqueness for trigonometric series. Because of Young's presentation [Y] we omit the proof.

THEOREM 3.7
Let $\{x_n : n \in \mathbb{Z}^d\}$ *be contained in* H. *The following are equivalent.*

(a) $\{x_n\}$ *is an exact frame for* H;

(b) $\{x_n\}$ *is a bounded unconditional basis of* H;

(c) $\{x_n\}$ *is a Riesz basis of* H.

Clearly, then, orthonormal bases are exact frames, which in turn are frames.

Example 3.8

(a) One of Duffin and Schaeffer's main results [DuSc, Theorem I] is the fact that $\{e_{-t_m}\}$ is a frame for $L^2[-\Omega, \Omega]$ in the case that $\{t_n : n \in \mathbb{Z}\} \subseteq \mathbb{R}$ is uniformly dense with uniform density greater that 2Ω. The notion of uniform density is defined in [Be5, Section 6], and the role of the Duffin and Schaeffer theorem in the modern theory of irregular sampling is also explained in [Be5, Section 6].

(b) If $\{e_{-t_m}\}$ is a frame for $L^2[-\Omega, \Omega]$ and $X_m \equiv \{e_{-t_n} : n \neq m\}$ is not a frame for $L^2[-\Omega, \Omega]$, then $\{e_{-t_n}\}$ is an exact frame for $L^2[-\Omega, \Omega]$ [DuSc, Lemma VII]. This result fails for arbitrary frames, cf. Theorem 3.5d.   ▯

REMARK 3.9

(*a*) Note that in light of Theorem 3.5a,b, we can assert that the map,

$$\ell^2(\mathbb{Z}^d) \to H$$
$$\{c_n\} \mapsto \sum c_n x_n,$$

is a continuous surjection in the case $\{x_n : n \in \mathbb{Z}^d\}$ is a frame.

(*b*) In light of Theorem 3.5a,b, and the importance of frame decompositions, it is natural to investigate the precise relationship between frames $\{x_n : n \in \mathbb{Z}^d\}$ for H and decompositions of the form $y = \sum c_n x_n$. In fact, a sequence $\{x_n : n \in \mathbb{Z}^d\} \subseteq H$ is a frame if and only if there is $C > 0$ such that for all $y \in H$,

$$\sum |\langle y, x_n \rangle|^2 < \infty, \qquad (3.10)$$

$$\exists c_y \equiv \{c_n\} \in \ell^2(\mathbb{Z}^d), \quad \text{such that } y = \sum c_n x_n \text{ in } H, \qquad (3.11)$$

and

$$\|c_y\|_{\ell^2(\mathbb{Z}^d)} \leq C\|y\|, \qquad (3.12)$$

cf. the definition and treatment of atoms in Section 3.6.

To prove this result, note that if $\{x_n\}$ is a frame then (3.10) follows by the upper frame bound, (3.11) is immediate by the frame decomposition and setting $c_y = \{\langle y, S^{-1} x_n \rangle\}$, and (3.12) again follows by the upper frame bound, where we can take $C = A^{-1/2}$ in (3.12).

Conversely, if we assume (3.10)–(3.12) then the upper frame bound is a consequence of (3.10) and (3.1). To obtain the lower frame bound we use (3.11) and (3.12) to make the following estimate:

$$|\langle y, x \rangle|^2 = \left| \left\langle y, \sum c_n(x) x_n \right\rangle \right|^2$$
$$= \left| \sum \overline{c_n(x)} \langle y, x_n \rangle \right|^2 \leq C^2 \|x\|^2 \sum |\langle y, x_n \rangle|^2.$$

Thus,

$$\|y\|^2 \equiv \sup_{\|x\|=1} |\langle y, x \rangle|^2 \leq C^2 \sum |\langle y, x_n \rangle|^2,$$

and so the lower frame bound is obtained with $A = 1/C^2$. ∎

3.4 Gabor frames for L^2

The point of view for this section goes back to the seminal ideas of Gabor (1946) [G] concerning time-frequency analysis in engineering problems, and even earlier in physics concerning the analysis of Weyl operators and coherent states, e.g., [vo, pages 405–407]. Gabor's approach and Duffin and Schaeffer's theory of frames lead naturally to the study of Gabor fames which is the subject of this section. Besides the references cited in Section 3.1, we also mention [H, Chapter 7].

Naturally, the Fourier transform plays a major role in this subject. The *Fourier transform* of $f \in L^1(\mathbb{R}^d)$ is

$$\hat{f}(\gamma) = \int f(t)e^{-2\pi it\cdot\gamma}\,dt, \qquad \gamma \in \hat{\mathbb{R}}^d,$$

where integration is over $\mathbb{R}^d$ and $\hat{\mathbb{R}}^d$ is $\mathbb{R}^d$ considered as the dual group. This allows us to distinguish notationally between "time" $t \in \mathbb{R}^d$ and "frequency" $\gamma \in \hat{\mathbb{R}}^d$. The *inverse Fourier transform* of F is denoted by $\check{F}$.

DEFINITION 3.10

(a) *Let $g \in L^2(\mathbb{R}^d)$ and let $a = (a_1,\ldots,a_d), b = (b_1,\ldots,b_d) \in \mathbb{R}^d$. Assume each $a_j, b_k > 0$ and define the translations and modulations,*

$$\tau_{na}f(t) \equiv f(t - na) \quad and \quad e_{mb} \equiv e^{2\pi it\cdot mb},$$

respectively, where $m,n \in \mathbb{Z}^d$, $f \in L^2(\mathbb{R}^d)$, $na = (n_1a_1,\ldots,n_da_d)$, and $mb = (m_1b_1,\ldots,m_db_d)$.

(b) *The Gabor or Weyl–Heisenberg system $\{\varphi_{m,n}: m, n \in \mathbb{Z}^d\}$ of coherent states is defined by*

$$\varphi_{m,n} = e_{mb}\tau_{na}g.$$

Clearly,

$$\hat{\varphi}_{m,n} = \tau_{mb}(e_{-na}\hat{g}).$$

If $\{\varphi_{m,n}\}$ is a frame for $L^2(\mathbb{R}^d)$, it is called a Gabor frame. Obviously, the Gabor system $\{\varphi_{m,n}\}$ is a Gabor frame for $L^2(\mathbb{R}^d)$ if and only if $\{\tau_{na}e_{mb}g\}$ is a frame for $L^2(\mathbb{R}^d)$.

(c) *The Zak transform of $f \in L^1_{\text{loc}}(\mathbb{R}^d)$ is formally defined as*

$$Zf(x,\omega) = a^{d/2} \sum_{k\in\mathbb{Z}^d} f(xa + ka)e^{2\pi ik\cdot\omega}, (x,\omega) \in \mathbb{R}^d \times \hat{\mathbb{R}}^d,$$

where multiplication is component-wise and $a^{d/2} = \prod a_j^{1/2}$, $a_j > 0$. There have been independent formulations by Weil–Brezin (1964, 1970), Igusa (1972), Auslander–Tolimieri (1975), and, of course, Zak (1967), cf. [J]. Formally, Zf is quasi-periodic in the sense that

$$Zf(x + n,\omega) = e^{-2\pi in\cdot\omega}Zf(x,\omega) \quad and \quad Zf(x,\omega + n) = Zf(x,\omega) \qquad (3.13)$$

for $(x,\omega) \in \mathbb{R}^d \times \hat{\mathbb{R}}^d$ and $n \in \mathbb{Z}^d$. In particular, Zf is completely determined on $Q \equiv [0,1)^d \times [0,1)^d$.

(d) *Let $a, b \in \mathbb{R}^d$, with each $a_j, b_k = 1$. As we shall see in Theorem 3.16, the Zak transform is useful in characterizing various closure properties of Gabor systems. This is a consequence of Theorem 3.15 and the calculation, for $a = b = 1$ and $g \in L^2(\mathbb{R}^d)$, that*

$$Z\varphi_{m,n}(x, \omega) = e_m(x)e_n(\omega)Zg(x, \omega) \equiv e_{m,n}(x, \omega)Zg(x, \omega), \ (x, \omega) \in \mathbb{R}^d \times \hat{\mathbb{R}}^d, \ (3.14)$$

see the normalization in [Be1, Section 2] and the discussion in [DGroMe], [Be5, Section 2.8].

The relationship between frames and dual frames was established in Theorem 3.2. The following result shows that the dual frame of a Gabor frame is a Gabor frame. An application is made in the proof of [Be5, Theorem 4.2].

THEOREM 3.11
Let $\{\varphi_{m,n} : m, n \in \mathbb{Z}^d\}$ be the Gabor frame for $L^2(\mathbb{R}^d)$ associated with $g \in L^2(\mathbb{R}^d)$ and $a, b \in \mathbb{R}^d$, and let S be the associated topological isomorphism, e.g., Definition/Remark 3.1 and Theorem 3.2. Then for all $m, n \in \mathbb{Z}^d$ and for all $f \in L^2(\mathbb{R}^d)$,

$$S^{-1}(e_{mb}\tau_{na}f) = e_{mb}\tau_{na}S^{-1}f. \tag{3.15}$$

PROOF First note that

$$S^{-1}(e_{mb}\tau_{na}f) = S^{-1}(e_{mb}\tau_{na}SS^{-1}f),$$

and so it is sufficient to verify that

$$S(e_{mb}\tau_{na}f) = e_{mb}\tau_{na}Sf. \tag{3.16}$$

To see this, we calculate

$$S(e_{mb}f_1) = \sum_{p,q}\langle e_{mb}f_1, e_{pb}\tau_{qa}g\rangle e_{pb}\tau_{qa}g$$

$$= \sum_{p,q}\langle f_1, e_{(p-m)b}\tau_{qa}g\rangle e_{pb}\tau_{qa}g$$

$$= e_{mb}\sum_{p,q}\langle f_1, e_{(p-m)b}\tau_{qa}g\rangle e_{(p-m)b}\tau_{qa}g = e_{mb}Sf_1,$$

and

$$S(\tau_{na}f_2) = \sum_{p,q}\langle \tau_{na}f_2, e_{pb}\tau_{qa}g\rangle e_{pb}\tau_{qa}g$$

$$= \sum_{p,q}\langle f_2, e_{pb}\tau_{(q-n)a}g\rangle e^{-2\pi i na \cdot pb}e_{pb}\tau_{qa}g$$

$$= \tau_{na}\left(\sum_{p,q}\langle f_2, e_{pb}\tau_{(q-n)a}g\rangle e_{pb}\tau_{(q-n)a}g\right) = \tau_{na}Sf_2.$$

Thus, if $f_1 = \tau_{na} f$, we have $S(e_{mb}\tau_{na}f) = e_{mb}S(\tau_{na}f) = e_{mb}\tau_{na}Sf$; and (3.16) is obtained. ∎

Let $g \in L^2(\mathbb{R}^d)$ and $a \in \mathbb{R}^d$, where each $a_j > 0$. We set

$$G(t) \equiv \sum_{n \in \mathbb{Z}^d} |g(t - na)|^2. \qquad (3.17)$$

G is defined a.e. since $g \in L^2(\mathbb{R}^d)$ and

$$\int_I G(t)\, dt = \int |g(t)|^2\, dt < \infty,$$

where $I \subseteq \mathbb{R}^d$ is the "rectangle" determined by the origin and $a \in \mathbb{R}^d$. Clearly, G is a-periodic on $\mathbb{R}^d$. The following result is a necessary condition for Gabor frames in terms of G.

THEOREM 3.12
Let $\{\varphi_{m,n}: m, n \in \mathbb{Z}^d\}$ be the Gabor frame for $L^2(\mathbb{R}^d)$ associated with $g \in L^2(\mathbb{R}^d)$ and $a, b \in \mathbb{R}^d$, and let A, B be frame bounds. If G is defined by (3.17) then

$$A \prod b_j \leq G(t) \leq B \prod b_j \quad a.e., \qquad (3.18)$$

and, in particular, $g \in L^\infty(\mathbb{R}^d) \cap L^2(\mathbb{R}^d)$.

PROOF Let $J \subseteq \mathbb{R}^d$ be the "rectangle" determined by the origin and $(1/b_1, \ldots, 1/b_d) \in \mathbb{R}^d$, and set $J_n = J + na$. It is easy to see that $\{b^{d/2}e_{mb}\}$ is an orthonormal sequence in $L^2(J_n)$, and, by a classical calculation, $\{b^{d/2}e_{mb}\}$ is an orthonormal basis of $L^2(J_n)$.

Now suppose ess inf $G(t) < A \prod b_j$. Thus, for $\epsilon > 0$ satisfying $A - \epsilon > 0$, there is $\Delta \subseteq J$ for which $|\Delta| > 0$ and

$$G(t) \leq (A - \epsilon) \prod b_j \quad \text{a.e. on } \Delta.$$

Letting $f \equiv \mathbf{1}_\Delta$, we use the aforementioned orthonormal basis with the Plancherel theorem to calculate

$$\sum_{m,n} |\langle f, e_{mb}\tau_{na}g \rangle|^2 = \sum_n \left(\sum_m |\langle f\tau_{na}\bar{g}, e_{mb} \rangle|^2 \right)$$

$$= \sum_n \frac{1}{\prod b_j} \int |f(t)|^2 |g(t - na)|^2\, dt$$

$$= \frac{1}{\prod b_j} \int |f(t)|^2 G(t)\, dt \leq A\|f\|_2^2 - \epsilon\|f\|_2^2,$$

thereby contradicting the frame hypothesis.

An analogous calculation yields the upper bound of (3.18). ∎

THEOREM 3.13

Let $\{\varphi_{m,n}: m, n \in \mathbb{Z}^d\}$ be the Gabor system associated with $g \in L^2(\mathbb{R}^d)$ and $a, b \in \mathbb{R}^d$, and assume there exist $A, B > 0$ such that G, defined by (3.17), satisfies the condition

$$A \leq G(t) \leq B \text{ a.e.}$$

If $\operatorname{supp} g \subseteq I$, where I is the "rectangle" determined by the origin and $(1/b_1, \dots, 1/b_d) \in \mathbb{R}^d$, then $\{\varphi_{m,n}\}$ is a Gabor frame for $L^2(\mathbb{R}^d)$ with frame bounds $A \prod b_j^{-1}$ and $B \prod b_j^{-1}$, and

$$\forall f \in L^2(\mathbb{R}^d), \ S^{-1}f = \frac{\prod b_j}{G} f. \tag{3.19}$$

PROOF Let $f \in L^2(\mathbb{R}^d)$. Then $\operatorname{supp} f \tau_{na} \bar{g} \subseteq I_n \equiv I + na$ for $n \in \mathbb{Z}^d$, $|I_n| = \prod b_j^{-1}$, and $f \tau_{na} \bar{g} \in L^2(\mathbb{R}^d)$ since $G \leq B$ a.e. Using the orthonormal basis $\{b^{d/2} e_{mb}\}$ and the Plancherel theorem, we compute

$$\sum_{m,n} |\langle f, e_{mb} \tau_{na} g \rangle|^2 = \prod b_j^{-1} \sum_n \int_{I_n} |f(t)|^2 |g(t - na)|^2 \, dt$$

$$= \prod b_j^{-1} \int |f(t)|^2 G(t) \, dt, \tag{3.20}$$

where the second equality utilizes the support hypothesis. Equations (3.20) and the necessary condition $A \leq G \leq B$ yield the first part of the result.

Since $\{b^{d/2} e_{mb}\}$ is an orthonormal basis of $L^2(I_n)$, we have the Fourier expansion

$$\forall f \in L^2(\mathbb{R}^d), \ \left(\prod b_j^{-1} \right) f \tau_{na} \bar{g} = \sum \langle f \tau_{na} \bar{g}, e_{mb} \rangle e_{mb} \tag{3.21}$$

in $L^2(\mathbb{R}^d)$. Because $\{\varphi_{m,n}\}$ is a frame, the associated operator S is a topological isomorphism; and, using (3.21), it can be written as

$$Sf = \sum_n \tau_{na} g \left(\prod b_j^{-1} \right) f \tau_{na} \bar{g} = \left(\prod b_j^{-1} \right) Gf. \tag{3.22}$$

Equation (3.19) is obtained by substituting $S^{-1}f$ for f in (3.22). $\blacksquare$

REMARK 3.14 Theorem 3.13 is the most elementary of the sufficient conditions for a Gabor frame. There is a generalization of the support constraint on g, in terms of an autocorrelation condition, which Daubechies used to study Gabor frames [D1]. This generalization led the second named author to use Wiener amalgam space criteria to obtain Gabor frames ([HW, Theorem 4.1.8, Theorem 4.2.1, Theorem 4.2.2], [W1], [W2]).

Autocorrelation conditions are required in Section 3.9 for Gabor decompositions. Fourier transforms of autocorrelations are power spectra; and we have used Wiener amalgam spaces for proving Wiener–Plancherel formulas. These are analogues of the Plancherel theorem for infinite energy signals, and they play a role in (power) spectral estimation, e.g., [Be3] and the reference there to work by the first named author with Benke and Evans. ∎

In the following result we mention the $p < 2$ cases because of our $p = 1$ result in Section 3.9. The most useful case is $p = 2$.

THEOREM 3.15
If $1 \leq p \leq 2$ then

$$Z: L^p(\mathbb{R}^d) \to L^p(Q)$$

is a well-defined continuous linear injection with operator norm $\|Z\| \leq 1$. If $p = 2$, then Z is a unitary map.

PROOF For simplicity let $a = (1, \ldots, 1)$. If $p = 2$, we see that $\{f(x+k)e^{2\pi i k \cdot \omega} : k \in \mathbb{Z}^d\}$ is orthogonal over Q, and, hence,

$$\|Zf\|_{L^2(Q)}^2 = \sum_k \int_Q |f(x+k)e^{2\pi i k \cdot \omega}|^2 \, dx \, d\omega = \|f\|_2^2.$$

Thus, the result is proved once we verify surjectivity.

Let $\{\varphi_{m,n}\}$ be the Gabor system corresponding to $g = \mathbf{1}_{[0,1)^d}$ and $a = b = 1$. $\{\varphi_{m,n}\}$ is an orthonormal basis of $L^2(\mathbb{R}^d)$ and $Z\varphi_{m,n}(x, \omega) = e_m(x)e_n(\omega)$ for $(x, \omega) \in Q$. Thus, Z maps the orthonormal basis $\{\varphi_{m,n}\}$ onto the orthonormal basis $\{e_m(x)e_n(\omega)\}$, and so Z is surjective, cf. the proof of unitarity in [Be1].

If $p = 1$ and $f \in L^1(\mathbb{R}^d)$ it is clear that $\|Zf\|_{L^1(Q)} \leq \|f\|_1$. Convexity theory yields the result for $p \in (1, 2)$, e.g., [StWe]. ∎

The following result has had partial formulations in the coherent states literature for many years going back to the von Neumann reference given at the beginning of the section. The proof is elementary, depending essentially on Theorem 3.15.

THEOREM 3.16
Let $\{\varphi_{m,n} : m, n \in \mathbb{Z}^d\}$ be the Gabor system associated with $g \in L^2(\mathbb{R}^d)$ and $a = b = (1, \ldots, 1) \in \mathbb{R}^d$.

(a) *$\{\varphi_{m,n}\}$ is complete in $L^2(\mathbb{R}^d)$ if and only if $Zg \neq 0$ a.e.*

(b) *$\{\varphi_{m,n}\}$ is minimal and complete in $L^2(\mathbb{R}^d)$ if and only if $1/Zg \in L^2(Q)$.*

(c) *$\{\varphi_{m,n}\}$ is an orthonormal basis of $L^2(\mathbb{R}^d)$ if and only if $|Zg| = 1$ a.e.*

(d) *$\{\varphi_{m,n}\}$ is a frame for $L^2(\mathbb{R}^d)$ with frame bounds A and B if and only if*

$$A \leq |Zg(x, \omega)|^2 \leq B \text{ a.e.} \tag{3.23}$$

In this case, $\{\varphi_{m,n}\}$ is an exact frame.

PROOF By unitarity and (3.14), $\{\varphi_{m,n}\}$ is complete in $L^2(\mathbb{R}^d)$ if and only if $\{e_{m,n}Zg\}$ is complete in $L^2(Q)$. If $\{e_{m,n}Zg\}$ is complete and $X \equiv$

$\{(x, \omega): Zg(x, \omega) = 0\}$, then we let $F \equiv \mathbf{1}_X$ and compute $\langle F, e_{m,n}Zg \rangle = \langle 0, e_{m,n} \rangle = 0$ for each $m, n \in \mathbb{Z}^d$. By the completeness hypothesis, $F = 0$ a.e. on Q and so $Zg \neq 0$ a.e.

Conversely, when $Zg \neq 0$ a.e. we must show that if F annihilates $\{e_{m,n}Zg\}$ then $F = 0$ a.e. Clearly, $F\overline{Zg} \in L^1(Q)$, and the annihilation hypothesis and completeness of $\{e_{m,n}\}$ imply $F\overline{Zg} = 0$ a.e. This implies $F = 0$ a.e. since $Zg \neq 0$ a.e. Part a is proved.

Part c follows from part d and the fact that orthonormal bases are exact frames with frame bounds $A = B = 1$, cf. Theorem 3.7 and the proof in [Be1, Theorem 3.5].

For part d, we first assume (3.23) and need only show $\{e_{m,n}Zg\}$ is a frame for $L^2(Q)$. Since $\{e_{m,n}\}$ is an orthonormal basis of $L^2(Q)$ we have

$$\forall F \in L^2(Q), \quad \sum_{m,n} |\langle F, e_{m,n}Zg \rangle|^2 = \|F\overline{Zg}\|^2_{L^2(Q)}. \tag{3.24}$$

The result is immediate from this equality, (3.23), and the definition of frame.

Conversely, by unitarity, $\{e_{m,n}Zg\}$ is a frame for $L^2(Q)$. Using (3.24) and the frame inequality for each $F \in L^2(Q)$, we obtain (3.23) by assuming (3.23) fails and obtaining a contradiction.

To finish part d, we invoke Theorem 3.5e, and shall verify that $X_{j,k} \equiv \{e_{m,n}Zg: (m, n) \neq (j, k)\}$ is incomplete. Without loss of generality, we take $j = k = 0 \in \mathbb{Z}^d$. By (3.23), $F \equiv 1/\overline{Zg} \in L^2(Q) \setminus \{0\}$; and if $(m, n) \neq (0, 0)$ then $\langle F, e_{m,n}Zg \rangle = 0$ so that $F \neq 0$ is orthogonal to $X_{0,0}$. Thus, $X_{0,0}$ is incomplete.

Part b follows from part a once we verify minimality assuming $1/Zg \in L^2(Q)$. This follows as in the previous paragraph. ∎

Example 3.17

(a) Let $\psi \in L^2(\mathbb{R})$. The *wavelet system* $\{\psi_{m,n}: m, n \in \mathbb{Z}\}$ is defined as

$$\psi_{m,n}(t) = 2^{m/2}\psi(2^m t - n).$$

There are *wavelet frames* for $L^2(\mathbb{R})$, and much of *wavelet theory* has centered around the construction of wavelet orthonormal bases, e.g., [D2], [M], [Me]. The function ψ is called a *wavelet* in the case $\{\psi_{m,n}\}$ is an orthonormal basis.

(b) By means of Theorem 3.16c, it is elementary to characterize those elements $g \in L^2(\mathbb{R}^d)$ for which $\{\varphi_{m,n}: m, n \in \mathbb{Z}\}$ is an orthonormal basis of $L^2(\mathbb{R}^d)$ in the case $a = b = (1, \ldots, 1) \in \mathbb{R}^d$, e.g., [Be1, Example 3.7b]. In particular, $g = \varphi_{0,0}$ need not have the property that $\int g(t)\, dt = 0$. On the other hand, if a wavelet system $\{\psi_{m,n}: m, n \in \mathbb{Z}\}$ is orthogonal and $\psi, \hat{\psi} \in L^1 \cap L^2$, then $\int \psi(t)\, dt = 0$, e.g., [Bat2]. We mention this vanishing moment property since it is a critical property of wavelets, and since it provides an insight into the distinction between Balian–Low phenomena for Gabor and wavelet systems, e.g., Example 3.25. The Balian–Low phenomenon is proved for Gabor systems in the following section.

(c) Since the vanishing moment property is sometimes proved under stronger hypotheses than $\psi, \hat{\psi} \in L^1 \cap L^2$, let us sketch the proof of the assertion in part b. If $\psi \in L^1(\mathbb{R}) \cap L^2(\mathbb{R}) \setminus \{0\}$ then $\hat{\psi}$ is continuous. Then, letting $\hat{\psi} \in L^1(\hat{\mathbb{R}})$, the L^1 Fourier inversion formula, e.g., [StWe], allows us to take ψ as a continuous function and to state that $(\hat{\psi})^{\vee} = \psi$. In particular, $\psi(-2^{m_0} n_0) \equiv \psi(-t_0) \neq 0$ for some $m_0, n_0 \in \mathbb{Z}$, where $m_0 < 0$. For any $m \leq m_0$, $t_0 = 2^m n$ where $n = 2^{m_0 - m} n_0$. The orthogonality hypothesis $\psi \perp \psi_{m,n}$, the continuity of $\hat{\psi}$, and the dominated convergence theorem allow us to write

$$0 = \hat{\psi}(0) \int \overline{\hat{\psi}(\gamma)} e^{2\pi i t_0 \gamma} \, d\gamma, \qquad (3.25)$$

when we let $m \to -\infty$. The right side of (3.25) is $\hat{\psi}(0)\bar{\psi}(-t_0)$ by the L^1 Fourier inversion formula, and so the proof is complete since $\psi(-t_0) \neq 0$. ▯

3.5   The Balian–Low theorem for Gabor frames

The Balian–Low theorem (BLT) reflects one of the most fundamental properties of the Fourier transform, and is intimately associated with topics such as uniqueness and uncertainty principle inequalities. These topics are treated in [Be5, Section 5, e.g., Examples 5.8 and 5.12a]. In this section we shall prove the Balian–Low theorem (Theorem 3.24) for Gabor frames.

This theorem and its verifications have led a full life since Balian's original statement (1981) [B], e.g., [Bat1], [D1], [D2, pages 108–112], [DJan], [DJaJo], [Lo]. In particular, there has been a full range of proofs. There are proofs which are really only true for the special case of orthonormal bases; there are persuasive arguments overshadowing technical flaws; there are formal calculations which we are unable to validate mathematically; and there are proofs which are true in the context of distributional but not classical differentiation. The proof we present is nondistributional and is due to Heil and the authors [BeHW]. Our reason for formulating it is related to expositions of the aforementioned verifications, and is fully explained in [BeHW].

The critical ideas for either the distributional or our nondistributional proof are both analytical and topological; the former are due to Coifman and Semmes, the latter are used by Balian and Low, and the synthesis is due to Daubechies [D1].

Our first step in proving the BLT for $g \in L^2(\mathbb{R})$ is to bound the variation of $G \equiv Zg$ over small squares $Q(x, \omega; r) \subseteq \mathbb{R} \times \hat{\mathbb{R}}$, where

$$Q(x, \omega; r) = \left\{ (y, \gamma) \in \mathbb{R} \times \hat{\mathbb{R}} : y \in [x - r/2, x + r/2), \ \gamma \in [\omega - r/2, \omega + r/2) \right\}.$$

This result, Lemma 3.18, is reminiscent of Hardy and Littlewood's (1928) characterization of one-dimensional bounded variation in terms of the L^1 norm.

LEMMA 3.18

Let f, $tf(t)$, and $\gamma \hat{f}(\gamma)$ be square integrable, set $F = Zf$, and let $\epsilon \in \mathbb{R}$ and $(x_0, \omega_0) \in [-3/2, 3/2] \times \hat{\mathbb{R}}$ be given. Then

$$\left(\iint_{Q(x_0,\omega_0;1)} |F(x - \epsilon, \omega) - F(x, \omega)|^2 \, dx \, d\omega \right)^{1/2} \leq 2\pi |\epsilon| \left(\int |\gamma \hat{f}(\gamma)|^2 \, d\gamma \right)^{1/2},$$

$$(3.26)$$

$$\left(\iint_{Q(x_0,\omega_0;1)} |F(x, \omega - \epsilon) - F(x, \omega)|^2 \, dx \, d\omega \right)^{1/2}$$

$$(3.27)$$

$$\leq 2\pi |\epsilon| \left[\left(\int |tf(t)|^2 \, dt \right)^{1/2} + 2 \left(\int |f(t)|^2 \, dt \right)^{1/2} \right].$$

PROOF   To prove (3.26) note that

$$F(x - \epsilon, \omega) = \sum_k f(x - \epsilon + k) e^{2\pi i k \omega} = Z(\tau_\epsilon f)(x, \omega).$$

Thus, by the unitarity of Z and the Plancherel theorem,

$$\left(\iint_{Q(x_0,\omega_0;1)} |F(x - \epsilon, \omega) - F(x, \omega)|^2 \, dx \, d\omega \right)^{1/2}$$

$$= \left(\iint_{Q(x_0,\omega_0;1)} |Z(\tau_\epsilon f - f)(x, \omega)|^2 \, dx \, d\omega \right)^{1/2}$$

$$= \left(\int |\tau_\epsilon f(t) - f(t)|^2 \, dt \right)^{1/2}$$

$$= \left(\int |e^{-2\pi i \epsilon \gamma} - 1|^2 |\hat{f}(\gamma)|^2 \, d\gamma \right)^{1/2}$$

$$= \left(\int |2 \sin(\pi \epsilon \gamma)|^2 |\hat{f}(\gamma)|^2 \, d\gamma \right)^{1/2}$$

$$\leq 2\pi |\epsilon| \left(\int |\gamma \hat{f}(\gamma)|^2 \, d\gamma \right)^{1/2}.$$

To prove (3.27), note that

$$F(x, \omega - \epsilon) = \sum_k f(x + k) e^{2\pi i k (\omega - \epsilon)}$$

$$= e^{2\pi i \epsilon x} \sum_k e^{-2\pi i \epsilon (x + k)} f(x + k) e^{2\pi i k \omega}$$

$$= e^{2\pi i \epsilon x} Z(e_{-\epsilon} f)(x, \omega).$$

By the unitarity of Z and the Plancherel theorem we therefore have

$$
\left(\iint\limits_{Q(x_0,\omega_0;1)} |F(x,\omega-\epsilon) - F(x,\omega)|^2 \, dx \, d\omega \right)^{1/2}
$$

$$
= \left(\iint\limits_{Q(x_0,\omega_0;1)} |e^{2\pi i\epsilon x} Z(e_{-\epsilon}f)(x,\omega) - Zf(x,\omega)|^2 \, dx \, d\omega \right)^{1/2}
$$

$$
\leq \left(\iint\limits_{Q(x_0,\omega_0;1)} |Z(e_{-\epsilon}f - f)(x,\omega)|^2 \, dx \, d\omega \right)^{1/2}
$$

$$
+ \left(\iint\limits_{Q(x_0,\omega_0;1)} |e^{2\pi i\epsilon x} - 1|^2 |Zf(x,\omega)|^2 \, dx \, d\omega \right)^{1/2}
$$

$$
= \left(\int |2\sin\pi\epsilon t|^2 |f(t)|^2 \, dt \right)^{1/2} + \left(\iint\limits_{Q(x_0,\omega_0;1)} |2\sin\pi\epsilon x|^2 |Zf(x,\omega)|^2 \, dx \, d\omega \right)^{1/2}
$$

$$
\leq 2\pi|\epsilon| \left[\left(\int |tf(t)|^2 \, dt \right)^{1/2} + 2 \left(\int |f(t)|^2 \, dt \right)^{1/2} \right]. \quad \blacksquare
$$

Next we shall prove a result (Lemma 3.20) similar to the above, obtaining a bound on the oscillation of F over a square of side $r < 1$. First, we fix some notation and prove some technicalities in the following lemma.

LEMMA 3.19
Let $f \in L^2(\mathbb{R})$, $s \in \mathbb{R}$, and $r > 0$ be given. Denote by $E(s,r)$ the set

$$
E(s,r) = \bigcup_{n\in\mathbb{Z}} [s - \frac{r}{2} + n, s + \frac{r}{2} + n].
$$

Then

(a) $Zf(x,\omega) = e^{-2\pi i x\omega} Z\hat{f}(-\omega,x) = e^{-2\pi i x\omega} Z\check{f}(\omega,-x)$,

(b) $Zf(x,\omega) \cdot \mathbf{1}_{E(s,r)}(x) = Z(f \cdot \mathbf{1}_{E(s,r)})(x,\omega)$ *for all* $(x,\omega) \in \mathbb{R} \times \hat{\mathbb{R}}$,

(c) $\lim_{r\to 0} \|f \cdot \mathbf{1}_{E(s,r)}\|_2 = 0$ *uniformly in s.*

PROOF To prove a, use the Poisson summation formula for $f \in \mathcal{S}(\mathbb{R})$, then use the density of $\mathcal{S}(\mathbb{R})$ in $L^2(\mathbb{R})$ to complete the proof. Statement b follows immediately from the definition of the Zak transform and the fact that $\mathbf{1}_{E(s,r)}$ is 1-periodic. To prove c, first fix any $\epsilon > 0$. Since $|f|^2$ is an integrable function, we can find an $N > 0$ such that $\int_{|t|>N} |f(t)|^2 \, dt \leq \epsilon/2$, and a $\delta > 0$ such that $\int_A |f(t)|^2 \, dt \leq \epsilon/2$ whenever $|A| \leq \delta$. Now, given any $s \in \mathbb{R}$, the number of $n \in \mathbb{Z}$

such that $[s - r/2 + n, s + r/2 + n] \cap [-N, N] \neq \emptyset$ is at most $2N + r + 1$, which is independent of s. Therefore, $|E(s,r) \cap [-N,N]| \leq (2N+2)r$ for all s when $r \leq 1$. Letting $r_0 = \delta/(2N+2)$, we therefore have

$$\|f \cdot \mathbf{1}_{E(s,r)}\|_2^2 \leq \int_{|t| > N} |f(t)|^2 \, dt + \int_{E(s,r) \cap [-N,N]} |f(t)|^2 \, dt \leq \epsilon$$

for all s and all $r \leq r_0$.   ∎

LEMMA 3.20
Let f, $tf(t)$, and $\gamma \hat{f}(\gamma)$ be square integrable, set $F = Zf$, and let $\epsilon \in \mathbb{R}$, $(x_0, \omega_0) \in [-3/2, 3/2] \times \hat{\mathbb{R}}$, and $0 < r < 1$ be given. Then

$$\left(\iint_{Q(x_0,\omega_0;r)} |F(x - \epsilon, \omega) - F(x,\omega)|^2 \, dx \, d\omega \right)^{1/2} \leq 2\pi |\epsilon| \|\gamma \hat{f}(\gamma) \cdot \mathbf{1}_{E(1-\omega_0,r)}(\gamma)\|_2,$$

$$(3.28)$$

$$\left(\iint_{Q(x_0,\omega_0;r)} |F(x, \omega - \epsilon) - F(x,\omega)|^2 \, dx \, d\omega \right)^{1/2}$$

$$\leq 2\pi |\epsilon| \left(\|tf(t) \cdot \mathbf{1}_{E(x_0,r)}(t)\|_2 + 2\|f \cdot \mathbf{1}_{E(x_0,r)}\|_2 \right).$$

$$(3.29)$$

PROOF   Using arguments similar to those in the proof of Lemma 3.18, and parts a and b of Lemma 3.19, we have

$$\left(\iint_{Q(x_0,\omega_0;r)} |F(x - \epsilon, \omega) - F(x,\omega)|^2 \, dx \, d\omega \right)^{1/2}$$

$$= \left(\iint_{Q(x_0,\omega_0;r)} |Z(\tau_\epsilon f - f)(x, \omega)|^2 \, dx \, d\omega \right)^{1/2}$$

$$= \left(\iint_{Q(x_0,\omega_0;r)} |Z(e_{-\epsilon} \hat{f} - \hat{f})(-\omega, x)|^2 \, dx \, d\omega \right)^{1/2}$$

$$= \left(\iint_{Q(x_0,-\omega_0;r)} |Z(e_{-\epsilon} \hat{f} - \hat{f})(\omega, x)|^2 \, d\omega \, dx \right)^{1/2}$$

$$= \left(\iint_{Q(x_0,1-\omega_0;r)} |Z(e_{-\epsilon} \hat{f} - \hat{f})(\omega, x)|^2 \, d\omega \, dx \right)^{1/2}$$

$$= \left(\iint_{Q(x_0,1-\omega_0;r)} |Z(e_{-\epsilon}\hat{f} - \hat{f})(\omega,x) \cdot \mathbf{1}_{E(1-\omega_0,r)}(\omega)|^2 \, d\omega \, dx \right)^{1/2}$$

$$\leq \left(\iint_{Q(x_0,1-\omega_0;1)} |Z((e_{-\epsilon}\hat{f} - \hat{f}) \cdot \mathbf{1}_{E(1-\omega_0,r)})(\omega,x)|^2 \, d\omega \, dx \right)^{1/2}$$

$$= \left(\int_{E(1-\omega_0,r)} |e_{-\epsilon}\hat{f}(\gamma) - \hat{f}(\gamma)|^2 \, d\gamma \right)^{1/2}$$

$$\leq 2\pi|\epsilon| \| \gamma \hat{f}(\gamma) \cdot \mathbf{1}_{E(1-\omega_0,r)}(\gamma) \|_2,$$

from which (3.28) follows.

For (3.29), we have that

$$\left(\iint_{Q(x_0,\omega_0;r)} |F(x,\omega - \epsilon) - F(x,\omega)|^2 \, dx \, d\omega \right)^{1/2}$$

$$\leq \left(\iint_{Q(x_0,\omega_0;r)} |Z(e_{-\epsilon}f - f)(x,\omega)|^2 \, dx \, d\omega \right)^{1/2}$$

$$+ \left(\iint_{Q(x_0,\omega_0;r)} |e^{2\pi i \epsilon x} - 1|^2 |Zf(x,\omega)|^2 \, dx \, d\omega \right)^{1/2}$$

$$\leq \left(\iint_{Q(x_0,\omega_0;r)} |Z(e_{-\epsilon}f - f)(x,\omega) \cdot \mathbf{1}_{E(x_0,r)}(x)|^2 \, dx \, d\omega \right)^{1/2}$$

$$+ 2\pi|\epsilon| \left(\iint_{Q(x_0,\omega_0;r)} |xZf(x,\omega) \cdot \mathbf{1}_{E(x_0,r)}(x)|^2 \, dx \, d\omega \right)^{1/2}$$

$$\leq \left(\iint_{Q(x_0,\omega_0;1)} |Z((e_{-\epsilon}f - f) \cdot \mathbf{1}_{E(x_0,r)})(x,\omega)|^2 \, dx \, d\omega \right)^{1/2}$$

$$+ 4\pi|\epsilon| \left(\iint_{Q(x_0,\omega_0;1)} |Z(f \cdot \mathbf{1}_{E(x_0,r)})(x,\omega)|^2 \, dx \, d\omega \right)^{1/2}$$

$$= \left(\int_{E(x_0,r)} |e_{-\epsilon}f(t) - f(t)|^2 \, dt \right)^{1/2} + 4\pi|\epsilon| \left(\int_{E(x_0,r)} |f(t)|^2 \, dt \right)^{1/2}.$$

This last term is bounded by the right side of (3.29).　　■

THEOREM 3.21
Let f, ϵ, r, and (x_0, ω_0) be as in Lemma 3.20 and in addition fix $0 < c < 1$. Then there are functions $C_1(r)$ and $C_2(r)$, depending on f alone, such that $\lim_{r \to 0} C_i(r) = 0$ for $i = 1, 2$ and

$$\iint\limits_{Q(x_0, \omega_0; cr)} |F(x - \epsilon, \omega) - F(x, \omega)| \, dx \, d\omega \leq cr|\epsilon|C_1(r), \tag{3.30}$$

$$\iint\limits_{Q(x_0, \omega_0; cr)} |F(x, \omega - \epsilon) - F(x, \omega)| \, dx \, d\omega \leq cr|\epsilon|C_2(r). \tag{3.31}$$

PROOF This result follows from Lemma 3.20 and a straightforward application of Hölder's inequality, where

$$C_1(r) = 2\pi \sup_{\omega \in \hat{\mathbb{R}}} \|\gamma \hat{f}(\gamma) \cdot \mathbf{1}_{E(\omega, r)}(\gamma)\|_2$$

and

$$C_2(r) = 2\pi \sup_{x \in \mathbb{R}} \left(\|tf(t) \cdot \mathbf{1}_{E(x,r)}(t)\|_2 + 2\|f \cdot \mathbf{1}_{E(x,r)}\|_2 \right).$$

The convergence to 0 of the C_i for $i = 1, 2$ follows from Lemma 3.19c. ∎

The next step in our proof of the BLT is to introduce a continuous approximation G_r of $G = Zg$. This idea is due to Coifman and Semmes. It turns out that G_r satisfies enough of a quasiperiodicity condition to allow a proof of the BLT similar to those which proceed by proving the continuity of the Zak transform.

Let $\rho = \mathbf{1}_Q$ and, for $r > 0$, let $\rho_r(x, \omega) = r^{-2} \rho(x/r, \omega/r)$. Then $\{\rho_r\}$ is an approximate identity on $\mathbb{R} \times \hat{\mathbb{R}}$ as $r \to 0$. We define the mean-value function,

$$G_r(x, \omega) = \iint\limits_Q G(x - y, \omega - \alpha)\rho_r(y, \alpha) \, dy \, d\alpha.$$

Some basic properties of the function G_r are the following, see, e.g., [D1]:

THEOREM 3.22
Let $g \in L^2(\mathbb{R})$ and $a = b = 1$ generate the Gabor frame $\{\varphi_{m,n}\}$ with frame bounds A and B, and let $G = Zg$. Then

(a) $|G_r(x_1, \omega_1) - G_r(x_2, \omega_2)| \leq 2r^{-1}B^{1/2}(|x_1 - x_2| + |\omega_1 - \omega_2|),$
(b) $\sup_{(x,\omega) \in Q} |G_r(x, \omega)| \leq B^{1/2},$
(c) $G_r(x, \omega + 1) = G_r(x, \omega),$
(d) $G_r(x+1, \omega) = e^{-2\pi i \omega} G_r(x, \omega) + \psi_r(x, \omega)$, where $\sup_{(x,\omega) \in Q} |\psi_r(x, \omega)| \leq \pi r B^{1/2}$,
(e) *for fixed (y, α), $(x, \omega) \in \mathbb{R} \times \hat{\mathbb{R}}$,*

$$|G(y, \alpha) - G_r(y, \alpha)| \geq A^{1/2} - 2r^{-1}B^{1/2}(|x - y| + |\omega - \alpha|) - |G_r(x, \omega)|$$

and, in particular, for fixed $(x, \omega) \in Q$ *and* $c < 1$,

$$\inf_{(y,\alpha)\in Q(x,\omega;cr)} |G(y,\alpha) - G_r(y,\alpha)| \geq A^{1/2} - 2cB^{1/2} - |G_r(x,\omega)|.$$

PROOF   To see a, note that

$$|G_r(x_1, \omega_1) - G_r(x_2, \omega_2)|$$

$$= \frac{1}{r^2} \left| \iint_{Q(z_1,\lambda_1;r)} G(y,\alpha) \, dy \, d\alpha - \iint_{Q(z_2,\lambda_2;r)} G(y,\alpha) \, dy \, d\alpha \right|$$

$$= \frac{1}{r^2} \left| \iint_{Q(z_1,\lambda_1;r)\Delta Q(z_2,\lambda_2;r)} G(y,\alpha) \, dy \, d\alpha \right|$$

$$\leq \frac{B^{1/2}}{r^2} |Q(z_1,\lambda_1;r)\Delta Q(z_2,\lambda_2;r)|$$

by (3.23), where Δ is the symmetric difference operator and $(z_j, \lambda_j) \equiv (x_j - \frac{r}{2}, \omega_j - \frac{r}{2})$ for $j = 1, 2$. An elementary geometrical argument yields the estimate,

$$|Q(z_1,\lambda_1;r)\Delta Q(z_2,\lambda_2;r)| \leq 2r(|x_1 - x_2| + |\omega_1 - \omega_2|),$$

and so part a is proved.

Conclusion b is immediate from (3.23), the approximate identity hypothesis, and the definition of G_r. Part c follows from the fact that G is periodic in the second variable. To see d, note that

$$G_r(x + 1, \omega) = \iint_Q G(x + 1 - y, \omega - \alpha)\rho_r(y,\alpha) \, dy \, d\alpha$$

$$= \iint_Q e^{-2\pi i(\omega-\alpha)} G(x - y, \omega - \alpha)\rho_r(y,\alpha) \, dy \, d\alpha$$

$$= e^{-2\pi i\omega} \iint_Q G(x - y, \omega - \alpha)\rho_r(y,\alpha) \, dy \, d\alpha + \psi_r(t,\omega),$$

where

$$\psi_r(x, \omega) = e^{-2\pi i\omega} \iint_Q (e^{2\pi i\alpha} - 1)G(x - y, \omega - \alpha)\rho_r(y,\alpha) \, dy \, d\alpha.$$

We estimate

$$|\psi_r(x, \omega)| \leq \iint_Q |e^{2\pi i\alpha} - 1||G(x - y, \omega - \alpha)|\rho_r(y,\alpha) \, dy \, d\alpha$$

$$\leq B^{1/2} \iint\limits_Q |2\pi\alpha| \rho_r(y,\alpha) \, dy \, d\alpha$$

$$\leq B^{1/2} \pi r.$$

Finally to see e observe from a that

$$|G(y,\alpha) - G_r(y,\alpha)| \geq |G(y,\alpha)| - |G_r(y,\alpha) - G_r(x,\omega)| - |G_r(x,\omega)|$$

$$\geq A^{1/2} - \frac{2}{r} B^{1/2} (|x-y| + |\omega - \alpha|) - |G_r(x,\omega)|.$$

Since $(y,\alpha) \in Q(x,\omega; cr)$ implies that $|x - y| \leq cr/2$ and $|\omega - \alpha| \leq cr/2$, it follows that

$$\inf_{(y,\alpha) \in Q(x,\omega;cr)} |G(y,\alpha) - G_r(y,\alpha)| \geq A^{1/2} - 2cB^{1/2} - |G_r(x,\omega)|. \quad \blacksquare$$

LEMMA 3.23
Suppose that g, $tg(t)$, and $\gamma\hat{g}(\gamma)$ are square integrable. Then for any $(x,\omega) \in Q$, $c < 1$, and $r < 1$, we have that

$$\iint\limits_{Q(x,\omega;cr)} |G(y,\alpha) - G_r(y,\alpha)| \, dy \, d\alpha \leq cr^2 C(r),$$

where $C(r)$ does not depend on (x,ω) and $\lim_{r\to 0} C(r) = 0$.

PROOF

$$\iint\limits_{Q(x,\omega;cr)} |G(y,\alpha) - G_r(y,\alpha)| \, dy \, d\alpha$$

$$\leq \iint\limits_{Q(x,\omega;cr)} \iint\limits_Q |\rho_r(v,\beta)||G(y,\alpha) - G(y-v, \alpha - \beta)| \, dv \, d\beta \, dy \, d\alpha$$

$$= \iint\limits_Q |\rho_r(v,\beta)| \iint\limits_{Q(x,\omega;cr)} |G(y,\alpha) - G(y-v, \alpha - \beta)| \, dy \, d\alpha \, dv \, d\beta$$

$$\leq \iint\limits_Q |\rho_r(v,\beta)| \iint\limits_{Q(x,\omega;cr)} |G(y,\alpha) - G(y-v, \alpha)| \, dy \, d\alpha \, dv \, d\beta$$

$$+ \iint\limits_Q |\rho_r(v,\beta)| \iint\limits_{Q(x-v,\omega;cr)} |G(z,\alpha) - G(z, \alpha - \beta)| \, dz \, d\alpha \, dv \, d\beta,$$

where we have applied the triangle inequality. Since ρ_r is supported in the square $Q(r/2, r/2; r)$ and since $r < 1$, we have that $x - v \in [-1/2, 1]$. Thus, by Theorem 3.21, the above quantity is bounded by

$$cr^2 (C_1(r) + C_2(r)).$$

Putting $C(r) = C_1(r) + C_2(r)$ completes the proof. $\quad \blacksquare$

We are now prepared to prove the Balian–Low theorem.

THEOREM 3.24   (BLT)

Let $g \in L^2(\mathbb{R})$ and $a = b = 1$. If the corresponding Gabor system $\{\varphi_{m,n}\}$ is a Gabor frame for $L^2(\mathbb{R})$ (with frame bounds A and B), then either $tg(t) \notin L^2(\mathbb{R})$ or $\gamma\hat{g}(\gamma) \notin L^2(\hat{\mathbb{R}})$.

PROOF   We shall assume that $tg(t) \in L^2(\mathbb{R})$, $\gamma\hat{g}(\gamma) \in L^2(\hat{\mathbb{R}})$, and that $\{\varphi_{m,n}\}$ is a Gabor frame for $L^2(\mathbb{R})$; and obtain a contradiction.

First, we claim that $|G_r(x,\omega)| \geq A^{1/2}/2$ for all $(x,\omega) \in Q$ and all r sufficiently small. To see this, note that by Theorem 3.22e and Lemma 3.23,

$$cr^2 C(r) \geq \iint\limits_{Q(x,\omega;cr)} |G(y,\alpha) - G_r(y,\alpha)| \, dy \, d\alpha \geq c^2 r^2 \left(A^{1/2} - 2cB^{1/2} - |G_r(t,\omega)| \right).$$

Fixing c so small that $A^{1/2} - 2cB^{1/2} > A^{1/2}/2$, we have that

$$|G_r(x,\omega)| \geq A^{1/2} - 2cB^{1/2} - \frac{C(r)}{c}.$$

Since $C(r) \to 0$ as $r \to 0$, independently of (x,ω), the claim is proved.

It is well known that every continuous, real or complex valued nonvanishing function F on $\bar{Q}$ can be written in the form $F(x,\omega) = |F(x,\omega)|e^{i\varphi(x,\omega)}$, where φ is a continuous real-valued function on $\bar{Q}$, see, e.g., [RRe, pages 377–385]. Thus, because of the estimate in the previous paragraph, for each $r > 0$, there is a continuous, real-valued function φ_r on $\bar{Q}$ such that

$$\forall (x,\omega) \in \bar{Q}, \ G_r(x,\omega) = |G_r(x,\omega)|e^{i\varphi_r(x,\omega)}. \tag{3.32}$$

Consequently, by Theorem 3.22c, for each $r > 0$ and for each $x \in [0,1]$, there is $k_{r,x} \in \mathbb{Z}$ for which $\varphi_r(x,1) = \varphi_r(x,0) + 2\pi k_{r,x}$. Since φ_r is continuous on $\bar{Q}$, we have that $\varphi_r(x,1) - \varphi_r(x,0)$ is continuous on $[0,1]$. Therefore, $k_{r,x} = k_r$, i.e.,

$$\forall r > 0, \ \exists k_r \in \mathbb{Z}, \quad \text{such that } \forall x \in [0,1], \ \varphi_r(x,1) = \varphi_r(x,0) + 2\pi k_r. \tag{3.33}$$

Next, let us consider the function

$$\delta_r(x,\omega) = 1 + \frac{\psi_r(x,\omega)}{e^{-2\pi i\omega}G_r(x,\omega)}.$$

Clearly δ_r is continuous. Also, using Theorem 3.22b, d, and the inequality $|G_r(x,\omega)| \geq A^{1/2}/2$ that we verified at the beginning of this proof, we see that δ_r is nonvanishing on Q for small enough r. Hence, as in (3.32), there is a continuous function $\theta_r(x,\omega)$ such that

$$\delta_r(x,\omega) = |\delta_r(x,\omega)|e^{i\theta_r(x,\omega)}. \tag{3.34}$$

Note that by Theorem 3.22d, δ_r converges uniformly to 1 on $\bar{Q}$ as $r \to 0$, and hence θ_r converges uniformly on $\bar{Q}$ to a multiple of 2π. Also note that by definition,

$$G_r(x+1,\omega) = e^{-2\pi i\omega}G_r(x,\omega)\delta_r(x,\omega) \tag{3.35}$$

for every x and ω.

Using (3.34) and (3.35), we compute

$$
\begin{aligned}
|G_r(1,\omega)|e^{i\varphi_r(1,\omega)} &= G_r(1,\omega) \\
&= e^{-2\pi i\omega}G_r(0,\omega)\delta_r(0,\omega) \\
&= |G_r(0,\omega)\delta_r(0,\omega)|e^{-2\pi i\omega+i\theta_r(0,\omega)+i\varphi_r(0,\omega)} \\
&= |G_r(1,\omega)|e^{-2\pi i\omega+i\theta_r(0,\omega)+i\varphi_r(0,\omega)}.
\end{aligned}
$$

Hence, as in (3.33), for each $r > 0$ there is an integer l_r such that for all $\omega \in [0,1]$,

$$\varphi_r(1,\omega) = -2\pi\omega + \theta_r(0,\omega) + \varphi_r(0,\omega) + 2\pi l_r. \tag{3.36}$$

Combining (3.33) and (3.36) we have

$$
\begin{aligned}
0 &= (\varphi_r(0,0) - \varphi_r(1,0)) + (\varphi_r(1,0) - \varphi_r(1,1)) \\
&\quad + (\varphi_r(1,1) - \varphi_r(0,1)) + (\varphi_r(0,1) - \varphi_r(0,0)) \\
&= (-\theta_r(0,0) - 2\pi l_r) + (-2\pi k_r) + (-2\pi + \theta_r(0,1) + 2\pi l_r) + (2\pi k_r) \\
&= -2\pi + \theta_r(0,1) - \theta_r(0,0).
\end{aligned} \tag{3.37}
$$

Letting $r \to 0$, we find that $0 = -2\pi$, a contradiction. ∎

Example 3.25

(a) In Example 3.17 we verified a vanishing moment property for orthogonal wavelet systems, and alluded to its relationship with a Balian–Low phenomenon for wavelet systems analogous to the Balian–Low phenomenon for Gabor systems. Now that we've quantified the latter statement in Theorem 3.24, we can be more precise about the former.

First, Meyer's initial example ψ of a wavelet satisfies the property that $\hat{\psi} \in C_c^\infty(\hat{\mathbb{R}})$, see, e.g., [Me]; and, hence,

$$\|p(t)\psi(t)\|_2\|q(\gamma)\hat{\psi}(\gamma)\|_2 < \infty$$

for all polynomials p and q, cf. [Be5, Theorem 5.6, Corollary 5.7]. Thus, the exact analogue of Theorem 3.24 for wavelet systems is not valid even in the case of orthonormal bases. On the other hand, Battle [Bat2] used a refinement of the vanishing moment property to prove that if $\{\psi_{m,n}\}$ is an orthogonal wavelet system, then

$$\|e^{|t|}\psi(t)\|_2\|e^{|\gamma|}\hat{\psi}(\gamma)\|_2 = \infty.$$

This fact, that if $\{\psi_{m,n}\}$ is an orthogonal wavelet system, then ψ does not possess "exponential localization" in *both* time and frequency, is the Balian–Low phenomenon for orthogonal wavelet systems. Heil and the authors have made an elementary extension of this fact for the case of exact frames.

(b) The "algebraic" parts of the proof of Theorem 3.24, e.g., (3.37), can be used to verify that if $f \in L^2(\mathbb{R})$ and Zf is continuous on $\mathbb{R} \times \hat{\mathbb{R}}$, then Zf has a zero, cf. [Be5, (2.5) and (2.6)], which deals with the Gaussian. ☐

3.6 Gabor decompositions of Bessel potential spaces

We have seen that in a Hilbert space, H, the frame inequality (Definition/Remark 3.1a) implies that the operator S is a topological isomorphism of H. In fact, these notions are equivalent (Theorem 3.2a). From this, it follows that there are stable decompositions of elements in H in terms of the frame (Theorem 3.2b). The notion of having a decomposition in terms of a fixed set of "atoms" is also an extremely powerful tool in the study of general Banach spaces B. In each case, one would like to have a "frame inequality" in the sense that the norm of $x \in B$ is equivalent to some expression depending only on the magnitude of the sequence of decomposition coefficients. In a Hilbert space, the frame inequality implies the existence of an atomic decomposition, but in a Banach space such an inequality is not sufficient. It is therefore convenient to extend the notion of a frame (cf. Section 3.9) and to consider the relationship between so-called "Banach frames" and atomic decompositions.

There has been much recent work studying atomic decompositions of Banach spaces in which the atoms are wavelets or Gabor functions, e.g., [FrJaw], [FrJawWe], [F2], [FGr1], [FGr2]. An interesting class of such spaces is the so-called Bessel potential spaces $\mathcal{L}_s^2(\mathbb{R}^d)$, e.g., [StWe]. The Bessel potential spaces are of interest since they include the classical Sobolev spaces (Definition/Remark 3.26d), and they are examples of spaces which admit strongly convergent atomic decompositions in wavelets as well as in Gabor functions. The spaces $\mathcal{L}_s^2(\mathbb{R}^d)$ are Hilbert spaces, and frames in the sense of Section 3.3 may exist for these spaces. However, the straightforward generalization of the results from Section 3.4 is not valid. The reasons for this are set down in Definition/Remark 3.26g–i. Therefore in order to study Gabor decompositions of $\mathcal{L}_s^2(\mathbb{R}^d)$, we shall ignore the Hilbert space structure, and view $\mathcal{L}_s^2(\mathbb{R}^d)$ as only a Banach space, that is, as a normed space embedded in $\mathcal{S}'(\mathbb{R}^d)$.

The purpose of this section is to consider the Bessel potential spaces and to study the properties of Banach frames and atomic decompositions of these spaces in terms of Gabor systems. The goal is to prove Theorem 3.32 which gives a sufficient condition on a Gabor system to be a Banach frame and a set of atoms for $\mathcal{L}_s^2(\mathbb{R}^d)$. The results in Section 3.7 characterizing Banach frames and sets of atoms rely heavily on the results proved here. The extensions of the notions of a frame and a set of atoms to the Banach space setting was done in [CRo] and generalized in [F2], [FGr1], and [FGr2]. A comprehensive study of frames and sets of atoms in Banach spaces with many interesting results not included here may be found in [Gr].

Definition/Remark 3.26

(a) Given $s \in \mathbb{R}$, we define the Bessel potential space $\mathcal{L}_s^2(\mathbb{R}^d)$ as

$$\mathcal{L}_s^2(\mathbb{R}^d) = \left\{ f : \int |\hat{f}(\gamma)|^2 (1 + |\gamma|)^{2s} \, d\gamma < \infty \right\}.$$

(b) The norm in $\mathcal{L}_s^2(\mathbb{R}^d)$ is given by

$$\|f\|_{\mathcal{L}_s^2} = \left(\int |\hat{f}(\gamma)|^2 (1 + |\gamma|)^{2s} \, d\gamma \right)^{1/2}. \tag{3.38}$$

(c) Note that $\mathcal{L}_s^2(\mathbb{R}^d)$ is a Hilbert space with respect to the inner product

$$\langle f, g \rangle_{\mathcal{L}_s^2} = \int \hat{f}(\gamma) \overline{\hat{g}(\gamma)} (1 + |\gamma|)^{2s} \, d\gamma.$$

The norm (3.38) may be thought of as a weighted L^2 norm with the weight on the Fourier transform side. The weight $(1 + |\gamma|)^s$ for $s \in \mathbb{R}$ satisfies the inequality

$$(1 + |\gamma_1 + \gamma_2|)^s \le (1 + |\gamma_1|)^{|s|}(1 + |\gamma_2|)^s.$$

When $s > 0$, this condition identifies $(1 + |\gamma|)^s$ as a *Beurling–Domar weight*.

(d) If $s = 0$, then $\mathcal{L}_s^2(\mathbb{R}^d) = L^2(\mathbb{R}^d)$. If $s = k \in \mathbb{Z}_+$, then a norm equivalent to (3.38) is given by

$$\|f\|_{\mathcal{L}_s^2} = \sum_{|\alpha| \le 2k} \|\partial^\alpha f\|_2,$$

where ∂^α represents the distributional derivative of order α. $\mathcal{L}_k^2(\mathbb{R}^d)$ can therefore be identified with the Sobolev space $L_{2k}^2(\mathbb{R}^d)$, cf. [St, Chapter V, Theorem 3].

(e) Given $s \in \mathbb{R}$, the dual space $(\mathcal{L}_s^2)'(\mathbb{R}^d)$ can be identified with the space $\mathcal{L}_{-s}^2(\mathbb{R}^d)$, with respect to the usual L^2 duality

$$\langle f, g \rangle = \int f(x) \overline{g(x)} \, dx.$$

(f) Given $s \in \mathbb{R}$, a Gabor system $\{\varphi_{m,n}\} = \{e_{mb} T_{na} g\}$, for some $a, b > 0$ and $g \in \mathcal{L}_{|s|}^2$, is a *Bessel sequence* for $\mathcal{L}_s^2(\mathbb{R}^d)$ provided that there exists $B > 0$ such that for all $f \in \mathcal{L}_s^2(\mathbb{R}^d)$,

$$\sum_n \sum_m |\langle f, \varphi_{m,n} \rangle|^2 (1 + |m|)^{2s} \le B \|f\|_{\mathcal{L}_s^2}^2. \tag{3.39}$$

(g) It is convenient here to discuss the difference between the notion of a *Bessel sequence* defined in part f and that defined in Definition/Remark 3.1b. Suppose that for $s \ge 0$, $g \in \mathcal{L}_{|s|}^2$ and $\hat{g}$ is supported in the cube $[-1/a, 1/a]^d$ for some $a > 0$. Fix $b \le 1/a$. Then $\{\varphi_{m,n}\} = \{e_{mb} T_{na} g\}$ is a frame for the

Hilbert space $\mathcal{L}_s^2(\mathbb{R}^d)$, that is, there exist constants $A, B > 0$ such that for all $f \in \mathcal{L}_s^2(\mathbb{R}^d)$,

$$A\|f\|_{\mathcal{L}_s^2}^2 \leq \sum_n \sum_m |\langle f, \varphi_{m,n} \rangle_{\mathcal{L}_s^2}|^2 \leq B\|f\|_{\mathcal{L}_s^2}^2,$$

if and only if there exist constants $c_1, c_2 > 0$ such that

$$c_1 \leq \sum_m |\hat{g}(\gamma - mb)|^2 (1 + |mb|)^{2s} \leq c_2 \text{ a.e.} \tag{3.40}$$

(h) We shall verify that (3.40) is a consequence of the frame hypothesis in part g. The opposite direction follows similarly. By a Fourier series argument we have

$$\sum_n \sum_m |\langle f, \varphi_{m,n} \rangle_{\mathcal{L}_s^2}|^2$$

$$= a^{-d} \sum_m \int_{[-1/a,1/a]^d + mb} |\hat{f}(\gamma)|^2 (1 + |\gamma|)^{4s} |\hat{g}(\gamma - mb)|^2 \, d\gamma \tag{3.41}$$

$$= a^{-d} \sum_m \int_{[-1/a,1/a]^d} |\hat{f}(\gamma + mb)|^2 (1 + |\gamma + mb|)^{4s} |\hat{g}(\gamma)|^2 \, d\gamma.$$

Now, since for all $s \in \mathbb{R}$,

$$(1 + |\gamma + mb|)^{2s} \leq (1 + |\gamma|)^{2|s|}(1 + |mb|)^{2s} \leq C(1 + |mb|)^{2s}$$

for $\gamma \in [-1/a, 1/a]^d$, the last term in (3.41) is

$$\leq Ca^{-d} \sum_m (1 + |mb|)^{2s} \int_{[-1/a,1/a]^d} |\hat{f}(\gamma + mb)|^2 (1 + |\gamma + mb|)^{2s} |\hat{g}(\gamma)|^2 \, d\gamma$$

$$= Ca^{-d} \int |\hat{f}(\gamma)|^2 (1 + |\gamma|)^{2s} \sum_m (1 + |mb|)^{2s} |\hat{g}(\gamma - mb)|^2 \, d\gamma.$$

Therefore, for all $f \in \mathcal{L}_s^2(\mathbb{R}^d)$, we have

$$A\|f\|_{\mathcal{L}_s^2} \leq Ca^{-d} \int |\hat{f}(\gamma)|^2 (1 + |\gamma|)^{2s} \sum_m (1 + |mb|)^{2s} |\hat{g}(\gamma - mb)|^2 \, d\gamma,$$

which implies that

$$\sum_m (1 + |mb|)^{2s} |\hat{g}(\gamma - mb)|^2 \, d\gamma \geq a^d A/C.$$

Since for $s \in \mathbb{R}$,

$$(1 + |\gamma - mb|)^{2s} \geq (1 + |\gamma|)^{-2|s|}(1 + |mb|)^{2s} \geq C(1 + |mb|)^{2|s|}$$

for $\gamma \in [-1/a, 1/a]^d$, the last term in (3.41) is

$$\geq Ca^{-d} \sum_m (1 + |mb|)^{2s} \int_{[-1/a,1/a]^d} |\hat{f}(\gamma - mb)|^2 (1 + |\gamma - mb|)^{2s} |\hat{g}(\gamma)|^2 \, d\gamma$$

$$= Ca^{-d} \int |\hat{f}(\gamma)|^2 (1+|\gamma|)^{2s} \sum_m (1+|mb|)^{2s} |\hat{g}(\gamma+mb)|^2 \, d\gamma.$$

Therefore, for all $f \in \mathcal{L}_s^2(\mathbb{R}^d)$, we have

$$B\|f\|_{\mathcal{L}_s^2} \geq Ca^{-d} \int |\hat{f}(\gamma)|^2 (1+|\gamma|)^{2s} \sum_m (1+|mb|)^{2s} |\hat{g}(\gamma+mb)|^2 \, d\gamma,$$

which implies that

$$\sum_m (1+|mb|)^{2s} |\hat{g}(\gamma+mb)|^2 \, d\gamma \leq a^d B/C.$$

This establishes (3.40).

(i) We now claim that (3.40) can never hold if $s \neq 0$. To see this, suppose that $s > 0$ and that $g \neq 0$ is given. Then there is a set $E \subseteq [-1/a, 1/a]^d$ with $|E| > 0$ and $|\hat{g}(\gamma)| > \alpha \geq 0$ on E. For a given $M > 0$, we can find m_0 such that $(1+|m_0 b|)^{2s} > M$, and so for $\gamma \in E + m_0 b$,

$$\sum_m |\hat{g}(\gamma-mb)|^2 (1+|mb|)^{2s} \geq |\hat{g}(\gamma-m_0 b)|^2 (1+|m_0 b|)^{2s} \geq \alpha M.$$

Thus,

$$\sum_m |\hat{g}(\gamma-mb)|^2 (1+|mb|)^{2s}$$

is unbounded, a contradiction.

For $s < 0$, note that if the upper bound in (3.40) holds then $\hat{g}$ must be bounded. Now, given $\epsilon > 0$, there is an m_0 such that $(1+|mb|)^{2s} < \epsilon$ for all $|m| \geq m_0$. Since $\hat{g}$ has compact support, for γ in any compact set, only finitely many terms in the sum are nonzero. Consider the sets $E_m \subseteq \hat{\mathbb{R}}^d$ defined by $E_m = [-1/a, 1/a]^d + mb$ for $m \in \mathbb{Z}^d$ and $F_m \subseteq \mathbb{Z}^d$ defined by $F_m = \{n \in \mathbb{Z}^d : E_n \cap E_m \neq \emptyset\}$. For any $m \in \mathbb{Z}^d$,

$$\operatorname*{ess\,sup}_{\gamma \in E_m} \sum_n |\hat{g}(\gamma-nb)|^2 (1+|nb|)^{2|s|}$$

$$\leq \sum_{n \in F_m} \operatorname*{ess\,sup}_{\hat{\mathbb{R}}^d} |\hat{g}(\gamma-nb)|^2 (1+|nb|)^{2|s|}$$

$$\leq \|g\|_\infty^2 \sum_{n \in F_m} (1+|nb|)^{2|s|}.$$

Now, since $\#F_m$ is bounded independently of m, and since for m large enough, $n \in F_m$ implies that $|m| \geq m_0$, we conclude that

$$\lim_{m \to \infty} \sum_{n \in F_m} (1+|nb|)^{2|s|} = 0.$$

Therefore, the lower bound in (3.40) fails in this case and hence (3.40) can never be satisfied if $s < 0$.

(j) The preceding remarks have shown that under the very reasonable conditions on g outlined in part g, $\{e_{mb}\tau_{na}g\}$ cannot satisfy the conditions to be a Gabor frame for the Hilbert space $\mathcal{L}_s^2(\mathbb{R}^d)$ unless $s = 0$. □

DEFINITION 3.27 *Given $s \in \mathbb{R}$, a Gabor system $\{\varphi_{m,n}\} = \{e_{mb}\tau_{na}g\}$, for some $a, b > 0$ and $g \in \mathcal{L}_{|s|}^2$, is a Banach frame for $\mathcal{L}_s^2(\mathbb{R}^d)$ provided there exist $A, B > 0$ such that for all $f \in \mathcal{L}_s^2(\mathbb{R}^d)$,*

$$A\|f\|_{\mathcal{L}_s^2}^2 \le \sum_n \sum_m |\langle f, \varphi_{m,n}\rangle|^2 (1 + |m|)^{2s} \le B\|f\|_{\mathcal{L}_s^2}^2.$$

The operator S is formally defined on $\mathcal{L}_s^2(\mathbb{R}^d)$ by

$$Sf = \sum_n \sum_m \langle f, \varphi_{m,n}\rangle \, \varphi_{m,n}.$$

DEFINITION 3.28 *Given $s \in \mathbb{R}$, a Gabor system $\{\varphi_{m,n}\} = \{e_{mb}\tau_{na}g\}$, for some $a, b > 0$ and $g \in \mathcal{L}_{|s|}^2$, is a set of atoms for $\mathcal{L}_s^2(\mathbb{R}^d)$ provided that there exists a collection of linear functionals $\{\lambda_{nm}\}$ on $\mathcal{L}_s^2(\mathbb{R}^d)$ such that*

(a) *for all $f \in \mathcal{L}_s^2(\mathbb{R}^d)$,*

$$f = \sum_n \sum_m \lambda_{nm}(f)\varphi_{m,n},$$

where the sum converges in norm to f, and

(b) *there exist $A, B > 0$ such that for all $f \in \mathcal{L}_s^2(\mathbb{R}^d)$,*

$$A\|f\|_{\mathcal{L}_s^2}^2 \le \sum_n \sum_m |\lambda_{nm}(f)|^2 (1 + |m|)^{2s} \le B\|f\|_{\mathcal{L}_s^2}^2.$$

THEOREM 3.29
Let $s \in \mathbb{R}$, and let $\{\varphi_{m,n}\} = \{e_{mb}\tau_{na}g\}$ be a Gabor system for some $a, b > 0$ and $g \in \mathcal{L}_{|s|}^2$. If $\{\varphi_{m,n}\}$ is a set of atoms for $\mathcal{L}_{-s}^2(\mathbb{R}^d)$ and a Bessel sequence for $\mathcal{L}_s^2(\mathbb{R}^d)$, then $\{\varphi_{m,n}\}$ is a Banach frame for $\mathcal{L}_s^2(\mathbb{R}^d)$.

PROOF We need only show the existence of the lower frame bound. To see this, let $f \in \mathcal{L}_s^2(\mathbb{R}^d)$, $h \in \mathcal{L}_{-s}^2(\mathbb{R}^d)$. Then

$$|\langle f, h\rangle|^2 = \left| \left\langle f, \sum_{n,m} \lambda_{nm}(h)\varphi_{m,n} \right\rangle \right|^2$$

$$= \left| \sum_{n,m} \langle f, \varphi_{m,n}\rangle \overline{\lambda_{nm}(h)} \right|^2$$

$$\le \sum_{n,m} |\langle f, \varphi_{m,n}\rangle|^2 (1 + |m|)^{2s} \sum_{n,m} |\lambda_{nm}(h)|^2 (1 + |m|)^{-2s}$$

$$\leq B\|h\|^2_{\mathcal{L}^2_{-s}} \sum_{n,m} |\langle f, \varphi_{m,n}\rangle|^2 (1+|m|)^{2s}.$$

Taking the supremum of both sides over all $h \in \mathcal{L}^2_{-s}$ with $\|h\|_{\mathcal{L}^2_{-s}} = 1$, we get

$$\|f\|^2_{\mathcal{L}^2_s} \leq B \sum_{n,m} |\langle f, \varphi_{m,n}\rangle|^2 (1+|m|)^{2s}. \quad \blacksquare$$

THEOREM 3.30
Let $s \in \mathbb{R}$, and let $\{\varphi_{m,n}\} = \{e_{mb}\tau_{na}g\}$ be a Gabor system for some $a, b > 0$ and $g \in \mathcal{L}^2_{|s|}$. If $\{\varphi_{m,n}\}$ is a Bessel sequence in $\mathcal{L}^2_s(\mathbb{R}^d)$ and $\mathcal{L}^2_{-s}(\mathbb{R}^d)$, then for each $f \in \mathcal{L}^2_s(\mathbb{R}^d)$, the sum

$$\sum_n \sum_m \langle f, \varphi_{m,n}\rangle \, \varphi_{m,n}$$

converges unconditionally in the norm of $\mathcal{L}^2_s(\mathbb{R}^d)$. Moreover the operator S is continuous on $\mathcal{L}^2_s(\mathbb{R}^d)$.

PROOF Let $F \subseteq \mathbb{Z}^d \times \mathbb{Z}^d$ be some finite index set. Then

$$\left| \left\langle \sum_{(n,m) \in F} \langle f, \varphi_{m,n}\rangle \varphi_{m,n}, \, h \right\rangle \right|^2 = \left| \sum_F \langle f, \varphi_{m,n}\rangle \overline{\langle h, \varphi_{m,n}\rangle} \right|^2$$

$$\leq \sum_F |\langle f, \varphi_{m,n}\rangle|^2 (1+|m|)^{2s} \sum_F |\langle h, \varphi_{m,n}\rangle|^2 (1+|m|)^{-2s}$$

$$\leq B\|h\|_{\mathcal{L}^2_{-s}} \sum_F |\langle f, \varphi_{m,n}\rangle|^2 (1+|m|)^{2s}.$$

Taking the supremum of both sides over all $h \in \mathcal{L}^2_{-s}$ with $\|h\|_{\mathcal{L}^2_{-s}} = 1$, we get

$$\left\| \sum_{(n,m)\in F} \langle f, \varphi_{m,n}\rangle \varphi_{m,n} \right\|^2_{\mathcal{L}^2_s} \leq B \sum_F |\langle f, \varphi_{m,n}\rangle|^2 (1+|m|)^{2s}.$$

The result now follows from a Cauchy criterion on the series

$$\sum_{n,m} |\langle f, \varphi_{m,n}\rangle|^2 (1+|m|)^{2s}$$

which converges since $\{\varphi_{m,n}\}$ is a Bessel sequence in $\mathcal{L}^2_s(\mathbb{R}^d)$. From this it also follows that S is continuous on $\mathcal{L}^2_s(\mathbb{R}^d)$. $\quad \blacksquare$

THEOREM 3.31
Let $s \in \mathbb{R}$, and let $\{\varphi_{m,n}\} = \{e_{mb}\tau_{na}g\}$ be a Gabor system for some $a, b > 0$ and $g \in \mathcal{L}^2_{|s|}$. If $\{\varphi_{m,n}\}$ is a Bessel sequence in $\mathcal{L}^2_s(\mathbb{R}^d)$ and $\mathcal{L}^2_{-s}(\mathbb{R}^d)$, and if S is a bijection onto $\mathcal{L}^2_s(\mathbb{R}^d)$, then $\{\varphi_{m,n}\}$ is a set of atoms for $\mathcal{L}^2_s(\mathbb{R}^d)$.

PROOF For $f \in \mathcal{L}_s^2(\mathbb{R}^d)$ and $n, m \in Z^d$, let

$$\lambda_{nm}(f) = \langle S^{-1}f, \varphi_{m,n} \rangle.$$

Then

$$f = SS^{-1}f = \sum_{n,m} \langle S^{-1}f, \varphi_{m,n} \rangle \varphi_{m,n},$$

the sum converging unconditionally and in norm in $\mathcal{L}_s^2(\mathbb{R}^d)$ by Theorem 3.30. Also, S is continuous on $\mathcal{L}_s^2(\mathbb{R}^d)$ so that by the open mapping theorem, S^{-1} is continuous. Using this continuity and the hypothesis that $\{\varphi_{m,n}\}$ is a Bessel sequence in $\mathcal{L}_s^2(\mathbb{R}^d)$ and $\mathcal{L}_{-s}^2(\mathbb{R}^d)$, we obtain

$$\sum_{n,m} |\langle S^{-1}f, \varphi_{m,n} \rangle|^2 (1 + |m|)^{2s} \leq B \|S^{-1}f\|_{\mathcal{L}_s^2}^2 \leq B' \|f\|_{\mathcal{L}_s^2}^2$$

and

$$\|f\|_{\mathcal{L}_s^2}^2 = \left\| \sum_{n,m} \langle S^{-1}f, \varphi_{m,n} \rangle \varphi_{m,n} \right\|_{\mathcal{L}_s^2}^2 \leq B \sum_{n,m} |\langle S^{-1}f, \varphi_{m,n} \rangle|^2 (1 + |m|)^{2s}. \quad \blacksquare$$

THEOREM 3.32

Let $s \in \mathbb{R}$, and let $\{\varphi_{m,n}\} = \{e_{mb}\tau_{na}g\}$ be a Gabor system for some $a, b > 0$ and $g \in \mathcal{L}_{|s|}^2$. If $\{\varphi_{m,n}\}$ is a Bessel sequence in $\mathcal{L}_s^2(\mathbb{R}^d)$ and $\mathcal{L}_{-s}^2(\mathbb{R}^d)$, and if S is a bijection onto $\mathcal{L}_s^2(\mathbb{R}^d)$ and $\mathcal{L}_{-s}^2(\mathbb{R}^d)$, then $\{\varphi_{m,n}\}$ is a set of atoms and a Banach frame for $\mathcal{L}_s^2(\mathbb{R}^d)$ and $\mathcal{L}_{-s}^2(\mathbb{R}^d)$.

PROOF By Theorem 3.31, we can conclude that $\{\varphi_{m,n}\}$ is a set of atoms for $\mathcal{L}_s^2(\mathbb{R}^d)$ and $\mathcal{L}_{-s}^2(\mathbb{R}^d)$. Combining this with Theorem 3.29 gives us that $\{\varphi_{m,n}\}$ is also a Banach frame for $\mathcal{L}_s^2(\mathbb{R}^d)$ and $\mathcal{L}_{-s}^2(\mathbb{R}^d)$. $\quad \blacksquare$

3.7 The Zak transform and Bessel potential spaces

The goal of this section is to prove a characterization of frames and sets of atoms for the Bessel potential spaces in terms of the Zak transform (Theorem 3.41). We first collect some results on the reservoir of basic functions which will generate the frames and sets of atoms. This set, the modulation space $M_{1,1}^{|s|}(\mathbb{R}^d)$, has been studied extensively by Feichtinger and Gröchenig, and has been used in this context in many papers, e.g., [FGr1], [FGr2], [F2], [Gr]. Our main result about $M_{1,1}^{|s|}(\mathbb{R}^d)$ is Theorem 3.36, which shows that if $g \in M_{1,1}^{|s|}(\mathbb{R}^d)$, then Zg is in a certain weighted Fourier algebra of functions on Q.

DEFINITION 3.33   *Given* $s \in \mathbb{R}$, *and* $\varphi \in \mathcal{S}(\mathbb{R}^d)$, *we define the space* $M_{1,1}^{|s|}(\mathbb{R}^d)$ *as*

$$M_{1,1}^{|s|}(\mathbb{R}^d) = \left\{ f \colon \|f\|_{M_{1,1}^{|s|}} \equiv \iint |\langle f, e_\gamma \tau_x \varphi \rangle| (1 + |\gamma|)^{|s|} \, d\gamma \, dx < \infty \right\}. \tag{3.42}$$

REMARK 3.34

(a) The space $M_{1,1}^{|s|}(\mathbb{R}^d)$ is a Banach space with norm given by the right side of (3.42). Different choices of $\varphi \in \mathcal{S}(\mathbb{R}^d)$ give equivalent norms [FGr1], [FGr2]. We shall briefly describe the reason for this equivalence and leave the details to the interested reader.

The Heisenberg group $\mathbb{H}^d$ is defined by

$$\mathbb{H}^d = \left\{ (z, x, \gamma) \colon z \in \mathbb{T}, \ x \in \mathbb{R}^d, \ \gamma \in \hat{\mathbb{R}}^d \right\}$$

with group multiplication defined as follows. Let $h_1 = (z_1, x_1, \gamma_1)$, $h_2 = (z_2, x_2, \gamma_2)$. Then

$$h_1 \cdot h_2 = (z_1 z_2 e^{2\pi i \gamma_1 \cdot x_2}, x_1 + x_2, \gamma_1 + \gamma_2).$$

Also,

$$h_1^{-1} = (z_1^{-1} e^{2\pi i x_1 \cdot \gamma_1}, -x_1, -\gamma_1).$$

Lebesgue measure $dh = dz \, dx \, d\gamma$ is the left and right invariant Haar measure on $\mathbb{H}^d$. The Schrödinger representation π of $\mathbb{H}^d$ on $L^2(\mathbb{R}^d)$ is defined by

$$\pi(h)f = z \tau_x e_\gamma f,$$

where $h \equiv (z, x, \gamma) \in \mathbb{H}^d$ and $f \in L^2(\mathbb{R}^d)$, cf. [Be5, Example 5.12b]. It is well known that π is an irreducible, unitary, square integrable representation, and consequently the following formula holds [GroMoPa]. For $f_1, f_2, g_1, g_2 \in L^2(\mathbb{R}^d)$,

$$\int_{\mathbb{H}^d} \langle f_1, \pi(h)g_1 \rangle \langle \pi(h)g_2, f_2 \rangle \, dh = \langle f_1, f_2 \rangle \langle g_1, g_2 \rangle. \tag{3.43}$$

We write $\langle f, \pi(h)g \rangle = V_g f(h)$ (cf. Definition 3.58). With this notation, we can define the space $M_{1,1}^{|s|}(\mathbb{R}^d)$ as follows. Let $\varphi \in \mathcal{S}(\mathbb{R}^d)$. Then

$$M_{1,1}^{|s|}(\mathbb{R}^d) = \left\{ f \colon \int_{\mathbb{H}^d} |V_\varphi f(h)| w_{|s|}(h) \, dh < \infty \right\},$$

where if $h = (z, x, \gamma)$ then $w_{|s|}(h) \equiv (1 + |\gamma|)^{|s|}$.

From (3.43) follows the reproducing formula,

$$\int_{\mathbb{H}^d} \langle f, \pi(u)g_1 \rangle \langle g_1, \pi(u^{-1}h)g_2 \rangle \, du$$

$$= \int_{\mathbb{H}^d} \langle f, \pi(u)g_1 \rangle \langle \pi(u)g_1, \pi(h)g_2 \rangle \, du$$

$$= \langle f, \pi(h)g_2 \rangle,$$

where $f, g_1, g_2 \in L^2(\mathbb{R}^d)$ and $\|g_1\|_2 = \|g_2\|_2 = 1$. In terms of group convolution, this identity can be written as

$$V_{g_1} f * V_{g_2} g_1 = V_{g_2} f.$$

Now, in order to show the equivalence of the norms (3.42) for different basic functions φ, we observe that

$$\int_{\mathbb{H}^d} |V_\psi \varphi(h)| w_{|s|}(h)\, dh \equiv C < \infty$$

for any $\psi, \varphi \in \mathcal{S}(\mathbb{R}^d)$. Since $w_{|s|}$ is a submultiplicative weight on $\mathbb{H}^d$, standard convolution inequalities give us that

$$\int_{\mathbb{H}^d} |V_\psi f(h)| w_{|s|}(h)\, dh = \int_{\mathbb{H}^d} |V_\varphi f * V_\psi \varphi(h)| w_{|s|}(h)\, dh$$

$$\leq \int_{\mathbb{H}^d} |V_\varphi f(h)| w_{|s|}(h)\, dh \int_{\mathbb{H}^d} |V_\psi \varphi(h)| w_{|s|}(h)\, dh$$

$$= C \int_{\mathbb{H}^d} |V_\varphi f(h)| w_{|s|}(h)\, dh.$$

The opposite inequality follows similarly.

(b) Given $r \in \mathbb{R}$, define the *weighted Fourier algebra* $A_r(\mathbb{R}^d)$ by

$$A_r(\mathbb{R}^d) = \left\{ f : \|f\|_{A_r(\mathbb{R}^d)} \equiv \int |\hat{f}(\gamma)|(1 + |\gamma|)^r\, d\gamma < \infty \right\}.$$

If $r = 0$, then $A_0(\mathbb{R}^d) = A(\mathbb{R}^d)$ is the *Fourier algebra*. We also define $A_r(\hat{\mathbb{R}}^d)$ by

$$A_r(\hat{\mathbb{R}}^d) = \left\{ f : \|f\|_{A_r(\hat{\mathbb{R}}^d)} \equiv \int |\check{f}(t)|(1 + |t|)^r\, dt < \infty \right\}.$$

If $r = 0$, we write $A_0(\hat{\mathbb{R}}^d) = A(\hat{\mathbb{R}}^d)$.

The $M_{1,1}^{|s|}(\mathbb{R}^d)$ norm of g, defined by (3.42), can be rewritten as

$$\|f\|_{M_{1,1}^{|s|}} \equiv \int \|f \tau_x \varphi\|_{A_{|s|}(\mathbb{R}^d)}\, dx.$$

To see why this is true, note that

$$\iint |\langle f, e_\gamma \tau_x \varphi \rangle|(1 + |\gamma|)^{|s|}\, d\gamma\, dx = \iint \left| \int f(t) \tau_x \overline{\varphi(t)} e^{-2\pi i \gamma \cdot t}\, dt \right|(1 + |\gamma|)^{|s|}\, d\gamma\, dx$$

$$= \iint |(f \tau_x \bar{\varphi})^\wedge(\gamma)|(1 + |\gamma|)^{|s|}\, d\gamma\, dx = \int \|f \tau_x \bar{\varphi}\|_{A_{|s|}(\mathbb{R}^d)}\, dx.$$

(c) From the last remark, we see that (3.42) defines a translation invariant norm on $M_{1,1}^{|s|}(\mathbb{R}^d)$, i.e.,

$$\|\tau_t f\|_{M_{1,1}^{|s|}} = \int \|\tau_t f \tau_x \varphi\|_{A_{|s|}(\mathbb{R}^d)}\, dx = \int \|f \tau_{x-t} \varphi\|_{A_{|s|}(\mathbb{R}^d)}\, dx = \int \|f \tau_x \varphi\|_{A_{|s|}(\mathbb{R}^d)}\, dx.$$

(d) Feichtinger has shown [F2] that a discrete norm equivalent to (3.42) can be given by the following definition. Let $\psi \in C_c^\infty(\mathbb{R}^d)$ have the properties that

1. $0 \leq \psi \leq 1$,

2. $\sum_n \tau_n \psi(t) = 1$.

Then the norm defined by

$$\|f\|_{M_{1,1}^{|s|}} \equiv \sum_n \|f\tau_n\psi\|_{A_{|s|}(\mathbb{R}^d)}$$

is equivalent to that defined by (3.42).

To see this, let $\varphi \in C_c^\infty(\mathbb{R}^d)$ with $\int \varphi = 1$. Since ψ has compact support, there is an $R > 0$ such that $\tau_x\bar\varphi\tau_n\psi = 0$ if $|x - n| > R$. We compute

$$\int \|f\tau_x\bar\varphi\|_{A_{|s|}(\mathbb{R}^d)}\, dx = \int \|f\tau_x\bar\varphi \sum_n \tau_n\psi\|_{A_{|s|}(\mathbb{R}^d)}\, dx$$

$$\leq \sum_n \int \|f\tau_x\bar\varphi\tau_n\psi\|_{A_{|s|}(\mathbb{R}^d)}\, dx \leq \sum_n \int_{|x-n|\leq R} \|f\tau_x\bar\varphi\tau_n\psi\|_{A_{|s|}(\mathbb{R}^d)}\, dx$$

$$\leq \|\varphi\|_{A_{|s|}(\mathbb{R}^d)} \sum_n \int_{|x-n|\leq R} \|f\tau_n\psi\|_{A_{|s|}(\mathbb{R}^d)}\, dx$$

$$= \|\varphi\|_{A_{|s|}(\mathbb{R}^d)} |B(R)| \sum_n \|f\tau_n\psi\|_{A_{|s|}(\mathbb{R}^d)}.$$

For the opposite inequality we observe that $\int \varphi = 1$ implies that $\int \tau_x\bar\varphi = 1$ for all $x \in \mathbb{R}^d$. Also, we note that for each n,

$$\left\|f\tau_n\psi \int \tau_x\bar\varphi\, dx\right\|_{A_{|s|}(\mathbb{R}^d)}$$

$$= \int \left|\int\int f(t)\tau_n\psi(t)\tau_x\bar\varphi(t)e^{-2\pi i\gamma\cdot t}\, dt\, dx\right| (1 + |\gamma|)^{|s|}\, d\gamma$$

$$\leq \int\int \left|\int f(t)\tau_n\psi(t)\tau_x\bar\varphi(t)e^{-2\pi i\gamma\cdot t}\, dt\right| (1 + |\gamma|)^{|s|}\, d\gamma\, dx$$

$$= \int \|f\tau_n\psi\tau_x\bar\varphi\|_{A_{|s|}(\mathbb{R}^d)}\, dx.$$

Therefore, we compute

$$\sum_n \|f\tau_n\psi\|_{A_{|s|}(\mathbb{R}^d)} = \sum_n \left\|f\tau_n\psi \int \tau_x\bar\varphi\, dx\right\|_{A_{|s|}(\mathbb{R}^d)}$$

$$\leq \int \sum_n \|f\tau_n\psi\tau_x\bar\varphi\|_{A_{|s|}(\mathbb{R}^d)}\, dx = \int \sum_{|n-x|\leq R} \|f\tau_n\psi\tau_x\bar\varphi\|_{A_{|s|}(\mathbb{R}^d)}\, dx$$

$$\leq \|\psi\|_{A_{|s|}(\mathbb{R}^d)}(R + 1)^d \int \|f\tau_x\bar\varphi\|_{A_{|s|}(\mathbb{R}^d)}\, dx.$$

(e) Feichtinger has also shown that $M_{1,1}^{|s|}(\mathbb{R}^d)$ can be characterized by the Fourier transform as follows. Let $\psi \in C_c^\infty(\hat{\mathbb{R}}^d)$ have the properties that

    1. $0 \le \psi \le 1$,

    2. $\sum_n \tau_n \psi(\gamma) = 1$.

Then the norm defined by

$$\|f\|_{M_{1,1}^{|s|}} \equiv \sum_n \|\hat{f}\tau_n\psi\|_{A(\hat{\mathbb{R}}^d)}(1 + |n|)^{|s|}$$

is equivalent to that defined by (3.42). For our purposes, we require only the following inequality. ∎

LEMMA 3.35

Let $\psi \in C_c^\infty(\hat{\mathbb{R}}^d)$ satisfy the conditions in Remark 3.34d. Then there is a constant C such that for all $f \in \mathcal{S}(\mathbb{R}^d)$,

$$\sum_n \|\hat{f}\tau_n\psi\|_{A(\hat{\mathbb{R}}^d)}(1 + |n|)^{|s|} \le C\|f\|_{M_{1,1}^{|s|}}.$$

PROOF   Assume without loss of generality that

$$\int \psi(\gamma)\,d\gamma = 1.$$

Then for each γ, $\int \tau_\lambda\psi(\gamma)\,d\lambda = 1$ and we may write

$$\|\hat{f}\tau_n\psi\|_{A(\hat{\mathbb{R}}^d)} = \left\| \int \hat{f}\tau_n\psi\tau_\lambda\psi\,d\lambda \right\|_{A(\hat{\mathbb{R}}^d)} \le \int \|\hat{f}\tau_n\psi\tau_\lambda\psi\|_{A(\hat{\mathbb{R}}^d)}\,d\lambda.$$

Since ψ is compactly supported, there exists a constant $R > 0$ such that $\tau_n\psi\tau_\lambda\psi = 0$ whenever $|\lambda - n| > R$. Note also that

$$(1 + |n|)^{|s|} \le (1 + |\lambda|)^{|s|}(1 + |\lambda - n|)^{|s|} \le (R+1)^{|s|}(1 + |\lambda|)^{|s|}$$

if $|\lambda - n| \le R$. Therefore, we have

$$\sum_n \|\hat{f}\tau_n\psi\|_{A(\hat{\mathbb{R}}^d)}(1 + |n|)^{|s|}$$

$$\le \sum_n \int \|\hat{f}\tau_n\psi\tau_\lambda\psi\|_{A(\hat{\mathbb{R}}^d)}(1 + |n|)^{|s|}\,d\lambda$$

$$= \sum_n \int_{|\lambda-n|\le R} \|\hat{f}\tau_n\psi\tau_\lambda\psi\|_{A(\hat{\mathbb{R}}^d)}(1 + |n|)^{|s|}\,d\lambda$$

$$\le C\sum_n \int_{|\lambda-n|\le R} \|\hat{f}\tau_\lambda\psi\|_{A(\hat{\mathbb{R}}^d)}(1 + |\lambda|)^{|s|}\,d\lambda$$

$$\le C\int \|\hat{f}\tau_\lambda\psi\|_{A(\hat{\mathbb{R}}^d)}(1 + |\lambda|)^{|s|}\,d\lambda$$

$$= C\iint |\langle f, e_x\tau_\lambda\bar{\psi}\rangle|(1 + |\lambda|)^{|s|}\,dx\,d\lambda$$

$$= C \iint |\langle f, \tau_x e_\lambda \check{\tilde{\psi}} \rangle|(1 + |\lambda|)^{|s|} \, dx \, d\lambda = C \|f\|_{M_{1,1}^{|s|}}$$

since $\check{\tilde{\psi}} \in \mathcal{S}(\mathbb{R}^d)$. ∎

Now we are ready to prove the following theorem.

THEOREM 3.36
Let $f \in M_{1,1}^{|s|}(\mathbb{R}^d)$, and define

$$c_{n,m} = \iint_Q |Zf(x,\omega)|^2 e^{2\pi i(m \cdot x + n \cdot \omega)} \, dx \, d\omega.$$

Then

$$\sum_n \sum_m |c_{n,m}|(1 + |m|)^{|s|} < \infty.$$

PROOF Note that by Theorem 3.15,

$$c_{n,m} = \iint_Q |Zf(x,\omega)|^2 e^{-2\pi i(m \cdot x + n \cdot \omega)} \, dx \, d\omega$$

$$= \iint_Q Zf(x,\omega) e^{-2\pi i(m \cdot x + n \cdot \omega)} \overline{Zf(x,\omega)} \, dx \, d\omega$$

$$= \iint_Q Z(\tau_n e_{-m} f)(x,\omega) \overline{Zf(x,\omega)} \, dx \, d\omega = \langle \tau_n e_{-m} f, f \rangle.$$

Now, let $\psi \in C_c^\infty(\hat{\mathbb{R}}^d)$ be such that for some $C > 0$, $C^{-1} \leq \inf_m \sum_k \tau_k \psi(m) \leq \sup_m \sum_k \tau_k \psi(m) \leq C$. Also, let $R > 0$ be such that $\tau_k \psi(m) = 0$ if $|m - k| > R$. Then

$$\sum_m |\langle f, \tau_n e_{-m} f \rangle|(1 + |m|)^{|s|}$$

$$= \sum_m \left| \int f(x) \tau_n \overline{f(x)} e^{-2\pi imx} \, dx \right| (1 + |m|)^{|s|}$$

$$= \sum_m |(f \tau_n \bar{f})^\wedge(m)|(1 + |m|)^{|s|}$$

$$\leq C \sum_m \sum_k |(f \tau_n \bar{f})^\wedge(m) \tau_k \psi(m)|(1 + |m|)^{|s|}$$

$$\leq C \sum_k \sum_{|m-k| \leq R} |(f \tau_n \bar{f})^\wedge(m) \tau_k \psi(m)|(1 + |m|)^{|s|}$$

$$\leq C(R+1)^d \sum_k \|(f \tau_n \bar{f})^\wedge \tau_k \psi\|_\infty (1 + |k|)^{|s|}$$

$$\leq C(R+1)^d \sum_k \|(f\tau_n\bar{f})^\wedge \tau_k\psi\|_{A(\hat{\mathbb{R}}^d)}(1+|k|)^{|s|} \leq C(R+1)^d \|f\tau_n\bar{f}\|_{M_{1,1}^{|s|}},$$

where we have used Lemma 3.35 and the fact that for any $f \in A(\hat{\mathbb{R}}^d)$, $\|f\|_\infty \leq \|f\|_{A(\hat{\mathbb{R}}^d)}$. Now, using the discrete norm characterization of $M_{1,1}^{|s|}(\mathbb{R}^d)$ of Remark 3.34d and with an appropriate choice of $\psi \in C_c^\infty(\mathbb{R}^d)$, we may write

$$\sum_n \|f\tau_n\bar{f}\|_{M_{1,1}^{|s|}} \leq C \sum_n \sum_k \|(f\tau_k\psi)(\tau_n\bar{f}\tau_k\psi)\|_{A_{|s|}(\mathbb{R}^d)}$$

$$\leq C \sum_k \|f\tau_k\psi\|_{A_{|s|}(\mathbb{R}^d)} \sum_n \|\tau_n f\tau_k\psi\|_{A_{|s|}(\mathbb{R}^d)}$$

$$\leq C \sum_k \|f\tau_k\psi\|_{A_{|s|}(\mathbb{R}^d)} \sum_n \|\tau_{-k}f\tau_{-n}\psi\|_{A_{|s|}(\mathbb{R}^d)}$$

$$\leq C \sum_k \|f\tau_k\psi\|_{A_{|s|}(\mathbb{R}^d)} \|\tau_{-k}f\|_{M_{1,1}^{|s|}} \leq C \|f\|_{M_{1,1}^{|s|}}^2. \quad \blacksquare$$

We are now ready to give a characterization of frames and sets of atoms in the Bessel potential spaces for certain lattice sizes in terms of the Zak transform (Theorem 3.41). This result is a generalization of one in [GrW]. A similar result appears in [W3].

LEMMA 3.37

Let $s \in \mathbb{R}$, $\psi \in \mathcal{S}(\mathbb{R}^d)$. Then there exist constants c_1 and $c_2 > 0$ such that for all $f \in \mathcal{L}_s^2(\mathbb{R}^d)$,

$$c_1\|f\|_{\mathcal{L}_s^2}^2 \leq \iint |\langle f, e_\gamma\tau_x\psi\rangle|^2(1+|\gamma|)^{2s}\, d\gamma\, dx \leq c_2\|f\|_{\mathcal{L}_s^2}^2.$$

PROOF   Note first that

$$|\langle f, e_\gamma\tau_x\psi\rangle| = |\langle f, \tau_x e_\gamma\psi\rangle|$$

and that

$$\langle f, \tau_x e_\gamma\psi\rangle = \langle \hat{f}, e_{-x}\tau_\gamma\hat{\psi}\rangle = (\hat{f}\tau_\gamma\overline{\hat{\psi}})^\wedge(-x).$$

Therefore, by Plancherel's theorem

$$\int |\langle f, e_\gamma\tau_x\psi\rangle|^2\, dx = \int |\hat{f}(\lambda)|^2|\hat{\psi}(\lambda-\gamma)|^2\, d\lambda.$$

Now,

$$\iint |\langle f, e_\gamma\tau_x\psi\rangle|^2(1+|\gamma|)^{2s}\, dx\, d\gamma$$

$$= \iint |\hat{f}(\lambda)|^2|\hat{\psi}(\lambda-\gamma)|^2(1+|\gamma|)^{2s}\, d\lambda\, d\gamma$$

$$= \int |\hat{\psi}(\lambda)|^2 \int |\hat{f}(\lambda+\gamma)|^2(1+|\gamma|)^{2s}\, d\gamma\, d\lambda$$

$$= \int |\hat{\psi}(\lambda)|^2 \int |\hat{f}(\gamma)|^2 (1 + |\gamma - \lambda|)^{2s} \, d\gamma \, d\lambda$$

$$\leq \int |\hat{\psi}(\lambda)|^2 (1 + |\lambda|)^{2|s|} \, d\lambda \int |\hat{f}(\gamma)|^2 (1 + |\gamma|)^{2s} \, d\gamma$$

$$= \|\psi\|^2_{\mathcal{L}^2_{|s|}} \|f\|^2_{\mathcal{L}^2_s} = c_1 \|f\|^2_{\mathcal{L}^2_s}.$$

Now, if $\psi \not\equiv 0$ then there is a ball B such that $|\hat{\psi}(\lambda)| > \epsilon > 0$ on B. Moreover, there is a constant C such that

$$(1 + |\gamma|)^{2s} \leq C(1 + |\gamma - \lambda|)^{2s}$$

for all $\lambda \in B$. Therefore, we have

$$\int |\hat{\psi}(\lambda)|^2 \int |\hat{f}(\gamma)|^2 (1 + |\gamma - \lambda|)^{2s} \, d\gamma \, d\lambda$$

$$\geq \int_B |\hat{\psi}(\lambda)|^2 \int |\hat{f}(\gamma)|^2 (1 + |\gamma - \lambda|)^{2s} \, d\gamma \, d\lambda$$

$$\geq C^{-1} \epsilon^2 |B| \|f\|^2_{\mathcal{L}^2_s} \equiv c_2 \|f\|^2_{\mathcal{L}^2_s}$$

and $c_2 > 0$.   ∎

LEMMA 3.38
Let $\psi, f \in \mathcal{S}(\mathbb{R}^d)$. Then $\|f\|_{\mathcal{L}^2_s}$ is equivalent to

$$\sum_{n,m} \int_Q \left| \int_Q e^{-2\pi i \gamma \cdot x} \overline{Z\psi(x - t, \omega - \gamma)} Zf(x, \omega) e^{2\pi i (m \cdot x - n \cdot \omega)} \, dx \, d\omega \right|^2 \tag{3.44}$$

$$\times (1 + |m|)^{2s} \, dt \, d\gamma.$$

PROOF   First observe that $Z(e_\gamma \tau_t \psi)(x, \omega) = e^{2\pi i \gamma \cdot x} Z\psi(x - t, \omega - \gamma)$. Since Z is unitary, we obtain

$$\iint |\langle f, e_\gamma \tau_t \psi \rangle|^2 (1 + |\gamma|)^{2s} \, d\gamma \, dt$$

$$= \iint |\langle Zf, Z(e_\gamma \tau_t \psi) \rangle|^2 (1 + |\gamma|)^{2s} \, d\gamma \, dt$$

$$= \iint \left| \iint_Q Zf(x, \omega) \overline{Z\psi(x - t, \omega - \gamma)} e^{-2\pi i \gamma \cdot x} \, dx \, d\omega \right|^2 (1 + |\gamma|)^{2s} \, d\gamma \, dt$$

$$= \sum_{n \in \mathbb{Z}^d} \sum_{m \in \mathbb{Z}^d} \iint_Q \left| \iint_Q Zf(x, \omega) \overline{Z\psi(x - t + n, \omega - \gamma + m)} e^{-2\pi i (\gamma - m) \cdot x} \, dx \, d\omega \right|^2$$

$$\times (1 + |\gamma - m|)^{2s} \, d\gamma \, dt.$$

From this and Lemma 3.37 it follows that $\|f\|_{\mathcal{L}_s^2}$ is equivalent to

$$
\sum_{n,m} \iint_Q \left| \iint_Q Zf(x,\omega)e^{-2\pi i\gamma \cdot x}\overline{Z\psi(x-t,\omega-\gamma)}e^{2\pi i(m\cdot x-n\cdot\omega)}\,dx\,d\omega \right|^2
$$
$$
\times (1+|m|)^{2s}\,dt\,d\gamma. \quad \blacksquare
$$

REMARK 3.39

(a) (3.44) can be interpreted as the weighted $\ell^2(\mathbb{Z}^d)$ norm of the Fourier coefficients of the function

$$
e^{-2\pi i\gamma \cdot x}\overline{Z\psi(x-t,\omega-\gamma)}Zf(x,\omega).
$$

(b) Given $r \geq 0$, we define the space $A_r(Q)$ as the space M of measurable functions on Q such that

$$
M(x,\omega) = \sum_{n\in\mathbb{Z}^d}\sum_{m\in\mathbb{Z}^d} \gamma_{n,m}e^{2\pi i(m\cdot x+n\cdot\omega)}
$$

and

$$
\|M\|_{A_r(Q)} = \sum_{n\in\mathbb{Z}^d}\sum_{m\in\mathbb{Z}^d} |\gamma_{n,m}|(1+|m|)^r < \infty.
$$

Since the weight $w(n,m) = (1+|m|)^r$ is a *Beurling–Domar weight* (Definition/Remark 3.26d), i.e.,

$$
w(n_1+n_2,m_1+m_2) \leq w(n_1,m_1)w(n_2,m_2),
$$

$A_r(Q)$ is a Banach algebra with unit under pointwise multiplication, see [K, Chapter VIII]. $\quad \blacksquare$

LEMMA 3.40
Given $s \in \mathbb{R}$, let $M(x,\omega) \in A_{|s|}(Q)$. For any $a > 0$, the operator $S_M = Z^{-1}MZ$ is continuous on $\mathcal{L}_s^2(\mathbb{R}^d)$ and

$$
\|S_M f\|_{\mathcal{L}_s^2} \leq \|M\|_{A_{|s|}(Q)}\|f\|_{\mathcal{L}_s^2}. \tag{3.45}
$$

PROOF Note first that since $\mathcal{L}_s^2(\mathbb{R}^d)$ is invariant under dilation, it suffices to consider only the case $a = 1$. Next observe that for each fixed $(t,\gamma) \in Q$ and $\psi \in \mathcal{S}(\mathbb{R}^d)$ the function

$$
e^{-2\pi i\gamma \cdot x}\overline{Z\psi(x-t,\omega-\gamma)}Zf(x,\omega)
$$

is periodic in (x,ω). Thus for each (t,γ) the inner integral in (3.44) is just the Fourier transform of this function at $(-n,m)$.

We first show (3.45) for $f \in \mathcal{S}(\mathbb{R}^d)$. Note that since $ZS_M f = MZf$, Lemma 3.38 implies that $\|S_M\|_{\mathcal{L}_s^2}$ is equivalent to

$$\sum_{n,m} \iint_Q \left| \iint_Q e^{-2\pi i \gamma \cdot x} M(x,\omega) Zf(x,\omega) \overline{Z\psi(x-t,\omega-\gamma)} e^{2\pi i(m\cdot x - n\cdot\omega)} \, dx \, d\omega \right|^2$$

$$\times (1 + |m|)^{2s} \, dt \, d\gamma.$$

Now, for each $(t,\gamma) \in Q$

$$\sum_{n,m} \left| \iint_Q e^{-2\pi i\gamma\cdot x} \overline{Z\psi(x-t,\omega-\gamma)} M(x,\omega) Zf(x,\omega) e^{2\pi i(m\cdot x - n\cdot\omega)} \, dx \, d\omega \right|^2$$

$$\times (1 + |m|)^{2s}$$

$$\leq \|M\|_{A_{|s|}(Q)}^2 \sum_{n,m} \left| \iint_Q e^{-2\pi i\gamma\cdot x} \overline{Z\psi(x-t,\omega-\gamma)} Zf(x,\omega) e^{2\pi i(m\cdot x - n\cdot\omega)} \, dx \, d\omega \right|^2$$

$$\times (1 + |m|)^{2s}. \tag{3.46}$$

This estimate holds because the left hand side is just the square of the weighted ℓ^2 norm of the Fourier coefficients of the function

$$M(x,\omega) e^{-2\pi i\gamma\cdot x} \overline{Z\psi(x-t,\omega-\gamma)} Zf(x,\omega),$$

which is the convolution of $\gamma_{n,m}$ with the Fourier coefficients of

$$e^{-2\pi i\gamma\cdot x} \overline{Z\psi(x-t,\omega-\gamma)} Zf(x,\omega).$$

By standard properties of convolutions, inequality (3.46) holds for each fixed $(t,\gamma) \in Q$.

Integrating both sides of (3.46) over $(t,\gamma) \in Q$ gives (3.45) for $f \in \mathcal{S}(\mathbb{R}^d)$. By the density of $\mathcal{S}(\mathbb{R}^d)$ in $\mathcal{L}_s^2(\mathbb{R}^d)$, (3.45) follows. ∎

THEOREM 3.41
Suppose $s \in \mathbb{R}$, $a,b > 0$, $ab = 1/N$, $N \in \mathbb{Z}_+$, and $g \in M_{1,1}^{|s|}(\mathbb{R}^d)$. Then the Gabor system $\{\varphi_{m,n}\} = \{e_{mb}T_{na}g\}$ is a frame and a set of atoms for $\mathcal{L}_r^2(\mathbb{R}^d)$, $|r| \leq |s|$, if and only if there exist constants $A,B > 0$ such that for a.e. $(x,\omega) \in Q$,

$$A \leq \sum_{j=0}^{N-1} |Zg(x, \omega + j/N)|^2 \leq B. \tag{3.47}$$

PROOF We observe first that if $g \in M_{1,1}^{|s|}(\mathbb{R}^d)$, then for any $a > 0$ and $b > 0$ there exists $C = C(a,b,g) > 0$ such that for all $f \in \mathcal{L}_r^2(\mathbb{R}^d)$ with $|r| \leq |s|$,

$$\sum_n \sum_m |\langle f, e_{mb}T_{na}g \rangle|^2 (1 + |m|)^{2r} \leq C\|f\|_{\mathcal{L}_r^2}^2,$$

cf. [W3] and Theorem 3.47, which is stronger than this inequality (see Remark 3.44c). Therefore $\{e_{mb}T_{na}g\}$ is always a Bessel sequence in $\mathcal{L}_r^2(\mathbb{R}^d)$ and $\mathcal{L}_{-r}^2(\mathbb{R}^d)$.

Suppose that condition (3.47) holds. Then by Theorem 3.32, it is sufficient to show that the operator S defined in Definition 3.27 is a topological isomorphism of $\mathcal{L}_r^2(\mathbb{R}^d)$. Using the notation of Definition 3.10, Lemma 3.40 and basic facts about the Zak transform, we compute

$$
\begin{aligned}
ZSf(x,\omega) &= \sum_{m,n} \langle Zf, Z\varphi_{m,n} \rangle Z\varphi_{m,n}(x,\omega) \\
&= \sum_{m,n} \langle Zf, e_{m/N}(x)e_{-n}(\omega)Zg(x,\omega - m/N) \rangle \\
&\quad \times e_{m/N}(x)e_{-n}(\omega)Zg(x,\omega - m/N) \\
&= \sum_{j=0}^{N-1} \sum_{k,n} \langle Zf\overline{Zg(x,\omega - j/N)}e_{-j/N}(x), e_k(x)e_{-n}(\omega) \rangle \\
&\quad \times e_k(x)e_{-n}(\omega)Zg(x,\omega - j/N)e_{j/N}(x) \\
&= \sum_{j=0}^{N-1} Zf(x,\omega)|Zg(x,\omega - j/N)|^2 = Zf(x,\omega)\sum_{j=0}^{N-1}|Zg(x,\omega - j/N)|^2.
\end{aligned}
$$

From this it follows that $S = S_M$ where

$$
M(x,\omega) = \sum_{j=0}^{N-1} |Zg(x,\omega + j/N)|^2.
$$

It follows from Theorem 3.36 that since $g \in M_{1,1}^{|s|}(\mathbb{R}^d)$, $M \in A_r(Q)$ for all $|r| \le |s|$. Therefore by Lemma 3.40, S is a continuous operator on $\mathcal{L}_r^2(\mathbb{R}^d)$.

Now, since (3.47) holds and $A_r(Q)$ is a Banach algebra with unit under pointwise multiplication, the theorem of Wiener–Levy can be applied ([K, page 209]) to conclude that $M^{-1} \in A_r(Q)$ for all $|r| \le |s|$. Hence S is a bijection of $\mathcal{L}_r^2(\mathbb{R}^d)$ with S^{-1} given by

$$
S^{-1} = Z^{-1}M^{-1}Z.
$$

Moreover by Lemma 3.12, S^{-1} is continuous and therefore, S is a topological isomorphism of $\mathcal{L}_r^2(\mathbb{R}^d)$.

Conversely, if $\{\varphi_{m,n}\}$ is a frame for $\mathcal{L}_r^2(\mathbb{R}^d)$ for all $|r| \le |s|$, then in particular, this holds for $\mathcal{L}_0^2(\mathbb{R}^d) = L^2(\mathbb{R}^d)$. Then by standard results on the Zak transform and Gabor frames for $L^2(\mathbb{R}^d)$ (e.g., [HW, Theorem 4.3.3]), (3.47) must hold. $\blacksquare$

REMARK 3.42 In fact a more general result than this holds. The Bessel potential spaces can be replaced by general modulation spaces $M_{p,q}^r(\mathbb{R}^d)$ with no change in the argument, cf. [W3], [GrW], [FGrW]. Also, in the special case of Sobolev spaces, Theorem 3.41 should be compared with [M, Theorem 3]. $\blacksquare$

Observe that in Theorem 3.41, the lattice density ab was restricted to $1/N$, $N \in \mathbb{Z}_+$. It is interesting to consider the case when the quantity ab is arbitrary. In what follows, we shall prove some results which address this situation. The reservoir of basic functions in this case is slightly larger than in the previous case. The results, however, are not sharp. To begin, we define our reservoir of basic functions.

DEFINITION 3.43 *Given $s \in \mathbb{R}$, $\varphi \in C_c^\infty(\hat{\mathbb{R}}^d)$, we define the space $W_{|s|}(\mathbb{R}^d)$ by*

$$W_{|s|}(\mathbb{R}^d) = \left\{f : \int \|\hat{f}\tau_\gamma\varphi\|_\infty(1+|\gamma|)^{|s|}\,d\gamma < \infty\right\}$$

with norm given by

$$\|f\|_{W_{|s|}} = \int \|\hat{f}\tau_\gamma\varphi\|_\infty(1+|\gamma|)^{|s|}\,d\gamma. \tag{3.48}$$

REMARK 3.44

(a) $W_{|s|}(\mathbb{R}^d)$ is a Banach space under the norm defined by (3.48) and a different choice of $\varphi \in C_c^\infty(\hat{\mathbb{R}}^d)$ defines an equivalent norm. The space $W_{|s|}(\mathbb{R}^d)$ has the property that

$$\|\tau_t f\|_{W_{|s|}} = \|f\|_{W_{|s|}}$$

and that

$$\|e_\lambda f\|_{W_{|s|}} = \int \|\tau_\lambda\hat{f}\tau_\gamma\varphi\|_\infty(1+|\gamma|)^{|s|}\,d\gamma = \int \|\hat{f}\tau_{\gamma-\lambda}\varphi\|_\infty(1+|\gamma|)^{|s|}\,d\gamma$$

$$= \int \|\hat{f}\tau_\gamma\varphi\|_\infty(1+|\gamma+\lambda|)^{|s|}\,d\gamma \leq (1+|\lambda|)^{|s|}\|f\|_{W_{|s|}}.$$

Also note that for any function h with $\hat{h}$ bounded,

$$\|hf\|_{W_{|s|}} \leq \|\hat{h}\|_\infty\|f\|_{W_{|s|}}.$$

(b) Let $\psi \in C_c^\infty(\hat{\mathbb{R}}^d)$ have the properties that
  1. $0 \leq \psi \leq 1$,
  2. $\sum_n \tau_n\psi(t) = 1$.
Then the norm defined by

$$\|f\|_{W_{|s|}} = \sum_{n\in\mathbb{Z}^d} \|\hat{f}\tau_n\psi\|_\infty(1+|n|)^s \tag{3.49}$$

is equivalent to that defined by (3.48).

(c) It is easy to see that $M_{1,1}^{|s|}(\mathbb{R}^d) \subseteq W_{|s|}(\mathbb{R}^d)$ and that the inclusion map is continuous since

$$\int \|\hat{f}\tau_\gamma\varphi\|_\infty(1+|\gamma|)^{|s|}\,d\gamma \leq \int \|\hat{f}\tau_\gamma\varphi\|_{A(\hat{\mathbb{R}}^d)}(1+|\gamma|)^{|s|}\,d\gamma.$$

(d) The space $W_{|s|}(\mathbb{R}^d)$ is an example of a Wiener amalgam space defined by Feichtinger in [F1]. The theory has been well developed by Feichtinger and others. For our purposes, we require only some basic facts about this space. The space $W_{|s|}(\mathbb{R}^d)$ will be the reservoir from which we choose basic functions for frames for $\mathcal{L}_r^2(\mathbb{R}^d)$, $|r| \leq |s|$. In the results which follow, we will show that under very nonrestrictive conditions on a function $g \in W_{|s|}(\mathbb{R}^d)$, the Gabor system $\{\varphi_{m,n}\}$ will be a frame and a set of atoms for $\mathcal{L}_r^2(\mathbb{R}^d)$, $|r| \leq |s|$ for a sufficiently small lattice. ∎

In what follows, assume $\varphi \in C_c^\infty(\hat{\mathbb{R}}^d)$ is fixed and $\varphi = 1$ on $B(0, \sqrt{d})$, $0 \leq \varphi \leq 1$, and $\varphi = 0$ off $B(0, 2\sqrt{d})$. For $a > 0$, define $\varphi_a(\gamma) = \varphi(\gamma/a)$ so that $\varphi_a = 1$ on $B(0, a\sqrt{d})$. Note that $B(0, \sqrt{d})$ contains the cube of side 1 centered at the origin.

LEMMA 3.45
Given $s \in \mathbb{R}$, $a > 0$, there exists a constant C depending only on $\mathrm{supp}(\varphi)$ such that for all functions h,

$$\sum_n \operatorname*{ess\,sup}_{\gamma \in \hat{\mathbb{R}}^d} |\hat{h}(\gamma)(1 + |\gamma|)^{|s|} \tau_{na}\varphi_a(\gamma)|$$

$$\leq C(1 + 2\sqrt{d}\max\{1, 1/a\})^d \int \|\hat{h}\tau_\lambda\varphi\|_\infty (1 + |\lambda|)^{|s|}\, d\lambda.$$

PROOF   Note first that $\int \varphi(\lambda)\, d\lambda = c \geq 1$ so that for each γ, $\int \tau_\lambda\varphi(\gamma)\, d\lambda = c \geq 1$. Therefore

$$\sum_n \operatorname*{ess\,sup}_{\gamma \in \hat{\mathbb{R}}^d} |\hat{h}(\gamma)(1 + |\gamma|)^{|s|} \tau_{na}\varphi_a(\gamma)|$$

$$\leq \sum_n \int \operatorname*{ess\,sup}_{\gamma \in \hat{\mathbb{R}}^d} |\hat{h}(\gamma)(1 + |\gamma|)^{|s|} \tau_{na}\varphi_a(\gamma)\tau_\lambda\varphi(\gamma)|\, d\lambda$$

$$= \int \sum_{n \in F_a(\lambda)} \operatorname*{ess\,sup}_{\gamma \in \hat{\mathbb{R}}^d} |\hat{h}(\gamma)(1 + |\gamma|)^{|s|} \tau_{na}\varphi_a(\gamma)\tau_\lambda\varphi(\gamma)|\, d\lambda,$$

where

$$F_a(\lambda) = \{n: |na - \lambda| \leq 2\sqrt{d}\max\{1, a\}\}$$

since $\tau_{na}\varphi_a\tau_\lambda\varphi = 0$ if $n \notin F_a(\lambda)$. Note that

$$\max_\lambda \#F_a(\lambda) \leq (1 + 2\sqrt{d}\max\{1, 1/a\})^d,$$

and that for some C depending only on $\mathrm{supp}\,\varphi$,

$$\sup_{\gamma \in \mathrm{supp}\,\tau_\lambda\varphi} (1 + |\gamma|)^{|s|} \leq C(1 + |\lambda|)^{|s|}.$$

Therefore,

$$\int \sum_{n \in F_a(\lambda)} \operatorname*{ess\,sup}_{\gamma \in \hat{\mathbb{R}}^d} |\hat{h}(\gamma)(1 + |\gamma|)^{|s|} \tau_{na}\varphi_a(\gamma)\tau_\lambda\varphi(\gamma)|\, d\lambda$$

$$\leq C(1 + 2\sqrt{d}\max\{1, 1/a\})^d \int \|\hat{h}\tau_\lambda\varphi\|_\infty (1 + |\lambda|)^{|s|} d\lambda. \quad \blacksquare$$

LEMMA 3.46
Let $s \in \mathbb{R}$, $g \in W_{|s|}(\mathbb{R}^d)$, $a, b > 0$. Define for each $j \in \mathbb{Z}^d$ the discrete autocorrelation function $\tilde{G}_j$ by

$$\tilde{G}_j \equiv \sum_m \tau_{mb}(\hat{g}\tau_{j/a}\overline{\hat{g}}).$$

Then

(a) $\sum_j (1 + |j/a|)^{|s|}\|\tilde{G}_j\|_\infty < \infty$, *and*

(b) *given $\eta > 0$, there exists $a_0 > 0$ depending on g and η such that for every $0 < a \leq a_0$, and $0 < b \leq 1$,*

$$\sum_{j \neq 0}(1 + |j/a|)^{|s|}\|\tilde{G}_j\|_\infty < b^{-d}\eta.$$

PROOF To see part a, note that since $\tilde{G}_j$ has period b and assuming $b \leq 1/a$, we have

$$\sum_j (1 + |j/a|)^{|s|}\|\tilde{G}_j\|_\infty$$

$$\leq \sum_j (1 + |j/a|)^{|s|}\|\varphi_b\varphi_{1/a}\sum_m \tau_{mb}(\hat{g}\tau_{j/a}\overline{\hat{g}})\|_\infty$$

$$\leq \sum_j (1 + |j/a|)^{|s|}\sum_m \operatorname*{ess\,sup}_{\gamma \in \hat{\mathbb{R}}^d} |\varphi_b(\gamma)\hat{g}(\gamma - mb)|$$

$$\times \operatorname*{ess\,sup}_{\gamma \in \hat{\mathbb{R}}^d} |\varphi_{1/a}(\gamma)\hat{g}(\gamma - mb - j/a)|$$

$$\leq \sum_j \sum_m \operatorname*{ess\,sup}_{\gamma \in \hat{\mathbb{R}}^d} |\varphi_b(\gamma)\hat{g}(\gamma - mb)(1 + |\gamma - mb|)^{|s|}|$$

$$\times \operatorname*{ess\,sup}_{\gamma \in \hat{\mathbb{R}}^d} |\varphi_{1/a}(\gamma)\hat{g}(\gamma - mb - j/a)(1 + |\gamma - mb - j/a|)^{|s|}|$$

$$= \sum_m \operatorname*{ess\,sup}_{\gamma \in \hat{\mathbb{R}}^d} |\tau_{mb}\varphi_b(\gamma)\hat{g}(\gamma)(1 + |\gamma|)^{|s|}|$$

$$\times \sum_j \operatorname*{ess\,sup}_{\gamma \in \hat{\mathbb{R}}^d} |\tau_{j/a}\varphi_{1/a}(\gamma)\hat{g}(\gamma - mb)(1 + |\gamma - mb|)^{|s|}|.$$

Applying Lemma 3.45 directly to the first sum, we have

$$\sum_m \operatorname*{ess\,sup}_{\gamma \in \hat{\mathbb{R}}^d} |\tau_{mb}\varphi_b(\gamma)\hat{g}(\gamma)(1 + |\gamma|)^{|s|}| \leq C(1 + 2\sqrt{d}\max\{1, 1/b\})^d\|g\|_{W_{|s|}}.$$

For the second sum, we apply Lemma 3.45 with $\hat{h}(\gamma)$ replaced by $\hat{g}(\gamma - mb)$ and $(1 + |\gamma|)^{|s|}$ replaced by $(1 + |\gamma - mb|)^{|s|}$. In this case, it is necessary to observe that the constant C in Lemma 3.45 does not depend on m. Therefore we have

$$\sum_j \operatorname*{ess\,sup}_{\gamma \in \hat{\mathbb{R}}^d} |\tau_{j/a}\varphi_{1/a}(\gamma)\hat{g}(\gamma - mb)(1 + |\gamma - mb|)^{|s|}|$$

$$\leq C(1 + 2\sqrt{d}\max\{1,a\})^d \int \|\tau_{mb}\hat{g}\tau_\lambda\varphi\|_\infty (1 + |\lambda - mb|)^{|s|}\,d\lambda$$

$$= C(1 + 2\sqrt{d}\max\{1,a\})^d \int \|\hat{g}\tau_\lambda\varphi\|_\infty (1 + |\lambda|)^{|s|}\,d\lambda$$

$$= C(1 + 2\sqrt{d}\max\{1,a\})^d \|g\|_{W_{|s|}}.$$

To see part b, note that for any $\epsilon > 0$, we may write $g = g_1 + g_2$ where

$$\operatorname{supp}\hat{g}_1 \subseteq B(0, M)$$

for some $M > 0$, and

$$\|g_2\|_{W_{|s|}} < \epsilon.$$

Hence, we may write

$$\tilde{G}_j = \sum_m \tau_{mb}(\hat{g}\tau_{j/a}\overline{\hat{g}})$$

$$= \sum_m \tau_{mb}(\hat{g}_1\tau_{j/a}\overline{\hat{g}_1}) + \sum_m \tau_{mb}(\hat{g}_1\tau_{j/a}\overline{\hat{g}_2})$$

$$+ \sum_m \tau_{mb}(\hat{g}_2\tau_{j/a}\overline{\hat{g}_1}) + \sum_m \tau_{mb}(\hat{g}_2\tau_{j/a}\overline{\hat{g}_2})$$

$$= \sum_m \tau_{mb}(\hat{g}_1\tau_{j/a}\overline{\hat{g}_2}) + \sum_m \tau_{mb}(\hat{g}_2\tau_{j/a}\overline{\hat{g}_1}) + \sum_m \tau_{mb}(\hat{g}_2\tau_{j/a}\overline{\hat{g}_2})$$

if $j \neq 0$ and $a < (2\sqrt{d}M)^{-1}$ since, in this case, the supports of $\hat{g}_1$ and $\tau_{j/a}\hat{g}_1$ are disjoint. Recalling that $b \leq 1$, and using the argument of part a, we have that

$$\sum_j (1 + |j/a|)^{|s|}\|\tilde{G}_j\|_\infty$$

$$\leq \sum_j (1 + |j/a|)^{|s|}\left\|\sum_m \tau_{mb}(\hat{g}_1\tau_{j/a}\overline{\hat{g}_2})\right\|_\infty$$

$$+ \sum_j (1 + |j/a|)^{|s|}\left\|\sum_m \tau_{mb}(\hat{g}_2\tau_{j/a}\overline{\hat{g}_1})\right\|_\infty$$

$$+ \sum_j (1 + |j/a|)^{|s|}\left\|\sum_m \tau_{mb}(\hat{g}_2\tau_{j/a}\overline{\hat{g}_2})\right\|_\infty$$

$$\leq C(1 + 2\sqrt{d}\max\{1,a\})^d(1 + 2\sqrt{d}\max\{1,1/b\})^d$$

$$\times (2\|g_1\|_{W_{|s|}}\|g_2\|_{W_{|s|}} + \|g_2\|_{W_{|s|}}^2)$$

$$< C(1 + 2\sqrt{d}\max\{1,a\})^d(1 + 2\sqrt{d}\max\{1,1/b\})^d(2\epsilon\|g_1\|_{W_{|s|}} + \epsilon^2) \leq b^{-d}\eta$$

since ϵ was arbitrary, $a < (2\sqrt{d}M)^{-1} < \infty$, and $b \le 1$. ∎

THEOREM 3.47
Let $s \in \mathbb{R}$, $g \in W_{|s|}(\mathbb{R}^d)$, $a, b > 0$. Then there exists a constant B such that for all $f \in \mathcal{L}^2_r(\mathbb{R}^d)$ with $|r| \le |s|$,

$$\sum_n \sum_m |\langle f, e_{mb}\tau_{na}g\rangle|^2 (1+|m|)^{2r} \le B\|f\|^2_{\mathcal{L}^2_r}.$$

In other words, for all $a, b > 0$, $\{\varphi_{m,n}\}$ is a Bessel sequence in $\mathcal{L}^2_r(\mathbb{R}^d)$ and $\mathcal{L}^2_{-r}(\mathbb{R}^d)$.

PROOF A standard Poisson summation formula argument, e.g., [HW], gives

$$\sum_n \sum_m |\langle f, e_{mb}\tau_{na}g\rangle|^2 (1+|m|)^{2r}$$

$$= a^{-d} \sum_k \int \overline{\hat{f}(\gamma)}\hat{f}(\gamma - k/a) \sum_m (1+|m|)^{2r} \hat{g}(\gamma - mb)\overline{\hat{g}(\gamma - mb - k/a)}\, d\gamma.$$

Now, for some constant c, and any $k \in \mathbb{Z}^d$, $\gamma \in \hat{\mathbb{R}}^d$, we have

$$(1+|m|)^{2r} \le c(1+|mb|)^{2r}$$
$$\le c(1+|\gamma|)^r (1+|\gamma - mb|)^{|r|}(1+|\gamma - k/a|)^r(1+|\gamma - mb - k/a|)^{|r|}$$
$$\le c(1+|\gamma|)^r (1+|\gamma - mb|)^{|s|}(1+|\gamma - k/a|)^r(1+|\gamma - mb - k/a|)^{|s|}.$$

Therefore,

$$\sum_n \sum_m |\langle f, e_{mb}\tau_{na}g\rangle|^2 (1+|m|)^{2r}$$

$$\le ca^{-d} \sum_k \int \overline{\hat{f}(\gamma)}(1+|\gamma|)^r \hat{f}(\gamma - k/a)(1+|\gamma - k/a|)^r$$

$$\sum_m \hat{g}(\gamma - mb)(1+|\gamma - mb|)^{|s|}\overline{\hat{g}(\gamma - mb - k/a)}(1+|\gamma - mb - k/a|)^{|s|}\, d\gamma$$

$$\le ca^{-d} \int |\hat{f}(\gamma)|^2(1+|\gamma|)^{2r}\, d\gamma$$

$$\sum_k \left\| \sum_m |\hat{g}(\gamma - mb)|(1+|\gamma - mb|)^{|s|}|\hat{g}(\gamma - mb - k/a)| \right.$$

$$\left. \times (1+|\gamma - mb - k/a|)^{|s|} \right\|_\infty$$

$$\le ca^{-d}\|f\|^2_{\mathcal{L}^2_r} \sum_m \operatorname*{ess\,sup}_{\gamma \in \hat{\mathbb{R}}^d} |\varphi_b(\gamma)\tau_{mb}(\hat{g}(\gamma)(1+|\gamma|)^{|s|})|$$

$$\times \sum_k \operatorname*{ess\,sup}_{\gamma \in \hat{\mathbb{R}}^d} |\varphi_{1/a}(\gamma)\tau_{k/a}(\tau_{mb}(\hat{g}(\gamma)(1+|\gamma|)^{|s|}))|$$

$$\leq ca^{-d}\|f\|_{\mathcal{L}_r^2}^2 \sum_m \operatorname*{ess\,sup}_{\gamma \in \hat{\mathbb{R}}^d} |\tau_{-mb}\varphi_b(\gamma)\hat{g}(\gamma)(1+|\gamma|)^{|s|}|$$

$$\times \sum_k \operatorname*{ess\,sup}_{\gamma \in \hat{\mathbb{R}}^d} |\tau_{-k/a}\varphi_{1/a}(\gamma)\hat{g}(\gamma - mb)(1+|\gamma - mb|)^{|s|}| \leq B\|f\|_{\mathcal{L}_r^2}^2 \|g\|_{W_{|s|}}^2$$

by Lemma 3.45. ∎

THEOREM 3.48

Let $s \in \mathbb{R}$, $g \in W_{|s|}(\mathbb{R}^d)$. *Assume that for some* $b > 0$, *there exist constants* $A, B > 0$ *such that*

$$A \leq \sum_m |\hat{g}(\gamma - mb)|^2 \leq B \text{ a.e.} \tag{3.50}$$

Then there is a constant $\tilde{C}(a, |s|)$ *depending also on* g *and* b *such that for each* $a > 0$,

$$\lim_{a \to 0} \tilde{C}(a, |s|) = 0, \tag{3.51}$$

and for each $f \in \mathcal{L}_r^2(\mathbb{R}^d)$, $|r| \leq |s|$,

$$\left\| f - \frac{2a^d}{B+A} Sf \right\|_{\mathcal{L}_r^2} \leq \frac{B - A + \tilde{C}(a, |s|)}{B+A} \|f\|_{\mathcal{L}_r^2}. \tag{3.52}$$

PROOF   A Poisson summation formula argument (e.g., [W2], [HW, Theorem 4.2.1]) shows that for $g \in W_{|s|}(\mathbb{R}^d)$, $f \in \mathcal{S}(\mathbb{R}^d)$,

$$(Sf)^\wedge(\gamma) = a^{-d} \sum_j \tau_{j/a}\hat{f}(\gamma)\tilde{G}_j(\gamma).$$

Combining this equation with Hölder's and Minkowski's inequalities, we compute

$$\left\| f - \frac{2a^d}{B+A} Sf \right\|_{\mathcal{L}_r^2}$$

$$= \left(\int \left| \hat{f}(\gamma) - \frac{2a^d}{B+A} (Sf)^\wedge(\gamma) \right|^2 (1 + |\gamma|)^{2r} \, d\gamma \right)^{1/2}$$

$$= \left(\int \left| \hat{f}(\gamma) - \frac{2}{B+A} \sum_j \tau_{j/a}\hat{f}(\gamma)\tilde{G}_j(\gamma) \right|^2 (1 + |\gamma|)^{2r} \, d\gamma \right)^{1/2}$$

$$\leq \left(\int \left| \hat{f}(\gamma)(1 - \frac{2}{B+A} \tilde{G}_0(\gamma)) \right|^2 (1 + |\gamma|)^{2r} \, d\gamma \right)^{1/2}$$

$$+ \left(\int \left| \frac{2}{B+A} \sum_{j \neq 0} \tau_{j/a}\hat{f}(\gamma)\tilde{G}_j(\gamma) \right|^2 (1 + |\gamma|)^{2r} \, d\gamma \right)^{1/2}$$

$$\leq \frac{B-A}{B+A} \left(\int |\hat{f}(\gamma)|^2 (1 + |\gamma|)^{2r} \, d\gamma \right)^{1/2}$$

$$+ \frac{2}{B+A} \sum_{j \neq 0} \|\tilde{G}_j\|_\infty \left(\int |\tau_{j/a}\hat{f}(\gamma)|^2 (1 + |\gamma|)^{2r} \, d\gamma \right)^{1/2}$$

$$\leq \frac{B-A}{B+A} \|f\|_{\mathcal{L}_r^2}^2 + \frac{2}{B+A} \sum_{j \neq 0} (1 + |j/a|)^{|s|} \|\tilde{G}_j\|_\infty \|f\|_{\mathcal{L}_r^2}^2$$

$$= \frac{B - A + \tilde{C}(a, |s|)}{B+A} \|f\|_{\mathcal{L}_r^2}^2,$$

where

$$\tilde{C}(a, |s|) = 2 \sum_{j \neq 0} (1 + |j/a|)^{|s|} \|\tilde{G}_j\|_\infty.$$

In fact, the second inequality requires the estimate

$$\left\| 1 - \frac{2}{B+A} G_0 \right\|_\infty = \max\left\{ 1 - \frac{2A}{B+A}, \frac{2B}{B+A} - 1 \right\}$$

$$= \frac{B-A}{B+A},$$

which follows from (3.50). By Lemma 3.46,

$$\lim_{a \to 0} \sum_{j \neq 0} (1 + |j/a|)^{|s|} \|\tilde{G}_j\|_\infty = 0,$$

and this is (3.51). ∎

We can now prove the following result.

COROLLARY 3.49
Let $s \in \mathbb{R}$, $g \in W_{|s|}(\mathbb{R}^d)$. Suppose that for some $b > 0$, (3.50) holds. Then for all $a > 0$ sufficiently small, $\{e_{mb}\tau_{na}g\}$ is a Banach frame and a set of atoms for $\mathcal{L}_r^2(\mathbb{R}^d)$ and $\mathcal{L}_{-r}^2(\mathbb{R}^d)$, $|r| \leq |s|$.

PROOF Fix $|r| \leq |s|$. By Theorem 3.47, the collection $\{e_{mb}\tau_{na}g\}$ is a Bessel sequence for $\mathcal{L}_r^2(\mathbb{R}^d)$ for any $a, b > 0$. By Theorem 3.32, it suffices to show that S is a topological isomorphism of $\mathcal{L}_r^2(\mathbb{R}^d)$. This follows from Theorem 3.48 and a Neumann series argument (cf. the proof of Theorem 3.2 and (3.9)). Specifically, we choose a so small that by (3.51),

$$\frac{B - A + \tilde{C}(a, |s|)}{B+A} < 1.$$

Then by (3.52) we have

$$\left\| I - \frac{2a^d}{B+A} S \right\| < 1,$$

where we consider S as an operator on $\mathcal{L}^2_r(\mathbb{R}^d)$. Therefore

$$S^{-1} = \frac{2a^d}{B+A} \sum_{k=0}^{\infty} \left(I - \frac{2a^d}{B+A} S \right)^k,$$

where the sum converges uniformly on bounded subsets of $\mathcal{L}^2_r(\mathbb{R}^d)$. Therefore, S is a topological isomorphism of $\mathcal{L}^2_r(\mathbb{R}^d)$ and by Theorem 3.32, $\{\varphi_{m,n}\}$ is a Banach frame and a set of atoms for $\mathcal{L}^2_r(\mathbb{R}^d)$. ∎

3.8 Frames of translates

Translation is the common geometrical feature of both Gabor and wavelet systems. In Theorem 3.56 we shall characterize when a sequence of integer translates generates a frame in terms of a boundedness condition reminiscent of the Zak transform characterization of frames given in Theorem 3.16. The simpler characterizations of orthonormal bases and exact frames of integer translates are proved in Theorems 3.50 and 3.51, respectively.

For $\varphi \in L^2(\mathbb{R}^d)$, we shall consider the periodic function

$$\Phi(\gamma) \equiv \sum |\hat{\varphi}(\gamma + m)|^2,$$

where summation ranges over $\mathbb{Z}^d$. Clearly, $\Phi \in L^1(\mathbb{T}^d)$ by the Beppo Levi and Plancherel theorems.

THEOREM 3.50
Let $\varphi \in L^2(\mathbb{R}^d)$ and let

$$V \equiv \overline{\text{Span}} \{\tau_k \varphi : k \in \mathbb{Z}^d\}$$

be a closed subspace of $L^2(\mathbb{R}^d)$. $\{\tau_k \varphi\}$ is an orthonormal basis of V if and only if $\Phi = 1$ a.e.

PROOF It suffices to prove that orthonormality is equivalent to the condition $\Phi = 1$ a.e. This is a consequence of the calculation

$$\int \varphi(t)\overline{\varphi(t-k)}\,dt = \int |\hat{\varphi}(\gamma)|^2 e^{2\pi i k \cdot \gamma}\,d\gamma$$
$$= \sum_n \int_{\mathbb{T}^d} |\hat{\varphi}(\gamma+n)|^2 e^{2\pi i k \cdot \gamma}\,d\gamma = \int_{\mathbb{T}^d} \Phi(\gamma) e^{2\pi i k \cdot \gamma}\,d\gamma, \tag{3.53}$$

and the fact that $\Phi = 1$ a.e. if and only if its Fourier coefficients are all 0 except for $k = 0$, in which case its Fourier coefficient is 1. ∎

THEOREM 3.51
Let $\varphi \in L^2(\mathbb{R}^d)$ and let

$$V \equiv \overline{\text{Span}}\{\tau_k\varphi \colon k \in \mathbb{Z}^d\}$$

be a closed subspace of $L^2(\mathbb{R}^d)$. $\{\tau_k\varphi\}$ is an exact frame for V if and only if

$$\exists A, B > 0 \quad \text{such that } A \leq \Phi(\gamma) \leq B \text{ a.e.} \tag{3.54}$$

PROOF Recall from Section 3.3 that $\{\tau_k\varphi\}$ is an exact frame if and only if it is a bounded unconditional basis, i.e., there exist $A, B > 0$ such that for all $\{c_n \colon n \in F \subseteq \mathbb{Z}^d\} \subseteq \mathbb{C}$,

$$A\sum_{n\in F}|c_n|^2 \leq \left\|\sum_{n\in F}c_n\tau_n\varphi\right\|_2^2 \leq B\sum_{n\in F}|c_n|^2, \tag{3.55}$$

where F is a finite set. Also, in a calculation similar to (3.53), the Plancherel theorem on $L^2(\mathbb{R}^d)$ allows us to write

$$\left\|\sum_{n\in F}c_n\tau_n\varphi\right\|_2^2 = \int_{\mathbb{T}^d}|\Theta(\gamma)|^2\Phi(\gamma)\,d\gamma, \tag{3.56}$$

where $\Theta(\gamma) \equiv \sum_{n\in F}c_ne^{-2\pi in\cdot\gamma}$. If (3.54) is valid then (3.55) is a consequence of (3.56) by the Plancherel theorem on $L^2(\mathbb{T}^d)$.

Conversely, suppose $\{\tau_k\varphi\}$ is an exact frame for V, i.e., suppose (3.55) holds. If $f = \sum_{n\in F}c_n\tau_n\varphi \in V$ and $\Theta(\gamma) = \sum_{n\in F}c_ne^{-2\pi in\cdot\gamma}$, then

$$\|f\|_2^2 = \int_{\mathbb{T}^d}|\Theta(\gamma)|^2\Phi(\gamma)\,d\gamma.$$

Thus, since $\|\Theta\|_{L^2(\mathbb{T}^d)}^2 = \sum_{n\in F}|c_n|^2$ and (3.55) holds, we can assert that $A \leq \Phi(\gamma) \leq B$ a.e. In fact, if $\Phi(\gamma) < A$ on a set E, $|E| > 0$, then there is a polynomial Θ, "concentrated" on E, so that $\int_{\mathbb{T}^d}|\Theta(\gamma)|^2(\Phi(\gamma) - A)\,d\gamma < 0$; this is the desired contradiction. ∎

Our characterization Theorem 3.56 of frames in terms of Φ is part of a general program developed in [BeLi]. This characterization and its proof in [BeLi] also make use of the following lemmas, some of which are of independent interest.

LEMMA 3.52
Let $\varphi \in L^2(\mathbb{R}^d)$ and let

$$V \equiv \overline{\text{Span}}\{\tau_k\varphi \colon k \in \mathbb{Z}^d\}$$

be a closed subspace of $L^2(\mathbb{R}^d)$. Consider the frame condition,

$$A\|f\|_2^2 \leq \sum|\langle f, \tau_k\varphi\rangle|^2 \leq B\|f\|_2^2. \tag{3.57}$$

(3.57) is valid for each $f \in V$ if and only if (3.57) is valid for each $f \in \text{span}\{\tau_k\varphi\}$.

PROOF   If (3.57) is valid for all $f \in V$ then it is true for all $f \in \operatorname{span}\{\tau_k \varphi\}$. For the converse, suppose (3.57) is valid for $\operatorname{span}\{\tau_k \varphi\}$, let $f \in V$, and let

$$\lim_{n\to\infty} \|f - f_n\|_2 = 0$$

for $\{f_n\} \subseteq \operatorname{span}\{\tau_k \varphi\}$.

(a) We begin this direction by noting that

$$\lim_{n\to\infty} |\langle f_n, \tau_k \varphi \rangle|^2 = |\langle f, \tau_k \varphi \rangle|^2 \tag{3.58}$$

since

$$\big| |\langle f_n, \tau_k \varphi \rangle|^2 - |\langle f, \tau_k \varphi \rangle|^2 \big| \le \|\varphi\|_2 (\|f_n\|_2 + \|f\|_2) |\langle f_n - f, \tau_k \varphi \rangle|$$
$$\le \|\varphi\|_2^2 (\|f_n\|_2 + \|f\|_2) \|f_n - f\|_2.$$

Because of (3.58) and Fatou's lemma applied to sums, we have by the right side of (3.57) for $\operatorname{span}\{\tau_k \varphi\}$ that

$$\sum |\langle f, \tau_k \varphi \rangle|^2 \le \liminf_{n\to\infty} \sum_k |\langle f_n, \tau_k \varphi \rangle|^2$$
$$\le B \liminf_{n\to\infty} \|f_n\|_2^2 = B\|f\|_2^2.$$

Thus, the right side of (3.57) for V is valid.

(b) To prove the first inequality of (3.57) for f, observe that for any n, we have by the triangle inequality that

$$\left(\sum |\langle f, \tau_k \varphi \rangle|^2 \right)^{1/2} \ge \left(\sum_k |\langle f_n, \tau_k \varphi \rangle|^2 \right)^{1/2} - \left(\sum_k |\langle f_n - f, \tau_k \varphi \rangle|^2 \right)^{1/2}.$$

Since the lower bound in (3.57) holds by assumption for f_n and by part a the upper bound holds for all f, we have that

$$\left(\sum |\langle f, \tau_k \varphi \rangle|^2 \right)^{1/2} \ge A^{1/2} \|f_n\|_2 - B^{1/2} \|f_n - f\|_2.$$

Letting $n \to \infty$ on the right hand side gives the first inequality of (3.57).   ∎

LEMMA 3.53
Let $\varphi \in L^2(\mathbb{R}^d)$, and define $f \equiv \sum_{k \in F} c_k \tau_k \varphi$ and $\Theta_F(\gamma) \equiv \sum_{k \in F} c_k e^{-2\pi i k \cdot \gamma}$, for a given finite set $F \subseteq \mathbb{Z}^d$. Then $f \in L^2(\mathbb{R}^d)$, $\Theta_F \in L^\infty(\mathbb{T}^d)$, and

$$\|f\|_2^2 = \int_{\mathbb{T}^d} |\Theta_F(\gamma)|^2 \Phi(\gamma)\, d\gamma < \infty. \tag{3.59}$$

PROOF   A direct calculation yields

$$\|f\|_2^2 = \sum_{m \in F} \sum_{n \in F} c_m \overline{c_n} \sum_k \int_{\mathbb{T}^d} |\hat{\varphi}(\gamma + k)|^2 e^{-2\pi i (m-n)\cdot\gamma}\, d\gamma. \tag{3.60}$$

We can write "$\sum_k \int_{\mathbb{T}^d} = \int_{\mathbb{T}^d} \sum_k$" in (3.60), using the dominated convergence theorem and the fact that $\Phi \in L^1(\mathbb{T}^d)$. (3.59) follows.   ∎

LEMMA 3.54

Let $\varphi \in L^2(\mathbb{R}^d)$, and define $f \equiv \sum_{k \in F} c_k T_k \varphi$ and $\Theta_F(\gamma) \equiv \sum_{k \in F} c_k e^{-2\pi i k \cdot \gamma}$ for a given finite set $F \subseteq \mathbb{Z}^d$. Assume $\Phi \in L^2(\mathbb{T}^d)$. Then $f \in L^2(\mathbb{R}^d)$, $\Theta_F \in L^\infty(\mathbb{T}^d)$, and

$$\sum |\langle f, T_k \varphi \rangle|^2 = \int_{\mathbb{T}^d} |\Theta_F(\gamma)|^2 \Phi(\gamma)^2 \, d\gamma < \infty. \tag{3.61}$$

PROOF

(a) A direct calculation yields

$$\sum_{k \in K} |\langle f, T_k \varphi \rangle|^2 = \sum_{m \in F} \sum_{n \in F} c_m \overline{c_n} \int |\hat{\varphi}(\gamma)|^2 e^{-2\pi i m \cdot \gamma}$$
$$\times \left(\int |\hat{\varphi}(\lambda)|^2 e^{2\pi i n \cdot \lambda} \sum_{k \in K} e^{2\pi i k \cdot (\gamma - \lambda)} \, d\lambda \right) d\gamma, \tag{3.62}$$

where $K \subseteq \mathbb{Z}^d$ is a finite set. Now, if

$$K = \prod_{j=1}^d [-K_j, K_j] \subseteq \mathbb{Z}^d,$$

we define the d-dimensional Dirichlet kernel

$$D_K(\gamma) = \prod_{j=1}^d \left(\sum_{k=-K_j}^{K_j} (e^{2\pi i \gamma_j})^k \right).$$

Thus, by elementary calculations, we have

$$D_K(\gamma) = \prod_{j=1}^d \frac{\sin \pi (2K_j + 1)\gamma_j}{\sin \pi \gamma_j},$$

$D_K(0) = \prod_{j=1}^d (2K_j + 1) = \|D_K\|_{L^\infty(\mathbb{T}^d)}$, $\int_{\mathbb{T}^d} D_K(\gamma) \, d\gamma = 1$, and

$$D_K(\gamma - \lambda) = \sum_{k \in K} e^{2\pi i k \cdot (\gamma - \lambda)}.$$

Hence, for this K, (3.62) can be rewritten as

$$\sum_{k \in K} |\langle f, T_k \varphi \rangle|^2 = \sum_{m \in F} \sum_{n \in F} c_m \overline{c_n} \int |\hat{\varphi}(\gamma)|^2 e^{-2\pi i m \cdot \gamma}$$
$$\times \left(\int |\hat{\varphi}(\lambda)|^2 e^{2\pi i n \cdot \lambda} D_K(\gamma - \lambda) \, d\lambda \right) d\gamma$$
$$= \sum_{m \in F} \sum_{n \in F} c_m \overline{c_n} \int |\hat{\varphi}(\gamma)|^2 e^{-2\pi i m \cdot \gamma} \tag{3.63}$$
$$\times \left(\int_{\mathbb{T}^d} \Phi(\lambda) e^{2\pi i n \cdot \lambda} D_K(\gamma - \lambda) \, d\lambda \right) d\gamma,$$

where the second equality is a consequence of the dominated convergence theorem.

(b) Note that

$$\int_{\mathbb{T}^d} \Phi(\lambda)e^{2\pi in\cdot\lambda}D_K(\gamma-\lambda)\,d\lambda = S_K(\Phi e_n)(\gamma) \equiv \sum_{k\in K}(\Phi e_n)^\wedge[k]e^{-2\pi ik\cdot\gamma},$$

where

$$(\Phi e_n)^\wedge[k] \equiv \int_{\mathbb{T}^d} \Phi(\lambda)e^{2\pi in\cdot\lambda}e^{2\pi ik\cdot\lambda}\,d\lambda.$$

Thus, (3.63) can be written as

$$\sum_{k\in K}|\langle f,\tau_k\varphi\rangle|^2 = \sum_{m\in F}\sum_{n\in F}c_m\overline{c_n}\int|\hat\varphi(\gamma)|^2 e^{-2\pi im\cdot\gamma}S_K(\Phi e_n)(\gamma)\,d\gamma. \qquad (3.64)$$

A dominated convergence argument, depending on the periodicity of e_{-m} and $S_K(\Phi e_n)$, the boundedness of $S_K(\Phi e_n)$ by $D_K(0)\|\Phi\|_{L^1(\mathbb{T}^d)}$, and the fact that $\Phi\in L^1(\mathbb{T}^d)$, can be used to prove that

$$\int|\hat\varphi(\gamma)|^2 e^{-2\pi im\cdot\gamma}S_K(\Phi e_n)(\gamma)\,d\gamma = \int_{\mathbb{T}^d}\Phi(\gamma)e^{-2\pi im\cdot\gamma}S_K(\Phi e_n)(\gamma)\,d\gamma. \qquad (3.65)$$

(c) For simplicity, let $K_1 = \cdots = K_d$. We now note that

$$\lim_{K_1\to\infty}\int_{\mathbb{T}^d}\Phi(\gamma)e^{-2\pi im\cdot\gamma}S_K(\Phi e_n)(\gamma)\,d\gamma = \int_{\mathbb{T}^d}\Phi(\gamma)^2 e^{-2\pi i(m-n)\cdot\gamma}\,d\gamma. \qquad (3.66)$$

In fact, the right side exists since $\Phi\in L^2(\mathbb{T}^d)$, and

$$\left|\int_{\mathbb{T}^d}\Phi(\gamma)e^{-2\pi im\cdot\gamma}\left(S_K(\Phi e_n)(\gamma)-\Phi(\gamma)e_n(\gamma)\right)d\gamma\right|$$
$$\leq \|\Phi\|_{L^2(\mathbb{T}^d)}\|S_K(\Phi e_n)-\Phi e_n\|_{L^2(\mathbb{T}^d)}.$$

The right side of the inequality goes to 0 as $K_1\to\infty$ by the Plancherel theorem.

(3.61) is a consequence of combining (3.64), (3.65), and (3.66).   ∎

LEMMA 3.55
Let $\varphi\in L^2(\mathbb{R}^d)$ and let

$$V \equiv \overline{\mathrm{Span}}\,\{\tau_k\varphi : k\in\mathbb{Z}^d\}$$

be a closed subspace of $L^2(\mathbb{R}^d)$. Assume $\Phi\in L^2(\mathbb{T}^d)$. The sequence $\{\tau_k\varphi\}$ is a frame for V with frame bounds A and B if and only if there are positive constants A and B such that, for all trigonometric polynomials

$$\Theta(\gamma) \equiv \Theta_F(\gamma) \equiv \sum_{k\in F}c_k e^{-2\pi ik\cdot\gamma}$$

on $\mathbb{T}^d$,

$$A \int_{\mathbb{T}^d} |\Theta(\gamma)|^2 \Phi(\gamma) \, d\gamma \leq \int_{\mathbb{T}^d} |\Theta(\gamma)|^2 \Phi(\gamma)^2 \, d\gamma \leq B \int_{\mathbb{T}^d} |\Theta(\gamma)|^2 \Phi(\gamma) \, d\gamma < \infty. \tag{3.67}$$

PROOF $(\Rightarrow)$ For a given $\Theta \equiv \Theta_F$, define $f(t) \equiv \sum_{k \in F} c_k \tau_k \varphi \in V$. By Lemma 3.53,

$$\|f\|_2^2 = \int_{\mathbb{T}^d} |\Theta(\gamma)|^2 \Phi(\gamma) \, d\gamma; \tag{3.68}$$

and by Lemma 3.54,

$$\sum |\langle f, \tau_k \varphi \rangle|^2 = \int_{\mathbb{T}^d} |\Theta(\gamma)|^2 \Phi(\gamma)^2 \, d\gamma. \tag{3.69}$$

By the hypotheses,

$$\exists A, B > 0 \quad \text{such that } \forall h \in V,$$
$$A\|h\|_2^2 \leq \sum |\langle h, \tau_k \varphi \rangle|^2 \leq B\|h\|_2^2 < \infty. \tag{3.70}$$

Substituting (3.68) and (3.69) into (3.70) for the case $h = f$ yields (3.67).

$(\Leftarrow)$ Now suppose (3.67) is valid for all trigonometric polynomials $\Theta = \Theta_F$. From Lemmas 3.53 and 3.54 we obtain

$$\forall f \in \text{span} \{\tau_k \varphi\}, \ A\|f\|_2^2 \leq \sum |\langle f, \tau_k \varphi \rangle|^2 \leq B\|f\|_2^2.$$

The result follows from Lemma 3.52. ∎

THEOREM 3.56 [BeLi]
Let $\varphi \in L^2(\mathbb{R}^d)$ and let

$$V \equiv \overline{\text{Span}} \{\tau_k \varphi : k \in \mathbb{Z}^d\}$$

be a closed subspace of $L^2(\mathbb{R}^d)$. The sequence $\{\tau_k \varphi\}$ is a frame for V if and only if there are positive constants A and B such that

$$A \leq \Phi(\gamma) \leq B \quad \text{a.e. on } \mathbb{T}^d \setminus N, \tag{3.71}$$

where $N \equiv \{\gamma \in \mathbb{T}^d : \Phi(\gamma) = 0\}$.

Note that the set N is defined only up to sets of measure zero, and a member of the equivalence class represented by N is given by the zero set of any member of the equivalence class in $L^2(\mathbb{T}^d)$ represented by Φ.

PROOF $(\Leftarrow)$ Assume (3.71) and let

$$\Theta(\gamma) \equiv \Theta_F(\gamma) \equiv \sum_{k \in F} c_k e^{-2\pi i k \cdot \gamma}$$

be a trigonometric polynomial on $\mathbb{T}^d$. We shall prove the inequalities (3.67) of Lemma 3.55.

$$
\begin{aligned}
A \int_{\mathbb{T}^d} |\Theta(\gamma)|^2 \Phi(\gamma) \, d\gamma &= A \int_{\mathbb{T}^d \setminus N} |\Theta(\gamma)|^2 \Phi(\gamma) \, d\gamma \\
&\leq \int_{\mathbb{T}^d \setminus N} |\Theta(\gamma)|^2 \Phi(\gamma)^2 \, d\gamma = \int_{\mathbb{T}^d} |\Theta(\gamma)|^2 \Phi(\gamma)^2 \, d\gamma \\
&= \int_{\mathbb{T}^d \setminus N} |\Theta(\gamma)|^2 \Phi(\gamma)^2 \, d\gamma \leq B \int_{\mathbb{T}^d \setminus N} |\Theta(\gamma)|^2 \Phi(\gamma) \, d\gamma \\
&= B \int_{\mathbb{T}^d} |\Theta(\gamma)|^2 \Phi(\gamma) \, d\gamma,
\end{aligned}
$$

where the two inequalities follow from (3.71).

$(\Rightarrow)$ **a.** Assume $\Phi(\gamma) < A$ on $E \subseteq \mathbb{T}^d \setminus N$, where $|E| > 0$. We shall obtain a contradiction to the first inequality of (3.67), thereby proving, by Lemma 3.55, that $\{\tau_k \varphi\}$ is not a frame for V. A similar contradiction will arise, in this case to the second inequality of (3.67), when we assume $\Phi(\gamma) > B$ on $E \subseteq \mathbb{T}^d \setminus N$, where $|E| > 0$.

b. By our hypothesis on Φ we can choose $\Theta \in L^\infty(\mathbb{T}^d) \subseteq L^2(\mathbb{T}^d)$ such that $\Theta = 0$ off E, $|\Theta| > 0$ on E, and

$$
A \int_{\mathbb{T}^d} |\Theta(\gamma)|^2 \Phi(\gamma) \, d\gamma > \int_{\mathbb{T}^d} |\Theta(\gamma)|^2 \Phi(\gamma)^2 \, d\gamma,
$$

e.g., $\Theta \equiv \mathbf{1}_E$. Thus,

$$
p \equiv \int_{\mathbb{T}^d} |\Theta(\gamma)|^2 (A\Phi(\gamma) - \Phi(\gamma)^2) \, d\gamma > 0. \tag{3.72}
$$

We shall find a trigonometric polynomial Ψ so that (3.72) is valid for Θ replaced by Ψ. This is the desired contradiction to the first inequality in (3.67). Note that if $\Phi \in L^2(\mathbb{T}^d) \setminus L^\infty(\mathbb{T}^d)$, then (3.72) is still valid for $\Theta \in L^\infty(\mathbb{T}^d)$; and the plan to choose Ψ does not a priori require $\Phi \in L^\infty(\mathbb{T}^d)$.

c. For any $\Psi \in L^\infty(\mathbb{T}^d)$ (still assuming only that $\Phi \in L^2(\mathbb{T}^d)$), we have

$$
\begin{aligned}
\int_{\mathbb{T}^d} &|\Psi(\gamma)|^2 (A\Phi(\gamma) - \Phi(\gamma)^2) \, d\gamma \\
&= \int_{\mathbb{T}^d \setminus E} |\Psi(\gamma)|^2 (A\Phi(\gamma) - \Phi(\gamma)^2) \, d\gamma \\
&\quad + \int_E |\Psi(\gamma) - \Theta(\gamma) + \Theta(\gamma)|^2 \Phi(\gamma)(A - \Phi(\gamma)) \, d\gamma. \tag{3.73}
\end{aligned}
$$

Since $A - \Phi > 0$ on E and $\Phi > 0$ a.e. on $\mathbb{T}^d \setminus N$, we have $\Phi(A - \Phi) > 0$ a.e. on E, and we consider $\Phi(A - \Phi)$ as a weight on a weighted L^2 space on

E. Thus, with $\Theta, \Psi \in L^\infty(\mathbb{T}^d)$ we have

$$\left(\int_E |\Psi(\gamma) - \Theta(\gamma) + \Theta(\gamma)|^2 \Phi(\gamma)(A - \Phi(\gamma)) \, d\gamma\right)^{1/2}$$

$$\geq \left(\int_E |\Psi(\gamma) - \Theta(\gamma)|^2 \Phi(\gamma)(A - \Phi(\gamma)) \, d\gamma\right)^{1/2}$$

$$- \left(\int_E |\Theta(\gamma)|^2 \Phi(\gamma)(A - \Phi(\gamma)) \, d\gamma\right)^{1/2}. \tag{3.74}$$

Combining (3.73) and (3.74) we have, by an easy calculation, that

$$\int_{\mathbb{T}^d} |\Psi(\gamma)|^2 (A\Phi(\gamma) - \Phi(\gamma)^2) \, d\gamma$$

$$\geq p - 2p^{1/2} \left(\int_E |\Psi(\gamma) - \Theta(\gamma)|^2 \Phi(\gamma)(A - \Phi(\gamma)) \, d\gamma\right)^{1/2} \tag{3.75}$$

$$+ \int_{\mathbb{T}^d} |\Psi(\gamma) - \Theta(\gamma)|^2 \Phi(\gamma)(A - \Phi(\gamma)) \, d\gamma.$$

d. Since the trigonometric polynomials are dense in $L^2(\mathbb{T}^d)$, we can choose a trigonometric polynomial Ψ so that, from (3.75), we obtain

$$\int_{\mathbb{T}^d} |\Psi(\gamma)|^2 (A\Phi(\gamma) - \Phi(\gamma)^2) \, d\gamma > \frac{p}{2} > 0,$$

e.g., [BeLi] for details. ∎

REMARK 3.57 Let $\varphi = \frac{1}{3} \mathbf{1}_{[-1,2)}$ on $\mathbb{R}$; in particular, $\{\tau_k \varphi\}$ is not orthogonal. It is easy to see that

$$\Phi(\gamma) = \frac{1}{9} e^{2\pi i \gamma} [1 + 2e^{-2\pi i \gamma} + 3e^{-2\pi i(2\gamma)} + 2e^{-2\pi i(3\gamma)} + e^{-2\pi i(4\gamma)}],$$

a trigonometric polynomial with two easily computable double roots. Thus, $\{\tau_k \varphi\}$ is not a frame for $V \equiv \overline{\text{Span}} \{\tau_k \varphi\}$. This function φ is generated by the quadrature mirror filter $H(\gamma) = \sqrt{2} e^{-\pi i \gamma} \cos 3\pi\gamma$ in the theory of multiresolution analysis, see, e.g., [M], [Me]. In [BeLi] there is a development of filter design based on Theorem 3.56, cf. [L]. ∎

3.9   Gabor decompositions of L^1 and the Fourier transform

In this section, we shall prove a Gabor decomposition for $f \in L^1(\mathbb{R}^d)$. Our result, Theorem 3.61, requires the notion of *autocorrelation* of bounded measurable
functions g. Although a discrete version of Theorem 3.61 is possible, our decomposition is in terms of the continuous Gabor transform. In this setting, the Gabor
transform reduces to the ordinary Fourier transform when $g \equiv 1$, and, in this case,
Theorem 3.61 becomes the usual L^1-inversion formula for the Fourier transform.

DEFINITION 3.58

(a) *Let $g \in L^1_{\text{loc}}(\mathbb{R}^d)$ and $c \in \mathbb{R}$. The Gabor or Weyl–Heisenberg system (of
coherent states) for g is the family of functions*

$$\varphi_{\gamma,x}(t) \equiv \varphi_g(t; x, \gamma) = e^{-2\pi i c x \cdot \gamma}(e_\gamma \tau_x g)(t), \ (x, \gamma) \in \mathbb{R}^d \times \hat{\mathbb{R}}^d,$$

cf. the definition of the Gabor system $\{\varphi_{m,n}\}$ (Definition 3.10b).

(b) *For a given Gabor coherent state φ_g the (continuous) Gabor transform of
$f \in L^1_{\text{loc}}(\mathbb{R}^d)$ is formally defined on $\mathbb{R}^d \times \hat{\mathbb{R}}^d$ as*

$$W_g(f)(x, \gamma) = \int f(t)\overline{\varphi_g}(t; x, \gamma) \, dt.$$

*W_g should be compared with the function V_g defined in Remark 3.34. $V_g(f)$
has the form*

$$\forall h \in \mathbb{H}^d, \ (V_g f)(h) = \int f(t)\bar{z}e^{2\pi i x \cdot \gamma}\overline{(e_\gamma \tau_x g)(t)} \, dt,$$

so that, vis-á-vis W_g, $c = 1$ and there is a toral component $\bar{z}$.

DEFINITION 3.59   *Let $g \in L^\infty(\mathbb{R}^d)$, and define*

$$\forall T > 0, \ P_T(t) = \frac{1}{|B(T)|} \int_{B(T)} g(t + x)\overline{g(x)} \, dx,$$

*where $B(T) \subseteq \mathbb{R}^d$ is the ball of radius T about $0 \in \mathbb{R}^d$. Suppose that there is a
continuous positive definite function P with $P(0) > 0$, for which $\lim_{T \to \infty} P_T = P$
pointwise on $\mathbb{R}^d$. Then $P \equiv P_g \in L^\infty(\mathbb{R}^d)$ is the autocorrelation of g, and the
bounded positive measure $\check{P}$ is the power spectrum of g. These notions, as
related to spectral estimation and Wiener's Generalized Harmonic Analysis, are
developed on $\mathbb{R}^d$ in [Be2], [Be3].*

Our aim in Theorems 3.60 and 3.61 is to give integral representations of $L^1(\mathbb{R}^d)$
in terms of the Gabor transform. To this end we let $\{\rho_n\} \subseteq L^1(\hat{\mathbb{R}}^d)$ have the
property that $\{\check{\rho}_n\} \subseteq L^1(\mathbb{R}^d)$ is an $L^1(\mathbb{R}^d)$ approximate identity, i.e., $\int \check{\rho}_n(t) \, dt = 1$,

$\sup_n \|\check{\rho}_n\|_1 < \infty$, and

$$\forall T > 0, \lim_{n \to \infty} \int_{B(T)^-} |\check{\rho}_n(t)| \, dt = 0.$$

THEOREM 3.60 [Be1]
Let $g \in (L^2(\mathbb{R}^d) \cap L^\infty(\mathbb{R}^d)) \setminus \{0\}$, and consider the corresponding Gabor system $\{\varphi_{\gamma,x} : (x, \gamma) \in \mathbb{R}^d \times \hat{\mathbb{R}}^d\}$ and Gabor transform W_g. Then, for each $f \in L^1(\mathbb{R}^d)$,

$$\lim_{n \to \infty} \frac{1}{\|g\|_2^2} \int\int W_g(f)\varphi_g(t; x, \gamma)\rho_n(\gamma) \, dx \, d\gamma = f(t)$$

in $L^1(\mathbb{R}^d)$ norm.

We omit the proof of Theorem 3.60. It is analogous to, but simpler than, the proof of Theorem 3.61, which we do give.

THEOREM 3.61 [Be1]
Let $g \in L^\infty(\mathbb{R}^d) \setminus \{0\}$, and consider the corresponding Gabor system $\{\varphi_{\gamma,x} : (x, \gamma) \in \mathbb{R}^d \times \hat{\mathbb{R}}^d\}$ and Gabor transform W_g. Assume g has an autocorrelation P_g. Then, for each $f \in L^1(\mathbb{R}^d)$,

$$\lim_{n \to \infty} \|f_n - f\|_1 = 0,$$

where

$$f_n(t) = \frac{1}{P(0)} \lim_{T \to \infty} \frac{1}{|B(T)|} \int\int_{B(T)} W_g(f)(x, \gamma)\varphi_g(t; x, \gamma)\rho_n(\gamma) \, dx \, d\gamma.$$

PROOF We shall prove the result for $d = 1$. The generalization to $\mathbb{R}^d$, for the case that balls $B(T)$ are replaced by cubes about the origin, is immediate because of the substitutions we make. The embedding of balls into cubes into balls is justified given the hypotheses of Theorem 3.61, and so the result as stated follows. A proof, dealing with $\{B(T) : T > 0\}$ directly, follows using a method developed in [Be2].

We first calculate

$$\frac{1}{2T} \int\int_{-T}^{T} W_g(f)(x, \gamma)\varphi_g(t; x, \gamma)\rho_n(\gamma) \, dx \, d\gamma = \int f(t-u)\check{\rho}_n(u)P_T(t, u) \, du, \quad (3.76)$$

where

$$P_T(t, u) \equiv \frac{1}{2T} \int_{-T+(t-u)}^{T+(t-u)} g(s + u)\overline{g(s)} \, ds$$

and where we can use Fubini's theorem in the verification of (3.76) since

$$\frac{1}{2T} \int\int\int_{-T}^{T} |f(v)||\rho_n(\gamma)||g(v - x)g(t - x)| \, dx \, d\gamma \, dv \leq \|g\|_\infty^2 \|f\|_1 \|\rho_n\|_1 < \infty.$$

Next, we make the estimate

$$\int \left| P_g(0)f(t) - \int f(t-u)\check{\rho}_n(u)P_T(t,u)\,du \right|\,dt$$

$$\leq \int |\check{\rho}_n(u)| \left(\int |P_g(0)f(t) - P_T(t,u)f(t-u)|\,dt \right) du$$

$$\leq \int |\check{\rho}_n(u)| \left(\int |P_T(t,u)||f(t) - f(t-u)|\,dt \right) du$$

$$+ \int |\check{\rho}_n(u)| \left(\int |f(t)||P_g(0) - P_T(t,u)|\,dt \right) du$$

$$\equiv I_1(n,T) + I_2(n,T). \tag{3.77}$$

Note that

$$\lim_{n\to\infty} \sup_{T>0} I_1(n,T) = 0. \tag{3.78}$$

In fact,

$$I_1(n,T) \leq \|g\|_\infty^2 \int |\check{\rho}_n(u)| \|f - \tau_u f\|_1\,du,$$

and

$$\lim_{n\to\infty} \int |\check{\rho}_n(u)| \|f - \tau_u f\|_1\,du = 0,$$

since $\{\check{\rho}_n\}$ is an $L^1(\mathbb{R})$ approximate identity (the property $\int \check{\rho}_n(u)\,du = 1$ is not used) and $\|f - \tau_u f\|_1$ is a bounded continuous function.

To estimate $I_2(n,T)$ we first compute

$$\lim_{T\to\infty} \int |\check{\rho}_n(u)||P_g(0) - P_T(t,u)|\,du = \int |\check{\rho}_n(u)||P_g(0) - P_g(u)|\,du$$

by the dominated convergence theorem which is applicable since

$$\lim_{T\to\infty} P_T(t,u) = P_g(u)$$

for each $t, u \in \mathbb{R}$ and since

$$|\check{\rho}_n(u)||P_g(0) - P_T(t,u)| \leq 2\|g\|_\infty^2 |\check{\rho}_n(u)| \in L^1(\mathbb{R}).$$

Consequently, because

$$|f(t)| \int |\check{\rho}_n(u)||P_g(0) - P_T(t,u)|\,du \leq 2\|g\|_\infty^2 (\sup \|\check{\rho}_n\|_1)|f(t)| \in L^1(\mathbb{R}), \tag{3.79}$$

we can apply the dominated convergence theorem again to compute

$$\lim_{T\to\infty} I_2(n,T) = \int |f(t)| \left(\int |\check{\rho}_n(u)||P_g(0) - P_g(u)|\,du \right) dt.$$

Now we have

$$\lim_{n \to \infty} \int |\check{\rho}_n(u)||P_g(0) - P_g(u)| \, du = 0$$

since $\{\check{\rho}_n\}$ is an $L^1(\mathbb{R})$ approximate identity and since $|P_g(0) - P_g(u)|$ is a bounded continuous function. Therefore, we can once again use the dominated convergence theorem, invoking the analogue of (3.79) with P_g replacing P_T, to state that

$$\lim_{n \to \infty} \lim_{T \to \infty} I_2(n, T) = 0. \qquad (3.80)$$

Substituting (3.76) into the left side of (3.77), and substituting (3.78) and (3.80) into the right side of (3.77), we obtain our result. ∎

REMARK 3.62 If $g \equiv 1$, $f \in L^1(\mathbb{R}^d)$, and $c = 0$ in the definition of φ_g, then $W_1(f)(x, \gamma) = \hat{f}(\gamma)$ for $(x, \gamma) \in \mathbb{R}^d \times \hat{\mathbb{R}}^d$. In this case, Theorem 3.61 asserts that

$$\lim_{n \to \infty} \|f(t) - \int \hat{f}(\gamma) e^{2\pi i t \cdot \gamma} \rho_n(\gamma) \, d\gamma\|_1 = 0,$$

noting that $f * \check{\rho}_n \in L^1(\mathbb{R}^d)$, see, e.g., [Be1]. ∎

Acknowledgement. The authors would like to thank Dr. Christopher Heil for sharing his expertise with us on many of the topics in this chapter.

Bibliography

[AT]   L. Auslander and R. Tolimieri. 1979. "Radar ambiguity functions and group theory," *SIAM J. Math. Anal.* **1**: 847–897.

[B]   R. Balian. 1981. "Un principe d'incertitude fort en théorie du signal ou en mécanique quantique," *C. R. Acad. Sci., Paris* **292**: 1357–1362.

[Ba]   N. Bari. 1951. "Biorthogonal systems and bases in Hilbert space," *Učen Zap. Mosk. Gos. Univ. 148, Mat.* **4**: 69–107.

[Bat1]   G. Battle. 1988. "Heisenberg proof of the Balian-Low theorem," *Lett. Math. Phys.* **15**: 175–177.

[Bat2]   G. Battle. 1989. "Phase space localization theorem for ondelettes," *J. Math. Phys.* **30**: 2195–2196.

[Be1]   J. Benedetto. 1989. "Gabor representations and wavelets," *Amer. Math. Soc. Contemp. Math.* **91**: 9–27.

[Be2]   J. Benedetto. 1991. "A multidimensional Wiener–Wintner theorem and spectrum estimation," *Trans. Amer. Math. Soc.* **327**: 833–852.

[Be3]   J. Benedetto. 1991. "The spherical Wiener–Plancherel formula and spectral estimation," *SIAM J. Math. Anal.* **22**: 1110–1130.

[Be4]   J. Benedetto. 1992. "Stationary frames and spectral estimation," *NATO–ASI–1991, Probabilistic and Stochastic Methods in Analysis*, J. Byrnes, ed. The Netherlands: Kluwer Academic Publishers, pp. 117–161.

[Be5]   J. Benedetto. 1993. "Frame decompositions, sampling, and uncertainty principle inequalities," this volume, Chap. 7.

[BeHW]   J. Benedetto, C. Heil, and D. Walnut. "Remarks on the proof of the Balian–Low theorem," *Ann. Scu. Norm. Sup., Pisa*, to appear.

[BeLi]   J. Benedetto and S. Li. "The theory of multiresolution analysis frames and applications to filter design," Preprint.

[BeuMa1] A. Beurling and P. Malliavin. 1962. "On Fourier transforms of measures with compact support," *Acta Math.* **107**: 291–309.

[BeuMa2] A. Beurling and P. Malliavin. 1967. "On the closure of characters and the zeros of entire functions," *Acta Math.* **118**: 79–93.

[CRo]   R. Coifman and R. Rochberg. 1980. "Representation theorems for holomorphic and harmonic functions," *Astérisque* **77**: 11–65.

[D1]   I. Daubechies. 1990. "The wavelet transform, time–frequency localization, and signal analysis," *IEEE Trans. Inform. Theory* **36**: 961–1005.

[D2]   I. Daubechies. 1992. *Ten Lectures on Wavelets*, CBMS-NSF Reg. Conf. Ser. Appl. Math. **61**, Philadelphia: Soc. Ind. Appl. Math,

[DGroMe] I. Daubechies, A. Grossmann, and Y. Meyer. 1986. "Painless nonorthogonal expansions," *J. Math. Phys.* **27**: 1271–1283.

[DJaJo] I. Daubechies, S. Jaffard, and J.-L. Journé. 1991. A simple Wilson orthonormal basis with exponential decay," *SIAM J. Math. Anal.* **22**: 554–572.

[DJan]   I. Daubechies and A. Janssen. "Two theorems on lattice expansions," *IEEE Trans. Inform. Theory*, to appear.

[DuSc]   R. Duffin and A. Schaeffer. 1952. "A class of nonharmonic Fourier series," *Trans. Amer. Math. Soc.* **72**: 341–366.

[F1]   H. Feichtinger. 1980. "Banach convolution algebras of Wiener type," *Functions, Series, Operators*, Proc. Conf. **38** Budapest: Colloq. Math. Soc. János Bolyai, pp. 509–524.

[F2]   H. Feichtinger. 1989. "Atomic decompositions of modulation spaces," *Rocky Mountain J. Math.* **19**: 113–126.

[FGr1]   H. Feichtinger and K. Gröchenig. 1988. "A unified approach to atomic decompositions through integrable group representations," *Function Spaces and Applications*, M. Cwikel et al., eds., Lect. Notes Math. **1302** Berlin: Springer-Verlag, pp. 52–73.

[FGr2]   H. Feichtinger and K. Gröchenig. 1989. "Banach spaces related to integrable group representations and their atomic decompositions, I," *J. Funct. Anal.* **86**: 307–340.

[FGr3]   H. Feichtinger and K. Gröchenig. 1989. "Banach spaces related to integrable group representations and their atomic decompositions, II," *Monat. fur Mat.* **108**: 129–148.

[FGrW]   H. Feichtinger, K. Gröchenig, and D. Walnut. 1992. "Wilson bases in modulation spaces," *Math. Nachr.* **155**: 7–17.

[FrJaw]   M. Frazier and B. Jawerth. 1990. "A discrete transform and decompositions of distribution spaces," *J. Funct. Anal.* **93**: 34–170.

[FrJawWe] M. Frazier, B. Jawerth and G. Weiss. 1991. *Littlewood–Paley Theory and the Study of Function Spaces*, CBMS-NSF, Providence: Amer. Math. Soc.

[G]   D. Gabor. 1946. "Theory of communications," *J. Inst. Elec. Eng. (London)* **93**: 429–457.

[GoGol]   I. Gohberg and S. Goldberg. 1981. *Basic Operator Theory*, Boston: Birkhäuser.

[Gr]   K. Gröchenig. 1991. "Describing functions: Atomic decompositions vs. frames," *Monat. fur Mat.* **112**: 1–41.

[GrW]   K. Gröchenig and D. Walnut. 1992. "A Riesz basis for Bargmann–Fock space related to sampling and interpolation," *Arkiv. fur Mat.* **30**: 283–295.

[GroMoPa] A. Grossmann, J. Morlet, and T. Paul. 1985. "Transforms associated to square integrable group representations, I, General results," *J. Math. Phys.* **26**: 2473–2479.

[H]   C. Heil. "Wiener amalgam spaces in generalized harmonic analysis and wavelet theory," Ph.D. thesis, University of Maryland, College Park, MD, 1990.

[HW]   C. Heil and D. Walnut. 1989. "Continuous and discrete wavelet transforms," *SIAM Review* **31**: 628–666.

[He]   H. Heuser. 1982. *Functional Analysis*, New York: John Wiley and Sons, Inc.

[Hö]   L. Hörmander. 1983. *The Analysis of Linear Partial Differential Operators, vols. I, II*, New York: Springer-Verlag.

[Hou]   C. Houdré. 1993. "Wavelets, probability, and statistics: some bridges," this volume, Chap. 9.

[J]   A. Janssen. 1988. "The Zak transform: a signal transform for sampled time-continuous signals," *Philips J. Res.* **43**: 23–69.

[K]   Y. Katznelson. 1976. *An Introduction to Harmonic Analysis*, New York: Dover Publications, Inc.

[Kö]   G. Köthe. 1936. "Das Trägheitsgesetz der quadratischen Formen im Hilbertschen Raum," *Math. Zeit.* **41**: 137–152.

[L]   W. Lawton. 1990. "Tight frames of compactly supported affine wavelets," *J. Math. Phys.* **31**: 1898–1901.

[Le]   N. Levinson. 1940. *Gap and Density Theorems*, Colloq. Publ **26**, Providence: Amer. Math. Soc.

[Lo]   F. Low. 1985. "Complete sets of wave packets," *A Passion for Physics—Essays in Honor of Geoffrey Chew*, C. DeTar et al., eds. Singapore: World Scientific, pp. 17–22.

[M]   S. Mallat. 1989. "Multiresolution approximations and wavelet orthonormal bases of $L^2(\mathbb{R})$," *Trans. Amer. Math. Soc.* **315**: 69–87.

[Me]   Y. Meyer. 1990. *Ondelettes et opérateurs*, vols. I, II, and III, Paris: Hermann.

[PWi]   R. Paley and N. Wiener. 1940. *Fourier Transforms in the Complex Domain*, Colloq. Publ. **19**, Providence: Amer. Math. Soc.

[RRe]   T. Rado and P. Reichelderfer. 1955. *Continuous Transformations in Analysis*, New York: Springer-Verlag.

[Ru]   W. Rudin. 1973. *Functional Analysis*, New York: McGraw–Hill, Inc.

[S]   L. Schwartz. 1966. *Théorie des distributions*, Paris: Hermann.

[St]   E. Stein. 1970. *Singular Integrals and Differentiability Properties of Functions*, Princeton: Princeton Univ. Press.

[StWe]   E. Stein and G. Weiss. 1971. *An Introduction to Fourier Analysis on Euclidean Spaces*, Princeton: Princeton Univ. Press.

[V]   G. Vitali. 1921. "Sulla condizione di chiusura di un sistema di funzioni ortogonali," *Atti R. Acad. Naz. Lincei, Rend. Cl. Sci. Fis. Mat. Nat.* **30**: 498–501.

[vo]   J. von Neumann. 1932, 1949, and 1955. *Mathematical Foundations of Quantum Mechanics*, Princeton: Princeton Univ. Press.

[W1]   D. Walnut. "Weyl–Heisenberg wavelet expansions: existence and stability in weighted spaces," Ph.D. thesis, University of Maryland, College Park, MD, 1989.

[W2]   D. Walnut. 1992. "Continuity properties of the Gabor frame operator," *J. Math. Anal. Appl.* **165**: 479–504.

[W3]   D. Walnut. 1993. "Lattice size estimates for Gabor decompositions," *Monat. fur Mat.* to appear.

[Y]   R. Young. 1980. *An Introduction to Nonharmonic Fourier Series*, New York: Academic Press.

4

Dilation equations and the smoothness of compactly supported wavelets

Christopher Heil* and David Colella

ABSTRACT   The construction of compactly supported wavelets with specified amounts of smoothness is an important problem in wavelet theory. This problem reduces to the construction of scaling functions, i.e., solutions f of dilation equations $f(t) = \sum_{k=0}^{N} c_k f(2t - k)$, with specified smoothness. This article characterizes all smooth, compactly supported scaling functions in terms of a joint spectral radius of two $N \times N$ matrices T_0, T_1 constructed from the coefficients $\{c_0, \ldots, c_N\}$ of the dilation equation, restricted to an appropriate subspace of $\mathbb{C}^N$. The number of continuous derivatives of the scaling function and the range of Hölder exponents of continuity of the last continuous derivative are determined by the value of this joint spectral radius. Numerous examples are provided to illustrate the results.

CONTENTS

*The first author is also Pure Mathematics Instructor at The Massachusetts Institute of Technology, Cambridge, Massachusetts 02139 and acknowledges partial support by National Science Foundation Grant DMS-9007212.

4.1 Introduction

Dilation equations play a crucial role in the construction and resulting properties of multiresolution analyses and wavelet orthonormal bases for $L^2(\mathbb{R})$, the
space of square-integrable functions on the real line [BW2], [ColHe1], [ColHe3],
[ColHe4], [DL1], [DL2], [E], [R1], [R2], [S], [V1], [V2], [W]. This connection is
one reason for the recent burst of activity in the study of such equations and their
solutions. There are also important applications of dilation equations to other
areas, most notably interpolating subdivision schemes, although those are outside
the scope of this chapter [CDaMi], [DeDu], [Du], [DyGLe], [DyLe], [MiP1],
[MiP2].

Multiresolution analyses are discussed in detail elsewhere in this volume and
therefore will not be defined here (see, for example, Chapter 1 by Strichartz or
references [D1], [D2], [He], [M], [Me]). Each multiresolution analysis determines
a *scaling function f*, i.e., a solution of a *dilation equation*

$$f(t) = \sum_{k=-\infty}^{\infty} c_k f(2t - k), \qquad (4.1)$$

which is square integrable. The coefficients $\{c_k\}$ are square-summable real or
complex numbers. This scaling function then determines a *wavelet*

$$g(t) = \sum_{k=-\infty}^{\infty} (-1)^k c_{1-k} f(2t - k), \qquad (4.2)$$

such that the collection $\{g_{n,k}\}_{n,k\in\mathbb{Z}}$ with $g_{n,k}(t) = 2^{n/2} g(2^n t - k)$ forms an orthonormal basis for $L^2(\mathbb{R})$ after suitable normalization of g. This *wavelet orthonormal
basis* is thus obtained by dilating and translating a single $L^2(\mathbb{R})$ function, and
therefore the properties of the basis elements are completely determined by the
corresponding properties of the wavelet.

Wavelet orthonormal bases are appealing for a variety of reasons, one being
that it is possible to find wavelets g which have good localization in both time and
frequency. For example, in this chapter we characterize all wavelets arising from
finite-coefficient dilation equations which are smooth and compactly supported–
such wavelets are necessarily well-localized in the time domain and have good
decay in their Fourier transforms. Moreover, we characterize the exact degree
of smoothness of these wavelets, by which we mean that we determine the total
number of continuous derivatives and the possible Hölder exponents of continuity
of the last derivative. As proved by Daubechies and Lagarias [DL1] (and discussed
in Section 4.7.1), compact support is incompatible with infinite differentiability.
A function h defined on the real line $\mathbb{R}$ is *Hölder continuous* if there exist constants
α, K such that $|h(x) - h(y)| \leq K|x - y|^\alpha$ for all $x, y \in \mathbb{R}$. The constants α and K
are referred to as a *Hölder exponent* and *Hölder constant* for h, respectively. If h
is differentiable then we can take $\alpha = 1$, but not conversely.

In order to ensure compact support for the wavelet g we assume that the number of nonzero coefficients c_k in Eq. (4.1) is finite. By translating the scaling function f if necessary, we therefore assume for the remainder of the chapter that the dilation equation has the form

$$f(t) = \sum_{k=0}^{N} c_k f(2t - k), \qquad (4.3)$$

i.e., we assume $c_k = 0$ for $k < 0$ or $k > N$. We seek square-integrable, compactly supported solutions of Eq. (4.3). Such scaling functions are necessarily integrable, and we show below that compactly supported, integrable solutions to Eq. (4.3) are unique up to multiplication by a constant (if they exist), and are supported in the finite interval $[0, N]$. If f is such a compactly supported scaling function then the wavelet g is obtained from f by a finite sum. Thus the smoothness of g is entirely determined by the corresponding smoothness of the scaling function f. We therefore concentrate for the remainder of the chapter on the properties of compactly supported scaling functions, rather than wavelets.

The class of dilation equations induced by multiresolution analyses is quite restrictive; in particular, the following two conditions must be satisfied:

$$\sum_{k} c_{2k} = \sum_{k} c_{2k+1} = 1 \qquad (4.4)$$

and

$$\sum_{k} c_k \bar{c}_{k+2j} = \begin{cases} 2, & \text{if } j = 0, \\ 0, & \text{if } j \neq 0. \end{cases} \qquad (4.5)$$

In fact, not only are these two conditions necessary for the existence of a multiresolution analysis, they are also almost sufficient. By this we mean the following. Assume that coefficients $\{c_0, \ldots, c_N\}$ are given which satisfy both Eqs. (4.4) and (4.5). Then it can be shown that the dilation equation will have an integrable and square-integrable solution f. If f is orthogonal to its integer translates, i.e., if $\int f(t)\overline{f(t-k)}\, dt = 0$ when $k \neq 0$, then f can be used to construct a multiresolution analysis [M]. Lawton [La] has shown that for each N, f will be orthogonal to its integer translates except for a set of coefficients $\{c_0, \ldots, c_N\}$ of measure zero in the set of all coefficients $\{c_0, \ldots, c_N\}$ satisfying Eqs. (4.4) and (4.5).

Let us illustrate the preceding remarks with a simple example. If we set $N = 1$ and define $c_0 = c_1 = 1$, then the dilation equation has the form

$$f(t) = f(2t) + f(2t - 1).$$

By inspection, one compactly supported solution is $f = \chi_{[0,1)}$, the characteristic function of the interval $[0, 1)$. Note that the coefficients satisfy both of Eqs. (4.4) and (4.5), and that f is (trivially) orthogonal to its integer translates. We therefore expect that the wavelet

$$g(t) = f(2t) - f(2t - 1) = \chi_{[0,1/2)}(t) - \chi_{[1/2,1)}(t)$$

will generate an orthonormal basis for $L^2(\mathbb{R})$, and in fact this is the case. This wavelet basis $\{g_{n,k}\}_{n,k \in \mathbb{Z}}$ was first constructed by Haar [H], and is today known as the *Haar system*. Although compactly supported, the Haar wavelet g is certainly not smooth! Smooth, compactly supported wavelets were first constructed by Daubechies [D1].

Eqs. (4.4) and (4.5) are necessary for the existence of multiresolution analyses. However, these equations do not play any significant role in many of the applications of dilation equations to other areas. Therefore, we will not impose any restrictions on the coefficients $\{c_0, \ldots, c_N\}$ in our characterization of smooth, compactly supported scaling functions, i.e., we will obtain all possible smooth, compactly supported solutions to dilation equations, without regard to their applicability to multiresolution analysis. There are several different methods for constructing such solutions, most of which can be classified as combinations of one or more of three techniques, which we shall refer to as Cascade Algorithm techniques, Fourier transform techniques, and dyadic interpolation techniques. We briefly outline these techniques in the remainder of this section. We then expand somewhat on the use of the Fourier transform in Section 4.2, and use dyadic interpolation thoughout the remainder of the chapter (Sections 4.3 through 4.8). We make minor use of the Cascade Algorithm in Section 4.7.2.

The Cascade Algorithm is based on the fact that a scaling function is a fixed point for the linear operator F defined by

$$Fh(t) = \sum_{k=0}^{N} c_k h(2t - k).$$

As usual, the fixed point is located by iterating, or "cascading", some reasonable starting function h_0 (typically, $h_0 = \chi_{[0,1)}$) through the operator F. That is, one hopes that the functions h_n defined by $h_n = Fh_{n-1}$ will converge to a scaling function f. However, convergence of the algorithm is not guaranteed or may occur only in a weak sense. This type of algorithm is typically used to study fractals; it is therefore not surprising that scaling functions or their last derivatives are often Hölder continuous with exponent strictly less than one. Articles in which the Cascade Algorithm is discussed include [BW2], [DL1], [DeDu], [DyGLe], [DyLe], [D1], [D2].

The use of the Fourier transform, on the other hand, is suggested by the convolutional nature of the dilation equation. We therefore take the Fourier transform of the dilation equation and obtain its equivalent Fourier transform version:

$$\hat{f}(\gamma) = m_0(\gamma/2)\hat{f}(\gamma/2), \tag{4.6}$$

where $m_0(\gamma) = \frac{1}{2}\sum c_k e^{ik\gamma}$ is the *symbol* of the coefficients $\{c_0, \ldots, c_N\}$. Iterating Eq. (4.6) we obtain

$$\hat{f}(\gamma) = m_0(\gamma/2)\hat{f}(\gamma/2) = m_0(\gamma/2)m_0(\gamma/4)\hat{f}(\gamma/4) = \cdots,$$

which suggests that properties of $\hat{f}$ will be determined by the convergence of an infinite product involving m_0. This technique can be used, for example, to prove uniqueness results or to study the smoothness of f via consideration of the decay of $\hat{f}$, but is usually not optimal for estimating time-domain properties of f such as the Hölder exponent of continuity. We discuss this technique in more detail in Section 4.2; articles in which Fourier transform techniques are used include [ColHe4], [DL1], [E], [V1], [V2], [D1], [D2], [M].

Finally, the dyadic interpolation technique is the method that we will use to construct and characterize all smooth, compactly supported scaling functions. It is based on the observation that if the values of a scaling function f are known at, say, the integer points, then the dilation equation immediately determines their values at the half-integer points. From this we obtain next the values at the quarter-integers, and so by recursion eventually finding the values at all *dyadic points*, i.e., all points of the form $k/2^n$. If f is uniformly continuous on this restricted set of points then it can be extended to all points in a continuous way. The implementation and execution of this method occupies us for the majority of the chapter, specifically Sections 4.3–4.8. We outline below the content of each of these sections.

In Section 4.3 we show how to implement dyadic interpolation in a matrix formulation. This requires an equivalent matrix form of the dilation equation, which we obtain as follows. First, the scaling function f, which is supported in the interval $[0, N]$, is converted into a vector-valued function v defined on the interval $[0, 1]$. Given $x \in [0, 1]$, the first component $v_1(x)$ of $v(x)$ is simply the value of $f(x)$; the second component $v_2(x)$ is then $f(x + 1)$, and so forth. We refer to this vector-valued function v as a *scaling vector*. Next, given $x \in [0, 1]$, we show that the collection of N equations obtained by evaluating the dilation equation at each of the points $x, x + 1, \ldots, x + N - 1$ can be expressed as a matrix equation involving the scaling vector v and an appropriate product of two matrices T_0, T_1 which are determined from the coefficients $\{c_0, \ldots, c_N\}$. The product has the form $T_{d_1} \cdots T_{d_m}$, where $d_1, \ldots, d_m$ are the first m digits of the binary (base two) expansion of x. We write this expansion as $x = .d_1 d_2 \cdots$; it is unique for nondyadic points. Dyadic points, on the other hand, have two binary expansions: a "terminating" expansion ending in infinitely many zeros, which we write for convenience as $x = .d_1 \cdots d_j$, and another expansion ending in infinitely many ones. Because the matrices T_0, T_1 satisfy a certain "consistency" condition, either expansion may be used to form the product $T_{d_1} \cdots T_{d_m}$ when x is dyadic.

Section 4.4 uses this matrix form of the dilation equation to find sufficient conditions under which a continuous scaling vector (and hence a continuous, compactly supported scaling function) can be constructed. This essentially amounts to finding conditions under which f, once defined at dyadic points by dyadic interpolation, will be uniformly continuous on this restricted set. The sufficient conditions are given in terms of a growth condition on all possible products of T_0 and T_1, restricted to an appropriate subspace W of $\mathbb{C}^N$. Moreover, a lower bound for the Hölder exponent of continuity is also obtained.

In Section 4.5 we show how these growth conditions on products of T_0 and T_1 can be viewed as a *joint spectral radius* of the two matrices. We therefore digress temporarily and discuss the basic properties of the joint spectral radius in Section 4.5.1, and methods of computing it in Section 4.5.2.

We return to considering the continuity of scaling vectors and associated scaling functions in Section 4.6. In particular, we prove that the results of Section 4.4 are sharp, i.e., the joint spectral radius can be used to precisely characterize all dilation equations which have continuous, compactly supported solutions, and to give precise bounds on the possible Hölder exponents of continuity.

The results of Sections 4.4 and 4.6 completely determine all possible continuous, compactly supported scaling functions. However, if f is a differentiable scaling function for the dilation equation with coefficients $\{c_0, \ldots, c_N\}$ then f' is a scaling function for the dilation equation with coefficients $\{2c_0, \ldots, 2c_N\}$. Since we know when this latter dilation equation will have a continuous solution, we have an implicit characterization of all dilation equations having continuously differentiable solutions, and, by iteration, of all n-times continuously differentiable solutions. In Section 4.7.1 we make this implicit characterization explicit, also in terms of an appropriate joint spectral radius. In Section 4.7.2 we show how the assumption of extra *sum rules* on the coefficients affects that joint spectral radius.

Finally, in Section 4.8 we give several examples illustrating the results of previous sections for specific dilation equations. In Section 4.8.1 we find all dilation equations with $N = 3$ which have smooth, compactly supported solutions. In Section 4.8.2 we consider the analogous problem for $N = 6$ with the additional assumptions that the coefficients $\{c_0, \ldots, c_6\}$ satisfy Eq. (4.4) and are symmetric, i.e., $c_0 = c_6$, $c_1 = c_5$, and $c_2 = c_4$.

Dyadic interpolation methods have been used by many authors, and we do not take credit for inventing them. In particular, Michelli and Prautzsch [MiP2] (motivated by interpolating subdivision problems) were the first to implement dyadic interpolation via products of two matrices. However, they did not make use of the joint spectral radius, nor did they obtain estimates for the Hölder exponent. Daubechies and Lagarias [DL1], [DL2] independently discovered the same implementation, and gave sufficient conditions for the existence of smooth, compactly supported scaling functions and lower bounds for the corresonding Hölder exponents of the last derivative in terms of a joint spectral radius. We improved on their sufficient conditions and proved that those improved conditions are also necessary [ColHe1], [ColHe3]. In particular, for the case of continuous, compactly supported scaling functions, the Daubechies and Lagarias theorem is essentially a restricted version of our Theorem 4.3 in Section 4.4. Our converse is Theorem 4.11 in Section 4.6. Equivalent results have also been obtained by Berger and Wang [BW2], [W]. We note that Rioul [R1], [R2] has a different implementation of dyadic interpolation, and obtains estimates for the Hölder exponent not expressed in terms of a joint spectral radius. Also, although we discuss smoothness only in global terms, the matrix methods we describe can be used to obtain detailed local information, e.g., smoothness at a point or at a class of points [DL2]. Finally, the techniques we discuss are also applicable to dilation equations with integer scale factors other than two, and to certain higher-dimensional dilation equations. Daubechies and Lagarias [DL2] give some indications of how this can be done.

4.2 The Fourier transform

Let us begin our study of dilation equations by assuming that f is an integrable scaling function—this includes, of course, the class of smooth, compactly supported scaling functions which is our primary interest. The Fourier transform $\hat{f}(\gamma) = \int f(t)e^{i\gamma t}\,dt$ of f is then a well-defined continuous function, so Eq. (4.6) gives an equivalent form for the dilation equation on the Fourier transform side. In particular, when $\gamma = 0$ we have $\hat{f}(0) = \Delta\hat{f}(0)$, where

$$\Delta = m_0(0) = \frac{1}{2}\sum c_k,$$

and therefore $\int f(t)\,dt = \hat{f}(0) = 0$ if $\Delta \neq 1$.

By iterating Eq. (4.6), we obtain that

$$\hat{f}(\gamma) = \hat{f}(\gamma/2^n)\prod_{j=1}^{n} m_0(\gamma/2^j) \tag{4.7}$$

for each $n > 0$. We are now tempted to let $n \to \infty$, but we may have convergence problems in the infinite product if $\Delta \neq 1$. Therefore, let us temporarily impose the restriction that $\Delta = 1$. Since m_0 is a trigonometric polynomial, it is not difficult to show that the infinite product

$$P(\gamma) = \prod_{j=1}^{\infty} m_0(\gamma/2^j) \tag{4.8}$$

then converges (uniformly on compact sets) to a continuous function, and therefore

$$\hat{f}(\gamma) = \hat{f}(0)P(\gamma).$$

In particular, note that $\hat{f}(0) \neq 0$ if f is nontrivial. Moreover, since P is independent of f, this shows that integrable solutions to dilation equations are unique up to scale when $\Delta = 1$, and we show next that such solutions are necessarily compactly supported. We will also show below that compactly supported, integrable scaling functions are likewise unique up to scale when $\Delta \neq 1$, although in that case there may also exist non-compactly supported, integrable scaling functions.

To calculate the support of f, we return to the time side. Let $\nu = \frac{1}{2}\sum c_k\delta_k$, where δ_k is the point mass at the point k. Then $\hat{\nu} = m_0$ and $\mathrm{supp}(\nu) = \{0,\dots,N\} \subset [0,N]$. Let ν_n be a compressed version of ν, i.e., $\nu_n = \frac{1}{2}\sum c_k\delta_{2^{-n}k}$, and let μ_n be the measure $\mu_n = \nu_1 * \cdots * \nu_n$, where the asterisk denotes convolution. Then

$\hat{\mu}_n(\gamma) = \prod_{j=1}^{n} m_0(\gamma/2^j)$ and $\text{supp}(\mu_n) \subset [0,N/2] + \cdots + [0,N/2^n] \subset [0,N]$. Since $\hat{f}(0)\hat{\mu}_n(\gamma) \to \hat{f}(\gamma)$ as $n \to \infty$, we therefore have $\hat{f}(0)\mu_n \to f$ in the weak* topology, and hence $\text{supp}(f) \subset [0,N]$.

Let us consider now what happens when $\Delta \neq 1$. In this case we adjust the terms of the infinite product in Eq. (4.8) in order to obtain convergence. Specifically, we rewrite Eq. (4.7) in the following form:

$$\hat{f}(\gamma) = \hat{f}(\gamma/2^n)\Delta^n \prod_{j=1}^{n} \frac{m_0(\gamma/2^j)}{\Delta}, \tag{4.9}$$

and note that

$$P_\Delta(\gamma) = \prod_{j=1}^{\infty} \frac{m_0(\gamma/2^j)}{\Delta}$$

converges to a continuous function. Since $\hat{f}(0) = 0$, it follows by letting $n \to \infty$ in Eq. (4.9) that $\hat{f} \equiv 0$ if $|\Delta| \leq 1$ with $\Delta \neq 1$. Thus there are no nontrivial integrable solutions of dilation equations when $|\Delta| \leq 1$ with $\Delta \neq 1$.

Assume therefore that $|\Delta| > 1$. Since $\int f(t)\,dt = \hat{f}(0) = 0$, it follows that if f is compactly supported then its primitive

$$f_1(t) = \int_{-\infty}^{t} f(s)\,ds$$

is a nontrivial integrable function with compact support. Moreover, f_1 is a solution of the dilation equation with coefficients $\{c_0/2, \ldots, c_N/2\}$, so we must either have $|\Delta/2| > 1$ or $\Delta/2 = 1$. If $|\Delta/2| > 1$ then we can repeat the process, forming the primitive f_2 of f_1, and conclude that either $|\Delta/4| > 1$ or $\Delta/4 = 1$. This process cannot continue indefinitely, i.e., we must have $\Delta = 2^n$ for some $n > 0$. Then f_n is an integrable solution to the dilation equation defined by the coefficients $\{2^{-n}c_0, \ldots, 2^{-n}c_N\}$ and hence (by our analysis for the case $\Delta = 1$) is the unique integrable solution of that dilation equation (up to scale) and has support contained in $[0,N]$. Thus f is the nth derivative of f_n, so $\text{supp}(f) \subset [0,N]$ as well, and we conclude that f is the unique compactly supported, integrable solution of its dilation equation (up to scale). On the other hand, if f is not compactly supported then its primitive f_1 need not be integrable, and the preceding analysis breaks down, i.e., we cannot prove uniqueness for integrable but non-compactly supported scaling functions when $|\Delta| > 1$. In fact, there exist dilation equations with $|\Delta| > 1$ which have infinitely many integrable solutions, and others which have none [DL1]. The above analysis does show that at most one integrable solution can be compactly supported.

The above characterization of compactly supported, integrable scaling functions was first proved by Daubechies and Lagarias [DL1] (some of our discussion was also inspired by Berger's excellent set of notes [B]). To summarize:

THEOREM 4.1

If there exists an integrable, compactly supported solution f to the dilation equation determined by the coefficients $\{c_0, \ldots, c_N\}$, then there exists an integer $n \geq 0$ such that the following statements hold (with uniqueness interpreted as holding up to scale).

(a) $\sum c_k = 2^{n+1}$.

(b) *The dilation equation determined by the coefficients $\{2^{-n}c_0, \ldots, 2^{-n}c_N\}$ has a unique integrable solution f_n, and $\text{supp}(f_n) \subset [0, N]$.*

(c) *f is the unique compactly supported, integrable solution to its dilation equation, and, with the proper choice of scale, f is the nth derivative of f_n.*

In particular, dilation equations satisfying $\sum c_k = 2$ (i.e., $\Delta = 1$) are in some sense "fundamental" (note that Eq. (4.4) is a special case). However, this condition is *not* sufficient to ensure the existence of an integrable solution! For example, the simplest dilation equation with $N = 0$ and $c_0 = 2$ is easily seen to have no integrable solutions. We have given some necessary or sufficient conditions for the existence of integrable solutions [ColHe4], and Wang [W] has conjectured a necessary and sufficient condition in terms of a type of joint spectral radius (similar to the joint spectral radius discussed in later sections of this chapter). On the other hand, distributional solutions for dilation equations satisfying $\sum c_k = 2$ can always be constructed—in fact, infinitely many distinct distributional solutions exist, but there is exactly one compactly supported distributional solution (up to scale) [ColHe4], [DL1]. Some of these non-compactly supported solutions may be realizable as functions; in fact, it is possible for a dilation equation to have both smooth, compactly supported and smooth, non-compactly supported solutions.

We have used the Fourier transform in this section only to obtain uniqueness results, but many other results are possible. For example, Daubechies [D1] estimates the decay of $\hat{f}$ by considering the infinite product in Eq. (4.8); see also the expostion in Chapter 1 by Strichartz. Estimates for the smoothness of f can be obtained from this, including even Hölder exponents of continuity, although the results are usually not optimal since the Hölder exponent is essentially a time domain rather than a frequency domain property. On the other hand, Eirola [E] and Villemoes [V1], [V2] obtain good estimates for the Sobolev exponent of continuity by using Fourier transforms, since the Sobolev exponent is essentially a frequency domain property.

4.3 Equivalence of scaling functions and scaling vectors

In the remainder of this chapter we use the dyadic interpolation method to characterize all smooth, compactly supported scaling functions. Our first step in this direction is to convert the dilation equation into an equivalent matrix formulation.

Assume, therefore, that coefficients $\{c_0, \ldots, c_N\}$ are given, and that a continuous, compactly supported solution f to the dilation equation exists. Then we know that $\text{supp}(f) \subset [0, N]$, so the vector-valued function

$$v(x) = \begin{pmatrix} f(x) \\ f(x+1) \\ \vdots \\ f(x+N-1) \end{pmatrix}, \qquad x \in [0, 1], \tag{4.10}$$

captures all information about f. Note that $v_1(0) = v_N(1) = 0$ since f is continuous, and $v_{i+1}(0) = v_i(1)$ for $i = 1, \ldots, N-1$ by construction, where $v_i(x)$ denotes the ith component of $v(x)$.

Now, if $0 \leq x \leq 1/2$ then it follows from the dilation equation and the fact that $\text{supp}(f) \subset [0, N]$ with $f(0) = f(N) = 0$ that the value of $v(x) = (f(x), \ldots, f(x+N-1))^{\text{t}}$ is completely determined from the value of $v(2x) = (f(2x), \ldots, f(2x+N-1))^{\text{t}}$ in a linear manner. Therefore, there exists an $N \times N$ matrix T_0 such that $v(x) = T_0 v(2x)$ when $0 \leq x \leq 1/2$. Similarly, if $1/2 \leq x \leq 1$ then $v(x)$ is determined from $v(2x-1)$ by a linear transformation, so $v(x) = T_1 v(2x-1)$ when $1/2 \leq x \leq 1$ for some matrix T_1. In fact, $(T_0)_{ij} = c_{2i-j-1}$ and $(T_1)_{ij} = c_{2i-j}$, i.e.,

$$T_0 = \begin{pmatrix} c_0 & 0 & 0 & \cdots & 0 & 0 \\ c_2 & c_1 & c_0 & \cdots & 0 & 0 \\ \vdots & \vdots & \vdots & \ddots & \vdots & \vdots \\ 0 & 0 & 0 & \cdots & c_N & c_{N-1} \end{pmatrix}$$

and

$$T_1 = \begin{pmatrix} c_1 & c_0 & 0 & \cdots & 0 & 0 \\ c_3 & c_2 & c_1 & \cdots & 0 & 0 \\ \vdots & \vdots & \vdots & \ddots & \vdots & \vdots \\ 0 & 0 & 0 & \cdots & 0 & c_N \end{pmatrix}.$$

Note that we have *consistency* at $x = 1/2$, i.e.,

$$v(1/2) = T_0 v(1) = T_1 v(0). \tag{4.11}$$

Thus, a continuous, compactly supported scaling function f determines a continuous, vector-valued function $v : [0, 1] \to \mathbb{C}^N$ which satisfies

$$v_1(0) = v_N(1) = 0, \tag{4.12}$$

$$v_{i+1}(0) = v_i(1), \quad i = 1, \ldots, N-1, \tag{4.13}$$

$$v(x) = T_{d_1} v(\tau x), \tag{4.14}$$

where $x = .d_1 d_2 \cdots$ is a binary expansion of x and

$$\tau x = 2x \bmod 1 = \begin{cases} 2x, & 0 \le x < 1/2, \\ 2x - 1, & 1/2 < x \le 1, \end{cases}$$

(so $\tau x = .d_2 d_3 \cdots$), and where Eq. (4.14) is interpreted for $x = 1/2$ to mean Eq. (4.11). If f is Hölder continuous with Hölder exponent α then it follows easily that v is also Hölder continuous with the same exponent, where Hölder continuity for vector-valued functions is defined by the inequality $\|v(x) - v(y)\| \le K|x - y|^\alpha$, where $\| \cdot \|$ is an arbitrary norm on $\mathbb{C}^N$.

If f is now an arbitrary compactly supported scaling function (i.e., not necessarily continuous) and v is defined as above by Eq. (4.10), then Eqs. (4.11)–(4.14) will still hold, *provided that $f(0) = f(N) = 0$*. Note that this must be the case if $c_0, c_N \ne 1$. If $|c_0| \ge 1$ or $|c_N| \ge 1$ then it is not difficult to show directly from the dilation equation that f must be discontinuous at 0 or N, respectively.

In light of the above discussion, we will call any vector-valued function $v :$ $[0, 1] \to \mathbb{C}^N$ satisfying Eqs. (4.11)–(4.14) for all $x \in [0, 1]$ a *scaling vector*.

The procedure described above is reversible. That is, assume that a scaling vector v is given, and define

$$f(x) = \begin{cases} 0, & x \le 0 \text{ or } x \ge N, \\ v_i(x), & i - 1 \le x \le i, \quad i = 1, \dots, N. \end{cases} \tag{4.15}$$

It follows immediately that f is a compactly supported solution to the dilation equation and is continuous if v is continuous, with the same Hölder exponent of continuity as v.

In summary, (Hölder) continuous, compactly supported scaling functions and (Hölder) continuous scaling vectors are completely equivalent, in the sense that one exists if and only if the other exists. Furthermore, as long as $c_0, c_N \ne 1$, a compactly supported scaling function exists if and only if a scaling vector exists. We therefore concentrate for the next several sections on the existence and properties of scaling vectors, rather than scaling functions.

4.4 Sufficient conditions for the existence of continuous scaling vectors

In this section we find sufficient conditions on the coefficients $\{c_0, \dots, c_N\}$ which allow construction of a continuous scaling vector v. Note first from Eq. (4.12) that if a scaling vector v exists then $v_1(0) = 0$. Since $T_0 v(0) = v(0)$ and only the first entry of the top row of T_0 is nonzero, we therefore have that $(v_2(0), \dots, v_N(0))^t$ is a right eigenvector for M for the eigenvalue 1, where M is the $(N - 1) \times (N - 1)$

submatrix of T_0 and T_1 defined by $M_{ij} = c_{2i-j}$, i.e.,

$$
M = \begin{pmatrix}
c_1 & c_0 & 0 & \cdots & 0 & 0 \\
c_3 & c_2 & c_1 & \cdots & 0 & 0 \\
\vdots & \vdots & \vdots & \ddots & \vdots & \vdots \\
0 & 0 & 0 & \cdots & c_N & c_{N-1}
\end{pmatrix}. \tag{4.16}
$$

Thus, a scaling vector can only exist if 1 is an element of the spectrum of M. This is the case, for example, when Eq. (4.4) is satisfied, for then $(1, \ldots, 1)$ must be a left eigenvector for M for the eigenvalue 1.

Assume, therefore, that 1 is an eigenvalue for M, and let us attempt to construct a scaling vector v. Let $a = (a_1, \ldots, a_{N-1})^t$ be any right eigenvector for M for the eigenvalue 1, and define $v(0) = (0, a_1, \ldots, a_{N-1})^t$ and $v(1) = (a_1, \ldots, a_{N-1}, 0)^t$. In terms of the scaling function, this gives the values of f at the integers. Equations (4.11)–(4.13) are then satisfied. Next, for each dyadic point $x = .d_1 \cdots d_m$, define

$$
v(x) = T_{d_1} \cdots T_{d_m} v(0). \tag{4.17}
$$

In terms of the scaling function, this amounts to plugging the values of f at the integers into the dilation equation to obtain the values of f at the half-integers, then plugging these into the dilation equation again to get the values at the quarter-integers, and so on recursively to all dyadic points.

Equation (4.17) immediately implies that Eq. (4.14) is satisfied *for dyadic* $x \in [0, 1]$, and we wish to extend the domain of v to *all* $x \in [0, 1]$ so that Eq. (4.14) holds for all x in this interval. For this it suffices to prove that v is uniformly continuous on its current domain of definition (the set of dyadic points in $[0, 1]$), for then there would exist a *unique* extension of v to all of $[0, 1]$ so that v is continuous, and Eq. (4.14) would follow for arbitrary $x \in [0, 1]$ upon taking limits. Moreover, if v satisfies a Hölder condition with exponent α for dyadic x, $y \in [0, 1]$ then it will also satisfy a Hölder condition with the same exponent for arbitrary $x, y \in [0, 1]$.

Consider, therefore, two dyadic points $x < y$ in $[0, 1]$. If x and y are close enough then they will share the first few digits in their binary expansions, e.g., if $x = .d_1 \cdots d_k$ and $2^{-m-1} \leq y - x < 2^{-m}$ with $m > k$ then $x = .d_1 \cdots d_m$ (with $d_i = 0$ for $i = k+1, \ldots, m$) and $y = .d_1 \cdots d_m d_{m+1} \cdots d_{m+j}$ for some j. Therefore,

$$
v(y) - v(x) = T_{d_1} \cdots T_{d_m} \left(v(\tau^m y) - v(0) \right). \tag{4.18}
$$

Thus, to prove continuity from the right on the set of dyadic points in $[0, 1]$ we must consider the behavior of all possible products $T_{d_1} \cdots T_{d_m}$ acting on all possible differences $v(z) - v(0)$ for dyadic $z \in [0, 1]$. Define, therefore,

$$
W = W(a) = \mathrm{span}\{v(z) - v(0): \text{ dyadic } z \in [0, 1]\}, \tag{4.19}
$$

and note that W is a subspace of $\mathbb{C}^N$ which is invariant under both T_0 and T_1. In general, W depends on the choice of a; however, for most choices of coefficients the eigenvalue 1 for M will be simple, in which case W is independent of a. By a slight abuse of terminology we therefore usually refer to W as if it was unique, and allow the dependence on a to be implicit.

If Eq. (4.4) holds then W is a subspace of

$$V = \{u \in \mathbb{C}^N : u_1 + \cdots + u_N = 0\},$$

which is also invariant in this case under both T_0 and T_1. In general, it is not difficult to determine W explicitly. For example, we have the following [ColHe3].

PROPOSITION 4.2
W is the smallest subspace of $\mathbb{C}^N$ which is invariant under both T_0 and T_1 and contains the vector $v(1) - v(0)$.

We have conjectured [ColHe3] that, for dilation equations satisfying Eq. (4.4), $W = V$ except for a set of coefficients $\{c_0, \ldots, c_N\}$ of measure zero. However, the distinction between W and V can be critical for determining whether a dilation equation has a continuous solution or not; some examples are given in Section 4.8.2.

Return now to consideration of Eq. (4.18); we want to determine the "size" of $v(y) - v(x)$. Fix, therefore, any norm $\| \cdot \|$ on $\mathbb{C}^N$, with corresponding operator norm

$$\|A\| = \sup_{u \neq 0} \frac{\|Au\|}{\|u\|}$$

defined for $N \times N$ matrices A. Then from Eq. (4.18) we obtain

$$\|v(y) - v(x)\| \leq C_1 \|(T_{d_1} \cdots T_{d_m})|_W\|, \tag{4.20}$$

where

$$C_1 = 2 \sup\{\|v(z)\| : \text{dyadic } z \in [0, 1]\}$$

(of course, we do not know *a priori* that $C_1 < \infty$, but let us assume this for the moment). The right hand side of Eq. (4.20) behaves geometrically with respect to m, so let $C, \theta > 0$ be such that

$$\max_{d_j = 0, 1} \|(T_{d_1} \cdots T_{d_m})|_W\| \leq C\theta^m \quad \text{for all } m > 0. \tag{4.21}$$

For example, $\theta = \max\{\|T_0|_W\|, \|T_1|_W\|\}$ is one possibility, but it is often possible to choose θ much smaller. Combining Eqs. (4.20) and (4.21) we have

$$\|v(y) - v(x)\| \leq C_1 C \theta^{-1} \theta^{m+1} = C_1 C \theta^{-1} (2^{-m-1})^\alpha \leq K|x - y|^\alpha,$$

where $K = C_1 C \theta^{-1}$ and $\alpha = -\log_2 \theta$. A symmetric argument establishes the same inequality for dyadic $y < x$ with $2^{-m-1} \leq x - y < 2^{-m}$. As K and α are independent of m, we have therefore nearly proved the following result.

THEOREM 4.3

Let $C, \theta > 0$ be such that Eq. (4.21) holds. If $\theta < 1$ then v determines a continuous, compactly supported scaling function f which is Hölder continuous with Hölder exponent $\alpha = -\log_2 \theta$.

A special case of Theorem 4.3, requiring the assumption of Eq. (4.4), was first proved by Daubechies and Lagarias [DL2].

The proof of Theorem 4.3 will be complete once we demonstrate that $C_1 < \infty$ whenever Eq. (4.21) holds with $\theta < 1$. This follows immediately from the calculation

$$\|T_{d_1} \cdots T_{d_m} v(0) - v(0)\|$$
$$\leq \|T_{d_1} \cdots T_{d_m} v(0) - T_{d_1} \cdots T_{d_{m-1}} v(0)\| + \cdots + \|T_{d_1} v(0) - v(0)\|$$
$$\leq \|(T_{d_1} \cdots T_{d_{m-1}})|_W\| \|T_{d_m} v(0) - v(0)\| + \cdots + \|T_{d_1} v(0) - v(0)\|$$
$$\leq (C\theta^{m-1} + \cdots + C)C_2$$
$$\leq \frac{CC_2}{1 - \theta},$$

where $C_2 = \max_{j=0,1} \|T_{d_j} v(0) - v(0)\|$.

4.5 The joint spectral radius

The hypothesis in Theorem 4.3 that Eq. (4.21) holds with $\theta < 1$ is essentially a spectral constraint on the two operators $T_0|_W$, $T_1|_W$. We devote this section to showing how this constraint can be formulated in terms of a "joint spectral radius" of $T_0|_W$ and $T_1|_W$.

4.5.1 Definition and properties

Recall that the usual spectral radius of a single matrix A is defined by

$$\rho(A) = \limsup_{m \to \infty} \|A^m\|^{1/m}. \tag{4.22}$$

This is independent of the choice of norm $\|\cdot\|$. The spectral radius can also be computed in an eigenvalue form, i.e.,

$$\rho(A) = \limsup_{m \to \infty} \sigma(A^m)^{1/m}, \tag{4.23}$$

where

$$\sigma(B) = \max\{|\lambda| : \lambda \text{ is an eigenvalue of } B\}.$$

Of course, Eq. (4.23) trivially implies

$$\rho(A) = \sigma(A) \tag{4.24}$$

since $\sigma(A^m) = \sigma(A)^m$, and in fact Eq. (4.24) is often used as the definition of the spectral radius.

The generalization of the usual spectral radius of a single matrix to the case of a set of elements of a normed algebra was first made by Rota and Strang [RoS]. For the specific case of two matrices, the definition is as follows.

DEFINITION 4.4 *The joint spectral radius $\hat{\rho}(A_0, A_1)$ of two $N \times N$ matrices A_0, A_1 is*

$$\hat{\rho}(A_0, A_1) = \limsup_{m \to \infty} \hat{\rho}_m,$$

where

$$\hat{\rho}_m = \hat{\rho}_m(A_0, A_1) = \max_{d_j = 0,1} \|A_{d_1} \cdots A_{d_m}\|^{1/m}.$$

The definition of the joint spectral radius for larger collections of matrices, or for matrices restricted to subspaces (e.g., $\hat{\rho}(T_0|_W, T_1|_W)$), is made in the obvious way.

The value of $\hat{\rho}(A_0, A_1)$ is independent of the choice of norm. Berger and Wang [BW1] proved the nontrivial analogy of Eq. (4.23), i.e.,

$$\hat{\rho}(A_0, A_1) = \limsup_{m \to \infty} \hat{\sigma}_m,$$

where

$$\hat{\sigma}_m = \hat{\sigma}_m(A_0, A_1) = \max_{d_j = 0,1} \rho(A_{d_1} \cdots A_{d_m})^{1/m}.$$

However, the analogue of Eq. (4.24) fails in general, i.e., $\hat{\rho}(A_0, A_1)$ may not equal the maximum of the absolute values of the eigenvalues of A_0 and A_1 (which is given by $\hat{\sigma}_1$). For example, consider $A_0 = \left(\begin{smallmatrix} 1 & 0 \\ 1 & 1 \end{smallmatrix}\right)$ and $A_1 = \left(\begin{smallmatrix} 1 & 1 \\ 0 & 1 \end{smallmatrix}\right)$; since $A_1 = A_0^t$ we have $\hat{\rho}(A_0, A_1) = \hat{\sigma}_2 = \sqrt{(3 + \sqrt{5})/2} > \hat{\sigma}_1 = 1$.

Given any $\theta > \hat{\rho}(A_0, A_1)$, it is easy to see that there must exist a constant $C > 0$ such that $\hat{\rho}_m^m \leq C\theta^m$ for every m. However, this need not be true with $\theta = \hat{\rho}(A_0, A_1)$ (if it is then we say that $A_0/\hat{\rho}(A_0, A_1)$, $A_1/\hat{\rho}(A_0, A_1)$ are *product bounded*, since all products of these two normalized matrices are then bounded in norm). In terms of the joint spectral radius, Theorem 4.3 may therefore be restated as follows.

THEOREM 4.5
If $\hat{\rho}(T_0|_W, T_1|_W) < 1$ then there exists a continuous scaling vector v which is Hölder continuous with Hölder exponent α for every $0 \leq \alpha < -\log_2 \hat{\rho}(T_0|_W, T_1|_W)$. If $\theta^{-1}T_0|_W$, $\theta^{-1}T_1|_W$ are product bounded with $\theta = \hat{\rho}(T_0|_W, T_1|_W)$ then v is also Hölder continuous with exponent $\alpha = -\log_2 \hat{\rho}(T_0|_W, T_1|_W)$.

4.5.2   Computing a joint spectral radius

Whereas the usual spectral radius of a single matrix is easily computed from the eigenvalues of that matrix, the joint spectral radius can be difficult to compute exactly. We therefore discuss in this subsection several methods of evaluating or approximating the joint spectral radius.

First, it is easy to see that the joint spectral radius is independent of the choice of basis, i.e., if B is an invertible matrix then $\hat{\rho}(BA_0B^{-1}, BA_1B^{-1}) = \hat{\rho}(A_0, A_1)$. Once a convenient basis is selected, the value of $\hat{\rho}(A_0, A_1)$ can be approximated to any desired degree of accuracy by the computation of the norms and eigenvalues of finitely many matrices. This follows from the next lemma and the fact that $\hat{\rho}(A_0, A_1) = \limsup \hat{\sigma}_m = \limsup \hat{\rho}_m$.

LEMMA 4.6
$\hat{\sigma}_m \le \hat{\rho}(A_0, A_1) \le \hat{\rho}_m$ *for every* $m > 0$.

PROOF   Given any $d_1, \ldots, d_m = 0, 1$ and any $n > 0$ we have

$$\rho(A_{d_1} \cdots A_{d_m})^{1/m} = \rho\big((A_{d_1} \cdots A_{d_m})^n\big)^{1/mn} \le \hat{\sigma}_{mn},$$

whence

$$\hat{\sigma}_m = \max_{d_j = 0,1} \rho(A_{d_1} \cdots A_{d_m})^{1/m} \le \limsup_{n \to \infty} \hat{\sigma}_{mn} \le \hat{\rho}(A_0, A_1).$$

For the second inequality, let

$$K = \max\{1, \hat{\rho}_m^{-1}, \ldots, \hat{\rho}_m^{-m+1}\} \quad \text{and} \quad L = \max\{1, \hat{\rho}_1, \ldots, \hat{\rho}_{m-1}^{m-1}\}.$$

Given any $n > 0$, write $n = mk + r$ with $0 \le r < m$. Then for any $d_1, \ldots, d_n = 0, 1$ we have

$$\|A_{d_1} \cdots A_{d_n}\| \le (\hat{\rho}_m^m)^k \hat{\rho}_r^r = \hat{\rho}_m^n \hat{\rho}_m^{-r} \hat{\rho}_r^r \le KL\hat{\rho}_m^n.$$

Taking the maximum over all $d_j = 0, 1$ we obtain $\hat{\rho}_n \le (KL)^{1/n} \hat{\rho}_m$, so $\hat{\rho}(A_0, A_1) = \limsup_{n \to \infty} \hat{\rho}_n \le \hat{\rho}_m$. ∎

As a consequence of Lemma 4.6, we have $\sup \hat{\sigma}_m = \hat{\rho}(A_0, A_1) = \lim \hat{\rho}_m = \inf \hat{\rho}_m$. Note that $\sup \hat{\sigma}_m$ is a lower-semicontinuous function of the matrices A_0, A_1 since each $\hat{\sigma}_m$ is a continuous function of those matrices. Similarly, $\inf \hat{\rho}_m$ is upper semicontinuous, so $\hat{\rho}(A_0, A_1)$ is a continuous function of the matrices A_0, A_1 [HeS].

Unfortunately, the number of matrix products involved in the computation of $\hat{\sigma}_m$ or $\hat{\rho}_m$ grows exponentially with m, and therefore it is usually impractical to achieve a good approximation to $\hat{\rho}(A_0, A_1)$ by simply computing $\hat{\sigma}_m$ and $\hat{\rho}_m$ for various m. On the other hand, if A_0, A_1 satisfy one of the hypotheses of the following lemma then $\hat{\rho}(A_0, A_1)$ can be computed exactly.

LEMMA 4.7

If A_0, A_1 can be simultaneously upper-triangularized or simultaneously Her-mitianized (i.e., there exists an invertible matrix B such that BA_0B^{-1} and BA_1B^{-1} are either both upper triangular or both Hermitian, respectively), then $\hat{\rho}(A_0, A_1) = \hat{\sigma}_1 = \max\{\rho(A_0), \rho(A_1)\}$.

PROOF The result for simultaneous upper-triangularization follows imme-diately from examination of the diagonal entries of the product of two upper-triangular matrices.

To prove the result for simultaneous Hermitianization, let $\| \cdot \|$ be any norm under which $\mathbb{C}^N$ is a Hilbert space, e.g., the Euclidean space norm, for then $\|A\| = \rho(A)$ for any Hermitian matrix A. Given $d_1, \ldots, d_m$, we therefore have

$$
\begin{aligned}
\|A_{d_1} \cdots A_{d_m}\| &= \|B^{-1}(BA_{d_1}B^{-1}) \cdots (BA_{d_m}B^{-1})B\| \\
&\leq \|B\|\|B^{-1}\|\|BA_{d_1}B^{-1}\| \cdots \|BA_{d_m}B^{-1}\| \\
&= \|B\|\|B^{-1}\|\rho(BA_{d_1}B^{-1}) \cdots \rho(BA_{d_m}B^{-1}) \\
&= \|B\|\|B^{-1}\|\rho(A_{d_1}) \cdots \rho(A_{d_m}) \\
&\leq \|B\|\|B^{-1}\|\hat{\sigma}_1^m,
\end{aligned}
$$

since each $BA_{d_j}B^{-1}$ is Hermitian. Thus $\hat{\rho}_m \leq \|B\|^{1/m}\|B^{-1}\|^{1/m}\hat{\sigma}_1$, so $\hat{\rho}(A_0, A_1) = \limsup \hat{\rho}_m \leq \hat{\sigma}_1 \leq \hat{\rho}(A_0, A_1)$. ∎

Note that $A_0 = \left(\begin{smallmatrix} 1 & 0 \\ 1 & 2 \end{smallmatrix}\right)$, $A_1 = \left(\begin{smallmatrix} 2 & 1 \\ 0 & 1 \end{smallmatrix}\right)$ can be simultaneously upper-triangularized but not simultaneously Hermitianized, while $A_0 = \left(\begin{smallmatrix} 3 & 0 \\ 3 & -2 \end{smallmatrix}\right)$, $A_1 = \left(\begin{smallmatrix} -2 & 3 \\ 0 & 3 \end{smallmatrix}\right)$ can be simultaneously Hermitianized but not simultaneously upper-triangularized.

The following lemma, a generalization of the first part of Lemma 4.7, can be useful for reducing the computational complexity of the calculation of $\hat{\rho}(A_0, A_1)$ when A_0, A_1 are large. Its proof follows easily using block matrix multiplication.

LEMMA 4.8

Assume A_0, A_1 can be simultaneously block upper-triangularized, i.e., there exists an invertible matrix B such that

$$
BA_iB^{-1} = \begin{pmatrix} C_i^1 & & * \\ & \ddots & \\ 0 & & C_i^k \end{pmatrix}, \qquad i = 0, 1,
$$

for some square submatrices $C_i^1, \ldots, C_i^k$. Then $\hat{\rho}(A_0, A_1) = \max_{j=1,\ldots,k}\{\hat{\rho}(C_0^j, C_1^j)\}$.

The above tools are not sufficient to exactly evaluate the joint spectral radius of all pairs of matrices, even when we consider only pairs of 2×2 matrices. For example, one pair which will be of interest in Section 4.8.1 is

$$S_0 = \frac{1}{5} \begin{pmatrix} 3 & 0 \\ 1 & 3 \end{pmatrix} \quad \text{and} \quad S_1 = \frac{1}{5} \begin{pmatrix} 3 & -3 \\ 0 & -1 \end{pmatrix}. \tag{4.25}$$

This pair is neither simultaneously upper-triangularizable nor simultaneously Hermitianizable. By direct brute-force calculation of $2^{31} - 2$ matrix products we find that

$$\min\{\hat{\rho}_1, \ldots, \hat{\rho}_{30}\} = \hat{\rho}_{30} = \|S_0^5 S_1 S_0^{11} S_1 S_0^{12}\|^{1/30} \approx 0.671271 \tag{4.26}$$

and

$$\max\{\hat{\sigma}_1, \ldots, \hat{\sigma}_{30}\} = \hat{\sigma}_{13} = \rho(S_1 S_0^{12})^{1/13} \approx 0.659679,$$

where we have chosen the norm $\|u\| = |u_1| + |u_2|$ on $\mathbb{C}^2$ to evaluate $\hat{\rho}_m$. This enormous calculation does not even specify $\hat{\rho}(S_0, S_1)$ to an accuracy of two decimal places!

Fortunately, there does exist a recursive algorithm which often will substantially reduce the computational complexity of approximating $\hat{\rho}(A_0, A_1)$ from above (unfortunately, we know of no similar method for reducing the computational burden of approximating from below). The algorithm is based on the following straightforward generalization of Lemma 4.6 [DL2].

LEMMA 4.9
Assume that $\{P_j\}$ is a set of building blocks of products of the matrices A_0, A_1. That is,

(a) *each P_j is a product of m_j of the matrices A_0, A_1, and*

(b) *there exists an $r \geq 0$ such that if P is any product of A_0, A_1 then $P = P_{j_1} \cdots P_{j_k} Q$ with Q a product of length at most r.*

Then $\hat{\rho}(A_0, A_1) \leq \max \|P_j\|^{1/m_j}$.

Given an initial "guess" θ for the joint spectral radius, the following algorithm [ColHe1] will construct a set of building blocks $\{P_j\}$ such that $\hat{\rho}(A_0, A_1) \leq \max \|P_j\|^{1/m_j} \leq \theta$ (if $\theta < \hat{\rho}(A_0, A_1)$ then the algorithm will not terminate).

Algorithm 4.10
Fix $\theta > \hat{\rho}(A_0, A_1)$. For each of $P = A_0$ and $P = A_1$ in turn, implement the following recursion.

> Let $P = A_{d_1} \cdots A_{d_m}$. If $\|P\|^{1/m} \leq \theta$ then keep P as a building block. Otherwise, repeat this step replacing $P = A_{d_1} \cdots A_{d_m}$ by each of $P = A_{d_1} \cdots A_{d_m} A_0$ and $P = A_{d_1} \cdots A_{d_m} A_1$ in turn. □

Algorithm 4.10 must terminate if the guess θ exceeds $\hat{\rho}(A_0, A_1)$ since there must be some m such that $\hat{\rho}(A_0, A_1) \leq \hat{\rho}_m \leq \theta$.

Returning now to the specific case of the matrices S_0, S_1 given in Eq. (4.25), we used Algorithm 4.10 to improve the estimate of $\hat{\rho}(S_0, S_1)$. In particular, Algorithm 4.10 did terminate after an initial guess of $\theta = 0.660025$ was input, so we conclude that $\hat{\rho}(S_0, S_1) \leq 0.660025$, a significant improvement over Eq. (4.26). Algorithm 4.10 output a set of 306473 building blocks after computing 612944 matrix products (including those not selected as building blocks). The longest product selected as a building block was of length 139. A direct calculation of $\hat{\rho}_{139}$ (even if it were possible!) could not improve this estimate since, for example,

$$\|S_0^3 S_1 S_0^{12} S_1 S_0^{13} S_1 S_0^{12} S_1 S_0^{11} S_1 S_0^9 S_1 S_0^{12} S_1 S_0^{10} S_1 S_0^{13} S_1 S_0^{12} S_1 S_0^{11} S_1 S_0^9 S_1 \|^{1/139}$$
$$= 0.661276,$$

whence $\hat{\rho}_{139} \geq 0.661276$. Based on the above numerical evidence, we conjecture that $\hat{\rho}(S_0, S_1) = \rho(S_1 S_0^{12})^{1/13}$.

4.6 Necessary conditions for the existence of continuous scaling vectors

We devote this section to proving that Theorems 4.3 and 4.5 are sharp, i.e., that we have the following converse [ColHe3].

THEOREM 4.11

If v is any continuous scaling vector and if W is defined by Eq. (4.19) then $\hat{\rho}(T_0|_W, T_1|_W) < 1$. Moreover, no Hölder exponent of continuity α for v can exceed $-\log_2 \hat{\rho}(T_0|_W, T_1|_W)$, and can equal $-\log_2 \hat{\rho}(T_0|_W, T_1|_W)$ if and only if $\theta^{-1} T_0|_W, \theta^{-1} T_1|_W$ are product bounded with $\theta = \hat{\rho}(T_0|_W, T_1|_W)$.

To begin the proof, let v be an arbitrary continuous scaling vector; v is then Hölder continuous for some Hölder exponent α in the range $0 \leq \alpha \leq 1$. Let K be any corresponding Hölder constant.

Choose now any fixed product $T = T_{d_1} \cdots T_{d_m}$ of T_0, T_1, and let $x \in [0, 1]$ be an arbitrary dyadic point (we will impose restrictions on x momentarily). Let $x = .e_1 e_2 \cdots$ be any binary expansion of x. Then by Eq. (4.14) and the continuity of v,

$$T^k v(x) = v(x_k) \rightarrow v(d) \quad \text{as } k \rightarrow \infty, \tag{4.27}$$

where

$$x_1 = .d_1 \cdots d_m e_1 e_2 \cdots,$$
$$x_2 = .d_1 \cdots d_m d_1 \cdots d_m e_1 e_2 \cdots,$$

$$\vdots$$

and $d \in [0, 1]$ is the rational (but not necessarily dyadic) point

$$d = .d_1 \cdots d_m d_1 \cdots d_m \cdots.$$

Intuitively, Eq. (4.27) should impose some constraint both on the spectral radius of T and on the Hölder exponent α. To make this precise, let $y_k \in [0, 1]$ be the dyadic points

$$y_1 = .d_1 \cdots d_m,$$
$$y_2 = .d_1 \cdots d_m d_1 \cdots d_m,$$

$$\vdots$$

We will use the y_k in place of d since d is not necessarily dyadic. Note that

$$v(x_k) - v(y_k) = T^k \big(v(x) - v(0) \big) \tag{4.28}$$

and

$$|x_k - y_k| = 2^{-mk} |x|. \tag{4.29}$$

If we define W as in Eq. (4.19) then certainly $v(x) - v(0) \in W$. So, let λ be that eigenvalue of $T|_W$ such that $|\lambda| = \rho(T|_W)$. If $v(x) - v(0)$ "has a component" in the λ eigenspace of $T|_W$, then Eq. (4.28) should imply that $\|v(x_k) - v(y_k)\|$ is on the order of $|\lambda|^k$; combining this with Eq. (4.29) then yields an estimate of the Hölder exponent. In general, the meaning of "component" can be precisely formulated in terms of the Jordan decomposition of W induced by $T|_W$; reference [ColHe3] contains an explicit description of how to do this. In any case, since W is spanned by vectors of the form $v(x) - v(0)$ for dyadic x, there must be *some* dyadic x such that $v(x) - v(0)$ has a component in the λ eigenspace of $T|_W$. Although we will omit the details, it is then straightforward to prove that there exists a constant $C > 0$ such that

$$\|T^k \big(v(x) - v(0) \big)\| \geq C |\lambda|^k \tag{4.30}$$

for all k [ColHe3]. In particular, since $v(x_k) - v(y_k) \to 0$ as $k \to \infty$, we must have $\rho(T|_W) = |\lambda| < 1$. Combining Eqs. (4.28)–(4.30), we find that

$$\|v(x_k) - v(y_k)\| \geq K_T |x_k - y_k|^{\alpha_T}, \tag{4.31}$$

where $\alpha_T = -\log_2 \rho(T)^{1/m}$ and $K_T = C|x|^{-\alpha_T}$. However, v is Hölder continuous with exponent α and constant K, so

$$\|v(x_k) - v(y_k)\| \leq K |x_k - y_k|^{\alpha}. \tag{4.32}$$

Since $|x_k - y_k| \to 0$ as $k \to \infty$, Eqs. (4.31) and (4.32) imply that $\alpha \leq \alpha_T$. Taking now the supremum over all $T = T_{d_1} \cdots T_{d_m}$ for $d_j = 0$, 1 we obtain $\hat{\sigma}_m < 1$ and $\alpha \leq -\log_2 \hat{\sigma}_m$. Thus $\hat{\rho}(T_0|_W, T_1|_W) = \sup \hat{\sigma}_m \leq 1$ and $\alpha \leq -\log_2 \hat{\rho}(T_0|_W, T_1|_W)$, i.e., we have proved the statement that the Hölder exponent of v cannot exceed $-\log_2 \hat{\rho}(T_0|_W, T_1|_W)$.

Our analysis has also nearly established that $\hat{\rho}(T_0|_W, T_1|_W) < 1$; in particular, we have shown that $\hat{\rho}(T_0|_W, T_1|_W) = \sup \hat{\sigma}_m \leq 1$ with $\hat{\sigma}_m < 1$ for every m. Thus, only the special case

$$\hat{\rho}(T_0|_W, T_1|_W) = \sup \hat{\sigma}_m = 1 \quad \text{and} \quad \hat{\sigma}_m < 1 \text{ for every } m \qquad (4.33)$$

remains. It is interesting to note that we do not know if Eq. (4.33) is possible. More generally, we know of no matrices A_0, A_1 for which we can demonstrate that the supremum $\sup \hat{\sigma}_m(A_0, A_1) = \hat{\rho}(A_0, A_1)$ is not achieved for some m. Lagarias and Wang [LW] have conjectured that no such matrices can exist, i.e., it is always the case that $\hat{\rho}(A_0, A_1) = \rho(A_{d_1} \cdots A_{d_m})^{1/m}$ for some finite matrix product $A_{d_1} \cdots A_{d_m}$.

We continue now with the proof of Theorem 4.11; in particular, we show now that the continuity of v implies $\hat{\rho}(T_0|_W, T_1|_W) < 1$ (here we use an argument due to Micchelli and Prautzsch [MiP2]; see [ColHe3] for another argument). Let $\{v(x_i) - v(0)\}_{i=1}^{J}$ be a basis for W, and define a norm on W by $\|w\| = \sum |a_i|$ for $w = \sum a_i(v(x_i) - v(0))$. Choose any $\varepsilon < 1$; since v is uniformly continuous there exists a $\delta > 0$ such that $\|v(x) - v(y)\| \leq \varepsilon$ whenever $|x - y| < \delta$. Let m be such that $2^{-m} < \delta$, and choose any product $T = T_{d_1} \cdots T_{d_m}$ of length m. Writing the binary expansion of x_i as $x_i = .e_1^i e_2^i \cdots$, define $X_i = .d_1 \cdots d_m e_1^i e_2^i \cdots$ and $Y = .d_1 \cdots d_m$. Then $|X_i - Y| \leq 2^{-m}$, so for an arbitrary element $w = \sum a_i(v(x_i) - v(0))$ of W we have

$$\|T_{d_1} \cdots T_{d_m} w\| \leq \sum_{i=1}^{J} |a_i| \, \|T_{d_1} \cdots T_{d_m}(v(x_i) - v(0))\|$$

$$= \sum_{i=1}^{J} |a_i| \, \|v(X_i) - v(Y)\|$$

$$\leq \varepsilon \sum_{i=1}^{J} |a_i|$$

$$= \varepsilon \|w\|.$$

Thus $\|T|_W\| \leq \varepsilon$, and therefore by taking the supremum over all such matrix products of length m we obtain $\hat{\rho}(T_0|_W, T_1|_W) \leq \hat{\rho}_m \leq \varepsilon^{1/m} < 1$.

The proof of the final statement of the theorem, that $\alpha = -\log_2 \hat{\rho}(T_0|_W, T_1|_W)$ implies product boundedness, follows easily by using a basis for W as above, and will be omitted [ColHe3].

4.7 Differentiability

Theorems 4.3, 4.5, and 4.11 characterize those dilation equations which have continuous, compactly supported solutions f. However, if f is a continuously differentiable, compactly supported scaling function for the dilation equation determined by the coefficients $\{c_0, \ldots, c_N\}$ then f' is a continuous, compactly supported scaling function for the dilation equation determined by the coefficients $\{2c_0, \ldots, 2c_N\}$. Thus, Theorems 4.3, 4.5, and 4.11 implicitly characterize those dilation equations having continuously differentiable, compactly supported solutions. We make this implicit characterization, and its extension to higher derivatives, explicit in this section.

4.7.1 Necessary and sufficient conditions

To begin, assume that f is an n-times continuously differentiable, compactly supported scaling function for a dilation equation determined by the coefficients $\{c_0, \ldots, c_N\}$. In light of Theorem 4.1, it suffices to consider coefficients satisfying the fundamental constraint $\sum c_k = 2$. Let the associated matrices M, T_0, T_1, scaling vector v, and subspace W be constructed as usual. Now, the jth derivative $f^{(j)}$ of f is a continuous, compactly supported scaling function for the dilation equation determined by the coefficients $\{2^j c_0, \ldots, 2^j c_N\}$ for each $j = 0, \ldots, n$, so let the matrices M^j, T_0^j, T_1^j, scaling vector v^j, and subspace W^j denote the obvious objects constructed with respect to $f^{(j)}$. Clearly, $M^j = 2^j M$, $T_0^j = 2^j T_0$, and $T_1^j = 2^j T_1$; however, the scaling vectors v^j and the subspaces W^j are quite distinct from v and W. In particular, since $1 \in \text{spectrum}(M^j)$ we have $2^{-j} \in \text{spectrum}(M)$ for $j = 0, \ldots, n$; this simply amounts to the realization that $(f^{(j)}(1), \ldots, f^{(j)}(N-1))^{\text{t}}$ is a right eigenvector for M for the eigenvalue 2^{-j}. Note that this implies $n < N-1$, and therefore f cannot be infinitely differentiable! From Theorem 4.11 we must have

$$\hat{\rho}(T_0^j|_{W^j}, T_1^j|_{W^j}) = 2^j \hat{\rho}(T_0|_{W^j}, T_1|_{W^j}) < 1 \quad \text{for each } j = 0, \ldots, n.$$

This supplies a necessary condition for the existence of an n-times continuously differentiable scaling vector.

   To make this more explicit, suppose we are given coefficients $\{c_0, \ldots, c_N\}$ such that $\sum c_k = 2$ and $1, 2^{-1}, \ldots, 2^{-n} \in \text{spectrum}(M)$, and let $a = (a_1, \ldots, a_{N-1})^{\text{t}}$ be a right eigenvector for M for the eigenvalue 2^{-n}. Define

$$v^n(0) = (0, a_1, \ldots, a_{N-1})^{\text{t}} \quad \text{and} \quad v^n(1) = (a_1, \ldots, a_{N-1}, 0)^{\text{t}}.$$

For dyadic $x = .d_1 \cdots d_m \in [0, 1]$ define

$$v^n(x) = 2^{mn} T_{d_1} \cdots T_{d_m} v^n(0),$$

and set

$$W^n = W^n(a) = \text{span}\{v^n(x) - v^n(0) : \text{dyadic } x \in [0, 1]\}.$$

Then we have the following theorem.

THEOREM 4.12
*Let coefficients $\{c_0,\ldots,c_N\}$ be given such that $\sum c_k = 2$ and $1, 2^{-1},\ldots,2^{-n} \in$
spectrum(M). Then there exists an n-times continuously differentiable, compactly supported scaling function f for the dilation equation determined by the coefficients $\{c_0,\ldots,c_N\}$ if and only if*

$$\hat{\rho}(T_0|_{W^n}, T_1|_{W^n}) < 2^{-n} \tag{4.34}$$

for some right eigenvector a for M for the eigenvalue 2^{-n}, where $W^n = W^n(a)$ is defined as above. In this case the nth derivative $f^{(n)}$ of f is Hölder continuous for each exponent

$$0 \le \alpha < \alpha_{\max} = -\log_2\left(2^n \hat{\rho}(T_0|_{W^n}, T_1|_{W^n})\right),$$

*but not for any $\alpha > \alpha_{\max}$, and $\alpha = \alpha_{\max}$ is allowed if and only if $\theta^{-1}T_0|_{W^n}$,
$\theta^{-1}T_1|_{W^n}$ are product bounded with $\theta = \hat{\rho}(T_0|_{W^n}, T_1|_{W^n})$.*

The sufficient condition in Theorem 4.12 is proved as follows. Assume $\hat{\rho}(T_0|_{W^n}, T_1|_{W^n}) < 2^{-n}$. Then v^n extends to a continuous scaling vector on $[0, 1]$ by Theorem 4.5, and therefore determines a continuous, compactly supported scaling function h for the dilation equation determined by the coefficients $\{2^n c_0,\ldots,2^n c_N\}$. Theorem 4.5 also implies that h is Hölder continuous for at least each $0 \le \alpha < \alpha_{\max}$, and Theorem 4.11 limits α to at most $\alpha_{\max}$. Together, Theorems 4.5 and 4.11 imply $\alpha = \alpha_{\max}$ is allowed if and only if the stated product boundedness condition holds. Finally, Theorem 4.1 implies that h is the nth derivative of an integrable, compactly supported scaling function f for the dilation equation determined by the coefficients $\{c_0,\ldots,c_N\}$.

4.7.2 Imposition of sum rules

Theorem 4.12 characterizes dilation equations having differentiable solutions without explicit restrictions on the coefficients $\{c_0,\ldots,c_N\}$. However, scaling functions useful in applications usually arise from coefficients satisfying a certain number of "equal-opportunity" restrictions on the even- and odd-indexed coefficients; Eq. (4.4) is one example of such a restriction. In particular, in addition to the fundamental constraint $\sum c_k = 2$, it is generally assumed that the coefficients satisfy the following $n + 1$ *sum rules*:

$$\sum_k (-1)^k k^j c_k = 0 \quad \text{for } j = 0,\ldots,n \tag{4.35}$$

(note that Eq. (4.35) is just Eq. (4.4) if $n = 0$). These sum rules impose certain conditions on the scaling function f and associated wavelet g which are related to, but distinct from, smoothness. For example, the $n + 1$ sum rules imply that the wavelet g will have $n + 1$ vanishing moments. For $j = 0$ this follows from the calculation

$$\int g(t)\, dt = \sum_k (-1)^k c_{N-k} \int f(2t - k)\, dt = 0,$$

and an easy induction then establishes that $\int t^j g(t)\, dt = 0$ for $j = 1, \dots, n$. We also have the following.

PROPOSITION 4.13
Assume the coefficients $\{c_0, \dots, c_N\}$ satisfy $\sum c_k = 2$ and $n + 1$ sum rules, and that an associated compactly supported scaling function f exists. Then $\sum k^j f(t - k)$ is a polynomial of degree j for each $j = 0, \dots, n$.

In particular, the polynomials $1, \dots, t^n$ can all be reproduced exactly from translates of f. Thus, an arbitrary smooth function can be approximated with error $\mathcal{O}(2^{-mn})$ by combinations of $\{f(2^m t - k)\}_{k \in \mathbb{Z}}$ for each scale 2^{-m} [S].

Proposition 4.13 can be proved by various methods; we sketch one version, restricted for clarity to the case $n = 1$. Assume f is normalized so that $\hat{f}(0) = 1$, and let θ_0 be any function such that

(a) $\hat{\theta}_0(0) = 1$,

(b) $\sum \theta_0(t - k) = 1$, and

(c) $\sum k \theta_0(t - k) = t + b$ for some constant b.

For example, this is true if θ_0 is the solution of a dilation equation satisfying two sum rules and such that θ_0 is orthogonal to its integer translates (admittedly, this is circular reasoning, but it is not difficult to prove Proposition 4.13 directly with this extra assumption of orthogonality). An example of such a θ_0 is the Daubechies scaling function D_4 discussed in Section 4.8.1. Now apply the Cascade Algorithm, i.e., define the functions θ_i for $i > 0$ by

$$\theta_i(t) = \sum_k c_k \theta_{i-1}(2t - k). \tag{4.36}$$

Then $\hat{\theta}_i(\gamma) \to \hat{f}(\gamma) = \prod_{j=1}^{\infty} m_0(\gamma/2^j)$ uniformly on compact sets since $\hat{\theta}_0(0) = 1$ and $\hat{\theta}_i(\gamma) = m_0(\gamma/2)\hat{\theta}_{i-1}(\gamma/2)$. In the time domain this implies $\theta_i \to f$ weakly in $L^2(\mathbb{R})$. Now, it follows from Eq. (4.36) and the two sum rules that

$$\sum_k \theta_i(t - k) = \sum_k \theta_{i-1}(2t - k) \tag{4.37}$$

and

$$\sum_k k \theta_i(t - k) = \frac{1}{2} \sum_k k \theta_{i-1}(2t - k) - \frac{A_1}{2} \sum_k \theta_{i-1}(2t - k), \tag{4.38}$$

where

$$A_1 = \sum (2k)c_{2k} = \sum (2k+1)c_{2k+1}.$$

Applying Eqs. (4.37) and (4.38) recursively and taking the limit as $i \to \infty$ we obtain $\sum f(t-k) = 1$ a.e. and $\sum kf(t-k) = t - A_1$ a.e., completing the proof.

As a corollary of Proposition 4.13 we have that if f is n-times differentiable and $n+1$ sum rules are satisfied then $\sum k^j f^{(n)}(t-k)$ is identically constant for each $j = 0, \ldots, n$. In the notation of Theorem 4.12, this implies

$$\sum_{k=1}^{N} k^j \left(v_k^n(x) - v_k^n(0) \right) = 0 \quad \text{for } j = 0, \ldots, n,$$

where $v_k^n(x)$ is the kth component of $v^n(x)$. Thus

$$W^n \subset V^n = \{u \in \mathbb{C}^N : e_j \cdot u = 0 \quad \text{for } j = 0, \ldots, n\},$$

where $e_j = (1^j, 2^j, \ldots, N^j)$. Moreover, as Daubechies and Lagarias proved [D1], the assumption of the sum rules implies the following facts about V^n.

LEMMA 4.14
Assume the coefficients $\{c_0, \ldots, c_N\}$ satisfy $\sum c_k = 2$ and $n+1$ sum rules. Then the following statements hold for each $j = 0, \ldots, n$.

- (a) *$U^j = \text{span}\{e_0, \ldots, e_j\}$ is left invariant under both T_0 and T_1 and $\text{spectrum}(T_0|_{U^j}) = \text{spectrum}(T_1|_{U^j}) = \{1, 2^{-1}, \ldots, 2^{-j}\}$.*

- (b) *V^j is the orthogonal complement of U^j in $\mathbb{C}^N$ and is right invariant under both T_0 and T_1.*

As a consequence, $1, 2^{-1}, \ldots, 2^{-n} \in \text{spectrum}(M)$.

PROOF Note that (b) follows immediately from (a) and the definition of V^j. To prove (a), it is sufficient to show that for each $i = 0, 1$ and for $j = 0, \ldots, n$ there exist constants a^i_{jl} such that

$$e_j T_i = 2^{-j} e_j + \sum_{l=0}^{j-1} a^i_{jl} e_l.$$

This is certainly true for $j = 0$ since $e_0 T_0 = e_0 T_1 = e_0$. For simplicity, we prove only the case $i = 0, j = 1$; the general case is similar.

Since $(T_0)_{pq} = c_{2p-q-1}$, we have

$$(e_1 T_0)_q = \sum_{p=1}^{N} p c_{2p-q-1}.$$

If q is even, say $q = 2r$, then

$$(e_1 T_0)_{2r} = \sum_p p c_{2p-2r-1}$$

$$= \sum_p (r + p + 1) c_{2p+1}$$

$$= \frac{2r+1}{2} \sum_p c_{2p+1} + \frac{1}{2} \sum_p (2p+1) c_{2p+1}$$

$$= \frac{1}{2} (2r + 1 + A_1)$$

$$= \frac{1}{2} (e_1)_{2r} + \frac{A_1 + 1}{2} (e_0)_{2r}.$$

A similar calculation applies if q is odd, so we conclude that

$$e_1 T_0 = \frac{1}{2} e_1 + \frac{A_1 + 1}{2} e_0,$$

as desired.

It follows from (a) that $1, 2^{-1}, \dots, 2^{-n}$ are eigenvalues of both T_0 and T_1, and it remiains only to show that these are also eigenvalues of M. However, given j we can find $u = (u_1, \dots, u_N) \in U^j$ which is a left eigenvector for T_0 for the eigenvalue 2^{-j}. Since u is a linear combination of $e_0, \dots, e_j$ we cannot have $u_2 = \cdots = u_N = 0$. Since $T_0 = \begin{pmatrix} c_0 & 0 \\ * & M \end{pmatrix}$, we conclude that $(u_2, \dots, u_N)$ is a left eigenvector for M for the eigenvalue 2^{-j}. ∎

As a consequence of Lemma 4.14 and Theorem 4.12, we find that

$$\hat{\rho}(T_0|_{V^n}, T_1|_{V^n}) < 2^{-n} \tag{4.39}$$

is sufficient to ensure the existence of an n-times continuously differentiable, compactly supported scaling function (although it need not be necessary). Since V^n is independent of the coefficients, this is often an easier condition to test than Eq. (4.34).

Note that Eq. (4.39) uses $n+1$ sum rules to conclude n-times differentiability. In practice, however, it is often the case that several more sum rules will be satisfied than the number of derivatives possessed by the scaling function. For example, the Daubechies scaling function D_4 satisfies two sum rules, yet is only continuous and not differentiable. Even so, we can use these "extra" sum rules to obtain extra simplification of the joint spectral radius calculations. This is made precise in the following proposition, which is a slightly weaker version of a result first proved by Daubechies and Lagarias [DL2].

PROPOSITION 4.15

Assume the coefficients $\{c_0, \dots, c_N\}$ satisfy $\sum c_k = 2$ and $n + 1$ sum rules. If

$$\hat{\rho}(T_0|_{V^n}, T_1|_{V^n}) < 2^{-l}$$

for some $0 \leq l \leq n$, then an l-times continuously differentiable, compactly supported scaling function exists, and its lth derivative is Hölder continuous for each exponent

$$0 \leq \alpha < \min\{1, -\log_2\left(2^l \hat{\rho}(T_0|_{V^n}, T_1|_{V^n})\right)\}.$$

PROOF Recall from Lemma 4.14 that $\{e_0, \dots, e_j\}$ is a basis for U^j for each $j = 0, \dots, n$. Since $U^0 \subset \cdots \subset U^n$ and each V^j is the orthogonal complement of U^j in $\mathbb{C}^N$, we have $V^0 \supset \cdots \supset V^n$. Moreover, $\dim(V^n) = N - n - 1$, so if $\{\tilde{e}_{n+1}, \dots, \tilde{e}_{N-1}\}$ is a basis for V^n then $\{e_{j+1}, \dots, e_n, \tilde{e}_{n+1}, \dots, \tilde{e}_{N-1}\}$ is a basis for V^j for each $j = 0, \dots, n$. By considering these complementary bases for U^j and V^j, we can easily construct a change-of-basis matrix B such that

$$BT_iB^{-1} = \begin{pmatrix} 1 & & & & * \\ & 2^{-1} & & & \\ & & \ddots & & \\ & & & 2^{-n} & \\ & 0 & & & C_i \end{pmatrix}, \qquad i = 0, 1,$$

and conclude then from Lemma 4.8 that

$$\hat{\rho}(T_0|_{V^j}, T_1|_{V^j}) = \begin{cases} \max\{2^{-j-1}, \dots, 2^{-n}, \hat{\rho}(C_0, C_1)\}, & j = 0, \dots, n-1, \\ \hat{\rho}(C_0, C_1), & j = n. \end{cases}$$

Thus,

$$\hat{\rho}(T_0|_{V^l}, T_1|_{V^l}) = \max\{2^{-l-1}, \hat{\rho}(T_0|_{V^n}, T_1|_{V^n})\},$$

so the hypothesis $\hat{\rho}(T_0|_{V^n}, T_1|_{V^n}) < 2^{-l}$ implies $\hat{\rho}(T_0|_{V^l}, T_1|_{V^l}) < 2^{-l}$. The result therefore follows from Theorem 4.12. ∎

In particular, note that the dimension of V^n can be significantly smaller than the dimension of V^l, resulting in a reduction in the size of the matrices involved in the computation of the joint spectral radius. Because W^l may be strictly smaller than V^l, the upper bound for the Hölder exponent of $f^{(l)}$ in Proposition 4.15 need not be sharp, as is the case for the upper bound in Theorem 4.12.

4.8 Examples

For simplicity, we will restrict our attention to real-valued coefficients $\{c_0, \dots, c_N\}$. By the uniqueness results of Theorem 4.1, the corresponding scaling functions must then also be real valued.

4.8.1   Four-coefficient dilation equations

There are no continuous, compactly supported solutions of dilation equations when $N = 0$ or 1. When $N = 2$ the assumptions $\sum c_k = 2$ and $1 \in \text{spectrum}(M)$ imply that Eq. (4.4) holds, whence $W \subset V$ is at most one dimensional and the computation of $\hat{\rho}(T_0|_W, T_1|_W)$ is trivial. We therefore move directly to the first nontrivial case, $N = 3$. Here we have four coefficients $\{c_0, c_1, c_2, c_3\}$; assume again that they satisfy $\sum c_k = 2$ and $1 \in \text{spectrum}(M)$. It follows then that either

(a) $c_0 + c_1 = c_2 + c_3 = 1$, or

(b) $c_0 + c_2 = c_1 + c_3 = 1$.

Note that case (b) is simply Eq. (4.4).

In case (a), it is easy to show that $\dim(W) = 3$ if $c_0 \neq c_3$, whence $\hat{\rho}(T_0|_W, T_1|_W) \geq \rho(T_0|_W) = \rho(T_0) \geq 1$. Thus there are no continuous solutions to case (a) when $c_0 \neq c_3$. Since case (a) with $c_0 = c_3$ is a special case of (b), we turn immediately to that case.

Assume, therefore, that Eq. (4.4) holds, and note that this class of dilation equations is a two-parameter family. We select the independent parameters to be c_0 and c_3, and therefore identify this family with the (c_0, c_3)-plane, i.e., each point (c_0, c_3) determines a dilation equation with coefficients $\{c_0, 1 - c_3, 1 - c_0, c_3\}$ and vice versa. Since Eq. (4.4) implies $W \subset V$ and $\dim(V) = 2$, it follows that either $W = V$ or W is one dimensional.

The (c_0, c_3)-plane is shown in Figure 4.1 along with several geometrical objects. It has been shown [ColHe4] that integrable scaling functions exist at least for all points in the shaded region, and cannot exist outside the ellipse. We show below that continuous scaling functions are restricted to a subregion of the interior of the triangle, and that differentiable scaling functions occur precisely on the solid portion of the dashed line. The circle $(c_0 - 1/2)^2 + (c_3 - 1/2)^2 = 1/2$ consists of those points which satisfy both Eqs. (4.4) and (4.5), which is a necessary condition for the existence of corresponding multiresolution analyses. In fact, every point on the circle with the single exception of $(c_0, c_3) = (1, 1)$ does determine a multiresolution analysis and hence a wavelet orthonormal basis for $L^2(\mathbb{R})$.

Now, a right eigenvector for M for the eigenvalue 1 is $(c_0, c_3)^t$; therefore $v(0) = (0, c_0, c_3)^t$ and $v(1) = (c_0, c_3, 0)^t$, up to normalization. Computing $v(1/2) = T_1 v(0)$, we find that $v(0)$, $v(1/2)$, and $v(1)$ are not collinear, and therefore $W = V$, if $1 - c_0 - c_3 \neq 0$. When $1 - c_0 - c_3 = 0$ it is not difficult to show that $\dim(W) = 1$, and therefore $\hat{\rho}(T_0|_W, T_1|_W)$ can be trivially computed in this case. However, $1 - c_0 - c_3 = 0$ is still interesting and we discuss it further below.

When $1 - c_0 - c_3 \neq 0$ we have $\dim(W) = 2$. Using the change-of-basis matrix

$$B = \begin{pmatrix} 1 & 1 & 1 \\ 1 & 0 & 0 \\ 0 & 0 & 1 \end{pmatrix},$$

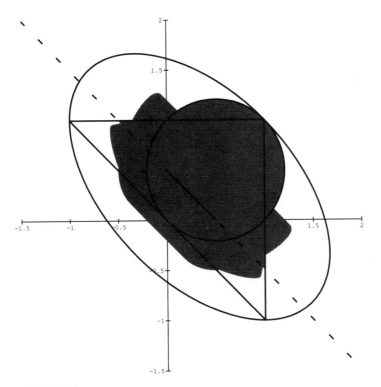

FIGURE 4.1
The (c_0, c_3)-plane, identified with real-valued, four-coefficient dilation equations satisfying
Eq. (4.4).

we find that

$$\hat{\rho}(T_0|_W, T_1|_W) = \hat{\rho}\big((BT_0B^{-1})|_{BW}, (BT_1B^{-1})|_{BW}\big) = \hat{\rho}(S_0, S_1),$$

where

$$S_0 = \begin{pmatrix} c_0 & 0 \\ -c_3 & 1 - c_0 - c_3 \end{pmatrix} \quad \text{and} \quad S_1 = \begin{pmatrix} 1 - c_0 - c_3 & -c_0 \\ 0 & c_3 \end{pmatrix}.$$

Therefore, a continuous, compactly supported solution to the dilation equation
exists if and only if $\hat{\rho}(S_0, S_1) < 1$. The "easy" cases for computing $\hat{\rho}(S_0, S_1)$ are
given in the following lemma, whose proof follows from tedious, but elementary,
algebra [ColHe2].

LEMMA 4.16

(a) S_0, S_1 *can be simultaneously upper-triangularized if and only if* $1 - c_0 - c_3 = 0$ *or* $1 - 2c_0 - 2c_3 = 0$.

(b) S_0, S_1 *can be simultaneously Hermitianized if and only if* $c_0 c_3 (1 - c_0 - c_3)(1 - 2c_0 - 2c_3) > 0$.

In either of these cases we have $\hat{\rho}(S_0, S_1) = \hat{\sigma}_1 = \max\{|c_0|, |c_3|, |1 - c_0 - c_3|\}$.

Only "half" of the (c_0, c_3)-plane satisfies one of the hypotheses of Lemma 4.16. Those points (c_0, c_3) satisfying one of those hypotheses and resulting in $\hat{\rho}(S_0, S_1) = \hat{\sigma}_1 < 1$ are shown in the shaded region of Figure 4.2. For the remaining points, $\hat{\rho}(S_0, S_1)$ must be approximated by some other method, e.g., by computing $\hat{\sigma}_m$, $\hat{\rho}_m$ or through the use of Algorithm 4.10 (see Section 4.5.2). The triangle shown in Figures 4.1–4.3 is the set $\{(c_0, c_3) : \hat{\sigma}_1 = 1\}$; continuous scaling functions are therefore restricted at least to points in the interior of the triangle. Combining Lemma 4.16 with brute-force calculations of $\hat{\sigma}_{18}$ and $\hat{\rho}_{18}$, we show in Figure 4.3 the following regions.

(a) The shaded region in Figure 4.3 is the set $\{(c_0, c_3) : \hat{\rho}_{18} < 1\}$ combined with the shaded region from Figure 4.2.

(b) The solid curve in Figure 4.3 is the set $\{(c_0, c_3) : \hat{\sigma}_{18} = 1\}$.

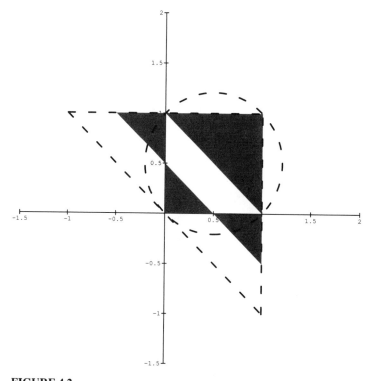

FIGURE 4.2
The subset of the (c_0, c_3)-plane where simultaneous Hermitianization occurs (shaded area).

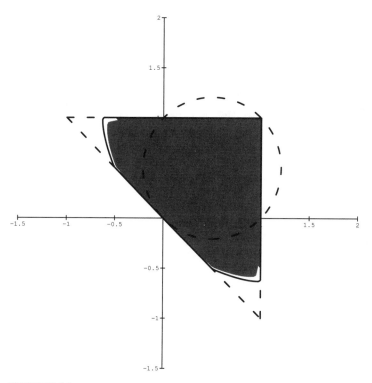

FIGURE 4.3
Numerical approximation of the region in the (c_0, c_3)-plane where continuous, compactly supported scaling functions exist (shaded area).

In particular, continuous, compactly supported scaling functions exist for all (c_0, c_3) lying within the shaded region, and cannot exist for any (c_0, c_3) lying on or outside the solid curve.

We consider now whether any of these scaling functions are differentiable. This requires that both 1 and 1/2 be eigenvalues of M, which implies $1 - c_0 - c_3 = 0$. In this case two sum rules are satisfied, so W^1 equals the one-dimensional space V^1. Therefore the scaling function f is differentiable if and only if $\hat{\rho}(T_0|_{W^1}, T_1|_{W^1}) = \max\{|c_0|, |c_3|\} < 1/2$, i.e., on the solid portion of the dashed line shown in Figure 4.1. No scaling function in the (c_0, c_3)-plane can be twice differentiable since this would require that 1, 1/2, and 1/4 all be eigenvalues of the 2×2 matrix M.

Finally, consider those dilation equations important for constructing multiresolution analyses, i.e., those lying on the circle shown in Figures 4.1–4.3. Only "half" of these have continuous solutions, and none can be differentiable. The "standard" example, the Daubechies scaling function D_4, is that unique point on the circle which satisfies two sum rules (that is, unique up to symmetry about the line $c_0 = c_3$, which corresponds to reversing the order of the coefficients in the dilation equation and time-reversing the corresponding scaling function). This point lies at the intersection of the circle and the dashed line $1 - c_0 - c_3 = 0$. The associated scaling function (shown in Figure 4.4) therefore has the highest regularity on the circle, in the sense of being able to reproduce exactly the greatest number of polynomials. However, it is not the smoothest scaling function on the circle in the sense of greatest Hölder exponent. In particular, D_4 satisfies the hypotheses of Lemma 4.16(a), so its maximum Hölder exponent is $-\log_2 \hat\rho(S_0, S_1) = -\log_2 \hat\sigma_1 \approx 0.550$ (since D_4 satisfies two sum rules, this can also be obtained more simply by applying Proposition 4.15 with $l = 0$, $n = 1$). On the circle, $-\log_2 \hat\sigma_1$ reaches its maximum value of $-\log_2 0.6 \approx 0.737$ at $(c_0, c_3) = (3/5, -1/5)$. Unfortunately, this point does not satisfy either of the hypotheses of Lemma 4.16, and this is not the maximum Hölder exponent of the corresponding scaling function (shown in Figure 4.5). This is, in fact, the example we studied in depth in Section 4.5.2, cf. Eq. (4.25). We determined there after extensive calculation that $0.659679 \leq \hat\rho(S_0, S_1) \leq 0.660025$, and therefore the maximum Hölder exponent of continuity of this scaling function is approximately $-\log_2 0.660 \approx 0.600$. We conjecture that this is the largest Hölder exponent occuring for points on the circle.

4.8.2 Symmetric seven-coefficient dilation equations

In general, the relationship between the smoothness of scaling functions and the coefficients of the dilation equation is much more complicated than is apparent from the discussion of four-coefficient dilation equations presented in the preceding subsection. This is a result of the fact that when $N = 3$ we have $\hat\rho(T_0|_W, T_1|_W) = \hat\rho(T_0|_V, T_1|_V)$ even when $W \neq V$. As V is independent of the co-

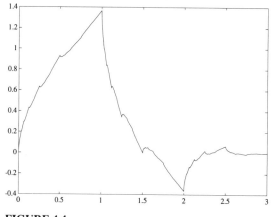

FIGURE 4.4
Daubechies scaling function D_4, corresponding to $(c_0, c_3) = ((1 + \sqrt{3})/4, (1 - \sqrt{3})/4)$.

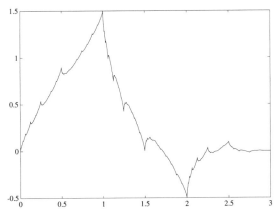

FIGURE 4.5
Scaling function corresponding to $(c_0, c_3) = (3/5, -1/5)$.

efficients, $\hat{\rho}(T_0|_W, T_1|_W)$ and the maximum Hölder exponent $-\log_2 \hat{\rho}(T_0|_W, T_1|_W)$ therefore become continuous as functions of the coefficients.

A markedly different situation can occur when $N > 3$; in particular, the space W and its dimensionality can change dramatically as the coefficients of the dilation equation vary. Since the joint spectral radius is a spectral condition and since each eigenvalue of each product of T_0, T_1 is a continuous function of the coefficients, what becomes important for the determination of $\hat{\rho}(T_0|_W, T_1|_W)$ is which eigenvalues of a product $T_{d_1} \cdots T_{d_m}$ are relevant when the product is restricted to W. In particular, when the dimension of W decreases certain eigenvalues can be rendered irrelevant, thereby changing the joint spectral radius discontinuously. The net effect is that $\hat{\rho}(T_0|_W, T_1|_W)$, and therefore also the maximum Hölder exponent, are "unstable" as functions of the coefficients. We illustrate some these ideas using symmetric seven-coefficient dilation equations.

Set $N = 6$ and suppose that the coefficients $\{c_0, c_1, c_2, c_3, c_4, c_5, c_6\}$ are real, satisfy Eq. (4.4), i.e.,

$$c_0 + c_2 + c_4 + c_6 = c_1 + c_3 + c_5 = 1,$$

and are symmetric, i.e.,

$$c_0 = c_6, \qquad c_1 = c_5, \qquad c_2 = c_4.$$

Together, these conditions reduce the number of free parameters to two, which we take as c_0 and c_1. We identify this class of dilation equation with the (c_0, c_1)-plane, i.e., each point (c_0, c_1) determines a dilation equation with coefficients $\{c_0, c_1, 1/2 - c_0, 1 - 2c_1, 1/2 - c_0, c_1, c_0\}$, and conversely. This (c_0, c_1)-plane is shown in Figure 4.6 along with several geometrical objects which we will discuss.

Since we have assumed that the coefficients to the dilation equation are symmetric, any associated scaling function will also be symmetric, i.e., $f(x) = f(6 - x)$.

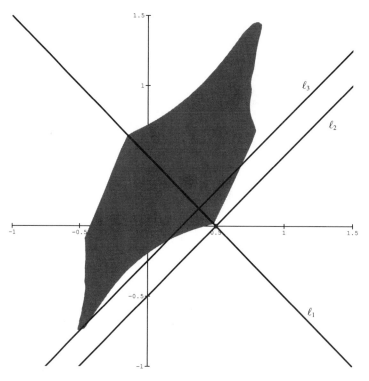

FIGURE 4.6
The (c_0, c_1)-plane, identified with symmetric, real-valued, seven-coefficient dilation
equations satisfying Eq. (4.4).

Daubechies [D1] has shown that, for real-valued coefficients, symmetry precludes
the possibility of an associated multiresolution analysis, with the single exception
of the Haar system. Thus these scaling functions cannot be used to construct
wavelet orthonormal bases for $L^2(\mathbb{R})$. However, symmetric functions can be used
to construct nonorthogonal wavelet bases [CoDF].

The shaded region in Figure 4.6 is a numerical approximation to the set of
points in the (c_0, c_1)-plane for which $\hat{\rho}(T_0|_V, T_1|_V) < 1$; in particular, the shaded
region consists of those points (c_0, c_1) such that $\hat{\rho}_{14}(T_0|_V, T_1|_V) < 1$, using the
norm $\|u\| = |u_1| + \cdots + |u_6|$. As $\hat{\rho}(T_0|_W, T_1|_W) \leq \hat{\rho}(T_0|_V, T_1|_V)$, each point in this
region has an associated continuous, compactly supported scaling function.

Consider now the point $(c_0, c_1) = (1/2, 0)$; the coefficients of the associated
dilation equation are

$$c_0 = 1/2, \quad c_1 = 0, \quad c_2 = 0, \quad c_3 = 1, \quad c_4 = 0, \quad c_5 = 0, \quad c_6 = 1/2. \quad (4.40)$$

Examination of the matrix M given by Eq. (4.16) shows that 1 is an eigenvalue with
multiplicity two. Two linearly independent right eigenvectors are, for example,

$(1, 2, 3, 2, 1)^t$ and $(0, 0, 1, 0, 0)^t$. As outlined in Section 4.4, any right eigenvector for M for the eigenvalue 1 is a possible candidate for defining a scaling vector; however, by uniqueness only one can possibly lead to an integrable solution. For example, setting $v(0) = (0, 0, 0, 1, 0, 0)^t$ we find that $W = V$, and therefore $\hat{\rho}(T_0|_W, T_1|_W) \geq 1$ since 1 is a multiple eigenvalue of the 6×6 matrices T_0 and T_1 and $\dim(V) = 5$. Therefore, by Theorem 4.11 this cannot lead to a continuous scaling vector. On the other hand, if we choose instead the vector $(0, 1, 2, 3, 2, 1)^t$ as the candidate for $v(0)$, we find that $v(x) = (x, x + 1, x + 2, 3 - x, 2 - x, 1 - x)^t$ for $x \in [0, 1]$. Thus W is one dimensional, from which one can determine that $\hat{\rho}(T_0|_W, T_1|_W) = 1/2$; by comparison, $\hat{\rho}(T_0|_V, T_1|_V) = 1$, and therefore it is not sufficient to examine only V to determine the existence of a continuous scaling vector. This scaling vector is therefore Hölder continuous with exponent 1; note this is obvious since the graph of v is a line segment. The associated scaling function is therefore also Hölder continuous with exponent 1; however, it is not differentiable. The scaling function is simply the triangle function with height 1 supported on the interval $[0, 6]$.

Note that this seven-coefficient dilation equation is simply a "stretched" version of a three-coefficient dilation equation, i.e., we obtain the coefficients in Eq. (4.40) by inserting two zeros between each of the coefficients of the $N = 2$ equation defined by $c_0 = 1/2$, $c_1 = 1$, $c_2 = 1/2$. This procedure effectively stretches the support of the solution of the original three-coefficient equation in the time domain by a factor of three. In particular, the solution to the three-coefficient equation is the symmetric triangle function with support $[0, 2]$ and height 1, so that our continuous solution to Eq. (4.40) is stretched out over the interval $[0, 6]$. Stretched dilation equations have been studied in some detail by Berger and Wang [BW2].

We show now that the stretched dilation equation corresponding to $(1/2, 0)$ lies on the boundary of the region where $\hat{\rho}(T_0|_W, T_1|_W) < 1$; this is due precisely to the fact that $\dim(W)$ changes as one moves away from this point in the (c_0, c_1)-plane. To see this, let us first examine the line ℓ_1 consisting of points of the form $(c_0, c_1) = (1/2 + \delta, -\delta)$ for $\delta \in \mathbb{R}$. This line is plotted in Figure 4.6; note that $\delta = 0$ corresponds to the stretched dilation equation $(1/2, 0)$ and therefore $\dim(W) = 1$ when $\delta = 0$. However, for $\delta \neq 0$ we can use Proposition 4.2 to show that $\dim(W) = 3$. Furthermore, in this case T_0 and T_1 can be simultaneously block upper-triangularized to the form

$$BT_iB^{-1} = \begin{pmatrix} C_i^1 & * & * & * \\ 0 & 1/2 & * & * \\ 0 & 0 & C_i^2 & * \\ 0 & 0 & 0 & 1 \end{pmatrix},$$

where each C_i^k is a 2×2 matrix, where the upper left 3×3 submatrix of BT_iB^{-1} corresponds to $T_i|_W$, and where the upper left 5×5 submatrix of BT_iB^{-1} corresponds to $T_i|_V$. Except for a small range of δ which will not affect us, each pair

C_0^k, C_1^k can be simultaneously Hermitianized; in particular, this simultaneous Hermitianization can be carried out for each δ in the interval $[-1/8, 1/8]$. Applying Lemmas 4.7 and 4.8 we find that

$$\hat{\rho}(T_0|_W, T_1|_W) = \max\left\{\hat{\rho}(C_0^1, C_1^1), \frac{1}{2}\right\} = \max\left\{\frac{1}{2}, \left|\frac{1}{2} + \delta\right|, \left|\frac{1}{2} + 2\delta\right|\right\} \quad (4.41)$$

and

$$\hat{\rho}(T_0|_V, T_1|_V) = \max\{\hat{\rho}(T_0|_W, T_1|_W), \hat{\rho}(C_0^2, C_1^2)\} = \max\{\hat{\rho}(T_0|_W, T_1|_W), |1 + 3\delta|\}. \quad (4.42)$$

In particular, note that $\hat{\rho}(T_0|_W, T_1|_W) = 1/2 < \hat{\rho}(T_0|_V, T_1|_V) < 1$ for $\delta \in [-1/8, 0)$ while $1/2 < \hat{\rho}(T_0|_W, T_1|_W) < 1 < \hat{\rho}(T_0|_V, T_1|_V)$ for $\delta \in (0, 1/8]$. Thus, although the points in the segment $[-1/8, 1/8]$ lie completely within the region $\hat{\rho}(T_0|_W, T_1|_W) < 1$ and have W as a proper subset of V, only half of the segment lies in the region $\hat{\rho}(T_0|_V, T_1|_V) < 1$. Note that the only stretched dilation equation along this segment occurs at the point $\delta = 0$.

Consider next the line ℓ_2 consisting of points of the form $(c_0, c_1) = (1/2 + \delta, \delta)$ for $\delta \in \mathbb{R}$. This line is plotted in Figure 4.6; we again have that $\delta = 0$ corresponds to the stretched dilation equation $(1/2, 0)$. Along this line, it can be shown [ColHe3] that $W = V$ whenever $\delta \neq 0$. Moreover, $\rho(T_0|_W) > 1$ for all $\delta \neq 0$ sufficiently close to zero. Therefore, there is an $\varepsilon > 0$ such that $\hat{\rho}(T_0|_W, T_1|_W) \geq \rho(T_0|_W) > 1$ for $0 < |\delta| < \varepsilon$, and so no dilation equation for these δ can have a continuous, compactly supported solution. Thus, although the point $(c_0, c_1) = (1/2, 0)$ has a continuous, compactly supported solution, an arbitrarily small change in the coefficients can result in a dilation equation without a continuous solution. In particular, the region where $\hat{\rho}(T_0|_W, T_1|_W) < 1$ is not an open set.

Now consider the line ℓ_3 consisting of those points of the form $(c_0, c_1) = (3/8 + \delta, 1/8 + \delta)$; this line is again plotted in Figure 4.6. The point $(3/8, 1/8)$, corresponding to $\delta = 0$, lies on the line ℓ_1, and so from Eqs. (4.41) and (4.42) has $\hat{\rho}(T_0|_W, T_1|_W) = 1/2$ and $\hat{\rho}(T_0|_V, T_1|_V) = 5/8$. Again it can be shown [ColHe3] that $W = V$ whenever $\delta \neq 0$. Since $\hat{\rho}(T_0|_V, T_1|_V)$ is a continuous function of the coefficients, there is an interval of points on ℓ_3 near $(3/8, 1/8)$ which have continuous, compactly supported solutions. For $\delta \neq 0$ in this interval, the maximum Hölder exponent of continuity is $-\log_2 \hat{\rho}(T_0|_V, T_1|_V)$. As δ converges to zero, this maximum Hölder exponent converges to the value of $-\log_2 \hat{\rho}(T_0|_V, T_1|_V)$ corresponding to the point $\delta = 0$, i.e., to $-\log_2 5/8$. However, the maximum Hölder exponent at the point $\delta = 0$ is $-\log_2 \hat{\rho}(T_0|_W, T_1|_W) = -\log_2 1/2 = 1$. Hence the maximum Hölder exponent is not continuous as a function of the coefficients even within the region $\hat{\rho}(T_0|_V, T_1|_V) < 1$.

As a final example, consider the point $(-1/16, 0)$, which was studied in detail in an early paper by Dubuc [Du]. This is the point in the (c_0, c_1)-plane satisfying the greatest number of sum rules (four), and will serve to illustrate the benefit of extra sum rules. In particular, since $\dim(V^3) = 2$ we can find 2×2 matrices C_0, C_1 such that $\hat{\rho}(T_0|_{V^3}, T_1|_{V^3}) = \hat{\rho}(C_0, C_1)$. Moreover, C_0, C_1 can be simultaneously Hermitianized, with the result that $\hat{\rho}(C_0, C_1) = 1/4$. It therefore follows from

Proposition 4.15 with $l = 1$, $n = 3$ that f is differentiable and f' is Hölder continuous for each exponent $0 \leq \alpha < 1$. We could prove this same result without using the two "extra" sum rules by appealing to Theorem 4.12 with $n = 1$, but this would require consideration of 4×4, rather than 2×2, matrices, since $W^1 = V^1$ is four dimensional in this case. However, this extra complexity is required if we want to determine whether f' is Hölder continuous with exponent $\alpha = 1$. In particular, after some calculation we find $\hat{\rho}(T_0|_{V^1}, T_1|_{V^1}) = 1/4$, and therefore Theorem 4.12 implies that $\alpha = 1$ is allowed if and only if $4T_0|_{V^1}, 4T_1|_{V^1}$ are product bounded. However, the eigenvalues of $T_0|_{V^1}$ are 1/4, 1/8, and $-1/16$, with 1/4 a degenerate eigenvalue of multiplicity two. Therefore, by an appropriate change of basis, $T_0|_{V^1}$ can be placed in Jordan canonical form

$$\begin{pmatrix} 1/4 & 1 & 0 & 0 \\ 0 & 1/4 & 0 & 0 \\ 0 & 0 & 1/8 & 0 \\ 0 & 0 & 0 & -1/16 \end{pmatrix},$$

whence $\|(4T_0|_{V^1})^k\| \to \infty$ as $k \to \infty$ for any choice of norm $\| \cdot \|$. Thus $4T_0|_{V^1}$, $4T_1|_{V^1}$ are not product bounded, so f' cannot be Hölder continuous for the exponent $\alpha = 1$ (in particular, f is not twice differentiable). This fact cannot be obtained by examining the smaller matrices $T_0|_{V^3}$, $T_1|_{V^3}$ since these ignore the degeneracy of the eigenvalue 1/4. In fact, $4T_0|_{V^3}$, $4T_1|_{V^3}$ are product bounded since they can be simultaneously Hermitianized. These results are the most we can obtain from the theorems we have presented. However, Dubuc [Du] proved directly that f' is "almost Lipschitz" in the sense that there exists a constant $C > 0$ such that

$$|f'(x) - f'(y)| \leq C |\log |x - y|| \, |x - y|.$$

Daubechies and Lagarias [DL2] proved that the same result follows from the product boundedness of the small matrices $4T_0|_{V^3}, 4T_1|_{V^3}$. More generally, they proved that if the hypotheses of Proposition 4.15 are fulfilled and if $\hat{\rho}(T_0|_{V^n}, T_1|_{V^n}) = 2^{-l-1}$ with $\theta^{-1}T_0|_{V^n}, \theta^{-1}T_1|_{V^n}$ product bounded for $\theta = \hat{\rho}(T_0|_{V^n}, T_1|_{V^n})$, then $f^{(l)}$ is almost Lipschitz. The results of Daubechies and Lagarias [DL2] on the local properties of functions can be used to refine even further the properties satisfied by this scaling function f.

Bibliography

[B]   M.A. Berger. 1993. "Lectures on wavelets," preprint.

[BW1]   M.A. Berger and Y. Wang. 1992. "Bounded semi-groups of matrices," *Lin. Alg. Appl.* **166**: 21–27.

[BW2]   M.A. Berger and Y. Wang. 1993. "Multi-scale dilation equations and iterated function systems," *Constr. Approx.*, to appear.

[CDaMi] A. Cavaretta, W. Dahmen, and C.A. Micchelli. 1991. "Stationary Subdivision," *Mem. Amer. Math. Soc.* **93**: 1–186.

[CoDF]   A. Cohen, I. Daubechies, and J.-C. Feauveau. 1992. "Biorthogonal bases of compactly supported wavelets," *Commun. Pure Appl. Math.* **65**: 485–560.

[ColHe1] D. Colella and C. Heil. 1992. "The characterization of continuous, four-coefficient scaling functions and wavelets," *IEEE Trans. Inf. Th., Special Issue on Wavelet Transforms and Multiresolution Signal Analysis* **38**: 876–881.

[ColHe2] D. Colella and C. Heil. 1992. "On the characterization of continuous scaling functions," Technical Report WP-91W00476, McLean, Va.: The MITRE Corporation.

[ColHe3] D. Colella and C. Heil. 1993. "Characterizations of scaling functions, I. Continuous solutions," *SIAM J. Matrix Anal. Appl.*, to appear.

[ColHe4] D. Colella and C. Heil. 1993. "Characterizations of scaling functions, II. Distributional and functional solutions," in preparation.

[D1]   I. Daubechies. 1988. "Orthonormal bases of compactly supported wavelets," *Commun. Pure Appl. Math.* **41**: 909–996.

[D2]   I. Daubechies. 1992. *Ten Lectures on Wavelets*, CBMS-NSF Reg. Conf. Ser. Appl. Math. **61**, Philadelphia: Soc. Ind. Appl. Math,

[DL1]   I. Daubechies and J. Lagarias. 1991. "Two-scale difference equations: I. Existence and global regularity of solutions," *SIAM J. Math. Anal.* **22**: 1388–1410.

[DL2]   I. Daubechies and J. Lagarias. 1992. "Two-scale difference equations: II. Local regularity, infinite products and fractals," *SIAM J. Math. Anal.* **23**: 1031–1079.

[DeDu]   G. Deslauriers and S. Dubuc. 1989. "Symmetric iterative interpolation processes," *Constr. Approx.* **5**: 49–68.

[Du]   S. Dubuc. 1986. "Interpolation through an iterative scheme," *J. Math. Anal. Appl.* **114**: 185–205.

[DyLe]   N. Dyn and D. Levin. 1990. "Interpolating subdivision schemes for the generation of curves and surfaces," in *International Series of Numerical Mathematics*, **94**, Boston: Birkhäuser, pp. 91–106.

[DyGLe] N. Dyn, J.A. Gregory, and D. Levin. 1991. "Analysis of uniform binary subdivision schemes for curve design," *Constr. Approx.* **7**: 127–147.

[E]   T. Eirola. 1992. "Sobolev characterization of solutions of dilation equations," *SIAM J. Math. Anal.* **23**: 1015–1030.

[H]   A. Haar. 1910. "Zur Theorie der orthogonalen Funktionensysteme," *Math. Ann.* **69**: 331–371.

[He]   C. Heil. 1993. "Methods of solving dilation equations," in *Proc. of the 1991 NATO ASI on Prob. and Stoch. Methods in Anal. with Appl.*, NATO Adv. Sci. Inst. Ser. C: Math. Phys. Sci., J.S. Byrnes, ed., Dordrecht: Kluwer Academic Publishers, to appear.

[HeS]   C. Heil and G. Strang. 1993. "Continuity of the joint spectral radius," in *Proc. of the Workshop on Linear Algebra for Signal Processing*, Institute for Mathematics and its Applications, to appear.

[LW]   J.C. Lagarias and Y. Wang. 1993. "The finiteness conjecture for the generalized spectral radius of a set of matrices," preprint.

[La]   W. Lawton. 1990. "Tight frames of compactly supported affine wavelets," *J. Math. Phys.* **31**: 1898–1901.

[M]   S.G. Mallat. 1989. "Multiresolution approximations and wavelet orthonormal bases for $L^2(\mathbb{R})$," *Trans. Amer. Math. Soc.* **315**: 69–87.

[Me]   Y. Meyer. 1985–1986. "Principe d'incertitude, bases hilbertiennes et algèbres d'opératers," *Séminaire Bourbaki* **662**.

[MiP1] C.A. Micchelli and H. Prautzsch. 1987. "Refinement and subdivision for spaces of integer translates of compactly supported functions," in *Numerical Analysis*, D.F. Griffith and G.A. Watson, eds., New York: Academic Press, pp. 192–222.

[MiP2] C.A. Micchelli and H. Prautzsch. 1989. "Uniform refinement of curves," *Lin. Alg. Appl.* **114/115**: 841–870.

[R1]   O. Rioul. 1993. "Simple regularity criteria for subdivision schemes," *SIAM J. Math. Anal.*, to appear.

[R2]   O. Rioul. 1993. "Simple, optimal regularity estimates for wavelets," preprint.

[RoS]   G.C. Rota and G. Strang. 1960. "A note on the joint spectral radius," *Kon. Nederl. Akad. Wet. Proc. A* **63**: 379–381.

[S]   G. Strang. 1989. "Wavelets and dilation equations: a brief introduction," *SIAM Rev.* **31**: 614–627.

[V1]   L. Villemoes. 1993. "Energy moments in time and frequency for two-scale difference equations," *SIAM J. Math. Anal.*, to appear.

[V2]   L. Villemoes. 1993. "Wavelet analysis of two-scale difference equations," preprint.

[W]   Y. Wang. 1993. "On two-scale dilation equations," preprint.

5

Remarks on the local Fourier bases

Pascal Auscher

ABSTRACT   We begin with a new, self-contained development of the theory of Wilson bases. Wilson bases are modifications of the Gabor system which can be constructed to have good time and frequency localization. An example of a Wilson basis with compact support in time is given. After a review of Malvar (or Coifman and Meyer) bases, a general notion of local Fourier bases is presented. We show that this notion includes Wilson and Malvar bases.

CONTENTS

5.1   Introduction

A famous theorem of R. Balian and F. Low ([B], [L]) tells us that no orthonormal basis for $L^2(\mathbb{R})$ of "nice" Gabor wavelets can be found. Let us recall that these functions are obtained by translation in phase-space (i.e., the time and frequency plane) of a single function g, often called a *window*:

$$g_{m,n}(x) = e^{2i\pi m\alpha x}g(x - n\beta), \quad m, n \in \mathbb{Z}, \tag{5.1}$$

where α and β are nonnegative numbers. More precisely, an orthonormal basis can only occur when $\alpha\beta = 1$ and either $xg(x) \notin L^2(\mathbb{R})$ or $\xi\hat{g}(\xi) \notin L^2(\mathbb{R})$. Here $\hat{g}(\xi)$ denotes the Fourier transform of g and is defined by $\int_{-\infty}^{+\infty} g(x)e^{2i\pi x\xi}\,dx$.

0-8493-8271-8/94/$0.00 + $.50

For example, choosing the window as the characteristic function of an interval with appropriate α and β furnishes a basis, but a smoothed version with compact support of this abrupt cut-off does not.

Several directions of research were undertaken to overcome the negative aspects of this theorem (see the introduction in [DJJo]): the theory of frames, use of irregular lattices, and the one we shall be concerned with in this note, which is probably the most striking. It can be summed up as follows.

Replace the exponential functions in $g_{m,n}$ by cosine or sine functions following some given rule so as to keep orthogonality and to allow "nice" windows.

This program has recently succeeded with two kinds of construction having distinct motivations.

The first kind arises from quantum mechanics following an idea of K. Wilson. He observed that for the study of the kinetic operator, one does not need basis functions that distinguish between positive and negative frequencies of the same order. Hence instead of using functions having one peak of localization in the Fourier domain, he used basis functions obtained from a combination of two functions each with one peak located symmetrically about the origin $\xi = 0$ of the other. In our context, this means that for each m, n, we pair $g_{m,n}$ with $g_{-m,n}$ and their combination leads us to replace the exponential function by a sine or a cosine. The construction in [Wi] and [Setal] has been dramatically simplified by I. Daubechies, S. Jaffard, and J.-L. Journé [DJJo] to yield what is now called a Wilson basis. In the first part of this work we shall reformulate the approach of the latter authors. This will bring simplifications of the proofs and yield different criteria to build such a basis. The examples obtained will present strong similarities with the constructions of the second kind.

The latter emerged from subband coding theory and are due to E. Malvar [M]. In signal processing, the block by block discrete cosine transform has been used for several years, one important problem being to cancel interactions between blocks when they are allowed to overlap. The solution proposed by Malvar is a modulated lapped transform for which the aliasing effect of subband coding is canceled and perfect reconstruction is allowed. The "Malvar bases" were independently discovered in a generalized formulation by R. Coifman and Y. Meyer [CMe].

Although these two kinds of constructions appear different, they show some "family ties;" they belong to a family of bases that we may call the "local Fourier bases." This led us to wonder whether they can be seen as particular cases of a general framework. Using the presentation in [AWeWic] of the local cosine and sine basis we show that this is actually the case and this is discussed in the second part of this chapter.

5.2 Wilson bases

5.2.1 The result of Daubechies, Jaffard, and Journé

We start by fixing the notations. From now on $\alpha = 1$ and $\beta = .5$ in Eqn. (5.1). Define a family of functions $\psi_{\ell,n}$ by

$$\begin{cases} \psi_{0,n} = g_{0,2n}, & n \in \mathbb{Z}, \\ \psi_{\ell,n} = \dfrac{1}{\sqrt{2}} (g_{\ell,n} + (-1)^{\ell+n} g_{-\ell,n}), & n \in \mathbb{Z}, \ell = 1, 2, \ldots. \end{cases} \tag{5.2}$$

Computing the resulting functions for arbitrary n leads to

$$\psi_{\ell,n}(x) = \begin{cases} g(x - n), & \text{if } \ell = 0; \\ \sqrt{2} \cos(2\pi\ell x) g \left(x - \dfrac{n}{2} \right), & \text{if } \ell > 0 \text{ and } \ell + n \text{ even}; \\ i\sqrt{2} \sin(2\pi\ell x) g \left(x - \dfrac{n}{2} \right), & \text{if } \ell > 0 \text{ and } \ell + n \text{ odd}. \end{cases} \tag{5.3}$$

Note that the number i in the third case is irrelevant; it will be omitted in the next result.

The main theorem in [DJJo] is the following.

THEOREM 5.1

Suppose that $\hat{g}$ is a real-valued function that satisfies

$$|\hat{g}(\xi)| \leq C(1 + |\xi|)^{-1-\varepsilon} \quad \text{for some } \varepsilon > 0.$$

Assume that

$$\sum_{\ell \in \mathbb{Z}} \hat{g}(\xi - \ell)\hat{g}(\xi - \ell - 2j) = \delta_{0,j}, \quad j \in \mathbb{Z}, \, \xi \in \mathbb{R}, \tag{5.4}$$

where $\delta_{p,q}$ denotes the Kronecker delta. Then, $\{\psi_{\ell,n}\}$ forms an orthonormal basis for $L^2(\mathbb{R})$, called a Wilson basis. Reciprocally, if $\{\psi_{\ell,n}\}$ forms an orthonormal basis for $L^2(\mathbb{R})$ then Eqn. (5.4) holds.

This theorem seems a little magical but has a nice interpretation in terms of frames. We first need to recall the notion of a *tight frame with redundancy* 2 as defined in [D]. The reader may find more on frames in the paper by Benedetto and Walnut in this book.

Let A be a nonnegative number. A family of vectors $\{e_k\}_{k \in \mathbb{Z}}$ in Hilbert space H is a tight frame with redundancy A if for all $f \in H$

$$\sum_{k \in \mathbb{Z}} |\langle f, e_k \rangle|^2 = A\|f\|^2.$$

Another way of saying this is that the *frame operator* defined by

$$Sf = \sum_{k \in \mathbb{Z}} \langle f, e_k \rangle e_k \tag{5.5}$$

is equal to $A I d$. If $A = 1$ with all e_k of unit length then one can easily check $\{e_k\}$ is actually an orthonormal basis of H.

In our concrete situation, it is shown in [DJJo] that $\{\psi_{\ell,n}\}$ forms an orthonormal basis for $L^2(\mathbb{R})$ if and only if $\{g_{m,n}\}$ is a tight frame with redundancy 2. The geometric interpretation of this result is that a tight frame with redundancy 2 contains twice as many vectors needed to form a basis and the linear combinations in Eqn. (5.2) eliminate this redundancy to actually yield an orthonormal basis.

We shall take this fact as a starting point for a simpler and more direct method to obtain Wilson bases. In particular Eqn. (5.4) will be replaced by an equivalent condition on g itself.

5.2.2   A simple approach

Let us denote by S the frame operator for the family $\{g_{m,n}\}$ as defined in Eqn. (5.5). We first need to make precise the sense of convergence in Eqn. (5.5) and prove the boundedness of S. We set $S_n f = \sum_{m \in \mathbb{Z}} \langle f, g_{m,n} \rangle g_{m,n}$.

PROPOSITION 5.2
Assume that $|g(x)| \leq C(1 + |x|)^{-1-\varepsilon}$ for some $\varepsilon > 0$. Then for f and $h \in L^2(\mathbb{R})$, the series $\sum_{n \in \mathbb{Z}} \langle S_n f, h \rangle$ converges absolutely. Moreover

$$\sum_{n \in \mathbb{Z}} |\langle S_n f, h \rangle| \leq C \|f\| \|h\|,$$

for a constant C independent of f and h.

This result implies the boundedness of S under this mild decay assumption on g. This result is classical and we include a proof for the convenience of the reader. Let f and h be L^2 functions and fix $n \in \mathbb{Z}$. We have

$$\int_0^1 \sum_{k=-\infty}^{+\infty} |f(x-k)\bar{g}\left(x - k - \frac{n}{2}\right)| \, dx = \int_{-\infty}^{+\infty} |f(x)\bar{g}\left(x - \frac{n}{2}\right)| \, dx < \infty.$$

Letting $g_n(x) = g\left(x - \frac{n}{2}\right)$ and $[u](x) = \sum_{k=-\infty}^{+\infty} u(x - k)$ a.e. for any $L^1(\mathbb{R})$-function $u(x)$, the above equality defines a.e. $[f\bar{g}_n](x)$ as a periodic $L^1(0,1)$-function and its mth Fourier coefficient is given by

$$\int_0^1 [f\bar{g}_n](x) e^{-2i\pi m x} \, dx = \langle f, g_{m,n} \rangle.$$

Thus, Parseval's formula gives us

$$\langle S_n f, h \rangle = \int_0^1 [f \bar{g}_n](x) \left[g_n \bar{h} \right](x) \, dx$$

$$= \int_0^1 \sum_{k,\ell} f(x-k) \bar{g} \left(x - k - \frac{n}{2} \right) g \left(x - \ell - \frac{n}{2} \right) \bar{h}(x - \ell) \, dx.$$

We now let n vary. We have not used the decay condition on g yet: it yields

$$\sum_{n=-\infty}^{+\infty} \left| \bar{g} \left(x - k - \frac{n}{2} \right) g \left((x - \ell - \frac{n}{2} \right) \right| \leq C(1 + |k - \ell|)^{-1-\varepsilon} \equiv \omega(k - \ell)$$

uniformly in $x \in [0, 1]$. It follows that

$$\Sigma = \sum_{n \in \mathbb{Z}} |\langle S_n f, h \rangle| \leq C \int_0^1 \sum_{k,\ell} |f(x-k)| \omega(k - \ell) |h(x - \ell)| \, dx.$$

Exchanging sums and integral and using the Cauchy–Schwarz inequality for each integral, we obtain

$$\Sigma \leq C \sum_{k,\ell} \left(\int_0^1 |f(x-k)|^2 \, dx \right)^{1/2} \omega(k - \ell) \left(\int_0^1 |h(x - \ell)|^2 \, dx \right)^{1/2}$$

$$\leq C \left(\sum_k \int_0^1 |f(x-k)|^2 \, dx \right)^{1/2} \left(\sum_\ell \int_0^1 |h(x - \ell)|^2 \, dx \right)^{1/2}$$

$$= C \|f\| \|h\|,$$

the last inequality being a consequence of the simple fact that the discrete convolution with the sequence $\{\omega(k)\}$ is bounded on $\ell^2(\mathbb{Z})$.

We can also introduce the frame operator for the family $\{\psi_{\ell,n}\}$, which we call R:

$$Rf = \sum_{\ell,n} \langle f, \psi_{\ell,n} \rangle \psi_{\ell,n},$$

where (ℓ, n) ranges through $\mathbb{N} \times \mathbb{Z}$. A similar, but slightly more involved argument, which we skip, applies to R under the same condition on g and shows that R is a well-defined operator acting boundedly on $L^2(\mathbb{R})$.

Our goal is to show that $R = Id$ whenever $S = 2Id$. It is, therefore, natural to compare $2R$ and S. Letting $T = 2R - S$, we have

LEMMA 5.3

$$Tf = \sum_{m,n} (-1)^{m+n} \langle f, g_{m,n} \rangle g_{-m,n}, \tag{5.6}$$

where (m, n) ranges through $\mathbb{Z} \times \mathbb{Z}$.

PROOF   Let $f \in L^2(\mathbb{R})$,

$$
2Rf = \sum_{n=-\infty}^{+\infty} \sum_{\ell=1}^{+\infty} \langle f, (g_{\ell,n} + (-1)^{\ell+n} g_{-\ell,n}) \rangle (g_{\ell,n} + (-1)^{\ell+n} g_{-\ell,n})
$$

$$
+ 2 \sum_{n=-\infty}^{+\infty} \langle f, g_{0,2n} \rangle g_{0,2n}
$$

$$
= \sum_{n=-\infty}^{+\infty} \sum_{\ell \neq 0} \langle f, g_{\ell,n} \rangle g_{\ell,n} + \sum_{n=-\infty}^{+\infty} \sum_{\ell \neq 0} (-1)^{\ell+n} \langle f, g_{\ell,n} \rangle g_{-\ell,n}
$$

$$
+ 2 \sum_{n=-\infty}^{+\infty} \langle f, g_{0,2n} \rangle g_{0,2n},
$$

and this equals $Sf + Tf$. Isolating Sf in the last expression yields the desired formula for Tf.   ∎

The next result establishes a way for computing S and T.

PROPOSITION 5.4
Assume that $|g(x)| \leq C(1 + |x|)^{-1-\varepsilon}$ for some $\varepsilon > 0$. Then, for $f, h \in L^2(\mathbb{R})$,

(i)
$$
\langle Sf, h \rangle = \sum_{k=-\infty}^{+\infty} \int_{-\infty}^{+\infty} f(x - k) \Delta_k(x) \bar{h}(x) \, dx,
$$

where $\Delta_k(x) = \sum_{n=-\infty}^{+\infty} \bar{g}(x - k - n/2) g(x - n/2)$;

(ii)
$$
\langle Tf, h \rangle = \sum_{k=-\infty}^{+\infty} \int_{-\infty}^{+\infty} f\left(x - k - \frac{1}{2}\right) E_k(x) \bar{h}(-x) \, dx,
$$

where $E_k(x) = \sum_{n=-\infty}^{+\infty} (-1)^n \bar{g}(x - k - n/2 - 1/2) g(-x - n/2)$.

Part (i) of this proposition was derived by D. Walnut [W] with a weaker hypothesis on g.

PROOF   Assume f and h to be Schwartz class functions so that all calculations make sense (the result for L^2 functions follows by density). The same argument as in Proposition 5.2 gives

$$
\langle Sf, h \rangle = \sum_{n=-\infty}^{+\infty} \int_0^1 \sum_{k,\ell} f(x - k) \bar{g}\left(x - k - \frac{n}{2}\right) g\left(x - \ell - \frac{n}{2}\right) \bar{h}(x - \ell) \, dx.
$$

Unfolding the sum over ℓ yields an integral over $\mathbb{R}$ and then exchanging the sum over k with the sum over n gives (i).

An examination of T and focusing on the factor $(-1)^{n+m}$ in Eqn. (5.6) leads us to separate the even m's from the odd ones; so this time it is natural to use a periodization argument with period 1/2. For u an $L^1(\mathbb{R})$-function we set $\{u\}(x) = \sum_{k=-\infty}^{+\infty} u(x - k/2)$.

Let

$$A_n = \sum_{m=-\infty}^{+\infty} (-1)^m \langle f, g_{m,n} \rangle \langle g_{-m,n}, h \rangle = \sum_{m \text{ even}} - \sum_{m \text{ odd}} \langle f, g_{m,n} \rangle \langle g_{-m,n}, h \rangle.$$

Writing $m = 2r$,

$$\sum_{m \text{ even}} = \sum_{r=-\infty}^{+\infty} \int_0^{1/2} \{ f \bar{g}_n \}(x) e^{-4i\pi r x} \, dx \int_0^{1/2} \{ g_n \bar{h} \}(y) e^{-4i\pi r y} \, dy.$$

Changing y to $-y$ and using Parseval's formula for periodic functions with period 1/2, we obtain

$$\sum_{m \text{ even}} = \frac{1}{2} \int_0^{1/2} \sum_{k,\ell} f\left(x - \frac{k}{2}\right) \bar{g}\left(x - \frac{k}{2} - \frac{n}{2}\right)$$

$$g\left(-x - \frac{\ell}{2} - \frac{n}{2}\right) \bar{h}\left(-x - \frac{\ell}{2}\right) dx.$$

Similarly, for $m = 2r + 1$,

$$\sum_{m \text{ odd}} = \frac{1}{2} \int_0^{1/2} \sum_{k,\ell} f\left(x - \frac{k}{2}\right) \bar{g}\left(x - \frac{k}{2} - \frac{n}{2}\right) e^{-2i\pi(x - k/2)}$$

$$\cdot e^{-2i\pi(-x - \ell/2)} g\left(-x - \frac{\ell}{2} - \frac{n}{2}\right) \bar{h}\left(-x - \frac{\ell}{2}\right) dx.$$

In each of these two integrals we unfold the sum over ℓ and adding up we obtain

$$A_n = \int_{-\infty}^{+\infty} \sum_{k \text{ odd}} f\left(x - \frac{k}{2}\right) \bar{g}\left(x - \frac{k}{2} - \frac{n}{2}\right) g\left(-x - \frac{n}{2}\right) \bar{h}(-x) \, dx.$$

We conclude by noticing that $\langle Tf, g \rangle = \sum_{-\infty}^{+\infty} (-1)^n A_n$ and letting $k = 2k' + 1$. This proves (ii). ∎

We now are ready to state the first result.

THEOREM 5.5
Assume that $|g(x)| \leq C(1 + |x|)^{-1-\varepsilon}$ for some $\varepsilon > 0$. Then, $\{\psi_{\ell,n}\}$ forms an orthonormal basis for $L^2(\mathbb{R})$ if and only if the two following conditions are fulfilled:

$$\Delta_k(x) = 2\delta_{0,k} \; a.e., \quad \forall k \in \mathbb{Z}; \tag{5.7a}$$

$$E_k(x) = 0 \; a.e., \quad \forall k \in \mathbb{Z}. \tag{5.7b}$$

REMARK Under the assumption for g, it can be checked that $E_k(x)$ and $\Delta_k(x)$ are bounded functions. Moreover, the latter is periodic with period 1/2. ∎

PROOF Suppose that Eqn. (5.7) holds. Lemma 5.3 and Proposition 5.4 yield

$$\langle Rf, h \rangle = \langle f, h \rangle,$$

which means that $R = Id$. Hence, $\{\psi_{\ell,n}\}$ forms an orthonormal basis for $L^2(\mathbb{R})$ provided the L^2 norm of each $\psi_{\ell,n}$ is 1. We get

$$\|\psi_{0,n}\|^2 = \|g\|^2 = \int_0^{1/2} \Delta_0(x)\,dx,$$

and, if $\ell \neq 0$,

$$\|\psi_{\ell,n}\|^2 = \|g\|^2 + (-1)^{\ell+n}\Re\langle g_{\ell,n}, g_{-\ell,n}\rangle = \|g\|^2 + (-1)^{\ell+n}\Re\int_0^{1/2} \Delta_0(x)e^{4i\pi\ell x}\,dx.$$

Therefore, they are unit vectors if $\Delta_0(x) = 2$, which is the case here.

Conversely, assume that $\{\psi_{\ell,n}\}$ forms an orthonormal basis for $L^2(\mathbb{R})$, we have $R = Id$ and, therefore,

$$2\langle f, h \rangle = \sum_{k=-\infty}^{+\infty} \int_{-\infty}^{+\infty} f(x-k)\Delta_k(x)\bar{h}(x)\,dx$$

$$+ \sum_{k=-\infty}^{+\infty} \int_{-\infty}^{+\infty} f\left(x - k - \frac{1}{2}\right) E_k(x)\bar{h}(-x)\,dx, \qquad (5.8)$$

for all f and h in the Schwartz class. Specific choices for these functions lead to Eqn. (5.7).

Since $\Delta_q(x)$ is 1/2-periodic, it suffices to prove $\Delta_q(x) = 2\delta_{0,q}$ a.e. for $x \in [q-1/4, q+1/4]$. Assume that h has support in this interval and that f is supported in $[-1/4, 1/4]$, but are otherwise arbitrary. Then Eqn. (5.8) reduces to the equality

$$\int_{q-1/4}^{q+1/4} f(x)\bar{h}(x)\,dx = \frac{1}{2} \int_{q-1/4}^{q+1/4} f(x-q)\Delta_q(x)\bar{h}(x)\,dx.$$

Notice that the integral on the left is 0 when $q \neq 0$ because of the support assumption on f and h. Letting these functions range over all the possible choices as above implies Eqn. (5.7a).

The proof of Eqn. (5.7b) is in the same spirit and makes use of the following relation:

$$E_{k-q}(x) = (-1)^q E_k\left(x + \frac{q}{2}\right), \qquad k, q \in \mathbb{Z}, x \in \mathbb{R}.$$

We leave the details to the reader. ∎

5.2.3 Examples with compact support

The examples in [DJJo] have exponential fall-off in both position space and momentum space. We shall derive here examples with compact support in position space directly from the above conditions.

The first remark is that if $g(-x) = \bar{g}(x)$, or, equivalently, if $\hat{g}$ is real valued, then E_k vanishes identically as easily seen from the change of index, $-n' = n + 2k + 1$, in the series defining E_k. No support condition is needed here and we are left with checking Eqn. (5.7a).

Assume that g is even and real valued and, furthermore, that g has support in $[-1/2, 1/2]$. Then, $\Delta_k = 0$ identically for $k \neq 0$. Moreover, the relation $\Delta_0 = 2$ reduces to

$$g^2 \left(x - \frac{1}{2} \right) + g^2(x) = 2, \quad x \in \left[0, \frac{1}{2} \right]. \tag{5.9}$$

It is then very easy to produce examples. Let $\theta(x)$ be an odd (increasing) function such that $\theta(x) = \pi/4$ for $x \geq 1/4$. Let $s(x) = \sin(\theta(x) + \pi/4)$ and $c(x) = \cos(\theta(x) + \pi/4)$. We define g by

$$g(x) = \sqrt{2} s \left(x + \frac{1}{4} \right) c \left(x - \frac{1}{4} \right). \tag{5.10}$$

We claim that all the required conditions are met for this choice. Indeed, we have the following properties:

1. $s(-x) = c(x)$ because of the oddness of $\theta(x)$;

2. $c(x)$ has support in $(-\infty, 1/4]$;

3. $s(x) = 1$ for $x \geq 1/4$.

Hence, properties 1 and 2 imply that g is an even real-valued function with support in $[-1/2, 1/2]$ and Eqn. (5.9) follows from properties 1 and 3 and the identity $\cos^2 u + \sin^2 u = 1$. Of course, g is a C^∞-function provided θ is such. Summarizing we have

THEOREM 5.6
There exists a real-valued compactly supported C^∞-function g such that the collection of functions

$$\psi_{\ell,n}(x) = \begin{cases} g(x - n), & \text{if } n \in \mathbb{Z} \text{ and } \ell = 0; \\ \sqrt{2} \cos(2\pi \ell x) g \left(x - \dfrac{n}{2} \right), & \text{if } \ell > 0, \ n \in \mathbb{Z} \text{ and } \ell + n \text{ even}; \\ \sqrt{2} \sin(2\pi \ell x) g \left(x - \dfrac{n}{2} \right), & \text{if } \ell > 0, \ n \in \mathbb{Z} \text{ and } \ell + n \text{ odd}, \end{cases}$$

forms an orthonormal basis for $L^2(\mathbb{R})$. Such a function is defined by Eqn. (5.10) and has support in $[-1/2, 1/2]$.

5.2.4 Symmetry properties

The statements of Theorem 5.1 and of Theorem 5.5 suggest that Eqns. (5.4) and (5.7a) should be equivalent. This is actually the case and this relates our method to the one used in [DJJo].

PROPOSITION 5.7

Let g be an L^2 function and $h = \hat{g}$. Then the following are equivalent:

$$\Delta_k(x) = \sum_{n=-\infty}^{+\infty} \bar{g}\left(x - k - \frac{n}{2}\right) g\left(x - \frac{n}{2}\right) = 2\delta_{0,k} \ a.e., \quad \forall k \in \mathbb{Z}; \quad (5.11a)$$

$$D_m(\xi) = \sum_{\ell=-\infty}^{+\infty} \bar{h}(\xi - \ell)h(\xi - \ell - 2m) = \delta_{0,m} \ a.e., \quad \forall m \in \mathbb{Z}. \quad (5.11b)$$

PROOF   Note that $\Delta_k(x)$ and $D_m(\xi)$ are periodic locally integrable functions. We compute their Fourier coefficients. Using the Parseval identity for Fourier integrals we obtain the following equalities:

$$\int_0^{1/2} \Delta_k(x)e^{-4i\pi mx} \, dx = \int_{-\infty}^{+\infty} \bar{g}(x - k)g(x)e^{-4i\pi mx} \, dx$$

$$= \int_{-\infty}^{+\infty} \bar{h}(\xi)e^{-2i\pi k\xi}h(\xi - 2m) \, d\xi = \int_0^1 D_m(\xi)e^{-2i\pi k\xi} \, d\xi.$$

This readily gives the desired equivalence.   ∎

This result also suggests an amusing symmetry property of Wilson bases.

PROPOSITION 5.8

Suppose that the functions g and $\hat{g}$ are real valued and satisfy

$$|g(x)| + |\hat{g}(x)| \le C(1 + |x|)^{-1-\varepsilon}.$$

Then, the following are equivalent:

i. *The collection of functions*

$$\begin{cases} g(x - n), & \text{if } n \in \mathbb{Z}; \\ \sqrt{2}\cos(2\pi\ell x)g\left(x - \dfrac{n}{2}\right), & \text{if } \ell > 0, \ n \in \mathbb{Z} \text{ and } \ell + n \text{ even}; \\ \sqrt{2}\sin(2\pi\ell x)g\left(x - \dfrac{n}{2}\right), & \text{if } \ell > 0, \ n \in \mathbb{Z} \text{ and } \ell + n \text{ odd} \end{cases}$$

forms an orthonormal basis for $L^2(\mathbb{R})$.

ii. *The collection of functions*

$$\begin{cases} \hat{g}(x - 2\ell), & \text{if } \ell \in \mathbb{Z}; \\ \sqrt{2}\cos(\pi nx)\hat{g}(x - \ell), & \text{if } n > 0, \ \ell \in \mathbb{Z} \text{ and } n + \ell \text{ even}; \\ \sqrt{2}\sin(\pi nx)\hat{g}(x - \ell), & \text{if } n > 0, \ \ell \in \mathbb{Z} \text{ and } n + \ell \text{ odd} \end{cases}$$

forms an orthonormal basis for $L^2(\mathbb{R})$.

The proof relies on the following observation. Since the Fourier transform is unitary (with our definition in the introduction, $\|f\| = \|\hat{f}\|$), a collection $\{e_k\}$ is a tight frame with redundancy A if and only if the same is true of $\{\hat{e}_k\}$.

Assume that (i) holds. Applying this remark to the family $\{g_{m,n}\}$, we see that the collection $\{e^{i\pi n\cdot}\hat{g}(\cdot - m)\}$ is a tight frame with redundancy 2. Then, we may use the same tricks to eliminate redundancy and Eqn. (5.11b) is exactly the condition that guarantees the collection (ii) to be an orthonormal basis. The converse is treated similarly.

5.3 A general framework for Wilson bases and Malvar bases

We shall quickly review the construction of the Malvar basis using the "smooth projections over an interval," as presented in great detail in [AWeWic], and the reader is referred there for detailed proofs. The only novelty is the observation that Wilson bases enter in that framework. This is presented at the end of this section. All functions here are real valued.

5.3.1 The family of smooth projections

i. Bell over an interval.

Let $\varepsilon > 0$ and $\theta_\varepsilon(x)$ be an odd (and increasing) function which takes the value $\pi/4$ for $x \geq \varepsilon$. Let

$$c_\varepsilon(x) = \cos\left(\theta_\varepsilon(x) + \frac{\pi}{4}\right) \tag{5.12}$$

and

$$s_\varepsilon(x) = \sin\left(\theta_\varepsilon(x) + \frac{\pi}{4}\right). \tag{5.13}$$

Let $I = [\alpha, \beta]$ be an interval and associated with the interval, two nonnegative numbers, ε and ε', subject to the condition $\alpha + \varepsilon \leq \beta - \varepsilon'$. Then, a *bell over I* is defined by

$$b_I(x) = s_\varepsilon(x - \alpha)c_{\varepsilon'}(x - \beta), \quad x \in \mathbb{R}. \tag{5.14}$$

This function depends on the particular choice of ε and ε' but we do not exhibit this dependence in our notation. Its support is the interval $[\alpha - \varepsilon, \beta + \varepsilon']$. It may be observed that $g(x)$, defined by Eqn. (5.10), is a bell over $[-1/4, 1/4]$ up to a multiplicative factor $\sqrt{2}$.

ii. The 4 smooth projections over I and polarities.

Let b_I be a bell over I. The operators defined by

$$(P_I f)(x) = b_I^2(x)f(x) \pm b_I(x)b_I(2\alpha - x)f(2\alpha - x)$$
$$\pm b_I(x)b_I(2\beta - x)f(2\beta - x) \tag{5.15}$$

are orthogonal projections. Roughly speaking, P_I is a smooth version of the operator of pointwise multiplication by the characteristic function of I. We have defined, in fact, 4 operators depending on the choice of signs at each endpoint, which we call the *polarities* of P_I or of I itself. If one of the endpoints is infinite then the corresponding term in Eqn. (5.15) disappears.

iii. *Compatible adjacent intervals and merging properties.*

Shortening a little the terminology in [AWeWic], we say that two adjacent intervals $I = [\alpha, \beta]$ and $J = [\beta, \gamma]$ are *compatible* if the bells are given by

$$b_I(x) = s_\varepsilon(x - \alpha)c_{\varepsilon'}(x - \beta), \qquad b_J(x) = s_{\varepsilon'}(x - \beta)c_{\varepsilon''}(x - \gamma),$$

with the same ε'.

Due to the properties of the functions $c_\varepsilon(x)$ and $s_\varepsilon(x)$ the bells of compatible intervals merge to yield a bell over the union:

$$b_I^2(x) + b_J^2(x) = b_{I \cup J}^2(x).$$

This implies the following results.

PROPOSITION 5.10
Suppose that $I = [\alpha, \beta]$ and $J = [\beta, \gamma]$ are compatible intervals and P_I, P_J have opposite polarity at β. Then, the projections are orthogonal to each other and $P_I + P_J = P_{I \cup J}$.

COROLLARY 5.11
Let $\mathcal{I} = \{I\}$ be a cover of the real line by adjacent compatible intervals. Let $\{P_I\}$ be the associated family of projections. If all pairs of projections associated with adjacent intervals in $\mathcal{I}$ have opposite polarities at their common endpoint, then $L^2(\mathbb{R})$ is the orthogonal direct sum of the spaces $\mathcal{H}_I = P_I L^2(\mathbb{R})$:

$$L^2(\mathbb{R}) = \bigoplus_{I \in \mathcal{I}} \mathcal{H}_I.$$

iv. *Description of the range spaces.*

Let $\alpha \in \mathbb{R}$. A function $f(x)$ has, by definition, polarity + (resp. $-$) at α if its graph is symmetric with respect to the vertical line $x = \alpha$ (resp. about the point $(\alpha, 0)$), i.e., $f(2\alpha - x) = f(x)$ (resp. $f(2\alpha - x) = -f(x)$).

PROPOSITION 5.12
Let I be an interval. Then, $f \in \mathcal{H}_I$ if and only if $f = b_I S$ where S is a function in $L^2(\mathbb{R})$ having the same polarity as I at its end points. Moreover, if $f_1 = b_I S_1 \in \mathcal{H}_I$ and $f_2 = b_I S_2 \in \mathcal{H}_I$, then

$$\int_{-\infty}^{+\infty} f_1(x) f_2(x) \, dx = \int_\alpha^\beta S_1(x) S_2(x) \, dx. \tag{5.16}$$

Equation (5.16) tells us that $S_I(f) = S|_I$, the restriction of S to I, is an isometry from $\mathcal{H}_I$ into $L^2(I)$; it is also onto. The inverse transformation is defined, for $S \in L^2(I)$, by $S_I^{-1}(S) = b_I S^I$, where S^I is the extension of S outside of I that has the same polarity as I at its end points.

This description is the key that allows a further analysis of $\mathcal{H}_I$ in terms of an orthonormal basis using orthogonal Fourier series expansions in $L^2(I)$.

5.3.2 Local Fourier bases

i. Fourier bases with polarity $(\pm, \pm)$ for $L^2(I)$.

For an interval $I = [\alpha, \beta]$, set $\hat{I}$ the interval of length $4|I|$ ($|I|$ denoting the length of I) and with center α. Since b_I has support contained in $\hat{I}$ we can assume, furthermore, that S^I (see above) is extended periodically outside of $\hat{I}$, with period $4|I|$: this does not change $b_I S^I$.

The function S^I can then be expanded as a Fourier series in the basis (for $L^2(\hat{I})$):

$$\mathcal{B} = \left\{ \sqrt{\frac{1}{|I|}}, \sqrt{\frac{2}{|I|}} \cos k \frac{\pi}{2|I|}(x - \alpha), \sqrt{\frac{2}{|I|}} \sin k \frac{\pi}{2|I|}(x - \alpha) \right\}, \quad k = 1, 2, \dots.$$

Notice that normalization has been so chosen that these functions are unit vectors in $L^2(I, dx)$. All these functions also have 1 of the 4 possible polarities at the endpoints of I. We split the collection $\mathcal{B}$ accordingly to obtain 4 subcollections, the *Fourier bases with polarity* $(\pm, \pm)$. Each is an orthonormal basis for $L^2(I)$.

ii. The local Fourier bases over an interval.

Let $I = [\alpha, \beta]$ be an interval with its polarity, $(-, +)$ say. The unitary operator S_I^{-1} carries the Fourier basis with polarity $(-, +)$ into an orthonormal basis for $\mathcal{H}_I$:

$$\left\{ \sqrt{\frac{2}{|I|}} b_I(x) \sin \frac{2k+1}{2} \frac{\pi}{|I|}(x - \alpha) \right\}, \quad k = 0, 1, 2, \dots.$$

This is the $(-, +)$-*local Fourier basis over I*. To obtain the others we consider the 3 different choices of polarity and, for each, the Fourier basis with the same polarity.

The $(+, -)$-local Fourier basis over I is:

$$\left\{ \sqrt{\frac{2}{|I|}} b_I(x) \cos \frac{2k+1}{2} \frac{\pi}{|I|}(x - \alpha) \right\}, \quad k = 0, 1, 2, \dots.$$

The $(+, +)$-local Fourier basis over I is:

$$\left\{ \sqrt{\frac{1}{|I|}} b_I(x), \sqrt{\frac{2}{|I|}} b_I(x) \cos k \frac{\pi}{|I|}(x - \alpha) \right\}, \quad k = 1, 2, \dots.$$

And the $(-,-)$-local Fourier basis over I is:

$$\left\{ \sqrt{\frac{2}{|I|}}\, b_I(x) \sin k \frac{\pi}{|I|}(x-\alpha) \right\}, \quad k = 1, 2, \dots.$$

5.3.3 Local Fourier bases for $L^2(\mathbb{R})$: Malvar basis and Wilson basis

Using Corollary 5.12 and the above description of local Fourier bases over an interval, we obtain a wide variety of *local Fourier bases for $L^2(\mathbb{R})$*. The ingredients are a covering of the real line by adjacent compatible intervals and, for each interval of the covering, the choice of a local Fourier basis with according polarity.

Let us distinguish two types of bases. The first type is when all intervals have different polarity at their end points. It may be $(-,+)$ for all, or $(+,-)$ for all. In the $(-,+)$-case, we obtain a local sine basis and in the $(+,-)$-case, we obtain a local cosine basis. These bases are the Malvar bases (in the formulation of Coifman and Meyer; Malvar constructs the discrete analogues where the basis functions are sampled at appropriate rates).

In the second type, all intervals have the same polarity at their end points. Then the sequence of pairs of polarity necessarily alternates $(+,+)$ and $(-,-)$. We claim that the Wilson basis of Theorem 5.6 is an example of this type of local Fourier basis.

Let $I_n = [-1/4 + n/2, 1/4 + n/2]$, $n \in \mathbb{Z}$. We associate the polarity $(+,+)$ to the I_n's with n even and the polarity $(-,-)$ to the I_n's with n odd. We obtain a suitable covering of $\mathbb{R}$ with adjacent compatible intervals.

The first observation is that the bells, b_n, over I_n are necessarily of the form $b_n(x) = b_0 \left(x - \frac{n}{2}\right)$. Let us now have a close look at the local Fourier basis over I_n. By translation invariance, we can restrict the study to the $n = 0$ and $n = 1$ cases.

For $n = 0$, we have the identities

$$2b_0(x) \cos 2k\pi \left(x + \frac{1}{4}\right) = 2b_0(x) \begin{cases} (-1)^{[k/2]} \cos 2k\pi x, & \text{if } k \text{ is even;} \\ (-1)^{[k/2]+1} \sin 2k\pi x, & \text{if } k \text{ is odd.} \end{cases}$$
$$(5.17)$$

And for $n = 1$:

$$2b_1(x) \sin 2k\pi \left(x - \frac{1}{4}\right) = 2b_0 \left(x - \frac{1}{2}\right) \begin{cases} (-1)^{[k/2]} \sin 2k\pi x, & \text{if } k \text{ is even;} \\ (-1)^{[k/2]+1} \cos 2k\pi x, & \text{if } k \text{ is odd.} \end{cases}$$
$$(5.18)$$

We have put $[u]$ the integer part of u.

In Eqns. (5.17) and (5.18) the term in the left hand side is an element of the Fourier local basis associated with the covering by the intervals I_n. Letting $g(x) = \sqrt{2}b_0(x)$ we recognize on the right hand side, up to powers of -1, elements of the Wilson basis in Theorem 5.6. A simple change of index in Eqn. (5.3) shows that, up to unimodular multiplicative constants, the two bases are the same.

Had we used the same covering with, this time, polarity $(-, -)$ for I_{2n} and polarity $(+, +)$ for I_{2n+1}, we would have obtained a Wilson type basis which can also be obtained by changing $(-1)^{\ell+n}$ to $(-1)^{\ell+n+1}$ and $g_{0,2n}$ to $g_{0,2n+1}$ in Eqn. (5.2).

5.4 Concluding remarks

We would like to point out that E. Laeng [La] constructed in his thesis a family of bases of the Wilson type. He imposed that the frequency domain is covered by a family of symmetric intervals not necessarily of the same size (observe that frequency domain and time domain can be interchanged in all these constructions; cf. Proposition 5.8, for example). It turns out that his construction can also be seen as what we just called a local Fourier basis of the second type.

Another remark is that this second type of local Fourier basis can also be seen as a refinement of the local cosine (or sine) basis; split each interval I of the accompanying covering in two pieces following the sign rule: $(+, -)$ for I gives $(+, +)$ for the left part and $(-, -)$ for the right part (or $(-, +)$ for I gives $(-, -)$ and $(+, +)$) and use Proposition 5.10. This means that local cosine (sine) bases are in some sense generic for all this collection of bases, the others being "variations on the theme."

Eventually, since this book is mostly devoted to the wavelet theory, we feel that some words on the connection of the topic described here to wavelets are necessary. There are strong relations between local Fourier bases and wavelet bases, more precisely the wavelet bases of Lemarié and Meyer can be obtained from some local Fourier bases. The calculations showing this are explicitly made in [AWeWic]. A recent work by A. Bonami, F. Soria and G. Weiss [BoSoWe] shows that in some sense the converse holds, i.e., the conditions to construct orthonormal bases of "band-limited" wavelets imply those satisfied by local Fourier bases.

Bibliography

[AWeWic] P. Auscher, G. Weiss, and M.V. Wickerhauser. 1992. "Local sine and cosine basis of Coifman and Meyer and the construction of smooth wavelets," in *Wavelets: A Tutorial in Theory and Applications*, C.K. Chui ed., Boston: Academic Press, pp. 237–256.

[B] R. Balian. 1981. Un principe d'incertitude fort en théorie du signal ou en mecanique quantique, *C. R. Acad. Sci.*, sér. 2 **292**(20): 1357–1362.

[BoSoWe] A. Bonami, F. Soria, and G. Weiss. 1993. "Band-limited wavelets," preprint, Washington University.

[CMe]   R.R. Coifman and Y. Meyer. 1991. "Remarques sur l'analyse de Fourier à fenêtre," *C. R. Acad. Sci.*, sér. 1 **312**: 259–261.

[D]   I. Daubechies. 1990. "The wavelet transform, time-frequency localization and signal analysis," *IEEE Trans. Inform. Theory* **36**: 961–1005.

[DJJo]   I. Daubechies, S. Jaffard, and J.-L. Journé. 1991. "A simple Wilson orthonormal basis with exponential decay," *SIAM J. Math. Anal.* **22**: 554–572.

[L]   F. Low. 1985. "Complete sets of wave-packets," in *A Passion for Physics: Essays in Honor of Geoffrey Chew*, Singapore: World Scientific, pp. 17–22.

[La]   E. Laeng. 1990. "Une base orthonormale de $L^2(\mathbb{R})$ dont les éléments sont bien localisés dans l'espace de phase et leurs supports adaptés à toute partition symétrique de l'espace des fréquences," *C. R. Acad. Sci.*, sér. 2 **311**: 677–680.

[M]   H. Malvar. 1990. "Lapped transform for efficient transform/subband coding," *IEEE Trans. Acoust. Speech Signal Process.* **38**: 969–978.

[Setal] D.J. Sullivan, J.J. Rehr, J.W. Wilkins, and K.G. Wilson. 1987. "Phase-space Wannier functions in electronic structure calculations," preprint, Cornell University.

[W]   D. Walnut. 1992. "Continuity properties of the Gabor frame operator," *J. Math. Anal. Appl.* **165**: 479–504.

[Wi]   K.G. Wilson. 1987. "Generalized Wannier functions," preprint, Cornell University.

Part II

Wavelets and Signal Processing

6

The sampling theorem, φ-transform, and Shannon wavelets for $\mathbb{R}$, $\mathbb{Z}$, $\mathbb{T}$, and $\mathbb{Z}_N$

Michael Frazier and Rodolfo Torres[*]

ABSTRACT   We present some sampling theorems for functions with Fourier transform supported in a B-transversal set. These results generalize the classical Shannon sampling theorem. As applications, we obtain the φ-transform identity and a simple orthonormal wavelet system called the Shannon wavelets. We obtain these results in each of the following settings: the real line, the integers, the circle, and the integers modulo N.

CONTENTS

[*]The first author was partially supported by NSF Grant DMS-8705935. The second author was partially supported by NSF Grant DMS-8806727.

6.1   Introduction

The classical Shannon sampling theorem ([S]) is a fundamental result in signal analysis. We develop some of its generalizations, based on the notion of transversality. From these sampling results, we obtain the φ-transform representation of functions ([FrJa1], [FrJa2]) and a simple example of an orthonormal wavelet basis, the Shannon basis ([FrJaW], [FrKu]). We shall present this sequence of ideas in the settings of main interest for engineering: the real line $\mathbb{R}$ (Section 6.2), the integers $\mathbb{Z}$ (Section 6.3), the circle $\mathbb{T} = \mathbb{R}/2\pi\mathbb{Z}$ which we identify with $[-\pi, \pi)$ as usual (Section 6.4), and the integers modulo N; $\mathbb{Z}_N = \{n \in \mathbb{Z} : -N/2 \leq n \leq N/2 - 1\}$ for N some power of 2 (Section 6.5).

The notation is adapted to each setting; e.g., the Fourier transform symbol "$\wedge$" has a different meaning in each section. The arguments in Sections 6.2–6.5, although parallel, are presented independently. However, some details in later sections, which are analogous to those earlier, are omitted. For simplicity we consider only one dimension, although there are straightforward generalizations to higher dimensions.

This chapter is designed partially as a tutorial introduction to sampling theory, giving background for the two subsequent chapters of this volume ([B2], [FGr]). Most (perhaps all) of the sampling formulae given here are known. An excellent survey article on sampling, giving much more information and history, is [H]. Several of our examples coincide with his (see [H], pp. 73–74). Another important reference which we have followed closely in our use of the key notion of transversality is [P]. Other good references for sampling theory include [J], [Bu], [K], [B1], and [G].

This chapter also elaborates on some of the remarks motivating wavelets in section II of the Introduction to this volume. Sampling theorems yield decomposition formulae for functions. As noted in the Introduction, it is desirable to have both space and frequency localization in such formulae. Using the sampling formulae we present, we develop the φ-transform identity in each setting, which is an almost orthogonal expansion with good space-frequency localization. These ideas have been fully developed for $\mathbb{R}$ previously ([FrJa1], [FrJa2]), but in some respects may be new in the other settings (however, see [Ku], [KuFu], and [T]).

The sampling results also yield the Shannon orthonormal wavelet basis in each setting (see [FrJaW] for $\mathbb{R}$ and [FrKu] for $\mathbb{Z}_N$). This basis is well localized in frequency but not in space. This motivates wavelet theory, which uses new methods to obtain orthonormality and, in some cases, dual localization. Wavelet methods for $\mathbb{R}$ are presented in Chapter 1 above ([St]), and for $\mathbb{Z}$ and $\mathbb{Z}_N$ in Chapter 2 ([FrKu]). For the unit circle $\mathbb{T}$, see the periodic wavelets in [M1] and [Ho]. We refer to the Introduction for background, history, and references on wavelets.

The reader familiar with locally compact abelian groups (LCAG) will not be surprised that there are similar results in the four settings given. We hope that our parallel presentation will help a general audience see how the same basic ideas apply in each setting. We do, however, formulate a generalization to the LCAG case in Section 6.6. It should be compared to the seminal LCAG result in [K], restated in [H], p. 73.

We have given our results for functions or signals in L^2, since that is the case of most interest in applications. However, these decompositions are useful in studying a wide range of function spaces. For L^p sampling, see [G]; for the φ-transform see [FrJa1] and [FrJa2] for $\mathbb{R}^n$ and [T] for $\mathbb{Z}$; for wavelets in $\mathbb{R}^n$ see [M2].

This chapter exhibits a bridge between sampling theory and orthonormal wavelets in each of the four settings. However, our basic approach, the use of sampling formulae to obtain decompositions with dual localization, has further importance. In other circumstances it is not known how mimicking the wavelet construction, because of the rigidity of the orthonormality requirement. Yet the sampling approach may work. An example of this is the expansion of radial functions in $\mathbb{R}^n$ in terms of radial components with appropriate properties in [EFr].

6.2   The real line $\mathbb{R}$

We say that a (Lebesgue measurable) function f belongs to $L^2 = L^2(\mathbb{R})$ if $\|f\|_{L^2} = (\int |f(x)|^2 dx)^{1/2} < +\infty$; the integral is understood to be over $\mathbb{R} = (-\infty, +\infty)$ if no domain is specified. The Fourier transform $^\wedge : L^2 \to L^2$ is defined by $\hat{f}(\xi) = \int f(x) e^{-ix\xi} dx$. The inverse transform $^\vee : L^2 \to L^2$ is then defined by $\check{f}(x) = (2\pi)^{-1} \int f(\xi) e^{ix\xi} d\xi$. It follows that $f = (\hat{f})^\vee = (\check{f})^\wedge$ and $\|\hat{f}\|_{L^2}^2 = 2\pi \|f\|_{L^2}^2$. For $f, g \in L^2$, $f * g(x) = \int f(x - y) g(y) dy$. Then $(f * g)^\wedge = \hat{f}\hat{g}$. Also let $\langle f, g \rangle = \int f(x)\overline{g(x)}dx$. We write $|E|$ for the Lebesgue measure of a (measurable) set $E \subseteq \mathbb{R}$, and χ_E for the function with value 1 on E and 0 elsewhere. The term "a.e. x" means for all x except on a set of measure 0. The support of a function f is the closure of $\{x : f(x) \neq 0\}$; it is denoted supp$f$.

For $B > 0$, we say that a set E is B-transversal if

$$\sum_{n \in \mathbb{Z}} \chi_E(x + 2Bn) = 1 \quad \text{for a.e. } x \in \mathbb{R}.$$

This means that the translates $\{E + 2Bn\}_{n \in \mathbb{Z}}$ form, except for a set of measure 0, a disjoint family whose union is $\mathbb{R}$ (for $\alpha \in \mathbb{R}$, $E + \alpha = \{x + \alpha : x \in E\}$). Examples that will be used later are $E = [-B, B]$, which is B-transversal, and $E = [-B, -B/2] \cup [B/2, B]$, which is $B/2$-transversal. If E is B-transversal, then by a change of variables,

$$|E| = \int \chi_E = \sum_{n \in \mathbb{Z}} \int_{(2n-1)B}^{(2n+1)B} \chi_E(y) dy$$

$$= \int_{-B}^{+B} \sum_{n \in \mathbb{Z}} \chi_E(x + 2Bn) dx = \int_{-B}^{B} dx = 2B.$$

(6.1)

The following lemma, which can be found in [P], shows the significance of the concept of B-transversality. We let $L^2(E)$ denote $\{f \in L^2(\mathbb{R}) : \operatorname{supp} f \subseteq E\}$.

LEMMA 6.1
Let $E \subseteq \mathbb{R}$ be B-transversal. For $k \in \mathbb{Z}$, define $e_k \in L^2(E)$ by

$$e_k(x) = (2B)^{-1/2} e^{-ik\pi x/B} \chi_E(x).$$

Then the family $\{e_k\}_{k \in \mathbb{Z}}$ is a complete orthonormal basis for $L^2(E)$. If $F \in L^2(E)$, then

$$\langle F, e_k \rangle = \sqrt{\frac{2}{B}} \, \pi \check{F}(k\pi/B) \tag{6.2}$$

and hence,

$$\|F\|_{L^2}^2 = \frac{2\pi^2}{B} \sum_{k \in \mathbb{Z}} |\check{F}(k\pi/B)|^2. \tag{6.3}$$

PROOF To prove the orthonormality, the same change of variables as in (6.1) and the B-transversality of E give

$$\langle e_j, e_k \rangle = \frac{1}{2B} \int \chi_E(y) e^{i(k-j)\pi y/B} dy$$

$$= \frac{1}{2B} \int_{-B}^{B} \sum_{n \in \mathbb{Z}} \chi_E(x + 2Bn) e^{i(k-j)\pi(x+2Bn)/B} dx$$

$$= \frac{1}{2B} \int_{-B}^{B} e^{i(k-j)\pi x/B} dx = \delta_{jk}.$$

For completeness, suppose $F \in L^2(E)$ and $\langle F, e_k \rangle = 0$ for all $k \in \mathbb{Z}$. Noting that $F\chi_E = F$ since $\operatorname{supp} F \subseteq E$, and changing variables as above, we obtain

$$\int_{-B}^{B} \sum_{n \in \mathbb{Z}} F(x + 2Bn) e^{ik\pi x/B} dx = 0 \quad \text{for all } k \in \mathbb{Z}.$$

By the standard completeness result for Fourier series, $\sum_{n \in \mathbb{Z}} F(x + 2Bn)$ is 0 in $L^2([-B,B])$, hence a.e. on $[-B,B]$. But this sum is $2B$-periodic and thus is 0 a.e. on $\mathbb{R}$. Since $\operatorname{supp} F \subseteq E$ and E is B-transversal, the functions $\{F(x + 2Bn)\}_{n \in \mathbb{Z}}$ have a.e. disjoint supports, hence each must be 0 a.e. on $\mathbb{R}$. Letting $n = 0$, we see that $F = 0$ a.e. on $\mathbb{R}$.

Equation (6.2) follows from the definitions and the assumption $F = F\chi_E$. Then (6.3) follows from (6.2) and the fact that the collection $\{e_k\}_{k \in \mathbb{Z}}$ is a complete orthonormal basis for $L^2(E)$. ∎

Lemma 6.1 easily yields the following sampling result. Its statement requires the following interpretation. Since $f \in L^2$, we have $\hat{f} \in L^2$. Since $|\operatorname{supp} \hat{f}| = |E| < +\infty$ by (6.1), we obtain $\hat{f} \in L^1$. Thus f agrees a.e. with the continuous function $(\hat{f})^\vee$. We identify f with this representative so that the sample values of f are well defined.

THEOREM 6.2
Suppose $E \subseteq \mathbb{R}$ is B-transversal, $f, g \in L^2$, supp $\hat{f} \subseteq E$, and supp $\hat{g} \subseteq E$. Then

$$f * g(x) = \frac{\pi}{B} \sum_{k\in\mathbb{Z}} f(k\pi/B)g(x - k\pi/B),$$ (6.4)

and

$$\|f\|_{L^2}^2 = \frac{\pi}{B} \sum_{k\in\mathbb{Z}} |f(k\pi/B)|^2.$$ (6.5)

PROOF   Let $\{e_k\}_{k\in\mathbb{Z}}$ be as in Lemma 6.1. Using this lemma with $F = \hat{f}$, we obtain

$$\hat{f} = \sum_{k\in\mathbb{Z}} \langle \hat{f}, e_k \rangle e_k = \sum_{k\in\mathbb{Z}} \sqrt{\frac{2}{B}} \pi f(k\pi/B) e_k.$$

Substituting this into $f * g = (\hat{f}\hat{g})^\vee$, we obtain

$$f * g = \sum_{k\in\mathbb{Z}} \sqrt{\frac{2}{B}} \pi f(k\pi/B)(\hat{g}e_k)^\vee.$$

The translation properties of the Fourier transform imply that $(g(x - k\pi/B))^\wedge = \sqrt{2B}\hat{g}e_k$. Using this above gives (6.4). Plancherel's formula $\|f\|_{L^2}^2 = \frac{1}{2\pi}\|\hat{f}\|_{L^2}^2$ and (6.3) with $F = \hat{f}$ imply (6.5). ∎

COROLLARY 6.3
Suppose E is a closed, B-transversal set, $f \in L^2$, and supp$\hat{f} \subseteq E$. Then

$$f(x) = \frac{\pi}{B} \sum_{k\in\mathbb{Z}} f(k\pi/B)\check{\chi}_E(x - k\pi/B).$$ (6.6)

PROOF   We apply Theorem 6.2 with $g = \check{\chi}_E$. Note that $|E| = 2B$ by (6.1), so χ_E, and hence g, belong to L^2. Since $\hat{g} = 1$ on supp$\hat{f}$, we have $\hat{f}\hat{g} = \hat{f}$, or $f * g = f$. Thus Theorem 6.2 gives the result. ∎

Example 6.4.   Shannon sampling formula, [S]
Let $E = [-B, B]$. Then E is B-transversal and $\check{\chi}_E(x) = B/\pi$, sinc Bx/π, where sinc x is $(\sin \pi x)/\pi x$ if $x \neq 0$, and sinc $0 = 1$. Suppose $f \in L^2$ and supp$\hat{f} \subseteq E$. (Such an f is said to be "band limited" in engineering terminology, or "of exponential type," in mathematics terminology.) Then (6.6) gives the classical Shannon sampling formula:

$$f(x) = \sum_{k\in\mathbb{Z}} f(k\pi/B) \text{ sinc} \left(\frac{Bx}{\pi} - k \right). □$$

Example 6.5. Annular sampling formula

Let $E = [-B, -B/2] \cup [B/2, B]$. Then E is $B/2$-transversal, and

$$\check{\chi}_E(x) = \frac{B}{2\pi} \operatorname{sinc}\left(\frac{Bx}{2\pi}\right)\left[2\cos\left(\frac{Bx}{2}\right) - 1\right].$$

Suppose $f \in L^2$ and $\operatorname{supp} \hat{f} \subseteq E$. Replacing B by $B/2$ in (6.6) and using the formula for $\check{\chi}_E$ gives the identity

$$f(x) = \sum_{k\in\mathbb{Z}} f(2k\pi/B) \operatorname{sinc}\left(\frac{Bx}{2\pi} - k\right)\left[2\cos\left(\frac{Bx}{2} - k\pi\right) - 1\right]. \tag{6.7}$$

Note that this f could also be represented by the Shannon formula above, but this would require sampling f at twice the rate required for (6.7).   ☐

We can use these results to represent a general $f \in L^2$ by partitioning $\hat{f}$ along transversal sets. We will describe two ways to do this. The first is the φ-transform identity. We begin by selecting functions φ and ψ in L^2 satisfying

$$\operatorname{supp}\hat{\varphi}, \operatorname{supp}\hat{\psi} \subseteq [-\pi, -\pi/4] \cup [\pi/4, \pi], \tag{6.8}$$

and

$$\sum_{v\in\mathbb{Z}} \overline{\hat{\varphi}(2^{-v}\xi)}\hat{\psi}(2^{-v}\xi) = 1 \quad \text{for } \xi \in \mathbb{R} \setminus \{0\}. \tag{6.9}$$

We also assume that $\hat{\varphi}$, $\hat{\psi}$ are bounded above, i.e., that there exists a constant $A_1 > 0$ such that

$$|\hat{\varphi}(\xi)|, |\hat{\psi}(\xi)| \le A_1 \quad \text{for all } \xi, \tag{6.10}$$

and that $\hat{\varphi}$ and $\hat{\psi}$ are bounded below on a certain subinterval of their support:

$$|\hat{\varphi}(\xi)|, |\hat{\psi}(\xi)| \ge A_2 \quad \text{for } \xi \in [-3\pi/4, -3\pi/8] \cup [3\pi/8, 3\pi/4], \tag{6.11}$$

for some $A_2 > 0$. It is not difficult to construct such functions; it is even possible to obtain such $\hat{\varphi}$ and $\hat{\psi}$ having bounded derivatives of arbitrarily high order (see, e.g., Lemma 6.9 in [FrJaW]). For $v, k \in \mathbb{Z}$, let $\varphi_{vk}(x) = 2^{v/2}\varphi(2^v x - k)$ and $\psi_{vk}(x) = 2^{v/2}\psi(2^v x - k)$.

THEOREM 6.6   (*φ-Transform Identity, [FrJa1], [FrJa2]*)
Suppose $f \in L^2$. Then

$$f = \sum_{v,k\in\mathbb{Z}} \langle f, \varphi_{vk}\rangle \psi_{vk}. \tag{6.12}$$

Also, there exist constants $c_1, c_2 > 0$ such that for all $f \in L^2$,

$$c_1\|f\|_{L^2}^2 \le \sum_{v,k\in\mathbb{Z}} |\langle f, \varphi_{v,k}\rangle|^2 \le c_2\|f\|_{L^2}^2. \tag{6.13}$$

PROOF   For $v \in \mathbb{Z}$, let $\varphi_v(x) = 2^v\varphi(2^v x)$ and $\psi_v(x) = 2^v\psi(2^v x)$. This is equivalent to $\hat{\varphi}_v(\xi) = \hat{\varphi}(2^{-v}\xi)$ and $\hat{\psi}_v(\xi) = \hat{\psi}(2^{-v}\xi)$. Also set $\tilde{\varphi}_v(x) = \overline{\varphi_v(-x)}$; this is equivalent to $\hat{\bar{\varphi}}_v = \hat{\tilde{\varphi}}_v$. By taking the Fourier transform of both sides, (6.9) implies that

$$f = \sum_{v \in \mathbb{Z}} \tilde{\varphi}_v * \psi_v * f.$$

Note that $\text{supp}(\tilde{\varphi}_v * f)^\wedge$, $\text{supp } \hat{\psi}_v \subseteq [-2^v\pi, -2^v\pi/4] \cup [2^v\pi/4, 2^v\pi] \subseteq [-2^v\pi, 2^v\pi]$. Hence for each v we can apply (6.4) with f replaced by $\tilde{\varphi}_v * f$, g replaced by ψ_v, and $E = [-B, B]$ where $B = 2^v\pi$. This gives

$$\tilde{\varphi}_v * \psi_v * f(x) = 2^{-v} \sum_{k \in \mathbb{Z}} \tilde{\varphi}_v * f(2^{-v}k) \psi_v(x - 2^{-v}k).$$

A simple computation shows that this last term is $\sum_k \langle f, \varphi_{vk} \rangle \psi_{vk}$. Hence, (6.12) follows.

To prove (6.13), first note that there exists $c > 0$ such that $\sum_{v \in \mathbb{Z}} |\hat{\varphi}_v(\xi)|^2 \leq c$. This holds because (6.8) and the identity $\hat{\varphi}_v(\xi) = \hat{\varphi}(2^{-v}\xi)$ imply that at most two terms in the sum are nonzero for any ξ, and hence (6.10) guarantees the boundedness. On the other hand, (6.11) implies that at each $\xi \neq 0$, there is at least one $v \in \mathbb{Z}$ such that $|\hat{\varphi}_v(\xi)| \geq A_2$, so $\sum_{v \in \mathbb{Z}} |\hat{\varphi}_v(\xi)|^2 \geq \tilde{c} > 0$ for some $\tilde{c}$, for all $\xi \neq 0$. Hence the ratio of $\|\hat{f}\|_{L^2}^2$ and

$$(*) = \int |\hat{f}(\xi)|^2 \sum_{v \in \mathbb{Z}} |\hat{\varphi}_v(\xi)|^2 d\xi$$

is bounded above and below (away from 0), with bounds independent of f. However, $\|\hat{f}\|_{L^2}^2 = 2\pi \|f\|_{L^2}^2$, and

$$(*) = \sum_{v \in \mathbb{Z}} \int |(\tilde{\varphi}_v * f)^\wedge(\xi)|^2 d\xi = 2\pi \sum_{v \in \mathbb{Z}} \|\tilde{\varphi}_v * f\|_{L^2}^2.$$

Moreover, by (6.5) with $B = 2^{-v}\pi$ and f replaced by $\tilde{\varphi}_v * f$,

$$\|\tilde{\varphi}_v * f\|_{L^2}^2 = 2^{-v} \sum_{k \in \mathbb{Z}} |\tilde{\varphi}_v * f(2^{-v}k)|^2 = \sum_{k \in \mathbb{Z}} |\langle f, \varphi_{vk} \rangle|^2.$$

Putting these facts together gives (6.13). ∎

If we choose $\hat{\varphi}$ to be very smooth, which is possible, as indicated above, then φ will decay rapidly at $\pm\infty$. More precisely, for the functions φ_{vk} (similarly ψ_{vk}), we can obtain

$$|\varphi_{vk}(x)| \leq c_M \left(1 + \frac{|x - 2^{-v}k|}{2^{-v}}\right)^{-M} \tag{6.14}$$

for any $M > 0$, where c_M is independent of v, k, and x. Roughly speaking, (6.14) says that φ_{vk} is concentrated near $2^{-v}k$, decaying rapidly on a scale of 2^{-v}. Also, by (6.8),

$$\text{supp } \hat{\varphi}_{vk} \subseteq [-2^v\pi, -2^v\pi/4] \cup [2^v\pi/4, 2^v\pi].$$

Thus the coefficient $\langle f, \varphi_{vk} \rangle$ measures the portion of f near $2^{-v}k$ carrying frequencies of magnitude between $2^v \pi/4$ and $2^v \pi$. In this way, the decomposition (6.12) gives a simultaneous space and frequency analysis of f.

It is possible to select φ and ψ to be identical; then (6.12) looks like an orthonormal decomposition. However, the φ_{vk}'s constructed in Theorem 6.6 can not be orthonormal (see, e.g., [FrJaW], §7). There is a way to use sampling methods to obtain an orthonormal basis $\{\varphi_{vk}\}_{v,k\in\mathbb{Z}}$, called the Shannon wavelet basis, obtained as above from a single φ. Its disadvantage is that the φ_{vk}'s do not decay rapidly; we obtain (6.14) only for $M \leq 1$. For the discussion of the Shannon wavelets, it is convenient to have the following simple lemma about general orthonormal systems.

LEMMA 6.7
Suppose $\{e_k\}_k$ is a collection of nonzero elements of a separable Hilbert space H such that every $x \in H$ can be written as $x = \sum_k \langle x, e_k \rangle e_k$. Then

$$\langle e_j, e_k \rangle = 0 \quad \text{whenever } j \neq k \tag{6.15}$$

if and only if

$$\|e_k\| = 1 \quad \text{for every } k. \tag{6.16}$$

PROOF For any $x \in H$, we have

$$\|x\|^2 = \langle x, x \rangle = \left\langle \sum_k \langle x, e_k \rangle e_k, x \right\rangle = \sum_k |\langle x, e_k \rangle|^2.$$

Applying this to $x = e_j$ gives

$$\|e_j\|^2 = \sum_k |\langle e_j, e_k \rangle|^2 = \|e_j\|^4 + \sum_{\{k, k \neq j\}} |\langle e_j, e_k \rangle|^2,$$

which implies the result. ∎

This lemma states that in the presence of the identity $x = \sum_k \langle x, e_k \rangle e_k$, we only have to check one of the two conditions defining an orthonormal set.

For $v \in \mathbb{Z}$, let $E_v = [-2^{v+1}\pi, -2^v\pi] \cup [2^v\pi, 2^{v+1}\pi]$ and $\varphi_v = \check{\chi}_{E_v}$. Note that $\varphi_v(x) = 2^v \varphi_0(2^v x)$. Let $\varphi_{vk}(x) = 2^{v/2} \varphi_0(2^v x - k) = 2^{-v/2} \varphi_v(x - 2^{-v}k)$. By Example 6.5, this gives

$$\varphi_{vk}(x) = 2^{v/2} \text{sinc}(2^v x - k)[2\cos(2^v \pi x - \pi k) - 1].$$

THEOREM 6.8 *(Shannon basis)*
The collection $\{\varphi_{vk}\}_{v,k\in\mathbb{Z}}$ forms a complete orthonormal basis for L^2.

PROOF For $f \in L^2$, we have $f = \sum_{v\in\mathbb{Z}} \varphi_v * f$, since $(\varphi_v * f)^\wedge = \hat{f}\chi_{E_v}$ and $\sum_{v\in\mathbb{Z}} \chi_{E_v} = 1$ a.e. For each $v \in \mathbb{Z}$, we apply (6.7) with $B = 2^{v+1}\pi$ and f replaced by $\varphi_v * f$ to obtain

$$f(x) = \sum_{v,k \in \mathbb{Z}} 2^{-v/2} \varphi_v * f(2^{-v}k) \varphi_{vk}(x) = \sum_{v,k \in \mathbb{Z}} \langle f, \varphi_{vk} \rangle \varphi_{vk}(x).$$

Note also that for each $v, k \in \mathbb{Z}$,

$$\|\varphi_{vk}\|_{L^2}^2 = 2^{-v} \|\varphi_v\|_{L^2}^2 = (2\pi)^{-1} 2^{-v} \|\chi_{E_v}\|_{L^2}^2 = 1.$$

Hence Lemma 6.7 completes the proof. ∎

6.3 The integers $\mathbb{Z}$

A function f defined on the integers, i.e., a complex-valued sequence $f = (\ldots, f(-1), f(0), f(1), \ldots) = \{f(n)\}_{n \in \mathbb{Z}}$, belongs to $\ell^2 = \ell^2(\mathbb{Z})$ if $\|f\|_{\ell^2} = (\sum_{n \in \mathbb{Z}} |f(n)|^2)^{1/2} < +\infty$. We define $L^2 = L^2(\mathbb{T})$ on the torus $\mathbb{T} = [-\pi, \pi)$ with respect to normalized (Haar) measure $dx/2\pi$: $F \in L^2$ if $\|F\|_{L^2} = (\frac{1}{2\pi} \int_{-\pi}^{\pi} |F(x)|^2 dx)^{1/2} < +\infty$. We sometimes identify F with its 2π-periodic extension F^* to $\mathbb{R}$; that is, the function F^* which agrees with F on $[-\pi, \pi)$ and satisfies $F^*(x + 2\pi) = F^*(x)$ for all $x \in \mathbb{R}$. The Fourier transform $\wedge : \ell^2 \to L^2$ is defined by $\hat{f}(x) = \sum_{n \in \mathbb{Z}} f(n) e^{inx}$. The inverse transform $\vee : L^2 \to \ell^2$ is then given by $\check{F}(n) = \frac{1}{2\pi} \int_{-\pi}^{\pi} F(x) e^{-inx} dx$. Then $(\hat{f})^{\vee} = f$ and $(\check{F})^{\wedge} = F$. For $f, g \in \ell^2$, let $f * g(n) = \sum_{m \in \mathbb{Z}} f(n - m) g(m)$. It follows that $(f * g)^{\wedge} = \hat{f}\hat{g}$. For $f, g \in \ell^2$, let $\langle f, g \rangle = \sum_{n \in \mathbb{Z}} f(n) \overline{g(n)}$, while for $F, G \in L^2$, let $\langle F, G \rangle = \frac{1}{2\pi} \int_{-\pi}^{\pi} F(x) \overline{G(x)} dx$. Then Parseval's relation is that $\langle \hat{f}, \hat{g} \rangle = \langle f, g \rangle$; in particular we have Plancherel's theorem $\|\hat{f}\|_{L^2} = \|f\|_{\ell^2}$. The notations $|E|$, χ_E, a.e. x, and supp$f$ are the same as in Section 6.2.

A (measurable) set $E \subseteq \mathbb{T}$ is said to be π/N-transversal ($N \in \mathbb{Z}, N \geq 1$) if

$$\sum_{n=0}^{N-1} \chi_E^*(x + 2\pi n/N) = 1 \quad \text{for a.e. } x \in \mathbb{T},$$

where χ_E^* is the 2π-periodic extension of χ_E, as described above. Examples are $E = [-\pi/N, \pi/N]$, which is π/N-transversal, and $E = [-\pi/N, -\pi/2N] \cup [\pi/2N, \pi/N]$, which is $\pi/2N$-transversal. If E is π/N-transversal, it follows that $|E| = 2\pi/N$.

We have formulated the definition of $E \subset \mathbb{T}$ being π/N-transversal in a way naturally adapted to the torus $\mathbb{T}$. However, this definition is equivalent to the statement that E, regarded now as a subset of $\mathbb{R}$, is π/N-transversal on $\mathbb{R}$ in the sense of Section 6.2. To see this, note that $\chi_E^*(x) = \sum_{m \in \mathbb{Z}} \chi_E(x + 2\pi m)$, substitute this in the above definition, observe that the result is $2\pi/N$-periodic, and compare to the definition in Section 6.2. This makes the following result, which could be obtained directly by the same arguments as for Lemma 6.1, a trivial consequence of that lemma.

LEMMA 6.9

Let $E \subseteq \mathbb{T}$ be π/N-transversal. For $k \in \mathbb{Z}$, define $e_k \in L^2$ by

$$e_k(x) = \sqrt{N} e^{ikNx} \chi_E(x).$$

Then the family $\{e_k\}_{k \in \mathbb{Z}}$ is a complete orthonormal basis for $L^2(E) \equiv \{F \in L^2 : \text{supp} F \subseteq E\}$. If $F \in L^2(E)$, then

$$\langle F, e_k \rangle = \sqrt{N} \check{F}(kN) \tag{6.17}$$

and hence,

$$\|F\|_{L^2}^2 = N \sum_{k \in \mathbb{Z}} |\check{F}(kN)|^2. \tag{6.18}$$

PROOF By the remarks preceding the theorem, the orthogonality and the completeness follow from Lemma 6.1. The normalization is different here because of the factor 2π in the definition of the $L^2(\mathbb{T})$ norm. Then (6.17) is a direct calculation, and these facts imply (6.18). ∎

From this, we obtain a version of the sampling theorem in the discrete context.

THEOREM 6.10

Suppose $E \subseteq \mathbb{T}$ is π/N-transversal, $f, g \in \ell^2$, $\text{supp} \hat{f} \subseteq E$, and $\text{supp} \hat{g} \subseteq E$. Then

$$f * g(n) = N \sum_{k \in \mathbb{Z}} f(kN) g(n - kN), \tag{6.19}$$

and

$$\|f\|_{\ell^2}^2 = N \sum_{k \in \mathbb{Z}} |f(kN)|^2. \tag{6.20}$$

PROOF Plancherel's theorem and (6.18) imply (6.20) immediately. From Lemma 6.9, we have that

$$\hat{f} = \sum_{k \in \mathbb{Z}} \langle \hat{f}, e_k \rangle e_k = \sqrt{N} \sum_{k \in \mathbb{Z}} f(kN) e_k.$$

Substituting this into $f * g = (\hat{f}\hat{g})^\vee$, using $\text{supp} \hat{g} \subseteq E$ and $(\hat{g}e_k)^\vee(n) = \sqrt{N} g(n-kN)$, we obtain (6.20). ∎

COROLLARY 6.11

Suppose E is closed and π/N-transversal. If $f \in \ell^2$ and $\text{supp} \hat{f} \subseteq E$, then

$$f(n) = N \sum_{k \in \mathbb{Z}} f(kN) \check{\chi}_E(n - kN). \tag{6.21}$$

PROOF Apply Theorem 6.10 with $g = \check{\chi}_E$, as in the proof of Corollary 6.3. ∎

Example 6.12. *Shannon sampling formula*
Suppose $f \in \ell^2$ with supp$\hat{f} \subseteq [-\pi/N, \pi/N]$, for some $N \geq 1$. Then

$$f(n) = \sum_{k \in \mathbb{Z}} f(kN) \operatorname{sinc}\left(\frac{n}{N} - k\right),$$

where $\operatorname{sinc} x$ is as in Example 6.4. This follows from Corollary 6.11 since $[-\pi/N, \pi/N] = E$ is π/N-transversal and $\check{\chi}_E(n) = \frac{1}{N}\operatorname{sinc}(n/N)$. $\Box$

Example 6.13. *Annular sampling formula*
Suppose $f \in \ell^2$ and supp$\hat{f} \subseteq [-\pi/N, -\pi/2N] \cup [\pi/2N, \pi/N] = E$, for some $N \geq 1$. Then

$$f(n) = \sum_{k \in \mathbb{Z}} f(2kN) \operatorname{sinc}\left(\frac{n}{2N} - k\right) \left[2\cos\left(\frac{\pi n}{2N} - \pi k\right) - 1\right]. \quad (6.22)$$

Again this is a consequence of Corollary 6.11, where here E is $\pi/2N$ transversal and $\check{\chi}_E(n) = \frac{1}{2N}\operatorname{sinc}(\frac{n}{2N})[2\cos(\frac{\pi n}{2N}) - 1]$. $\Box$

We can formulate an ℓ^2 version of the φ-transform as follows. We say that a pair of sequences $\{\varphi_v\}_{v=0}^{\infty}$, $\{\psi_v\}_{v=0}^{\infty}$, where $\varphi_v, \psi_v \in \ell^2$ for each v, is an admissible pair if

$$\operatorname{supp}\hat{\varphi}_v, \operatorname{supp}\hat{\psi}_v \subseteq R_v \equiv [-2^{-v}\pi, -2^{-v}\pi/4] \cup [2^{-v}\pi/4, 2^{-v}\pi] \quad \text{for each } v, (6.23)$$

$$\sum_{v=0}^{\infty} \bar{\hat{\varphi}}_v(x)\hat{\psi}_v(x) = 1 \quad \text{for } x \in \mathbb{T} \setminus \{0\}, \quad (6.24)$$

and there exist constants $A_1, A_2 > 0$ (independent of v) such that

$$|\hat{\varphi}_v(x)|, |\hat{\psi}_v(x)| \leq A_1 \quad \text{for all } x, v \quad (6.25)$$

and

$$|\hat{\varphi}_v(x)|, |\hat{\psi}_v(x)| \geq A_2 \quad \text{if } x \in T_v, \quad (6.26)$$

where $T_0 = [-\pi, -3\pi/8] \cup [3\pi/8, \pi]$ and, for $v \geq 1$, $T_v = [-3 \cdot 2^{-v-2}\pi, -3 \cdot 2^{-v-3}\pi] \cup [3 \cdot 2^{-v-3}\pi, 3 \cdot 2^{-v-2}\pi]$. It is not difficult to construct an admissible pair. For each $v \geq 0$ and $k \in \mathbb{Z}$, we define functions $\varphi_{vk}, \psi_{vk} \in \ell^2$ by

$$\varphi_{vk}(n) = 2^{v/2}\varphi_v(n - 2^v k) \quad \text{and} \quad \psi_{vk}(n) = 2^{v/2}\psi_v(n - 2^v k).$$

For comparison with Section 6.2, note that the indexing is reversed: here the points in supp$\hat{\varphi}_v, \hat{\psi}_v$ go to 0 as v goes to ∞, whereas in Section 6.2 these points go to ∞ with v.

THEOREM 6.14 (φ-*Transform Identity for* ℓ^2, *[KuFu] and [T]*)
Suppose that $\{\varphi_v\}_{v=0}^{\infty}$, $\{\psi_v\}_{v=0}^{\infty}$ is an admissible pair of sequences in ℓ^2. Suppose $f \in \ell^2$. Then

$$f = \sum_{v=0}^{\infty} \sum_{k \in \mathbb{Z}} \langle f, \varphi_{vk} \rangle \psi_{vk}. \quad (6.27)$$

Also there exist constants $c_1, c_2 > 0$ depending only on A_1, A_2 in (6.25) and (6.26) such that

$$c_1 \|f\|_{\ell^2}^2 \leq \sum_{v=0}^{\infty} \sum_{k \in \mathbb{Z}} |\langle f, \varphi_{vk} \rangle|^2 \leq c_2 \|f\|_{\ell^2}^2. \tag{6.28}$$

PROOF For each v, let $\tilde{\varphi}_v(n) = \overline{\varphi_v(-n)}$, which is equivalent to $\hat{\tilde{\varphi}}_v = \bar{\hat{\varphi}}_v$. Then, by Fourier inversion and (6.24), we have

$$f = \sum_{v=0}^{\infty} \sum_{k \in \mathbb{Z}} \tilde{\varphi}_v * \psi_v * f.$$

By (6.23), $\operatorname{supp}(\tilde{\varphi}_v * f)^\wedge$, $\operatorname{supp} \hat{\psi}_v \subseteq R_v \subseteq [-2^{-v}\pi, 2^{-v}\pi]$, and $[-2^{-v}\pi, 2^{-v}\pi]$ is π/N-transversal for $N = 2^v$. Note that $(\tilde{\varphi}_v * f)^\wedge \in L^2$ and hence $\tilde{\varphi}_v * f \in \ell^2$, because $\hat{f} \in L^2$ and (6.25) holds. Hence we can apply Theorem 6.10 to obtain, for each v,

$$\tilde{\varphi}_v * \psi_v * f(n) = 2^v \sum_{k \in \mathbb{Z}} \tilde{\varphi}_v * f(2^v k) \psi_v(n - 2^v k) = \sum_{k \in \mathbb{Z}} \langle f, \varphi_{vk} \rangle \psi_{vk}.$$

Substituting this above gives (6.27). Using (6.20), (6.25), and (6.26), we obtain (6.28) in exactly the same way that (6.13) was obtained in Theorem 6.6. We omit the details. ∎

REMARK 6.15 In Section 6.2 we assumed that the functions $\hat{\varphi}_v$ were dilates of each other. Similarly, here it is possible to obtain

$$\hat{\varphi}_v(x) = \hat{\varphi}_1(2^{v-1}x) \quad \text{and} \quad \hat{\psi}_v(x) = \hat{\psi}_1(2^{v-1}x) \quad \text{for } v \geq 1 \tag{6.29}$$

by setting

$$\varphi_v(n) = 2^{-v+1} \sum_{k \in \mathbb{Z}} \varphi_1(k) \operatorname{sinc}(2^{-v+1}n - k), \tag{6.30}$$

and similarly for ψ_v. (The case $v = 0$ is different since (6.23) and (6.24) imply that $\bar{\hat{\varphi}}_0 \hat{\psi}_0 = 1$ on $[-\pi, -\pi/2] \cup [\pi/2, \pi]$, while the other $\hat{\varphi}_v$'s and $\hat{\psi}_v$'s can go smoothly to 0 at the endpoints of their supporting intervals.) To see (6.30) from (6.29), write

$$\varphi_v(n) = (\hat{\varphi}_v)^\vee(n) = (2\pi)^{-1} \int_{-2^{-v}\pi}^{2^{-v}\pi} \hat{\varphi}_1(2^{v-1}y) e^{-iny} dy$$

$$= (2^v \pi)^{-1} \int_{-\pi}^{\pi} \hat{\varphi}_1(x) e^{-i2^{-v+1}nx} dx.$$

Substituting $\hat{\varphi}_1(x) = \sum_{k \in \mathbb{Z}} \varphi_1(k) e^{ikx}$, a simple computation gives (6.30). We can select $\varphi_v = \psi_v$ for each v, if we like. Also we can select $\hat{\varphi}_v$ and $\hat{\psi}_v$ to be smooth (infinitely differentiable), which guarantees that the functions φ_{vk} and ψ_{vk} decrease rapidly to 0 away from $2^v k$ on a scale of 2^v (similar to (6.14), with v

replaced by $-v$). Then (6.27) gives a simultaneous time and frequency analysis of $f \in \ell^2(\mathbb{T})$, as (6.12) does for $f \in L^2(\mathbb{R})$. ∎

THEOREM 6.16 *(Shannon basis)*
For $v \in \mathbb{Z}$, $v \geq 1$, and $k \in \mathbb{Z}$, define

$$\varphi_{vk}(n) = 2^{-v/2} \operatorname{sinc}(2^{-v}n - k)[2\cos(2^{-v}\pi n - \pi k) - 1].$$

Then the collection $\{\varphi_{vk}\}_{v,k \in \mathbb{Z}, v \geq 1}$ forms a complete orthonormal basis for ℓ^2.

PROOF   For $v \in \mathbb{Z}$, $v \geq 1$, let $E_v = [-2^{-v+1}\pi, -2^{-v}\pi] \cup [2^{-v}\pi, 2^{-v+1}\pi]$. Let $\varphi_v = \check{\chi}_{E_v}$, which gives, by Example 6.13 with $N = 2^{v-1}$,

$$\varphi_v(n) = 2^{-v} \operatorname{sinc}(2^{-v}n)[2\cos(2^{-v}\pi n) - 1].$$

Note that $\varphi_{vk}(n) = 2^{v/2}\varphi_v(n - 2^v k)$. Since $\sum_{v=1}^{\infty} \hat{\varphi}_v(x) = 1$ for $x \in \mathbb{T} \setminus \{0\}$, we have, for $f \in \ell^2$, $f = \sum_{v=1}^{\infty} \varphi_v * f$. Example 6.13 then gives

$$\varphi_v * f(n) = 2^v \sum_{k \in \mathbb{Z}} \varphi_v * f(2^v k)\varphi_v(n - 2^v k),$$

which implies $\varphi_v * f = \sum_{v=1}^{\infty} \sum_{k \in \mathbb{Z}} \langle f, \varphi_{vk} \rangle \varphi_{vk}$. Since

$$\|\varphi_{vk}\|_{\ell^2}^2 = 2^v \|\varphi_v\|_{\ell^2}^2 = 2^v \|\hat{\varphi}_v\|_{L^2}^2 = 1 \quad \text{for all } v, k,$$

the result follows from Lemma 6.7. ∎

As in the case of $\mathbb{R}$, the Shannon wavelets are orthonormal but only decay like $1/n$ at ∞, while the φ_{vk}'s for the φ-transform can be chosen to decay rapidly at ∞, but are not orthonormal. As noted in the introduction, wavelet theory allows us to gain both advantages.

6.4   The torus $\mathbb{T}$

The case of $L^2 = L^2(\mathbb{T})$ is dual to the case of $\ell^2 = \ell^2(\mathbb{Z})$ considered in Section 6.3. We use the same definitions for L^2 and ℓ^2, and their norms and inner products, as in the previous section. The notation for the Fourier transform is reversed: $\wedge : L^2 \to \ell^2$ is defined for $f \in L^2$ by $\hat{f}(n) = (2\pi)^{-1} \int_{-\pi}^{\pi} f(x)e^{-inx} dx$ (the usual Fourier coefficients of f) while $\vee : \ell^2 \to L^2$ is defined for $F = \{F(n)\}_{n \in \mathbb{Z}} \in \ell^2$ by $\check{F}(x) = \sum_{n \in \mathbb{Z}} F(n)e^{inx}$. Here we always identify $f \in L^2$ with its 2π-periodic extension to $\mathbb{R}$ (denoted f^* in Section 6.3). We define convolution for $f, g \in L^2$ by $f * g(x) = (2\pi)^{-1} \int_{-\pi}^{\pi} f(x - y)g(y)dy$. Then $(f * g)^{\wedge} = \hat{f}\hat{g}$. Plancherel's theorem is that $\|f\|_{L^2} = \|\hat{f}\|_{\ell^2}$. For $E \subseteq \mathbb{Z}$, χ_E is as usual the function with value 1 on E and 0 elsewhere, and card E is the number of elements in E. For $F \in \ell^2$, $\operatorname{supp} F = \{n \in \mathbb{Z} : F(n) \neq 0\}$.

A set $E \subseteq \mathbb{Z}$ is said to be N-transversal ($N \in \mathbb{Z}, N \geq 1$) if

$$\sum_{n \in \mathbb{Z}} \chi_E(k + 2Nn) = 1 \quad \text{for all } k \in \mathbb{Z}.$$

Let $B_N = \{m \in \mathbb{Z} : -N \leq m \leq N - 1\}$. Then B_N is N-transversal. Suppose E is any N-transversal set. Then for each $m \in B_N$, there is exactly one $n = n(m) \in \mathbb{Z}$ such that $m + 2Nn(m) \in E$; the map $m \to m + 2Nn(m)$ provides a one-to-one map of B_N onto E. In particular, card $E = 2N$.

LEMMA 6.17
Let $E \subseteq \mathbb{Z}$ be N-transversal. For $k \in E$, define $e_k \in \ell^2$ by

$$e_k(n) = (2N)^{-1/2} e^{-ikn\pi/N} \chi_E(n).$$

Then the family $\{e_k\}_{k \in E}$ is a complete orthonormal basis for $\ell^2(E) \equiv \{F \in \ell^2 : \text{supp} F \subseteq E\}$. For $F \in \ell^2(E)$, we have

$$\langle F, e_k \rangle = (2N)^{-1/2} \check{F}(k\pi/N) \tag{6.31}$$

and

$$\|F\|_{\ell^2}^2 = (2N)^{-1} \sum_{k \in E} |\check{F}(k\pi/N)|^2. \tag{6.32}$$

PROOF By the N-transversality of E, for each $\ell \in E$ there exists a unique $m \in B_N$ such that $\ell = m + 2Nn(m)$. Hence

$$\langle e_j, e_k \rangle = \frac{1}{2N} \sum_{m \in B_N} e^{i(j-k)(m+2Nn(m))\pi/N}$$

$$= \frac{1}{2N} \sum_{m=-N}^{N-1} e^{i(j-k)m\pi/N} = \delta_{jk}.$$

Hence the family $\{e_k\}_{k \in E}$ is orthonormal. Since it has cardinality $2N$, the dimension of $\ell^2(E)$, it must be complete also. Then (6.31) follows from the definitions, and (6.32) from the above results. ∎

THEOREM 6.18
Suppose $E \subseteq \mathbb{Z}$ is N-transversal, $f, g \in L^2$, $\text{supp}\hat{f} \subseteq E$, and $\text{supp}\hat{g} \subseteq E$ (in particular, f and g are trigonometric polynomials). Then

$$f * g(x) = \frac{1}{2N} \sum_{k \in E} f(k\pi/N) g(x - k\pi/N) = \frac{1}{2N} \sum_{m \in B_N} f(m\pi/N) g(x - m\pi/N), \tag{6.33}$$

and

$$\|f\|_{L^2}^2 = \sum_{k \in E} |f(k\pi/N)|^2 = \sum_{m \in B_N} |f(m\pi/N)|^2. \tag{6.34}$$

PROOF   Applying Plancherel's formula and (6.32) for $F = \hat{f}$ gives the first part of (6.34). By Lemma 6.17,

$$\hat{f} = \sum_{k \in E} \langle \hat{f}, e_k \rangle e_k = (2N)^{-1/2} \sum_{k \in E} f(k\pi/N) e_k.$$

We substitute this into $f * g = (\hat{f}\hat{g})^\vee$ and use $(\hat{g}e_k)^\vee(x) = (2N)^{-1/2} g(x - k\pi/N)$ to obtain the first part of (6.33). Writing each $k \in E$ as $k = m + 2Nn(m)$ for $m \in B_N$, as described above, and using the 2π-periodicity of f and g, gives the second parts of (6.33) and (6.34).   ∎

We remark that with the assumptions $\mathrm{supp}\hat{f}$, $\mathrm{supp}\,\hat{g} \subseteq E$, the sum on $k \in E$ in (6.33) and (6.34) can be replaced by the sum over $k \in F$, for any N-transversal set F. This follows because the argument at the end of the proof shows that in each case the expression involving the sum on $k \in F$ agrees with the far right hand side of (6.33) or (6.34), respectively.

COROLLARY 6.19
Suppose E is N-transversal, $f \in L^2$, and $\mathrm{supp}\hat{f} \subseteq E$. Then

$$f(x) = \frac{1}{2N} \sum_{k \in E} f(k\pi/N)\check{\chi}_E(x - k\pi/N) = \frac{1}{2N} \sum_{k \in B_N} f(k\pi/N)\check{\chi}_E(x - k\pi/N). \quad (6.35)$$

PROOF   Let $g = \check{\chi}_E$ in Theorem 6.18, cf. to the proof of Corollary 6.3.   ∎

Example 6.20.   Shannon sampling formula
Suppose $f \in L^2$ and $\mathrm{supp}\hat{f} \subseteq B_N$, for some $N \geq 1$. Note that B_N is N-transversal. For $x \in \mathbb{T}$ and $x \neq 0$, let

$$h_N(x) = e^{-ix/2}(\sin Nx)/\sin(x/2),$$

and let $h_N(0) = 2N$. Then

$$f(x) = \frac{1}{2N} \sum_{k=-N}^{N-1} f(k\pi/N)h_N(x - k\pi/N).$$

This follows from Corollary 6.19 and a computation showing that $\check{\chi}_{B_N} = h_N$.   ⬚

Example 6.21.   Annular sampling formula
Suppose $f \in L^2$ and

$$\mathrm{supp}\hat{f} \subseteq R_N \equiv \{n \in \mathbb{Z} : -N \leq n \leq -\frac{N}{2} - 1 \quad \text{or} \quad \frac{N}{2} \leq n \leq N - 1\},$$

for some even N, $N \geq 2$. Then R_N is $N/2$-transversal, so by Corollary 6.19,

$$f(x) = \frac{1}{N} \sum_{k \in R_N} f(2k\pi/N)\check{\chi}_{R_n}(x - 2k\pi/N) = \frac{1}{N} \sum_{k \in B_{N/2}} f(2k\pi/N)\check{\chi}_{R_N}(x - 2k\pi/N),$$

$$(6.36)$$

where a computation shows that

$$\check{\chi}_{R_N}(x) = \frac{e^{-ix/2}}{\sin(x/2)} \sin(Nx/2)[2\cos(Nx/2) - 1], \quad \text{for } x \neq 0,$$

and $\check{\chi}_{R_N}(0) = N$. ☐

A pair of sequences $\{\varphi_v\}_{v=0}^{\infty}$, $\{\psi_v\}_{v=0}^{\infty}$, where $\varphi_v, \psi_v \in L^2$ for each v, is said to be admissible if

$$\operatorname{supp}\hat{\varphi}_v, \operatorname{supp}\hat{\psi}_v \subseteq S_v \quad \text{for all } v, \tag{6.37}$$

where $S_0 = \{-2, -1, 0, 1\}$ and $S_v = \{n \in \mathbb{Z} : -2^{v+1} \leq n \leq -2^{v-1} - 1 \text{ or } 2^{v-1} \leq n \leq 2^{v+1} - 1\}$ for $v \geq 1$,

$$\sum_{v=0}^{\infty} \bar{\hat{\varphi}}_v(n)\hat{\psi}_v(n) = 1 \quad \text{for all } n \in \mathbb{Z}, \tag{6.38}$$

and there exist $A_1, A_2 > 0$ such that for all $v \in \mathbb{Z}$,

$$|\hat{\varphi}_v(n)|, |\hat{\psi}_v(n)| \leq A_1 \quad \text{for all } n \in \mathbb{Z}, \tag{6.39}$$

and

$$|\hat{\varphi}_v(n)|, |\hat{\psi}_v(n)| \geq A_2 \quad \text{for all } n \in T_v, \tag{6.40}$$

where $T_0 = S_0$, $T_1 = S_1$, and $T_v = \{n \in \mathbb{Z} : -2^v - 2^{v-1} - 1 \leq n \leq 2^v + 2^{v-2} - 1 \text{ or } 2^v - 2^{v-2} \leq n \leq 2^v + 2^{v-1}\}$ for $v \geq 2$. For each $v \in \mathbb{Z}$, $v \geq 0$, and $k \in \mathbb{Z}$ with $-2^{v+1} \leq k \leq 2^{v+1} - 1$, let

$$\varphi_{vk}(x) = 2^{-v/2-1}\varphi_v(x - 2^{-v-1}k\pi) \quad \text{and} \quad \psi_{vk}(x) = 2^{-v/2-1}\psi_v(x - 2^{-v-1}k\pi),$$

where, as usual, φ_v and ψ_v are identified with their 2π-periodic extensions to $\mathbb{R}$.

THEOREM 6.22   (φ-*Transform Identity for* $L^2(\mathbb{T})$)
Suppose that $\{\varphi_v\}_{v=0}^{\infty}, \{\psi_v\}_{v=0}^{\infty}$ *is an admissible pair of sequences in* L^2. *Suppose* $f \in L^2$. *Then*

$$f = \sum_{v=0}^{\infty} \sum_{k=-2^{v+1}}^{2^{v+1}-1} \langle f, \varphi_{vk} \rangle \psi_{vk}. \tag{6.41}$$

There exist constants $c_1, c_2 > 0$ *(depending only on* A_1, A_2 *in (6.39) and (6.40))* *such that*

$$c_1\|f\|_{L^2}^2 \leq \sum_{v=0}^{\infty} \sum_{k=-2^{v+1}}^{2^{v+1}-1} |\langle f, \varphi_{vk} \rangle|^2 \leq c_2\|f\|_{L^2}^2. \tag{6.42}$$

PROOF   By (6.38) and Fourier inversion, we have $f = \sum_{v=0}^{\infty} \bar{\varphi}_v * \psi_v * f$, where as usual $\bar{\varphi}_v(x) = \overline{\varphi_v(-x)}$ so that $\hat{\bar{\varphi}}_v = \bar{\hat{\varphi}}_v$. By (6.37) we have $\operatorname{supp}(\bar{\varphi}_v * f)^{\wedge}$,

$\operatorname{supp} \hat{\psi}_v \subseteq B_{2^{v+1}}$. Applying Theorem 6.18 with $N = 2^{v+1}$ to $(\tilde{\varphi}_v * f)^\wedge$ and $\hat{\psi}_v$, with $E = B_N$, we obtain

$$\tilde{\varphi}_v * \psi_v * f = \sum_{k=-2^{v+1}}^{2^{v+1}-1} 2^{-v-2} \tilde{\varphi}_v * f(2^{-v-1}k\pi)\psi_v(x - 2^{-v-1}k\pi)$$

$$= \sum_{k=-2^{v+1}}^{2^{v+1}-1} \langle f, \varphi_{vk} \rangle \psi_{vk}.$$

This gives (6.41), and an argument similar to the one above establishing (6.13) (using the far right-hand side of (6.34)) gives (6.42). ∎

Since $\hat{\varphi}_{vk}$, $\hat{\psi}_{vk}$ are supported in the annular regions S_v, (6.41) gives a frequency analysis of f. If we select φ_v and ψ_v strongly concentrated near the origin in some appropriate sense, then φ_{vk} and ψ_{vk} are concentrated near $2^{-v-1}k\pi$, and (6.41) also gives a spatial analysis of f. However, in this case it is not clear how to construct admissible $\{\varphi_v\}_{v=0}^\infty$, $\{\psi_v\}_{v=0}^\infty$ having optimal concentration, or even what optimality criterion should be used.

THEOREM 6.23 *(Shannon basis)*
Let $C_0 = \{-1, 0\}$ and, for $v \geq 1$, $v \in \mathbb{Z}$, let $C_v = B_{2^{v-1}} = \{n \in \mathbb{Z} : -2^{v-1} \leq n \leq 2^{v-1} - 1\}$. Let $\varphi_{00}(x) = 2^{-1/2}(1 + e^{-ix})$ and $\varphi_{0-1}(x) = 2^{-1/2}(1 - e^{-ix})$. For $v \geq 1$ and $k \in C_v$, let

$$\varphi_{vk}(x) = 2^{-v/2} \check{\chi}_{R_{2^v}}(x - 2^{-v+1}k\pi)$$

$$= \frac{2^{-v/2} e^{i(\frac{1}{2}x - 2^{-v}k\pi)}}{\sin(\frac{1}{2}x - 2^{-v}k\pi)} (-1)^k \sin(2^{v-1}x)[(-1)^k \cos(2^{v-1}x) - 1],$$

for $x \neq 2^{-v+1}k\pi$, and $\varphi_{vk}(2^{-v+1}k\pi) = 2^{v/2}$, where R_{2^v} is as in Example 6.21. Then the collection $\{\varphi_{vk}\}_{v \geq 0, k \in C_v}$ is a complete orthonormal basis for L^2.

PROOF Let $\varphi_0(x) = \check{\chi}_{C_0}(x) = 1 + e^{-ix}$ and, for $v \geq 1$, $\varphi_v = \check{\chi}_{R_{2^v}}$. Then $\sum_{v=0}^\infty \hat{\varphi}_v \equiv 1$ on $\mathbb{Z}$, so by Fourier inversion, for $f \in L^2$, $f = \sum_{v=0}^\infty \varphi_v * f$. Example 6.20 shows that

$$\varphi_0 * f(x) = \frac{1}{2} \sum_{k=-1}^{0} \varphi_0 * f(k\pi)\varphi_0(x - k\pi) = \sum_{k \in C_0} \langle f, \varphi_{0k} \rangle \varphi_{0k}(x).$$

For each $v \geq 1$ we apply Example 6.21 with $N = 2^v$. The right-hand side of (6.36) gives

$$\varphi_v * f(x) = 2^{-v} \sum_{k \in C_v} \varphi_v * f(2^{-v+1}k\pi)\varphi_v(x - 2^{-v+1}k\pi) = \sum_{k \in C_v} \langle f, \varphi_{vk} \rangle \varphi_{vk}(x).$$

It is easy to check that $\|\varphi_{0k}\|_{L^2} = 1$ for $k = -1, 0$, and, for $v \geq 1$, $k \in C_v$, $\|\varphi_{vk}\|_{L^2}^2 = 2^{-v}\|\varphi_v\|_{L^2}^2 = 2^{-v}\|\chi_{R_{2^v}}\|_{\ell^2}^2 = 1$. The conclusion now follows from Lemma 6.7. ∎

6.5   The integers modulo N

In this chapter we consider a setting which is both discrete, like $\mathbb{Z}$, and bounded, like $\mathbb{T}$. We fix the following notation: we assume $L \in \mathbb{Z}, L \geq 1, N = 2^L$, and $M = N/2$. Then $\mathbb{Z}_N = \{n \in \mathbb{Z} : -M \leq n \leq M - 1\}$. We consider finite sequences $f = \{f(n)\}_{n \in \mathbb{Z}_N}$, which can be regarded as elements of $\mathbb{C}^N$, or, equivalently, $\ell^2(\mathbb{Z}_N)$. The discrete Fourier transform (DFT) of $f \in \mathbb{C}^N$ is defined by

$$\hat{f}(m) = \sum_{n=-M}^{M-1} f(n)e^{-2\pi inm/N}, \quad \text{for } m \in \mathbb{Z}_N,$$

and the inverse DFT by $\check{F}(n) = \frac{1}{N}\sum_{m=-M}^{M-1} f(m)e^{2\pi inm/N}$, for $n \in \mathbb{Z}_N$. Then $(\hat{f})^\vee = f = (\check{f})^\wedge$. (We choose the more symmetric definition of $\mathbb{Z}_N$ above, rather than the traditional choice of $\{0, 1, \ldots, N-1\}$, for convenience. For the DFT, the periodicity of the complex exponentials shows that these choices are equivalent.) We identify $f \in \ell^2 \equiv \ell^2(\mathbb{Z}_N)$ with its extension to $\mathbb{Z}$ which is periodic with period N. We then define $f * g(n) = \sum_{m \in \mathbb{Z}_N} f(n-m)g(m)$; it follows that $(f*g)^\wedge = \hat{f}\hat{g}$. We set $\langle f, g \rangle = \sum_{n \in \mathbb{Z}_N} f(n)\overline{g(n)}$ and $\|f\|_{\ell^2} = (\sum_{n \in \mathbb{Z}_N} |f(n)|^2)^{1/2}$. Parseval's relation and Plancherel's formula are $\langle \hat{f}, \hat{g} \rangle = N\langle f, g \rangle$ and (hence) $\|\hat{f}\|_{\ell^2}^2 = N\|f\|_{\ell^2}^2$, respectively. The notations χ_E, card E, and supp f, for $E \subseteq \mathbb{Z}_N$ and $f \in \ell^2(\mathbb{Z}_N)$, are as in Section 6.4.

Our assumption that $N = 2^L$ is used in Theorems 6.29 and 6.30 below. Until then it suffices to assume $N = 2M$, where $M \in \mathbb{Z}$, $M > 1$, is not prime. The notation $m \mid M$ means that $M = \ell m$ for some $\ell \in \mathbb{Z}$.

For $m \in \mathbb{Z}$, $m \geq 1$, such that $m \mid M$, we say that $E \subseteq \mathbb{Z}_N$ is m-transversal if

$$\sum_{k=0}^{M/m} \chi_E(n + 2mk) = 1 \quad \text{for all } n \in \mathbb{Z}_N,$$

where χ_E is extended to be N-periodic on $\mathbb{Z}$, as usual. This implies that card $E = 2m$. As in the corresponding case in Section 6.3, the definition of E being m-transversal for $\mathbb{Z}_N$ is equivalent to the statement that E, regarded now as a subset of $\mathbb{Z}$, is m-transversal on $\mathbb{Z}$ in the sense of Section 6.4. This leads to the following result.

LEMMA 6.24
Suppose $m \mid M$ and $E \subseteq \mathbb{Z}_N$ is m-transversal. For $k \in E$, define $e_k \in \ell^2$ by

$$e_k(n) = (2m)^{-1/2}e^{-ikn\pi/m}\chi_E(n).$$

Then the family $\{e_k\}_{k \in E}$ is a complete orthonormal basis for $\ell^2(E) = \{F \in \ell^2 : \text{supp } F \subseteq E\}$. For $F \in \ell^2(E)$, we have

$$\langle F, e_k \rangle = \frac{N}{\sqrt{2m}}\check{F}(kN/m), \tag{6.43}$$

and hence

$$\|F\|_{\ell^2}^2 = \frac{N^2}{2m} \sum_{k \in E} |\check{F}(kN/2m)|^2. \tag{6.44}$$

PROOF The orthonormality and completeness of the set $\{e_k\}_{k \in E}$ follow from Lemma 6.17 and the remarks above. Then (6.43) is a simple calculation, and (6.44) follows from these results. ∎

For m such that $m \mid M$, let $B_m = \{n \in \mathbb{Z} : -m \le n \le m - 1\}$. Then B_m is m-transversal. As in Section 6.4, any m-transversal set E is in one-to-one correspondence with B_m via a map taking $k \in E$ to $\ell \in B_m$, where $k = \ell + 2mn(\ell)$ for a unique integer $n(\ell)$.

THEOREM 6.25
Suppose $m \mid M$; $E \subseteq \mathbb{Z}_N$ is m-transversal; $f, g \in \ell^2$; $\operatorname{supp} \hat{f} \subseteq E$; and $\operatorname{supp} \hat{g} \subseteq E$. Then

$$f * g(n) = \frac{N}{2m} \sum_{k \in E} f(kN/2m) g(n - kN/2m) = \frac{N}{2m} \sum_{k \in B_m} f(kN/2m) g(n - kN/2m), \tag{6.45}$$

and

$$\|f\|_{\ell^2}^2 = \frac{N}{2m} \sum_{k \in E} |f(kN/2m)|^2 = \frac{N}{2m} \sum_{k \in B_m} |f(kN/2m)|^2. \tag{6.46}$$

PROOF The proof is similar to the proof of Theorem 6.18. Expanding $\hat{f}$ in terms of the orthonormal basis $\{e_k\}_{k \in E}$ in Lemma 6.24 and using $f * g = (\hat{f} \hat{g})^{\vee}$ gives

$$f * g = \sum_{k \in E} \frac{N}{\sqrt{2m}} f(kN/2m)(\hat{g} e_k)^{\vee}.$$

Then the translation properties of the DFT give that $(\hat{g} e_k)^{\vee}(n) = (2m)^{-1/2} g(n - kN/2m)$. This gives the left equality in (6.45). The left side of (6.46) follows from Plancherel's formula and (6.44). The right sides of (6.45) and (6.46) follow then from the correspondence between E and B_m described above and the N-periodicity of f and g. ∎

COROLLARY 6.26
Suppose $m \mid M$, E is m-transversal, $f \in \ell^2$ and $\operatorname{supp} \hat{f} \subseteq E$. Then

$$f(n) = \frac{N}{2m} \sum_{k \in E} f(kN/2m) \check{\chi}_E(n - kN/2m) = \frac{N}{2m} \sum_{k \in B_m} f(kN/2m) \check{\chi}_E(n - kN/2m). \tag{6.47}$$

PROOF Let $g = \check{\chi}_E$ in Theorem 6.25, which implies $\hat{f} \hat{g} = \hat{f}$, hence $f = f * g$. ∎

Example 6.27.   Shannon sampling formula
Suppose $m \mid M, f \in \ell^2$, and $\mathrm{supp}\hat{f} \subseteq B_m$. For $n \in \mathbb{Z}_N$, $n \neq 0$, let

$$h_m(n) = \frac{1}{N} e^{-i\pi n/N} \sin(2\pi nm/N)/\sin(\pi n/N),$$

and $h_m(0) = 2m/N$. Then

$$f(n) = \frac{N}{2m} \sum_{k=-m}^{m-1} f(kN/2m)h_m(n - kN/2m).$$

This follows by Corollary 6.26 and a computation showing that $\check{\chi}_{B_m} = h_m$.   □

Example 6.28.   Annular sampling formula
Suppose $m \mid M$ and m is even. Let $R_m = \{n \in \mathbb{Z} : -m \leq n \leq -m/2 - 1$ or $m/2 \leq n \leq m - 1\}$. Then R_m is $(m/2)$-transversal. Hence by Corollary 6.26, if $f \in \ell^2$ and $\mathrm{supp}\hat{f} \subset R_m$,

$$f(n) = \frac{N}{m} \sum_{k \in R_m} f(kN/m)\check{\chi}_{R_m}(n - kN/m) = \frac{N}{m} \sum_{k \in B_{m/2}} f(kN/m)\check{\chi}_{R_m}(n - kN/m),$$
(6.48)

where a computation shows that

$$\check{\chi}_{R_m}(n) = \frac{e^{-i\pi n/N}}{N\sin(\pi n/N)} \sin(\pi nm/N)[2\cos(\pi nm/N) - 1], \quad \text{for } n \neq 0$$

and $\check{\chi}_{R_m}(0) = m/N$.   □

The version of the φ-transform for $\ell^2(\mathbb{Z}_N)$ that we consider next will allow a little more flexibility (compared to the versions in Sections 6.2, 6.3, and 6.4) by introducing a parameter v_0. The functions $\psi_{v_0 k}$ will be the lowest frequency terms in the decomposition. Selecting v_0 amounts to choosing how finely to subdivide the low end of the frequency spectrum. To keep the notation simpler, we have not treated this variation in the cases of $\mathbb{R}$, $\mathbb{Z}$, and $\mathbb{T}$, but arguments similar to those we will give for $\mathbb{Z}_N$ lead to natural analogues in these three previous cases.

Recall that we assume $N = 2^L$ and set $M = N/2$. Fix $v_0 \in \mathbb{Z}$ with $0 \leq v_0 < L - 2$. We say that a pair of sequences $\{\varphi_v\}_{v=v_0}^{L-2}$, $\{\psi_v\}_{v=v_0}^{L-2}$, where $\varphi_v, \psi_v \in \ell^2$ for each v, is admissible if

$$\mathrm{supp}\,\hat{\varphi}_v, \mathrm{supp}\,\hat{\psi}_v \subseteq S_v \quad \text{for each } v,$$
(6.49)

where $S_{v_0} = \{n \in \mathbb{Z}_N : -2^{v_0+1} \leq n \leq 2^{v_0+1} - 1\}$, and, for $v_0 < v \leq L - 2$, $S_v = \{n \in \mathbb{Z}_N : -2^{v+1} \leq n \leq -2^{v-1} - 1$ or $2^{v-1} \leq n \leq 2^{v+1} - 1\}$ (as in Section 6.4),

$$\sum_{v=v_0}^{L-2} \bar{\hat{\varphi}}_v(n)\hat{\psi}_v(n) = 1 \quad \text{for all } n \in \mathbb{Z}_N,$$
(6.50)

there exist constants $A_1, A_2 > 0$ (independent of v and N) such that

$$|\hat{\varphi}_v(n)|, |\hat{\psi}_v(n)| \leq A_1 \quad \text{for all } n \in \mathbb{Z},$$
(6.51)

and

$$|\hat{\varphi}_v(n)|, |\hat{\psi}_v(n)| \geq A_2 \quad \text{for all } n \in T_v. \tag{6.52}$$

Here if $2 \leq v_0 < L - 2, T_{v_0} = \{n \in \mathbb{Z} : -2^{v_0} - 2^{v_0-1} - 1 \leq n \leq 2^{v_0} + 2^{v_0-1}\}$; in this case for $v_0 < v < L - 2, T_v = \{n \in \mathbb{Z} : -2^v - 2^{v-1} - 1 \leq n \leq -2^v + 2^{v-2} - 1$ or $2^v - 2^{v-2} \leq n \leq 2^v + 2^{v-1}\}$, and for $v = L-2, T_v = \{n \in \mathbb{Z} : -M \leq n \leq -2^v + 2^{v-2} - 1$ or $2^v - 2^{v-2} - 1 \leq n \leq M - 1\}$. If $v_0 = 0$, then $T_0 = S_0 = \{-2, -1, 0, 1\}$, $T_1 = S_1 = \{-4, -3, -2, 1, 2, 3\}$, and T_v is as above for $v > 1$, while if $v_0 = 1$, $T_1 = S_1 = \{n \in \mathbb{Z} : -4 \leq n \leq 3\}$ and T_v is as above for $v > 1$. For each $v \in \mathbb{Z}$ with $v_0 \leq v \leq L - 2$, and $k \in \mathbb{Z}$ with $-2^{v+1} \leq k \leq 2^{v+1} - 1$, let

$$\varphi_{vk}(n) = \sqrt{N} 2^{-v/2-1} \varphi_v(n - 2^{-v-2}kN)$$

and

$$\psi_{vk}(n) = \sqrt{N} 2^{-v/2-1} \psi_v(n - 2^{-v-2}kN),$$

where as usual φ_{vk} and ψ_{vk} are extended to $\mathbb{Z}$ to be N-periodic.

THEOREM 6.29   (*φ-Transform Identity for $\ell^2(\mathbb{Z}_N)$, [Ku]*)
Suppose $N = 2^L$, $0 \leq v_0 < L - 2$, and $\{\varphi_v\}_{v=v_0}^{L-2}$, $\{\psi_v\}_{v=v_0}^{L-2}$ is an admissible pair of sequences in $\ell^2(\mathbb{Z}_N)$. Suppose $f \in \ell^2$. Then

$$f = \sum_{v=v_0}^{L-2} \sum_{k=-2^{v+1}}^{2^{v+1}-1} \langle f, \varphi_{vk} \rangle \psi_{vk}. \tag{6.53}$$

There exist constants $c_1, c_2 > 0$ depending only on A_1, A_2 in (6.51) and (6.52) (hence independent of f and N) such that

$$c_1 \|f\|_{\ell^2}^2 \leq \sum_{v=v_0}^{L-2} \sum_{k=-2^{v+1}}^{2^{v+1}-1} |\langle f, \varphi_{vk} \rangle|^2 \leq c_2 \|f\|_{\ell^2}^2. \tag{6.54}$$

PROOF   As usual, we set $\tilde{\varphi}_v(n) = \overline{\varphi_v(-n)}$ so that $\hat{\tilde{\varphi}}_v = \overline{\hat{\varphi}_v}$. Then Fourier inversion and (6.50) imply that $f = \sum_{v=v_0}^{L-2} \tilde{\varphi}_v * \psi_v * f$. Noting that $\text{supp}(\tilde{\varphi}_v * f)^\wedge$, $\text{supp } \hat{\psi}_v \subseteq B_{2^{v+1}}$, an application of Theorem 6.25 leads to (6.53), as for Theorems 6.6, 6.14, and 6.22. We have chosen the T_v's in (6.52) to overlap, so (6.51) and (6.52) imply that there exist constants B_1, B_2 independent of N such that

$$B_2 \leq \sum_{v=v_0}^{L-2} |\hat{\varphi}_v(n)|^2 \leq B_1 \quad \text{for all } n \in \mathbb{Z}_N.$$

This and an argument similar to the one in the proof of Theorem 6.6 give (6.54).   ∎

One would like to pick the functions φ_v and ψ_v highly concentrated near the origin, so that (6.53) gives a well-localized spatial analysis of f (it is well localized in frequency automatically by (6.49)). However, as in Section 6.4, it is not clear how to do this optimally.

THEOREM 6.30   **(Shannon basis)**
Let $N = 2^L$ and $0 \le v_0 < L - 2$. Let $C_{v_0} = \{n \in \mathbb{Z} : -2^{v_0} \le n \le 2^{v_0} - 1\}$, and for $v_0 < v \le L - 1$, let $C_v = \{n \in \mathbb{Z} : -2^{v-1} \le n \le 2^{v-1} - 1\}$. Let

$$\varphi_{v_0}(n) = \check{\chi}_{B_{2^{v_0}}}(n) = \frac{1}{N} e^{-i\pi n/N} \sin(2^{v_0+1} n\pi/N) / \sin(n\pi/N), \quad \text{for } n \ne 0,$$

and $\varphi_{v_0}(0) = 2^{v_0+1}/N$. For $k \in C_{v_0}$, let

$$\varphi_{v_0 k}(n) = 2^{-(v_0+1)/2} \sqrt{N} \varphi_{v_0}(n - 2^{-v_0-1} kN).$$

For $v_0 < v \le L - 1$, let

$$\varphi_v(n) = \check{\chi}_{R_{2^v}}(n) = \frac{e^{-i\pi n/N}}{N \sin(\pi n/N)} \sin(2^v n\pi/N)[2\cos(2^v n\pi/N) - 1], \quad \text{for } n \ne 0,$$

and $\varphi_v(0) = m/N$. For $k \in C_v$, $v_0 < v \le L - 1$, let

$$\varphi_{vk}(n) = 2^{-v/2} \sqrt{N} \varphi_v(n - 2^{-v} kN).$$

Then the collection $\{\varphi_{vk}\}_{v_0 \le v \le L-1, k \in C_v}$ is a complete orthonormal basis for $\ell^2(\mathbb{Z}_N)$.

PROOF   The formulae for $\check{\chi}_{B_{2^{v_0}}}$ and $\check{\chi}_{R_{2^v}}$ come from Example 6.27 and 6.28. We have $\sum_{v=v_0}^{L-1} \hat{\varphi}_v \equiv 1$ on $\mathbb{Z}_N$, and hence for $f \in \ell^2(\mathbb{Z}_N)$, $f = \sum_{v=v_0}^{L-1} \varphi_v * f$. Using Example 6.27 for $\varphi_{v_0} * f$ and Example 6.28 for each $\varphi_v * f$, $v_0 < v \le L - 1$, we obtain

$$f = \sum_{v=v_0}^{L-1} \sum_{k \in C_v} \langle f, \varphi_{vk} \rangle \varphi_{vk}.$$

Also for each $k \in C_{v_0}$, we have, using Plancherel's theorem,

$$\|\varphi_{v_0 k}\|_{\ell^2}^2 = 2^{-v_0-1} N \|\varphi_{v_0}\|_{\ell^2}^2 = 2^{-v_0-1} \|\chi_{B_{2^{v_0}}}\|_{\ell^2}^2 = 1,$$

and for $k \in C_v$, $v_0 < v \le L - 1$,

$$\|\varphi_{vk}\|_{\ell^2}^2 = 2^{-v} N \|\varphi_v\|_{\ell^2}^2 = 2^{-v} \|\chi_{R_{2^v}}\|_{\ell^2}^2 = 1.$$

Hence Lemma 6.7 completes the proof.   ∎

6.6   Group theoretic setting

As a reader familiar with the theory of locally compact Abelian groups should expect, there is a general version of the basic sampling result (Theorems 6.2, 6.10, 6.18, and 6.25). Each case considered above is then an example of this general result.

Let G be a locally compact Abelian group with given Haar measure μ. (We refer to [R] for the definitions and background needed here.) Let $\hat{G}$ be the dual group of G. For a character $\xi \in \hat{G}$ and an element $x \in G$, denote the action of ξ on x (and vice versa) by $e_x(\xi)$. Define the Fourier transform for $\xi \in \hat{G}$ by

$$\hat{f}(\xi) = \int_G f(x)e_{-x}(\xi)d\mu(x).$$

Let the Haar measure $\hat{\mu}$ on $\hat{G}$ be normalized so that the Fourier inversion formula takes the form

$$f(x) = (\hat{f})^\vee(x) = \int_{\hat{G}} \hat{f}(\xi)e_x(\xi)d\hat{\mu}(\xi) \quad \text{for } x \in G.$$

A set $E \subseteq \hat{G}$ is said to be L-transversal if $\hat{\mu}(E) = 2L < +\infty$ and there exists a set $\{x_i\}_{i \in A} \subseteq G$ such that the family $\{(2L)^{-1/2}e_{-x_i}(\xi)\chi_E(\xi)\}_{i \in A}$ is an orthonormal basis for $L^2(E, d\hat{\mu})$.

THEOREM 6.31
Suppose $E \subseteq \hat{G}$ is L-transversal; $f, g \in L^2(G)$; $\mathrm{supp}\hat{f} \subseteq E$; and $\mathrm{supp}\,\hat{g} \subseteq E$. Then

$$f * g(x) = \frac{1}{2L} \sum_{i \in A} f(x_i)g(x - x_i), \quad x \in G. \tag{6.55}$$

PROOF We have

$$\langle \hat{f}, e_{-x_i}\chi_E \rangle = \int_{\hat{G}} \hat{f}(\xi)\overline{e_{-x_i}(\xi)}d\hat{\mu}(\xi)$$

$$= \int_{\hat{G}} \hat{f}(\xi)e_{x_i}(\xi)d\hat{\mu}(\xi) = f(x_i)$$

by Fourier inversion. So expanding $\hat{f}$ in the above orthonormal basis gives

$$\hat{f} = \frac{1}{2L} \sum_{i \in A} f(x_i)e_{-x_i}\chi_E.$$

Also we have

$$(\hat{g}(\cdot - x_i))^\wedge(\xi) = \int_G g(x - x_i)e_{-x}(\xi)d\mu(x)$$

$$= \int_G g(y)e_{-x_i-y}(\xi)d\mu(y) = e_{-x_i}(\xi)\hat{g}(\xi),$$

or $(\hat{g}e_{-x_i})^\vee(x) = g(x - x_i)$. Putting these two facts into $f * g = (\hat{f}\hat{g})^\vee$ and using $\hat{g}\chi_E = \hat{g}$ gives (6.55). $\blacksquare$

This result should be compared with Kluvánek's sampling result for locally compact Abelian groups in [K] or [H].

COROLLARY 6.32

Suppose $E \subseteq \hat{G}$ is closed and L-transversal. If $f \in L^2(G)$ and supp$\hat{f} \subseteq E$, then

$$f(x) = \frac{1}{2L} \sum_{i \in A} f(x_i) \check{\chi}_E(x - x_i), \quad x \in G.$$

PROOF Let $g = \check{\chi}_E$ in Theorem 6.31 and note that $f * g = f$ since $\hat{f}\hat{g} = \hat{f}$. ∎

If we can find transversal sets that are balls or annuli around the origin, then natural analogues of the Shannon and annular sampling theorems can be formulated as examples. If $\hat{G}$ can then be partitioned according to the "size" of ξ, using transversal sets appropriately, then analogues of the φ-transform identity and Shannon wavelets can also be found. Results of this type have been obtained in the context of Vilenkin groups in [OWe].

Acknowledgments. The first author would like to thank Dr. Arun Kumar some helpful conversations. Both authors would like to thank anonymous referees for pointing out some important references.

Bibliography

[B1] J. Benedetto. 1992. "Irregular sampling and frames," in *Wavelets: A Tutorial in Theory and Applications*, C.K. Chui, ed., Boston: Academic Press, pp. 445–507.

[B2] J. Benedetto. 1993. "Frame decompositions, sampling, and uncertainty principle inequalities," this volume, Chap. 7.

[Bu] P.L. Butzer. 1983. "A survey of the Whittaker–Shannon sampling theorem and some of its extensions," *J. Math. Res. Exposition* **3**: 185–212.

[EFr] J. Epperson and M. Frazier. 1992. "An almost orthogonal radial wavelet expansion for radial distributions," preprint.

[FGr] H. Feichtinger and C. Gröchenig. 1993. "Theory and practice of irregular sampling," this volume, Chap. 8.

[FrJa1] M. Frazier and B. Jawerth. 1985. "Decomposition of Besov spaces," *Ind. Univ. Math. J.* **34**: 777–799.

[FrJa2] M. Frazier and B. Jawerth. 1990. "A discrete transform and decompositions of distribution spaces," *J. Funct. Anal.* **93**: 34–170.

[FrJaW] M. Frazier, B. Jawerth, and G. Weiss. 1991. *Littlewood–Paley Theory and the Study of Function Spaces*, CBMS Conf. Lect. Notes **79**, Providence: Amer. Math. Soc.

[FrKu] M. Frazier and A. Kumar. 1993. "An introduction to the orthonormal wavelet transform on discrete sets," this volume, Chap. 2.

[G]   R.P. Gosselin. 1963. "On the L^p theory of cardinal series," *Ann. Math.* **78**: 567–581.

[H]   J.R. Higgins. 1985. "Five short stories about the cardinal series," *Bull. Amer. Math. Soc.* **12**: 45–89.

[Ho]   M. Holschneider. 1990. "Wavelet analysis on the circle," *J. Math. Phys.* **31**(1): 39–44.

[J]   A.J. Jerri. 1977. "The Shannon sampling theorem—Its various extensions and applications: A tutorial review," *Proc. IEEE* **65**: 1565–1596.

[K]   I. Kluvánek. 1965. "Sampling theorem in abstract harmonic analysis," *Mat.-Fyz. Časopis Sloven. Akad. Vied* **15**: 43–48.

[Ku]   A. Kumar. 1990. "Time-frequency analysis and transform coding," Ph.D. dissertation, Electrical Eng. Dept., Washington Univ., St. Louis.

[KuFu] A. Kumar and D. Fuhrmann. 1990. "The Frazier–Jawerth transform," *Proc. Int. Conf. Acoust. Speech Signal Process.*, **5**: 2483–2486, Albuquerque, N.M.

[M1]   Y. Meyer. 1989. "Wavelets and operators," in *Analysis at Urbana*, E. Berkson et al., eds., London Math. Soc. Lect. Note **137**, Cambridge: Cambridge Univ. Press, pp. 256–364.

[M2]   Y. Meyer. 1990. *Ondelettes et Opérateurs*, Paris: Hermann.

[OWe]   C. Onneweer and S. Weiyi. 1989. "Homogeneous Besov spaces on locally compact Vilenkin groups," *Studia Math.* **93**: 17–39.

[P]   J. Price. 1985. "Uncertainty principles and sampling theorems," in *Fourier Techniques and Applications*, J. Price, ed., New York: Plenum Press, pp. 25–44.

[R]   W. Rudin. 1962. *Fourier Analysis on Groups*, New York: Interscience Publishers.

[S]   C.E. Shannon. 1949. "Communication in the presence of noise," *Proc. IRE* **37**: 10–21.

[St]   R. Strichartz. 1993. "Construction of orthonormal wavelets," this volume, Chap. 1.

[T]   R. Torres. 1991. "Spaces of sequences, sampling theorem, and functions of exponential type," *Studia Math.* **100**: 51–74.

7

Frame decompositions, sampling, and uncertainty principle inequalities

John J. Benedetto

ABSTRACT   Gabor's approach to signal analysis has been formalized and generalized in several ways. The approach in this chapter integrates signal reconstruction and the uncertainty principle. The signal reconstruction is in terms of frame decompositions and sampling theory, with an emphasis on local sampling and effective irregular sampling theorems and algorithms. The uncertainty principle is formulated in terms of the uncertainty principle for variances in the context of Gabor and wavelet systems; and is developed to include general weighted uncertainty principle inequalities.

Examples include dissections of the information plane, real-time sampling, auditory models, stationary systems, the Plancherel–Pólya and Boas theory, aliasing, zero-crossings, entropy inequalities, uniqueness theory and logarithmic integrals, local uncertainty principle inequalities, and the role of Riesz transforms and A_p-spaces. The theory unifies some of these notions in the parlance of wavelets and coherent states. The methods are from harmonic and functional analysis.

CONTENTS

0-8493-8271-8/94/$0.00 + $.50

7.1   Introduction

We shall pose a Fundamental Problem in Signal Analysis based on Gabor's decomposition formula (1946) and on his use of the classical uncertainty principle inequality in choosing modulates and translates of the Gaussian as elementary signals in this formula. This Fundamental Problem and its solution depend on sampling formulas and uncertainty principle inequalities; and these topics are the subject matter of this chapter.

The Fourier transform $\hat{f}$ of a complex-valued function $f : \mathbb{R} \to \mathbb{C}$ defined on the real line $\mathbb{R}$ is formally written as

$$\hat{f}(\gamma) = \int f(t)e^{-2\pi it\gamma}\, dt,$$

for $\gamma \in \hat{\mathbb{R}}(=\mathbb{R})$, where integration is over $\mathbb{R}$. The Fourier transform is well defined on

$$L^1(\mathbb{R}) = \left\{ f : \|f\|_1 = \int |f(t)|\, dt < \infty \right\}$$

and on

$$L^2(\mathbb{R}) = \left\{ f : \|f\|_2 = \left(\int |f(t)|^2\, dt \right)^{1/2} < \infty \right\}.$$

The *classical sampling theorem* asserts that if $0 < 2T\Omega \leq 1$ then

$$\forall f \in PW_\Omega, \quad f(t) = T\sum f(nT)s(t - nT), \tag{7.1}$$

where the convergence is uniform on $\mathbb{R}$ and in $L^2(\mathbb{R})$ norm, and where

$$PW_\Omega = \left\{ f \in L^2(\mathbb{R}) : \mathrm{supp}\hat{f} \subseteq [-\Omega, \Omega] \right\}$$

is the *Paley–Wiener space* and

$$s(t) = \frac{\sin(2\pi\Omega t)}{\pi t} \equiv d_{2\pi\Omega}(t).$$

The *classical uncertainty principle inequality* asserts that

$$\forall f \in L^2(\mathbb{R}) \quad \text{and} \quad \forall(t_0, \gamma_0) \in \hat{\mathbb{R}} \times \mathbb{R}, \tag{7.2}$$
$$\|f\|_2^2 \leq 4\pi\|(t - t_0)f(t)\|_2\|(\gamma - \gamma_0)\hat{f}(\gamma)\|_2,$$

and there is equality in (7.2) if and only if f is of the form

$$\varphi(t) = Ce^{2\pi it\gamma_0}e^{-s(t-t_0)^2} \tag{7.3}$$

for $C \in \mathbb{C}$ and $s > 0$.

These two particular formulas, (7.1) and (7.2), are at the foundation of our formulation of the Fundamental Problem, and are apposite for the understanding and development of some of the basic ideas in engineering, mathematics, and the sciences. They are also what we call *creative formulas*, as well as being intimately related to each other in certain critical applications. The developments associated with (7.1) and (7.2) as creative formulas, and the theory associated with the relations between (7.1) and (7.2), are the theme of the chapter, written as it is in the context of the Fundamental Problem.

What do we mean by a creative formula? The Pythagorean theorem is an example of such a formula. It leads to the proper definition of distance between points in Euclidean space. It is fundamental in defining lengths of curves, areas of surfaces, etc. It extends to defining the notion of distance by means of the L^2 norm in spaces of square integrable functions. It provides a fundamental guideline in the geometry of Hilbert space. It is a backdrop from which various non-Euclidean geometries are assessed. It inspires new developments in the sciences, with concepts such as *energy* or fields of study such as the *special theory of relativity*. Similarly, sampling formulas and uncertainty principle inequalities have given rise to profound generalizations and extensions, and new points of view.

There is a third formula which is at the theoretical basis of the Fundamental Problem. The formula, which is well known and elementary to verify, asserts that

$$S = L^*L, \tag{7.4}$$

where L is the map

$$L : H \to l^2(\mathbb{Z}^d)$$
$$y \mapsto \{\langle y, x_n \rangle : n \in \mathbb{Z}^d\},$$

$\langle .,. \rangle$ is the inner product on the Hilbert space H, L^* is the adjoint of L, and S is the frame operator for the frame $\{x_n\}$. These notions are explained in Sections 7.3 and 7.4; to fix ideas, H will always be a complex infinite-dimensional separable Hilbert space.

The Fundamental Problem is described in Section 7.3. Section 7.4 is a functional analytic excursion through various types of decompositions. Frames and exact frames are characterized in terms of L in Theorem 7.15 and Theorem 7.19; and the role of topics such as stationarity and the Plancherel–Pólya–Boas theory is examined.

Sections 7.5–7.8 can be read independently from the point of view developed in Sections 7.3 and 7.4. These sections exhibit useful sampling formulas and algorithms, as well as uncertainty principle inequalities.

In Section 7.5, Theorem 7.22 is a *general* and *implementable* regular sampling theorem proved in terms of frames; and Theorem 7.26 is the first of our constructive local regular sampling results, which have many natural applications. Section 7.7 provides analogous effective results for the case of irregular sampling. In particular, Theorem 7.45 is a general and implementable irregular sampling theorem proved in terms of frames; and Algorithm 7.51a is a successful implementation of Theorem 7.45. Algorithm 7.51b briefly outlines an important application of the general point of view of this chapter. This application is based on our wavelet auditory model (WAM) for data compression and signal reconstruction in terms of irregular sampling.

In Section 7.6, the classical uncertainty principle inequality is analyzed in terms of classical as well as functional analysis, i.e., Theorem 7.30 and Remark 7.31 as well as Theorem 7.32. The remainder of Section 7.6 studies the role of the uncertainty principle in wavelet and Gabor systems. In Section 7.8, we give a taste of the power of harmonic analysis to formulate and prove weighted uncertainty principle inequalities.

Besides the usual notation in analysis as found in the books by Hörmander [Hö1], Schwartz [S], and Stein and Weiss [StWei], we use the conventions and notation described in Section 7.2.

7.2 Notation

The integral over $\mathbb{R}^d$, resp., the sum over $\mathbb{Z}^d$, is designated by "$\int$," resp., "$\sum$." The d-dimensional torus $\mathbb{T}^d$ is the compact group $\mathbb{R}^d/\mathbb{Z}^d$, and we write $\mathbb{T}_r \equiv \mathbb{R}/(r\mathbb{Z})$ for $r > 0$. The characteristic function of a set S is $\mathbf{1}_S$, and we write $\mathbf{1}_{(\Omega)}$ for $S = [-\Omega, \Omega)$; and the Lebesgue measure of S is $|S|$. "card" designates cardinality.

The L^1 dilation of f is $f_\lambda(t) \equiv \lambda f(t\lambda)$, so that the Fourier transform of $\mathbf{1}_{(\Omega)}$ is $d_{2\pi\Omega}$, where $d(t) \equiv (\sin t)/(\pi t)$. The involution of g is $\tilde{g}(t) \equiv \bar{g}(-t)$, the inverse Fourier transform of F is $F^\vee$, the support of F is $\operatorname{supp} F$, and $\delta_{m,n}$ is 1 if $m = n$ and is 0 if $m \neq n$.

The conjugate exponents for the L^p spaces are p and p'. The Schwartz space is $\mathcal{S}$, the space of bounded Radon measures is M_b, the space of absolutely convergent Fourier transforms is A, and its dual space, the space of pseudo-measures, is A'.

The exponential $e_b(t)$ is $e^{2\pi i t b}$ and translation $\tau_a g(t)$ is $g(t - a)$. We write $\varphi_{m,n}(t) = e_{mb}(t)\tau_{na}g(t)$ and $\psi_{m,n}(t) = 2^{m/2}\psi(2^m t - n)$ for $t \in \mathbb{R}$. Finally, if $\{\theta_n\}$ is contained in a Banach space, then $sp\{\theta_n\}$ is the linear span of $\{\theta_n\}$, and $\overline{sp}\{\theta_n\}$ is the closed linear span of $\{\theta_n\}$.

7.3 A fundamental problem in signal analysis

In 1946, Dennis Gabor formulated a fundamental approach for signal decomposition in terms of elementary signals. His approach has become a paradigm for the spectral analysis associated with time-frequency methods such as short time Fourier transforms (STFT), Wigner transforms and ambiguity functions, and wavelet theory. Gabor's idea is encapsulated in his remark that

> "The elementary signals ... assure the best utilization of the information area in the sense that they possess the smallest product of effective duration by effective frequency width" [Ga, p. 437].

We shall explain this remark, and use it as the basis of our point of view relating the topics of frame decompositions, sampling, and uncertainty principle inequalities. (In 1947, Gabor conceived of holography, and in 1971 he was awarded the Nobel Prize in Physics.)

DEFINITION 7.1

(a) *Let $g \in L^2(\mathbb{R})$ and let $a, b > 0$.*

    i. *The Weyl–Heisenberg system or Gabor system (of coherent states) is the sequence $\{\varphi_{m,n} : (m, n) \in \mathbb{Z} \times \mathbb{Z}\}$, where*

$$\varphi_{m,n}(t) = e^{2\pi itmb} g(t - na) = e_{mb}(t)\tau_{na}g(t).$$

Clearly,

$$\hat{\varphi}_{m,n}(\gamma) = e^{2\pi inamb} e^{-2\pi ina\gamma} \hat{g}(\gamma - mb) = \tau_{mb}(e_{-na}\hat{g})(\gamma).$$

    ii. *The Zak transform Zg of g is*

$$Zg(x, \omega) = a^{1/2} \sum_k g(xa + ka)e^{2\pi ik\omega}.$$

Its properties are developed in [BenWa] and we shall use it in Theorem 7.8.

(b) *Let $\psi \in L^2(\mathbb{R})$. The affine system or wavelet system is the sequence $\{\psi_{m,n} : (m, n) \in \mathbb{Z} \times \mathbb{Z}\}$, where*

$$\psi_{m,n}(t) = 2^{m/2}\psi(2^m t - n).$$

Clearly,

$$\hat{\psi}_{m,n}(\gamma) = 2^{-m/2} e^{-2\pi in(\gamma/2^m)} \hat{\psi}(\gamma/2^m) = 2^{-m/2}(e_{-n}\hat{\psi})(\gamma/2^m).$$

The major treatise on wavelet systems, especially wavelet orthonormal bases, is due to Meyer [Me].

Example 7.2. The Gaussian

Let $C \in \mathbb{C}, s > 0$, and $(t_0, \gamma_0) \in \mathbb{R} \times \hat{\mathbb{R}}$. Define

$$\varphi(t) = Ce^{2\pi it\gamma_0} e^{-s(t-t_0)^2},$$

so that

$$\hat{\varphi}(\gamma) = C\sqrt{\frac{\pi}{s}} e^{-2\pi it_0(\gamma-\gamma_0)} e^{-(\pi^2/s)(\gamma-\gamma_0)^2}.$$

In light of equation (7.2), we compute

$$\|\varphi\|_2^2 = \frac{|C|^2\sqrt{\pi}}{\sqrt{2s}},$$

$$\|(t - t_0)\varphi(t)\|_2^2 = \frac{|C|^2\sqrt{\pi}}{2(2s)^{3/2}},$$

and

$$\|(\gamma - \gamma_0)\hat{\varphi}(\gamma)\|_2^2 = \frac{|C|^2\sqrt{s}}{4\sqrt{2}\pi^{3/2}}.$$

As a consequence, we obtain the equality asserted in (7.3). ☐

The Gabor decomposition 7.3

For a given $s > 0$, choose any $C \in \mathbb{C}$ for which

$$|C|^2 = \sqrt{\frac{2s}{\pi}}.$$

Then the modulated and translated Gaussians φ, defined in Example 7.2, have the property that $\|\varphi\|_2 = 1$. For these L^2-normalized Gaussians, we have

$$\|(t - t_0)\varphi(t)\|_2^2 = \frac{1}{4s} \quad \text{and} \quad \|(\gamma - \gamma_0)\hat{\varphi}(\gamma)\|_2^2 = \frac{1}{4\pi^2}s.$$

As such, $\Delta^2 t_s \equiv 4\pi(1/4s) = \pi/s$ and $\Delta^2 \gamma_s \equiv 4\pi(s/4\pi^2) = s/\pi$ are the *variances* associated with φ and $\hat{\varphi}$, respectively; and we have

$$\Delta^2 t_s \Delta^2 \gamma_s = 1,$$

which is just another way of writing the equality asserted in (7.3), cf. Example 7.33.

Gabor's idea required a tiling of the time-frequency domain by nonoverlapping (half-open) rectangles $R_{m,n}$ centered at $(n\Delta^2 t_s, m\Delta^2 \gamma_s) \in \mathbb{R} \times \hat{\mathbb{R}}$, with length $\Delta^2 t_s$ and width (height) $\Delta^2 \gamma_s$. His goal, and the goal of any "spectral analysis," was to write discrete decompositions of signals f in terms of "optimal" elementary signals.

Gabor's notion of optimality stemmed from the classical uncertainty principle inequality, thereby leading to a "time-frequency analysis." He reasoned that optimal elementary signals should provide an efficient decomposition of the *information plane* (or *area*), i.e., the *time-frequency domain* or *phase space*. This efficient decomposition should determine countably many components of the information plane. Each component should be sufficiently localized in time and frequency so that a coefficient c associated with a component R would characterize the amount of information from R in the signal f; and it would not require smaller subcomponents in order to distinguish different types of information from R. Thus, it became desirable to find a countable collection of elementary signals with small associated areas in the information plane.

The product of temporal and frequency variances is a natural notion of "information area" to implement. In fact, if the product $\|(t - t_0)g(t)\|_2 \|(\gamma - \gamma_0)\hat{g}(\gamma)\|_2$ is small relative to other signals, then g and its spectrum $\hat{g}$ are (relatively) localized near (t_0, γ_0); i.e., the information characterizing g is relegated to a (relatively) small component R. This observation, coupled with the classical uncertainty principle inequality and the assertion that equality is achieved in (7.2) by modulated and translated Gaussians, lead Gabor to seek signal decompositions of the form

$$f(t) = \sum_{m,n} c_{m,n} \left(\frac{2s}{\pi} \right)^{1/4} e^{2\pi itm\Delta^2 \gamma_s} e^{-s(t-n\Delta^2 t_s)^2}, \tag{7.5}$$

since φ (from Example 7.2), with $C = (2s/\pi)^{1/4}$ and $(t_0, \gamma_0) = (n\Delta^2 t_s, m\Delta^2 \gamma_s)$, is associated with $R_{m,n}$.

Gabor also provided a means of estimating the coefficients $c_{m,n}$ associated with the components $R_{m,n}$. His method has been analyzed, and we shall not pursue it here. ⬚

DEFINITION/DISCUSSION 7.4

(a) *A sequence $\{x_n\}$ in an Hilbert space H is a frame if there exist $A, B > 0$ such that*

$$\forall y \in H, A\|y\|^2 \le \sum |\langle y, x_n \rangle|^2 \le B\|y\|^2, \tag{7.6}$$

where the norm of $y \in H$ is $\|y\| = \langle y, y \rangle^{1/2}$. A and B are the frame bounds, and a frame is tight if $A = B$. A frame is exact if it is no longer a frame whenever any one of its elements is removed. This definition and its consequences are the subject of the chapter [BenWa], found earlier in this volume. The theory of frames is due to Duffin and Schaeffer [DuSc], cf. [Y], [DGrMe], [HWa], [Ho].

(b) *The frame operator of the frame $\{x_n\}$ is the function $S : H \rightarrow H$ defined as $Sy = \sum \langle y, x_n \rangle x_n$.*

(c) *Let $\{x_n\} \subseteq H$ be a frame with frame bounds A and B. The frame decomposition theorem asserts that S is a topological isomorphism (Definition 7.14a) with inverse $S^{-1} : H \rightarrow H$, $\{S^{-1} x_n\}$ is a frame with frame bounds B^{-1} and A^{-1}, and*

$$\forall y \in H \quad y = \sum \langle y, S^{-1} x_n \rangle x_n$$
$$= \sum \langle y, x_n \rangle S^{-1} x_n. \tag{7.7}$$

(d) *Let $H = L^2(\mathbb{R}), g \in L^2(\mathbb{R})$, and $a, b > 0$. If the Gabor system $\{\varphi_{m,n}\}$ defined in Definition 7.1 is a frame, then it is a Gabor frame.*

Let $H = L^2(\mathbb{R})$ and $\psi \in L^2(\mathbb{R})$. If the wavelet system $\{\psi_{m,n}\}$ defined in Definition 7.1 is a frame, then it is a wavelet frame.

Fundamental problem 7.5

Let f be a complex signal. To fix a setting, suppose f is defined on $\mathbb{R}$, although we could just as well consider discrete signals, or signals defined on a variety of other domains.

The Fundamental Problem is to find frame decompositions of f of the form

$$f = \sum c_n s_n, \tag{7.8}$$

where the *coefficients* c_n and *harmonics* (or *elementary signals*) s_n provide a "best possible" time-frequency analysis in the following sense.

The coefficients in (7.8) should be *computable* and *relevant* to the goals of the signal analysis at hand. As such, computable sampled values of the signal are a natural type of coefficient initially to consider. The classical sampling theorem (7.1) provides a decomposition formula in the case of band-limited signals and regularly spaced samples. There are important variants of this theorem, where regularly spaced samples are not available, or are not the best information to use for signal reconstruction when they are available. In fact, there are a host of sampling theorems in the spirit of (7.1), but with different types of accessible sampled values depending on the physical problem. Such theorems are discussed in Section 7.7.

The elementary signals in (7.8) should be *elementary* and *relevant* to the goals of the signal analysis at hand. As such, Gabor chose modulates and translates of Gaussians as elementary signals since the product of their time and frequency variances are optimal vis-à-vis the classical uncertainty principle inequality (7.2); details concerning this choice were described in Subsection 7.3. Other notions of minimal area in the information plane, determined by other criteria for time and frequency durations besides variances, can also serve as constraints for signal reconstruction, depending on the given problem. Many such notions are a consequence of uncertainty principle inequalities in the spirit of (7.2), but with different physical consequences. Such inequalities are discussed in Section 7.8.

Besides the role of sampling and uncertainty for the coefficients and harmonics, respectively, in (7.8), there must also be a *compatibility* between these coefficients and harmonics. In fact, certain classes of desirable coefficients and harmonics might not combine to yield reconstructions of the form (7.8). Quantitatively, we translate this compatibility issue into our use of frame decompositions and the discussion of topological isomorphisms in Section 7.4.

Our theme, then, is to view the Fundamental Problem in terms of generalizations of the classical sampling theorem and classical uncertainty principle inequality, with reconstructions based on frame decompositions. The time-frequency methods mentioned in the opening paragraph of this Section 7.3 can be viewed in this way, and it seems a natural format with which to model apparently ad hoc problems, see, e.g., Remark 7.7; cf. [AU]. ☐

Example 7.6. *Information plane for the Dirichlet kernel*

(a) Let $g = d_{2\pi\Omega} \in L^2(\mathbb{R})$ be the *Dirichlet kernel* defined as

$$d_{2\pi\Omega}(t) = \frac{\sin 2\pi\Omega t}{\pi t}.$$

The function d_λ is the L^1 dilation $d_\lambda(t) = \lambda d(\lambda t)$ where $d(t) = (\sin t)/(\pi t)$. Clearly, $\hat{d}_{2\pi\Omega} = \mathbf{1}_{(\Omega)}$; and by the criteria for exact frames in terms of Fourier transforms, e.g., [BenWa, Theorem 3.51], we see that $\{\tau_k g : k \in \mathbb{Z}\}$ is an exact frame for $V_0 \equiv \overline{\mathrm{sp}}\{\tau_k g\}$ if and only if $\Omega \geq 1/2$. Using Parseval's theorem, we see that $\{\tau_k g : k \in \mathbb{Z}\}$ is an orthonormal basis for V_0 if and only if $\Omega = 1/2$. In the case $\Omega > 1/2$, $\{\tau_k \varphi\}$ is an orthonormal basis for V_0, where

$$\hat{\varphi}(\gamma) = \frac{\hat{g}(\gamma)}{(\sum |\hat{g}(\gamma+n)|^2)^{1/2}}.$$

Suppose $\Omega < 1/2$. Then by our characterization of frames in [BenWa, Theorem 3.56], we see that $\{\tau_k g\}$ is a frame for V_0.

(b) Let $a, b > 0$ and consider the Gabor system $\{\varphi_{m,n}\}$ defined by

$$\varphi_{m,n}(t) = e^{2\pi itmb} d_{2\pi\Omega}(t - na),$$

so that

$$\hat{\varphi}_{m,n}(\gamma) = \tau_{mb}(e_{-na}\mathbf{1}_{(\Omega)})(\gamma).$$

Thus,

$$\|(t - na)\varphi_{m,n}(t)\|_2 = \infty,$$

thereby precluding a decomposition of the information plane in terms of variance as described in Subsection 7.3, cf. Theorem 7.35a.

On the other hand, since $d_{2\pi\Omega}(t) = 0$ at $t = \pm 1/(2\Omega)$, the interval

$$I_{m,n} = \left[na - s\left(\frac{1}{2\Omega}\right), na + s\left(\frac{1}{2\Omega}\right)\right),$$

where $0 < s \leq 1$ is fixed, is contained in the domain of the major lobe of $\varphi_{m,n}$. As such, $I_{m,n}$ accounts for a significant portion of the set where we could reasonably assert that $d_{2\pi\Omega}$ is concentrated. The interval $I_{m,n}$ has length s/Ω. Also, it is reasonable to assert that $\hat{\varphi}_{m,n}$ is concentrated in the interval

$$J_{m,n} = [mb - \Omega, mb + \Omega).$$

We define the half-open rectangles $R_{m,n} = I_{m,n} \times J_{m,n}$. The area $2s$ of these rectangles reflects time and frequency "concentration," and, hence, fits in with the program of Subsection 7.3.

If $a \leq s/\Omega$ and $b \leq 2\Omega$, i.e., $na + s/(2\Omega) \geq (n+1)a - s/(2\Omega)$ and $mb + \Omega \geq (m+1)b - \Omega$, then we obtain a tiling of the information plane by $\{R_{m,n}\}$; and the tiling is non-overlapping if and only if $a = s/\Omega$ and $b = 2\Omega$. If $a > s/\Omega$ or $b > 2\Omega$ then $\{R_{m,n}\}$ is not a tiling of the information plane.

(c) Let $\psi = d_{2\pi\Omega}$, and consider the wavelet system $\{\psi_{m,n}\}$ defined by

$$\psi_{m,n}(t) = 2^{m/2} d_{2\pi\Omega}(2^m t - n),$$

so that

$$\hat{\psi}_{m,n}(\gamma) = 2^{-m/2}(e_{-n}\mathbf{1}_{(\Omega)})\left(\frac{\gamma}{2^m}\right).$$

Since $d_{2\pi\Omega}(t) = 0$ at $t = \pm 1/(2\Omega)$, it is reasonable to assert that $d_{2\pi\Omega}$ is concentrated in the interval

$$\left[-s\left(\frac{1}{2\Omega}\right), s\left(\frac{1}{2\Omega}\right)\right),$$

where $0 < s \le 1$ is fixed. Thus, $\psi_{m,n}$ is concentrated in the interval

$$I_{m,n} = \left[\frac{n}{2^m} - \frac{s}{2^m}\left(\frac{1}{2\Omega}\right), \frac{n}{2^m} + \frac{s}{2^m}\left(\frac{1}{2\Omega}\right)\right)$$

centered at $n/2^m$ and of length

$$|I_{m,n}| = \frac{s}{2^{m-1}}\left(\frac{1}{2\Omega}\right).$$

Also, since ψ is concentrated in $[-\Omega, \Omega)$, $\hat{\psi}_{m,n}$ is concentrated in the interval

$$J_{m,n} = [-2^m\Omega, 2^m\Omega),$$

centered at 0 and of length $|J_{m,n}| = 2\Omega2^m$. We define the half-open rectangles $R_{m,n} = I_{m,n} \times J_{m,n}$, with areas $2s$.

If $s/(2\Omega) \ge 1$ then, for every fixed m,

$$\bigcup_n R_{m,n} = \mathbb{R} \times [-2^m\Omega, 2^m\Omega),$$

and the union is disjoint if $s/(2\Omega) = 1$. Thus, if $s/(2\Omega) \ge 1$, then $\{R_{m,n}\}$ is an overlapping tiling of the information plane, cf. [D2, p. 965], [RioVe, pp. 15–18]. ▯

REMARK 7.7 Our approach to the Fundamental Problem posed in Subsection 7.5 is decidedly mathematical, and is presented in this chapter in the spirit of an emerging theory. However, the notions we've combined in Subsection 7.5 for our extension of Gabor's original theory (in Subsection 7.3) are the ingredients unifying a host of seemingly ad hoc, mundane, and difficult engineering problems, cf. [Ma]. We now give two such examples.

1. *Computation problem.* Let g be a discrete signal, e.g., let g be the impulse response of a cochlear filter. Let x be a (speech) signal to be processed by g, so that $x * g$ is to be computed, cf. Section 7.2. Then g and x must be sampled in such a way that $x * g$ can be computed with *speed* and *accuracy*. The computation of $x * g$ corresponds to the decomposition aspect of the Fundamental Problem, where accuracy (minimal loss of information) is reflected by proper sampling, and speed is reflected by the proper uncertainty principle constraint in conjunction with this sampling.

2. *Real-time sampling problem.* Suppose a sensing system, e.g., a group of antennas, is receiving data-intensive signals. The problem is to sample the

incoming data in real time, and to determine whether or not certain desirable (or undesirable!) signals are embedded in the data. Thus, there is a problem of signal determination (the decomposition aspect of the Fundamental Problem), minimizing loss of information by means of proper sampling, in conjunction with minimizing processing time by means of the proper choice of uncertainty principle criteria. The uncertainty principle arises in "resource allocation processing," i.e., optimizing "cost" by sampling only when necessary. For example, in analyzing lightning, there is generally no need to sample when there is no lightning, but there might be a need to sample extensively during a storm. ∎

In light of Gabor's use of the Gaussian in Subsection 7.3, we state the following result which summarizes some of the completeness and decomposition properties of the Gabor system of Gaussians, cf. [Fo, pp. 166–169].

THEOREM 7.8
Let $g(t) = (2s/\pi)^{1/4}e^{-st^2}$, for some $s > 0$, and let $a, b > 0$. Define the Gabor system $\{\varphi_{m,n}\}$ for g as in Definition 7.1.

(a) *The span $sp\{\varphi_{m,m}\}$ is dense in $L^2(\mathbb{R})$ if and only if $ab \leq 1$.*

(b) *The Gabor system $\{\varphi_{m,n}\}$ is a frame for $L^2(\mathbb{R})$ if and only if $ab < 1$. (Naturally, part b implies that if $ab < 1$ then $\overline{sp}\{\varphi_{m,n}\} = L^2(\mathbb{R})$.)*

This type of result for the Gaussian was initiated by von Neumann in the early 1930s [v, Section 5.4, pp. 504 ff.]. The proof of Theorem 7.8a can be found in [Baretal], [Pe] (1971). The proof in part *b* that if $ab < 1$ then $\{\varphi_{m,n}\}$ is a frame is due to [SeWal], cf. [DGrMe], [D2]. The material in [SeWal] and the related work [Se] deals with *Beurling density*, e.g., Definition 7.42. The proof that $\{\varphi_{m,n}\}$ is not a frame when $ab = 1$ follows from the following facts:

$$Zg\left(\frac{1}{2}, \frac{1}{2}\right) = 0, \tag{7.9}$$

$$Zg \text{ is continuous on } \mathbb{R} \times \hat{\mathbb{R}}, \tag{7.10}$$

and the Gabor system for a given L^2 function is a frame (in fact, an exact frame in this case that $ab = 1$) if and only if its Zak transform Z satisfies the inequalities $0 < A \leq Z \leq B < \infty$ a.e. ([BenWa, Section 3.4]). To verify (7.9), note that

$$Zg\left(\frac{1}{2}, \frac{1}{2}\right) = a^{1/2}\left(\frac{2s}{\pi}\right)^{1/4}\sum_k(-1)^k e^{-sa^2((1/2)+k)^2} \equiv C\sum_k a_k,$$

where

$$a_0 = e^{-s(a/2)^2}, a_{-1} = -e^{-s(a/2)^2}, a_{-2} = e^{-s(3a/2)^2}, a_1 = -e^{-s(3a/2)^2},$$

etc. Thus, since the series $Zg(1/2, 1/2)$ converges absolutely, we obtain (7.9). Further, to verify (7.10) we need only prove that

$$\forall K \in \mathbb{N}, \quad \sum e^{-sa^2(x+k)^2}$$

converges uniformly on $[-K, K]$. This is valid since $(x+k)^2 > (k-K)^2 > 0$ for $x \in [-K, K]$ and $k > K$, and $(x+k)^2 > (k+K)^2 > 0$ for $x \in [-K, K]$ and $k < -K$.

The $ab = 1$ case can also be proved by computing Zg directly in terms of certain Jacobi theta functions, see, e.g., [DGrMe, pp. 1274–1275], cf. [Pe].

Motivated by Subsection 7.3, we could take $a = \Delta t_s$ and $b = \Delta \gamma_s$ in the $ab = 1$ case of Theorem 7.8, but, in fact, the result is true for any other values of $a, b > 0$. In any case, Gabor's original idea of implementing the Gaussian and its variances does not yield a frame for $L^2(\mathbb{R})$.

On the other hand, each element $f \in L^2(\mathbb{R})$ does have a Gaussian Gabor system representation $f = \sum c_{m,n} \varphi_{m,n}$ for a computable system $\{c_{m,n}\}$ in the case $ab = 1$. The weaknesses of this decomposition are that $\{c_{m,n}\}$ can be an unbounded sequence and the convergence is much weaker than the $L^2(\mathbb{R})$-norm topology [Ja1, Theorems 4.6 and 4.7; Ja2, Section 4.4], cf. [Bas].

REMARK 7.9

(a) Time-frequency analysis has a long history in physics, music, mathematics, and other subjects, see, e.g., [C], [Fo], [Ri]. Gabor [Ga] was the first to make a systematic study in engineering; and we have emphasized his contribution because the natural development of his ideas is the basis for our approach, and because signal processing is our ultimate application.

(b) Recently, the information plane has been the playing field of wavelet theory, as in e.g., [D2]. In this case, if $\psi \in L^2(\mathbb{R})$ is "concentrated" in $[0, 1)$ and $\hat{\psi}$ is "concentrated" in $[1, 2) \cup (-2, -1]$, then the wavelet system provides a nonoverlapping covering of the information plane exclusive of the time axis. In fact, by our concentration hypotheses on ψ, we see that $\psi_{m,n}$ is "concentrated" in $[n/2^m, (n+1)/2^m)$, and $\hat{\psi}_{m,n}$ is concentrated in $[2^m, 2^{m+1}) \cup (-2^{m+1}, -2^m]$. Other dissections of the information plane can be made in terms of the bases described in [CMW2] and by Theorem 7.40; and the information plane provides a useful device for feasibility studies of desired decompositions. In fact, there is a cottage industry emerging to provide imaginative and sometimes applicable dissections of the information plane that go beyond Gabor and wavelet systems and their kin.

(c) Our definition of the information plane, before dissections motivated by physical problems, is simply $\mathbb{R} \times \hat{\mathbb{R}}$, where $\hat{\mathbb{R}}$ is the domain of Fourier transforms of functions defined on $\mathbb{R}$. We mention this since there is a natural inclination to compare scale and frequency in wavelet theory, and there are various points of view and subtleties associated with this comparison. ∎

7.4   Frame decompositions

In the context of Subsection 7.5, decompositions of signals f are sought so that the coefficients have computable and relevant properties, and so that the elementary signals ("harmonics") have optimal properties in terms of the information plane. Because of accessibility, it is natural to deal with some type of sampling of f when analyzing computability and relevance; and, as indicated in Section 7.3, it is reasonable to implement some type of uncertainty principle inequality when dealing with optimality in the information plane. Finally, there must be a compatibility between the sampling procedures and uncertainty principle inequalities in order to obtain effective decompositions. In this section we shall formulate the structure of these decompositions in general terms by means of the theory of frames and mappings between signals and coefficients. Analytic results in sampling and uncertainty principle inequalities appear in later sections.

H will denote a complex infinite-dimensional separable Hilbert space. It is well known that a complex infinite-dimensional Hilbert space contains an orthonormal basis if and only if it is separable, (e.g., [GoGol, pp. 34–35]).

THEOREM 7.10

Let $\{x_n : n \in \mathbb{Z}^d\}$ be an orthonormal sequence in H.

(**a**) *Bessel's inequality. The mapping*

$$L : H \to \ell^2(\mathbb{Z}^d)$$
$$y \mapsto \{\langle y, x_n \rangle\}$$

(7.11)

is continuous; in fact,

$$\forall y \in H, \quad \sum |\langle y, x_n \rangle|^2 \leq \|y\|^2.$$

(**b**) *For each* $y \in H, \sum \langle y, x_n \rangle x_n$ *converges in* H.

(**c**) $\sum c[n] x_n$ *converges in* H *if and only if* $c = \{c[n] \in \mathbb{C} : n \in \mathbb{Z}^d\} \in \ell^2(\mathbb{Z}^d)$.

(**d**) *If* $y = \sum c[n] x_n$ *converges in* H *then each* $c[n] = \langle y, x_n \rangle$.

This result on *orthonormality and convergence* is classical and elementary (cf. [GoGol, pp. 25–26]). Bessel's inequality gives way to the *Parseval–Plancherel equality* and to an isometry in the case of orthonormal bases, see, e.g., [GoGol, pp. 27–28]:

THEOREM 7.11

Let $\{x_n : n \in \mathbb{Z}^d\}$ be an orthonormal sequence in H. The following are equivalent.

i. $\{x_n\}$ *is an orthonormal basis of* H.

ii. *If* $\langle y, x_n \rangle = 0$ *for each* n *then* $y = 0$.

iii. $\overline{\text{sp}}\{x_n\} = H$.

iv. Parseval–Plancherel equality. The mapping L of (7.11) is a linear surjective isometry; in fact,

$$\forall y \in H, \quad \|y\| = \left(\sum |\langle y, x_n \rangle|^2\right)^{1/2}.$$

v. Parseval–Plancherel formula.

$$\forall x, y \in H, \quad \langle x, y \rangle = \sum \langle x, x_n \rangle \overline{\langle y, x_n \rangle}.$$

The surjectivity in part iv of Theorem 7.11 means that if $c \in \ell^2(\mathbb{Z}^d)$, then by parts c and d of Theorem 7.10 there is $y \in H$ for which $Ly = c$.

COROLLARY 7.12

Let $\{x_n : n \in \mathbb{Z}^d\}$ be an orthonormal basis of H. Then

$$\forall y \in H, \quad y = \sum \langle y, x_n \rangle x_n.$$

PROOF If $y \in H$ then $\sum \langle y, x_n \rangle x_n = x$ in H by Theorem 7.10b, and, hence, $\langle x, x_n \rangle = \langle y, x_n \rangle$ for each n by Theorem 7.10d. The result follows by part iv of Theorem 7.11. ∎

The following result is really the same as part iv of Theorem 7.11.

THEOREM 7.13

Let $\{x_n : n \in \mathbb{Z}^d\}$ and $\{e_n : n \in \mathbb{Z}^d\}$ be orthonormal bases of H and $\ell^2(\mathbb{Z}^d)$, respectively. The mapping

$$L : H \to \ell^2(\mathbb{Z}^d)$$
$$y \mapsto \sum \langle y, x_n \rangle e_n$$

is a well defined linear surjective isometry.

PROOF The linear map L is well-defined by Theorem 7.10a,c. The isometry, and in particular the injectivity of L, is a consequence of the Parseval–Plancherel equality, Corollary 7.12, and Theorem 7.10d:

$$\|Ly\|^2 = \sum |\langle Ly, e_n \rangle|^2 = \sum |\langle y, x_n \rangle|^2 = \|y\|^2.$$

For the surjectivity, let $c \in \ell^2(\mathbb{Z}^d)$, note that $Lx_m = e_m$, and calculate

$$c = \sum \langle c, e_n \rangle e_n = \sum \langle c, e_n \rangle Lx_n = L\left(\sum \langle c, e_n \rangle x_n\right),$$

where, because of Theorem 7.10c, $y = \sum \langle c, e_n \rangle x_n \in H$. Thus, $Ly = c$. ∎

DEFINITION/REMARK 7.14

(a) *A linear map $L : H_1 \to H_2$ is a topological isomorphism if it is bijective and continuous, in which case its inverse L^{-1} is also continuous.*

(b) *A sequence $\{x_n : n \in \mathbb{Z}^d\}$ in H is stationary if the inner product*

$$R_n = \langle x_{n+k}, x_k \rangle, \quad n \in \mathbb{Z}^d$$

is independent of $k \in \mathbb{Z}^d$.

(c) *A sequence $\{x_n : n \in \mathbb{Z}^d\} \subseteq H$ is minimal if*

$$\forall n \in \mathbb{Z}^d, \quad x_n \notin \overline{\mathrm{sp}}\{x_k : k \neq n\}.$$

(d) *The sequences $\{x_n : n \in \mathbb{Z}^d\}, \{y_n : n \in \mathbb{Z}^d\} \subseteq H$ are biorthonormal if*

$$\forall m, n \in \mathbb{Z}^d, \quad \langle x_m, y_n \rangle = \delta_{m,n}.$$

For a given sequence $\{x_n\} \subseteq H$, a Hahn–Banach argument shows that there is a sequence $\{y_n\} \subseteq H$ so that $\{x_n\}, \{y_n\}$ are biorthonormal if and only if $\{x_n\}$ is minimal. Furthermore, $\{y_n\}$ is uniquely determined if and only if $\{x_n\}$ is not only minimal in H but also $\overline{\mathrm{sp}}\{x_n\} = H$.

(e) *A sequence $\{x_n : n \in \mathbb{Z}^d\} \subseteq H$ is a Bessel sequence if the Bessel map*

$$L : H \to \ell^2(\mathbb{Z}^d)$$

$$y \mapsto \{\langle y, x_n \rangle\}$$

is a well-defined linear map. $\{x_n\}$ is a Hilbert sequence if

$$\forall c = \{c[n]\} \in \ell^2(\mathbb{Z}^d), \quad \exists y_c \in H,$$

such that

$$\forall n \in \mathbb{Z}^d, \quad c[n] = \langle y_c, x_n \rangle.$$

Clearly, a Bessel sequence is a Hilbert sequence if and only if the Bessel map L is surjective.

(f) *Let $\{x_n : n \in \mathbb{Z}^d\} \subseteq H$ be a Bessel sequence with Bessel map L. Then L is continuous, i.e.,*

$$\exists B > 0 \quad \text{such that } \forall y \in H, \tag{7.12}$$

$$\sum |\langle y, x_n \rangle|^2 \leq B \|y\|^2;$$

and

$$\forall c = \{c[n]\} \in \ell^2(\mathbb{Z}^d), \quad \exists y \equiv L^* c = \sum c[n] x_n \text{ in } H, \tag{7.13}$$

where $L^* : \ell^2(\mathbb{Z}^d) \to H$ is the adjoint of L, and the "geometry" of the norm convergence of the sum in (7.13) is specified by s_n (below), cf. (7.4).

To prove (7.12), we first define the linear maps

$$L_N : H \to \ell^2(\mathbb{Z}^d)$$

$$y \mapsto \{\langle y, x_n \rangle : |n| \leq N\},$$

where $n = (n_1, \ldots, n_d)$ and $|n|^2 \equiv \sum |n_j|^2$. Each L_N is obviously continuous, and since $Ly \in \ell^2(\mathbb{Z}^d)$ for $y \in H$, we have

$$\|L_N y - Ly\|_2^2 = \sum_{|n|>N} |\langle y, x_n \rangle|^2,$$

which goes to 0 as $N \to \infty$ by the definition of L. Thus,

$$\forall y \in H, \quad \lim_{N \to \infty} L_N y = Ly \text{ in } \ell^2(\mathbb{Z}^d),$$

and so L is bounded by the uniform boundedness principle. In fact, we have

$$\sup_N \|L_N\| \equiv B^{1/2} < \infty.$$

To prove (7.13) we use (7.12) in the following way. Let $c \in \ell^2(\mathbb{Z}^d)$ and let

$$s_N = \sum_{|n| \leq N} c[n] x_n.$$

Then, for $N > M$, the definition of the norm and Hölder's inequality allow us to write

$$\|s_N - s_M\|^2 = \sup_{\|y\|=1} \left| \sum_{M<|n|\leq N} c[n] \langle x_n, y \rangle \right|^2$$

$$\leq \sup_{\|y\|=1} \sum_{M<|n|\leq N} |c[n]|^2 \sum_{M<|n|\leq N} |\langle x_n, y \rangle|^2$$

$$\leq B \sup_{\|y\|=1} \|y\|^2 \sum_{M<|n|\leq N} |c[n]|^2.$$

Therefore, $\{s_N\} \subseteq H$ is a Cauchy sequence and (7.13) follows. Letting $y \equiv \lim s_N$, the calculation that $L^* c = y$ is immediate.

(g) *The notions we have just formulated have a long and related history, treated in, e.g., [Ben3, Definition 2.14–Theorem 2.17], where care must be taken since Bessel and Hilbert sequences are defined in terms of $\{y_n\}$ for the case that $\{x_n\}$ is minimal and complete.*

If $\{x_n : n \in \mathbb{Z}^d\} \subseteq H$ is minimal and complete, and $\{x_n\}, \{y_n\} \subseteq H$ are biorthonormal, then it is not necessarily true that $\overline{\mathrm{sp}}\{y_n\} = H$; an old example of Kaczmarz and Steinhaus (1935) for the case $d = 1$ provides a counterexample ([Hil, pp. 19–20]). On the other hand, Masani and Rozanov, ([Mas]) have proven that $\overline{\mathrm{sp}}\{y_n\} = H$ in the case that $\{x_n : n \in \mathbb{Z}\}$ is not only minimal and complete, but also stationary. This result is also true for $d > 1$, and depends on the linear isometric isomorphism,

$$U : L_\mu^2(\mathbb{T}^d) \to \overline{\mathrm{sp}}\{x_n\},$$

defined by

$$\forall n \in \mathbb{Z}^d, \quad U(e^{2\pi i n \cdot \gamma}) \equiv x_n,$$

where $\{x_n\}$ is stationary and μ is the power spectrum measure defined by $\mu^{\vee}[n] = R_n$, see [Ben3, Theorem 2.5].

THEOREM 7.15
Let $\{x_n : n \in \mathbb{Z}^d\} \subseteq H$. The sequence $\{x_n\}$ is a frame for H with frame bounds A and B if and only if the mapping

$$L : H \rightarrow \ell^2(\mathbb{Z}^d)$$
$$y \mapsto \{\langle y, x_n \rangle\} \tag{7.14}$$

is a well-defined topological isomorphism onto a closed subspace of $\ell^2(\mathbb{Z}^d)$. In this case

$$\|L\| \leq B^{1/2} \quad and \quad \|L^{-1}\| \leq A^{-1/2},$$

where L^{-1} is defined on the range $L(H)$.

PROOF Assume $\{x_n : n \in \mathbb{Z}^d\}$ is a frame. Then

$$\forall y \in H, \quad \|Ly\|^2_{\ell^2(\mathbb{Z}^d)} = \sum |\langle y, x_n \rangle|^2 \leq B\|y\|^2.$$

In particular, L is a well-defined linear map, L is a bounded operator, and $\|L\| \leq B^{1/2}$. If $\{\langle y, x_n \rangle\} = 0 \in \ell^2(\mathbb{Z}^d)$ for some $y \in H$, then $y = 0$ by the lower frame bound condition, and so L is injective so that L^{-1} exists as a linear injection on $L(H)$. Then

$$\forall c \in L(H), \quad \|L^{-1}c\|^2 = \|y_c\|^2 \leq \frac{1}{A} \sum |\langle y_c, x_n \rangle|^2,$$

where $Ly_c = c$ and $c = \{\langle y_c, x_n \rangle\}$. Thus, L^{-1} is continuous, and $\|L^{-1}\| \leq A^{-1/2}$. $L(H)$ is closed in $\ell^2(\mathbb{Z}^d)$ by general principles, but we see it explicitly as follows. Let $\{c_n : n \in \mathbb{N}\} \subseteq L(H)$ converge to $c \in \ell^2(\mathbb{Z}^d)$, and let $Ly_n = c_n$ for each $n \in \mathbb{N}$. Then $\{c_n : n \in \mathbb{N}\}$ is Cauchy in $\ell^2(\mathbb{Z}^d)$, so that $\{y_n : n \in \mathbb{N}\}$ is Cauchy in H by the (uniform) continuity of L^{-1}; and hence $\lim_{n \to \infty} y_n = y$ exists in H. Finally, $c = Ly$ since $Ly \in L(H)$, by the continuity of L, and since $\lim_{n \to \infty} \|c_n - c\|_{\ell^2(\mathbb{Z}^d)} = 0$.

Conversely, if L is a topological isomorphism onto a closed subspace $L(H)$ of $\ell^2(\mathbb{Z}^d)$, then

$$\forall y \in H, \quad \|Ly\|_{\ell^2(\mathbb{Z}^d)} \leq \|L\|\|y\| \tag{7.15}$$

and

$$\forall c = Ly_c \in L(H), \quad \|L^{-1}c\| \leq \|L^{-1}\|\|c\|_{\ell^2(\mathbb{Z}^d)}. \tag{7.16}$$

Since $\|Ly\|^2_{\ell^2(\mathbb{Z}^d)} = \sum |\langle y, x_n \rangle|^2$, (7.15) and (7.16) imply that

$$\forall y \in H, \quad \|L^{-1}\|^{-2}\|y\|^2 \leq \sum |\langle y, x_n \rangle|^2 \leq \|L\|^2\|y\|^2.$$

Hence, $\{x_n\}$ is a frame with frame bounds A and B, where $A \leq \|L^{-1}\|^{-2}$ and $\|L\|^2 \leq B$. ∎

THEOREM 7.16

(a) Let $\{x_n : n \in \mathbb{Z}^d\} \subseteq H$ be a frame for H. $\{x_n\}$ is an exact frame if and only if it is minimal, see, e.g., [Ben3, Theorem 4.6] for a proof.

(b) Let $\{x_n : n \in \mathbb{Z}^d\} \subseteq H$ be a stationary, minimal, complete sequence with injective Bessel map L. If $\{x_n\}$ is a Hilbert sequence then $\{x_n\}$ is an exact frame for H.

PROOF   To prove part b, consider that, if $\{x_n\}$ is a Hilbert sequence, then $L^{-1} : \ell^2(\mathbb{Z}^d) \to H$ is well defined, and we may apply the open mapping theorem since L is continuous. Thus, $\{x_n\}$ is a frame, and we obtain exactness by the minimality hypothesis and part a. ∎

Stationarity can be used to distinguish Gabor and wavelet systems as follows.

Example 7.17. *Stationarity and Gabor systems*

(a) Let $g \in L^2(\mathbb{R})$ and let $a, b > 0$, and consider the Gabor system $\{\varphi_{m,n}\}$. If $ab = 1$ then the sequence $x : \mathbb{Z}^2 \to H = L^2(\mathbb{R})$, defined by $x(m, n) = \varphi_{m,n}$ $((m, n) \in \mathbb{Z} \times \mathbb{Z} = \mathbb{Z}^2)$, is a stationary sequence; see, e.g., [Ben3, Section 4] for further properties of stationary Gabor systems and their relevance to other problems in analysis. In the case of frames, recall that $\{\varphi_{m,n}\}$ is exact for the case $ab = 1$.

(b) If $\psi \in L^2(\mathbb{R})$, the wavelet system $\{\psi_{m,n}\}$ is not stationary. ⬜

THEOREM 7.18
Let $L : H \to \ell^2(\mathbb{Z}^d)$ be a topological isomorphism of H onto the closed subspace $L(H)$ of $\ell^2(\mathbb{Z}^d)$, and let $\{e_n : n \in \mathbb{Z}^d\} \subseteq L(H)$.

(a) The sequence $\{e_n\}$ is a frame for the Hilbert space $L(H)$ if and only if the sequence $\{L^{-1}e_n\}$ is a frame for H. In the case $\{e_n\}$ is a frame with frame bounds A and B, then

$$A_H = A\|L\|^{-2} \quad and \quad B_H = B\|L^{-1}\|^2$$

can be taken as frame bounds for $\{L^{-1}e_n\}$.

(b) If $\{e_n\}$ is an orthonormal basis of $L(H)$ then, for all $y \in H$,

$$Ly = \sum \langle Ly, e_n \rangle e_n \text{ in } L(H), \quad \{\langle Ly, e_n \rangle\} \in \ell^2(\mathbb{Z}^d),$$

and

$$y = \sum \langle Ly, e_n \rangle L^{-1}e_n \text{ in } H. \tag{7.17}$$

(c) *If $H \subseteq \ell^2(\mathbb{Z}^d)$ and $\{e_n\}$ is an orthonormal basis of $L(H)$, then $\{L^{-1}e_n\}$ is an exact frame for H.*

PROOF Part c follows from the definition of Riesz basis, and part b follows immediately by hypothesis, Theorem 7.10, and Corollary 7.12. We shall only prove one direction of part a since a similar calculation works in the other direction.

Let $\{e_n\} \subseteq L(H)$ be a frame with frame constants A and B. Then, for $y \in H$,

$$\sum |\langle y, L^{-1}e_n\rangle|^2 = \sum |\langle e_n, (L^{-1})^*y\rangle|^2$$

and

$$A\|(L^{-1})^*y\|^2_{\ell^2(\mathbb{Z}^d)} \le \sum |\langle (L^{-1})^*y, e_n\rangle|^2 \le B\|(L^{-1})^*y\|^2_{\ell^2(\mathbb{Z}^d)}.$$

We have

$$\|(L^{-1})^*y\|_{\ell^2(\mathbb{Z}^d)} \le \|(L^{-1})^*\|\|y\| = \|L^{-1}\|\|y\|,$$

and

$$\|y\| = \|L^*(L^*)^{-1}y\| \le \|L^*\|\|(L^*)^{-1}y\|_{\ell^2(\mathbb{Z}^d)} = \|L\|\|(L^{-1})^*y\|_{\ell^2(\mathbb{Z}^d)}.$$

Combining these calculations establishes that $\{L^{-1}e_n\}$ is a frame with frame bounds A_H and B_H. ∎

Thus, our general problem, described in Subsection 7.5 and the beginning of Section 7.4, is to construct topological isomorphisms $L : H \to L(H) \subseteq \ell^2(\mathbb{Z}^d)$ and (orthonormal) bases $\{e_n\}$ of $L(H)$ so that the sequence $\{\langle Ly, e_n\rangle\}$ of coefficients in (7.17) reflects accessible and desired properties, and so that the sequence $\{L^{-1}e_n\}$ of elementary signals in (7.17) reflects desired duration properties in the information plane.

In the spirit of Theorem 7.15 we have the following result for exact frames.

THEOREM 7.19
Let $\{x_n : n \in \mathbb{Z}^d\} \subseteq H$ be a frame for H. The sequence $\{x_n\}$ is an exact frame for H if and only if the mapping

$$L : H \to \ell^2(\mathbb{Z}^d) \tag{7.18}$$
$$y \mapsto \{\langle y, x_n\rangle\}$$

is a well-defined surjective topological isomorphism.

PROOF

(a) Let $\{x_n : n \in \mathbb{Z}^d\}$ be an exact frame. Then $\{x_n\}$ is a Riesz basis for H ([BenWa, Section 3.3]), and so it possesses a unique biorthonormal sequence $\{y_n : n \in \mathbb{Z}^d\}$ which, by duality, is also a Riesz basis for H. Then, by (7.7),

$$\forall y \in H, \quad y = \sum \langle y, x_n\rangle y_n \text{ in } H, \tag{7.19}$$

where $\{\langle y, x_n \rangle\} \in \ell^2(\mathbb{Z}^d)$.

Since $\{x_n\}$ is a frame, we proceed to obtain L as a well-defined topological isomorphism onto a closed subspace of $\ell^2(\mathbb{Z}^d)$, just as we did in Theorem 7.15.

It remains to show that L is surjective. Let $c = \{c[n]\} \in \ell^2(\mathbb{Z}^d)$. Then, by (7.13),

$$y_c \equiv \sum c[n] y_n \text{ in } H. \tag{7.20}$$

Since $\{y_n\}$ is a basis, the representation in (7.20) is unique, and hence, $c[n] = \langle y_c, x_n \rangle$ for all n by (7.19). Thus, $Ly_c \equiv \{\langle y_c, x_n \rangle\} = c$.

(b) Assume L is surjective, cf. Theorem 7.15. Thus, L^* is injective and so $L^* c = 0$ if and only if $c = 0$. Pick any $m \in \mathbb{Z}^d$ and consider $S_m = \{x_n : n \in \mathbb{Z}^d \setminus \{m\}\}$. If this sequence is a frame then there is $\{d[n] : n \in \mathbb{Z}^d \setminus \{m\}\}$ for which $\sum d[n] x_n = x_m$; and so we obtain a contradiction to the injectivity for $c = \{c[n] : c[n] = d[n], \text{ for } n \neq m, \text{ and } c[m] = -1\}$. Thus, S_m is not a frame, and so $\{x_n\}$ is a exact frame. ∎

Example/Remark 7.20.   Bessel mappings on the Paley–Wiener space

(a) A classical theorem of Plancherel–Pólya asserts that if $f \in PW_\Omega$ and $\{t_n : n \in \mathbb{Z}\}$ is uniformly discrete, i.e., $\inf\{|t_m - t_n| : m \neq n\} = d > 0$, then

$$\sum |f(t_n)|^2 \leq B \|f\|_2^2, \tag{7.21}$$

where B can be taken as

$$B = \frac{4}{\pi^2 \Omega d^2} (e^{\pi \Omega d} - 1),$$

see, e.g., [Ben2, Lemma 42], cf., [Bo2, pp. 97–103]. Thus, the mapping $PW_\Omega \to \ell^2(\mathbb{Z})$, taking $f \in PW_\Omega$ into $\{\langle f, \tau_{-t_n} d_{2\pi\Omega}\rangle\} = \{f(t_n)\}$ is the Bessel mapping corresponding to the Bessel sequence $\{\tau_{-t_n} d_{2\pi\Omega}\} \subseteq PW_\Omega$. Boas [Bo1, Theorems 1,2,3] (1940) has provided a proof of (7.21) different from the argument of Plancherel and Pólya. An attractive feature of Boas' approach is that it leads us to rewrite (7.21) as a *weighted Fourier transform norm inequality*, viz.,

$$\|f\|_{2,\nu}^2 \leq B \|\hat{f}\|_{2,\mu}^2.$$

In the case of (7.21), ν is the discrete measure $\sum \delta_{t_n}$, μ is the absolutely continuous measure $\mu = \mathbf{1}_{(\Omega)}$ (usually more comfortably written as "$d\mu(\gamma) = \mathbf{1}_{(\Omega)}(\gamma) d\gamma$"), and PW_Ω is the weighted L^2-space, $L_\mu^2(\hat{\mathbb{R}})$. Weights such as $\mu = \mathbf{1}_{(\Omega)}$ play a role in the Bell Labs uncertainty principle theory from the early 1960s ([L2]), and measure weights play a significant role in general uncertainty principle theory, see, e.g., ([Ben1], [Bo2, Chapter 10], and Section 7.8.

Boas and Plancherel–Pólya are also responsible for the opposite inequalities to (7.21), cf. [Bo2, pp. 196–199]. Together, one obtains frame-like inequalities for exponentials $\{e_{t_n}(\gamma)\}$ in the spirit of nonharmonic Fourier series and the work of Duffin and Schaeffer, see, e.g., [DuSc], [Le], [PWie].

(b) If $T, \Omega > 0$ and $2T\Omega \leq 1$, then the mapping

$$L : PW_\Omega \to \ell^2(\mathbb{Z})$$
$$f \mapsto \{\sqrt{T}f(nT)\} \tag{7.22}$$

is a well-defined (injective) isometric isomorphism onto a closed subspace $L(PW_\Omega)$ of $\ell^2(\mathbb{Z})$. L is surjective in the case $2T\Omega = 1$.

The isometry follows by the following calculation for $f \in PW_\Omega$.

$$\sqrt{T}\|f\|_2 = \left(T \int_{-\Omega}^{\Omega} |\hat{f}(\gamma)|^2 \, d\gamma \right)^{1/2}$$
$$= \left(T \int_{-1/(2T)}^{1/(2T)} |G(\gamma)|^2 \, d\gamma \right)^{1/2}$$
$$= \left(\sum |\hat{G}[n]|^2 \right)^{1/2} = T \left(\sum |f(nT)|^2 \right)^{1/2},$$

where G is defined as the $(1/T)$-periodic function equal to $\hat{f}$ on $[-\Omega, \Omega)$ and vanishing on $[-1/2T, 1/2T) \backslash [-\Omega, \Omega)$, the Fourier series of G is $\sum \hat{G}[n]e^{-\pi in(2T)\gamma}$, and $\hat{G}[n] = Tf(nT)$.

Further, $\{\sqrt{T}\tau_{nT}d_{2\pi\Omega}\} \subseteq PW_\Omega$ is a Bessel sequence with Bessel map L (defined by (7.22)), since

$$\forall f \in PW_\Omega, \quad \langle f, \sqrt{T}\tau_{nT}d_{2\pi\Omega}\rangle$$
$$= \sqrt{T} \int_{-\Omega}^{\Omega} \hat{f}(\gamma)e^{2\pi inT\gamma} \, d\gamma = \sqrt{T}f(nT) = (Lf)(n).$$

Note that $\|\sqrt{T}\tau_{nT}d_{2\pi\Omega}\|_2 = (2T\Omega)^{1/2}$. Now if $c = \{c[n]\} \in \ell^2(\mathbb{Z})$, then

$$f(t) = \sum c[n]\sqrt{T}d_{2\pi\Omega}(t - nT) \text{ in } PW_\Omega$$

by (7.13). This equation combined with the classical sampling theorem (7.1) allows us to assert that

$$\sum \left(\sqrt{T}c[n] - Tf(nT) \right) d_{2\pi\Omega} = 0 \text{ in } PW_\Omega. \tag{7.23}$$

Using (7.23) we obtain the surjectivity of L when $2T\Omega = 1$ since $\{(1/2\Omega)e^{\pi in\gamma/\Omega}\}$ is an orthonormal basis of $L^2(\mathbb{T}_{2\Omega})$ in this case.

For a computational verification of the surjectivity in the case $2T\Omega = 1$, we take $c = \{c[n]\} \in \ell^2(\mathbb{Z})$; and define the $(1/T)$-periodic function $F \in L^2(\mathbb{T}_{1/T})$ by its Fourier series $\sum c[n]e^{-\pi in\gamma(2T)}$. We define $G \in L^2(\hat{\mathbb{R}})$ to be F on

$[-\Omega, \Omega) = [-1/2T, 1/2T)$ and 0 elsewhere; and we see that there is $f \in PW_\Omega$ for which $\hat{f} = G$. Then

$$c[n] = \hat{F}[n] = \int_{\mathbb{T}_{1/T}} F(\gamma) e^{\pi i n \gamma / \Omega} \, d\gamma$$

$$= \frac{1}{2\Omega} \int_\Omega^\Omega G(\gamma) e^{\pi i n \gamma / \Omega} \, d\gamma = \frac{1}{2\Omega} f(\frac{n}{2\Omega});$$

and so L, with the factor $\sqrt{T}$, is a surjection.

This calculation fails for $2T\Omega < 1$, essentially because $\{e^{2\pi i n T \gamma}\}$ is not a basis for $L^2[-\Omega, \Omega]$. □

7.5   The classical sampling theorem

This section will be devoted to sampling theorems for regularly spaced *sampling sequences* $\{nT\}$. There are spectacular expositions and research tutorials by Bützer et al. [BuSpSte] and Higgins [Hi2], some remarkable singular research contributions such as Kluvánek's sampling theorem for locally compact abelian groups [Kl], [Hi2], [Pr2], and, alas, the natural and inevitable nexus with Gabor and wavelet systems, as in [Ben2], [BenHel1], [D2], [DGrMe], [FeGrö], [FrTo], [W].

In Theorem 7.22 we shall use Gabor frames to prove an efficient and general sampling theorem [BenHel1]. Then we shall pose a local sampling problem which arises in applications and solve it in Theorem 7.26 for a special but useful case [BenKar].

We begin by restating the *classical sampling theorem* (7.1).

THEOREM 7.21
Let $T, \Omega > 0$ be constants for which $2T\Omega \leq 1$. Then

$$\forall f \in PW_\Omega, \quad f = T \sum f(nT) \tau_{nT} \, d_{2\pi\Omega}, \tag{7.24}$$

where the convergence is uniform on $\mathbb{R}$ and in $L^2(\mathbb{R})$.

This theorem and its generalizations, Theorem 7.22 and Theorem 7.24, have several proofs, each ultimately depending on the relation between Fourier series and Fourier transforms. We shall give a frame theoretic proof of Theorem 7.22 to illustrate a natural means of dealing with aliasing and with non-band-limited sampling formulas, cf. [Ben2, Section 2, Section 10, and Definition/Discussion 28] for aliasing and classical proofs, as well as [BuSte].

The sampling function $d_{2\pi\Omega}$ of Theorem 7.21 is replaced by rather general sampling functions s in Theorem 7.22. These force oversampling but can give rise to faster convergence in applications. Although the hypotheses seem technical, the critical hypotheses are that $2T\Omega \leq 1$ and $\hat{s} = 1$ on $[-\Omega, \Omega]$.

THEOREM 7.22   [BenHell]
Let $T, \Omega > 0$ be constants for which $0 < 2T\Omega \leq 1$, and let $g \in PW_{1/(2T)}$ have the properties that $\hat{g} \in L^\infty(\hat{\mathbb{R}})$,

$$\hat{g} = 1 \ on \ [-\Omega, \Omega],$$

and, in case $2T\Omega < 1$, $\hat{g}$ is continuous and

$$|\hat{g}| > 0 \ on \ \left(-\frac{1}{2T}, -\Omega\right] \cup \left[\Omega, \frac{1}{2T}\right).$$

Set

$$G(\gamma) \equiv \sum |\hat{g}(\gamma - mb)|^2 \quad and \quad s(t) \equiv \left(\frac{\hat{g}}{G}\right)^{\vee}(t),$$

where $\Omega + 1/2T \leq b < 1/T$ in case $2T\Omega < 1$ and $\Omega + 1/2T = b$ if $2T\Omega = 1$. Then

$$\exists A, B > 0 \quad such \ that \ A \leq G(\gamma) \leq B \ a.e.,$$

$s \in PW_{1/(2T)}$, $\hat{s} \in L^\infty(\hat{\mathbb{R}})$, $\hat{s} = 1$ on $[-\Omega, \Omega]$,

$$\forall f \in L^2(\mathbb{R}), \quad f = T \sum \langle \hat{f}, e_{nT}\tau_{mb}\hat{g} \rangle \tau_{-nT}(e_{mb}s) \ in \ L^2(\mathbb{R}), \tag{7.25}$$

and

$$\forall f \in PW_\Omega, \quad f = T \sum f(nT) \tau_{nT} s \ in \ L^2(\mathbb{R}). \tag{7.26}$$

PROOF

  (a) Let $2T\Omega < 1$ so that

$$\Omega = \Omega + \frac{1}{2T} - \frac{1}{2T} \leq b - \frac{1}{2T} < \frac{1}{T} - \frac{1}{2T} = \frac{1}{2T}.$$

Thus, the left hand "endpoint" of supp $\tau_b\hat{g}$ is in the half-open interval $[\Omega, \frac{1}{2T})$; and so, by the hypothesis on $\hat{g}$ in this case, we have $A \leq G \leq B$ a.e. In case $2T\Omega = 1$, we have $G = 1$ a.e.

By definition of s and the properties of G we have $\hat{s} = \hat{g}/G \in L^\infty(\hat{\mathbb{R}})$, and so $s \in PW_{1/(2T)}$. By our choice of b, $G = 1$ on $[-\Omega, \Omega]$, and hence $\hat{s} = 1$ on $[-\Omega, \Omega]$. We require $\hat{g} \in L^\infty(\hat{\mathbb{R}})$ for the upper bound $G \leq B$, and we use the positivity hypothesis on $\hat{g}$ for the lower bound $A \leq G$, cf. [Hel] where this latter hypothesis is weakened.

In any case, the assertions about G and s follow from our choice of b.

  (b) Let $a = T$ so that $|\operatorname{supp}\hat{g}| = \frac{1}{T} = \frac{1}{a}$. Then, using the fact $A \leq G \leq B$ a.e., and the elementary sufficient conditions for Gabor frames ([BenWa], Section 3.4]), we see that $\{e_{na}\tau_{mb}\hat{g}\}$ is a frame for $L^2(\hat{\mathbb{R}})$. Since supp $\hat{g}$ is compact, we have

$$\forall h \in L^2(\mathbb{R}), \quad S^{-1}\hat{h} = T\frac{\hat{h}}{G}$$

by basic properties of the inverse frame operator ([BenWa, Section 3.4]). Hence, we have the frame decomposition

$$\forall f \in L^2(\mathbb{R}), \quad \hat{f} = T \sum \langle \hat{f}, e_{nT} T_{mb} \hat{g} \rangle e_{na} T_{mb} \hat{s} \text{ in } L^2(\hat{\mathbb{R}}),$$

where we have used the fact that $S^{-1}(e_{na} T_{mb} \hat{g}) = e_{na} T_{mb} S^{-1} \hat{g}$ ([BenWa, Section 3.4]). Thus,

$$\forall f \in L^2(\mathbb{R}), \quad f = T \sum \langle \hat{f}, e_{nT} T_{mb} \hat{g} \rangle \tau_{-na} (e_{mb} s) \text{ in } L^2(\mathbb{R})$$

by the basic frame-decomposition formula (7.7). This is (7.25).

If $f \in PW_\Omega$ then

$$\langle \hat{f}, e_{nT} T_{mb} \hat{g} \rangle = \begin{cases} f(-nT), & \text{if } m = 0 \\ 0, & \text{if } m \neq 0. \end{cases}$$

In this case, (7.25) reduces to (7.26). ∎

Technically, Theorem 7.22 yields Theorem 7.21 by making the special choice $b = 2\Omega$.

Example 7.23. *Smooth sampling functions*

The functions $s \in PW_{1/(2T)}$ of Theorem 7.22 can be chosen in the Schwartz space $\mathcal{S}(\mathbb{R})$ when $2T\Omega < 1$. We begin the construction by defining

$$\psi_\epsilon(\gamma) = \frac{\varphi(\epsilon - |\gamma|^2)}{\int \varphi(\epsilon - |\lambda|^2) d\lambda},$$

where $\varphi \in C^\infty(\hat{\mathbb{R}})$ vanishes on $(-\infty, 0]$ and equals $e^{-1/\gamma}$ on $[0, \infty)$. Thus $\psi_\epsilon \in C_c^\infty(\hat{\mathbb{R}})$ is an even function satisfying the conditions $\operatorname{supp} \psi_\epsilon = [-\epsilon, \epsilon]$ and $\int \psi_\epsilon(\gamma) d\gamma = 1$. Next set

$$\psi_{U,V} = \frac{1}{|V|} \mathbf{1}_V * \mathbf{1}_{U-V}, \quad U, V \subseteq \hat{\mathbb{R}},$$

where $0 < |V| < \infty$ and U is compact. Thus, $\psi_{U,V}$ is 1 on U and vanishes off of $U + V - V$. The function $g \in \mathcal{S}(\mathbb{R})$ of Theorem 7.22 is defined in terms of $\hat{g}$ as $\hat{g} = \psi_{U,V} * \psi_\epsilon$, where we specify ϵ, U, and V as follows. Let $U = [-u, u]$, where $u \in (\Omega, 1/2T)$ is arbitrary, and let $\epsilon = u - \Omega$. Choose $V = [-v, v]$ by setting $v = (w - u)/2$, where $w = 1/2T - \epsilon$. These choices are necessitated by a simple geometric argument, and the resulting function $\hat{g}$ satisfies the conditions of Theorem 7.22 and $\hat{g} \in C_c^\infty(\hat{\mathbb{R}})$. The definition of s in Theorem 7.22 completes the construction. ☐

The following is an elementary but aesthetically pleasing generalization of Theorem 7.21, cf. [Bo2, p. 220], [K], [Pl].

THEOREM 7.24
*Let $E \subseteq \hat{\mathbb{R}}$ have positive finite Lebesgue measure, let $PW_E \equiv \{f \in L^2(\mathbb{R}) : \hat{f} = 0$
a.e. off $E\}$, and let $\{g_n : n \in \mathbb{Z}\}$ be an orthonormal basis of the Hilbert space
PW_E. For any $f \in L^2(\mathbb{R}), f \in PW_E$ if and only if*

$$f = \sum c[n]g_n \quad \text{converges uniformly on } \mathbb{R},$$

where $c = \{c[n]\} \in \ell^2(\mathbb{Z})$. In this case,

$$\forall n, \quad c[n] = \int_E \hat{f}(\gamma)\overline{\hat{g}_n(\gamma)}\,d\gamma \equiv \langle \hat{f}, \hat{g}_n \rangle;$$

cf. Theorem 7.10 and Corollary 7.12.

PROOF

 (a) Let $f \in PW_E$. We obtain the uniform convergence from the formal estimate,

$$\left| f(t) - \sum_{-M}^{N} c[n]g_n(t) \right| = \left| \int_E \hat{f}(\gamma)e^{2\pi it\gamma}\,d\gamma - \sum_{-M}^{N} c[n]\int_E \hat{g}_n(\gamma)e^{2\pi it\gamma}\,d\gamma \right|$$

$$\leq |E|^{1/2} \left\| \hat{f} - \sum_{-M}^{N} c[n]\hat{g}_n \right\|_{L^2(E)},$$

and this last term tends to 0 as $M, N \to \infty$ when $c[n] = \langle \hat{f}, \hat{g}_n \rangle$.

 This calculation is legitimate once we show that the elements of PW_E are
continuous and that the pointwise inversion formula is true.

 To this end, first note that $PW_E^{\wedge} = L^2(E) \subseteq L^1(E)$, where the inclusion is
a consequence of Hölder's inequality and the hypothesis $|E| < \infty$. Thus, if
$f \in PW_E$ then

$$g(t) = \int \hat{f}(\gamma)e^{2\pi it\gamma}\,d\gamma \in A(\mathbb{R}) \cap L^2(\mathbb{R}).$$

It remains to show that $f = g$ a.e. If $h \in \mathcal{S}(\mathbb{R})$ then, distributionally,

$$\langle \hat{g}, \hat{h} \rangle = \langle g, h \rangle = \int\!\!\int_E \hat{f}(\gamma)e^{2\pi it\gamma}\bar{h}(t)\,d\gamma dt$$

$$= \int \hat{f}(\gamma)\left(\int \bar{h}(t)e^{2\pi it\gamma}\,dt \right)d\gamma = \langle \hat{f}, \hat{h} \rangle,$$

where the change in order of integration follows since $\hat{f}\bar{h} \in L^1(\hat{\mathbb{R}} \times \mathbb{R})$. Thus,
$\hat{g} = \hat{f}$ and so $g = f$ a.e. by the Fourier uniqueness theorem for tempered
distributions. (It should be noted that Fourier inversion is valid for $f, \hat{f} \in L^1$
or for f integrable and of bounded variation. There is an L^2 version of this
latter theorem, see, e.g., [Ti2, p. 83].)

 (b) In the opposite direction, we let $f \in L^2(\mathbb{R})$, and use the convergence of
$\sum |c[n]|^2$ and the Riesz–Fischer theorem (Theorem 7.10 c,d) to assert the

existence of $g \in PW_E$ for which

$$g = \sum c[n]g_n \text{ in } PW_E,$$

where each $c[n] = \langle g, g_n \rangle$. By hypothesis, $f = \sum c[n]g_n$ uniformly on $\mathbb{R}$. Thus,

$$|f(t) - g(t)| \leq \left| f(t) - \sum_{-M}^{N} c[n]g_n \right| + |E|^{1/2} \left\| \hat{g} - \sum_{-M}^{N} c[n]\hat{g}_n \right\|_{L^2(E)}$$

by part a; and so $f = g$. $\blacksquare$

Theorem 7.24 is obviously true for $\mathbb{R}^d$. Also, if $E = [-\Omega, \Omega]$ and $g_n = \tau_{n/(2\Omega)}d_{2\pi\Omega}$, then Theorem 7.24 falls into the setting of Theorem 7.21 for sampling at the Nyquist rate, i.e., the *sampling period* is $T = 1/2\Omega$. It is well-known that if $f \in PW_E$ and

$$|\{t : f(t) \neq 0\}| < \infty,$$

then $f = 0$ a.e. [Bene1].

Local problem 7.25

(a) The *local sampling problem* is to represent a signal f on an interval I in terms of samples $\{Tf(nT)\}$, where each nT is in a neighborhood of I.

There are several ways of quantifying this general statement. For example, suppose we are given a frequency band $\Omega > 0$ and an error margin or threshold $\epsilon > 0$. The problem is to construct a sampling function s and the smallest possible rim $r > 0$, such that

$$\forall I \quad \text{and} \quad \forall f \in PW_\Omega, \quad \left\| f - T \sum_{nT \in I_r} f(nT)\tau_{nT}s \right\| < \epsilon, \tag{7.27}$$

where $2T\Omega = 1, \| \cdot \|$ is a preordained norm, $I \subseteq \mathbb{R}$ is an interval, and

$$I_r \equiv \{t \in \mathbb{R} : \text{dist}(t, I) < r\}.$$

(b) In light of Theorem 7.22 we shall consider sampling functions s of the form

$$s = \frac{1}{\|g\|_2^2} (g * \tilde{g}) d_{2\pi(\Gamma + \Omega)}, \tag{7.28}$$

where $g \in PW_\Gamma$. Then $\hat{s}$ is 1 on $[-\Omega, \Omega]$ and vanishes off of $[-(2\Gamma + \Omega), 2\Gamma + \Omega]$.

We can make the connection between Γ and the *Nyquist rate* (*sampling frequency*) $1/T = 2\Omega$ (or for the case $2T\Omega < 1$ when *oversampling* occurs) in the following way. Because of supp $\hat{s}$, we let

$$\frac{1}{2T} = 2\Gamma + \Omega.$$

Thus, if we're given $T, \Omega > 0$ for which $2T\Omega \leq 1$, then we can define $\Gamma = 1/2(1/2T - \Omega)$ which can be as large as we like in the case of extensive oversampling. Similarly, in the case $\Gamma \geq 0, \Omega > 0$ are given then we can define $T = [2(2\Gamma + \Omega)]^{-1}$ so that $2T\Omega \leq 1$. Of course, if $\Gamma = 0$ then $PW_0 = \{0\}$ since $\delta^{(k)} \notin L^2(\hat{\mathbb{R}})$. ▯

The following is typical of the type of result we can prove using Gabor systems. We have taken s as in (7.28) and $\hat{g} = \mathbf{1}_{(\Omega)}$ in Theorem 7.26.

THEOREM 7.26 [BenKar]
Let $\epsilon > 0$ and $\Omega > 0$. If $r > 0$ is defined as

$$r \equiv \left(\frac{1}{\sqrt{6\pi^2 \epsilon \Omega}} \right)^{2/3},$$

then for all $f \in PW_\Omega$, for which $\|f\|_2 = 1$, and for each bounded interval I,

$$\left| f(t) - \frac{1}{2\Omega} \sum_{\frac{n}{2\Omega} \in I_r} f(\frac{n}{2\Omega}) \tau_{\frac{n}{2\Omega}} s(t) \right| < \epsilon$$

uniformly in I. In this case of $\hat{g} = \mathbf{1}_{(\Omega)}$ and s defined by (7.28), we have

$$\tau_{\frac{n}{2\Omega}} s(t) = \frac{(-1)^n \cos 2\pi t \Omega}{\Omega} d_{2\pi\Omega}^2 \left(t - \frac{n}{2\Omega} \right).$$

REMARK 7.27 The local sampling problem provides a natural illustration of the uncertainty principle in action.

In the case Γ is very large and $\hat{g}$ is very smooth, then we can think of $g \in PW_\Gamma$ as close to δ and having rapid decay. Thus, by the form of the sampling function s in (7.28), local sampling formulas such as (7.27) yield excellent local reconstruction (on the t-axis), but implement very high frequencies (due to Γ) in the process.

In Theorem 7.26, where $\Gamma = \Omega$, the sampling intervals can not be refined significantly outside of mollifying $\hat{g}$, but the bandwidth does not go beyond a modest factor of Ω which is important when limited bandwidth is available. ▌

Example 7.28. **Aliasing and linear time invariant systems**

(a) In the setting of Theorem 7.22, if $f \notin PW_\Omega$ we say that *aliasing* occurs, see, e.g., [Ben2]. This means that phase information contributed by $m \neq 0$ terms in (7.25) is required in the frame decomposition of a signal. To quantify this observation, we define the *aliasing pseudomeasure* $\alpha_{t,\Omega}$ on $\mathbb{R}$ (for the low pass impulse response $s = d_{2\pi\Omega}$ and for $2T\Omega = 1$) to be the distributional Fourier transform, $\alpha_{t,\Omega} = A_{t,\Omega}^\vee$, where each t is fixed and

$$A_{t,\Omega} \equiv \sum \left(e^{2\pi i (2m\Omega) t} - 1 \right) \tau_{2m\Omega} \mathbf{1}_{(\Omega)} \in L^\infty(\hat{\mathbb{R}}).$$

If $f \in L^2(\mathbb{R}) \cap A(\mathbb{R})$ then (7.25) yields

$$f(t) = T \sum f(nT) \tau_{nT} d_{2\pi\Omega}(t)$$

$$+ T \sum (f * \alpha_{t,\Omega})(nT) \tau_{nT} d_{2\pi\Omega}(t) \text{ in } L^2(\mathbb{R}),$$

where the first term is the usual sampling formula and the second is the aliasing term.

(b) Let X and Y be Banach subalgebras of the convolution algebra $M_b(\mathbb{R})$ of bounded Radon measures, let $\delta \in X$, assume $X \subseteq Y$, and let $L : X \to Y$ be a continuous linear map. L is a *linear time-invariant system* if the *impulse response* $L\delta = h$ has the property that

$$\forall f \in X, \quad Lf = h * f.$$

The Fourier transform $\hat{h}$ is the *frequency response* or *transfer function* of the system, and L or h is called a *filter* if the inclusion $\operatorname{supp} \hat{h} \subseteq \hat{\mathbb{R}}$ is proper. The proof of the following formal statement is along the lines of the classical proofs of Theorem 7.21 or our proof of Theorem 7.22 when natural hypotheses are made. Let $L : X \to Y$ be a linear time-invariant system with impulse response $h \in Y$ and frequency response $\hat{h}$. Assume there are constants $T, \Omega > 0$ such that $2T\Omega \le 1$, and define the function

$$s(t) = \int_{-\Omega}^{\Omega} \frac{1}{\hat{h}(\gamma)} e^{2\pi i t \gamma} d\gamma.$$

If $f \in X$ satisfies the support condition $\operatorname{supp} \hat{f} \subseteq [-\Omega, \Omega]$, then

$$f = T \sum (Lf)(nT) \tau_{nT} s \text{ in } Y. \quad \square$$

Example 7.29. Zero-crossings
Elements of PW_Ω are entire functions of exponential type. It is an important feature of such functions f that they are completely determined by their zeros. For example, Titchmarsh [Ti1] proved the following result.

Let $F \in L^1(\hat{\mathbb{R}})$ have compact support $\operatorname{supp} F = [-\Omega, \Omega]$, and assume this support is "tight," i.e.,

$$\forall \epsilon > 0, \quad \int_{-\Omega}^{-\Omega+\epsilon} |F(\gamma)| d\gamma \ne 0 \quad \text{and} \quad \int_{\Omega-\epsilon}^{\Omega} |F(\gamma)| d\gamma \ne 0.$$

Define the entire function $f(z) \equiv F^\vee(z)$, $z \in \mathbb{C}$, and assume, without loss of generality, that $f(0) \ne 0$. If $\{z_n : n = 1, \ldots\} \subseteq \mathbb{C}$ is the zero set of f, "ordered" by increasing amplitude, then

$$f(z) = f(0) \prod_{n=1}^{\infty} \left(1 - \frac{z}{z_n}\right),$$

where the infinite product is not absolutely convergent.

A consequence of this result is that reconstruction of $f \in PW_\Omega$ in terms of its zeros requires a knowledge of all the zeros. The real zeros of $f \in PW_\Omega$ are the *zero-crossings* of f, and perfect reconstruction of f can not be made in terms of zero-crossings in the case that f has complex zeros. This is not to say that zero-crossing algorithms have not been practically successful, see, e.g., [Ke]. ☐

7.6 The classical uncertainty principle inequality

This section will be devoted to classical uncertainty principle inequalities such as (7.2), which depends essentially on variances.

An instructive musical analogue of such classical inequalities is the following idealized piano experiment. The standard for concert pitch is that the *A* above middle *C* should have 440 vibrations per second. Thus, the *A* four octaves down (and the last key on the piano) should have 27.5 vibrations per second. Suppose we could strike this last key for a time interval of 1/30 seconds, i.e., the hammer strikes the string and 1/30 seconds later the damper returns to the string, thereby stopping the sound. We have very precise time information but correspondingly imprecise frequency information since the emitted sound is anything but the desired pure periodic pitch of this low *A*. Thus, if a sound f is emitted at time t_0 and lasts a very short time, then (by means of a slightly fanciful quantification) (7.2) asserts that the frequency range for f is quite broad. In particular, f is not close to a pure tone of frequency γ_0, for if it were, then $\|(\gamma - \gamma_0)\hat{f}(\gamma)\|_2$, as well as $\|(t - t_0)f(t)\|_2$, would be small in contrast to the "loudness" $\|f\|_2$.

For perspective, note that (7.2) does not provide useful information for certain signals, e.g., the right side of (7.2) is infinite for the signals f in Example 7.6b or for signals $f \in L^2(\mathbb{R})$ behaving like $|t|^a$ as $|t| \to \infty$, where $a \in [-3/2, -1/2)$. Besides our weighted approach in Section 7.8, there have been many contributions to deal with this limitation of (7.2). A "sampling" of recent work is [DoSta], [F], [L2], [Pr2].

THEOREM 7.30
Let $(t_0, \gamma_0) \in \mathbb{R} \times \hat{\mathbb{R}}$. Then

$$\forall f \in \mathcal{S}(\mathbb{R}), \quad \|f\|_2^2 \le 4\pi \|(t - t_0)f(t)\|_2 \|(\gamma - \gamma_0)\hat{f}(\gamma)\|_2, \tag{7.29}$$

and there is equality in (7.29) if and only if

$$f(t) = Ce^{2\pi i t \gamma_0} e^{-s(t - t_0)^2} \tag{7.30}$$

for $C \in \mathbb{C}$ and $s > 0$.

PROOF The mapping $f(t) \mapsto f(t + t_0)e^{-2\pi i t \gamma_0}$ shows that it is sufficient to verify (7.29) for $(t_0, \gamma_0) = (0, 0)$.

The following calculation gives (7.29) for $(t_0, \gamma_0) = (0, 0)$:

$$\|f\|_2^4 = \left(\int t |f(t)^2|' \, dt \right)^2 \leq \left(\int |t| |f(t)^2|' \, dt \right)^2$$

$$\leq 4 \left(\int |t \bar{f}(t) f'(t)| \, dt \right)^2 \leq 4 \|t f(t)\|_2^2 \|f'(t)\|_2^2 \qquad (7.31)$$

$$= 16 \pi^2 \|t f(t)\|_2^2 \|\gamma \hat{f}(\gamma)\|_2^2.$$

If f is defined by (7.30), then equality is obtained in (7.29) by direct calculation, see, e.g., Example 7.2. ∎

The main ingredients in (7.31) are integration by parts, Hölder's inequality, and the Plancherel theorem. We can extend and refine Theorem 7.30 in several ways, cf. Section 7.8. The main ingredients of such proofs are the same: integration by parts or conceptually similar ideas such as generalizations of Hardy's inequality; Hölder's inequality; and weighted norm inequalities for the Fourier transform, of which the Plancherel theorem is a special case.

Theorem 7.30 is not stated for all $f \in L^2(\mathbb{R})$, but the result is in fact valid for $L^2(\mathbb{R})$. The proof requires closure arguments and results described in Remark 7.31. Our description is fairly general in light of the direction we pursue in Section 7.8.

REMARK 7.31.   Closure theory

(a) Let $m, n \geq 0$ be integers and let $1 \leq p \leq \infty$. The *Sobolev space* L_m^p is the Banach space of functions $f \in L^p(\mathbb{R}^d)$ for which

$$\|f\|_{m,p} \equiv \sum_{|\alpha| \leq m} \|\partial^\alpha f\|_p < \infty.$$

The *weighted space* $L_{0,n}^p$ is the Banach space of functions $f \in L^p(\mathbb{R}^d)$ for which

$$\|f\|_{p,n} \equiv \sum_{|\beta| \leq n} \|t^\beta f(t)\|_p < \infty.$$

The *Bi-Sobolev space* $L_{m,n}^p$ is the Banach space of functions $f \in L_m^p \cap L_{0,n}^p$ for which

$$\|f\|_{m,p,n} \equiv \|f\|_{m,p} + \|f\|_{p,n} < \infty.$$

(b) Let $m, n \geq 0$. $C_c^\infty(\mathbb{R}^d)$ is dense in the Hilbert space $(L_{m,n}^2, \| \cdot \|_{m,2,n})$ with *inner product*

$$[f, g] = \sum_{|\alpha| \leq m} \langle \partial^\alpha f, \partial^\alpha g \rangle + \sum_{|\beta| \leq n} \langle t^\beta f, t^\beta g \rangle,$$

where $\langle \cdot, \cdot \rangle$ is the usual inner product on $L^2(\mathbb{R}^d)$. The proof involves classical techniques including Young's theorem, Leibniz's formula, and an equicontinuity argument ([Ben1, Theorem A.1.2]).

(c) Let $\mathcal{S}_0 \equiv \{f \in \mathcal{S}(\mathbb{R}^d) : \hat{f}(0) = 0\}$, let $v \in L^1_{loc}(\mathbb{R}^d)$ be positive a.e., and choose $p \in (1, \infty)$. If $v^{1-p'} \notin L^1(\mathbb{R}^d)$, then $\mathcal{S}_0 \cap L^p_v(\mathbb{R}^d)$ is dense in $L^p_v(\mathbb{R}^d)$, where $L^p_v(\mathbb{R}^d)$ consists of the Borel measurable functions f on $\mathbb{R}^d$ for which $\|fv^{1/p}\|_p < \infty$, cf. Section 7.8 and [BenHe]. This result is not true for $p = 1$; in fact, for $v = 1$ the closure of $\mathcal{S}_0$ in $L^1(\mathbb{R}^d)$ is a (proper closed) maximal ideal in $L^1(\mathbb{R}^d)$. The closure of $\mathcal{S}_0$ in $L^2_{1,1}$ is $\{f \in L^2_{1,1} : \hat{f}(0) = 0\}$. ∎

THEOREM 7.32

Let A, B be self-adjoint operators on a complex Hilbert space H (A and B need not be continuous). Define the commutator $[A, B] \equiv AB - BA$, the expectation or expected value $E_x(A) \equiv \langle Ax, x \rangle$ of A at $x \in D(A)$ ($D(A)$ is the domain of A), and the variance $\Delta_x^2(A) \equiv E_x(A^2) - \{E_x(A)\}^2$ of A at $x \in D(A^2)$. If $x \in D(A^2) \cap D(B^2) \cap D(i[A, B])$ and $\|x\| \le 1$, then

$$\{E_x(i[A, B])\}^2 \le 4\Delta_x^2(A)\Delta_x^2(B). \tag{7.32}$$

PROOF By self-adjointness, we first compute

$$
\begin{aligned}
E_x(i[A, B]) &= i\{\langle Bx, Ax \rangle - \langle Ax, Bx \rangle\} \\
&= 2\,\mathrm{Im}\langle Ax, Bx \rangle.
\end{aligned}
\tag{7.33}
$$

Also note that $D(A^2) \subseteq D(A)$.

Next, since $\|x\| \le 1$ and $\langle Ax, x \rangle, \langle Bx, x \rangle \in \mathbb{R}$ (by self-adjointness), we have

$$\|(B + iA)x\|^2 - |\langle (B + iA)x, x \rangle|^2 \ge 0 \tag{7.34}$$

and

$$|\langle (B + iA)x, x \rangle|^2 = \langle Bx, x \rangle^2 + \langle Ax, x \rangle^2. \tag{7.35}$$

Also, by definition of $\|\cdot\|$, we compute

$$\|(B + iA)x\|^2 = \|Bx\|^2 + \|Ax\|^2 - 2Im\langle Ax, Bx \rangle. \tag{7.36}$$

Substituting (7.35) and (7.36) into (7.34) yields the inequality,

$$
\begin{aligned}
\|Ax\|^2 - \langle Ax, x \rangle^2 + \|Bx\|^2 - \langle Bx, x \rangle^2 \\
\ge 2\,\mathrm{Im}\langle Ax, Bx \rangle.
\end{aligned}
\tag{7.37}
$$

Letting $r, s \in \mathbb{R}$, so that rA and sB are also self-adjoint, inequality (7.37) becomes

$$
\begin{aligned}
r^2\left(\|Ax\|^2 - \langle Ax, x \rangle^2\right) + s^2\left(\|Bx\|^2 - \langle Bx, x \rangle^2\right) \\
\ge 2rs\,\mathrm{Im}\langle Ax, Bx \rangle.
\end{aligned}
\tag{7.38}
$$

Setting $r^2 = \|Bx\|^2 - \langle Bx, x \rangle^2$ and $s^2 = \|Ax\|^2 - \langle Ax, x \rangle^2$, substituting into (7.38), squaring both sides and dividing, we obtain

$$\left(\|Ax\|^2 - \langle Ax, x \rangle^2 \right) \left(\|Bx\|^2 - \langle Bx, x \rangle^2 \right) \geq (\mathrm{Im} \langle Ax, Bx \rangle)^2 \,.$$

The uncertainty principle inequality (7.32) follows from this inequality and (7.33). ∎

Example 7.33. *The classical uncertainty principle inequality and entropy*

(a) Equation (7.29) is a corollary of Theorem 7.32 for the case $H = L^2(\mathbb{R})$, where the operators A and B are defined as

$$A(f)(t) \equiv (t - t_0)f(t)$$

and

$$B(f)(t) \equiv i(2\pi i \left(\gamma - \gamma_0 \right) \hat{f}(\gamma))^{\vee}(t).$$

Straightforward calculations show that A and B are self-adjoint, and that

$$E_f(A) = \int (t - t_0)|f(t)|^2 \, dt,$$

$$E_f(B) = -2\pi \int (\gamma - \gamma_0)|\hat{f}(\gamma)|^2 \, d\gamma,$$

$$\Delta_f^2(A) = \int |t - t_0|^2 |f(t)|^2 \, dt - \left(\int (t - t_0)|f(t)|^2 \, dt \right)^2,$$

$$\Delta_f^2(B) = 4\pi^2 \left(\int |\gamma - \gamma_0|^2 |\hat{f}(\gamma)|^2 \, d\gamma - \left(\int (\gamma - \gamma_0)|\hat{f}(\gamma)|^2 \, d\gamma \right)^2 \right),$$

$$\{E_f(i[A, B])\}^2 = \|f\|_2^4,$$

cf. the variances in Subsection 7.3, where the expected values of the modulated and translated Gaussians vanish since they are symmetric about t_0 and γ_0.

(b) The Fast Wavelet Packet Transform (FWPT) was developed by Coifman, Meyer and Wickerhauser ([CoMeWi]), and has been applied to data compression problems. A critical step, and one which demands deeper study for specific applications, is to choose "best bases." In the spirit of the Fundamental Problem, this "best basis" is chosen in [CoMeWi] as the orthonormal basis $\{x_n\}$ of $H = \mathbb{R}^{2^d}$ for which a given signal $c \in \mathbb{R}^{2^d}$ has the least entropy,

$$- \sum_n |\langle c, x_n \rangle|^2 \log |\langle c, x_n \rangle|^2,$$

relative to $\{x_n\}$.

To make our tie-in with (7.29), we define the entropy of $f \in H = L^2(\mathbb{R})$ to be

$$\epsilon(f) = - \int |f(t)| \log |f(t)| \, dt.$$

Hirschman [Hir] proved that

$$\epsilon(|f|^2) + \epsilon(|\hat{f}|^2) \geq 0$$

for $f \in L^2(\mathbb{R})$ for which $\|f\|_2 = 1$, cf. [DeCovTh]. He also noted the plausibility of the inequality

$$\epsilon(|f|^2) + \epsilon(|\hat{f}|^2) \geq 1 - \log 2; \tag{7.39}$$

and pointed out that (7.39) includes (7.29) by means of an argument found in [ShWe, pp. 55–57].

It turns out that (7.39) is true [F]; and depends on the following sharp form of the Hausdorff–Young–Titchmarsh inequality, viz., if $1 \leq p \leq 2$ and $1/p + 1/p' = 1$, then

$$\forall f \in \mathcal{S}(\mathbb{R}^d), \quad \|\hat{f}\|_{p'} \leq B_d(p)\|f\|_p, \tag{7.40}$$

where $B_d(p)$ is the Babenko–Beckner constant,

$$B_d(p) = \left(p^{1/p}(p')^{-1/p'}\right)^{d/2}, \tag{7.41}$$

cf. [Be]. □

Notation 7.34

"σ^2" is the usual notation for *variance*, but we are using "Δ^2" in light of some classical conventions with the uncertainty principle. Specifically, we shall use the notation

$$\Delta^2(f, t_0) \equiv \int |t - t_0|^2 |f(t)|^2 \, dt - \left(\int (t - t_0)|f(t)|^2 \, dt\right)^2,$$

and

$$\Delta^2(\hat{f}, \gamma_0) \equiv 4\pi^2 \left(\int |\gamma - \gamma_0|^2 |\hat{f}(\gamma)|^2 \, d\gamma - \left(\int (\gamma - \gamma_0)|\hat{f}(\gamma)|^2 \, d\gamma\right)^2\right),$$

cf. Example 7.33a.

Further, for a given function $\theta_{m,n}$, we shall use the notation,

$$t_{m,n} = \int t|\theta_{m,n}(t)|^2 \, dt \quad \text{and} \quad \gamma_{m,n} = \int \gamma|\hat{\theta}_{m,n}(\gamma)|^2 \, d\gamma$$

for the *expected values*. □

The proof of the following result is a straightforward calculation.

THEOREM 7.35

(a) Let $g \in L^2(\mathbb{R})$ satisfy the conditions that $\|g\|_2 = 1$, $tg(t) \in L^2(\mathbb{R})$, and $\gamma\hat{g}(\gamma) \in L^2(\mathbb{R})$; and assume that $\{\varphi_{m,n} : (m,n) \in \mathbb{Z} \times \mathbb{Z}\}$ is the corresponding

Gabor system. Then $t_{m,n}, \gamma_{m,n} \in \mathbb{C}$, *and*

$$\forall (t_0, \gamma_0) \in \mathbb{R} \times \hat{\mathbb{R}} \quad and \quad \forall (m, n) \in \mathbb{Z} \times \mathbb{Z},$$
$$\Delta^2(\varphi_{m,n}, t_0) = \Delta^2(\varphi_{0,0}, 0) = \|(t - t_{m,n})\varphi_{m,n}\|_2^2 < \infty \tag{7.42}$$

and

$$\Delta^2(\hat{\varphi}_{m,n}, \gamma_0) = \Delta^2(\hat{\varphi}_{0,0}, 0) = 4\pi^2 \|(\gamma - \gamma_{m,n})\hat{\varphi}_{m,n}\|_2^2 < \infty$$

$(\varphi_{0,0} = g)$.

(b) *Let* $\psi \in L^2(\mathbb{R})$ *satisfy the conditions that* $\|\psi\|_2 = 1$, $t\psi(t) \in L^2(\mathbb{R})$, *and* $\gamma\hat{\psi}(\gamma) \in L^2(\hat{\mathbb{R}})$; *and assume that* $\{\psi_{m,n} : (m, n) \in \mathbb{Z} \times \mathbb{Z}\}$ *is the corresponding wavelet system. Then* $t_{m,n}, \gamma_{m,n} \in \mathbb{C}$, *and*

$$\forall (t_0, \gamma_0) \in \mathbb{R} \times \hat{\mathbb{R}} \quad and \quad \forall (m, n) \in \mathbb{Z} \times \mathbb{Z},$$
$$\Delta^2(\psi_{m,n}, t_0) = 2^{-2m}\Delta^2(\psi_{0,0}, 0) = \|(t - t_{m,n})\psi_{m,n}(t)\|_2^2 < \infty \tag{7.43}$$

and

$$\Delta^2(\hat{\psi}_{m,n}, \gamma_0) = 2^{2m}\Delta^2(\hat{\psi}_{0,0}, 0) = 4\pi^2 \left\|(\gamma - \gamma_{m,n})\hat{\psi}_{m,n}(\gamma)\right\|_2^2 < \infty$$

$(\psi_{0,0} = \psi)$.

COROLLARY 7.36

Suppose the hypotheses of Theorem 7.35 are satisfied. Then

$$\forall n, \quad \sup_m \|(t - t_{m,n})\psi_{m,n}(t)\|_2 = \infty \tag{7.44}$$

and

$$\forall n, \quad \sup_m 2\pi \left\|(\gamma - \gamma_{m,n})\hat{\psi}_{m,n}(\gamma)\right\|_2 = \infty, \tag{7.45}$$

whereas

$$\sup_{m,n} 2\pi \|(t - t_{m,n})\psi_{m,n}(t)\|_2 \left\|(\gamma - \gamma_{m,n})\hat{\psi}_{m,n}(\gamma)\right\|_2 < \infty. \tag{7.46}$$

Example 7.37. **Uniqueness and the classical uncertainty principle inequality**

(a) Even though (7.42) is valid, it is easy to see that

$$\forall (t_0, \gamma_0) \in \mathbb{R} \times \hat{\mathbb{R}}, \quad \sup_{m,n} 2\pi \|(t - t_0)\varphi_{m,n}(t)\|_2 \|(\gamma - \gamma_0)\hat{\varphi}_{m,n}(\gamma)\|_2 = \infty, \tag{7.47}$$

because of translation from the origin.

Because of open questions concerning the Balian–Low Theorem (Theorem 7.38), we shall give a complicated proof of (7.47) which allows for dealing with refinements of (7.47) in which the sup is replaced by various paths in $\mathbb{Z} \times \mathbb{Z}$.

First, if f is analytic in the tube $\mathbb{R} \times (0, b)$, f has nontangential limits a.e. on $\mathbb{R}$, and $f = 0$ on $K \subseteq \mathbb{R}$ for which $|K| > 0$, then f is the zero function. In this generality, David Hamilton has pointed out that topological arguments à la Privalov's theorem can be used in the proof. Alternatively, we could proceed in the usual way by Jensen's theorem after setting up the following mapping. Choose $C \subseteq K, |C| > 0$, such that $|f(z)| \leq A$ on $\cup T_\gamma$, $\gamma \in C$, where $T_\gamma \subseteq \mathbb{R} \times [0,b)$ is a triangle and $T_\gamma \backslash \{\gamma\} \subseteq \mathbb{R} \times (0,b)$. Then for an appropriate set $B \supseteq \cup T_\gamma$ choose a bi-absolutely continuous map from B onto the unit disc, and apply the Jensen's theorem argument found in the books of Koosis [Ko, pp. 76–77] or Levinson [Le, page 74].

Second, if $F \in L^2(\hat{\mathbb{R}}) \backslash \{0\}$ and $\text{supp} F$ is contained in a half-line then $\hat{F} = f \in H^2(\mathbb{R}) \backslash \{0\}$, and, in particular, f is an analytic function with $\mathbb{R}$ as a boundary. This fact is the easy direction of the Paley–Wiener theorem.

Using these two facts, our complicated proof of (7.37) proceeds as follows. We write

$$v(m,n) \equiv \|(t - t_0)\varphi_{m,n}(t)\|_2^2 \|(\gamma - \gamma_0)\hat{\varphi}_{m,n}(\gamma)\|_2^2$$

$$= \int |(u - t_0 + na)g(u)|^2 \, du \int |(\gamma - \gamma_0 + mb)\hat{g}(\gamma)|^2 \, d\gamma.$$

For case 1, suppose $\text{supp} \, g$ is not contained in a half-line so that for all $n \geq 0$,

$$v(m,n) \geq (na)^2 \int_{t_0}^\infty |g(u)|^2 \, du \int |(\gamma - \gamma_0 + mb)\hat{g}(\gamma)|^2 \, d\gamma.$$

Thus,

$$\forall m, \quad \overline{\lim_{n \to \infty}} \, v(m,n) = \infty.$$

For case 2, suppose $\text{supp} \, g$ is contained in a half-line. Then, for all $m \geq 0$,

$$v(m,n) \geq (mb)^2 \int |(u - t_0 + na)g(u)|^2 \, du \int_{\gamma_0}^\infty |\hat{g}(\gamma)|^2 \, d\gamma.$$

By our two facts, $\int_{\gamma_0}^\infty |\hat{g}(\gamma)|^2 \, d\gamma > 0$, and so

$$\forall n, \quad \overline{\lim_{m \to \infty}} \, v(m,n) = \infty.$$

(b) The uniqueness result, which we invoked in part a, asserts that if f vanishes on a "large enough" set and if $\hat{f}$ is supported by a "small enough" set then $f \equiv 0$; or, in terms of decay, f and $\hat{f}$ cannot both decay too rapidly. This interplay is central to the *uncertainty principle phenomenon in harmonic analysis*. □

The following is an "$H^1(\mathbb{R})$" version of an early uniqueness theorem due to F. and M. Riesz: If $f \in L^1(\mathbb{R}), |\{t : f(t) = 0\}| > 0, \hat{f} \in A(\hat{\mathbb{R}})$ is a bounded measure, and $\text{supp} \hat{f} \subseteq [-\Omega, \infty)$, then $f \equiv 0$. This result can be generalized by replacing the support condition by the condition

$$\forall \gamma \leq 0, \quad |\hat{f}(\gamma)| \leq Be^{2\pi b\gamma},$$

for some $B, b > 0$. To prove uniqueness in this case, simple estimates on the integrals formally defining $f(z)$ and $f'(z)$ for $z \in \mathbb{C}$ show that f is analytic in the tube $\mathbb{R} \times (0, b)$. Because $f \in L^1(\mathbb{R})$, a Jensen's theorem argument, more direct than that outlined in part a, allows us to conclude that $f \equiv 0$.

The role of logarithmic integrals in this subject is illustrated by the following theorem, see, e.g., [Le, Chapter V], [Beu, Volume 1, pp. 405–412]: let the weight $u \geq 1$ be a continuous function on $\hat{\mathbb{R}}$ for which $\log u$ is uniformly continuous on $\hat{\mathbb{R}}$ and

$$\int \frac{\log u(\gamma)}{1 + \gamma^2} \, d\gamma = \infty;$$

if f is the inverse Fourier transform of a bounded measure $\hat{f} \equiv \mu$ subject to the constraint

$$\int u(\gamma) \, d\mu(\gamma) < \infty,$$

and $|\{t : f(t) = 0\}| > 0$, then $f \equiv 0$. Beurling's contributions were recorded in his Stanford lectures (1961), and there are comparable results due to deBranges. More recent uniqueness theorems have been proven by Benedicks [Bene2] and Gabardo [Ga].

Hardy made the following contribution concerning *uncertainty principle phenomena in harmonic analysis*: if $a, b > 0$, $ab > \pi^2$, and

$$|f(t)| \leq Ae^{-at^2} \quad \text{and} \quad |\hat{f}(\gamma)| \leq Be^{-b\gamma^2},$$

then $f \equiv 0$, cf. [PWie, pp. 64–67]. In this spirit Beurling [Beu, Volume 2, p. 372] (1942) proved that if $f \in L^1(\mathbb{R})$ and

$$\iint |f(t)\hat{f}(\gamma)| e^{|t\gamma|} \, dt d\gamma < \infty,$$

then $f \equiv 0$, e.g., [Hö2] for details. As a corollary, he proved that if φ and ψ are conjugate convex functions and

$$\int |f(t)| e^{\varphi(t)} \, dt < \infty \quad \text{and} \quad \int |\hat{f}(\gamma)| e^{\psi(\gamma)} \, d\gamma < \infty,$$

then $f \equiv 0$, cf. (7.51).

It is natural to investigate the frame properties of Gabor systems $\{\varphi_{m,n}\}$ in light of our Fundamental Problem and Theorem 7.35a. The most striking result in this area is the Balian–Low theorem, which was first stated by Balian [Bal] (1981) and, independently, by Low [Lo] (1985), and which is discussed and proved in [BenWa].

THEOREM 7.38

Let $g \in L^2(\mathbb{R})$ and let $ab = 1$ where $a, b > 0$. If the Gabor system $\{\varphi_{m,n}\}$ is a frame (and thus an exact frame since $ab = 1$), then

$$\sup_{m,n} 4\pi^2 \|(t - t_{m,n})\varphi_{m,n}(t)\|_2^2 \|(\gamma - \gamma_{m,n})\hat{\varphi}_{m,n}(\gamma)\|_2^2$$

$$= \Delta^2(\varphi_{0,0}, 0)\Delta^2(\hat{\varphi}_{0,0}, 0) = \infty,$$

i.e., either $tg(t) \notin L^2(\mathbb{R})$ or $\gamma\hat{g}(\gamma) \notin L^2(\hat{\mathbb{R}})$.

Problem 7.39

In contrast to Theorem 7.38, even the earliest examples of wavelet orthonormal bases $\{\psi_{m,n}\}$ were noted to satisfy (7.46), see, e.g., [Me] and Meyer's Seminaire Bourbaki, 1985–1986, number 662. In this regard, and taking Theorem 7.38 as well as (7.44) and (7.45) into account, there is the problem of determining whether or not there are orthonormal bases $\{\theta_{m,n}\}$ of $L^2(\mathbb{R})$ for which

$$\sup_{m,n} \|(t - t_{m,n})\theta_{m,n}(t)\|_2 < \infty \quad \text{and} \quad \sup_{m,n} \|(\gamma - \gamma_{m,n})\hat{\theta}_{m,n}(\gamma)\|_2 < \infty, \quad (7.48)$$

where $t_{m,n}, \gamma_{m,n}$ are the expected values defined in Subsection 7.34. This particular problem was posed by Balian [Bal] in conjunction with his statement of Theorem 7.38. Clearly, there are related problems which can be posed. This particular problem, with the constraints imposed by (7.48), was answered by Bourgain [Bou] in the following way, cf. [By], [Sha]. $\square$

THEOREM 7.40

For every $\epsilon > 0$ there is an orthonormal basis $\{\theta_{m,n}\}$ of $L^2(\mathbb{R})$ having expected values $(t_{m,n}, \gamma_{m,n}) \in \mathbb{R} \times \hat{\mathbb{R}}$ and satisfying the inequalities

$$\sup_{m,n} \|(t - t_{m,n})\theta_{m,n}(t)\|_2 < \frac{1}{2\sqrt{\pi}} + \epsilon \qquad (7.49)$$

and

$$\sup_{m,n} \|(\gamma - \gamma_{m,n})\hat{\theta}_{m,n}(\gamma)\|_2 < \frac{1}{2\sqrt{\pi}} + \epsilon.$$

Thus, invoking (7.29),

$$\forall m, n, \quad 1 = \|\theta_{m,n}\|_2^2 \leq 4\pi \|(t - t_{m,n})\theta_{m,n}(t)\|_2 \|(\gamma - \gamma_{m,n})\hat{\theta}_{m,n}(\gamma)\|_2$$

$$< 4\pi \left(\frac{1}{2\sqrt{\pi}} + \epsilon\right)^2. \qquad (7.50)$$

Consider the tiling of information plane in terms of Bourgain's functions and their variances. Because of (7.49), (7.50), and the classical uncertainty principle inequality, the "tiles" are essentially smallest possible squares.

Examples 7.41.   Balian–Low phenomenon for wavelet systems and incompleteness for Gabor systems

(a) As mentioned at the beginning of Problem 7.39, the analogue of Theorem 7.38 fails for the case of wavelet systems; and Meyer's first example of a wavelet orthonormal basis $\{\psi_{m,n}\}$ has the property that $\psi \in \mathcal{S}(\mathbb{R})$ and $\hat{\psi} \in C_c^{\infty}(\hat{\mathbb{R}})$. Thus, a problem for wavelet systems, in the spirit of the Balian–Low phenomenon, is to determine whether or not there are functions $\psi \in L^2(\mathbb{R})$ for which the elements of the wavelet system $\{\psi_{m,n}\}$ are pairwise orthogonal, and for which

$$\||e^{|t|}\psi(t)\|_2 \|e^{|\gamma|}\hat{\psi}(\gamma)\|_2 < \infty. \tag{7.51}$$

Battle [Bat] proved that this can not happen; and, recalling the hypothesis $ab = 1$ of Theorem 7.38, Heil, Walnut, and I made a simple extension to an exact frame setting. Battle's result does *not* depend on the completeness of $\{\psi_{m,n}\}$. His proof uses the following elementary but useful fact: if $\psi, \hat{\psi} \in L^1 \cap L^2$ and the elements of the wavelet system $\{\psi_{m,n}\}$ are mutually orthogonal, then $\hat{\psi}(0) = 0$. The hypotheses for this fact are interesting in light of the role the conclusion plays in basic wavelet theory, see, e.g., [Mal], [Me].

(b) In Theorem 7.8a we noted that if $ab > 1$ then $\overline{\mathrm{sp}}\{\varphi_{m,n}\} \neq L^2(\mathbb{R})$ for the case of the Gabor system generated by a Gaussian. Because of the gaps in the information plane, this result, as well as the analogous result for any Gabor system, is expected. In fact, it is true that if $ab > 1$, $g \in L^2(\mathbb{R})$, and $\{\varphi_{m,n}\}$ is the Gabor system associated with a, b, and g, then $\overline{\mathrm{sp}}\{\varphi_{m,n}\} \neq L^2(\mathbb{R})$. The proof due to Rieffel [R] (1981) is surprisingly abstract. Baggett [B] obtained the result by studying representations $\rho^{a,b}$ of the discrete Heisenberg group Γ of 3×3 integer matrices. Elements of Γ have the form

$$\begin{pmatrix} 1 & n & p \\ 0 & 1 & m \\ 0 & 0 & 1 \end{pmatrix},$$

and, for a given g and corresponding Gabor system $\{\varphi_{m,n}\}$, the representation $\rho^{a,b}$ of Γ on $L^2(\mathbb{R})$ is defined as

$$\rho_{m,n,p}^{a,b} g(t) = e^{2\pi i abp} \varphi_{m,n}(t).$$

Thus, $\overline{\mathrm{sp}}\{\varphi_{m,n}\} = L^2(\mathbb{R})$ if and only if g is a cyclic vector for the representation $\rho^{a,b}$. ☐

The most significant contribution in classical analysis related to Rieffel's theorem is due to H. Landau [L3]. Using some of his previous work [L1] and a theorem of Daubechies [D2, Theorem 3.1], he proved that if g and $\hat{g}$ have sufficient decay and $\{\varphi_{m,n}\}$ is a Gabor frame for $L^2(\mathbb{R})$, then $ab \leq 1$. This is, of course, weaker than Rieffel's theorem, but the method, although intricate, is constructive. The following are sufficient decay conditions to implement Daubechies' theorem: there are constants $C > 0$ and $\alpha > 1/2$ such that

$$\forall (t, \gamma) \in \mathbb{R} \times \hat{\mathbb{R}}, \quad |g(t)| \leq \frac{C}{(1 + t^2)^\alpha} \tag{7.52}$$

and

$$|\hat{g}(\gamma)| \leq \frac{C}{(1 + \gamma^2)^\alpha}.$$

Of course, if $\alpha > 3/4$ and $ab = 1$ then, by Theorem 7.38, there are no such g for which $\{\varphi_{m,n}\}$ is a Gabor frame. Note that $g(t) \equiv H(t)\sqrt{t}e^{-t}$ ($H \equiv$ Heaviside function) satisfies (7.52) for $\alpha = 3/4$ as well as the conclusion of Theorem 7.38, since

$$\hat{g}(\gamma) = Cu(\gamma)/(1 + \gamma^2)^{3/4}, \quad |u(\gamma)| \equiv 1.$$

7.7 Sampling theorems

This section will be devoted to sampling theorems for irregularly spaced *sampling sequences* $\{t_n\}$. There is an extensive tutorial on the subject in [Ben2], which features new results and formulations in terms of frames, as well as a guide to the literature. Our goal here is to present a general and constructive irregular sampling formula (Theorem 7.45) and an implementable algorithm (Algorithm 7.51), cf. [Ben2, Algorithm 50]. The discussion is rounded off by examples and a comparison with other methods.

DEFINITION 7.42 *Let $\{t_n : n \in \mathbb{Z}\} \subseteq \mathbb{R}$ be a strictly increasing sequence for which $\lim_{n \to \pm\infty} t_n = \pm\infty$, and for which*

$$\exists d > 0 \quad \text{such that } \forall m \neq n, \quad |t_m - t_n| \geq d. \tag{7.53}$$

Sequences satisfying (7.53) are uniformly discrete, cf. Example/Remark 7.20.

(a) The sequence $\{t_n\}$ is uniformly dense with uniform density $\Delta > 0$ if

$$\exists L > 0 \quad \text{such that } \forall n \in \mathbb{Z}, \quad |t_n - \frac{n}{\Delta}| \leq L.$$

(b) The sequence $\{t_n\}$ has natural density $\Delta \geq 0$ if

$$\exists \lim_{r \to \infty} \frac{n_0(r)}{r} = \Delta,$$

where

$$n_0(r) \equiv \text{card}\left\{t_n \in \left[-\frac{r}{2}, \frac{r}{2}\right]\right\}.$$

Even if a sequence fails to have a natural density, it always has (finite or infinite) upper and lower natural densities defined in terms of $\overline{\lim}$ *and* $\underline{\lim}$, *respectively, ([Bo2, Section 1.5]). The comparison between uniform density and natural density is made in [Ben2, Example 37].*

(c) *The sequence* $\{t_n\}$ *has lower, resp., upper, uniform Beurling density* $\Delta^- \geq 0$, *resp.,* $\Delta^+ \geq 0$, *if*

$$\Delta^- = \lim_{r \to \infty} \frac{n^-(r)}{r}, \ resp., \Delta^+ = \lim_{r \to \infty} \frac{n^+(r)}{r}, \tag{7.54}$$

where

$$n^-(r) \equiv \inf_I n_I(r), \ n^+(r) \equiv \sup_I n_I(r),$$

$$n_I(r) \equiv \mathrm{card}\{t_n \in I : |I| = r\}$$

for all intervals $I \subseteq \mathbb{R}$ *and* $r > 0$. *The limits exists in (7.54) since* $n^-(r_1+r_2) \geq n^-(r_1) + n^-(r_2)$ *and* $n^+(r_1 + r_2) \leq n^+(r_1) + n^+(r_2)$.

The relation between Beurling density and the other topics of this section is discussed in [Ben2, Section 9]. For Beurling's basic theorem in this area we refer to [Beu, Volume2, pp. 341 ff.], cf. [L1].

(d) *The frame radius of* $\{t_n\}$ *is*

$$\Omega_r \equiv \sup \{\Omega \geq 0 : \{e_{-t_n}\} \ \textit{is a frame for } L^2[-\Omega, \Omega]\} \in [0, \infty].$$

A central result for the modern theory of irregular sampling is the remarkable Duffin–Schaeffer theorem [DuSc].

THEOREM 7.43
Let $\{t_n : n \in \mathbb{Z}\} \subseteq \mathbb{R}$ *be a uniformly dense sequence with uniform density* Δ. *If* $0 < 2\Omega < \Delta$ *then* $\{e_{-t_n}(\gamma)\}$ *is a frame for* $L^2[-\Omega, \Omega]$.

After 40 years, Theorem 7.43 is still difficult to prove. Among other notions and estimates, its proof involves fundamental properties of entire functions of exponential type associated with the work of Plancherel–Pólya and Boas, referred to in Example/Remark 7.20. It is easy to see that Theorem 7.43 can be translated into the assertion that if $\{t_n\} \subseteq \mathbb{R}$ has uniform density $\Delta > 0$, then $2\Omega_r \geq \Delta$.

Using the Duffin–Schaeffer theorem (Theorem 7.43) and (7.21) for one direction, Jaffard [J] has provided the following characterization of frames $\{e_{-t_n}\}$ for $L^2[-\Omega, \Omega]$.

THEOREM 7.44
Let $\{t_n\} \subseteq \mathbb{R}$ *be a strictly increasing sequence for which* $\lim_{n \to \pm\infty} t_n = \pm\infty$, *and let* $I \subseteq \mathbb{R}$ *denote an interval.*

(a) *The following two assertions are equivalent:*

    i. There is $I \subseteq \mathbb{R}$ for which $\{e_{-t_n}\}$ is a frame for $L^2(I)$;

    ii. The sequence $\{t_n\}$ is a disjoint union of a uniformly dense sequence with uniform density Δ and a finite number of uniformly discrete sequences.

(b) *In the case assertion ii of part a holds, then $\{e_{-t_n}\}$ is a frame for $L^2(I)$ for each $I \subseteq \mathbb{R}$ for which $|I| < \Delta$. (Hence, $2\Omega_r \geq \Delta$.)*

The following general irregular sampling formula originated in [BenHel1]. We shall deal with implementation in Algorithm 7.51, cf. [BenHel1], [Hel].

THEOREM 7.45
Suppose $\Omega > 0$ and $\Omega_1 \succ \Omega$, and let $\{t_n\} \subseteq \mathbb{R}$ be a strictly increasing sequence for which $\lim_{n \to \pm\infty} t_n = \pm\infty$. Assume the sequence $\{t_n\}$ is a disjoint union of a uniformly dense sequence with uniform density $\Delta > 2\Omega_1$ and a finite number of uniformly discrete sequences. Let $s \in L^2(\mathbb{R})$ have the properties that $\hat{s} \in L^\infty(\hat{\mathbb{R}})$, $\operatorname{supp} \hat{s} \subseteq [-\Omega_1, \Omega_1]$, and $\hat{s} = 1$ on $[-\Omega, \Omega]$.

(a) *The sequence $\{e_{-t_n}\}$ is a frame for $L^2[-\Omega_1, \Omega_1]$ with frame operator S.*

(b) *Each $f \in PW_\Omega$ has the representation*

$$f = \sum c_f[n] \tau_{t_n} s \text{ in } L^2(\mathbb{R}),$$

where

$$c_f[n] \equiv \langle S^{-1}(\hat{f} \mathbf{1}_{(\Omega_1)}), e_{-t_n} \rangle_{[-\Omega_1, \Omega_1]}$$

and

$$\{c_f[n]\} \in \ell^2(\mathbb{Z}).$$

PROOF   Part a is a restatement of Theorem 7.44.

For part b, since $\{e_{-t_n}\}$ is a frame for $L^2[-\Omega_1, \Omega_1]$ and $\operatorname{supp} \hat{f} \subseteq [-\Omega, \Omega]$, we have

$$\hat{f} = \hat{f} \mathbf{1}_{(\Omega_1)} = \sum \langle S^{-1}(\hat{f} \mathbf{1}_{(\Omega_1)}), e_{-t_n} \rangle_{[-\Omega_1, \Omega_1]} e_{-t_n} \text{ in } L^2[-\Omega_1, \Omega_1]. \tag{7.55}$$

Equation (7.55) is a direct consequence of (7.7) and the fact that S^{-1}, being a positive operator, is self-adjoint. Using the hypothesis, $f \in PW_\Omega$, we can rewrite (7.55) as

$$\hat{f} = \sum c_f[n](e_{-t_n} \mathbf{1}_{(\Omega_1)}) \text{ in } L^2(\hat{\mathbb{R}}). \tag{7.56}$$

In fact,

$$\left\| \hat{f} - \sum_{-M}^{N} c_f[n](e_{-t_n} \mathbf{1}_{(\Omega_1)}) \right\|_2^2$$

$$= \int_{-\Omega_1}^{\Omega_1} \left| \hat{f}(\gamma) - \sum_{-M}^{N} c_f[n] e_{-t_n}(\gamma) \right|^2 d\gamma$$

$$= \left\| \hat{f} - \sum_{-M}^{N} c_f[n] e_{-t_n} \right\|^2_{L^2[-\Omega_1, \Omega_1]}$$

Next, we note that $\hat{f} = \hat{f}\hat{s}$, and hence,

$$\left\| \hat{f} - \sum_{-M}^{N} c_f[n](e_{-t_n} \hat{s}) \right\|^2_2$$

$$= \left\| \hat{f}\hat{s} - \sum_{-M}^{N} c_f[n](e_{-t_n} \mathbf{1}_{(\Omega_1)}) \hat{s} \right\|^2_2$$

$$\leq \|\hat{s}\|^2_\infty \left\| \hat{f} - \sum_{-M}^{N} c_f[n](e_{-t_n} \mathbf{1}_{(\Omega_1)}) \right\|^2_2 .$$

Using this estimate, equation (7.56), and the hypotheses on s, we obtain

$$\hat{f} = \sum c_f[n](e_{-t_n} \hat{s}) \text{ in } L^2(\hat{\mathbb{R}}). \tag{7.57}$$

The proof is completed by taking the Fourier transform in equation (7.57). ∎

Theorem 7.45 is a *sampling theorem* in the conventional sense that the coefficients $c_f[n]$ can be described in terms of values of f on the sampling set $\{t_m\}$. However, because a given sampling set is irregularly spaced in Theorem 7.45, we can not, in general, write $c_f[n] = f(t_n)$. Certain special cases and natural approximations have been calculated in [Ben2], [Hel]; and a more complete analysis is contained in [BenHel2].

REMARK 7.46. Comparisons and implementation One expects, and can prove, implementable irregular sampling theorems for the case

$$T \equiv \sup_n (t_{n+1} - t_n) < \frac{1}{2\Omega_1}, \tag{7.58}$$

see, e.g., [Ben2, Lemma 47, Theorem 48, Theorem 51], [BenHel1], [FeGrö], [Hel], cf. [OSe]. On the other hand, Theorem 7.45 allows gaps as large as

$$t_{n+1} - t_n = 2L + \frac{1}{\Delta} < 2L + \frac{1}{2\Omega_1},$$

when $(t_{n+1} - (n+1)/\Delta), (n/\Delta - t_n) \geq 0$ and where L in Definition 7.42a can be arbitrarily large (but fixed). In either case, the computation of explicit frame constants, a quantitative error analysis, and an effective local analysis of the sampling distribution are critical issues for implementation. In Algorithm 7.51, we outline an iterative procedure for dealing with these criteria in terms of adaptive sampling functions and factorization of associated frame operators [BenT], [T]. ∎

The choice of rapidly decreasing sampling functions s in Theorem 7.45 allows for better convergence in the sampling formula than one can expect with classical functions such as $d_{2\pi\Omega}$, cf. Example 7.47b which deals with control of the coefficients $c_f[n]$ in the sampling formula.

Example 7.47.   For Theorem 7.45

(a) Suppose $\{t_{1,n} : n \in \mathbb{Z}\}$ has uniform density $\Delta = 1$, upper bound $L \in \mathbb{N}$, and lower uniformly discrete bound d (from Definition 7.42.a). Let $\{t_n\} = \{t_{1,n}\} \cup \{t_{2,n}\}$, where $\{t_{2,n}\}$ is uniformly discrete. There are two phenomena we wish to mention about such sequences.

First, if d is small enough, then $\{t_{1,n}\}$ can have the property that there is a subsequence of $\mathbb{Z}$ each of whose elements can have as many as $2L + 1$ elements of $\{t_{1,n}\}$ close to it. For example, let $L = 3$ and $d = 1/10$, and for simplicity of exposition, let $\{u_n\} \equiv \{t_{1,n}\}$ be symmetric. Set $u_1 = 1, u_2 = 5 - 3d, u_3 = 5 - 2d, u_4 = 5 - d, u_5 = 5, u_6 = 5 + d, u_7 = 5 + 2d$, and $u_8 = 5 + 3d$. Hence, $2L + 1 = 7$ elements of $\{u_n\}$ are within a distance of $Ld = 3d = 3/10$ from $n = 5$. In this case, of course, it is necessary that $u_9 \geq 6$.

Second, for a properly chosen sequence $\{t_{2,n}\}$, the sequence $\{t_n\}$ can have the property that $\lim_{|n| \to \infty}(t_{n+1} - t_n) = 0$.

(b) Notwithstanding oversampling, the additional sampling points provided by adding uniformly discrete sequences to a uniformly dense sequence can provide finer control of the coefficients $c_f[n]$ in Theorem 7.45 in the following way.

In terms of Theorem 7.43, let A_1 be the lower frame bound associated with a given uniformly dense sequence $T_1 \equiv \{t_{1,n}\}$. Then the lower frame bound A associated with $T \equiv T_1 \cup T_2$ by Theorems 7.54 and 7.55, where $T_2 \equiv \{t_{2,n}\}$ is uniformly discrete, can be expected to dominate A_1 since

$$A_1\|f\|_2^2 \leq A_1\|f\|_2^2 + \sum_{t \in T_2}|\langle \hat{f}, e_{-t}\rangle|^2 \leq \sum_{t \in T}|\langle \hat{f}, e_{-t}\rangle|^2,$$

i.e., one expects that there is $A \geq A_1$ for which

$$A_1\|f\|_2^2 \leq A\|f\|_2^2 \leq A_1\|f\|_2^2 + \sum_{t \in T_2}|\langle \hat{f}, e_{-t}\rangle|^2.$$

Because of (7.21), one can also assert that the upper frame bounds B and B_1 for the sequences T and T_1 satisfy the inequality $B \geq B_1$.

The control of the coefficients $c_f[n]$ is then ruled by the inequality,

$$\frac{B}{B_1} < \frac{A}{A_1} \tag{7.59}$$

for a properly chosen sequence T_2. (Several such sequences might be chosen.) To see this, let S be the frame operator associated with T and with frame bounds A, B as in Theorems 7.54 and 7.55. It is easy to see that

$$\left\| I - \frac{2}{A+B} S \right\| \le \frac{B-A}{A+B} < 1, \tag{7.60}$$

so that by the Neumann expansion we have

$$S^{-1} = \frac{2}{A+B} \sum_{k=0}^{\infty} \left(I - \frac{2}{A+B} S \right)^k, \tag{7.61}$$

see, e.g., [Ben2, Algorithm 50]. Thus, convergence of S^{-1}, and hence computation of $c_f[n]$, is controlled by (7.60). Clearly, the convergence of S^{-1} will be better than that of the inverse frame operator associated with the sequence T_1 in case $(B - A)/(A + B) < (B_1 - A_1)/(A_1 + B_1)$; and this inequality is a consequence of (7.59). □

Equation (7.4), i.e., $S = L^*L$, and its ramifications provided perspective for earlier work in signal reconstruction, ([DGrMe], [D2]); and it is natural and productive to isolate and analyze by functional analysis methods the fundamental operators that arise in many applications, see, e.g., [Hö1], [v]. The following result, which is in the spirit of Section 7.4, collects some elementary facts about (7.4) and general continuous linear operators, with an eye towards implementation of our approach to irregular sampling.

THEOREM 7.48
Let $\{x_n : n \in \mathbb{Z}^d\} \subseteq H$ be a Bessel sequence with Bessel map $L : H \to \ell^2(\mathbb{Z}^d)$, and let $L^* : \ell^2(\mathbb{Z}^d) \to H$ be the adjoint map, cf. Definition/Remark 7.14f for mode of convergence in the following assertions.

(a) $L, L^*, R \equiv LL^*$, *and* $S \equiv L^*L$ *are continuous linear operators. R and S are self-adjoint,* $\|L\| = \|L^*\|$, *and* $\|R\| = \|S\|$.

(b) $R : \ell^2(\mathbb{Z}^d) \to \ell^2(\mathbb{Z}^d)$ *is given by the formula,*

$$\forall c = \{c[n]\} \in \ell^2(\mathbb{Z}^d), \quad Rc[m] = \sum_n c[n] \langle x_n, x_m \rangle.$$

(c) $S : H \to H$ *is given by the formula,*

$$\forall y \in H, \quad Sy = \sum \langle y, x_n \rangle x_n.$$

(d) L^{-1} *exists and is continuous on the range of L if and only if L^* is surjective. $(L^*)^{-1}$ exists and is continuous on the range of L^* if and only if L is surjective.*

(e) L^{-1} *exists and is continuous on $\ell^2(\mathbb{Z}^d)$ if and only if $(L^*)^{-1}$ exists and is continuous on H, in which case,*

$$(L^*)^{-1} = (L^{-1})^*.$$

(f) *If $\{x_n\}$ is a frame for H then L is injective, and this is equivalent to $L^*(\ell^2(\mathbb{Z}^d))$ being dense in H. Of course, by Theorem 7.15, L^{-1} is continuous on the closed range of L; and if L is surjective, then $\{x_n\}$ is an exact frame by Theorem 7.19.*

(g) *If $\{x_n\}$ is a basis of H, then L^* is injective, and this is equivalent to $L(H)$ being dense in H. In this case, if L^* is surjective, then $(L^*)^{-1}$ is continuous.*

If $L : H_1 \to H_2$ is any continuous linear operator then parts a, d, and e of Theorem 7.48 are still valid, and we have the following result.

THEOREM 7.49
Let $L : H_1 \to H_2$ be a continuous linear operator on H_1 and let $R \equiv LL^*$. Assume L^{-1} exists and is continuous on the range of L.

(a) *L^* is surjective and $Range\ L = Range\ R$.*

(b) *If $(L^*)^{-1}$ exists and is continuous on H_1, then L is surjective, and $R^{-1} \equiv (L^*)^{-1}L^{-1}$ exists and is continuous on H_2.*

(c) *If R^{-1} exists and is continuous on $Range\ R \subseteq H_2$ then $(L^*)^{-1} \equiv R^{-1}L$ exists and is continuous on H_1, and L is surjective.*

PROOF Using the previous result, the only thing that has to be checked is that Range $L \subseteq$ Range R, and this is a consequence of the surjectivity of L^*. ∎

THEOREM 7.50
Suppose $\Omega > 0$ and $\Omega_1 \geq \Omega$, and let $\{t_n\} \subseteq \mathbb{R}$ be a uniformly discrete sequence.

(a) *If $g \in A'(\mathbb{R})$, i.e., the distributional Fourier transform $\hat{g}$ is a element of $L^\infty(\hat{\mathbb{R}})$, then the mapping*

$$L_g : PW_\Omega \to \ell^2(\mathbb{Z})$$

$$f \mapsto \{f * \tilde{g}(t_n)\} \equiv \{\langle f, \tau_{t_n}g\rangle\}$$

is a continuous linear operator, and

$$\forall f \in PW_\Omega, \quad \|Lf\|_{\ell^2(\mathbb{Z})} \leq B\|\hat{g}\|_\infty\|f\|_2,$$

where B was defined after (7.21).

(b) *If $f, g \in L^2(\mathbb{R})$ then $f * \tilde{g} \in A(\mathbb{R})$ and $\langle \hat{f}, \hat{g}e_{-t}\rangle = f * \tilde{g}(t)$ for each $t \in \mathbb{R}$.*

(c) *If $g \in A'(\mathbb{R}) \cap PW_{\Omega_1}$ and $\{\tau_{t_n}g\}$ is complete in PW_{Ω_1}, or if $g \in A'(\mathbb{R}) \cap L^2(\mathbb{R})$ and $\{\hat{g}e_{-t_n}\}$ is complete in $L^2[-\Omega, \Omega]$, then L_g is injective.*

(d) *Let $\{e_{-t_n}\}$ be a frame for $L^2[-\Omega_1, \Omega_1]$ with frame bounds A_1, B_1. If $g \in A'(\mathbb{R}) \cap PW_\Omega$ and $C \geq \hat{g} \geq c > 0$ a.e. on $[-\Omega, \Omega]$ (therefore $\hat{g}$ is not continuous on $\hat{\mathbb{R}}$), or if $g \in A'(\mathbb{R}) \cap L^2(\mathbb{R})$ and $C \geq \hat{g} \geq c > 0$ a.e. on $[-\Omega, \Omega]$, then $\{\hat{g}e_{-t_n}\}$ is a frame for $L^2[-\Omega, \Omega]$ with frame bounds $A = c^2A_1, B = C^2B_1$.*

PROOF Part a is a consequence of the Plancherel–Pólya theorem (7.21). Part b is an elementary consequence of the Parseval formula, see [Ti2, p. 90], cf. Theorem 7.11.

For Part c, let $\{\tau_{t_n} g\}$ be complete in PW_{Ω_1}. If $L_g f = 0$ then $\langle \hat{f}, \hat{g} e_{-t_n} \rangle = 0$ for each n, so that $\hat{f} = 0$ since $\Omega_1 \geq \Omega$ and by the completeness. If $\{\hat{g} e_{-t_n}\}$ is complete in $L^2[-\Omega, \Omega]$, then the same argument works.

Part d results from making the obvious lower and upper bounds on $\sum |\langle \hat{f}, \hat{g} e_{-t_n} \rangle|^2$ where the inner product is an integral over $[-\Omega, \Omega]$. ∎

The following algorithm is designed for signal reconstruction in terms of irregularly spaced data. It is meant to be an implementable version of Theorem 7.45. Some of the derivations in the algorithm are not rigorous in the generality in which they are stated. However, our goal is to provide an algorithm for a broad range of applications. Our own applications of the algorithm in speech processing and data compression, using our wavelet auditory model, *WAM*, have produced excellent results [BenT].

Algorithm 7.51

(a) Using the setting of Theorem 7.50, we define $R = L_g L_g^*$ and $S = L_g^* L_g$. We have

$$L_g f = \{f * \tilde{g}(t_n)\}, \qquad L_g^* c(t) = \sum c[n] \tau_{t_n} g(t),$$

$$Sf(t) = \sum \langle f, \tau_{t_n} g \rangle \tau_{t_n} g,$$

and R has the matrix representation

$$R = (\langle \tau_{t_n} g, \tau_{t_m} g \rangle) = (g * \tilde{g}(t_m - t_n)) \equiv (r_{m,n}).$$

A signal $f \in PW_\Omega$ is written as

$$f = S^{-1} Sf = L_g^* R^{-1} L_g f. \tag{7.62}$$

As in (7.61),

$$f = \sum_{k=0}^{\infty} (I - \lambda S)^k (\lambda S) f$$

when $\|I - \lambda S\| < 1$, cf. (7.60). By an elementary calculation, we check that

$$\lambda \sum_{k=0}^{\infty} L_g^* (I - \lambda L_g L_g^*)^k L_g f = \lambda \sum_{k=0}^{\infty} (I - \lambda L_g^* L_g)^k L_g^* L_g f.$$

Hence,

$$f = \lambda L_g^* \left(\sum_{k=0}^{\infty} (I - \lambda L_g L_g^*)^k \right) L_g f, \tag{7.63}$$

cf. (7.62).

The implementation of (7.63) proceeds as follows. For a given signal $f \in PW_\Omega$ we set $f_0 = 0$ and $c_0 = L_g f$. We then define $h_n = L^* c_n$, $c_{n+1} = c_n - \lambda L_g h_n$, and $f_{n+1} = f_n + h_n$.

At the first stage we compute c_0, h_0, and f_1; at the second stage we compute c_1, h_1, and f_2; and at the nth stage we compute c_{n-1}, h_{n-1}, and f_n. Thus,

$$f_{n+1} = L_g^* \left(\sum_{k=0}^{n} (I - \lambda L_g L_g^*)^k \right) c_0, \qquad (7.64)$$

and so

$$\lim_{n \to \infty} \lambda f_n = f. \qquad (7.65)$$

(b) Without going into detail about the nonlinear and physiologically–based structure of WAM, let us point out how the Algorithm is applied in this context. A (speech) signal f is processed through WAM and produces an irregularly spaced sequence $\{t_{m,n}(f)\}$ of points in an auditory time-scale plane; this plane for auditory systems corresponds to the notion of information plane described in Section 7.3. Along with $\{t_{m,n}(f)\}$, there is a sequence $\{\theta_{m,n}\}$, which is based on the cochlear process; and there is WAM output $w_{m,n}(f)$ at $\{t_{m,n}(f)\}$, which is based on the physiologically realistic lateral inhibitory network. $\{\theta_{m,n}\}$ is a frame candidate.

It is a remarkable fact that the WAM output $\{w_{m,n}\}$ is precisely $\{\langle f, \theta_{m,n} \rangle\} \equiv Lf$.

Notwithstanding the redundancy in the early stages of the auditory system, it is reasonable to expect that the brain processes very little WAM output while obtaining high resolution. As such, after choosing a threshold $\epsilon > 0$, we discard WAM output below the threshold, and define a frame operator $Sf = \sum w_{m,n} \theta_{m,n}$ over the remaining points $t_{m,n}(f, \epsilon)$.

Typically, one can think of transmitting the small sequence $\{w_{m,n}\}$ corresponding to $\{t_{m,n}(f, \epsilon)\}$. As such, the receiver reconstructs f from this data by the Algorithm, i.e., by (7.64) and (7.65). A full description of WAM and its implementation is found in [BenT]. ⬜

REMARK 7.52 Daubechies [D1, p. 911] observed that the higher the redundancy of a frame $\{\theta_{m,n}\}$ for $L^2(\mathbb{R})$, the smaller is the subspace $L(L^2(\mathbb{R}))$, where $LF = \{\langle f, \theta_{m,n} \rangle\}$; and, importantly, that this is a desirable feature for purposes such as the reduction of calculational noise, cf. Example/Remark 7.20b for the case $2T\Omega < 1$. In a corresponding analysis involving pseudo-inverse operators, Teolis [T] drew the analogous conclusion that the larger the kernel of L^* the more stable the signal reconstruction will be. ∎

7.8   Uncertainty principle inequalities

In this section we lay the groundwork for generalizing Gabor's idea, described in Subsection 7.3. Instead of the classical uncertainty principle inequality, we prove weighted uncertainty principle inequalities in Theorems 7.55 and 7.62. It is natural to prove such inequalities since there are many criteria besides minimization of variance that are relevant in applications. Once we have a weighted uncertainty principle inequality, the goal is to determine a minimizer for this inequality, just as the Gaussian is a minimizer for the classical uncertainty principle inequality. Then, as motivated in Section 7.3 in terms of dissections of the information plane, the program is to develop the decomposition and sampling theory associated with these minimizers.

DEFINITION 7.53   *Let u and v be nonnegative Borel measurable functions on $\hat{\mathbb{R}}$ and $\mathbb{R}$, respectively. We refer to u and v as weights. If $1 < p, q < \infty$ and if there is a constant $K > 0$ such that*

$$\sup_{s>0} \left(\int_0^{1/s} u(\gamma)\, d\gamma \right)^{1/q} \left(\int_0^{s} v(t)^{-p'/p}\, dt \right)^{1/p'} = K, \tag{7.66}$$

then we write $(u, v) \in F(p, q)$. Also, $L_v^p(\mathbb{R})$ is the set of Borel measurable functions f for which

$$\|f\|_{p,v} \equiv \left(\int |f(t)|^p v(t)\, dt \right)^{1/p} < \infty.$$

The following is a weighted Hausdorff–Young–Titchmarsh inequality, cf. (7.40) and (7.41); H. Heinig and I proved Theorem 7.54 in 1982, see [BenHe] for more recent references and generalizations.

THEOREM 7.54
Let $1 < p \le q < \infty$, and let u and v be even weights on $\hat{\mathbb{R}}$ and $\mathbb{R}$ for which $(u, v) \in F(p, q)$ with constant K (as in (7.66)). Assume $1/u$ and v are increasing on $(0, \infty)$. Then there is a constant $C(K)$ such that

$$\forall f \in \mathcal{S}(\mathbb{R}) \cap L_v^p(\mathbb{R}), \quad \|\hat{f}\|_{q,u} \le C(K)\|f\|_{p,v}. \tag{7.67}$$

The closure theory in Remark 7.31 can be used to verify the inequality in (7.67) for a larger class of functions. There are sometimes subtle problems for such extensions even so far as the proper definition of the Fourier transform is concerned ([BenLa]).

If $p = 1$ and $q > 1$, then Theorem 7.54 is true for any weight u. In this case the proof is routine and the constant $C(K)$ is explicit. If $p > 1$ the constant $C(K)$ is less explicit, but it can be estimated by examining the proof of Calderón's rearrangement inequality (*Studia Math.*, 26 (1966), 273–299), which we used in the proof of Theorem 7.54.

After completing Theorem 7.54, Heinig and I realized we could prove the following weighted uncertainty principle inequality.

THEOREM 7.55

Let $1 < p \leq q < \infty$, and let u and v be even weights on $\hat{\mathbb{R}}$ and $\mathbb{R}$ for which $(u, v) \in F(p, q)$ with constant K (as in (7.66)). Assume $1/u$ and v are increasing on $(0, \infty)$. Then there is a constant $C(K)$ (the same as Theorem 7.54) such that

$$\forall f \in \mathcal{S}(\mathbb{R}), \quad \|f\|_2^2 \leq 4\pi C(K)\|tf(t)\|_{p,v}\|\gamma\hat{f}(\gamma)\|_{q',u^{-q'/q}}. \tag{7.68}$$

PROOF   By means of the first part of (7.31) (for $\hat{f}$ instead of f), Hölder's inequality, and Theorem 7.54, we have the estimate,

$$\|f\|_2^2 = \|\hat{f}\|_2^2 \leq \int |\gamma(\bar{\hat{f}})(\gamma)(\hat{f})'(\gamma)|\, d\gamma$$

$$= 2 \int \left|\gamma(\bar{\hat{f}})(\gamma)u(\gamma)^{-1/q}\right| \left|(\hat{f})'(\gamma)u(\gamma)^{1/q}\right| d\gamma$$

$$\leq 2 \left(\int \left|\gamma(\bar{\hat{f}})(\gamma)\right|^{q'} u(\gamma)^{-q'/q}\, d\gamma\right)^{1/q'} \left(\int \left|(\hat{f})'(\gamma)\right|^{q} u(\gamma)\, d\gamma\right)^{1/q}$$

$$\leq 2C(K)\|((\hat{f})')^{\vee}(t)\|_{p,v}\|\gamma\hat{f}(\gamma)\|_{q',u^{-q'/q}},$$

and the result is obtained since $((\hat{f})')^{\vee}(t) = 2\pi i t f(t)$.   ∎

As pointed out after Theorem 7.30, the main ingredients for such inequalities are integration by parts, Hölder's inequality, the Plancherel theorem, and their generalizations. In the case of Theorem 7.55, we have used Theorem 7.54 instead of Plancherel's theorem. Our next goal, Theorem 7.62, is a bit more sophisticated and is also proved in $\mathbb{R}^d$.

THEOREM 7.56   [Si, Theorem 4.1]

Let $1 < q < \infty$, and let u and v be weights on $\mathbb{R}^d$.

   (a) There is a constant $C > 0$ such that

$$\forall g \in C_c^{\infty}(\mathbb{R}^d), \quad \|g\|_{q,u} \leq C\|t \cdot \nabla g(t)\|_{q,v} \tag{7.69}$$

   if and only if

$$\sup_{s \in \mathbb{R}^d} \left(\int_0^1 u(xs)x^{d-1}\, dx\right)^{1/q} \left(\int_1^{\infty} (v(xs)x^d)^{-q'/q}x^{-1}\, dx\right)^{1/q'} \equiv K < \infty.$$

   The constants C and K satisfy the inequalities, $K \leq C \leq Kq^{1/q}(q')^{1/q'}$.
   (b) There is a constant $C > 0$ such that

$$\forall g \in C_c^{\infty}(\mathbb{R}^d) \quad \text{for which } g(0) = 0, \quad \|g\|_{q,u} \leq C\|t \cdot \nabla g(t)\|_{q,v}$$

if and only if

$$\sup_{s \in \mathbb{R}^d} \left(\int_1^\infty u(xs) x^{d-1} \, dx \right)^{1/q} \left(\int_0^1 (v(xs) x^d)^{-q'/q} x^{-1} \, dx \right)^{1/q'} = K < \infty.$$

DEFINITION 7.57   Let $1 < p < \infty$, and let w be a weight on $\mathbb{R}^d$. w is an A_p-weight, written $w \in A_p$, if

$$\sup_Q \left(\frac{1}{|Q|} \int_Q w(x) \, dx \right) \left(\frac{1}{|Q|} \int_Q w(x)^{-p'/p} \, dx \right)^{p/p'} \equiv K < \infty,$$

where Q is a compact cube with sides parallel to the axes and having nonempty interior.

THEOREM 7.58

Let $1 < p \leq q \leq p' < \infty$, and let w be a weight on $\mathbb{R}^d$. Assume $w(|t|)$ is increasing on $(0, \infty)$. There is a constant $C > 0$ such that

$$\forall f \in C_c^\infty(\mathbb{R}^d), \quad \left(\int |\hat{f}(\gamma)|^q |\gamma|^{d((q/p')-1)} w \left(\frac{1}{|\gamma|} \right)^{q/p} d\gamma \right)^{1/q} \leq C \|f\|_{p,w} \quad (7.70)$$

if and only if $w \in A_p$.

(The reference for the case $d = 1$ is [BenHeJo]. The $d > 1$ version stated above is in [HeSi, Theorem 2.10].)

REMARK 7.59   We view Theorem 7.58 as the culmination of some interesting classical analysis. Take $d = 1$. In case $p = q$ and $w = 1$, (7.70) is the Hardy, Littlewood, Paley theorem (1931),

$$\left(\int |\hat{f}(\gamma)|^p |\gamma|^{p-2} \, d\gamma \right)^{1/p} \leq C \|f\|_p.$$

If $q = p'$ and $w = 1$, (7.70) is the Hausdorff–Young theorem. If $w(t) = |t|^\alpha, 0 \leq \alpha < p - 1$, then (7.70) reduces to Pitt's theorem (1937),

$$\left(\int |\hat{f}(\gamma)|^q |\gamma|^{-\beta} \, d\gamma \right)^{1/q} \leq C \left(\int |f(t)|^p |t|^\alpha \, dt \right)^{1/p},$$

where $\beta = q/p(\alpha + 1) + 1 - q$. The fact that Fourier transform inequalities are characterized in terms of A_p-weights was initially surprising since the A_p-condition was associated with maximal function and singular integral norm inequalities. $\blacksquare$

DEFINITION 7.60   *The d-dimensional Riesz transforms are the d singular integral operators $R_1, \ldots, R_d$ defined by the odd kernels $k_j(x) = \Omega_j(x)/|x|^d, j = 1, \ldots, d$, where $\Omega_j(x) = c_d x_j/|x_j|$ and $c_d = \Gamma((d + 1)/2)/\pi^{(d+1)/2}$. In fact,*

$$(R_j f)(x) = \lim_{T^{-1}, \epsilon \to 0} \int_{\epsilon \leq |t| \leq T} f(x - t) k_j(t) \, dt$$

exists a.e. for each $f \in L^p(\mathbb{R}^d)$, $1 < p < \infty$, *and there is* $C = C(p)$ *such that*

$$\forall f \in L^p(\mathbb{R}^d), \quad \|R_j f\|_p \leq C \|f\|_p,$$

$j = 1, \dots, d$. $C = C(p)$ *does not depend on* d *[GarFra, p. 223]. Also, we compute*

$$\hat{k}_j(\gamma) = -i \frac{\gamma_j}{|\gamma|}, \quad j = 1, \dots, d.$$

THEOREM 7.61   **(Hunt, Muckenhoupt, and Wheeden, 1973)**
Let $1 < p < \infty$, *and suppose* $w \in A_p$. *Then*

$$R_j : L_w^p(\mathbb{R}^d) \rightarrow L_w^p(\mathbb{R}^d)$$

is a continuous linear map for $j = 1, \dots, d$, *see [GarFra, pp. 196, 204, 411–413].*

THEOREM 7.62
Given $1 < r \leq 2$ *and a nonnegative radial weight* $w \in A_r$ *on* $\mathbb{R}^d$ *for which* $w(|t|)$
is increasing on $(0, \infty)$. *Assume*

$$\sup_{s \in \mathbb{R}^d} \left(\int_0^1 \frac{w(xs)^{-r'/r}}{|xs|^{r'}} x^{d-1} dx \right)^{1/r'}$$
$$\times \left(\int_1^\infty w \left(\frac{1}{|xs|} \right)^{-1} |xs|^r x^{-\frac{dr}{r}-1} dx \right)^{1/r} \equiv K < \infty. \tag{7.71}$$

Then there is a constant $C = C(K) > 0$ *such that*

$$\forall f \in C_c^\infty(\mathbb{R}^d), \quad \|f\|_2^2 \leq C \||t|f(t)\|_{r,w} \||\gamma|\hat{f}(\gamma)\|_{r,w}.$$

PROOF

(a) For $1 < r < \infty$ we have

$$\|f\|_2^2 \leq \||t|f(t)\|_{r,w} \|f\|_{r',u}, \tag{7.72}$$

where

$$u(t) = |t|^{-r'} w(t)^{-r'/r}.$$

(b) The second factor on the right side of (7.72) is estimated by means of
Theorem 7.56a, where q and v in (7.69) are $q = r'$ and

$$v(t) = |t|^{-r'} w \left(\frac{1}{|t|} \right)^{r'/r},$$

respectively. Thus,

$$\|f\|_{r',u} \leq C_1 \|t \cdot \nabla f(t)\|_{r',v} \tag{7.73}$$

if and only if (7.71) holds.

(c) By Minkowski's inequality the right side of (7.73) is bounded by

$$C_1 \sum_{j=1}^{d} \left(\int |(R_j G_j)^\vee(t)|^{r'} w \left(\frac{1}{|t|} \right)^{r'/r} dt \right)^{1/r'}, \tag{7.74}$$

where $G_j^\vee(t) = \partial_j f(t)$. Combining (7.73) and (7.74), and applying Theorem 7.58 for the case $p = r$ and $q = p'$ (so that $1 < r \leq 2$), we obtain

$$\|f\|_{r',u} \leq C_1 C_2 \sum_{j=1}^{d} \|R_j G_j\|_{r,w}. \tag{7.75}$$

(d) Finally, combining (7.72) and (7.75) and applying Theorem 7.61 to the right side of (7.75), we have the estimate

$$\|f\|_2^2 \leq C_1 C_2 C_3 \||t|f(t)\|_{r,w} \sum_{j=1}^{d} \|(\partial_j f)^\wedge(\gamma)\|_{r,w}$$

$$\leq 2\pi d^{1/r'} C_1 C_2 C_3 \||t|f(t)\|_{r,w} \left(\int |\hat{f}(\gamma)|^r \left(\sum_{j=1}^{d} |\gamma_j|^r \right) w(\gamma) d\gamma \right)^{1/r} \tag{7.76}$$

$$\leq 2\pi d^{1/2} C_1 C_2 C_3 \||t|f(t)\|_{r,w} \||\gamma|\hat{f}(\gamma)\|_{r,w}. \quad \blacksquare$$

COROLLARY 7.63
Given $1 < r \leq 2$ and $d > r'$, there is $C > 0$ such that

$$\forall f \in \mathcal{S}(\mathbb{R}^d), \quad \|f\|_2^2 \leq C \||t|f(t)\|_r \||\gamma|\hat{f}(\gamma)\|_r. \tag{7.77}$$

REMARK 7.64

(a) In the notation of (7.76) the constant C in Corollary 7.63 is of the form

$$C = 2\pi d^{1/2} C_1(r,d) B_d(r) C_3(r).$$

Since it is of interest to measure the growth of C as d increases, we note that $C_1(r,d)$ can be estimated in terms of K in (7.71) for any w.

(b) Theorem 7.56b gives rise to an analogue of Theorem 7.62 which, for $w = 1$, yields (7.77) for $d < r'$. $\blacksquare$

Example 7.65. Local uncertainty principles inequalities
Analogous to the local sampling results discussed in Section 7.5, there are also local uncertainty principles inequalities. To understand what they mean, recall that the classical uncertainty principle inequality asserts that if f is concentrated then $\hat{f}$ is spread out. Local uncertainty principle inequalities quantify the folklore that this spread is smooth and is not the result of $\hat{f}$ having isolated peaks far from

the origin. For example, Faris [F, (3.2)] proved that

$$\forall f \in \mathcal{S}(\mathbb{R}), \quad \int_E |\hat{f}(\gamma)|^2 \, d\gamma \le 2\pi |E| \|tf(t)\|_2 \|f\|_2 \qquad (7.78)$$

for any fixed Borel set $E \subset \hat{\mathbb{R}}$. Thus, if $\hat{f}$ has a peak about γ_0, then $\|\hat{f} \mathbf{1}_{E_j}\|_2$ would be more or less constant as $|E_j| \to 0$, where $\gamma_0 \in E_j$, thereby contradicting (7.78). Subsequent local uncertainty principle inequalities were proved by Price [P1], and then by Heinig and the author, see [Ben1] for references and some details. □

Acknowledgments. This work was supported by Prometheus Inc., under DARPA Contract DAA-H01-91- CR212. The author is also Professor of Mathematics at the University of Maryland, College Park. I would like to thank Dr. Sam Earp, Sandra Saliani, and Georg Zimmermann for sharing their expertise and providing several valuable insights on this material.

Bibliography

[AU]   A. Aldroubi and M. Unser. "Families of multiresolution and wavelet spaces with optimal properties," preprint, 1992.

[B]   L. Baggett. 1990. "Processing a radar signal and representations of the discrete Heisenberg group," *Coll. Math.* **60/61**: 195–203.

[Bal]   R. Balian. 1981. "Un principe d'incertitude fort en théorie du signal ou en mécanique quantique," *C. R. Acad. Sci.* **292**: 1357–1362.

[Baretal] V. Bargmann, P. Butero, L. Girardello, and J. Klauder. "On the completeness of coherent states," *Rep. Mod. Phys.*, 1971. **2**: 221–228.

[Bas]   M. Bastiaans. "Gabor's expansion of a signal into Gaussian elementary signals," *Proc. IEEE* 1980. **68**: 538–539.

[Bat]   G. Battle. 1989. "Phase space localization theorem for ondelettes," *J. Math. Phys.* **30**: 2195–2196.

[Be]   W. Beckner. 1975. "Inequalities in Fourier analysis," *Ann. Math.* **102**: 159–182.

[Ben1]   J. Benedetto. 1990. "Uncertainty principle inequalities and spectrum estimation," in *Recent Advances in Fourier Analysis and Its Applications*, NATO-ASI Series C **315**, J.S. and J.L. Byrnes, eds., Dordrecht: Kluwer Academic Publishers, pp. 143–182.

[Ben2]   J. Benedetto. 1992. "Irregular sampling and frames," in *Wavelets: A Tutorial in Theory and Applications*, C.K. Chui, ed., Boston: Academic Press, pp. 445–507.

[Ben3]   J. Benedetto. 1992. "Stationary frames and spectral estimation," in *Probabilistic and Stochastic Methods in Analysis*, NATO-ASI-1991,

J.S. Byrnes, ed., Dordrecht: Kluwer Academic Publishers, pp. 117–161.

[BenHe]   J. Benedetto and H. Heinig. 1992. "Fourier transform inequalities with measure weights," *Adv. Math.* **96**: 194–225.

[BenHeJo]   J. Benedetto, H. Heinig, and R. Johnson. 1987. "Fourier inequalities with A_p-weights" *ISNM* **80**: 217–232.

[BenHel1]   J. Benedetto and W. Heller. 1990. "Irregular sampling and the theory of frames," *Mat. Note* **10**(Suppl. 1): 103–125.

[BenHel2]   J. Benedetto and W. Heller. "Irregular sampling and the theory for frames, II," preprint, 1993.

[BenKar]   J. Benedetto and C. Karanikas. "Local sampling theory and the Gabor transform," preprint, 1993.

[BenLa]   J. Benedetto and J. Lakey. 1993. "The definition of the Fourier transform for weighted inequalities," *J. Funct. Anal.*, to appear.

[BenT]   J. Benedetto and A. Teolis. 1993. "Wavelet auditory models and data compression," *Appl. Comp. Harmonic Anal.*, to appear.

[BenWa]   J. Benedetto and D. Walnut. 1993. "Gabor frames for L^2 and related spaces," this volume, Chap. 3.

[Bene1]   M. Benedicks. 1985. "On Fourier transforms of functions supported on sets of finite Lebesgue measure," *J. Math. Anal. Appl.* **106**: 180–183.

[Bene2]   M. Benedicks. 1984. "The support of functions and distributions with a spectral gap," *Math. Scand.* **55**: 285–309.

[BerWa]   E. Bernstein and D. Walnut. "A wavelet based phoneme discriminator," preprint, 1992.

[Beu]   A. Beurling. 1989. *Collected Works*, vols. 1 and 2, Boston: Birkhäuser.

[Bo1]   R. Boas. 1940. "Entire functions bounded on a line," *Duke J. Math.* **6**: 148–169.

[Bo2]   R. Boas. 1954. *Entire Functions*, New York: Academic Press.

[Bou]   J. Bourgain. 1988. "A remark on the uncertainty principle for Hilbertian basis," *J. Funct. Anal.* **79**: 136–143.

[BuSpSte]   P. Butzer, W. Splettstösser, and R. Stens. 1988. "The sampling theorem and linear prediction in signal analysis," *Jber. d. Dt. Math.-Verein.* **90**: 1–70.

[BuSte]   P. Butzer and R. Stens. 1992. "Sampling theory for not necessarily band-limited functions: a historical overview," *SIAM Rev.* **34**: 40–53.

[By]   J. Byrnes. "A low crest factor complete orthonormal set," preprint, 1992.

[C]   L. Cohen. 1989. "Time-frequency distributions-a review," *Proc. IEEE* 77: 941–981.

[CoMeWi]   R. Coifman, Y. Meyer, and V. Wickerhauser. 1992. "Wavelet analysis and signal processing," in *Wavelets and Their Applications*, M. Ruskai, ed., Boston: Jones and Bartlett, pp. 153–178.

[D1]   I. Daubechies. 1988. "Orthonormal bases of compactly supported wavelets," *Comm. Pure Appl. Math.* 41: 909–996.

[D2]   I. Daubechies. 1990. "The wavelet transform, time-frequency localization, and signal analysis," *IEEE Trans. Inform. Theory* 36: 961–1005.

[DGrMe]   I. Daubechies, A. Grossmann, and Y. Meyer. 1986. "Painless nonorthogonal expansions," *J. Math. Phys.* 27: 1271–1283.

[DeCovTh] A. Dembo, T. Cover, and J. Thomas. 1991. "Information theoretic inequalities," *IEEE Trans. Inform. Theory* 37: 1501–1518.

[DoSta]   D. Donoho and P. Stark. 1989. "Uncertainty principles and signal recovery," *SIAM J. Appl. Math.* 49: 906–931.

[DuSc]   R. Duffin and A. Schaeffer. 1952. "A class of non-harmonic Fourier series," *Trans. Amer. Math. Soc.* 72: 341–366.

[F]   W. Faris. 1978. "Inequalities and uncertainty principles," *J. Math. Phys.* 19: 461–466.

[FeGrö]   H. Feichtinger and K. Gröchenig. 1993. "Theory and practice of irregular sampling," this volume, Chap. 8.

[Fo]   G. Folland. 1989. *Harmonic Analysis in Phase Space*, Princeton: Princeton Univ. Press.

[FrTo]   M. Frazier and R. Torres. 1993. "The sampling theorem, φ-transform, and Shannon wavelets for $\mathbb{R}$, $\mathbb{Z}$, $\mathbb{T}$, and $\mathbb{Z}_N$," this volume, Chap. 6.

[Ga]   J.-P. Gabardo. 1989. "Tempered distributions with spectral gaps," *Math. Proc. Cambr. Philos. Soc.* 106: 143–162.

[Ga]   D. Gabor. 1946. "Theory of communication," *J. IEE London* 93: 429–457.

[GarFra]   J. Garcia-Cuerva and J. Rubio de Francia. 1985. *Weighted Norm Inequalities and Related Topics*, Amsterdam: North-Holland.

[GoGol]   I. Gohberg and S. Goldberg. 1981. *Basic Operator Theory*, Boston: Birkhäuser.

[HWa]   C. Heil and D. Walnut. 1989. "Continuous and discrete wavelet transforms," *SIAM Rev.* 31: 628–666.

[HeSi]   H. Heinig and G. Sinnamon. 1989. "Fourier inequalities and integral representations of functions in weighted Bergman spaces over tube domains," *Ind. Univ. Math. J.* 38: 603–628.

[Hel]   W. Heller. 1991. "Frames of exponentials and applications," Ph.D. thesis, University of Maryland, College Park, Md.

[Hi1]   J. Higgins. 1977. *Completeness and Basis Properties of Sets of Special Functions*, Cambridge: Cambridge Univ. Press.

[Hi2]   J. Higgins. 1985. "Five short stories about the cardinal series," *Bull. Amer. Math. Soc.* **12**: 45–89.

[Hir]   I. Hirschman. 1957. "A note on entropy," *Amer. J. Math.* **79**: 152–156.

[Höl]   L. Hörmander. 1983. *The Analysis of Linear Partial Differential Operators*, vols. I and II. New York: Springer-Verlag.

[Hö2]   L. Hörmander. 1991. "A uniqueness theorem of Beurling for Fourier transform pairs," *Ark. Mat.* **29**: 237–240.

[Ho]   C. Houdrè. 1993. "Wavelets, probability, and statistics: Some bridges," this volume, Chap. 9.

[J]   S. Jaffard. 1991. "A density criterion for frames of complex exponentials," *Mich. Math. J.* **38**: 339–348.

[Ja1]   A. Janssen. 1981. "Gabor representation of generalized functions," *J. Math. Anal. Appl.* **83**: 377–394.

[Ja2]   A. Janssen. 1982. "Bargmann transform, Zak transform, and coherent states," *J. Math. Phys.* **23**: 720–731.

[K]   J. Kampé de Fériet. 1950–1951. "Mathematical methods used in the statistical theory of turbulence: harmonic analysis," Inst. Fluid Dynamics, U. of Maryland **1**: 1–108.

[Ke]   B. Kedem. 1986. "Spectral analysis and discrimination by zero-crossings," *Proc. IEEE* **74**: 1477–1493.

[Kl]   I. Kluvánek. 1965. "Sampling theorem in abstract harmonic analysis," *Mat. Fyz. Casopsis Sloven. Akad. Vied.* **15**: 43–48.

[Ko]   P. Koosis. 1980. *Introduction to H_p Spaces*, Cambridge: Cambridge Univ. Press.

[L1]   H. Landau. 1967. "Necessary density condition for sampling and interpolation of certain entire functions," *Acta Math.* **117**: 37–52.

[L2]   H. Landau. 1985. "An overview of time and frequency limiting," in *Fourier Techniques and Applications*, J. Price, ed., New York: Plenum Press, pp. 201–220.

[L3]   H. Landau. 1989. "On the density of phase-space expansions," preprint.

[Le]   N. Levinson. 1940. *Gap and Density Theorems*, Colloq. Publ. **26**, Providence: Amer. Math. Soc.

[Lo]   F. Low. 1985. "Complete sets of wavepackets," in *A Passion for Physics: Essays in Honor of Geoffrey Chew*, C. DeTar et al., eds., Singapore: World Scientific, pp. 17–22.

[Mal] S. Mallat. 1989. "Multiresolution approximations and wavelet orthonormal bases of $L^2(\mathbb{R})$," *Trans. Amer. Math. Soc.* **315**: 69–87.

[Ma] F. Marvasti. "A unified approach to zero-crossings and non-uniform sampling," ISBN 09618167-08, 1987.

[Mas] P. Masani. 1960. "The prediction theory of multivariate stochastic processes, III," *Acta Math.* **104**: 141–162.

[Me] Y. Meyer. 1990. *Ondelettes et Opérateurs*, vols. I, II, and III, Paris: Hermann.

[OSe] P. Olsen and K. Seip. 1992. "A note on irregular discrete wavelet transform," *IEEE Trans. Inform. Theory* **38**: 861–863.

[OpWil] A. Opperheim and A. Willsky. 1983. *Signals and Systems*, Englewood Cliffs, N.J.: Prentice-Hall, Inc.

[PWie] R. Paley and N. Wiener. 1934. *Fourier Transforms in the Complex Domain*, Colloq. Publ. **19**, Providence: Am. Math. Soc.

[Pa] A. Papoulis. 1977. *Signal Analysis*, New York: McGraw-Hill.

[Pe] A. Perelomov. 1971. "On the completeness of the system of coherent states," *Theor. Math. Phys.* **6**: 156–164.

[Pl] M. Plancherel. 1949. "Intégrales de Fourier et fonctions entières," *Colloq. sur l'analyse harmonique, Nancy*: 31–43.

[Pr1] J. Price. 1983. "Inequalities and local uncertainty principles," *J. Math. Phys.* **24**: 1711–1714.

[Pr2] J. Price. 1985. "Uncertainty principles and sampling theorems," in *Fourier Techniques and Applications*, J. Price, ed., New York: Plenum Press, pp. 25–44.

[R] M. Rieffel. 1981. "Von Neumann algebras associated with pairs of lattices in Lie groups," *Math. Ann.* **257**: 403–418.

[Ri] A. Rihaczek. 1985. *Principles of High-Resolution Radar*, Los Altos, Calif: Peninsula Publishing.

[RioVe] O. Rioul and M. Vetterli. 1991. "Wavelets and signal processing," *IEEE Signal Process. Mag.* October: 14–38.

[S] L. Schwartz. 1966. *Théorie des Distributions*, Paris: Hermann.

[Se] K. Seip. 1992. "Density theorems for sampling and interpolation in the Bargmann–Fock space," *Bull. Amer. Math. Soc.* **26**: 322–328.

[SeWal] K. Seip and R. Wallstén. 1990. *Sampling and Interpolation in the Bargmann–Fock Space*, Djursholm: Report Mittag-Leffler.

[ShWe] C. Shannon and W. Weaver. 1949. *A Mathematical Theory of Communication*, Univ. of Illinois Press.

[Sha] H. Shapiro. 1991. "Uncertainty principles for bases in $L^2(\mathbb{R})$," preprint.

[Si]   G. Sinnamon. 1989. "A weighted gradient inequality," *Proc. Roy. Soc. Edinburgh* **111**: 329–335.

[StWei]   E. Stein and G. Weiss. 1971. *An Introduction to Fourier Analysis on Euclidean Spaces*, Princeton: Princeton Univ. Press.

[T]   A. Teolis. 1992. "Implementation of irregular sampling algorithms," Ph.D. thesis, University of Maryland, College Park, Md.

[Ti1]   E. Titchmarsh. 1926. "The zeros of certain integral functions," *Proc. London Math. Soc.* **25**: 283–302.

[Ti2]   E. Titchmarsh. 1948. *Theory of Fourier Integrals*, 2nd ed., Oxford: Oxford Univ. Press.

[v]   J. von Neumann. 1932, 1949, and 1955. *Mathematical Foundations of Quantum Mechanics*, Princeton: Princeton Univ. Press.

[W]   G. Walter. 1992. "A sampling theorem for wavelet subspaces, Part II," *IEEE Trans. Inform. Theory* **38**: 881–884.

[Y]   R. Young. 1980. *An Introduction to Nonharmonic Fourier Series*, New York: Academic Press.

8

Theory and practice of
irregular sampling

Hans G. Feichtinger and Karlheinz Gröchenig[*]

ABSTRACT This chapter presents methods for the reconstruction of band-limited functions from irregular samples. The first part discusses algorithms for a constructive solution of the irregular sampling problem. An important aspect of these algorithms is the explicit knowledge of the constants involved and efficient error estimates. The second part discusses the numerical implementation of these algorithms and compares the performance of various reconstruction methods.

Although most of the material has already appeared in print, several new results are included: (a) a new method to estimate frame bounds, (b) a reconstruction of band-limited functions from partial information, which is not just the samples; (c) a new result on the complete reconstruction of band-limited functions from local averages; (d) a systematic exposition of recent experimental results.

CONTENTS

[*]The first author was partially supported by the Austrian National Bank through the project "Experimental Signal Analysis" and the Kammer der Gewerblichen Wirtschaft von Wien through the "Hochschuljubiläumspreis." The final draft was prepared as part of project PO8784 of FWF, the Austrian Science Foundation. The second author acknowledges the partial support by a DARPA Grant AFOSR-90-0311.

0-8493-8271-8/94/$0.00 + $.50

8.1 Introduction: The irregular sampling problem

The sampling problem is one of the standard problems in signal analysis. Since a signal $f(x)$ cannot be recorded in its entirety, it is sampled at a sequence $\{x_n, n \in \mathbb{Z}\}$. Then the question arises how f can be reconstructed or at least approximated from the samples $f(x_n)$.

In practice only a finite number of samples can be measured and stored. In this case the irregular sampling problem becomes a finite-dimensional problem and can be approached with methods of linear algebra—see Part II of this chapter.

For the theoretical model we will assume that an infinite sequence of samples is given. Then the reconstruction of arbitrary signals is clearly an ill-defined problem. In most applications, however, it is reasonable to assume that the signal is band limited, i.e., that it does not contain high frequencies.

If the samples are equally spaced, then the famous sampling theorem by Shannon, Whittaker, Kotel'nikov, and others provides an explicit reconstruction. Assume that $f \in L^2(\mathbb{R})$ and that $\text{supp}\hat{f} \subset [-\omega, \omega]$. Then for any $\delta \leq \pi/\omega$ the signal f has the representation

$$f(x) = \sum_{n=-\infty}^{\infty} f(n\delta) \frac{\sin \omega(x - n\delta)}{\omega(x - n\delta)}, \tag{8.1}$$

where the series converges in $L^2(\mathbb{R})$ and uniformly on compact sets. The case of regular sampling is well understood and has led to countless variants. We refer to the reviews [BuSpSte], [Hi2], [Je].

The problem for irregular samples x_n, $n \in \mathbb{Z}$, is more difficult. Some difficulties stem from the fact that Poisson's formula is no longer applicable; also, for very irregular sampling sets much more attention has to be paid to the convergence of the infinite series involved.

The irregular sampling problem has inspired many difficult and deep theorems which culminate in the work of A. Beurling, P. Malliavin, and H. Landau [Be], [BeMal], [L]. We apologize in advance that in this chapter many theoretical developments will not receive the treatment they deserve. Instead we refer to J. Benedetto's excellent survey [B].

The objective of this article is quite different. Our goal is the development of efficient reconstruction algorithms and ultimately their numerical implementation. This requires a *quantitative* theory of irregular sampling. Merely qualitative statements of the form "There exists a constant, such that ... " may be insufficient in concrete applications. Even so-called explicit reconstruction formulas may be of little use, when their ingredients are very complicated.

Instead we would like to find algorithms that are *simple* enough to be used in numerical computations. Furthermore, in order to make a step beyond the trial-and-error method in numerical computations, all numerical constants should be *explicit* and should yield reasonable *error estimates.* For iterative algorithms, which are our main concern, this should also include estimates for the number of iterations required to achieve a desired accuracy of the reconstruction.

In view of this program the chapter is divided into two parts. In the first part, written by the second-named author, a class of iterative algorithms is developed which satisfies the requirements listed above.

Section 8.2 starts with an abstract description of iterative reconstruction algorithms and discusses the concept of frames. A related approach has been taken by Benedetto and Heller [BHe]. We introduce a new method for the estimation of frame bounds. In Section 8.3 we briefly outline results in the literature which have been or could be used in numerical work with irregular sampling. In Section 8.4 we derive the basic reconstruction algorithm, the so-called "adaptive weights method." Since this algorithm has proved to be the most efficient method in the numerical experiments, we give a detailed proof for this theorem. In Section 8.5 we present several variations of this method. We discuss: (a) a discrete model for irregular sampling. The adaptive weights method in this situation can be implemented directly. (b) We touch briefly on the problem of irregular sampling in higher dimensions, which is a very important problem in image processing. (c) We improve a recent result of Donoho and Stark [DoSt] on the uncertainty principle and the problem of missing segments. In Section 8.6 we give a recent application of sampling results for band-limited functions to wavelet theory. We state irregular sampling results for the wavelet transform and the short time Fourier transform.

The second part, written by the first-named author, is devoted to the numerical implementations of these algorithms and provides empirical results which either support or complement the theoretical results given in Part I.

Section 8.7 gives a short description of the discrete model underlying the numerical experiments. It is followed by a summary of the hard- and software environment (MATLAB on PC and Sun workstations), on which the collection "IRSATOL" of MATLAB *M*-files is built (cf. [FStr]). The main section of this part is Section 8.9. It gives a general comparison of the five most important methods studied so far in terms of iterations and time. We also discuss the sensitivity of different methods with respect to irregularities in the sampling set (cf. Subsection 8.9.4) and the role of relaxation parameters (Subsection 8.9.7). More practical questions are treated in Subsection 8.9.3 (how to measure errors). In the last subsection theoretical and practical convergence rates are compared. Section 8.10 deals among others with questions of speed. In Subsection 8.10.3 we indicate how to modify the algorithms for repeated reconstruction tasks with fixed spectrum and sampling set. Subsection 8.10.4 shows that the reconstruction from averages works surprisingly well. The last section summarizes the arguments for the use of ADPW, the Adaptive Weights Method, as proposed by the authors.

I A quantitative theory of irregular sampling

8.2 Iterative reconstruction algorithms

In this section we discuss the generic form of many reconstruction methods that are currently used in signal analysis as well as in other areas of mathematics. This class of algorithms is distinguished (a) by their linearity, (b) by their being iterative, and (c) by the geometric convergence of successive approximations. From the functional analytic point of view they are just the inversion of a linear operator by a Neumann series.

A particular case of these methods are the algorithms based on frames. We outline a new method for constructing frames which will be very useful in obtaining explicit estimates in the following section.

PROPOSITION 8.1
Let A be a bounded operator on a Banach space $(B, \| \cdot \|_B)$ that satisfies for some positive constant $\gamma < 1$

$$\|f - Af\|_B \leq \gamma \|f\|_B \quad \text{for all } f \in B. \tag{8.2}$$

Then A is invertible on B and f can be recovered from Af by the following iteration algorithm. Setting $f_0 = Af$ and

$$f_{n+1} = f_n + A(f - f_n) \tag{8.3}$$

for $n \geq 0$, we have

$$\lim_{n \to \infty} f_n = f \tag{8.4}$$

with the error estimate after n iterations

$$\|f - f_n\|_B \leq \gamma^{n+1} \|f\|_B. \tag{8.5}$$

PROOF By inequality (8.2) the operator norm of $Id - A$ is less than γ. This implies that A is invertible and that the inverse can be represented as a Neumann series:

$$A^{-1} = \sum_{n=0}^{\infty} (Id - A)^n \tag{8.6}$$

and any $f \in B$ is determined by Af and the norm-convergent series $f = A^{-1}Af = \sum_{n=0}^{\infty} (Id - A)^n Af$.

The reconstruction (8.4) and the error estimate (8.5) follow easily after we have shown that the nth approximation f_n as defined in (8.3) coincides with the nth partial sum $\sum_{k=0}^{n} (Id - A)^k Af$. This is clear for $n = 0$, since $f_0 = Af$ by definition. Next assume that we know already that

$$f_n = \sum_{k=0}^{n} (Id - A)^k Af. \tag{8.7}$$

Then we obtain for $n + 1$

$$\sum_{k=0}^{n+1} (Id - A)^k Af = Af + \sum_{k=1}^{n+1} (Id - A)Af$$

$$= Af + (Id - A) \sum_{k=0}^{n} (Id - A)^k Af \quad \text{(by induction)}$$

$$= Af + (Id - A)f_n = f_n + A(f - f_n).$$

Now clearly $f = \lim_{n \to \infty} f_n$ and since $\sum_{k=n+1}^{\infty} (Id - A)^k = (Id - A)^{n+1} A^{-1}$, we obtain

$$\|f - f_n\|_B = \left\| \sum_{k=n+1}^{\infty} (Id - A)^k Af \right\|_B$$

$$= \|(Id - A)^{n+1} A^{-1} Af\|_B \leq \gamma^{n+1} \|f\|_B. \quad \blacksquare$$

It is a noticeable strength of this reconstruction algorithm that it allows determination *a priori* of the number of iterations required to achieve a given precision of the approximation f_n.

To prove a sampling theorem combined with an efficient reconstruction algorithm, it is thus necessary to "assemble" a linear approximation operator A which contains only the sampled values of f and then prove a norm estimate of the form (8.2). This strategy has been applied successfully to prove irregular sampling theorems in spaces of analytic functions [CoR], for short time Fourier transforms and wavelet transforms [G1], and for a general class of spaces of band-limited functions [FG1], [FG2], [FG3].

An important instance of this algorithm occurs in the presence of *frames* in a Hilbert space. Frames were introduced by R. Duffin and A. Schaeffer [DuSc] as a generalization of Riesz bases in their fundamental work on irregular sampling of band-limited functions. The concept was revived in wavelet theory through the work of I. Daubechies and A. Grossmann [D], [DGr], [DGrMe].

DEFINITION 8.2   *A sequence* $\{e_n\}_{n\in\mathbb{N}}$ *in a (separable) Hilbert space* $\mathcal{H}$ *with inner product* $\langle \cdot, \cdot \rangle$ *is called a frame if there exist constants* $0 < A \le B$ *such that*

$$A\|f\|^2 \le \sum_{n=1}^{\infty} |\langle f, e_n \rangle|^2 \le B\|f\|^2 \quad \text{for all } f \in \mathcal{H}. \tag{8.8}$$

The left-hand inequality implies not only the uniqueness of f, but also that f depends continuously on the frame coefficients $\langle f, e_n \rangle$. With Proposition 8.1 and (8.8) we can effectively reconstruct f from the sequence $\langle f, e_n \rangle$. For this we introduce the so-called *frame-operator*

$$Sf = \frac{2}{A+B} \sum_{n=1}^{\infty} \langle f, e_n \rangle e_n. \tag{8.9}$$

Then S is a positive operator, and (8.8) implies the inequalities

$$-\frac{B-A}{B+A}\|f\|^2 \le \|f\|^2 - \frac{2}{A+B}\sum_n |\langle f, e_n \rangle|^2 \tag{8.10}$$

$$= \langle (Id - S)f, f \rangle \le \frac{B-A}{B+A}\|f\|^2.$$

Since $Id - S$ is self-adjoint, (8.10) entails that the operator norm of $Id - S$ on $\mathcal{H}$ is smaller than $(B-A)/(B+A) < 1$. Now Proposition 8.1 applies, and yields an iterative reconstruction of f from Sf with a rate of convergence determined by $\gamma = (B-A)/(B+A)$. Since the input for Sf consists of the frame coefficients $\langle f, e_n \rangle$ only, this is indeed the desired reconstruction.

After setting $g_n = S^{-1}e_n$, we may write the reconstruction of f from $\langle f, e_n \rangle$ in "closed" form

$$f = S^{-1}Sf = \sum_n \langle f, e_n \rangle g_n. \tag{8.11}$$

We get immediately a nonorthogonal expansion of f with respect to the "basis" functions e_n:

$$f = SS^{-1}f = \sum_n \langle S^{-1}f, \ e_n \rangle e_n = \sum_n \langle f, \ g_n \rangle e_n. \tag{8.12}$$

Note that the two *frame bounds* A and B play a vital role in the algorithm and determine the speed of convergence of the algorithm. It is therefore of practical importance to estimate A, B as sharply as possible.

The upper estimate in (8.8) is rarely a problem. It expresses the continuity of $f \longmapsto \langle f, e_n \rangle_{n\in\mathbb{N}}$ from $\mathcal{H}$ into l^2 and it is not difficult to derive reasonable estimates for B. On the other hand, the lower estimate is usually the difficult and deep part of the argument, and good estimates for the lower frame bound A are mostly beyond reach.

This problem is often dealt with in the following way: one is content with the mere existence of frame bounds A and B and considers the family of operators

$$S_\lambda f = \lambda \sum_n \langle f, e_n \rangle e_n$$

where λ is the so-called *relaxation parameter*. With an estimate similar to (8.10) one obtains

$$\| f - S_\lambda f \| \leq \gamma(\lambda) \| f \|$$

where $\gamma(\lambda) = \max\{|1 - \lambda A|, |1 - \lambda B|\}$ and $\gamma(\lambda) < 1$ for small values of λ. As long as numerical efficacy is not an issue, this trick is good enough. Let us mention that the choice $\lambda_0 = 2/(A + B)$ gives the optimal value $\gamma(\lambda_0) = (B - A)/(B + A)$, and that the actual rate of convergence is better the closer λ is to λ_0.

Here we present a *new* method for obtaining frame bounds. In the applications to irregular sampling of band-limited functions this alternative strategy leads to better, explicit estimates of the frame bounds.

This new method is based on the observation that in most cases it is more natural to work with an approximation operator of the form

$$Af = \sum_n \langle f, e_n \rangle h_n \tag{8.13}$$

for two sequences $\{e_n\}$ and $\{h_n\}$ in $\mathcal{H}$.

PROPOSITION 8.3
Suppose that the sequences $\{e_n\}_{n \in \mathbb{N}}$ and $\{h_n\}_{n \in \mathbb{N}}$ have the following properties: there exist constants $C_1, C_2 > 0, 0 \leq \gamma < 1$, so that

$$\sum_n |\langle f, e_n \rangle|^2 \leq C_1 \| f \|^2 \tag{8.14}$$

$$\left\| \sum_n \lambda_n h_n \right\|^2 \leq C_2 \sum_n |\lambda_n|^2 \tag{8.15}$$

and

$$\left\| f - \sum_n \langle f, e_n \rangle h_n \right\| \leq \gamma \| f \| \tag{8.16}$$

for all $f \in \mathcal{H}$ and $(\lambda_n)_{n \in \mathbb{N}} \in l^2$. Then $\{e_n\}$ is a frame with frame bounds $(1 - \gamma)^2/C_2$ and C_1, and $\{h_n\}$ is a frame with bounds $(1 - \gamma)^2/C_1$ and C_2.

PROOF Let A be defined as in (8.13). Then A is a bounded operator on $\mathcal{H}$ by (8.14) and (8.15), A is invertible by (8.16) with $A^{-1} = \sum_{n=0}^\infty (Id - A)^n$, and the operator norm of A^{-1} is less than $(1 - \gamma)^{-1}$. Using (8.14) and (8.15) we obtain the inequality

$$\|f\|^2 = \|A^{-1}Af\|^2 \le (1-\gamma)^{-2} \left\| \sum_n \langle f, e_n \rangle h_n \right\|^2 \tag{8.17}$$

$$\le (1-\gamma)^{-2} C_2 \sum_n |\langle f, e_n \rangle|^2 \le (1-\gamma)^{-2} C_1 C_2 \|f\|^2.$$

Next we verify two inequalities which are dual to (8.14) and (8.15):

$$\left\| \sum_n \lambda_n e_n \right\|^2 \le C_1 \sum_n |\lambda_n|^2 \tag{8.18}$$

$$\sum_n |\langle f, h_n \rangle|^2 \le C_2 \|f\|^2 \tag{8.19}$$

for all $(\lambda_n) \in l^2$ and $f \in \mathcal{H}$. Since $\|\sum_n \lambda_n e_n\| = \sup_{f:\|f\|=1} |\langle \sum_n \lambda_n e_n, f \rangle|$, (8.18) follows from

$$\left| \left\langle \sum_n \lambda_n e_n, f \right\rangle \right|^2 = \left| \sum_n \lambda_n \langle e_n, f \rangle \right|^2$$

$$\le \left(\sum_n |\lambda_n|^2 \right) \left(\sum_n |\langle f, e_n \rangle|^2 \right) \le C_1 \|f\|^2 \sum_n |\lambda_n|^2.$$

The inequality (8.19) follows similarly. To see that $\{h_n\}_{n \in \mathbb{N}}$ is a frame, we can apply the same argument as in (8.17) to the adjoint operator

$$A^* f = \sum_n \langle f, h_n \rangle e_n. \quad \blacksquare$$

It follows that f can be reconstructed from the frame coefficients $\langle f, e_n \rangle$ in two ways: (a) we use the approximation operator A, which requires the use of both frames $\{e_n\}$ and $\{h_n\}$, and the iteration defined in Proposition 8.1, or (b) we use the frame operator $Sf = \lambda \sum_n \langle f, e_n \rangle e_n$ for a suitable relaxation parameter λ, e.g., $\lambda = 2C_2((1-\gamma)^2 + C_1 C_2)^{-1}$.

Clearly, in its abstract form, Proposition 8.3 does not have much substance, but in the following sections it will be a valuable guide to prove the existence of frames, irregular sampling theorems, and explicit numerical constants.

8.3 Qualitative sampling theory

This section provides an overview of constructive, but qualitative solutions of the irregular sampling problem. A detailed discussion of these results and their applicability and numerical efficiency will be carried out in Part II.

To fix notation, let B_ω be the space of band-limited functions

$$B_\omega = \left\{ f : \|f\| = \left(\int_{\mathbb{R}} |f(x)|^2 dx \right)^{1/2} < \infty \text{ and } \operatorname{supp}\hat{f} \subseteq [-\omega, \omega] \right\}. \quad (8.20)$$

Here the L^2 norm corresponds to the *energy* of the signal f, and $2\omega > 0$ is the *bandwidth* of f. The Fourier transform $f \to \mathcal{F}f = \hat{f}$ is normalized by

$$\mathcal{F}f(\xi) = \hat{f}(\xi) = \int_{\mathbb{R}} f(x)e^{-ix\xi} dx. \quad (8.21)$$

Its inverse is $f(x) = 1/2\pi \int_{\mathbb{R}} \hat{f}(\xi)e^{ix\xi} d\xi$, so that we have

$$\langle \hat{f}, \hat{g} \rangle = 2\pi \langle f, g \rangle. \quad (8.22)$$

B_ω is a closed subspace of $L^2(\mathbb{R})$ with some remarkable properties.

1. A multiple of the Fourier transform, $1/\sqrt{2\pi}\mathcal{F}$ is a unitary transformation from B_ω onto $L^2(-\omega, \omega)$.

   If χ_A is the characteristic function of a set A and $L_x f(t) = f(t - x)$ is translation by x, then for $x \in \mathbb{R}$

$$\mathcal{F}^{-1}\left(e^{-ix\xi}\chi_{[-\omega,\omega]}(\xi) \right)(t) = \frac{\sin \omega(t - x)}{\pi(t - x)} = \frac{\omega}{\pi} L_x \operatorname{sinc}_\omega t, \quad (8.23)$$

   where $\operatorname{sinc}_\omega x = \sin \omega x / \omega x$ for $x \neq 0$ and $\operatorname{sinc}_\omega 0 = 1$. From (8.23) one concludes immediately the following.

2. Since $\{1/\sqrt{2\omega} e^{i\pi n\xi/\omega}, n \in \mathbb{Z}\}$ is an orthonormal basis for $L^2(-\omega, \omega)$ the collection, $\{\sqrt{\omega/2\pi} L_{\pi n/\omega} \operatorname{sinc}_\omega, n \in \mathbb{Z}\}$ is an orthonormal basis of B_ω.

3. B_ω is a reproduction kernel Hilbert space, i.e., point evaluations $f \to f(x)$ are continuous on B_ω. More precisely, for $f \in B_\omega$,

$$f(x) = \frac{1}{2\pi} \int_{\mathbb{R}} \hat{f}(\xi)e^{ix\xi}\chi_{[-\omega,\omega]}(\xi)d\xi \quad (8.24)$$

$$= \int f(t)\frac{\sin \omega(t - x)}{\pi(t - x)} dx = \frac{\omega}{\pi} \langle f, L_x \operatorname{sinc}_\omega \rangle = \frac{\pi}{\omega} f * \operatorname{sinc}_\omega(x).$$

   The regular sampling theorem (8.1) for the Nyquist rate $\delta = \pi/\omega$ follows immediately from 2 and 3.

4. Every $f \in B_\omega$ extends to an entire function $f(x + iy)$ of exponential growth ω. By a famous theorem of Paley–Wiener, the converse is also true [PWi]. Each of these properties of B_ω can be taken as the foundation of a sampling theory.

The most profound results are based on the identification of band-limited functions with entire functions, see [PWi], [Be], [BeMal], [Le]. The work of Nashed and Walter [NWa] and part of our own work [FG2], [FG3] are based on the reproducing kernel property (8.24). Finally, any statement about a nonharmonic Fourier series $\varphi(\xi) = \sum_n C_n e^{ix_n\xi}$ in $L^2(-\omega, \omega)$ is equivalent to a sampling theorem by (8.23) and (8.24). Indeed, many well-known irregular sampling theorems are just reformulations about earlier results on nonharmonic Fourier series, cf. [You1] for an introduction.

8.3.1   Theorems following Paley–Wiener

Paley and Wiener [PWi] were interested in the question of which perturbations $\{e^{ix_n\xi}, n \in \mathbb{Z}\}$ of the orthonormal basis $\{e^{i\pi n\xi/\omega}, n \in \mathbb{Z}\}$ are still a Riesz basis for $L^2(-\omega, \omega)$. Recall that a Riesz basis of a Hilbert space is the image of an orthonormal basis under an invertible operator.

In the context of irregular sampling the sharp form of their result can be stated as follows:

THEOREM 8.4
Suppose that the irregular sampling set $\{x_n, n \in \mathbb{Z}\}$ satisfies

$$\sup_{n\in\mathbb{Z}} \left| x_n - \frac{\pi n}{\omega} \right| \leq L < \frac{\pi}{4\omega}. \tag{8.25}$$

Then there exists a unique sequence $\{g_n\} \subseteq B_\omega$, so that every $f \in B_\omega$ has the representations

$$f(x) = \sum_{n\in\mathbb{Z}} f(x_n)g_n(x) = \frac{\omega}{\pi} \sum_{n\in\mathbb{Z}} \langle f, g_n \rangle L_{x_n} \operatorname{sinc}_\omega x. \tag{8.26}$$

The series converges in $L^2(\mathbb{R})$ and uniformly on compact sets. The collections $\{g_n, n \in \mathbb{Z}\}$ and $\{L_{x_n} \operatorname{sinc}_\omega, n \in \mathbb{Z}\}$ are Riesz bases of B_ω, and $\omega/\pi \langle L_{x_n} \operatorname{sinc}_\omega, g_n \rangle = \delta_{kn}$ for $k, n \in \mathbb{Z}$.

REMARKS 8.5

1. A proof of a slightly weaker statement can be based on Propositions 8.1 and 8.3. For this, one considers the approximation

$$Af = \sum f\left(\frac{n\pi}{\omega}\right) L_{x_n} \operatorname{sinc}_\omega \tag{8.27}$$

on B_ω. Since $f = \sum f(n\pi/\omega) L_{n\pi/\omega} \operatorname{sinc}_\omega$, it is not difficult to show that

$$\|f - Af\| \leq \Theta\|f\| \quad \text{for all } f \in B_\omega \text{ and some } \Theta < 1,$$

provided that $|x_n - \pi n/\omega| < \ln 2/\omega$. See [DuE], [You2] for details. Using the adjoint operator $A^*f = \sum f(x_n) L_{n\pi/\omega} \operatorname{sinc}_\omega$, Proposition 8.1 provides an iterative reconstruction of f from A^*f, in other words, a reconstruction from the irregular samples $f(x_n)$.

2. It is more difficult to find the optimal constant in (8.25). The precise bound $\pi/4\omega$ in (8.25) is due to Kadec [Ka].

3. Levinson [Le] has found explicit formulas for the g_n's in terms of Lagrange interpolation functions. Set

$$g(x) = (x - x_0) \prod_{n=1}^{\infty} \left(1 - \frac{x}{x_n}\right) \left(1 - \frac{x}{x_{-n}}\right), \quad \text{then } g_n(x) = \frac{g(x)}{(x - x_n)g'(x_n)}.$$

In this way one obtains an "explicit" reconstruction of $f \in B_\omega$ from its samples $f(x_n)$.

4. The theorem of Paley–Wiener and Kadec has been reformulated and interpreted many times, see, for example, [Beu1], [Beu2], [YTh], [Hi2].

5. Finally, the sampling density is the same that is required for regular sampling. If $n(r)$ denotes the minimum number of samples found in an interval of length r, then $\lim_{r \to \infty} n(r)/r = \omega/\pi$ is exactly the Nyquist density. ∎

8.3.2 Theorems following Duffin–Schaeffer

Theorem 8.4 permits only a slight perturbation of the regular sampling geometry. Weakening the notion of basis, Duffin and Schaeffer in their very influential paper [DuSc] found the following Theorem.

THEOREM 8.6
Suppose that there exist constants $0 < \varepsilon < 1$, $\alpha, L > 0$, so that the sampling set $\{x_n, n \in \mathbb{Z}\}$ satisfies $|x_n - x_m| \geq \alpha > 0$ for $n \neq m$ and

$$\sup_{n \in \mathbb{Z}} \left| x_n - \varepsilon \frac{\pi n}{\omega} \right| \leq L < \infty. \tag{8.28}$$

Then one can find constants $0 < A \leq B$, depending only on ε, α, L, so that for any $f \in B_\omega$

$$A\|f\|^2 \leq \sum_{n \in \mathbb{Z}} |f(x_n)|^2 \leq B\|f\|^2. \tag{8.29}$$

REMARKS 8.7

1. Since $f(x_n) = \omega/\pi \langle f, L_{x_n} \text{sinc}_\omega \rangle$, (8.29) is equivalent to the statement that the *collection* $\{L_{x_n} \text{sinc}_\omega, n \in \mathbb{Z}\}$ *is a frame for* B_ω. Proposition 8.1 now provides an *iterative reconstruction* of f from $f(x_n)$ by means of the frame operator

$$Sf = \sum f(x_n)L_{x_n} \text{sinc}_\omega.$$

2. The theorem of Duffin–Schaeffer is very powerful and its proof is still considered difficult. Many applications and several early numerical treatments of irregular sampling are just reformulations of this theorem, see, e.g., [M3], [Wi], [YTh], [BHe].

From a practical and numerical point of view the range of this theorem is limited, because the constants A and B are only implicit. Whereas it is not difficult to see that the upper bound B in (8.29) behaves like L/ε, of the lower bound A only existence is known.

As long as there is no concern about the numerical efficacy or the required number of iterations, one can work with an appropriate relaxation parameter determined by trial and error (!), see [MA], and achieve a reasonable reconstruction from irregular samples. Since the frame reconstruction has been used implicitly by Marvasti and previously in a very similar form by Wiley, we well refer to this method in Part II as the Wiley–Marvasti (for short WIL/MAR), or frame method.

3. Let $n(r)$ be the minimal number of samples in an interval of length r. Then we obtain from (8.28)

$$\lim_{r\to\infty} \frac{n(r)}{r} = \frac{\omega}{\varepsilon\pi} > \frac{\omega}{\pi}$$

for the sampling density. Compared to the regular sampling theorem (8.1), Theorem 8.6 always requires oversampling. ∎

In 1967, H. Landau proved a converse [L].

THEOREM 8.8

If an irregular sampling set $\{x_n, n \in \mathbb{Z}\}$ generates a frame $\{L_{x_n}\,\mathrm{sinc}_\omega, n \in \mathbb{Z}\}$ for B_ω, then necessarily

$$\liminf_{r\to\infty} \frac{n(r)}{r} > \frac{\omega}{\pi}.$$

Recently, S. Jaffard has completely characterized all sampling sequences that generate frames [J]. The proof requires the theorems of Duffin–Schaeffer and of Landau. For the formulation we recall that a set $\{x_n, n \in \mathbb{Z}\}$ is called *relatively separated* [FG2] if it can be divided into a finite number of subsets so that $|x_n - x_m| \geq \alpha > 0$ for a fixed $\alpha > 0$, $n \neq m$, and x_n, x_m in the same subset.

THEOREM 8.9

A sampling sequence $\{x_n, n \in \mathbb{Z}\}$ generates a frame $\{L_{x_n}\,\mathrm{sinc}_\omega, n \in \mathbb{Z}\}$ for B_ω if and only if $\{x_n, n \in \mathbb{Z}\}$ is relatively separated and

$$\liminf_{r\to\infty} \frac{n(r)}{r} > \frac{\omega}{\pi}.$$

This is sometimes regarded as the most general sampling theorem. As explained above, its applicability is marred by the absence of estimates for the frame bounds. The relaxation parameter in the frame reconstruction has to be determined by trial-and-error. For very irregular sampling sets the convergence of the frame algorithm can be very slow.

8.3.3 The POCS method

The POCS method (=projection onto convex sets) was used in [YeSt] to interpolate band-limited functions from a *finite* number of irregular samples. More precisely, given N samples (x_n, y_n), $n = 1, 2, \ldots, N$, the task is to find a band-limited $f \in B_\omega$, such that $f(x_n) = y_n$. An easy argument shows that this problem is always solvable and that the set of solutions is a closed convex subset of B_ω contained in a subspace of codimension N.

Although this is a very weak form of the irregular sampling problem, we discuss its treatment briefly because the POCS method has brought a new idea into the field of irregular sampling.

The abstract background of the POCS method is described as follows [YoWe] (it goes back to a classical result by von Neumann [Ne, p. 55] and Kaczmarcz [K]; cf. [St] and [Hu] for more recent references): Let C_n, $n = 1, \ldots, N$, be closed convex subsets of a Hilbert space $\mathcal{H}$ and let P_n be the (nonlinear) projection operators from $\mathcal{H}$ onto C_n, uniquely defined by $\|f - Pf\| = \min_{g \in C_n} \|f - g\|$. The problem is to find $C := \bigcap_{n=1}^N C_n$ and the corresponding projection in terms of P_n.

The theory of the POCS method [YoWe] guarantees that the projection Pf of a given f onto C can be computed iteratively by $f_0 = f$,

$$f_{k+1} = P_N P_{N-1} \ldots P_2 P_1 f_k, \quad k \geq 0, \tag{8.30}$$

and $f = \lim_{k \to \infty} f_k$ in $\mathcal{H}$.

More generally, with appropriate relaxation parameters λ_n, $0 < \lambda_n < 2$, $n = 1, \ldots, N$, and the operators $T_n = f + \lambda_n(P_n f - f)$, the projection onto $\bigcap_{n=1}^N C_n = C$ can be approximated by $g_0 = f$,

$$g_{k+1} = T_N T_{N-1} \ldots T_1 g_k, \quad k \geq 0,$$

and $f = \lim_{k \to \infty} g_k$.

In the application of the POCS method to the interpolation of a band-limited function f from the samples $\{(x_n, y_n), n = 1, \ldots, N\}$, one looks at the closed convex sets of B_ω

$$C_n = \{f \in B_\omega : f(x_n) = y_n\}$$

and has to find an $f \in \bigcap_{n=1}^N C_n$ by means of the iteration (8.30). In this problem the projection P_n onto C_n is calculated explicitly in [YeSt] as

$$P_n f = \frac{\omega}{\pi} f * \text{sinc}_\omega + \frac{\pi}{\omega} \left(y_n - \frac{\pi}{\omega} f * \text{sinc}_\omega(x_n) \right) L_{x_n} \text{sinc}_\omega. \tag{8.31}$$

The iteration (8.30) then produces a band-limited function $f \in B_\omega$, $f(x_n) = y_n$, $n = 1, \ldots, N$. This algorithm extends immediately to higher dimensions. It works reasonably well as long as the number of samples is small, less than 100, say. However, when the number of sampling points becomes larger, both the POCS method described above, and the so-called single step method, which requires the calculation of a (pseudo-) inverse matrix whose size is the number of sampling points, become very slow or not feasible at all. We will report about this, but also about ways to enhance its performance in a simple way, in Subsection 8.9.2 (see also [Fetal]).

8.4   The adaptive weights method

In this section, we discuss a new method for the reconstruction of band-limited functions from irregular samples. It is practical and it has produced the best numerical results so far. Whereas Duffin–Schaeffer's theorem is too hard to be proved in this review, the adaptive weights method is easy enough to be proved completely. Additional benefits of this method are explicit frame bounds and estimates for the rate of convergence.

We assume that the sampling points x_n, $n \in \mathbb{Z}$, are ordered by magnitude, i.e., $\cdots < x_{n-1} < x_n < x_{n+1} < \cdots$ and that $\lim_{n \to \pm\infty} x_n = \pm\infty$. The sampling density is measured by the maximal length of the gaps between the samples, i.e.,

$$\delta = \sup_{n \in \mathbb{Z}} (x_{n+1} - x_n). \tag{8.32}$$

We denote the midpoints between the samples by y_n, i.e., $y_n = (x_n + x_{n+1})/2$ and the distance between the midpoints by $w_n = (x_{n+1} - x_{n-1})/2$. If the maximal gap length is δ, then we obviously have $y_n - x_n \leq \delta/2$ and $x_n - y_{n-1} \leq \delta/2$.

According to Proposition 8.1, we have to construct an approximation of the identity operator on B_ω that requires only the samples $f(x_n)$ as input in order to obtain an iterative reconstruction of f.

An obvious and simple approximation procedure is to interpolate f by a step function first, followed by the projection onto B_ω.

Denoting the characteristic function of the interval $[y_{n-1}, y_n)$ by χ_n and the orthogonal projection from $L^2(\mathbb{R})$ onto B_ω by P, where P is given by $(Pf)^\wedge = \hat{f} \chi_{[-\omega,\omega]}, f \in L^2(\mathbb{R})$, then our first guess for an approximation of f by its samples is the operator

$$Af = P \left(\sum_{n \in \mathbb{Z}} f(x_n) \chi_n \right). \tag{8.33}$$

It is easy to verify that A maps B_ω into B_ω (see also below), so the question is how well A approximates the identity on B_ω.

For this we need two simple inequalities.

LEMMA 8.10   (Bernstein's inequality)
If $f \in B_\omega$, then $f' \in B_\omega$ and

$$\|f'\| \leq \omega \|f\|. \tag{8.34}$$

PROOF Using (8.22) we have,

$$\|f'\|^2 = \frac{1}{2\pi} \|(f')^\wedge\|^2 = \frac{1}{2\pi} \|i\xi\hat{f}(\xi)\|^2 \leq \frac{\omega^2}{2\pi} \|\hat{f}\|^2 = \omega^2\|f\|^2. \quad \blacksquare$$

LEMMA 8.11 (Wirtinger's inequality)
If $f, f' \in L^2(a, b)$ and either $f(a) = 0$ or $f(b) = 0$, then

$$\int_a^b |f(x)|^2 dx \leq \frac{4}{\pi^2}(b-a)^2 \int_a^b |f'(x)|^2 dx. \tag{8.35}$$

Lemma 8.11 follows from [HaLiPó], p. 184, by a change of variables. It is interesting to note that (8.35) has also been used to obtain uniqueness results for band-limited functions [W].

PROPOSITION 8.12
If $\delta = \sup_{n\in\mathbb{Z}}(x_{n+1} - x_n) < \pi/\omega$, then for all $f \in B_\omega$

$$\|f - Af\| < \frac{\delta\omega}{\pi}\|f\|. \tag{8.36}$$

Consequently, A is bounded and invertible on B_ω with bounds

$$\|Af\| \leq \left(1 + \frac{\delta\omega}{\pi}\right)\|f\| \tag{8.37}$$

and

$$\|A^{-1}f\| \leq \left(1 - \frac{\delta\omega}{\pi}\right)^{-1}\|f\|. \tag{8.38}$$

PROOF Since $f = Pf = P(\sum_{n\in\mathbb{Z}} f\chi_n)$ for $f \in B_\omega$, and since the characteristic functions χ_n have mutually disjoint support, we may write

$$\|f - Af\|^2 = \left\|Pf - P\left(\sum_{n\in\mathbb{Z}} f(x_n)\chi_n\right)\right\|^2 \tag{8.39}$$

$$\leq \left\|\sum_{n\in\mathbb{Z}} (f - f(x_n))\chi_n\right\|^2 = \sum_{n\in\mathbb{Z}} \int_{y_{n-1}}^{y_n} |f(x) - f(x_n)|^2 dx.$$

Now split each integral into two parts, $\int_{y_{n-1}}^{y_n} \ldots = \int_{y_{n-1}}^{x_n} + \int_{x_n}^{y_n}$, and apply Wirtinger's inequality:

$$\int_{y_{n-1}}^{y_n} |f(x) - f(x_n)|^2 dx$$

$$\leq \frac{4}{\pi^2}(x_n - y_{n-1})^2 \int_{y_{n-1}}^{x_n} |f'(x)|^2 dx + \frac{4}{\pi^2}(y_n - x_n)^2 \int_{x_n}^{y_n} |f'(x)|^2 dx$$

$$\leq \frac{\delta^2}{\pi^2} \int_{y_{n-1}}^{y_n} |f'(x)|^2 dx.$$

In the last step we sum over n and use Bernstein's inequality to obtain

$$\sum_{n\in\mathbb{Z}}\int_{y_{n-1}}^{y_n}|f(x)-f(x_n)|^2\,dx \le \frac{\delta^2}{\pi^2}\sum_{n\in\mathbb{Z}}\int_{y_{n-1}}^{y_n}|f'(x)|^2\,dx \tag{8.40}$$

$$= \frac{\delta^2}{\pi^2}\|f'\|^2 \le \frac{\delta^2\omega^2}{\pi^2}\|f\|^2.$$

A combination of these estimates yields (8.36). The inequalities (8.37) and (8.38) are obvious:

$$\|Af\| < \|f\| + \|Af - f\| \le \left(1+\frac{\delta\omega}{\pi}\right)\|f\|$$

and

$$\|A^{-1}f\| = \left\|\sum_{k=0}^{\infty}(Id-A)^k f\right\| \le \sum_{k=0}^{\infty}\left(\frac{\delta\omega}{\pi}\right)^k\|f\|. \quad\blacksquare$$

To obtain a *quantitative sampling theory*, we apply Propositions 8.1, 8.3, and 8.12 in various ways.

THEOREM 8.13 *(Iterative reconstruction—Voronoi method)*
If $\delta = \sup_{n\in\mathbb{Z}}(x_{n+1}-x_n) < \pi/\omega$, then $f \in B_\omega$ is uniquely determined by its samples $f(x_n)$ and can be reconstructed iteratively as follows: Set

$$f_0 = Af = P\left(\sum_{n\in\mathbb{Z}}f(x_n)\chi_n\right), \tag{8.41}$$

$$f_{k+1} = f_k + A(f - f_k), \quad for\ k \ge 0.$$

Then $\lim_{k\to\infty}f_k = f$ in $L^2(\mathbb{R})$ and

$$\|f - f_k\| \le \left(\frac{\delta\omega}{\pi}\right)^{k+1}\|f\|. \tag{8.42}$$

PROOF   Combine Propositions 8.1 and 8.12. $\blacksquare$

To deduce further results, we look once more at the approximation operator A. With (8.17) and $w_n = \int \chi_n(x)dx = (x_{n+1}-x_{n-1})/2$, A becomes

$$Af = \frac{\omega}{\pi}\sum_{n\in\mathbb{Z}}\langle f, \sqrt{w_n}L_{x_n}\,\mathrm{sinc}_\omega\rangle\frac{1}{\sqrt{w_n}}P\chi_n. \tag{8.43}$$

This normalization is natural, because $\|1/\sqrt{w_n}P\chi_n\| \le 1$. It is now easy to verify the assumptions of Proposition 8.3. Therefore the collections $\{\sqrt{w_n}L_{x_n}\,\mathrm{sinc}_\omega, n \in \mathbb{Z}\}$ and $\{1/(\sqrt{w_n})P\chi_n, n \in \mathbb{Z}\}$ are frames for B_ω.

THEOREM 8.14
Let $\{x_n, n \in \mathbb{Z}\}$ be a sampling set with $\delta = \sup_{n\in\mathbb{Z}}(x_{n+1}-x_n) < \pi/\omega$. Then:

(a) *Weighted frames. For any* $f \in B_\omega$

$$\left(1 - \frac{\delta\omega}{\pi}\right)^2 \|f\|^2 \leq \sum_{n \in \mathbb{Z}} w_n |f(x_n)|^2 \leq \left(1 + \frac{\delta\omega}{\pi}\right)^2 \|f\|^2. \qquad (8.44)$$

Consequently, the collection $\{\sqrt{w_n}(\omega/\pi)L_{x_n} \text{sinc}_\omega, n \in \mathbb{Z}\}$ *is a frame with frame bounds*

$$A = \left(1 - \frac{\delta\omega}{\pi}\right)^2 \quad and \quad B = \left(1 + \frac{\delta\omega}{\pi}\right)^2. \qquad (8.45)$$

(b) *The adaptive weights method. Any* $f \in B_\omega$ *can be reconstructed from its samples* $f(x_n)$ *by the adapted frame algorithm:*

$$f_0 = \frac{\pi^2}{\pi^2 + \delta^2\omega^2} \sum_{n \in \mathbb{Z}} f(x_n) w_n \frac{\omega}{\pi} L_{x_n} \text{sinc}_\omega := Sf \qquad (8.46)$$

$$f_{k+1} = f_k + S(f - f_k), \quad k \geq 0,$$

with $f = \lim_{k \to \infty} f_k$ *in* $L^2(\mathbb{R})$ *and*

$$\|f - f_k\| \leq \gamma^{k+1} \|f\| \qquad (8.47)$$

where $\gamma = 2\pi\delta\omega/\pi^2 + \delta^2\omega^2$.

PROOF OF (a) For the left hand inequality of (8.44) we use Proposition 8.12 and obtain

$$\|f\|^2 = \|A^{-1}Af\|^2 \leq \left(1 - \frac{\delta\omega}{\pi}\right)^{-2} \left\|P\left(\sum_{n \in \mathbb{Z}} f(x_n)\chi_n\right)\right\|^2 \qquad (8.48)$$

$$\leq \left(1 - \frac{\delta\omega}{\pi}\right)^{-2} \left\|\sum_{n \in \mathbb{Z}} f(x_n)\chi_n\right\|^2.$$

Since the supports of χ_n are mutually disjoint, we have

$$\left\|\sum_{n \in \mathbb{Z}} f(x_n)\chi_n\right\|^2 = \int \left|\sum_n f(x_n)\chi_n(x)\right|^2 dx$$

$$= \sum_n \int_{y_{n-1}}^{y_n} |f(x_n)|^2 \chi_n(x)dx = \sum_{n \in \mathbb{Z}} |f(x_n)|^2 w_n,$$

and (8.44) follows. For the right side we use

$$\sum_n |f(x_n)|^2 w_n = \|\sum_n f(x_n)\chi_n\|^2 \leq \left(\|f\| + \|\sum_n f(x_n)\chi_n - f\|\right)^2 \qquad (8.49)$$

and the estimates (8.39) and (8.40). Since $f(x_n) = \omega/\pi\langle f, L_{x_n} \text{sinc}_\omega\rangle$, the second statement is clear.

(b) now follows from Proposition 8.1, (8.9) and (8.10), when we use the relaxation parameter

$$\lambda = \frac{2}{A+B} = \frac{1}{1 + \delta^2\omega^2/\pi^2},$$

which gives as rate of convergence

$$\gamma = \frac{B-A}{B+A} = \frac{2\pi\delta\omega}{\pi^2 + \delta^2\omega^2} < 1. \quad \blacksquare$$

The dual theorem is new and comes as a big surprise. Since for $f \in B_\omega$ the inner product

$$\left\langle f, \frac{1}{\sqrt{w_n}} P\chi_n \right\rangle = \left\langle Pf, \frac{1}{\sqrt{w_n}} \chi_n \right\rangle = \sqrt{w_n} \frac{1}{w_n} \int_{y_{n-1}}^{y_n} f(x)dx \qquad (8.50)$$

is just a multiple of the *average* of f over the interval $[y_{n-1}, y_n]$, the fact that the collection $\{1/(\sqrt{w_n})P\chi_n, n \in \mathbb{Z}\}$ is a frame for B_ω implies that *any band-limited function is completely determined by its local averages.* Since it is impossible to measure a sample of f at a precise instant x_n as a consequence of the uncertainty principle and technical limitations, the reconstruction from averages is a more realistic approach to the sampling problem. In this sense the reconstruction from averages can be considered as an extreme form of *stability* of sampling theorems and an important addition to the common error analysis [Pa], [FG4].

The precise statement is the content of the next theorem:

THEOREM 8.15   *(Reconstruction from local averages)*
Let $\{x_n, n \in \mathbb{Z}\}$ be a sampling sequence with $\delta = \sup_{n\in\mathbb{Z}}(x_{n+1} - x_n) < \pi/\omega$, y_n the midpoints $y_n = (x_n + x_{n+1})/2$, and $w_n = y_n - y_{n-1}$. Let φ_n be the local average $\varphi_n = 1/w_n \int_{y_{n-1}}^{y_n} f(x)dx$ for a function f.

Then every $f \in B_\omega$ is uniquely determined by its averages φ_n and can be reconstructed by the following algorithm: Set

$$f_0 = \sum_{n\in\mathbb{Z}} \varphi_n w_n \frac{\omega}{\pi} L_{x_n} \operatorname{sinc}_\omega \qquad (8.51)$$

$$f_{k+1} = f_k + f_0 - \sum_{n\in\mathbb{Z}} \left(\int_{y_{n-1}}^{y_n} f_k(x)dx \right) \frac{\omega}{\pi} L_{x_n} \operatorname{sinc}_\omega, \quad k \geq 0.$$

Then $f = \lim_{k\to\infty} f_k$ in $L^2(\mathbb{R})$ and

$$\|f - f_k\| \leq \left(\frac{\delta\omega}{\pi} \right)^{k+1} \|f\|.$$

PROOF   We first compute the adjoint of the approximation operator A from (8.33): for $f, g \in B_\omega$ we find

$$\langle A^*f, g \rangle = \langle f, Ag \rangle = \left\langle f, \sum_n g(x_n)P\chi_n \right\rangle$$

$$= \sum_n \langle Pf, \chi_n \rangle \frac{\omega}{\pi} \langle L_{x_n} \operatorname{sinc}_\omega, g \rangle,$$

or, in brief,

$$A^*f = \sum_{n \in \mathbb{Z}} \left(\frac{1}{w_n} \int_{y_{n-1}}^{y_n} f(x)dx \right) \frac{\omega}{\pi} w_n L_{x_n} \operatorname{sinc}_\omega. \tag{8.52}$$

Under the hypothesis stated, Proposition 8.12 implies $\|f - A^*f\| \le (\delta\omega/\pi)\|f\|$ for all $f \in B_\omega$. The equation (8.51) is just a concrete formulation of the iteration of Proposition 8.1. Since the input for this algorithm is $f_0 = A^*f$ and requires only the local averages, this is the desired reconstruction. $\blacksquare$

According to the remarks after Proposition 8.3, a reconstruction can also be executed by means of the corresponding frame operator

$$Sf = \sum_{n \in \mathbb{Z}} \left(\frac{1}{w_n} \int_{y_{n-1}}^{y_n} f(x)dx \right) P\chi_n \tag{8.53}$$

and an appropriate relaxation parameter.

REMARKS 8.16

1. The theorems of this section require a slightly more restrictive concept of density than [L] and [DuSc] used. This seems to be the price for explicit constants and a numerically efficient theory.

2. The proof of Theorem 8.14 contains a new idea to compute frame bounds explicitly. Compared with the difficulty of Duffin–Schaeffer's theorem [DuSc] it is almost disappointing that with a new method this should be so simple. On the other hand, it is exactly this simplicity which allows to apply the method to other situations which are intractable by entire function theory. Some variations are discussed in the next section; we refer to [G2] for further sampling theorems.

3. It is notable that all estimates are independent of the distribution of the sampling points. This is relevant when the local sampling density varies and when the sampling points are not separated by a minimal distance. By contrast the frame bounds in the unweighted case of [DuSc] seem to be more sensitive to irregularities of the sampling set.

This is seen dramatically in examples when Jaffard's frame criterion fails, but the adaptive weights method still applies: choose $\omega = \pi$ as the bandwidth and let the sampling set consist of the points

$$\{k/2, |k| \le 20\} \cup \bigcup_{n=1}^{\infty} \{j + k/10^n \quad \text{for } 10^n \le j < 10^{n+1}, 0 \le k < 10^n\}.$$

The maximal gap for this set is 1/2, so that Theorem 8.14 gives a weighted frame with estimates for the frame bounds $A = 1/4$, $B = 9/4$ and a rate of convergence $\gamma = 4/5$. On the other hand, the unweighted sequence $L_{x_n} \text{sinc}_\omega$ fails to be a frame!

Thus one obtains the following qualitative picture: adding more points, in other words, more information, to the sampling set makes the ordinary frame reconstruction behave worse. The ratio B/A of the frame bounds will increase, and the reconstruction from the samples will converge more slowly with more information—a rather counter-intuitive behavior! On the other hand, the adaptive weights methods is not affected, since the frame bounds are uniform. An increase of information is likely to improve the performance of this algorithm. See Part II for a numerical demonstration of this effect.

4. The algorithms of Theorems 8.13 and 8.14 are very special cases of the sampling theorems in [FG2], [FG3] adapted to the L^2 norm. Their quantitative analysis was carried out in [G2].

   The first form of Theorem 8.15 also appeared in [G2]. There a higher sampling rate is required, but very general averaging procedures are allowed.

5. The adaptive weights method is also used in wavelet theory. It serves the construction of irregular wavelet and Gabor frames and allows to prove irregular sampling theorems for wavelet and short time Fourier transforms, cf., [G1], Thm. U. ∎

8.5 Variations

In this section we present several variations of the adaptive weights method. Since the proofs are similar to those of the previous section, we give only brief indications and refer to the original publications for details.

8.5.1 A discrete model for irregular sampling

The irregular sampling problem rarely arises in concrete applications and in numerical implementations in the form discussed so far. In digital signal processing a signal s consists of a finite sequence of data $s(n)$, $n = 0, 1, \ldots, N - 1$, where N is usually large. The signal is sampled at a subsequence $0 \le n_1 < n_2 < \cdots < n_r < N$. Thus the problem of irregular sampling is the question under which conditions and how s can be completely reconstructed from its samples $s(n_i)$, $i = 1, \ldots, r$.

This problem can be treated in complete analogy with band-limited functions on $\mathbb{R}$. For this we extend the finite signals periodically, i.e., $s(j + lN) = s(j)$ for $j = 0, 1, \ldots, N - 1$ and $l \in \mathbb{Z}$, and interpret s as a function on the finite cyclic group $\mathbb{Z}_N$. The discrete Fourier transform on $\mathbb{Z}_N$ is defined as

$$\hat{s}(k) = N^{-1/2} \sum_{n=0}^{N-1} s(n) e^{-2\pi i k n / N}; \tag{8.54}$$

$s \rightarrow \hat{s}$ is a unitary operator on $l^2(\mathbb{Z}_N) = \mathbb{C}^N$ and preserves the energy norm $\|s\| = (\sum_{n=0}^{N-1} |s(n)|^2)^{1/2} = \|\hat{s}\|$.

Using the periodicity of s and $\hat{s}$ and negative indices, the space of *discrete band-limited signals* of bandwidth $M \leq N/2$ is

$$\mathcal{B}_M = \{s \in \mathbb{C}^N | \hat{s}(k) = 0 \quad \text{for } |k| > M\}.$$

The orthogonal projection from $\mathbb{C}^N$ onto $\mathcal{B}_M$ is given by $(Ps)^\wedge(k) = \hat{s}(k)$ for $|k| \leq M$ and $(Ps)^\wedge(k) = 0$ for $M < |k| \leq N/2$. With proper care and the discrete versions of Bernstein's and Wirtinger's inequalities [KyTTo], the previous reconstruction Theorems 8.13–8.15 can all be translated into this discrete model of irregular sampling.

We use the notation $[q]$ for the largest integer $\leq q \in \mathbb{R}$, $l_i = [(n_i + n_{i+1})/2]$ for the midpoints between the sampling points, $\Delta_i = l_{i+1} - l_i$ for the weights. To be consistent with periodicity, we have to define $n_{r+1} = n_1 + N$, $n_0 = n_r - N$, and $l_{r+1} = l_1$.

If $\varepsilon_l \in \mathbb{C}^N$ is the sequence $\varepsilon_l(l) = 1$, $\varepsilon_l(k) = 0$ for $k \neq l$, then the substitute of sinc_ω is

$$P\varepsilon_l(n) = \frac{\sin(2M+1)\pi(n-l)/N}{N \sin \pi(n-l)/N}. \tag{8.55}$$

We can now state the *discrete version of the adaptive weights method* in the exact form in which it is implemented.

THEOREM 8.17 **[G3]**

Let $0 \leq n_1 < n_2 < \ldots < n_r < N$ be a subset of $\mathbb{Z}_N$ and $d = \max_{i=1,\ldots,r}(n_{i+1} - n_i)$. If

$$2M \left(2\left[\frac{d}{2}\right] + 1\right) < N \tag{8.56}$$

and $\lambda > 0$ is a sufficiently small relaxation parameter, then every $s \in \mathcal{B}_M$ can be reconstructed iteratively from its samples $s(n_i)$, $i = 1, \ldots, r$, by

$$s_0 = \lambda \sum_{i=1}^r s(n_i) \Delta_i P\varepsilon_{n_i}, \tag{8.57}$$

$$s_{k+1} = s_k + \lambda \sum_{i=1}^r (s(n_i) - s_k(n_i)) \Delta_i P\varepsilon_{n_i}.$$

If

$$\gamma = \frac{\sin \frac{\pi M}{N}}{\sin \frac{\pi}{4[d/2]-2}},$$

the algorithm converges to s for any λ, $0 < \lambda < 2/(1+\gamma)^2$. The rate of convergence is

$$\|s - s_k\| \leq \gamma(\lambda)^{k+1} \|s\| \tag{8.58}$$

where $\gamma(\lambda) = \max\{|1 - \lambda(1 + \gamma)^2|, |1 - \lambda(1 - \gamma)^2|\}.$

The proof follows the same steps as the proofs of Theorems 8.13 and 8.14. Again all constants are explicit and depend only on the bandwidth M and the maximal gap length d.

A critical evaluation of this theorem requires consideration of two aspects.

1. The irregular sampling problem for discrete band-limited signals is finite dimensional and amounts to solving the following system of linear equations for $\hat{s}(k)$, $|k| \leq M$:

$$N^{-1/2} \sum_{k=-M}^{M} \hat{s}(k)e^{2\pi i n_j k/N} = s(n_j) \quad j = 1, \ldots, r. \tag{8.59}$$

Here both M and r are usually large, typically of order 100–1000, and the system is overdetermined.

Since there is an extensive literature on the (iterative) solution of large linear systems, e.g., [HYou], the question is rather to determine which solution method is appropriate for the particular system (8.59). The iteration (8.57) can be written in terms of matrix multiplication and is known as *Richardson's method* for the solution of linear systems. The merit of Theorem 8.17 is the explicit range for the relaxation parameter and the rate of convergence. From the point of view of linear algebra we have proved that the condition number of Richardson's method is small and that this algorithm is well adapted to the irregular sampling problem.

2. If we identify a band-limited signal $s \in \mathcal{B}_M$ with the trigonometric polynomial $p(x) = \sum_{k=-M}^{M} \hat{s}(k)e^{2\pi i k x}$, then $s(n_i) = p(n_i/N)$, and the discrete problem seems to be entirely trivial. The polynomial p is of order M and therefore uniquely determined by its values at any $2M + 1$ distinct points. The well-known Lagrange interpolation formula then provides an explicit reconstruction of p from the $p(n_i/N)$. However, Lagrange interpolation tends to be fairly ill-conditioned and high-order Lagrange polynomials are cumbersome to use. In this sense Theorem 8.17 provides a practical and computable solution with error estimates and a good condition number.

For more on the discrete theory of irregular sampling see [G4].

8.5.2 Irregular sampling in higher dimensions

Whereas irregular sampling in one dimension is well-developed both in theory and in its numerical exploitation, we are only at the beginning of a coherent and explicit theory in higher dimensions. This is an active field of current research which offers many difficult and interesting problems.

We present first some easy extensions of the one-dimensional theory which are certainly well known, but apparently not in print. To keep notation simple, we restrict it to dimension 2.

Given a compact set $\Omega \subseteq \mathbb{R}^2$ let

$$B_\Omega(\mathbb{R}^2) = \{f \in L^2(\mathbb{R}^2) : \operatorname{supp}\hat{f} \subseteq \Omega\}$$

be the subspace of band-limited functions f with spectrum in Ω and with finite energy.

If Ω is a rectangle, $\Omega = [-\omega_1, \omega_1] \times [-\omega_2, \omega_2]$, then B_Ω is the tensor product of the corresponding subspaces of $L^2(\mathbb{R})$:

$$B_\Omega(\mathbb{R}^2) = B_{\omega_1} \hat{\otimes} B_{\omega_2}. \tag{8.60}$$

This means that $B_\Omega(\mathbb{R}^2)$ is the closed subspace of $L^2(\mathbb{R}^2)$ generated by all products $f(x)g(y)$ with $f \in B_{\omega_1}$, $g \in B_{\omega_2}$.

The following abstract lemma allows to extend one-dimensional results to higher dimensions.

LEMMA 8.18
Let $\mathcal{H} = \mathcal{H}_1 \hat{\otimes} \mathcal{H}_2$ and suppose that $\{e_m, m \in \mathbb{Z}\} \subseteq \mathcal{H}_1$ is a frame (a Riesz basis) for $\mathcal{H}_1$, and $\{f_n, n \in \mathbb{Z}\} \subseteq \mathcal{H}_2$ a frame (a Riesz basis) for $\mathcal{H}_2$. Then the collection $\{e_m \otimes f_n, m, n \in \mathbb{Z}\}$ is a frame (a Riesz basis) for $\mathcal{H}_1 \otimes \mathcal{H}_2$.

PROOF Let S_i be the frame operators for $\mathcal{H}_i$, i.e., $S_1 g = \sum_m \langle g, e_m \rangle e_m$, $S_2 h = \sum_n \langle h, f_n \rangle f_n$, and S the frame operator on $\mathcal{H}_1 \otimes \mathcal{H}_2$. We test S on an element of the form $g \otimes h \in \mathcal{H}_1 \otimes \mathcal{H}_2$ and find

$$S(g \otimes h) = \sum_m \sum_n \langle g \otimes h, e_m \otimes f_n \rangle e_m \otimes f_n$$

$$= \sum_m \sum_n \langle g, e_m \rangle \langle h, f_n \rangle e_m \otimes f_n$$

$$= \left(\sum_m \langle g, e_m \rangle e_m \right) \otimes \left(\sum_n \langle h, f_n \rangle f_n \right) = S_1 g \otimes S_2 h.$$

In other words, $S = S_1 \otimes S_2$, therefore its spectrum is the product of the spectra of S_1 and S_2. In particular, S is invertible and thus $e_m \otimes f_n$ is a frame. Its frame bounds are $A_1 A_2$ and $B_1 B_2$ if A_i, B_i are the frame bounds of $\{e_m\}$ and $\{f_n\}$, respectively.

The case of a Riesz basis is treated similarly. ∎

THEOREM 8.19
Assume that $\Omega = [-\omega_1, \omega_1] \times [-\omega_2, \omega_2]$ and that the sampling set is of the form $(x_m, y_n), m, n \in \mathbb{Z}$. Then any $f \in B_\Omega$ is uniquely determined in a stable way under any of the following conditions:

(a) Paley–Wiener:

$$\left| x_m - \frac{\pi m}{\omega_1} \right| \leq L_1 < \frac{\pi}{4\omega_1},$$

and

$$\left| y_n - \frac{\pi n}{\omega_2} \right| \leq L_2 < \frac{\pi}{4\omega_2}$$

for all $m, n \in \mathbb{Z}$.

(b) *Duffin–Schaeffer:*

$$|x_m - x_n| \geq \alpha > 0, \quad |y_m - y_n| \geq \alpha > 0 \quad \text{for } m \neq n$$

and

$$\left| x_m - \frac{\gamma_1 m\pi}{\omega_1} \right| \leq L_1 < \infty, \left| y_n - \gamma_2 \frac{\pi n}{\omega_2} \right| \leq L_2 < \infty$$

for some constants $\alpha > 0$, $L_i > 0$, $0 < \gamma_1, \gamma_2 < 1$.

(c) *Adaptive Weights Method: Both sequences are increasing and*

$$\delta_1 = \sup_{m \in \mathbb{Z}} (x_{m+1} - x_m) < \frac{\pi}{\omega_1}, \delta_2 = \sup_{n \in \mathbb{Z}} (y_{n+1} - y_n) < \frac{\pi}{\omega_2}.$$

In the first case the collection

$$\left\{ \frac{\sin \omega_1 (x - x_m)}{\pi (x - x_m)} \cdot \frac{\sin \omega_2 (y - y_n)}{\pi (y - y_n)}, n, m \in \mathbb{Z} \right\}$$

is a Riesz basis for B_Ω, in the second case it is a frame, in the third case is a frame after multiplying with the weights $1/4(x_{m+1} - x_{m-1})(y_{n+1} - y_{n-1})$. Its frame bounds are

$$A = \left(1 - \frac{\delta_1 \omega_1}{\pi} \right)^2 \left(1 - \frac{\delta_2 \omega_2}{\pi} \right)^2 \quad \text{and} \quad B = \left(1 + \frac{\delta_1 \omega_1}{\pi} \right)^2 \left(1 + \frac{\delta_2 \omega_2}{\pi} \right)^2.$$

The proof follows from the one-dimensional theorems and Lemma 8.18.

The theorem of Paley–Wiener–Levinson (case a) has been extended to sampling sets of the form $\{(x_{mn}, y_n), m, n \in \mathbb{Z}\}$ in [BuHin].

It has been shown that appropriate versions of the reconstruction algorithms of Theorems 8.13–8.15 converge for arbitrary spectrum Ω and arbitrary sampling sets $\{\boldsymbol{x}_n, n \in \mathbb{Z}\} \subseteq \mathbb{R}^2$, *provided that the maximal distance to the next neighbors is small enough* [FG2], [FG3]. A two-dimensional POCS method was proposed in [SAl].

To summarize, the known reconstruction methods for two-dimensional irregular sampling require either a very restrictive product structure of the sampling set or the strong faith that the used algorithm will converge somehow, despite its qualitative nature.

The following theorem is a first and moderate step towards a quantitative sampling theory in $\mathbb{R}^2$. Its proof is similar to the proof of Theorem 8.13 and can be found in [G2].

For the following theorem we measure the density of a sampling set $\{\boldsymbol{x}_i, i \in I\}$ by the size of rectangles centered at $\boldsymbol{x}_i$ that is required to cover $\mathbb{R}^2$. For $\boldsymbol{x} = (x_1, x_2) \in \mathbb{R}^2$ and $\boldsymbol{\delta} = (\delta_1, \delta_2) \in \mathbb{R}^2$ let $R(\boldsymbol{x}, \boldsymbol{\delta})$ denote the rectangle

$$R(\boldsymbol{x}, \boldsymbol{\delta}) = \left[x_1 - \frac{\delta_1}{2}, x_1 + \frac{\delta_1}{2} \right] \times \left[x_2 - \frac{\delta_2}{2}, x_2 + \frac{\delta_2}{2} \right].$$

We say that a set $\{x_i, i \in I\} \subseteq \mathbb{R}^2$ is δ-dense if

$$\bigcup_{i \in I} R(x_i, \delta) = \mathbb{R}^2.$$

For a δ-dense set $\{x_i\}$ we choose a partition of unity $\{\Psi_i\}$ with the properties $0 \leq \Psi_i \leq 1$, supp $\Psi_i \subseteq R(x_i, \delta)$, and $\sum_i \Psi_i \equiv 1$. Finally P_Ω is the orthogonal projection from $L^2(\mathbb{R}^2) \to B_\Omega$.

THEOREM 8.20
Assume that $\Omega = [-\omega_1, \omega_1] \times [-\omega_2, \omega_2]$ and that $\{x_i, i \in I\}$ is any δ-dense subset of $\mathbb{R}^2$ satisfying

$$\delta_1 \omega_1 + \delta_2 \omega_2 < \ln 2. \tag{8.61}$$

Then every $f \in B_\Omega(\mathbb{R}^2)$ is uniquely determined by its samples $f(x_i)$, $i \in I$, and it can be reconstructed iteratively by

$$f_0 = P_\Omega \left(\sum_{i \in I} f(x_i) \Psi_i \right), \tag{8.62}$$

$$f_{k+1} = f_k + P_\Omega \left(\sum_{i \in I} (f(x_i) - f_k(x_i)) \Psi_i \right), \quad k \geq 0.$$

Then $f = \lim_{k \to \infty} f_k$ in $L^2(\mathbb{R}^2)$ and

$$\|f - f_k\| \leq \left(e^{\delta_1 \omega_1 + \delta_2 \omega_2} - 1 \right)^{k+1} \|f\|. \tag{8.63}$$

The proof consists of the fact that the operator $Af = P_\Omega(\sum_{i \in I} f(x_i) \Psi_i)$ satisfies

$$\|Af - f\| \leq (e^{\delta_1 \omega_1 + \delta_2 \omega_2} - 1) \|f\| \tag{8.64}$$

for all $f \in B_\Omega$, see [G2].

If the partition of unity consists of characteristic functions $\Psi_i = \chi_{V_i}$ with $V_i \subseteq R(x_i, \delta)$, then one obtains by Theorem 8.20 a *two-dimensional adaptive weights method*.

COROLLARY 8.21
Set $w_i = \int \chi_{V_i}(x) dx$. Then any $f \in B_\Omega$ satisfies

$$\left(2 - e^{\delta_1 \omega_1 + \delta_2 \omega_2} \right)^2 \|f\|^2 \leq \sum_{i \in I} w_i |f(x_i)|^2 \leq e^{2\delta_1 \omega_1 + 2\delta_2 \omega_2} \|f\|^2. \tag{8.65}$$

REMARKS 8.22

1. The corresponding reconstruction algorithm differs from the adaptive weights method in Theorem 8.14(b) only in the details of notation. We leave its formulation to the reader.

2. Once the estimate (8.64) is accepted, the corollary is proved in exactly the same way as Theorem 8.14. ∎

This corollary is of theoretical interest because it is the first genuine generalization of Duffin and Schaeffer's theorem to higher dimensions. It is of practical interest because it is the basis for the efficient implementations of Part II.

8.5.3 The uncertainty principle and recovery of missing segments

In [DoSt] D. Donoho and P. Stark consider versions of the uncertainty principle of the following form: if $f \in L^2(\mathbb{R})$ is essentially concentrated on a measurable set $T \subseteq \mathbb{R}$ and $\hat{f}$ is essentially concentrated on a measurable set W, then

$$|T||W| \geq 2\pi - \delta,$$

where $|T|$ is the Lebesgue measure of T and where $\delta > 0$ is small and depends on the precise definition of "essentially concentrated."

They motivate one of their applications in the following way: "Often the uncertainty principle is used to show that certain things are impossible.... We present ... examples where the generalized uncertainty principle shows something unexpected is *possible*; specifically, the recovery of a signal or image despite significant amounts of missing information." [DoSt], p. 912. This is the content of the following theorem.

THEOREM 8.23

[DoSt]: Let W and $T \subseteq \mathbb{R}$ be arbitrary measurable sets with $|W||T| < 2\pi$. Suppose that $\mathrm{supp}\hat{f} \subseteq W$ and that f is known on the complement of T. Then f is uniquely determined and can be reconstructed in a stable way from $f \cdot (1 - \chi_T)$ by the algorithm

$$f_0 = (1 - \chi_T)f,$$
$$f_{k+1} = (1 - \chi_T)f + \chi_T P_W f_k, \quad k \geq 0,$$

and

$$f = \lim_{k \to \infty} f_k \quad in\ L^2(\mathbb{R}),$$

where P_W from $L^2(\mathbb{R})$ onto $\{f \in L^2(\mathbb{R}) : \mathrm{supp}\hat{f} \subseteq W\}$ is the orthogonal projection.

The point of the theorem is that the spectrum can be an *arbitrary* measurable set.

If the spectrum W is an interval or contained in an interval $W = [-\omega, \omega]$, then the theorem of Donoho and Stark is not so surprising. Information on $f \in B_\omega$ is missing only on a subset T of measure $|T| < 2\pi/|W| = \pi/\omega$. If we cover T by disjoint intervals I_l, so that $|T| \leq \sum_l |I_l| < \pi/\omega$, the gaps of missing information are at most of length $|I_l| \leq |T| < \pi/\omega$. Therefore we may select a discrete set of samples $f(x_n)$ of f, so that $x_{n+1} - x_n \leq |T| < \pi/\omega$ for all n and so that $x_n \notin T$. Theorems 8.13 and 8.14 now provide a complete reconstruction of f. For this argument the size of T is completely irrelevant; what matters are the gaps between the given information on f. Thus, at least in the case $W = [-\omega, \omega]$ one can do "infinitely better." The missing segment T can have infinite measure or even be the complement of a set of measure zero!

In order to apply one of the Theorems 8.13 or 8.14, one has to discard even more information and preserve only samples of f on a discrete set. In most applications it is quite unreasonable to throw away any given information on f. In the following we give the modifications of the algorithms in order to use the full information $f(1 - \chi_T)$.

Since Lebesgue measure is regular, we can cover T for any $\varepsilon > 0$ by an at most countable union of open intervals, i.e., $T \subseteq \bigcup_{n \in \mathbb{Z}}(\xi_n, \eta_n)$, so that $\xi_n < \eta_n \leq \xi_{n+1}$ and $|\bigcup_{n \in \mathbb{Z}}(\xi_n, \eta_n) \setminus T| < \varepsilon$.

In order to obtain an approximation of f in B_ω, interpolate f linearly between the gaps and then project onto B_ω, i.e., define

$$If(x) = \begin{cases} f(x) & \text{for } x \notin T \\ f(\xi_n) + \dfrac{f(\eta_n) - f(\xi_n)}{\eta_n - \xi_n}(x - \xi_n) & \text{for } x \in (\xi_n, \eta_n) \end{cases} \tag{8.66}$$

and

$$A_1 f = PIf. \tag{8.67}$$

Finally we need another version of Wirtinger's inequality [KyTTo];

LEMMA 8.24
If $f(a) = f(b) = 0$ and $f, f' \in L^2(a, b)$, then

$$\int_a^b |f(x)|^2 dx \leq \left(\frac{b-a}{\pi}\right)^4 \int_a^b |f''(x)|^2 dx. \tag{8.68}$$

The next theorem contains a considerable improvement of Theorem 8.23 and uses the full information $f(1 - \chi_T)$.

THEOREM 8.25
Assume that $T \subseteq \bigcup_{n \in \mathbb{Z}}(\xi_n, \eta_n)$ and that

$$\delta := \sup_{n \in \mathbb{Z}}(\eta_n - \xi_n) < \frac{\pi}{\omega}. \tag{8.69}$$

Then $f \in B_\omega$ is uniquely determined by $f(1 - \chi_T)$ and can be reconstructed as follows:

$$f_0 = A_1 f, \quad f_{k+1} = f_k + A_1(f - f_k) \quad \text{for } k \geq 0,$$

and $f = \lim_{k \to \infty} f_k$ *with the error estimate*

$$\|f - f_k\| \leq \left(\frac{\delta\omega}{\pi}\right)^{2(k+1)} \|f\|.$$

PROOF We have to show that $\|f - A_1 f\| \leq (\delta\omega/\pi)^2 \|f\|$ and then apply Proposition 8.1. For $f \in B_\omega$ we obtain by the same argument as in (8.39):

$$\|f - A_1 f\|^2 = \|P(f - If)\|^2 \leq \|f - If\|^2$$

$$= \sum_{n \in \mathbb{Z}} \int_{\xi_n}^{\eta_n} \left| f(x) - f(\xi_n) - \frac{f(\eta_n) - f(\xi_n)}{\eta_n - \xi_n} (x - \xi_n) \right|^2 dx.$$

Since the integrand vanishes at both end points ξ_n and η_n, Lemma 8.24 implies

$$\|f - A_1 f\|^2 \leq \sum_{n \in \mathbb{Z}} \left(\frac{\eta_n - \xi_n}{\pi}\right)^4 \int_{\xi_n}^{\eta_n} |f''(x)|^2 dx$$

$$\leq \left(\frac{\delta}{\pi}\right)^4 \|f''\|^2 \leq \left(\frac{\delta\omega}{\pi}\right)^4 \|f\|^2.$$

In the last two inequalities we have used the hypothesis (8.69) and Bernstein's inequality. ∎

   The following corollary which specializes the theorem to sampling sets $\mathbb{R} \setminus T = \{x_n, n \in \mathbb{Z}\}$ has already been obtained in [G2]. It gives an improved rate of convergence of the reconstruction algorithm.

COROLLARY 8.26
Let $\{x_n, n \in \mathbb{Z}\}$ *be a sampling set so that* $\delta := \sup(x_{n+1} - x_n) < \pi/\omega$. *Then every* $f \in B_\omega$ *can be reconstructed iteratively from its samples* $f(x_n)$ *by* $f_0 = A_1 f$, $f_{k+1} = f_k + A_1(f - f_k)$ *for* $k \geq 0$, *and* $f = \lim_{k \to \infty} f_k$ *with the error estimate*

$$\|f - f_k\| \leq \left(\frac{\delta\omega}{\pi}\right)^{2(k+1)} \|f\|.$$

Here $A_1 f = PIf$, *where*

$$If(x) = f(x_n) + \frac{f(x_{n+1}) - f(x_n)}{x_{n+1} - x_n} (x - x_n)$$

for $x \in [x_n, x_{n+1}]$.

8.6   Irregular sampling of the wavelet and the short time Fourier transform

In this section we discuss how the well-known results of I. Daubechies, A. Gross-mann, and Y. Meyer [DGrMe] and [D] on the construction of wavelet frames can be extended to irregularly spaced wavelet frames.

Let $g \in L^2(\mathbb{R})$ be a fixed function (a "wavelet"). Then the *continuous wavelet transform* of a function f on $\mathbb{R}$ is defined as

$$W_g f(x,s) = \frac{1}{s} \int_{\mathbb{R}} \bar{g}\left(\frac{t-x}{s}\right) f(t)dt \quad \text{for } x, s \in \mathbb{R}, s > 0, \qquad (8.70)$$

and the short time Fourier transform $S_g f$ of a function f is

$$S_g f(x,y) = \int_{\mathbb{R}} e^{-iyt} \bar{g}(t-x) f(t)dt \quad \text{for } x, y \in \mathbb{R}. \qquad (8.71)$$

Given g, the function f is uniquely determined by both $W_g f$ and $S_g f$ and can be recovered by explicit inversion formulas, see, e.g., [GrMoPau]. However, a representation of f by one of these transforms is highly redundant, and one seeks to reduce the redundancy by sampling.

The case of regular sampling is well understood through the work of I. Daubechies [D]. She has found nearly optimal estimates on the density (α, β), so that any f can be reconstructed in a stable way from $W_g f(\alpha\beta^j k, \beta^j), j, k \in \mathbb{Z}$, or from $S_g f(\alpha j, \beta k)$.

For irregular sampling only qualitative results are known. $W_g f$ and $S_g f$ are completely determined by their samples on a "sufficiently dense" set in the plane [FrJa], [G1].

In the following we consider band-limited wavelets $g \in B_\omega$ and sampling sets of the form $(x_{jk}, s_j), j, k \in \mathbb{Z}$, with the usual assumptions that $\lim_{k \to \pm\infty} x_{jk} = \pm\infty$ and $\lim_{j \to \pm\infty} \ln s_j = \pm\infty$, and that all sequences $\{x_{jk}, k \in \mathbb{Z}\}$ and $\{s_j\}$ are increasing. This amounts to sampling the wavelet transform first irregularly in the scaling parameter s, and then on each scale s_j at the points x_{jk}.

THEOREM 8.27   [G3]
Let $g \in B_\omega \cap L^1(\mathbb{R})$ and $\hat{g}(\xi) = 0$ for $|\xi| < \varepsilon < \omega$ with some (small) $\varepsilon > 0$. Suppose that the sequence $\{s_j, j \in \mathbb{Z}\} \subseteq \mathbb{R}^+$ satisfies, for two constants $a, b > 0$,

$$a \le \sum_{j \in \mathbb{Z}} |\hat{g}(s_j\xi)|^2 \le b \quad \text{for a.a. } \xi. \qquad (8.72)$$

Let $\{x_{jk}, j, k \in \mathbb{Z}\} \subseteq \mathbb{R}$ be a double sequence satisfying

$$\delta := \sup_{j,k \in \mathbb{Z}} \frac{x_{jk+1} - x_{jk}}{s_j} < \frac{\pi}{\omega}. \qquad (8.73)$$

Then we have for all $f \in L^2(\mathbb{R})$

$$a\left(1 - \frac{\delta\omega}{\pi}\right)^2 \|f\|^2 \leq \sum_{j,k\in\mathbb{Z}} \frac{1}{2}(x_{j,k+1} - x_{j,k-1})|W_g f(x_{jk}, s_j)|^2 \qquad (8.74)$$

$$\leq b\left(1 + \frac{\delta\omega}{\pi}\right)^2 \|f\|^2.$$

A reformulation of (8.74)

The collection of functions

$$\left\{\frac{1}{s_j} g\left(\frac{x - x_{jk}}{s_j}\right), j, k \in \mathbb{Z}\right\} \qquad (8.75)$$

is a frame for $L^2(\mathbb{R})$ with frame bounds

$$A = a\left(1 - \frac{\delta\omega}{\pi}\right)^2 \quad \text{and} \quad B = b\left(1 + \frac{\delta\omega}{\pi}\right)^2. \qquad (8.76)$$

The reconstruction of f from the irregular samples of $W_g f$ can be carried out explicitly as follows. Let

$$\lambda = \frac{2}{A + B} \quad \text{and} \quad \gamma = \frac{B - A}{B + A} = \frac{b(\pi + \delta\omega)^2 - a(\pi - \delta\omega)^2}{b(\pi + \delta\omega)^2 + a(\pi - \delta\omega)^2}.$$

Let S be the wavelet frame operator

$$Sf(x) = \lambda \sum_{j,k\in\mathbb{Z}} \frac{1}{2}(x_{j,k+1} - x_{j,k-1})W_g f(x_{jk}, s_j)\frac{1}{s_j} g\left(\frac{x - x_{jk}}{s_j}\right). \qquad (8.77)$$

The sequence f_k, where

$$f_0 = Sf, f_{k+1} = f_k + S(f - f_k) \quad \text{for } k \geq 0, \qquad (8.78)$$

converges to f in $L^2(\mathbb{R})$ and

$$\|f - f_k\| \leq \gamma^{k+1}\|f\|.$$

By duality, every $f \in L^2(\mathbb{R})$ has a wavelet expansion of the form

$$f(x) = \sum_{j,k\in\mathbb{Z}} c_{jk} \frac{1}{2}(x_{j,k+1} - x_{j,k-1})\frac{1}{s_j} g\left(\frac{x - x_{jk}}{s_j}\right) \qquad (8.79)$$

with coefficients satisfying

$$\left[b\left(1 + \frac{\delta\omega}{\pi}\right)\right]^{-2} \|f\|^2 \leq \sum_{j,k\in\mathbb{Z}} |c_{jk}|^2 \leq \left[a\left(1 - \frac{\delta\omega}{\pi}\right)\right]^{-2} \|f\|^2. \quad \square$$

REMARKS 8.28

1. The conditions on the wavelet g are identical to those in [DGrMe], and (8.72) is just a paraphrase of the known condition for regular sampling.

2. The left side of (8.72) implies that $s_{j+1}/s_j \leq \omega/\varepsilon$ for all j. Otherwise the dilates of supp $\hat{g}$ would not cover $\mathbb{R}$ and (8.72) is violated. The right side of (8.72) requires $s_{j+1}/s_j \geq \alpha > 0$ for all j.

   The condition that the s_j be separated becomes unnecessary if appropriate weights for the scaling parameter s are used ([G3]).

3. Since g is band limited, the frequency resolution of $W_g f$ is the best possible. In order to obtain good spatial resolution as well, the wavelet g must have rapid decay. This is achieved by a choice $g \in B_\omega$, $\hat{g} \in C^\infty$.

4. A similar theorem without explicit frame bounds is obtained in the article of Benedetto and Heller [BHe].

5. A different reconstruction method that uses a deconvolution argument is indicated in [G3]. ∎

PROOF OF THEOREM 8.27   In the light of Section 8.2 and Theorem 8.14 the proof is much easier than the task of formulating the theorem.

We observe first that for fixed scale $s > 0$,

$$W_g f(x, s) = \int f(t) \frac{1}{s} \bar{g}\left(-\frac{x-t}{s}\right) dt = f * g_s^\sim(x) \tag{8.80}$$

where $g^\sim(t) = \overline{g(-t)}$ and $g_s(t) = 1/s g(t/s)$. Taking Fourier transform of $f * g_s^\sim$, we find

$$\mathrm{supp}(f * g_s^\sim)^\wedge = \mathrm{supp} \hat{f} g_s^{\sim\vee} = \mathrm{supp} \hat{f}(\xi) \overline{\hat{g}(s\xi)} \subseteq \left[-\frac{\omega}{s}, \frac{\omega}{s}\right].$$

Thus at each scale $s > 0$, $W_g f(\cdot, s)$ is a band-limited function with band-width ω/s.

Step 1: Reconstruction of $W_g f(x, s_j)$ from $W_g f(x_{jk}, s_j)$.

Since $W_g(\cdot, s_j) = f * g_{s_j}^\sim \in B_{\omega/s_j}$ and by assumption for each $j \in \mathbb{Z}$, $\sup_{k \in \mathbb{Z}}(x_{j,k+1} - x_{j,k}) \leq s_j \delta$ and $(s_j \delta) \cdot (\omega/s_j) = \delta\omega$, Theorem 8.14, (8.44) yields for each j

$$\left(1 - \frac{\delta\omega}{\pi}\right)^2 \|f * g_{s_j}^\sim\|^2 \leq \sum_k \frac{1}{2}(x_{j,k+1} - x_{j,k-1}) |W_g f(x_{jk}, s_j)|^2$$

$$\leq \left(1 + \frac{\delta\omega}{\pi}\right)^2 \|f * g_{s_j}^\sim\|^2. \tag{8.81}$$

Step 2: Reconstruction of $W_g f$ from $W_g f(\cdot, s_j)$. For this we use Plancherel's theorem (8.15) and write $\|f * g_{s_j}^\sim\|^2 = 1/2\pi \|\hat{f} g_{s_j}^{\sim\wedge}\|^2 = 1/2\pi \int_\mathbb{R} |\hat{f}(\xi)|^2 |\hat{g}(s_j\xi)|^2 d\xi$. Now we take the sum over j and obtain with (8.72)

$$\sum_{j \in \mathbb{Z}} \|f * g_{s_j}^\sim\|^2 = \frac{1}{2\pi} \int |\hat{f}(\xi)|^2 \left(\sum_{j \in \mathbb{Z}} |\hat{g}(s_j\xi)|^2\right) d\xi$$

$$\leq b \frac{1}{2\pi} \|\hat{f}\|^2 = b\|f\|^2 \quad \text{and} \quad \geq a\|f\|^2.$$

Combined with (8.81) this proves the theorem. Since $W_g f(x, s)$ is the inner product of f with the function $s^{-1} g(t - x/s)$, the other statements are obvious. ∎

A parallel statement for the short time Fourier transform reads as follows.

THEOREM 8.29
Suppose $g \in B_\omega \cap L^1(\mathbb{R})$ and that for two constants $a, b > 0$ the sequence $\{y_j, j \in \mathbb{Z}\}$ satisfies

$$0 < a \leq \sum_{j \in \mathbb{Z}} |\hat{g}(\xi - y_j)|^2 \leq b. \tag{8.82}$$

If $\{x_{jk}, j, k \in \mathbb{Z}\}$ is a double sequence satisfying

$$\delta := \sup_{j,k} (x_{j,k+1} - x_{j,k}) < \frac{\pi}{\omega},$$

then for any $f \in L^2(\mathbb{R})$

$$a \left(1 - \frac{\delta\omega}{\pi}\right)^2 \|f\|^2 \leq \sum_{j,k \in \mathbb{Z}} \frac{1}{2} (x_{j,k+1} - x_{j,k-1}) |S_g f(x_{j,k}, y_j)|^2$$

$$\leq b \left(1 + \frac{\delta\omega}{\pi}\right)^2 \|f\|^2. \tag{8.83}$$

Thus the collection $\{e^{iy_j t} g(t - x_{j,k}), j, k \in \mathbb{Z}\}$ is a frame for $L^2(\mathbb{R})$ with frame bounds $A = a(1 - \delta\omega/\pi)^2$ and $B = b(1 + \delta\omega/\pi)^2$.

The proof of Theorem 8.29 is almost identical to the previous proof, and the necessary modifications are left to the reader.

It is worth noticing that the (8.83) contains implicitly a sampling theorem for the short time Fourier transform with a compactly support wavelet g. This follows from the symmetry relation

$$2\pi S_g f(x, y) = e^{-ixy} S_{\hat{g}} \hat{f}(y, -x).$$

In particular, the collection $\{(1/2)(x_{j,k+1} - x_{j,k-1}) e^{ix_{j,k} t} \hat{g}(t - y_j), j, k \in \mathbb{Z}\}$ is now a frame for $L^2(\mathbb{R})$ based on the compactly supported wavelet $\hat{g}$.

II The practice of irregular sampling

The aim of this part is to support some of the general statements made in Part I through numerical evidence. In particular, we aim at a verification of the fact that the Adaptive Weights Method (ADPW) is in many respects the method of choice due to its several advantages, such as good guaranteed rates of convergence and low computational complexity. Furthermore we will provide some evidence that there are additional features which do not follow from the theory but which have to be taken into account if one is to judge the usefulness of different methods. Such observations have more to do with statistical features and are thus not directly related to the theoretical results which have to be considered—at the present state—as worst case analysis.

8.7   Setup for numerical calculations

Restricting ourselves to the discussion of one-dimensional (1-D) signals, we consider synthetic signals which are obtained through discrete Fourier transform, with known spectrum. Thus, depending on our point of view we are dealing with functions on the cyclic group of order n (typically we take $n = 512$, or some other power of two, in order to take advantage of the existence of the FFT, the fast Fourier transform, in this case), or trigonometric polynomials. From the first point of view we assume that the spectrum (the set of coordinates for which we allow nonzero Fourier coefficients) is given, and it is small compared to the full signal length. Using the second, equivalent description, we are assuming that the frequencies which constitute the trigonometric polynomial form a finite subset of an appropriate lattice in the frequency domain. Practically, we define the signal as a superposition of those admissible frequencies with random amplitudes. We understand the irregular sampling problem as the question for constructive and efficient methods to recover signals from sufficiently many sampling values. Theory guarantees that several iterative methods will converge with a geometric rate provided that the maximal gap between successive samples is not larger than the Nyquist distance, i.e., the maximal gap size admissible for regular sampling sets according to Shannon's theorem.

Let us give next a commented version of the MATLAB code containing the essential parts of an experiment, which establishes a random signal with a given spectrum from the sampled coordinates and then apply the ADPW method to solve it. At the beginning the filter sequence filt of a length n (taking the value 1 over the spectrum yp, and 0 elsewhere) and a subsequence xp of the integer sequence $1, \ldots, n$, representing the sampling positions are given. The variable iter gives the number of iterations.

```
% APDW.M H.G. Feichtinger, 9/91, ADAPTIVE WEIGHTS METHOD
n = length(filt);          % n = signal length
yp = (filt == 1);          % yp = spectrum
sp = length(yp);           % sp = spectral dimension
y = randc(1,n)             % generating complex random spectrum
y = y.*filt;               % setting high frequencies to zero
x = ifft(y);x = x/norm(x); % generating normalized signal
xs = x(xp);                % determining sampling values
```

```
lxp = length(xp);                    % lxp: number of sampling points
xp1 = [xp(2:lxp),n+xp(1)]; xp2 = [(xp(lxp)-n), xp(1:(lxp-1))];
w = 0.5*(xp1-xp2);                   % calculating "adaptive" weights
xa = zeros(x); xd = zeros(x);        % initialization...

for j = 1 : iter                     % ITERATIONS
 xd(xp) = (xs-xa(xp)).*w;            % establishing auxiliary signal
 xa = xa+ifft((filt.*fft(xd)));      % filtering auxiliary signal
end;
```

Figure 8.1 shows a typical signal (real and imaginary part) and the sampling values and coordinates (marked by crosses), which are the data from which the reconstruction process has to start. The number of spectral values (non-vanishing DFT coefficients) is 51, the signal length 512. We have a total of 93 sampling points, and there are three gaps of length 10 (which is about the Nyquist rate for this spectrum) in the sampling set.

Figure 8.2 shows the corresponding "sinc" function, i.e., the signal obtained by means of IFFT from the filter function. It is the discrete analogue of the classical (cardinal) sinc function. It has the property of reproducing any band-limited signal with appropriate spectrum through convolution.

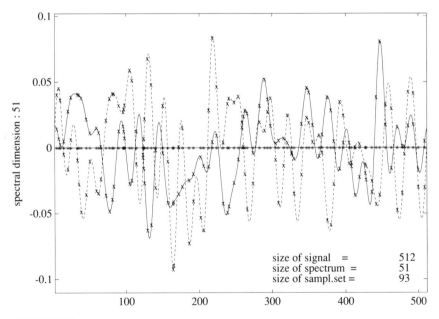

FIGURE 8.1
A typical signal and the sampling set.

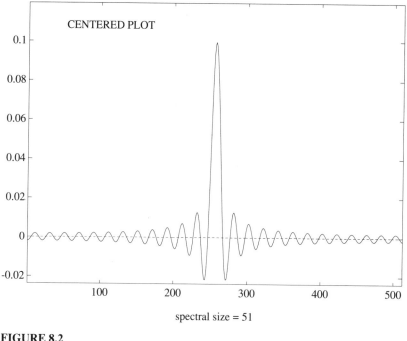

FIGURE 8.2
Symmetric sinc function.

8.8 Hard and software environment: PC386/486 and Sun, IRSATOL/MATLAB

Let us only give a brief description of the practical setup on which the numerical experiments described below have been carried out by a group of mathematicians at the Mathematics Department of the University of Vienna under the guidance of the first-named author.* The most systematic contributions over the time came from Thomas Strohmer, without whose constant engagement and cooperation the package would not exist in its present form. We also gratefully acknowledge his help in producing the plots, with technical assistance from B. van Thannen, who also took care of setting up an appropriate environment on the Sun workstations. H. Steier deserves to be mentioned separately, because he has designed an independent collection of M-files.

---

*Among the colleagues that have helped with the project were C. Cenker, M. Herrmann, M. Mayer, and F. Stettinger.

We have chosen MATLAB as the basis for our experiments, because it is a widespread mathematical software package by "MathWorks," running on a large variety of platforms. Most of our experiments were done on a PC386 and PC486 (timing given below was taken on a 33 MHz PC486), and to some extent on a Sun workstation. Extending the basic collection of M-files (i.e., modular program pieces, such as function, auxiliary utilities, and script files, containing a sequence of commands) by a collection of several hundred additional M-files, we have produced the program package IRSATOL (for IRregular SAmpling TOoLbox). A more detailed report about some general features, more from the project management and the programmer's viewpoint is given in the report [FStr].

Some of the MATLAB M-files which are of general interest also to other users can be obtained, at least for some limited period, through anonymous ftp from the InterNet address TYCHE.MAT.UNIVIE.AC.AT ($=$ 131.130.22.36). It is assumed that colleagues making use of those M-files or find improvements will notify us (FEI@TYCHE.MAT.UNIVIE.AC.AT). More details can be obtained from the same address.

8.9 Results and observations from the experiments

8.9.1 Comparison of different methods (efficiency of methods in terms of iterations or time)

We are mainly comparing the following *five* methods in this article:

1. The Wiley/Marvasti or (ordinary/unweighted) frame method, for short WIL/MAR method. It is based on the fact that the collection of shifted sinc functions forms a frame (cf. Theorem 8.6).

2. The VOR(onoi) method as described in Theorem 8.13 which requires step functions, constant in the nearest neighborhood of any of the sampling points, followed by convolution with a sinc function as approximation step.

3. The PLN (piecewise-linear) method, which make use of piecewise linear interpolations, given the sampling values, again followed by low pass filtering for the approximation operator A. (cf. Cor. 8.26).

4. The (single step) POCS method, as described in Subsection 8.3.3, where at each step the given approximate signal xa is updated by adding a shifted sinc function to correct the sampling value of the approximate signal at the sampling point under consideration.

5. The main role will be played by the ADPW method discovered by the authors, cf. Theorems 8.14 and 8.19. This method is also labelled as "dps," following the older notation for the relevant approximation operator.

Those methods have already been described at several occasions and proofs for their guaranteed rate of convergence are known, at least for the VOR and

ADPW methods, if the maximal gap does not exceed the Nyquist rate (cf. [CFHer], [CFStei], [G2], [FGHer] for details).

Figure 8.3 shows the convergence behavior for the signal of Figure 8.1, described through the natural logarithm of the ℓ^2 norm of the error versus the number of iterations (we have taken 32 iterations for each method in this example). Note that -35 corresponds to a numerical value of the error of about 10^{-16} which is of the order of machine precision for MATLAB. The methods are ordered by efficiency. Thus, in this particular example the piecewise linear method (labelled "pln") is slightly more efficient in terms of iterations than the ADPW method (labelled "dps"). We have chosen this example despite the fact that we have found that on the average ADPW did perform better even in terms of this form of efficiency (for timing see below).

One iteration is one step of any of the iterative methods discussed. We call the gain at each step (which actually comes close to its limit value after few steps), i.e., $\|xx - xa_{n+1}\|_2/\|xx - xa_n\|_2$, the *rate of convergence*. With the exception of the POCS method (labelled as "pcs") for all methods under discussion one step consists of establishing an auxiliary function and subsequent filtering. Thus each such step makes use of all the sampling values at once. In contrast, a single step for the POCS method makes use of only one sampling value. We have therefore

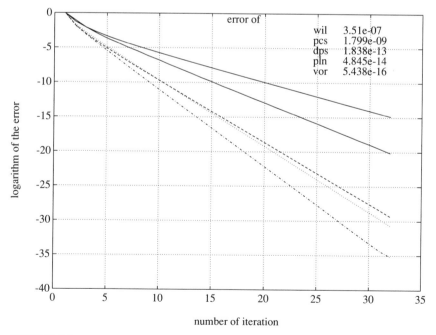

FIGURE 8.3
Convergence behavior of several methods.

decided to count one cycle, made up by a sequence of single steps and using all the sampling values once, as one iteration in our comparison. This way of counting clearly disguises the much higher effort to run through a single cycle as can be seen from Figure 8.4, which shows the convergence behavior in terms of the natural logarithm of approximation error versus time.

The time-plot Figure 8.4 describes the time needed for the same reconstruction task in order to achieve machine precision. It shows that the ADPW method and the frame method are fast compared to VOR and PLN (because the auxiliary signals are easier to build). Since one step of the frame method takes about the same time as one step of ADPW, the fact that WIL/MAR takes about twice the time of ADPW is only due to the fact that about twice as many iterations are needed to obtain the prescribed degree of approximation. Figure 8.4 also shows that in this particular example the POCS method is by far the slowest, despite an efficient implementation, where a matrix of shifted sinc functions is stored at first. Of course, it takes some time to do this, and in this particular case about the same time as to run the complete ADPW algorithm (to full precision). Without this trick the rate of convergence of POCS would be even slower.

In general the Voronoi method as well as the adaptive weights method show very good performance in terms of the degree of approximation within a given number of iterations. On the other hand the computational costs for each iteration

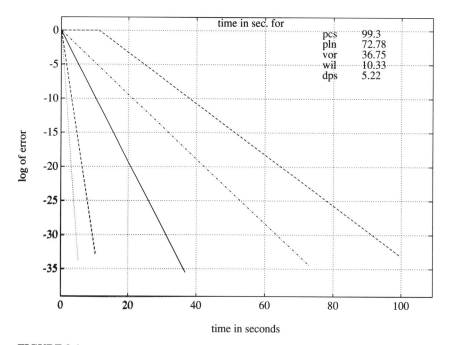

FIGURE 8.4
Comparison of time to reach precision 1.11×10^{-14}.

of the frame method or the ADPW method are low (essentially a double FFT) compared to that of the Voronoi or PLN method, because those methods require the construction of an auxiliary function (step or piecewise linear). Actually, the computation of the auxiliary functions gets more expensive as more sampling points are available. Thus there is a trade-off between the gain in performance through improved rate of convergence and increasing computational effort in order to perform a single iteration for the last two methods. In contrast, the computational costs of a single step of the WIL/MAR or ADPW method are *independent* of the number of sampling points.

We have to point out here that usually there are several different ways to describe each of the algorithms and to implement it. Typically a linear algorithm involving some Fourier transform may be described as a linear mapping on a given vector space (using a basis and matrix representations) or by direct use of the FFT. A discussion about the choice of the implementation for a given method and situation will be given elsewhere. For the present comparison we haven chosen implementations for the Voronoi iteration (or PLN) which are based on the following observation: Since step functions or piecewise-linear interpolants over a fixed geometry can always be written as linear combinations of box or triangular functions, respectively one may store those primitive auxiliary function first and then build only linear combinations with the given coefficients at the moment when the new step function (PLN function) is needed. In most cases it is even better to "filter" those functions and store the low-pass versions of those functions for later use during the iterations.

This method is *much faster* than the naive one, where at *each iteration* a new step (or PLN) function has to be built from a sequence of coefficients.

Summary

Among the discussed methods the Voronoi and the Adaptive Weights Method are the most efficient ones. But VOR is rather slow, while PLN is both slow and often not very efficient, in contrast to theoretical predictions (more about this point in Subsection 8.9.8). Similar observations apply to methods which use higher order spline interpolations for the iterative process. ADPW and the WIL/MAR method are both fast, but WIL/MAR tends to be very sensitive towards irregularities of the sampling geometry (see below). Therefore ADPW can be seen as the best approach, both in terms of speed and robustness.

8.9.2 Special features of the POCS method

There are several positive features of the POCS method. First of all, it is a relatively simple method which permits the inclusion of different kinds of convex conditions, including positivity (e.g., of the Fourier coefficients), or the knowledge of several values of derivatives or averages, because all these conditions can be formulated as membership in convex sets with easily described projection operators. In most cases a single POCS condition is just the membership of the signal in a hyperplane of the L^2 space under consideration, or rather B_Ω. Nevertheless the practical value of the POCS method for reconstruction of band-limited signals from irregular sampling appears to be limited compared to the other methods available.

The main reason why POCS is slow for our example is the fact that the number of sampling points is not very small. Increasing the number of sampling points makes each cycle computationally more expensive, while the rate of convergence (per cycle) does not necessarily improve strongly enough to compensate this effect. For a small number of sampling points the situation might be different, but for such a case many different efficient methods are available. Actually, it may be the best solution just to solve an appropriate and comparatively small linear system of equations. Since for small problems such direct methods are most efficient anyway the lack of speed is one of the major drawbacks of the POCS method.

Nevertheless, there are several strategies of improving the straightforward method for POCS (which consists in going through the sampling set from left to right) by an appropriate choice of the order or points. Some of those strategies will be discussed in M. Mayer's thesis ([Ma]); see also [Fetal] for a very recent account of this approach. Despite the fact that it appears at first sight a "natural" order, e.g., from left to right, this turns out to be a fairly inefficient way of using the given information. In fact, a second look tells us that the improvement from a pair of adjacent points (the second being very close to the first) is almost the same as of just *one* of those points, since after applying the correction based on the first sampling value, the value at the second will be almost correct. Therefore a correction at the second point will only give minimal further improvement. On the other hand, if two points are far from each other, this phenomenon does *not* arise.

It turns out that one may either randomly choose the next point at each single step, or—this is easier from a practical point of view—to fix a random enumeration of the given sampling set, and follow that fixed sequence within each cycle. In both cases the outcome is similar. Figure 8.5 exactly demonstrates this. Whereas the rate of convergence for the direct POCS method is quite poor, 6 out of 7 random orders have achieved machine precision within 22 iterations. Of course the efficiency of this trick depends on the density of sampling points. Usually a similar effect can be obtained by applying the so-called red/black method. It consists of using the odd indices first and the even afterwards, i.e., in the following order: $x_1, x_3, x_5, \ldots, x_{2l+1}, x_{2l}, x_{2l-2}, x_{2l-4}, \ldots, x_4, x_2$.

Another idea is to choose the order of points which are used within the series of single POCS steps according to the size of the errors at the corresponding sampling points (use those points first where the maximal error occurs). Unfortunately, determining those points becomes a computationally intensive extra task if the number of sampling points is large.

Since from those arguments it appears advantageous to use only the minimal number of sampling points, one might be tempted to try to reduce the number of points by ignoring some of the sampling values in order to perform the recon-

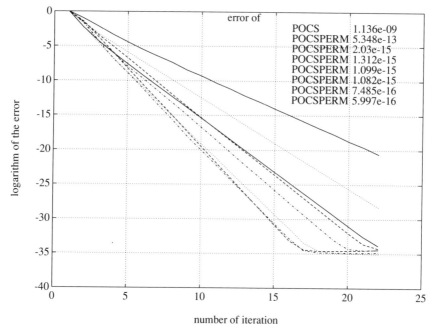

FIGURE 8.5
POCS with and without permutation of sampling points.

struction. This will do as long as exactness prevails (e.g., if we are dealing with synthetic signals), but becomes a serious problem in a more practical situation, where data are contaminated by noise (which even might vary from point to point, and it might have been the reason for sampling the signal at a variable rate). Thus, in fact, ignoring available information does not improve the algorithm.

The following example indicates that the POCS method shows more sensitivity towards noise than most of the other methods: Assume that there is a single disturbed sampling value. It is obvious that this error is fully transmitted to the reconstructed signal if the POCS method happens to use this point last. Since one usually does *not* know which one of a sequence of given data points is corrupted most, and also subsequent correct values do not help to correct the error, this is a serious problem when dealing with noisy data using the POCS method.

8.9.3   Measuring the error

Plots such as Figure 8.3 or Figure 8.4 also allow to read off easily how many iterations or how many units of time are needed in order to reach a certain precision with a given method. For both plots the choice of the logarithmic scale turns convergent curves into almost straight lines for all of the methods under discussion. Typically, after an initial phase of few iterations with steeper decay a certain fixed quotient is obtained at each of the additional iterations.

Since we have worked with synthetic signals we are able to compute the error (band-limited signal minus approximation) in an exact way. Figure 8.6 describes the error behavior measured with respect to different norms. Note that we have used normalized signals xx of ℓ^2 norm equal to 1, so that absolute and relative error coincide in that case. The ℓ^1 error (absolute sum of coordinate deviations) represents the top line of that graph, and the lowest line describes the maximal error (sup-norm, or ℓ^∞ norm). However, in practical situations we cannot measure the true deviation, because only the sampling values are given. If we just try to avoid the use of the limit element (in case it is unknown) we might use the size of the relative error as a criterion. In the example this value appears in line two.

On the other hand we can measure the error between the given sampling values and sampling values of the reconstructed signal. The second line in the graph from the bottom represents the ℓ^2 sum (sometimes this is called the sampling energy), divided by the sampling energy of the given signal. We also look at the relative change of the sampling energy (third line from the top).

It is the frame estimate (cf. (8.29)) which tells us that it is sufficient to know that the deviation of sampling values (in the ℓ^2 norm) tends to zero at the same speed as the global deviation. This general equivalence of two norms implies that the curve for the sampling error has to stay within a strip of fixed height, parallel to

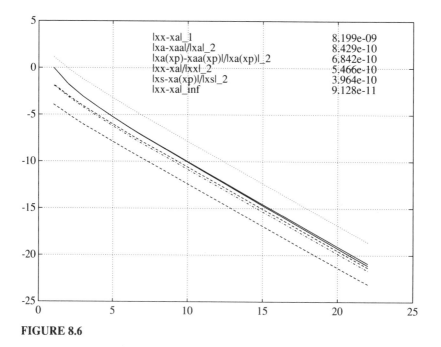

FIGURE 8.6

the "true" error curve. However, it does not directly imply that the curve obtained by plotting the natural log of the ℓ^2-sampling error is itself almost a straight line, as can indeed be seen from Figure 8.6.

This is very useful for practical applications, as it gives a fairly precise estimate of the number of required steps (hence time needed) in order to reach a given precision (e.g., the machine precision, which is 10^{-16} for MATLAB). This is particularly useful for those implementations where the approximate signal is not fully calculated at each step, but only certain coefficients or parameters, which are sufficient to continue the iteration (cf. [CFStei] for explanations in this direction).

8.9.4   Sampling geometry and spectral size

It is of great practical importance to know the extent to which the sampling geometry or the spectral size have an influence on the performance of a given method (see Figure 8.7). A detailed discussion of this topic has to be left for a separate note, because it depends excessively on the implementation of the method under consideration.

It is not surprising that the size of the maximal gap in a sampling set has considerable influence on the speed of convergence for *any* of the methods under

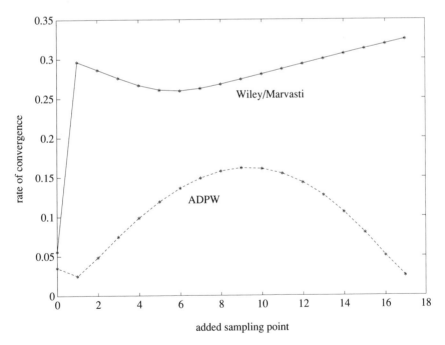

FIGURE 8.7
Rate of convergence vs. added sample point for ADPW and Wiley/Marvasti.

consideration. If the maximal gap is too large (the critical size depends also on the size of the spectrum Ω in use), then practically all methods under consideration will diverge, at least for the natural relaxation parameters (see section 7 for details). Very small relaxation parameters will in principle enforce convergence, but often enough at a rate which is so poor that it is not useful from a practical point of view.

Of course there are many other "geometric" features that a sampling set may have. For example, the degree of irregularity plays an important role. We cannot give a formal definition here. A sampling family may have slowly varying sampling density (e.g., many sampling points on the left side, fewer sampling points on the right), or may be a set with almost uniform density, but "spoiled" by few clusters of points, i.e., regions of higher density. In any case the quality of an algorithm should include statements about its sensitivity with respect to those different forms of irregularity.

Among all methods under discussion the Wiley/Marvasti method (the only one which compares with ADPW in terms of computational effort per step of the iteration) has shown the highest sensitivity in this respect. This is no surprise if we understand the choice of a relaxation parameter as the choice of a constant weight. Therefore it cannot compensate the irregularities of the sampling set as ADPW does. Only for sampling sets with uniform density, such as sampling sets which are generated by a uniform random number generator, is the performance of the simple frame comparable to that of the ADPW method. This can be explained by the fact that "on the average" the weights can be assumed to be equal in this case. Since many authors in the field restrict their attention too much on such sampling sets, we have to insist here on making the difference between "arbitrary" and "random" sampling sets. In the first case the sampling geometry is given and may have any kind of irregularity. In the second case, randomness might have played a role in the process generating the sampling set.

We also found that not only the ADPW method, but also the PLN and VOR methods are not very sensitive towards irregularities. One possible explanation for this is the fact that the "local mass balance" of the auxiliary signal built up before filtering (which is a piecewise-linear or step function respectively, and a discrete measure for ADPW) is not far from that of the full signal for those methods, whereas it will substantially deviate for Wiley/Marvasti.

Figure 8.7 shows the (negative) effect of adding points to a given sampling set. We start with a "quasiregular" sampling sequence, i.e., an arithmetic progression with step width 17, consisting of 15 points. Since we work on the cyclic group of order 256 there is one gap of size 18, so the sampling set is *not* a subgroup of the full group. Then we add successively single points at any of the 17 possible positions of that largest gap and compare the (natural) ADPW method with the frame method with optimal relaxation parameter (although it appears unrealistic that one can obtain that for practical applications), to make things fair.

The plot shows the actual rate of convergence (factor by which the error decreases per iteration) for both methods, for different positions of the point added.

At the far left end we have the situation where *no* point is added. For both methods the error decreases by a factor of about 0.05 per iteration. One can explain this by observing that the original sampling set is very close to a regular one. Adding now extra (single) points results in a weaker performance. For the WIL/MAR method the effect is very strong, in any case the rate of convergence is less than 0.25. Even for the ADPW method (with two exceptions) performance degrades, but only to a rate of about 0.15 in the worst case. However, in any of the cases the actual rate of convergence of ADPW is much better than that of WIL/MAR. Moreover, the worst case for the frame method is that of double points (positions 1 and 17), whereas those situations are the best ones for the ADPW method. As explanation we offer the fact that those positions are the ones where adding a point causes the strongest irregularity.

For other, less dramatic, situations we have found, that usually the performance of the simple, unweighted frame method decreases quickly if one increases the irregularity of a sampling set by adding further points. On the contrary, the added information (through knowledge of additional sampling values) will result in *improved* rates of convergence (at constant costs in terms of computational expenses) in the case of the ADPW method. Furthermore (see section 9.7 below) the knowledge of the optimal relaxation parameter for the frame method cannot be used to make a guess for the new optimal relaxation parameter, even if we add only few sampling points (so in a practical situation, where the optimal relaxation parameter is *not* known, the discrepancy between WIL/MAR and ADPW will be even larger than in Figure 8.7). In contrast, the weights for the ADPW method adjust automatically and no relaxation parameter has to be calculated. Furthermore, if only a few points are added the old weight can be used, and only those weight factors have to be changed which correspond to points not far from the new points.

8.9.5   Actual spectrum and known spectrum

In this section we deal with the question of lacking knowledge of the actual spectrum of the signal that should be reconstructed. Even if the sampled signal is a low frequency signal, i.e., if it has a small spectrum, there might not be enough knowledge about its spectrum, so that for the reconstructive procedure a larger filter than necessary has to be used. It then shows that the rate of convergence depends more on the size of the filger used than on the actual maximal frequency of the signal. Putting this into other words we can say that higher smoothness of a signal does not speed up the reconstruction process if this information is not incorporated as a priori knowledge. To illustrate this, think of a pure frequency (e.g., with the 8th Fourier coefficient different from zero), and assume that we know that we have to look for frequencies up to order 8 only. Then the rate of convergence will be—for all the methods under comparison—much better than for the reconstruction of a pure frequency 3, if we assume to know only that it is smaller than 11. In fact, the auxiliary signals (such as the VOR step function) will always contain significant contributions within the full spectrum under discussion which spoil the efficiency of the iterations in a significant way. Of course, taking the filter too small prevents us from complete reconstruction, and therefore this case has to be avoided under all circumstances. Figure 8.8 shows the actual (logarithmic) decay for a fixed low-pass signal, ADPW method (left), and the frame method with natural relaxation (right). The best approximation is obtained using the true spectrum, and the figure shows the effect of adding more and more points to the spectrum. For the two methods the degradation effect due to the growing spectrum is comparable, although in each case ADPW was performing much better than the frame method.

8.9.6   Choice of weight sequence

The description of the ADPW method includes the choice of the weights, which in higher dimensions are to be taken as the areas/volumes of the Voronoi domains (sets of nearest neighbors of a given sampling point). Here we cannot address the problem of actual calculation of those Voronoi weights for more than one dimension. The 1-D case is trivial: we just take half the distance between the left and the right neighbor as the length of the Voronoi interval. From our numerical

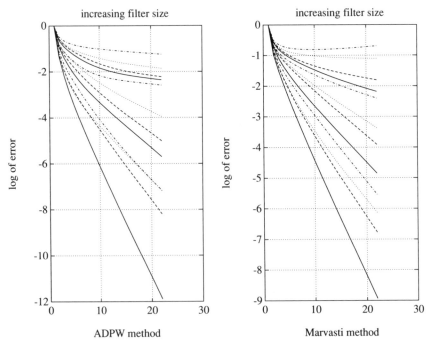

FIGURE 8.8
Log of error vs. increasing filter size. Left, ADPW method; right, Marvasti method.

experiments we can say the following concerning the dependence of the rate of convergence on the choice of the weight:

1. In most cases the speed of convergence does not depend much on the specific weights in use, as long as they compensate the irregularities of the sampling geometry. In particular, it is important to have smaller weight factors for areas of higher sampling density. On the other hand, for sampling sets with gaps larger than that for which theory gives a guaranteed rate of convergence (maximal Nyquist gaps) we have found that one should not take the formula for weights too "literal." In fact, points which are very isolated would gain too much weight, and the effect would be negative. Instead of combining such a weight sequence with a global relaxation parameter $\lambda < 1$ (which helps to guarantee convergence in this case), it is better to determine some threshold. Numerical experiments indicate that the weight which corresponds to a Nyquist gap appears as a good recommendation for the threshold.

2. As a rule the total sum of weights should be equal to the signal length. For the 1-D case this comes up in a natural way if one follows the Voronoi rule (neighborhoods are taken in the cyclic sense). For alternative methods it makes sense to normalize the weights (which might roughly represent the variable density of the sampling set) to make their total sum equal to the signal length. If we apply this rule to a constant, we end up with the recommendation to choose the weight factor n/lxp (number of sampling points divided by signal length). Actually it is just a relaxation parameter for the unweighted frame reconstruction. For sampling sets which roughly show a "uniform distribution" of sampling points this is a good choice. In fact, the Wiley/Marvasti reconstruction method with this relaxation parameter is more or less the same as the ADPW method!

The choice of weights, i.e., the ADPW method, can also be applied in a situation as described by Donoho/Stark in [DoSt], where a function with multiband spectrum (the support of its Fourier transform is contained in a collection of intervals) is given over a sequence of intervals with relatively small gaps between them. Figure 8.9 compares the rate of convergence for the frame method with different relaxation parameters. In this example the optimal relaxation parameter 1.25 is more efficient than the natural one (using n/lxp as reference value). Since we had 90 sampling points for n = 128, the direct application of the Donoho/Stark algorithm (which simply replaces the given data at each step over the intervals by the given ones) corresponds to the frame method with a relaxation parameter of 1.422 in our terminology. It only ranks third after 32 iterations in our example. Evidently, the ADPW method performs much better than any of the relaxed frame methods (about twice the gain of the frame method per iteration) even for such a sampling geometry.

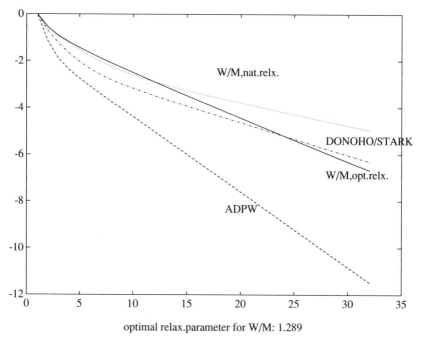

optimal relax.parameter for W/M: 1.289

FIGURE 8.9
Comparing Donoho/Stark, frames with opt.relax, and ADPW.

8.9.7   Relevance and choice of relaxation parameters

For different methods the choice of relaxation parameters plays a completely different role. In our experiments we have found that the effect of choosing relaxation parameters different from 1 does not bring any significant improvement to the Voronoi or PLN method on the average, although there are few exceptional cases. In contrast, for the WIL/MAR method a good choice of the good relaxation parameter very often has significant influence on the actual speed of convergence. Figure 8.10 displays the rate of convergence for the signal shown in Figure 8.1 for the frame method, using different relaxation parameters.

From the frame-theory point of view one should just take $\lambda = 2/(A + B)$, given the frame constants A and B. In practice, however, A and B cannot be obtained numerically from the sampling geometry and the filter itself, and even a good guess appears to be difficult by simple methods. For small problems it is possible to calculate the two values A and B as the largest and smallest eigenvalues of the matrix describing the positive and invertible frame operator. Of course, the computational complexity for this analysis is greater than that for solving the irregular sampling problem directly. Therefore this approach helps us with the analysis, but does not provide an efficient practical method. In general we can

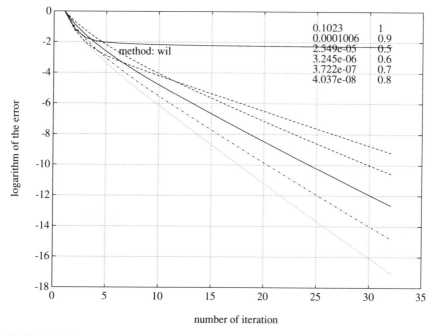

FIGURE 8.10
Convergence and error for different relaxation parameters.

only say that B tends to be large if there are accumulations of sampling points, and that for this reason λ has to be a small value, which in turn is responsible for slow convergence in this case.

On the other hand, in contrast to statements in the literature (cf. [MA]) we can give a simple recipe, derived from the ADPW approach, to give a fair rule of thumb for sampling sets with mild irregularities by setting the relaxation parameter to `n/lxp`. Although this is a reasonably good first choice as a relaxation parameter for the frame method there are two arguments why one should take a slightly smaller value (we suggest `0.8 * n/lxp`).

Argument (a): In general it is true that a small relaxation parameter improves stability of the method, at the cost of speed. For the frame method this means that the additional factor 0.8 prevents divergence for sampling geometries with more than just slight irregularities, while not sacrificing speed of convergence for sampling situations which are fairly regular. A smaller value would slow down for most cases and help to overcome situations which are somewhat more irregular.

Argument (b) is an immediate consequence of the two histograms (Figure 8.11) of optimal relaxation parameters for a large number of random geometries having a fixed maximal gap. We have carried out a number of similar experiments and have found that one can make the following general observations:

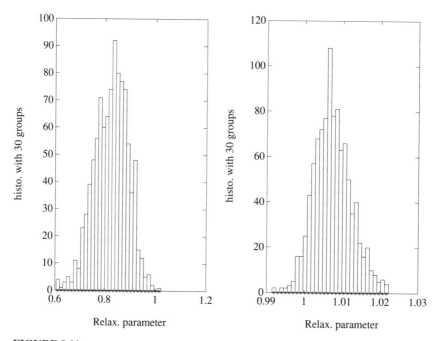

FIGURE 8.11
30-group histograms vs. relaxation parameter. Left, Wiley; right, ADPW.

whereas the variation of optimal relaxation parameters is relatively large for the ordinary frame method, (Wiley/Marvasti method) optimal relaxation parameters are very well concentrated near 1 in the case of the ADPW method. The first is due to the fact that it is likely that an accumulation of points occurs, which makes the upper frame constant large, and therefore requires a small relaxation parameter for optimal relaxation. In the case of ADPW the small weights for areas of high sampling density compensate this automatically.

The optimal relaxation parameter for ADPW is usually very close to one, often within 2 decimals. Therefore optimal adjustment of the relaxation parameter is already part of the method and there is no further need to obtain optimal relaxation through trial and error or other more sophisticated methods.

The histogram for unweighted frames shows another reason to make use of the additional factor 0.8 as relaxation parameter (thus $\lambda = 0.8 \times n/\mathtt{lxp}$ altogether): the expectation of obtaining a suboptimal relaxation parameter is highest in that case.

Warning

Improved frame estimates do not guarantee better convergence!   []

From a practical point of view we have to mention that a given recipe (how to calculate the optimal relaxation parameter from the frame bounds) only gives an optimal rate of convergence if the frame bounds are close to the optimal ones. It is *not* true in general that knowledge of better frame bounds *per se* gives a better relaxation parameter. As an example we consider a frame which is simply an ONS, hence has frame bounds $A = 1 = B$. Assume now we only know that $B \leq 1.5$ and $A = 1$. The general recipe would suggest using the relaxation parameter $2/(A + B) = 4/5$, which results in a convergence rate of 0.2. On the other hand, if we would only have had the estimate $A \geq 0.5$, we would have been lead to the choice $2/(A + B) = 1$, hence perfect reconstruction after one step. It is true, however, that an improvement of an estimate that is close to optimal will also result in an improved guess of the optimal relaxation parameter.

8.9.8 Theory and practice

One of the surprising observations made over and over again is the stunning efficiency of the Voronoi method in terms of iterations, in comparison to the PLN method. Whereas in theory (not unexpected), if we take sophistication as

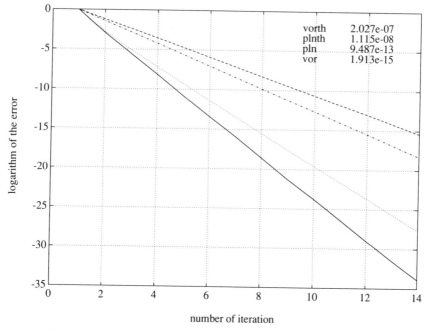

FIGURE 8.12
Comparison of theoretical and practical convergence.

a measure for the expected performance of a method, the PLN method should show better decay per iteration, we found that in practice the Voronoi method is astonishingly efficient, comparable to ADPW in this respect. Simple comparisons between theoretical and actual rates of convergence show that for PLN the theory comes quite close to the true decay. Since the present estimates just give the rate of decay (i.e., the steepness of the error curve in the logarithmic scale), one cannot expect to have sharp absolute values. On the other hand those results can be used to calculate the number of iterations needed to obtain a given precision. For the PLN method such a prediction will not be too far from the number of steps actually required in a given situation. For the VOR method the situation is quite different, and numerical error curves are substantially better than the predicted ones. Whether this is a statistical phenomenon (for a high percentage of cases this is true; few exceptional cases show the predicted rate) or a topic where theory has to be adjusted to actual observations is a question that has to be analyzed in the future.

8.10 Additional results and observations

8.10.1 Different implementations

For the comparisons presented we have used the standard implementations, using FFT to carry out convolution for all methods except the POCS method (which does not require the smoothing step). However, we have to mention that there are several different implementations for most of the methods, and which one (for the same mathematical algorithm) is the best it differs from case to case. We have developed certain heuristic rules for a good choice (depending on the signal length, spectral dimension, number of sampling points, ...). A detailed discussion of these questions requires a separate report. Typically in alternative implementations one replaces a single step of the iteration by a matrix multiplication. Depending on the size of that matrix (either number of spectral components or sampling points) and the speed of the FFT for the given signal length, one or the other implementation is better. Note that the signal may be determined by the problem (in the worst case it is a prime number); this prohibits the use of an efficient FFT.

8.10.2 2-D situations

Although we already have some experience with most of the 2-D versions of the methods described, we do not have enough space here to describe the special tricks required in order to implement those 2-D algorithms efficiently. For example, it is nontrivial to establish the Voronoi step functions, and even an efficient calculation of the area of the Voronoi domains of each of the sampling points of an irregular sampling set in the plane is not an obvious task. We will give a comprehensive discussion of the 2-D case elsewhere. We note that for 2-D signal reconstruction the advantages of the ADPW method over other methods are even greater than in the 1-D situation. In fact, sampling sets in the plane tend to be quite irregular. Furthermore, the number of variables and spectral points become quite large, so that it is a computationally intensive (if not impossible) task to build auxiliary functions (such as piecewise-linear ones over triangulations) or to describe the iterations using matrix representations of the underlying linear mapping. On the other hand, it was no problem to run the ADPW method in one of our 2-D examples (which is a 256×256 image) with about $25,000$ variables and $40,000$ sampling values. In fact, it is possible to carry out one iterative step within less than 1 minute on a Sun Sparc-station.

8.10.3 Performance, repeated reconstruction tasks

For questions of speed one has to take different situations into account. For real-time applications it might be necessary to obtain a fair partial reconstruction within a given time rather than a better one much later. In other practical situations a method will be considered "good" if it is good on the average, i.e., if the expected waiting time for a series of reconstructions is minimal among the available methods. The experimental evidence is overwhelming that among the considered methods the ADPW method is by far the most efficient reconstruction method.

There are also situations where different signals with the same spectrum are reconstructed from samples taken at a fixed family of sampling points. In that case preprocessing the information about the spectrum and the sampling geometry can be used to considerably shorten the reconstruction time needed for a series of signals. In principle this is possible by means of matrix methods. The basic idea is to make use of the fact that the operators involved are linear or at least affine mappings (for POCS). Therefore it suffices to run the algorithm first using appropriate unit vectors. If the problem under discussion are not too large direct matrix methods might be the best solution.

A typical application is the reconstruction of a 2-D band-limited signal with rectangular spectrum from a sampling set which arises as the product of sampling sets in each of the coordinates. Thus the sampling coordinates are of the form $(k_j, n_l)_{1 \leq j \leq j_0, 1 \leq l \leq l_0}$. Since the restriction of such a band-limited image to a single row (or column) is again band limited (the spectrum is just the projected rectangle on the corresponding axis), we can reconstruct the rows $(k_j)_{1 \leq j \leq j_0}$ from the given sampling values. Having done this we have to reconstruct (many!) columns (vertical sections of the image) from their values at the coordinates $(k_j)_{1 \leq j \leq j_0}$. Obviously all those sections have the *same* spectrum and have to be reconstructed from their sampling values over a *fixed* geometry. Since the number of column vectors is usually large compared to the number j_0 of sampling points in each of those columns, the application of the above-mentioned algorithms will be much more efficient than the application of the standard versions of any of those algorithms (not taking into account the special situation). We have successfully used such a scheme for the reconstruction of band-limited images of size 256×256, where a number of horizontal and vertical lines has been removed.

8.10.4   Signal reconstruction from local averages

We have tested the reconstruction of signals from averages as described in Theorem 7 for the case that averages are taken over a sequence of intervals which are not longer than the Nyquist rate (see Figure 8.13). It has been observed that the rate of convergence (per iteration) is approximately the same as that for the Voronoi method (using the same intervals). This is surprising since one might expect that the knowledge of average values of a function contains less information than that of exact sampling values. However, one has to admit that, at least for the natural implementation of the "reconstruction from averages" method, the computational effort is increased, because at each step the local averages for the new approximate signal have to be calculated (instead of sampling the signal).

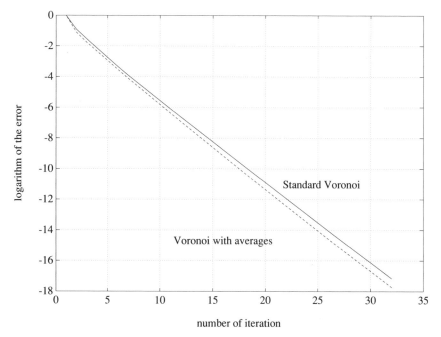

FIGURE 8.13
Convergence behavior of Voronoi method.

8.10.5 Further improvements, recent numerical experiments

We only mention that there are of course more advanced methods to improve the rate of convergence, e.g., by an adaptive choice of the value of a correction term at each step (variable over different iterations, contrary to the *fixed* rate used by the idea of the relaxation method). A typical representative of this approach is the steepest descent method. Combining the ADPW approach with the steepest descent method sometimes gives a significant improvement over unweighted steepest descent and ordinary ADPW iterations. Furthermore the idea of Chebyshev acceleration and in particular the application of conjugate gradient methods has turned out to be extremely efficient. Again it is often possible to reconstruct the signal from sampling values successfully, even if the Nyquist criterion is not satisfied. Also in this sense practical evidence shows that the performance of the ADPW method is even better than in theory.

8.11 Conclusion

Among the five methods discussed in this note only the ADPW method is both simple (low computational effort per step) and efficient, and at the same time fairly stable with respect to irregularities of the sampling set. The Voronoi method comes next: it is also stable, but at much higher computational expenses. Similarly, POCS is efficient for a small number of sampling points, but as in VOR and PLN the computational load per step increases with the number of sampling points. This is not the case for the ordinary frame method and the ADPW method. However, unweighted frames show strong sensitivity to irregularities in the sampling geometry, such as accumulations of sampling points. Furthermore, the WIL/MAR method often performs slowly if one does not choose the optimal relaxation parameter, which in turn cannot be obtained from the given data (sampling set and spectrum) in a simple way. In contrast, the optimal relaxation parameter for ADPW is in most cases very close to 1.

Bibliography

[B]   J.J. Benedetto. 1992. "Irregular sampling and frames," in *Wavelets: A Tutorial in Theory and Applications*, C.K. Chui, ed., Boston: Academic Press, pp. 445–508.

[BHe]   J. Benedetto and W. Heller. 1990. "Frames and irregular sampling," *Math. Note* **10**(Suppl. 1): 181–194.

[Be]   A. Beurling. 1963–1964. "Local harmonic analysis with some applications to differential operators." In *Some Recent Advances in the Basic Sciences*, vol. I, Belfer Grad. School Sci. Annu. Sci. Conf. Proc., A. Gelbart, ed., pp. 109–125,

[BeMal]   A. Beurling and P. Malliavin. 1967. "On the closure of characters and the zeros of entire functions," *Acta Math.* **118**: 79–95.

[Beu1]   F. Beutler. 1961. "Sampling theorems and bases in Hilbert space," *Inform. Contr.* **4**: 97–117.

[Beu2]   F.J. Beutler. 1966. "Error-free recovery of signals from irregularly spaced samples," *SIAM Rev.* **8**: 328–335.

[BuHin]   P.L. Butzer and L. Hinsen. 1989. "Two-dimensional nonuniform sampling expansions, I, II," *Appl. Anal.* **32**: 53–68, 69–85.

[BuSpSte]   P.L. Butzer, W. Splettstößer, and R. Stens. 1987. "The sampling theorem and linear prediction in signal analysis," *Jber. d. Dt. Math.-Verein.* **90**: 1–70.

[CFHer]   C. Cenker, H.G. Feichtinger, and M. Herrmann. 1991. "Iterative methods in irregular sampling: A first comparison of methods," in *Proc. 10th IPCCC*, March '91, Scottsdale, Ariz., New York: IEEE Comput. Soc. Press, pp. 483–489.

[CFStei]   C. Cenker, H.G. Feichtinger, and H. Steier. 1991. "Fast iterative and non-iterative reconstruction of band-limited functions from irregular sampling values," *Proc. Int. Conf. Acoust. Speech Signal Process. 1991*, Toronto, vol. 3, New York: IEEE Signal Process. Soc., pp. 1773–1776.

[CoR]   R.R. Coifman and R. Rochberg. 1980. "Representation theorems for holomorphic and harmonic functions in L^p," *Astérisque* **77**: 11–66.

[D]   I. Daubechies. 1990. "The wavelet transform, time-frequency localization and signal analysis," *IEEE Trans. Inf. Theory*, **36**: 961–1005.

[DGr]   I. Daubechies and A. Grossmann. 1985. "Frames in the Bargmann space of entire functions," *Commun. Pure Appl. Math.* **41**: 151–164.

[DGrMe]   I. Daubechies, A. Grossmann, and Y. Meyer. 1986. "Painless nonorthogonal expansions," *J. Math. Phys.* **27**: 1271–1283.

[DoSt]   D.L. Donoho and P.B. Stark. 1989. "Uncertainty principles and signal recovery," *SIAM J. Appl. Math.* **49**: 906–931.

[DuE]   R.J. Duffin and J.J. Eachus. 1942. "Some notes on an expansion theorem of Paley and Wiener," *Bull. Amer. Math. Soc.* **48**: 850–855.

[DuSc]   R. Duffin and A. Schaeffer. 1952. "A class of nonharmonic Fourier series," *Trans. Amer. Math. Soc.* **72**: 341–366.

[Fetal]   H.G. Feichtinger, C. Cenker, M. Mayer, H. Steier, and T. Strohmer. 1992. "New variants of the POCS method using affine subspaces of finite codimension, with applications to irregular sampling," in *SPIE92 Conf., Visual Communications and Image Processing*, Boston, Nov. 1992. pp. 299–310.

[FG1]   H.G. Feichtinger and K. Gröchenig. 1989. "Multidimensional irregular sampling of band-limited functions in L^p-spaces," in *Proc. Conf. Oberwolfach*, Feb. 1989, ISNM **90**, Boston: Birkhäuser, pp. 135–142.

[FG2]   H.G. Feichtinger and K. Gröchenig. 1992. "Iterative reconstruction of multivariate band-limited functions from irregular sampling values," *SIAM J. Math. Anal.* **23**: 244–261.

[FG3]   H.G. Feichtinger and K. Gröchenig. 1992. "Irregular sampling theorems and series expansions of band-limited functions," *J. Math. Anal. Appl.* **167**: 530–556.

[FG4]   H.G. Feichtinger and K. Gröchenig. 1992. "Error analysis in regular and irregular sampling theory," *Appl. Anal.*, to appear.

[FGHer]   H.G. Feichtinger, K. Gröchenig, and M. Hermann. 1990. "Iterative methods in irregular sampling theory. Numerical Results," in *7th Aachener Symposium für Signaltheorie*, ASST 1990, **253**, Aachen, Informatik Fachber, Berlin: Springer, pp. 160–166.

[FStr]   H.G. Feichtinger and T. Strohmer. 1992. "IRSATOL: Irregular sampling of band-limited signals toolbox," in *Conference on Computers for Teaching*, CIP-92, Berlin, Sept. 1992 (available through anonymous ftp from 131.130.22.36 (=tyche.mat.univie.ac.at)).

[FrJa]   M. Frazier and B. Jawerth. 1990. "A discrete transform and decomposition of distribution spaces," *J. Functional Anal.* **93**: 34–170.

[G1]   K. Gröchenig. 1991. "Describing functions: atomic decompositions versus frames," *Monatsh. Math.* **112**: 1–42.

[G2]   K. Gröchenig. 1992. "Reconstruction algorithms in irregular sampling," *Math. Comp.* **59**: 103–125.

[G3]   K. Gröchenig. 1993. "Irregular sampling of wavelet and short time Fourier transforms," *Constr. Approx.* **9**: 283–297 (special issue on wavelet theory).

[G4]   K. Gröchenig. 1993. "A discrete theory of irregular sampling," *Lin. Alg. Appl.*, to appear.

[GrMoPau] A. Grossmann, J. Morlet, and T. Paul. 1985. "Transforms associated to square integrable group representations I: general results," *J. Math. Phys.* **26**: 2473–2479.

[HYou]   L.A. Hageman and D.M. Young. 1981. *Applied Iterative Methods*, New York: Academic Press.

[HaLiPó]   G. Hardy, J.E. Littlewood, and G. Pólya. 1952. *Inequalities*, 2nd Ed., Cambridge: Cambridge Univ. Press.

[Hi1]   J.R. Higgins. 1976. "A sampling theorem for irregularly sampled points," *IEEE Trans. Inform. Theory*, Sept., pp. 621–622.

[Hi2]   J.R. Higgins. 1985. "Five short stories about the cardinal series," *Bull. Amer. Math. Soc.* **12**: 45–89.

[Hu]   N.E. Hurt. 1989. *Phase Retrieval and Zero Crossings: Mathematical methods in image reconstruction*, Dordrecht: Kluwer Academic Publ.

[J]   S. Jaffard. 1991. "A density criterion for frames of complex exponentials," *Michigan Math. J.* **38**: 339–348.

[Je]   A.J. Jerri. 1977. "The Shannon sampling theorem—Its various extensions and applications. A tutorial review," *Proc. IEEE* **65**: 1565–1596.

[K]   S. Kaczmarcz. 1972. "Angenäherte Auflösung von Systemen von linearen Gleichungen," *Bull. Acad. Polon. Sci. Lett.* **40**: 596–599.

[Ka]   M.I. Kadec. 1964. "The exact value of the Paley–Wiener constant," *Sov. Math. Dokl.* **5**: 559–561.

[KyTTo]   Ky Fan, O. Taussky, and J. Todd. 1955. "Discrete analogs of inequalities of Wirtinger," *Monatsh. Math.* **59**: 73–90.

[L]   H. Landau. 1967. "Necessary density conditions for sampling and interpolation of certain entire functions," *Acta Math.* **117**: 37–52.

[Le]   N. Levinson. 1940. *Gap and Density Theorems*, Coll. Publ. **26**, New York: Amer. Math. Soc.

[M1]   F.A. Marvasti. 1986. "Spectral analysis of random sampling and error free recovery by an iterative method," *Trans. Inst. Elect. Commun. Eng. Jpn* **E69**(2).

[M2]   F.A. Marvasti. 1987. "A unified approach to zero-crossing and nonuniform sampling of single and multidimensional systems," Nonuniform, P.O. Box 1505, Oak Park, IL 60304.

[M3]   F.A. Marvasti. 1989. "An iterative method to compensation for interpolation distortion," *IEEE Trans. Acoust. Speech Signal Process.* **37/1**, pp. 1617–1621.

[MA]   F. Marvasti and M. Analoui. 1989. "Recovery of signals from nonuniform samples using iterative methods," in *Proc. Int. Symp. Circuits Syst.*, May 1989, Portland, Ore.

[Ma]   M. Mayer. 1993. "Variants of the POCS method applied to irregular sampling," Thesis, Univ. Vienna.

[NWa]   M.Z. Nashed and G.G. Walter. 1991. "General sampling theorems for functions in reproducing kernel Hilbert spaces," *Math. Control Signals Syst.*, **4**: 373–412.

[Ne]   J. von Neumann. 1950. *Functional Operators*, vol. II, Ann. Math. Studies **22**, Princeton: Princeton Univ. Press.

[PWi]   R.E.A.C. Paley and N. Wiener. 1934. *Fourier Transforms in the Complex Domain*, Coll. Publ. **19**, Providence: Amer. Math. Soc.

[Pa]   A. Papoulis. 1966. "Error analysis in sampling theory," *Proc. IEEE* **54**/7: 947–955.

[SAl]   K.D. Sauer and J.P. Allebach. 1987. "Iterative reconstruction of band-limited images from nonuniformly spaced samples," *IEEE Trans.* **CAS-34**/2: 1497–1506.

[St]   H. Stark. 1987. *Image Recovery: Theory and Application*, San Diego: Academic Press.

[W]   W.J. Walker. 1987. "The separation of zeros for entire functions of exponential type," *J. Math. Anal. Appl.* **122**: 257–259.

[Wi]   R.G. Wiley. 1978. "Recovery of band-limited signals from unequally spaced samples," *IEEE Trans. Commun.* **COMM-26**: 135–137.

[YTh]   K. Yao and J.O. Thomas. 1967. "On some stability and interpolatory properties of nonuniform sampling expansions," *IEEE Trans. Circuit Theory* **14**: 404–408.

[YeSt]   S. Yeh and H. Stark. 1990. "Iterative and one-step reconstruction from nonuniform samples by convex projections," *J. Opt. Soc. Amer. A*, **7**: 491–499.

[YoWe]   D.C. Youla and H. Webb. 1982. "Image restoration by the method of convex projections: Part I," *IEEE Trans. Med. Imag.* **MI-1**: 81–94.

[You1]   R. Young. 1980. *An Introduction to Nonharmonic Fourier Series*, New York: Academic Press.

[You2]   R. Young. 1987. "On the stability of exponential bases in $L^2(-\pi, \pi)$," Proc. Amer. Math. Soc. **100**: 117–122.

9

Wavelets, probability, and statistics:
Some bridges

Christian Houdré

ABSTRACT   The rôle of some wavelet methods in probability and statistics is illustrated via a sample of three problems: We show how properties of processes can be read from properties of their wavelet transform. We discuss how the missing data problem can be approached via frames of complex exponentials. We explain how wavelets can be used to span classes of admissible estimators in nonparametric function estimation. It is also the purpose of this paper to show that bridges can be crossed in the other directions: Random products of matrices determine the smoothness of compactly supported wavelets. Nonstationary prediction theory gives new results on frames in Hilbert space.

CONTENTS

*This research was supported by the Office of Naval Research Contract No. N00014-91-J-1003 while the author was visiting the Department of Mathematics, University of Maryland, College Park, Md., as well as the Center for Computational Statistics, George Mason University, Fairfax, Va. This research has also been supported in part by the Air Force Office of Scientific Research Contract No. F49620-92-J-0154 while the author was visiting the Center for Stochastic Processes, Department of Statistics, University of North Carolina, Chapel Hill, N.C. This paper was also revised while the author was supported by an NSF Mathematical Sciences Post-Doctoral Fellowship.

0-8493-8271-8/94/$0.00 + $.50

9.1   The wavelet transform of second-order random processes

In this first section, we wish to understand the effects of the wavelet transform on the second-order properties of stochastic processes. The wavelet transform provides a time-scale decomposition of processes from which their properties can be read off. We show in this section how the second-order statistics of a process are characterized by the second-order statistics of its wavelet transform.

The results presented here are particular cases of results obtained in [CHou]. The starting point of [CHou] was the observation due to Flandrin [F] that (the nonstationary) fractional Brownian motion could be stationarized via the wavelet transform. Flandrin's observation was in turn taken up by Ramanathan and Zeitouni [RZ], who showed that in fact the form of the wavelet transform did characterize fractional Brownian motion. They also showed that no nontrivial Gaussian process could have a wavelet transform invariant with respect to translations and dilations.

Although nonstationary, fractional Brownian motion has *stationary increments* and in fact is the only mean square continuous, Gaussian process whose increments are stationary and self-similar. This property of the increments of fractional Brownian motion is a first crucial ingredient which stationarizes its wavelet transform. A second crucial ingredient is the existence of zero mean wavelets. It is shown here that as long as the wavelet has zero mean, the wavelet transform of a process is stationary if and only if the process has stationary increments. A similar result holds for self-similarity (of different orders). A combination of these two facts, with the characterizing property of fractional Brownian motion, solves the mystery.

Let (Ω, B, P) be a probability space and let $X = \{X(t, \omega); t \in \mathbb{R}, \omega \in \Omega\}$ be a random process, i.e., a jointly measurable map $X : \mathbb{R} \times \Omega \to \mathbb{C}$. Let E denote expectation, i.e., integration with respect to P; then X is a second-order process if $E|X(t)|^2 = \int_\Omega |X(t, \omega)|^2 dP(\omega) < +\infty$, for all $t \in \mathbb{R}$.

Given a process X, and a function $\psi : \mathbb{R} \to \mathbb{C}$, the continuous wavelet transform of X is the random field defined by

$$W = \left\{ W(t, a, \omega) = \frac{1}{\sqrt{a}} \int_\mathbb{R} X(u, \omega) \psi \left(\frac{t - u}{a} \right) du; \ t \in \mathbb{R}, \ a > 0, \ \omega \in \Omega \right\}$$
(9.1)

provided the above path integral exists with probability one. At a given scale $a > 0$, the wavelet transform is the stochastic process

$$W_a = \{W_a(t, \omega) = W(t, a, \omega); \ t \in \mathbb{R}, \ \omega \in \Omega\}.$$

Here, we are only interested in the second-order properties of the wavelet transform. We thus wish to impose conditions on the wavelet ψ and the process X to ensure that W has also finite second moments. Throughout, let X be a second-order process with covariance function R_X, i.e., $R_X(u, v) = EX(u)\overline{X}(v)$, for all $u, v \in \mathbb{R}$. Now ψ and X are said to satisfy *condition* (C) whenever:

(C) $(1 + t^2)^n \psi(t) \in L^1(\mathbb{R})$ for all $n \in \mathbb{Z}$, R_X is continuous, and
$(1 + t^2 + s^2)^{-N} R_X(t, s) \in L^\infty(\mathbb{R}^2)$ for some $N > 0$.

Condition (C) simply represents a nice workable sufficient condition under which

$$E\left\{ \int_{\mathbb{R}} \left| X(u)\psi\left(\frac{t-u}{a}\right) \right| du \right\}^2 \leq \left\{ \int_{\mathbb{R}} \{R_X(u,u)\}^{1/2} \left| \psi\left(\frac{t-u}{a}\right) \right| du \right\}^2 < +\infty.$$
(9.2)

In other words, the path integral defining W exists with probability one as a Lebesgue integral and W_a is a second-order process. When (9.2) is satisfied, then

$$R_W(t, a, s, b) = E W_a(t) \overline{W}_b(s)$$
(9.3)

$$= \frac{1}{\sqrt{ab}} \iint_{\mathbb{R}^2} R_X(u, v)\psi\left(\frac{t-u}{a}\right) \overline{\psi}\left(\frac{s-v}{b}\right) du\,dv.$$

Now that conditions ensuring the existence of the wavelet transform of second-order processes have been found, let us recall some properties of second-order processes. A process X with covariance function $EX(t)\overline{X}(s) = R_X(t, s)$ is called *weakly stationary* if

$$R_X(t, s) = R_X(t - s), \quad \forall\, t, s \in \mathbb{R}.$$
(9.4)

Two second-order processes X and Y with joint second moments $EX(t)\overline{Y}(s) = R_{X,Y}(t, s)$ are *cross weakly stationary* if

$$R_{X,Y}(t, s) = R_{X,Y}(t - s, 0), \quad \forall\, t, s \in \mathbb{R}.$$
(9.5)

Similarly, X has *weakly stationary increments* if the second moments of the increments

$$R_X(t_1, \tau_1; t_2, \tau_2) = E(X(t_1 + \tau_1) - X(t_1))(\overline{X}(t_2 + \tau_2) - \overline{X}(t_2))$$

depends on t_1, t_2 only through $t_1 - t_2$, i.e.,

$$R_X(t_1, \tau_1; t_2, \tau_2) = R_X(t_1 - t_2, \tau_1; 0, \tau_2), \quad \forall\, t_1, t_2, \tau_1, \tau_2 \in \mathbb{R}.$$
(9.6)

Finally X and Y have *cross weakly stationary increments* if

$$R_{X,Y}(t_1, \tau_1; t_2, \tau_2) = R_{X,Y}(t_1 - t_2, \tau_1; 0, \tau_2), \quad \forall\, t_1, t_2, \tau_1, \tau_2 \in \mathbb{R}.$$
(9.7)

The first result presented now relates the stationary properties of the increments of a process and of its wavelet transform, whenever the wavelet is "all pass", i.e., $\hat{\psi}(\lambda) \neq 0$, for all $\lambda \in \mathbb{R}$. In particular, this last condition prevents the wavelet from having any vanishing moment.

THEOREM 9.1

Let X and ψ satisfy condition (C). Let also $\hat{\psi}(\lambda) \neq 0$ for all $\lambda \in \mathbb{R}$. Then X is weakly stationary (resp. has weakly stationary increments) if and only if its *wavelet transforms W_a, W_b at any two scales $a, b > 0$ (with possibly $a = b$) are cross weakly stationary (resp. have cross weakly stationary increments).*

PROOF We only prove the result for the stationary case. The "stationary increments" part of the statement can be proved in a similar way. Of course, when X is stationary the polynomial growth condition on R_X is always satisfied.

If X is stationary, then $R_X(u, v) = R_X(u + T, v + T)$ for all $u, v, T \in \mathbb{R}$. Hence by (9.3), we have for any $a, b > 0$

$$R_{W_a, W_b}(t + T, s + T) = \frac{1}{\sqrt{ab}} \iint_{\mathbb{R}^2} R_X(u, v) \psi \left(\frac{t + T - u}{a} \right) \overline{\psi} \left(\frac{s + T - v}{b} \right) du dv$$

$$= \frac{1}{\sqrt{ab}} \iint_{\mathbb{R}^2} R_X(u + T, v + T) \psi \left(\frac{t - u}{a} \right) \overline{\psi} \left(\frac{s - v}{b} \right) du dv$$

$$= R_{W_a, W_b}(t, s) \quad \text{for all } t, s, T \in \mathbb{R}.$$

Hence W_a and W_b are cross weakly stationary. Conversely, if for any $a, b > 0$,

$$R_{W_a, W_b}(t + T, s + T) = R_{W_a, W_b}(t, s) \quad \text{for all } t, s, T,$$

then

$$\iint_{\mathbb{R}^2} \{R_X(u + T, v + T) - R_X(u, v)\} \psi \left(\frac{t - u}{a} \right) \overline{\psi} \left(\frac{s - v}{b} \right) du dv = 0.$$

Now, (essentially) by Wiener's Tauberian theorem, it follows since $\hat{\psi}(\lambda) \neq 0$, for all $\lambda \in \mathbb{R}$, that $R_X(u + T, v + T) = R_X(u, v)$, for all $u, v, T \in \mathbb{R}$. X is thus stationary. ∎

The proof of Theorem 9.1 is a simple consequence of Wiener's Tauberian theorem and so is a result on spans of translates. The condition $\hat{\psi}(\lambda) \neq 0$ for all $\lambda \in \mathbb{R}$ is essential as our next result shows. If for example $\hat{\psi}(0) = 0$, then the characterization given above becomes very different.

THEOREM 9.2
Let (C) be satisfied with also $\int_{\mathbb{R}} \psi(t) dt = 0 \neq \int_{\mathbb{R}} t \psi(t) dt$ and moreover $\hat{\psi}(\lambda) \neq 0$ for all $\lambda \neq 0$. Then X has weakly stationary increments if and only if its wavelet transforms W_a, W_b at any two scales $a, b > 0$ are cross weakly stationary.

PROOF First, let X have weakly stationary increments. From (9.3) we have

$$R_{W_a, W_b}(t, s) = \frac{1}{\sqrt{ab}} \iint_{\mathbb{R}^2} R_X(u, v) \psi \left(\frac{t - u}{a} \right) \overline{\psi} \left(\frac{s - v}{b} \right) du dv.$$

Hence for any $T, x, y \in \mathbb{R}$, by (9.6) and since $\int_{\mathbb{R}} \psi(u) du = 0$, it follows that

$$
R_{W_a, W_b}(t + T, s + T)
$$

$$
= \frac{1}{\sqrt{ab}} \iint_{\mathbb{R}^2} \{R_X(u + T, v + T) - R_X(x, v + T) - R_X(u + T, y) + R_X(x, y)\}
$$

$$
\psi \left(\frac{t - u}{a} \right) \overline{\psi} \left(\frac{s - v}{b} \right) du dv
$$

$$
= \frac{1}{\sqrt{ab}} \iint_{\mathbb{R}^2} E(X(u) - X(x - T))(\overline{X}(v) - \overline{X}(y - T)) \psi \left(\frac{t - u}{a} \right)
$$

$$
\overline{\psi} \left(\frac{s - v}{b} \right) du dv.
$$

But again $\int_{\mathbb{R}} \psi(u) du = 0$, hence $R_{W_a, W_b}(t + T, s + T) = R_{W_a, W_b}(t, s)$ for all t, s, T. In other words, W_a, W_b are cross weakly stationary. For the converse, let

$$
R_{W_a, W_b}(t + T, s + T) = R_{W_a, W_b}(t, s) \quad \text{for all } t, s, T \in \mathbb{R}.
$$

Then it again follows from (9.3) that

$$
\iint_{\mathbb{R}^2} \{R_X(u + T, v + T) - R_X(u, v)\} \psi \left(\frac{t - u}{a} \right) \overline{\psi} \left(\frac{s - v}{b} \right) du dv = 0, \tag{9.8}
$$

$$
\forall \, t, s, T \in \mathbb{R}.
$$

Now

$$
\int_{\mathbb{R}} \psi \left(\frac{t - u}{a} \right) \int_{\mathbb{R}} \{R_X(u + T, v + T) - R_X(u, v)\} \overline{\psi} \left(\frac{s - v}{b} \right) dv du = 0.
$$

Hence for any $s \in \mathbb{R}$ it (essentially) follows from Wiener's Tauberian theorem that if we denote by f_s the inner integral, the support of $\hat{f}_s$ is included in the zero set of $\hat{\psi}$. Hence,

$$
\hat{f}_s = \sum_{n=0}^{N_s} c_n(s) \delta_0^{(n)} \quad \text{and} \quad f_s(u) = \sum_{n=0}^{N_s} d_n(s) u^n \tag{9.9}
$$

for almost all u (Lebesgue). Now, for any $t, s \in \mathbb{R}$

$$
\int_{\mathbb{R}} \psi \left(\frac{t - u}{a} \right) f_s(u) du = 0.
$$

Using (9.9), this becomes

$$
\sum_{n=0}^{N_s} d_n(s) \sum_{m=0}^{n} \binom{n}{m} t^m (-1)^{n-m} F_{n-m} = 0
$$

where $F_k = \int_{\mathbb{R}} t^k \psi(t) dt$. Since $F_0 = 0$ and $F_1 \neq 0$ we conclude that $f_s(u) = d_0(s)$ for almost all $u \in \mathbb{R}$.

Thus, for each s we have

$$\int_{\mathbb{R}} \{R_X(u+T, v+T) - R_X(u, v)\} \overline{\psi} \left(\frac{s-v}{b} \right) dv = d_0(s)$$

From which we conclude that for some u_0 and almost all u (Lebesgue)

$$\int_{\mathbb{R}} \{R_X(u+T, v+T) - R_X(u, v) - R_X(u_0+T, v+T) + R_X(u_0, v)\}$$
$$\overline{\psi} \left(\frac{s-v}{b} \right) dv = 0.$$

An argument similar to the one used for f_s as well as the continuity and antisymmetry of R_X finally gives

$$R_X(u+T, v+T) - R_X(u, v) = h_1(u) + \overline{h}_1(v) \tag{9.10}$$

for some function h; all u, v and T. To conclude, putting $u = u_0, v = v_0$ in (9.10) we get

$$E(X(u+T) - X(u_0+T))(\overline{X}(v+T) - \overline{X}(v_0+T)) = E(X(u) - X(u_0))(\overline{X}(v) - \overline{X}(v_0))$$

for all $u, v, T \in \mathbb{R}$, i.e., X has weakly stationary increments. ∎

Stationarity corresponds to translation invariance. We now wish to tackle the scale parameter in the wavelet transform and to study scale invariance, i.e., self-similarity.

A second-order random field $X = \{X(t); t \in \mathbb{R}^n\}$, $n \geq 1$, is called weakly *H-self-similar* if

$$EX(ct)\overline{X}(cs) = c^{2H} EX(t)\overline{X}(s), \quad \forall t, s \in \mathbb{R}^n, \quad \forall c > 0. \tag{9.11}$$

When X is replaced by its increments in (9.11), the process has *weakly H-self-similar increments*. For all-pass and zero mean wavelets, the results below are again quite different. First an unsurprising result whose proof is similar to the proof of Theorem 9.1 and is omitted.

THEOREM 9.3
Let X and ψ satisfy condition (C) with moreover $\hat{\psi}(\lambda) \neq 0$, for all $\lambda \in \mathbb{R}$. Then X is weakly H-self-similar (resp. has weakly H-self-similar increments) if and only if its wavelet transform field W is weakly $(H + 1/2)$-self-similar (resp. has weakly $(H + 1/2)$-self-similar increments).

When the wavelet has zero mean, properties of the increments do again translate to properties of the process.

THEOREM 9.4
Let (C) be satisfied and let also $\int_{\mathbb{R}} \psi(t)dt = 0 \neq \int_{\mathbb{R}} t\psi(t)dt$ with moreover $\hat{\psi}(\lambda) \neq 0$ for all $\lambda \neq 0$. Then, X has weakly H-self-similar increments if and only if its wavelet transform field W is weakly $(H + 1/2)$-self-similar.

PROOF (SKETCH) Since $\int_{\mathbb{R}} \psi(t)dt = 0$, we have

$$EW(ct, ca)\overline{W}(cs, cb)$$

$$= \frac{c}{\sqrt{ab}} \iint_{\mathbb{R}^2} R_X(cu, cv)\psi\left(\frac{t-u}{a}\right)\overline{\psi}\left(\frac{s-v}{b}\right) dudv$$

$$= \frac{c}{\sqrt{ab}} \iint_{\mathbb{R}^2} \{R_X(cu, cv) - R_X(cu, cv_0) - R_X(cu_0, cv) + R_X(cu_0, cv_0)\}$$

$$\psi\left(\frac{t-u}{a}\right)\overline{\psi}\left(\frac{s-v}{b}\right) dudv. \tag{9.12}$$

Now since X has weakly H-self-similar increments, (9.12) is equal to

$$\frac{c^{1+2H}}{\sqrt{ab}} \iint_{\mathbb{R}^2} \{R_X(u, v) - R_X(u, v_0) - R_X(u_0, v) + R_X(u_0, v_0)\}$$

$$\psi\left(\frac{t-u}{a}\right)\overline{\psi}\left(\frac{s-v}{b}\right) dudv$$

$$= \frac{c^{1+2H}}{\sqrt{ab}} \int_{\mathbb{R}^2} \int R_X(u, v)\psi\left(\frac{t-u}{a}\right)\overline{\psi}\left(\frac{s-v}{b}\right) dudv$$

$$= c^{1+2H}EW(t, a)\overline{W}(s, b).$$

Conversely, if W is weakly $(H + 1/2)$-self-similar, then proceeding as in the proof of Theorem 9.2, we get that

$$f(u, v) = E(X(cu) - X(cu_0))(\overline{X}(cv) - \overline{X}(cv_0))$$
$$- c^{2H} E(X(u) - X(u_0))(\overline{X}(v) - \overline{X}(v_0))$$
$$= g(u) + h(v), \quad \text{for all } u, v \in \mathbb{R}, \ c > 0.$$

Since $f(u_0, v) = f(u, v_0) = f(u_0, v_0) = 0$ it follows that $f(u, v) = 0$, i.e., X has weakly self-similar increments. ∎

To finish this section, we specialized Theorems 9.2 and 9.4 to the Gaussian case and thus recover the results in [F] and [RZ].

COROLLARY 9.5

Let (C) be satisfied, let $\int_{\mathbb{R}} \psi(t)dt = 0 \neq \int_{\mathbb{R}} t\psi(t)dt$, and $\hat{\psi}(\lambda) \neq 0$ for all $\lambda \neq 0$. Let also X be Gaussian with $X(0) = 0$. Then X is fractional Brownian motion of order H if and only if at any scale $a > 0$, W_a is weakly stationary and W is weakly $H + 1/2$-self-similar.

PROOF The only mean square continuous Gaussian process with self-similar and stationary increments is fractional Brownian motion (see Mandelbrot and Van Ness [MaV]). Since $X(0) = 0$, self-similarity of the process and of its increments are equivalent. Apply Theorems 9.2 and 9.4. ∎

Wavelets can have vanishing moments of arbitrary order (Daubechies [D], Meyer [Me]). When instead of having a nonvanishing first moment we have

$$\int_{\mathbb{R}} \psi(t)dt = \int_{\mathbb{R}} t\psi(t)dt = 0 \neq \int_{\mathbb{R}} t^2 \psi(t)dt$$

with again $\hat{\psi}(\lambda) \neq 0$ for all $\lambda \neq 0$. Then Theorem 9.2 becomes (see [CHou] for more details):

THEOREM 9.6

X has weakly stationary (resp. H-self-similar) increments of order 2 if and only if its wavelet transforms W_a, W_b at scales $a, b > 0$ are cross weakly stationary (resp. W is weakly $(H + 1/2)$-self-similar).

Only the continuous wavelet transform has been considered in this section. Results on the discrete wavelet transform of second-order processes and its connections with multiresolution analysis (Mallat [M], Meyer [Me]) are presented in [Hou4].

9.2   Random matrices and dilation equations

At the outset of the section, we must say that none of the results presented below should be attributed in any way or form to the author. Our task in this section is not to present new results, but rather to draw bridges between dilation equations and a part of the probability literature.

It has been shown by Daubechies and Lagarias [DLa1], [DLa2] that the smoothness of compactly supported wavelets is determined by the "arbitrary" products of two matrices. To date, the methods developed to understand these products have much to do with linear algebra, e.g., the notion of joint spectral radius. We wish below to complement this approach with a probabilistic point of view. By replacing "arbitrary" by "random", much becomes known on the asymptotics of products of matrices. This will be explained below. As with the other sections we try to be rather self-contained. Thus, we start by recalling some elements of Daubechies' construction of compactly supported wavelets and explaining how products of matrices enter the picture. The article by Heil and Colella [HCo] in this volume should be consulted for further details. Strang [St] also presents a nice review of dilation equations.

Let $\{\psi_{j,k}\}_{j,k \in \mathbb{Z}}$ be an orthonormal basis of $L^2(\mathbb{R})$ obtained by translating and dilating a single function ψ, namely, $\psi_{j,k}(t) = 2^{-j/2}\psi(2^{-j}t - k)$. Such a family $\{\psi_{j,k}\}_{j,k \in \mathbb{Z}}$ is called a *wavelet basis* of $L^2(\mathbb{R})$ and ψ is the *mother wavelet*. Since the $\psi_{j,k}$ are obtained from a single ψ, it is clear that the characteristics (continuity, smoothness, decay, etc.) of ψ determine the characteristics of the elements of the basis. Wavelets can be obtained via

$$\psi(t) = \sum_k (-1)^k c_{1-k} \phi(2t - k), \tag{9.13}$$

where the $\{c_k\}_{k \in \mathbb{Z}}$ are square summable and where ϕ is a *scaling function*. The function ϕ is itself the solution of a *dilation equation*

$$\phi(t) = \sum_k c_k \phi(2t - k). \tag{9.14}$$

Since we are interested in compactly supported scaling functions and hence compactly supported mother wavelets, the sums in (9.13) and (9.14) are taken to be finite, i.e., k varies from 0 to N.

Let us now explain a way of solving equation (9.14). First, we are looking for a solution which is normalized, say $\int_{\mathbb{R}} \phi(t)dt = 1$. Then by taking Fourier transform (in the sense of distribution) in (9.14) (as motivated by the convolution form of the equation), it follows that

$$\hat{\phi}(\lambda) = \int_{\mathbb{R}} e^{-it\lambda} \sum_{k=0}^N c_k \phi(2t - k) dt$$

$$= \sum_{k=0}^N \frac{c_k}{2} \int_{\mathbb{R}} e^{-i(t+k/2)\lambda} \phi(t) dt$$

$$= \sum_{k=0}^N \frac{c_k}{2} e^{-ik\lambda/2} \hat{\phi}\left(\frac{\lambda}{2}\right)$$

$$= m_0\left(\frac{\lambda}{2}\right) \hat{\phi}\left(\frac{\lambda}{2}\right) \tag{9.15}$$

where $m_0(\lambda) = 1/2 \sum_{k=0}^N c_k e^{-ik\lambda}$.

Now we iterate (9.15) and get

$$\hat{\phi}(\lambda) = m_0\left(\frac{\lambda}{2}\right) m_0\left(\frac{\lambda}{4}\right) \hat{\phi}\left(\frac{\lambda}{4}\right). \tag{9.16}$$

In turn (9.16) can be iterated and so on .... Since $\int_{\mathbb{R}} \phi(t)dt = \hat{\phi}(0) = 1$, the iterative process leads to an infinite (convergent) product and

$$\hat{\phi}(\lambda) = \prod_{n=1}^{\infty} m_0\left(\frac{\lambda}{2^n}\right). \tag{9.17}$$

Although (9.17) exists, nothing in the construction guarantees that ϕ (hence ψ) is, say, continuous or differentiable.

To obtain smooth solutions, the dilation equation (9.14) is first transformed into an equivalent matrix problem: Let ϕ be a compactly supported (in $[0, N]$), continuous solution to (9.14). Let also $V : [0,1] \to \mathbb{C}^N$ be defined via

$$V(t) = \begin{pmatrix} \phi(t) \\ \phi(t+1) \\ \vdots \\ \phi(t+N-1) \end{pmatrix}. \tag{9.18}$$

Now since ϕ has compact support, it follows from (9.14) that

$$V(t) = T_0 V(2t), \quad 0 \le t \le \frac{1}{2}$$

and

$$V(t) = T_1 V(2t-1), \quad \frac{1}{2} \le t \le 1, \tag{9.19}$$

where T_0 and T_1 are $\mathbb{C}^{N \times N}$-matrices given by $(T_0)_{n,k} = c_{2n-k-1}$ and $(T_1)_{n,k} = c_{2n-k}$. So given a continuous compactly supported ϕ satisfying (9.14), there is a continuous V satisfying

$$V(t) = T_{d_1} V(\tau t), \tag{9.20}$$

where $t = .d_1 d_2 d_3 \ldots$ is a binary expansion and where $\tau t = .d_2 d_3 d_4 \ldots$. Moreover (see for example [HCo]) the converse also holds. Now by induction (9.20) becomes

$$V(t) = T_{d_1} T_{d_2} \ldots T_{d_m} V(\tau^m t) \tag{9.21}$$

where again $t = .d_1 d_2 d_3 \ldots$ and $\tau^m t = .d_{m+1} d_{m+2} \ldots$. We are now in business since continuity and smoothness can be quantified as "closeness" in binary expansions. So if $2^{-m-1} \le t - s \le 2^{-m}$, then

$$V(t) - V(s) = T_{d_1} T_{d_2} \ldots T_{d_m} (V(\tau^m t) - V(0)), \tag{9.22}$$

where $t = .d_1 \ldots d_m d_{m+1} \ldots d_{m+k}$, $s = .d_1 \ldots d_m$. Now continuity means finer and finer dyadic approximation, hence the need to study products $T_{d_1} \cdots T_{d_m}$ of arbitrary length and order.

A key word in understanding the above products is "random matrices". At any position in the product, either T_0 or T_1 can be present (of course with equal probability 1/2). We can thus model these independent equally likely choices of T_0 and T_1 with a sequence of matrix-valued, independent, identically distributed (i.i.d.) Bernoulli random variables $\{X_i\}_{i=1}^{\infty}$. Namely, let $\{X_i\}_{i=1}^{\infty}$ be a sequence of mutually independent measurable functions on some probability space (Ω, B, P), $X_i : \Omega \to \mathbb{C}^{N \times N}$, such that

$$P\{X_i = T_0\} = P\{X_i = T_1\} = \frac{1}{2}.$$

With this model in hand, it is clear that studying the asymptotics of the products $T_{d_1} \cdots T_{d_m}$ is equivalent to studying the asymptotics of the products $X_1 \cdots X_m$. The basic result on matrix products of this type is due to Furstenberg and Kesten [FuKe]. A specialized (to our framework) version of their result is stated below ($\| \cdot \|_0$ is the operator norm of the corresponding matrix).

THEOREM 9.7

Let $\{X_i\}_{i=1}^{\infty}$ be i.i.d. Bernoulli random variables taking for values the invertible matrices T_0, T_1. Then, with probability one,

$$\lim_{n \to \infty} \frac{1}{n} \log \|X_1 X_2 \ldots X_n\|_0 = \Lambda$$

where Λ (the maximal Lyapunov exponent) of $\{X_i\}_{i=1}^{\infty}$ is almost surely constant.

Stated differently, Theorem 9.7 says that for almost all orderings of T_0 and T_1's the products do grow *exponentially*. Of course, Theorem 9.7 does not give any information about the convergence of a particular product since it is a global quantitative result.

Much more is known on products of random matrices, and their connections to ergodic theory (e.g., the Oseledec multiplicative ergodic theorem). We refer to Furstenberg [Fu], Le Page [Le], Slud [S], as well as Goldsheid and Guivarc'h [GoGu] for more details and recent results. We do not pursue here a detailed discussion of Theorem 9.7 in relation to dilation equations. This is left for future work and we just close this section hoping that we have shown that bridges can be crossed in either direction.

9.3 Frames, nonstationary processes and prediction

In connection with wavelet theory, the notion of frame introduced by Duffin and Schaeffer [DuSc] has also experienced renewed interest (see Meyer [Me], Heil and Walnut [HWa]). Except for the original definition, few ways are actually available to characterize them. We wish in this section to slightly remedy this situation and to provide some identification criteria (more or less practical). The basic ideas involved in proving these criteria are borrowed from the study of some classes of nonstationary second-order processes.

Let us first establish some notations and recall some definitions which will be used throughout the rest of this paper. H is always a separable complex Hilbert space with inner product $\langle \cdot, \cdot \rangle$ and norm $\| \cdot \|$. H is equipped with the strong topology (and in the statements below, $\sum_{-\infty}^{\infty} = \lim \sum_{-N}^{N}$). $\mathcal{B}(\Pi), \mathcal{B}(\Pi^2)$ are respectively the Borel σ-algebra on the circle $\Pi =]-\pi, \pi]$ and the torus Π^2. A *Hilbertian measure* Z is a σ-additive set function $Z : \mathcal{B}(\Pi) \to H$; it is called *orthogonally scattered* whenever $\langle Z(F), Z(G) \rangle = 0$, $F, G \in \mathcal{B}(\Pi)$, $F \cap G = \phi$. We need below to integrate functions with respect to Hilbertian measures. This is taken in the sense of Bartle, Dunford and Schwartz (Dunford and Schwartz

[DunSch, IV]). In particular, Borel bounded functions are integrable in that sense (see [DunSch] for more details).

Given a Hilbertian measure Z, the function β defined on the Cartesian product $\mathcal{B}(\Pi) \times \mathcal{B}(\Pi)$ via $\beta(\cdot, \cdot) = \langle Z(\cdot), Z(\cdot) \rangle$ is separately σ-additive, i.e., β is a *bimeasure* and in general does not extend to a measure on $\mathcal{B}(\Pi^2)$. However, when β takes nonnegative values it does extend to a measure on $\mathcal{B}(\Pi^2)$ (Horowitz [Ho]). In particular, when Z is orthogonally scattered, $\langle Z(A), Z(B) \rangle = \|Z(A \cap B)\|^2 = \lambda(A \cap B)$ defines a bounded positive measure concentrated on the diagonal of Π^2. Although β is not of bounded variation, a (nonabsolute) integral with respect to a bimeasure β can be defined. We refer to [Hou2] for more details on this restricted Morse–Transue integral. Let us just mention that whenever a bimeasure is induced by a Hilbertian measure, i.e., whenever $\beta(\cdot, \cdot) = \langle Z(\cdot), Z(\cdot) \rangle$, and whenever the functions f and g are Z-integrable, the pair (f, g) is also β-integrable with

$$\left\langle \int_{-\pi}^{\pi} f(\theta) dZ(\theta), \int_{-\pi}^{\pi} g(\psi) dZ(\psi) \right\rangle = \int_{-\pi}^{\pi} \int_{-\pi}^{\pi} f(\theta) \overline{g}(\psi) d\beta(\theta, \psi).$$

Now, recall (see [DuSc]) that a sequence $\{x_n\}_{n \in \mathbb{Z}} \subset H$ is a *frame* if there exist $A, B > 0$ (the frame constants) such that

$$A\|x\|^2 \leq \sum_{-\infty}^{\infty} |\langle x, x_n \rangle|^2 \leq B\|x\|^2, \quad \forall x \in H. \tag{9.23}$$

The frame is *tight* whenever $A = B$ and is *exact* if it fails to remain a frame by the deletion of any one of its elements.

Equation (9.23) is made of two inequalities of very different nature. We first deal with boundedness, i.e., the right hand side inequality. Sequences satisfying such a condition are also known as *Bessel sequences* (see Young [Y]). With the help of this preliminary material we can now present our first result.

THEOREM 9.8
The following are equivalent:

  i. $\sum_{-\infty}^{\infty} |\langle x, x_n \rangle|^2 \leq B\|x\|^2, \forall x \in H.$

  ii. $\|\sum_{-\infty}^{\infty} \alpha_n x_n\|^2 \leq B \sum_{-\infty}^{\infty} |\alpha_n|^2, \forall \{\alpha_n\}_{n \in \mathbb{Z}} \in \ell^2(\mathbb{Z}) = \{\{\alpha_n\}_{n \in \mathbb{Z}} : \sum_{-\infty}^{\infty} |\alpha_n|^2 < +\infty\}.$

  iii. *There exists a (unique) Hilbertian measure Z such that $x_n = \int_{-\pi}^{\pi} e^{in\theta} dZ(\theta)$, $n \in \mathbb{Z}$ and $\|\int_{-\pi}^{\pi} f(\theta) dZ(\theta)\|^2 \leq B \int_{-\pi}^{\pi} |f(\theta)|^2 d\theta, \forall f \in L^2(\Pi) = \{f :] - \pi, \pi] \to \mathbb{C} : \int_{-\pi}^{\pi} |f(\theta)|^2 d\theta < +\infty\}.$*

  iv. *There exists a (unique positive definite) bimeasure β such that $\langle x_n, x_m \rangle = \int_{-\pi}^{\pi} \int_{-\pi}^{\pi} e^{in\theta} e^{-im\psi} d\beta(\theta, \psi), n, m \in \mathbb{Z}$ and*

$$\int_{-\pi}^{\pi} \int_{-\pi}^{\pi} f(\theta) \overline{f}(\psi) d\beta(\theta, \psi) \leq B \int_{-\pi}^{\pi} |f(\theta)|^2 d\theta, \quad \forall f \in L^2(\Pi).$$

v. *There exists a Hilbert space $K \supset H$ and an orthogonal sequence $\{y_n\}_{n \in \mathbb{Z}} \subset K$ with $\|y_n\|_K^2 = B$ such that $x_n = Qy_n$, where Q is the orthogonal projection from K onto H.*

PROOF The above equivalences are not new. The first one is very well known (see, e.g., [Y]). The others are included in greater generality in [Hou1] where the reader will find precise references and further details. Briefly, (iii) follows from the fact that $T : \sum_{-N}^{N} \alpha_n e^{in\theta} \rightarrow \sum_{-N}^{N} \alpha_n x_n$ extends to a bounded linear operator from $C(\Pi)$ (the continuous functions on Π) to H, hence can be represented via a Hilbertian measure. In turn, (iii) clearly implies (iv). It is also clear that

$$\left\| \sum_{-N}^{N} \alpha_n Q y_n \right\|_K^2 \leq \|P\|^2 \left\| \sum_{-N}^{N} \alpha_n y_n \right\|^2 \leq \sum_{-N}^{N} |\alpha_n|^2 \|y_n\|^2 = B \sum_{-N}^{N} |\alpha_n|^2$$

and (v) $\Rightarrow$ (ii). To conclude let us indicate the proof of (iv) $\Rightarrow$ (v). First, $[f,g] = B \int_{-\pi}^{\pi} f(\theta) \overline{g}(\theta) d\theta - \int_{-\pi}^{\pi} \int_{-\pi}^{\pi} f(\theta) \overline{g}(\psi) d\beta(\theta, \psi)$ defines a semi-inner product on $C(\Pi)$. Taking equivalent classes and completion $(C(\Pi), [\cdot, \cdot])$ becomes a Hilbert space H_1. Now, let $K = H \oplus H_1$, let $\{w_n\}_{n \in \mathbb{Z}} \subset H_1$ be (the image of) $\{e^{in \cdot}\}_{n \in \mathbb{Z}}$ and let $y_n = x_n + w_n$. It is clear that $x_n = Qy_n$; furthermore

$$\langle y_n, y_m \rangle_K = \langle x_n, x_m \rangle + [w_n, w_m] = \int_{-\pi}^{\pi} \int_{-\pi}^{\pi} e^{in\theta} e^{-im\psi} d\beta(\theta, \psi) + B \int_{-\pi}^{\pi} e^{i(n-m)\theta} d\theta$$

$$- \int_{-\pi}^{\pi} \int_{-\pi}^{\pi} e^{in\theta} e^{-im\psi} d\beta(\theta, \psi)$$

$$= B \delta_{n,m},$$

where $\delta_{n,m}$ is Kronecker's symbol. ∎

REMARK 9.9 In (iv), the bimeasure β is necessarily positive definite (p.d.), namely, $\sum_{i=1}^{N} \sum_{j=1}^{N} a_i \overline{a}_j \beta(A_i, A_j) \geq 0$, for any $a_1, \ldots, a_N \in \mathbb{C}, A_1, A_2, \ldots, A_N \in \mathcal{B}(\Pi)$. Fourier coefficients of bimeasures are not pathological objects but rather typical: let the sequence $\{x_n\}_{n \in \mathbb{Z}} \in H$ be such that $\langle x_n, x_n \rangle = a_n \delta_{n,m}, 0 \leq a_n \leq C$, (throughout C is an absolute constant whose value might change from one expression to another) with $\lim_{n \to \infty} a_n \neq \lim_{n \to -\infty} a_n$. Then $\langle x_n, x_n \rangle = \hat{\beta}(n, m)$, but β is *not* a measure on Π^2. The statement (iv) is of interest since it allows a transfer of properties from Bessel sequences to exponentials in $L^2(\beta) = \{f : \Pi \rightarrow \mathbb{C} : \int_{-\pi}^{\pi} \int_{-\pi}^{\pi} f(\theta) \overline{f}(\psi) d\beta(\theta, \psi) < +\infty\}$. ∎

A way of paraphrasing the condition (ii) is to say that the doubly infinite Gramian matrix $G = (G_{n,m} = \langle x_n, x_m \rangle)_{n,m \in \mathbb{Z}}$ is a positive bounded linear operator on $\ell^2(\mathbb{Z})$. We wish now to slightly improve on (ii) by presenting an equivalent spectral radius type condition.

PROPOSITION 9.10
Let $G = (G_{n,m} = \langle x_n, x_m \rangle)_{n,m \in \mathbb{Z}}$. Then the following are equivalent:

    i. $\{x_n\}_{n\in\mathbb{Z}}$ *is a Bessel sequence*

    ii. $((G^{2^p})_{n,n})^{1/2^p} \leq B, \forall p, n \in \mathbb{Z}, p \geq 0.$

PROOF First (see [Y, IV]), G is a bounded linear operator on $\ell^2(\mathbb{Z})$ if and only if its sections $G_N = (G_{n,m})_{-N \leq n,m \leq N}$ are uniformly bounded, i.e., $\|G_N\|_0 \leq B$, where $\|\cdot\|_0$ is the operator norm. So let $\{x_n\}_{n\in\mathbb{Z}}$ be a Bessel sequence, i.e., let G be bounded. Then, for any positive integers N and p, since the matrices G_N are positive semi-definite, we have $((G^{2^p})_{N,N})^{1/2^p} \leq \|G_N^{2^p}\|_0^{1/2^p} = \|G_N\|_0 \leq B$. On the other hand, since again the G_N are $N \times N$ positive semi-definite matrices, we have

$$\|G_N\|_0 = \|G_N^{2^p}\|_0^{1/2^p} \leq N^{1/2^p} \sup_{-N \leq n \leq N} \{(G^{2^p})_{n,n}\}^{1/2^p} \leq N^{1/2^p} B.$$

Letting $p \to +\infty$ gives the rest of the result. ∎

By a repeated application of Hölder's inequality, it follows that $\{(G^{2^p})_{n,n}\}^{1/2^p} \leq \sum_{k=-\infty}^{\infty} |\langle x_n, x_k \rangle|$ and Proposition 9.10 recovers the so-called Schur's Lemma (see [Y, IV]).

We wish now to investigate the left hand side inequality in (9.23). First, it is clear taking (for example) an incomplete orthonormal sequence that the left hand side of (9.23) cannot be replaced by:

$$A \sum_{-\infty}^{\infty} |\alpha_n|^2 \leq \left\| \sum_{-\infty}^{\infty} \alpha_n x_n \right\|^2. \tag{9.24}$$

Hence frames are not necessarily *Riesz–Fischer* sequences or vice versa. However, when the sequence is complete more can be said. Before doing so, we recall that a sequence $\{x_n\}_{n\in\mathbb{Z}}$ in H is *stationary* whenever $\langle x_n, x_m \rangle = \langle x_{n+1}, x_{m+1} \rangle$, for all $n, m \in \mathbb{Z}$. Hence $\langle x_n, x_m \rangle = \langle x_{n-m}, x_0 \rangle$, and by Herglotz's theorem $\langle x_n, x_m \rangle = \int_{-\pi}^{\pi} e^{i(n-m)\theta} d\nu(\theta)$, for a (unique) positive finite measure ν on Π (ν is the so-called *spectrum* of the sequence $x = \{x_n\}_{n\in\mathbb{Z}}$). Equivalently, $x_n = \int_{-\pi}^{\pi} e^{in\theta} dZ(\theta), n \in \mathbb{Z}$, where the Hilbertian measure Z is orthogonally scattered.

THEOREM 9.11
For a complete sequence $\{x_n\}_{n\in\mathbb{Z}}$, *the following are equivalent.*

    i. $\{x_n\}_{n\in\mathbb{Z}}$ *is an exact frame with bounds A and B.*

    ii.

$$A \sum_{-\infty}^{\infty} |\alpha_n|^2 \leq \left\| \sum_{-\infty}^{\infty} \alpha_n x_n \right\|^2 \leq B \sum_{-\infty}^{\infty} |\alpha_n|^2, \quad \forall \alpha = \{\alpha_n\} \in \ell^2(\mathbb{Z}). \tag{9.25}$$

    iii. *There exist an orthonormal sequence* $\{e_n\}_{n\in\mathbb{Z}}$ *and a bounded invertible linear operator* $T : H \to H$ *such that* $x_n = Te_n, n \in \mathbb{Z}$.

    iv. *There exist a stationary sequence* $\{y_n\}_{n\in\mathbb{Z}}$, *whose spectrum and its inverse are essentially bounded, i.e.,* $d\nu(\theta) = f(\theta)d\theta, f, f^{-1} \in L^\infty(\Pi)$, *and a bounded invertible linear operator* $T : H \to H$ *such that* $x_n = Ty_n, n \in \mathbb{Z}$.

v. There exist two bounded invertible linear operators T and S such that $\sup_{n \in \mathbb{Z}} \|S^n\|_0 < +\infty$; $x_n = S^n x_0$, $n \in \mathbb{Z}$; and $\{Tx_n\}_{n \in \mathbb{Z}}$ is a stationary sequence whose spectrum and its inverse are essentially bounded.

PROOF   The equivalence of (i), (ii), and (iii) is again well-known and present in Young [Y]. Since an orthonormal sequence is a stationary sequence with constant spectrum, i.e., $d\nu(\theta) = d\theta$, (iii) clearly implies (iv). Let us now assume that (iv) holds. Then, since $\{y_n\}_{n \in \mathbb{Z}}$ is stationary, its shift is a unitary operator on H ($\{y_n\}$ is complete). Hence,

$$x_n = Ty_n = TU^n y_0 = TU^n T^{-1} Ty_0 = TU^n T^{-1} x_0.$$

Now, $S = TUT^{-1}$ and $y_n = T^{-1} x_n$ satisfy the requirements of (v). Finally, let $y_n = Tx_n$ be stationary with spectrum and inverse in $L^\infty(\Pi)$. Then,

$$\frac{1}{\|T\|} \left\| \sum_{-\infty}^{\infty} \alpha_n y_n \right\| \le \left\| \sum_{-\infty}^{\infty} \alpha_n x_n \right\| = \left\| T^{-1} \sum_{-\infty}^{\infty} \alpha_n y_n \right\| \le \|T^{-1}\| \left\| \sum_{-\infty}^{\infty} \alpha_n y_n \right\|$$

and since $\| \sum_{-\infty}^{\infty} \alpha_n y_n \|^2 = \int_{-\pi}^{\pi} |\sum_{-\infty}^{\infty} \alpha_n e^{in\theta}|^2 f(\theta) d\theta$ with $f, f^{-1} \in L^\infty(\Pi)$, (v) implies (ii).   ∎

It is clear from (ii) and (iii) of Theorem 9.8, that (ii) above can also be rewritten as:

$$A \int_{-\pi}^{\pi} |f(\theta)|^2 d\theta \le \left\| \int_{-\pi}^{\pi} f(\theta) dZ(\theta) \right\|^2 \le B \int_{-\pi}^{\pi} |f(\theta)|^2 d\theta$$

or

$$A \int_{-\pi}^{\pi} |f(\theta)|^2 d\theta \le \int_{-\pi}^{\pi} \int_{-\pi}^{\pi} f(\theta) \overline{f}(\psi) d\beta(\theta, \psi) \le B \int_{-\pi}^{\pi} |f(\theta)|^2 d\theta.$$

Let us now present another simple condition necessarily satisfied by frames in Hilbert space. Associated to a frame $\{x_n\}_{n \in \mathbb{Z}} \subset H$ is the well-defined frame operator T given via $Tx = \sum_{-\infty}^{\infty} \langle x, x_n \rangle x_n$, $x \in H$ (again see [DuSc]). T is positive and invertible with $A \le T \le B$, i.e., $A\langle x, x \rangle \le \langle Tx, x \rangle \le B\langle x, x \rangle$, for all $x \in H$. With the help of the frame operator, we have the following simple proposition which says that the frame operator cannot be of trace class.

PROPOSITION 9.12
Let $\{x_n\}_{n \in \mathbb{Z}}$ be a frame. Then $\sum_{-\infty}^{\infty} \|x_n\|^2 = +\infty$.

PROOF   Let $\sum_{-\infty}^{\infty} \|x_n\|^2 < +\infty$, and let $\{e_n\}_{n \in \mathbb{Z}}$ be an orthonormal basis of H. Then,

$$+\infty > \sum_{-\infty}^{\infty} \|x_n\|^2 = \sum_{n=-\infty}^{\infty} \sum_{k=-\infty}^{\infty} |\langle x_n, e_k \rangle|^2$$

$$= \sum_k \sum_n |\langle x_n, e_k \rangle|^2 = \sum_k \langle Te_k, e_k \rangle \ge \sum_{-\infty}^{\infty} A \|e_k\|^2 = +\infty.$$ ∎

To finish this section, we present a last set of ideas originating in prediction theory allowing other interpretations of the conditions of Theorem 9.11. Another motivation for introducing these ideas is that they are reminiscent of the multiresolution analysis (Mallat [M], Meyer [Me]) or the other way around.

Associated to a sequence $\{x_n\}_{n \in \mathbb{Z}} \in H$ is a set of spaces embodying the "information" about this sequence: The *present and past* subspace $H_n = \overline{sp}\{x_k : k \leq n\}$ where $\overline{sp}\{\cdot\}$ is the closed linear span of the corresponding set. The *remote past* $H_{-\infty} = \cap_{n \in \mathbb{Z}} H_n$ and the time domain $H_{+\infty} = \overline{sp}\{x_n : n \in \mathbb{Z}\} = \overline{\cup_{n \in \mathbb{Z}} H_n}$.

Using these spaces, classification of sequences can be performed, namely, $\{x_n\}_{n \in \mathbb{Z}}$ is *purely deterministic* if $H_{-\infty} = H_{+\infty}$; *purely nondeterministic* if $H_{-\infty} = \{0\}$; and *nondeterministic* if $H_{-\infty} \not\subset H_{+\infty}$. Since for each n, $H_{n-1} \subset H_n$, we have $x_n = (x_n|H_{n-1}) + e_n$ where $(x_n|H_{n-1})$ is the orthogonal projection of x_n onto H_{n-1}. The sequence $e = \{e_n\}_{n \in \mathbb{Z}} = \{x_n - (x_n|H_{n-1})\}_{n \in \mathbb{Z}}$ which is clearly orthogonal is called the *innovation* of $x = \{x_n\}_{n \in \mathbb{Z}}$.

The importance of the above time domain analysis is illustrated by the following well known Wold–Cramér decomposition (see Cramér [Cr]).

THEOREM 9.13

A sequence $x = \{x_n\}_{n \subset \mathbb{Z}} \subset H$ can be uniquely decomposed as $x = u + v$, the sequences $u = \{u_n\}_{n \in \mathbb{Z}}$ and $v = \{v_n\}_{n \in \mathbb{Z}}$ having the following properties:

   i. $\langle u_n, v_m \rangle = 0$, *for all $n, m \in \mathbb{Z}$*

   ii. *u is purely nondeterministic and v is purely deterministic.*
     In addition u can be decomposed as

   iii. $u_n = \sum_{k=-\infty}^{n} a_{n,k} e_k$, $n \in \mathbb{Z}$, *where $e = \{e_n\}_{n \in \mathbb{Z}}$ is the innovation sequence.*

The sequences u and v are respectively given by $u_n = (x_n|H_n \ominus H_{-\infty})$ and $v_n = (x_n|H_{-\infty})$, $n \in \mathbb{Z}$. Let $\sigma_k = \langle e_k, e_k \rangle$. Then $\|u_n\|^2 = \sum_{k=-\infty}^{n} |a_{n,k}|^2 \sigma_k < +\infty$. The coefficients $a_{n,k}$ are uniquely determined if and only if $\sigma_k > 0$, $k \in \mathbb{Z}$, while the $a_{n,k}\sigma_k$ are uniquely determined by $a_{n,k}\sigma_k = \langle u_n, e_k \rangle$, $k < n$, and $a_{n,n}\sigma_n = \sigma_n = \bar{a}_n\sigma_n$, $n \in \mathbb{Z}$.

It is not difficult to see ([Hou2]) that a Bessel sequence is purely nondeterministic, i.e., that $x_n = \sum_{k=-\infty}^{n} a_{n,k} e_k$, $n \in \mathbb{Z}$. This last observation leads to

THEOREM 9.14

Let $\{x_n\}_{n \in \mathbb{Z}} \subset H$ be a complete sequence. Then $\{x_n\}_{n \in \mathbb{Z}}$ is an exact frame (with bounds A and B) if and only if there exist two constants A and $B > 0$ such that

$$A \sum_{-\infty}^{\infty} |\alpha_n|^2 \leq \sum_{-\infty}^{\infty} \left| \sum_{k=-\infty}^{n} a_{n,k} \sqrt{\sigma_k} \alpha_k \right|^2 \leq B \sum_{-\infty}^{\infty} |\alpha_n|^2, \quad \forall \alpha = \{\alpha_n\} \in \ell^2(\mathbb{Z}).$$

PROOF $\{x_n\}_{n\in\mathbb{Z}}$ is an exact frame if and only if its Gramian matrix G is a bounded linear invertible operator on $\ell^2(\mathbb{Z})$. But, $G = \Gamma\Gamma^*$ where the doubly infinite matrix Γ is upper triangular and given via

$$(\Gamma)_{n,k} = \begin{cases} a_{n,k}, \sqrt{\sigma_k} & k \leq n \\ 0 & k > n. \end{cases}$$

Now, boundedness and invertibility can be expressed (since G is positive) as

$$A\langle\alpha,\alpha\rangle_{\ell^2(\mathbb{Z})} \leq \langle\Gamma\Gamma^*\alpha,\alpha\rangle_{\ell^2(\mathbb{Z})} \leq B\langle\alpha,\alpha\rangle_{\ell^2(\mathbb{Z})}, \quad \alpha \in \ell^2(\mathbb{Z}).$$

The result follows. ∎

This last result ends the section. Further, as well as different, connections between stationary sequences, frames, prediction theory, and weighted norm inequalities can be found in Benedetto [B1].

9.4 Frames of exponentials and irregular sampling

In this section we study the path reconstruction of a weakly stationary process from a set of its samples. The results presented here are taken from [Hou3], where statistical irregular sampling is studied. The article on frames and irregular sampling by Benedetto [B2] will provide the reader with a good introduction to the subject (in a deterministic setting) as well as more complete references.

Let us introduce the framework and show the relevance of frames of exponentials to the study of the "missing data" problem.

Let (Ω, B, P) be a probability space, and let $L^2(P)$ be the corresponding space of complex random variables equipped with the norm $(E|\cdot|^2)^{1/2} = \|\cdot\|_2$ (E is the expectation). The class of processes considered here is the class of (mean square continuous, weakly) stationary processes. It is well known that a stationary process $X = \{X_t\}_{t\in\mathbb{R}}$ has an integral representation $X_t = \int_{\mathbb{R}} e^{it\lambda} dZ(\lambda)$, where the random measure Z is orthogonally scattered; equivalently, the covariance function $R(t) = EX_t\overline{X}_0 = \int_{\mathbb{R}} e^{it\lambda} dF(\lambda)$, where F is a bounded positive measure. A stationary process X is *band-limited* to $(-\gamma, \gamma)$, $\gamma > 0$, if $Z = 0$ a.s. P outside of $(-\gamma, \gamma)$.

We recall from the previous section that, given a frame $\{x_k\}_{k\in\mathbb{Z}}$ in a Hilbert space H, the frame operator defined by $Tx = \sum_{-\infty}^{\infty} \langle x, x_n\rangle x_n$ is bounded and invertible with moreover $x = \sum_{-\infty}^{\infty} \langle x, T^{-1}x_n\rangle x_n = \sum_{-\infty}^{\infty} \langle x, x_n\rangle T^{-1}x_n$, for all $x \in H$. Furthermore, since $A \leq T \leq B$ (A and B are the frame bounds), we have $T^{-1} = B^{-1} \sum_{n=0}^{\infty} (I - B^{-1}T)^n$ in the uniform operator topology (This last fact simply follows from the Neumann expansion of $B^{-1}T$).

Recall (again see [DuSc]) that a sequence of reals $\{t_k\}_{k\in\mathbb{Z}}$ has *uniform density one*, if $D = \sup_{k\in\mathbb{Z}} |t_k - k| < +\infty$ and $\delta = \inf_{k\neq n} |t_k - t_n| > 0$. For this class of sequences, Duffin and Schaeffer developed a theory of non harmonic Fourier series following Levinson [Le, IV] and Paley and Wiener [PWi]. Their most important results can be summarized as:

LEMMA 9.15
Let $\{t_k\}_{k\in\mathbb{Z}}$ have uniform density one and let $0 < \gamma < \pi$. Then $\{e^{it_k \cdot}\}_{k\in\mathbb{Z}}$ is a frame for $L^2(-\gamma,\gamma)$, and there exists a unique (see Remark 9.16) sequence $\{h_k\}_{k\in\mathbb{Z}} \subset L^2(-\gamma,\gamma)$ such that for any $g \in L^2(-\gamma,\gamma)$, $g = \lim_{n\to\infty}\sum_{-n}^{n}(1/2\gamma\int_{-\gamma}^{\gamma}\overline{h}_k(x)g(x)dx)e^{it_k \cdot}$ in $L^2(-\gamma,\gamma)$. Moreover

$$\sum_{-\infty}^{\infty}\left(\frac{1}{2\gamma}\int_{-\gamma}^{\gamma}\overline{h}_k(x)g(x)dx\right)e^{it_k \cdot}$$

converges in $L^2(-\pi,\pi)$, and the corresponding ordinary and nonharmonic Fourier series are uniformly equiconvergent over any $[-\pi+\epsilon, \pi-\epsilon]$, $\epsilon > 0$.

REMARK 9.16 The uniqueness part of the result is the fact that if $\sum_{-\infty}^{\infty} a_k e^{it_k \cdot}$ is another nonharmonic Fourier expansion of g, then

$$\sum_{-\infty}^{\infty}|a_k|^2 \geq \sum_{-\infty}^{\infty}\left|\frac{1}{2\gamma}\int_{-\gamma}^{\gamma}\overline{h}_k(x)g(x)dx\right|^2.$$

Furthermore, let h denote the limit (in $L^2(-\pi,\pi)$) of the partial sums

$$\sum_{-n}^{n}\left(\frac{1}{2\gamma}\int_{-\gamma}^{\gamma}\overline{h}_k(x)g(x)dx\right)e^{it_k\lambda}.$$

Then

$$\lim_{n\to\infty}\left\{\sum_{-n}^{n}\hat{h}(k)e^{ik\lambda} - \sum_{-n}^{n}\left(\frac{1}{2\gamma}\int_{-\gamma}^{\gamma}\overline{h}_k(x)g(x)dx\right)e^{it_k\lambda}\right\} = 0,$$

uniformly for $\lambda \in [-\pi+\epsilon, \pi-\epsilon]$, $\epsilon > 0$. It is clear that a sequence with uniform density one retains this property after the removal of finitely many of its elements. Hence the conclusion of Lemma 9.15 is unaffected by such a change. ∎

For sequences $\{t_k\}_{k\in\mathbb{Z}}$ such that $D = \sup_{k\in\mathbb{Z}}|t_k - k| < 1/4$, in which case $\delta = \inf_{k\neq n}|t_k - t_n| > 0$, γ can be replaced by π. Moreover, the h_k are then given via $\Psi_k(t) = 1/2\pi\int_{-\pi}^{\pi}\overline{h}_k(x)e^{ixt}dx = G(t)/G'(t_k)(t - t_k)$, where $G(t) = (t-t_o)\prod_{n=1}^{\infty}(1-t/t_n)(1-t/t_{-n}), t \in \mathbb{R}$. The functions Ψ_k are called Lagrange interpolating functions since

$$\Psi_k(t_n) = \begin{cases} 0 & \text{for } k \neq n \\ 1 & \text{for } k = n. \end{cases}$$

Furthermore, when $t_k = k$, $G(t) = \sin \pi t / \pi$ and $\Psi_k(t) = \sin \pi(t - k)/\pi(t - k)$. The bound 1/4 is tight and is a sufficient condition for $\{e^{it_k \cdot}\}_{k \in \mathbb{Z}}$ to form a bounded unconditional basis of $L^2(-\pi, \pi)$ (this is known as Kadec's 1/4-condition [K]). It is already present in Levinson [Le, IV] and originates in Paley–Wiener. For $D < 1/4$, the series expansion fails to hold by the removal of a single sampling point.

Our next lemma gives a more explicit knowledge of the sampling coefficients $1/2\gamma \int_{-\gamma}^{\gamma} \bar{h}_k(x)g(x)dx$. This result already present in [Hou3] has its origin in Benedetto and Heller [BHe] (see also [B2]).

LEMMA 9.17
Let $\{t_k\}_{k \in \mathbb{Z}}$ have uniform density one, and let $B \geq (1 + |e^{\gamma D} - 1|)^2$, $0 < \gamma \leq \pi$.
Then, $1/2\gamma \int_{-\gamma}^{\gamma} \bar{h}_k(x)g(x)dx = B^{-1} \sum_{n=0}^{\infty} \sum_{p=0}^{n} (-B)^{p-n} \binom{n}{p} \Phi(n - p, t_k, \gamma)$, where
$\Phi(0, t_k, \gamma) = 1/2\gamma \int_{-\gamma}^{\gamma} e^{-it_k x}g(x)dx$ and where for $q \geq 1$,

$$\Phi(q, t_k, \gamma) = \sum_{m_1=-\infty}^{\infty} \sum_{m_2=-\infty}^{\infty} \cdots \sum_{m_{q-1}=-\infty}^{\infty} \sum_{m_q=-\infty}^{\infty} \frac{1}{2\gamma} \int_{-\gamma}^{\gamma} e^{-it_{m_1}x}g(x)dx$$

$$\times \frac{\sin \gamma(t_{m_1} - t_{m_2}) \cdots \sin \gamma(t_{m_{q-1}} - t_{m_q}) \sin \gamma(t_{m_q} - t_k)}{\gamma(t_{m_1} - t_{m_2}) \cdots \gamma(t_{m_{q-1}} - t_{m_q})\gamma(t_{m_q} - t_k)}.$$

PROOF The frame operator is given by

$$Tg(\lambda) = \sum_{k=-\infty}^{\infty} \frac{1}{2\lambda} \int_{-\gamma}^{\gamma} e^{-it_k x}g(x)dx e^{it_k \lambda}, \quad g \in L^2(-\gamma, \gamma).$$

Moreover, in $L^2(-\gamma, \gamma)$,

$$g(\lambda) = \sum_{k=-\infty}^{\infty} \frac{1}{2\gamma} \int_{-\gamma}^{\gamma} \bar{h}_k(x)g(x)dx e^{it_k \lambda},$$

and

$$\int_{-\gamma}^{\gamma} \bar{h}_k(x)g(x)dx = \int_{-\gamma}^{\gamma} e^{it_k x}T^{-1}g(x)dx$$

(stated differently, $h_k = T^{-1}e^{it_k \cdot}$). By inverting $B^{-1}T$, we get

$$\int_{-\gamma}^{\gamma} \bar{h}_k(x)g(x)dx = B^{-1} \sum_{n=0}^{\infty} \int_{-\gamma}^{\gamma} e^{-it_k x}(I - B^{-1}T)^n g(x)dx$$

$$= B^{-1} \sum_{n=0}^{\infty} \sum_{p=0}^{n} (-B^{-1})^{n-p} \binom{n}{p} \int_{-\gamma}^{\gamma} e^{-it_k x}T^{n-p}g(x)dx.$$

The defining property of T gives $\Phi(0, t_k, \gamma) = 1/2\gamma \int_{-\gamma}^{\gamma} e^{-it_k x} g(x) dx$, and for $n-p \geq 1$,

$$\frac{1}{2\gamma} \int_{-\gamma}^{\gamma} e^{it_k x} T^{n-p} g(x) dx$$

$$= \sum_{m_1=-\infty}^{\infty} \sum_{m_2=-\infty}^{\infty} \cdots \sum_{m_{n-p-1}=-\infty}^{\infty} \sum_{m_{n-p}=-\infty}^{\infty}$$

$$\frac{1}{2\gamma} \int_{-\gamma}^{\gamma} e^{-it_{m_1} x} g(x) dx \frac{1}{2\gamma} \int_{-\gamma}^{\gamma} e^{i(t_{m_1} - t_{m_2}) x} dx \cdots \frac{1}{2\gamma} \int_{-\gamma}^{\gamma} e^{i(t_{m_{n-p-1}} - t_{m_{n-p}}) x} dx$$

$$\frac{1}{2\gamma} \int_{-\gamma}^{\gamma} e^{i(t_{m_{n-p}} - t_k) x} dx.$$

Integrating the exponentials and using the estimates $B = (1 + |e^{\gamma D} - 1|)^2$ (see Lemma 2 in [DuSc]) gives the result. ∎

With the help of the results presented above, our task in the forthcoming pages is mainly to digress on the theme of the following

THEOREM 9.18

Let X be stationary, band-limited to $(-\gamma, \gamma)$, $0 < \gamma < \pi$, and let $\{t_k\} = \{k, s_1, \ldots s_\ell\}_{k \in \mathbb{Z} - \{k_1, \ldots, k_r\}}$. Then, $X(t) = \sum_{k=-\infty}^{\infty} \Psi_k(t) X(t_k)$ in $L^2(P)$, uniformly on compact subsets of $\mathbb{R}$. The interpolating series converges almost surely if and only if $\lim_{p \to \infty} e^{i\gamma t} Z(\gamma - 2^{-p}, \gamma) + e^{-i\gamma t} Z(-\gamma, -\gamma + 2^{-p}) = 0$ a.s. P.
The Ψ_ks are given by

$$\Psi_k(t) = \frac{\gamma}{\pi} \left\{ \frac{\sin \gamma(t_k - t)}{\gamma(t_k - t)} + \frac{\gamma}{\pi} \sum_{p=1}^{r} \sum_{q=1}^{r} \frac{\sin \gamma(k_p - t)}{\gamma(k_p - t)} \frac{\sin \gamma(t_k - k_q)}{\gamma(t_k - t_q)} V_{q,p} \right.$$

$$\left. - \frac{\gamma}{\pi} \sum_{p=1}^{\ell} \sum_{q=1}^{\ell} \frac{\sin \gamma(s_p - t)}{\gamma(s_p - t)} \frac{\sin \gamma(t_k - s_q)}{\gamma(t_k - s_q)} V_{q,p} \right\},$$

where $V_{q,p}$ is the (q,p)-entry of the inverse of

$$V = \left(I - \frac{\gamma}{\pi} \frac{\sin(f_p - f_q)}{\gamma(f_p - f_q)} \right)_{p,q=1,\ldots,r+\ell},$$

where $t \in \mathbb{R}$, and

$$f_p = \begin{cases} k_p & p = 1, \ldots, r, \\ s_p & p = r+1, \ldots, \ell. \end{cases}$$

This theorem, which is a special case of a result in [Hou3], can be interpreted as statistical "missing data" results. Under a uniform density d condition and for data sets containing mixtures of regular and irregular points with gaps, the part of the process whose spectrum is contained $(-\gamma d, \gamma d)$, $\gamma < \pi$, can be exactly recovered (in $L^2(P)$). The mean square error committed with the band-limited assumption is majorized by $E|\int_{|\lambda|\geq\gamma d} e^{it\lambda}dZ(\lambda)|^2 = \int_{|\lambda|\geq\gamma d} dF(\lambda)$ (see Parzen [Pa] for more details on the "missing data" problem).

We first wish to present some computations which will give the form of the Ψ_k's in Theorem 9.18. Replacing $g(x)$ by e^{itx} in Lemma 9.15 provides in a more clumsy way other expressions. Again, the statement and the full proof can be found in [Hou3].

LEMMA 9.19
Let $\{t_k\}_{k\in\mathbb{Z}} = \{k, s_1, s_2, \ldots, s_\ell\}_{k\in\mathbb{Z}-\{k_1,\ldots,k_r\}}$, *and let* $0 < \gamma < \pi$. *The statements in Lemma 9.15 continue to hold with*

$$\frac{1}{2\gamma}\int_{-\gamma}^{\gamma}\overline{h}_k(x)g(x)dx = \frac{\gamma}{\pi}\langle g(\cdot), e^{it_k\cdot}\rangle_\gamma$$

$$+ \left(\frac{\gamma}{\pi}\right)^2\sum_{p=1}^{r}\sum_{q=1}^{r}\langle g(\cdot), e^{ik_p\cdot}\rangle_\gamma\langle e^{ik_q\cdot}, e^{it_k\cdot}\rangle_\gamma V_{q,p}$$

$$- \left(\frac{\gamma}{\pi}\right)^2\sum_{p=1}^{\ell}\sum_{q=1}^{\ell}\langle g(\cdot), e^{is_p\cdot}\rangle_\gamma\langle e^{is_q\cdot}, e^{it_k\cdot}\rangle_\gamma V_{q,p},$$

where $V_{q,p}$ *is the* (q,p)-*entry of the inverse of the matrix*

$$\left(I - \frac{\gamma\sin\gamma(f_p - f_q)}{\pi\gamma(f_p - f_q)}\right)_{p,q=1,\ldots,r+\ell},$$

with

$$f_p = \begin{cases} k_p & p = 1,\ldots,r \\ s_p & p = r+1,\ldots,r+\ell. \end{cases}$$

PROOF Appealing to Lemma 9.15, it is clear that the exponentials $\{e^{it_k\cdot}\}$ form a frame for $L^2(-\gamma,\gamma)$, $0 < \gamma < \pi$, which is neither tight nor exact. Now, for any $g \in (L^2(-\gamma,\gamma), \langle\cdot,\cdot\rangle_\gamma)$,

$$\sum_{k\neq k_1,\ldots,k_r}|\langle g(\cdot), e^{ik\cdot}\rangle_\gamma|^2 + \sum_{j=1}^{\ell}|\langle g(\cdot), e^{is_j\cdot}\rangle_\gamma|^2$$

$$\leq \sum_{k=-\infty}^{\infty}|\langle g(\cdot), e^{ik\cdot}\rangle_\gamma|^2 + \ell\|g\|_\gamma^2$$

$$= \left(\frac{\pi}{\gamma}\right)^2\sum_{k=-\infty}^{\infty}\left|\frac{1}{2\pi}\int_{-\gamma}^{\gamma}e^{-ikx}g(x)dx\right|^2 + \ell\|g\|_\gamma^2$$

$$= \left(\left(\frac{\pi}{\gamma}\right)^2 + \ell\right)\|g\|_\gamma^2,$$

by the Plancherel theorem. Hence we can invert the frame operator and get $T^{-1} = C \sum_{n=0}^{\infty} (I - CT)^n$ where $C = ((\pi/\gamma)^2 + \ell)^{1/2}$. But the frame operator T is also given by $T = \pi/\gamma I - S$ (unfortunately it is not clear that $\|\gamma/\pi S\| < 1$), where $Sg(\lambda) = \sum_{j=1}^{r} \langle g(\cdot), e^{ik_j \cdot} \rangle_\gamma e^{ik_j \lambda} - \sum_{j=1}^{\ell} \langle g(\cdot), e^{is_j \cdot} \rangle_\gamma e^{is_j \lambda}$. Hence, in the uniform operator topology, $T^{-1} = C \sum_{n=0}^{\infty} ((1 - C\pi/\gamma)I + CS^n)$. Now in $L^2(-\gamma, \gamma)$,

$$g(\lambda) = \sum_{k=-\infty}^{\infty} \frac{1}{2\gamma} \int_{-\gamma}^{\gamma} \overline{h}_k(x) g(x) dx e^{it_k \lambda}$$

$$= \sum_{k=-\infty}^{\infty} \langle T^{-1} g(\cdot), e^{it_k \cdot} \rangle_\gamma e^{it_k \lambda}.$$

Hence,

$$\frac{1}{2\gamma} \int_{-\gamma}^{\gamma} \overline{h}_k(x) g(x) dx$$

$$= C \sum_{n=0}^{\infty} \left\langle \left(\left(1 - C\frac{\pi}{\gamma} \right) I + CS \right)^n g(\cdot), e^{it_k \cdot} \right\rangle_\gamma$$

$$= \frac{\gamma}{\pi} \langle g(\cdot), e^{it_k \cdot} \rangle_\gamma + C \sum_{n=0}^{\infty} \sum_{p=1}^{n} \binom{n}{p} \left(1 - C\frac{\pi}{\gamma} \right)^{n-p} C^p \langle S^p g(\cdot), e^{it_k \cdot} \rangle_\gamma. \quad (9.26)$$

To conclude we now need to evaluate $\langle S^n g(\cdot), e^{it_k \cdot} \rangle_\gamma$, $n \geq 1$. We claim (see [Hou3] for details) that $\langle S^n g(\cdot), e^{it_k \cdot} \rangle_\gamma = {}^k x M^{n-1} y$, $n \geq 1$, where ${}^k x$ is a $1 \times (r + \ell)^2$ matrix with entries

$$^k x_{(p-1)(r+\ell)+q} = \langle \pm g(\cdot), e^{if_p \cdot} \rangle_\gamma \langle e^{if_q \cdot}, e^{it_k \cdot} \rangle_\gamma, p = 1, \ldots, r + \ell$$

and $q = 1, \ldots, r + \ell$, ($+g$ when $q = 1, \ldots, r$ and $-g$ when $q = r + 1, \ldots, r + \ell$); where M is a $(r + \ell)^2 \times (r + \ell)^2$ block diagonal matrix with identical $(r + \ell) \times (r + \ell)$ blocks $(\langle e^{if_p \cdot}, e^{if_q \cdot} \rangle_\gamma)_{p,q=1,\ldots,r+\ell}$, and where y is a $(r + \ell)^2 \times 1$ matrix with entries

$$y_q = \begin{cases} 1 & \text{if } q = (p - 1)(r + \ell) + p, \ p = 1, \ldots, r + \ell \\ 0 & \text{elsewhere.} \end{cases}$$

It thus follows from (9.26) and the form of $\langle S^p g(\cdot), e^{it_k \cdot} \rangle_\gamma$ that

$$\frac{1}{2\gamma} \int_{-\gamma}^{\gamma} \overline{h}_k(x) g(x) dx$$

$$= \frac{\gamma}{\pi} \langle g(\cdot), e^{it_k \cdot} \rangle_\gamma + C \, {}^k x \sum_{n=0}^{\infty} \sum_{p=1}^{n} \binom{n}{p} \left(1 - C\frac{\pi}{\gamma} \right)^{n-p} C^p M^{p-1} y$$

$$= \frac{\gamma}{\pi} \langle g(\cdot), e^{it_k \cdot} \rangle_\gamma + C \, {}^k x M^{-1} \left\{ \sum_{n=0}^{\infty} \left(\left(1 - C\frac{\pi}{\gamma} \right) I + CM \right)^n - \left(1 - C\frac{\pi}{\gamma} \right)^n \right\} y$$

$$= \frac{\gamma}{\pi} \langle g(\cdot), e^{it_k \cdot} \rangle_\gamma + C \, {}^k x M^{-1} \left\{ \left(C\frac{\pi}{\gamma} I - CM \right)^{-1} - \left(C\frac{\pi}{\gamma} \right)^{-1} \right\} y$$

$$= \frac{\gamma}{\pi} \langle g(\cdot), e^{it_k \cdot} \rangle_\gamma + \frac{\gamma}{\pi} {}^k x M^{-1} \left\{ \left(I - \frac{\gamma}{\pi} M \right)^{-1} - I \right\} y$$

$$= \frac{\gamma}{\pi} \langle g(\cdot), e^{it_k \cdot} \rangle_\gamma + \left(\frac{\gamma}{\pi} \right)^2 {}^k x \left(I - \frac{\gamma}{\pi} M \right)^{-1} y. \tag{9.27}$$

Finally, from (9.26), (9.27) and the form of M we get

$$\frac{1}{2\gamma} \int_{-\gamma}^{\gamma} \overline{h}_k(x) g(x) dx$$

$$= \frac{\gamma}{\pi} \langle g(\cdot), e^{it_k \cdot} \rangle_\gamma + \left(\frac{\gamma}{\pi} \right)^2 \sum_{p=1}^{r} \sum_{q=1}^{r} \langle g(\cdot), e^{ik_p \cdot} \rangle_\gamma \langle e^{ik_q \cdot}, e^{it_k \cdot} \rangle_\gamma V_{q,p}$$

$$- \left(\frac{\gamma}{\pi} \right)^2 \sum_{p=1}^{\ell} \sum_{q=1}^{\ell} \langle g(\cdot), e^{it_p \cdot} \rangle_\gamma \langle e^{it_q \cdot}, e^{it_k \cdot} \rangle_\gamma V_{q,p}. \quad \blacksquare$$

Now that the Ψ_k are explicitly given (replacing $g(\cdot)$ by $e^{it \cdot}$ in Lemma 9.19), let us indicate some of the elements of the proof of the stated theorem. First, the equiconvergence result (Lemma 9.15) or the boundedness of the kernels $\sum_{-n}^{n} \Psi_k(t) e^{it_k \lambda}$ give the L^2-result.

THEOREM 9.20
Let X be stationary, band-limited to $(-\lambda, \lambda)$, $0 < \lambda < \pi$, and let $\{t_k\}_{k \in \mathbb{Z}} = \{k, s_1, \ldots, s_\ell\}_{k \in \mathbb{Z} - \{k_1, \ldots, k_r\}}$. Then $X(t) = \sum_{k=-\infty}^{\infty} \Psi_k(t) X(t_k)$ in $L^2(P)$, uniformly on any compact subset of $\mathbb{R}$.

PROOF   Applying the previous lemma to the functions $g_t(\lambda) = e^{it\lambda}$, $t \in \mathbb{R}$, $-\gamma < \lambda < \gamma$, we get

$$X(t) = \int_{|\lambda| < \gamma} e^{it\lambda} dZ(\lambda) = \int_{|\lambda| < \gamma} \lim_{n \to \infty} \sum_{k=-n}^{n} \Psi_k(t) e^{it_k \lambda} dZ(\lambda).$$

We now wish to interchange the limit and the integral. This can be done as follows: In $L^2(-\pi, \pi)$, $e^{it\lambda} \chi_{(-\gamma, \gamma)}(\lambda) = \sum_{k=-\infty}^{\infty} \Psi_k(t) e^{it_k \lambda}$. It thus follows from the equiconvergence result (Lemma 9.15 or 9.19 and Remark 9.16) that

$$\lim_{n \to \infty} \sup_{-\gamma < \lambda < \gamma} \left| \sum_{k=-n}^{n} \frac{\sin \gamma(t-k)}{\gamma(t-k)} e^{ik\lambda} - \Psi_k(t) e^{it_k \lambda} \right| = 0.$$

But, it is classical that $\sup_{-\gamma < \lambda < \gamma} | \sum_{k=-n}^{n} \sin \gamma(t-k)/\gamma(t-k) e^{ik\lambda} | \leq C, t \in K \subset \mathbb{R}$, K compact and $n \geq 0$. Since the spectral measure F of X is finite, we can appeal to Bounded Convergence to get the result. $\blacksquare$

L^2-convergence is established, we can now look for the a.s. result. The next lemma (given without proof) provides some useful estimates on the kernels $\sum_{k=-n}^{n} \Psi_k(t) e^{it_k \lambda}$.

LEMMA 9.21

Let $S_n(\lambda) = \sum_{k=-n}^{n} \Psi_k e^{it_k \lambda}$, $\lambda \in (-\gamma, \gamma)$, $t \in K \subset \mathbb{R}$, K compact.

   i. $|S_n(\lambda) - S_m(\lambda)| \leq C(m-n)/m$, $2^p \leq n < m < 2^{p+1}$.

   ii. $|S_n(\pm\lambda) - S_m(\pm\lambda)| \leq C\{(m-n)|\gamma \mp \lambda| + 1/n\}$, $1 \leq n < m$, $\gamma - 1/n \leq \pm\lambda < \gamma$, $\lambda \geq 0$.

   iii. $|S_n(\pm\lambda) - e^{\pm i\lambda t} + e^{\pm i\gamma t}| \leq C\{n|\gamma \mp \lambda| + 1/n\}$, $n \geq 1$, $\gamma - 1/n \leq \pm\lambda < \gamma$, $\lambda \geq 0$.

   iv. $|S_n(\pm\lambda) - e^{\pm i\lambda t}| \leq C\{1/n|\gamma \mp \lambda| + 1/n\}$, $n \geq 1$, $0 \leq \pm\lambda \leq \gamma - 1/n$, $\lambda \geq 0$.

We now have all the ingredients for the proof by combining the above estimates with a strategy of proof due to Gaposhkin [G].

SKETCH OF PROOF OF THEOREM 9.18   First, it is shown that

$$\lim_{p \to +\infty} \max_{2^p < n \leq 2^{p+1}} |S_n X - S_{2^p} X| = 0,$$

i.e., that the problem can be reduced to studying the kernels $S_n X(t) = \sum_{-n}^{n} \Psi_k(t) X(t_k)$ along the dyadic integers. This is done by using the dyadic decomposition of n and the estimates (i), (ii) and (iv) above. Once the first reduction of the problem is achieved, we write

$$X(t) - S_n X(t) = \{-S_n X(t) + S_{2^p} X(t)\}$$

$$+ \{X(t) - S_{2^p} X(t) - e^{i\gamma t} Z(\gamma - 2^{-p}, \gamma) - e^{-i\gamma t} Z(-\gamma, -\gamma + 2^{-p})\}$$

$$+ \{e^{i\gamma t} Z(\gamma - 2^{-p}, \gamma) + e^{-i\gamma t} Z(-\gamma, -\gamma + 2^{-p})\}.$$

Proving Theorem 9.18 is equivalent to showing that with probability one

$$\lim_{p \to \infty} \{X(t) - S_{2^p} X(t) - e^{-i\gamma t} Z(-\gamma, -\gamma + 2^{-p}) - e^{i\gamma t} Z(\gamma - 2^{-p}, \gamma)\} = 0,$$

uniformly for t in a compact of $\mathbb{R}$. It is in turn enough to show that:

$$\sum_{p=1}^{\infty} \int_{-\gamma}^{\gamma} |e^{i\lambda t} - S_{2^p}(\lambda) - e^{i\gamma t} \chi_{(\gamma - 2^{-p}, \gamma)}(\lambda) - e^{-i\gamma t} \chi_{(-\gamma, -\gamma + 2^{-p})}(\lambda)|^2 dF(\lambda) < +\infty.$$

Dividing $|\lambda| < \gamma$ into $0 < |\lambda \mp \gamma| < 2^{-p}$ and $|\lambda \mp \gamma| \geq 2^{-p}$, and providing estimates over these two regions the result follows. For example, for $|\lambda \mp \gamma| < 2^{-p}$, and if

$A_k = \{2^{-k-1} \le |\lambda \pm \gamma| < 2^{-k}\}$, Lemma 9.21 (iii) gives

$$\sum_{p=1}^{\infty} \int_{0<|\lambda \mp \gamma|<2^{-p}} |e^{\pm i\lambda t} - S_n(\pm \lambda) - e^{\pm i\gamma t}|^2 dF$$

$$\le \sum_{p=1}^{\infty} 2^{2p} \sum_{k=p}^{\infty} \int_{A_k} |\lambda \mp \gamma|^2 dF + C \sum_{p=1}^{\infty} \frac{1}{2^{2p}}$$

$$\le C \sum_{p=1}^{\infty} 2^{2p} \sum_{k=p}^{\infty} 2^{-2k} F(A_k) + C \sum_{p=1}^{\infty} \frac{1}{2^{2p}}$$

$$\le C \sum_{p=1}^{\infty} F(A_p) + C \sum_{p=1}^{\infty} \frac{F(A_p)}{2^p} + C < +\infty.$$

For $|\lambda \mp \gamma| \ge 2^{-p}$, the estimates are essentially similar to the ones given above using Lemma 9.21 (iv) instead of Lemma 9.21 (iii). ∎

The essential moral of Theorem 9.18 is that in a random environment and as far as sampling and reconstruction are concerned, a realization of a process is representative if the end points of the band are filtered out or if the spectrum of the process is smooth enough.

We now give two easy corollaries which state that for particular cases of X or $\{t_k\}_{k \in \mathbb{Z}}$, simplifications occur. The first corollary applies in particular to stationary Gaussian processes.

COROLLARY 9.22
Let X and $\{t_k\}_{k \in \mathbb{Z}}$ satisfy the hypothesis of Theorem 9.18 and let Z have independent increments. Then, with probability one, $X(t) = \sum_{k=-\infty}^{\infty} \Psi_k(t) X(t_k)$, $t \in \mathbb{R}$.

PROOF By Theorem 9.20, and in $L^2(P)$

$$\lim_{p \to \infty} e^{i\gamma t} Z(\gamma - 2^{-p}, \gamma) + e^{-i\gamma t} Z(-\gamma, -\gamma + 2^{-p})$$

$$= \lim_{p \to \infty} \sum_{k=p}^{\infty} e^{i\gamma t} Z(\gamma - 2^{-k}, \gamma - 2^{-k-1}] + e^{-i\gamma t} Z[-\gamma + 2^{-k-1}, -\gamma + 2^{-k})$$

$$= 0$$

Independence gives the a.s. convergence. ∎

COROLLARY 9.23
Let X and $\{t_k\}_{k \in \mathbb{Z}}$ satisfy the hypothesis of Theorem 9.18 and let $dF(\theta) = f d\theta$, $f \in L^{1+\epsilon}(-\gamma, \gamma)$, $\epsilon > 0$. Then, with probability one $X(t) = \sum_{k=-\infty}^{\infty} \Psi_k(t) X(t_k)$.

PROOF

$$\sum_{p=1}^{\infty} E \left| e^{i\gamma t} Z(\gamma - 2^{-p}, \gamma) + e^{-i\gamma t} Z(-\gamma, -\gamma + 2^{-p}) \right|^2$$

$$\leq C \sum_{p=1}^{\infty} \int_{0 < |\lambda \pm \gamma| < 2^{-p}} f(\lambda) d\lambda$$

$$\leq C \sum_{p=1}^{\infty} 2^{-p\epsilon/(1+\epsilon)} \left(\int_{|\lambda| < \gamma} f^{(1+\epsilon)}(\lambda) d\lambda \right)^{1/1+\epsilon} < +\infty. \quad \blacksquare$$

We will not pursue this approach any longer, and once more refer the reader to [Hou3], to get the best possible "behavior" of the spectral density f in order to get convergence with probability one. The reader will also find there optimal conditions on the covariance function, sampling results for general sequences with uniform density one, as well as for more general classes of processes.

9.5 Nonparametric function estimation via wavelets

To conclude this chapter, we present some ideas on applications of wavelets methods to nonparametric statistics. These ideas are presently being developed in "work in progress" with Ed Wegman.

Consider a general function f to be estimated based on some sampled data, say $x_1, x_2, \ldots, x_n$. This is, in fact, the most elementary estimation problem in statistical inference. Often the function, f, in question is the probability distribution function or the probability density function. Most frequently the approach taken is to place the function within a parametric family indexed by some parameter, say θ. Rather than estimate f directly, the parameter θ is estimated with f_θ then being estimated by $\hat{f}_\theta = f_{\hat{\theta}}$. Under a variety of circumstances, it is much more desirable to take a nonparametric approach so as to avoid problems associated with misspecification of parametric family. This is particularly the case when data is relatively plentiful and the information captured by the parametric model is not needed for statistical efficiency.

Probability density estimation and nonparametric, nonlinear regression are probably the two most widely studied nonparametric function estimation problems. However, other problems of interest coming to mind are spectral density estimation, transfer function estimation, impulse response function estimation, all in the time series setting. In the reliability/biometry setting are corresponding problems, failure rate function estimation and survival function estimation. While it may be the case that we simply want an unconstrained estimate of the function, it is more often the case that we wish to impose one or more constraints, for example, positivity, smoothness, isotonicity, convexity, transience, and fixed discontinuities, to name but a few. By far, the most common assumption is smoothness and frequently the estimation is via a kernel or convolution smoother. We would like to formulate a general optimal nonparametric framework.

We formulate the optimization problem as follows. Let H be a Hilbert space of functions (say $\mathbb{R}$-valued) with inner product $\langle \cdot, \cdot \rangle$ and norm $\| \cdot \|$. Let $\mathcal{L} : H \to \mathbb{R}$ be a linear *functional*, i.e., $\mathcal{L}(\alpha f + \beta g) = \alpha \mathcal{L}(f) + \beta \mathcal{L}(g)$ for every $f, g \in H$ and $\alpha, \beta \in \mathbb{R}$. In addition, $\mathcal{L}$ is *convex* on $S \subseteq \mathcal{H}$ if $\mathcal{L}(tf + (1-t)g) \leq t\mathcal{L}(f) + (1-t)\mathcal{L}(g)$, for every $f, g \in S$ with $f, g \in S$ and $0 \leq t \leq 1$. $\mathcal{L}$ is *concave* if the inequality is reversed. $\mathcal{L}$ is *strictly convex (concave)* on S if the inequality is strict. $\mathcal{L}$ is *uniformly convex* on S if $t\mathcal{L}(f) + (1+t)\mathcal{L}(g) - \mathcal{L}(tf + (1-t)g) \geq ct(1-t)\|f - g\|^2$ for every $f, g \in S$ and $0 \leq t \leq 1$.

We wish to use $\mathcal{L}$ as the general objective functional in our optimization framework. For example, if we are concerned with likelihood, we may consider the log likelihood,

$$\mathcal{L}(f) = \sum_{i=1}^{n} \log f(x_i) \quad \text{with } x_i \text{ a random sample from } f.$$

If we have censored samples, we may wish to consider

$$\mathcal{L}(g) = \sum_{i=1}^{n} \delta_i \log g(x_i) + \sum_{i=1}^{n} (1 - \delta_i) \log \overline{G}(x_i);$$

x_i is again a random sample, δ_i a censoring random variable, $\overline{G} = 1 - G$, and $G(x) = \int_{-\infty}^{x} g(u) du$. This is the censored log likelihood. Another example is the generalized least squares. In this case

$$\mathcal{L}(g) = \sum_{i=1}^{n} (y_i - g(x_i))^2 + \lambda \int_{a}^{b} (Lg(u))^2 \, du.$$

Here L is a differential operator and the solution of this optimization problem over appropriate domains is called a penalized smoothing L-spline. If $L = D^2$, then the solution is the familiar cubic spline.

The basic idea is to construct $S \subseteq H$ where S is the collection of functions g, which satisfy our desired constraints such as smoothness or isotonicity. We wish to optimize $\mathcal{L}(g)$ over S. The optimized estimator will be an element of S and hence will inherit whatever properties we choose for S. The estimator will optimize $\mathcal{L}(g)$ and hence will be chosen according to whatever optimization criterion appeals to the investigator. In this sense we can construct designer estimators, i.e., estimators that are designed by the investigator to suit the specifics of the problem at hand.

Of course, in a wide variety of rather disparate contexts, many of these estimators are already known. However, they may be proven to exist in a general framework according to the following theorem due to Wegman [W].

THEOREM 9.24

Consider the following optimization problem: Minimize (maximize) $\mathcal{L}(f)$ *subject to* $f \in S \subseteq H$. *Then*

    i. *If* H *is finite dimensional;* $\mathcal{L}$ *is continuous and convex (concave); and* S *is closed and bounded, then there exists at least one solution.*

    ii. *If* H *is infinite dimensional,* $\mathcal{L}$ *is continuous and convex (concave) and* S *is closed, bounded and convex, then there exists at least one solution.*

    iii. *If* $\mathcal{L}$ *in (i) or (ii) is strictly convex (concave), the solution is unique.*

    iv. *If* H *is infinite dimensional,* $\mathcal{L}$ *is continuous and uniformly convex (concave), and* S *is closed and convex, then there exists a unique solution.*

This theorem gives a unified framework for the construction of optimal non-parametric function estimators. It does not, however, give us a definitive method for construction of nonparametric function estimators. We give a constructive framework in the next several sections and refer the reader to Wegman [W] for the proof of Theorem 9.24 and more examples of the use of this result.

Suppose now that we have a basis for H, call it $\{e_k\}_{k=1}^{\infty}$. This basis obviously also spans any subset S of H and hence any of our "designer" functions in S can be written in terms of the basis, $\{e_k\}_{k=1}^{\infty}$. The unnecessary basis elements will simply have coefficients 0. In a sense, however, this basis is too rich and in a noisy estimation setting superfluous basis elements will only contribute to estimating noise. As part of our "designer" set, S, philosophy, we would like to have a minimal basis set for S. Let us give some construction methods. Consider a basis $\{e_k\}_{k=1}^{\infty}$ for H. Form B_S which is to be the basis for S by the following routine: if there is a $g \in S$ such that $\langle g, e_k \rangle \neq 0$, then let $e_k \in B_S$. If on the other hand there is a $g \in S^{\perp}$ such that $\langle g, e_k \rangle \neq 0$, then let $e_k \in B_{S^{\perp}}$. Unfortunately, it may not be that $B_S \cap B_{S^{\perp}} = \phi$. But this algorithm yields $\{e_k\} = B_S \cup B_{S^{\perp}}$. Moreover $S \subseteq \text{span}(B_S)$. Thus we may be able to eliminate unnecessary basis elements. Usually if we know the properties of the set S we desire and the nature of the basis set $\{e_k\}$, it will be straightforward to construct a test function g to construct the basis set B_S. If S is a subspace, then $S = \text{span}(B_S)$. In any case we can carry out our estimation by

$$\hat{f} = \sum_{e_k \in B_S} \hat{c}_k e_k. \tag{9.28}$$

In a completely noiseless setting (9.28) is really an equality norm, i.e., $\|f - \sum_k c_k e_k\| = 0$. If $H = L^2(\mathbb{R})$, then (9.28) is really

$$f = \sum_k c_k e_k, \quad \text{almost everywhere Lebesgue with } c_k = \langle f, e_k \rangle. \tag{9.29}$$

This choice of c_k is a minimum norm choice. Nevertheless, in a noisy setting where we do not know f exactly, we cannot compute c_k directly. However, we may be able to estimate c_k by standard inference techniques.

Example 9.25.   Norm Estimate
The minimum norm estimate of c_k is the choice which minimizes $\|f - \sum_k c_k e_k\|$, i.e., $c_k = \langle f, e_k \rangle$. In the $L^2(\mathbb{R})$ context,

$$\langle f, e_k \rangle = \int_{\mathbb{R}} f(x) e_k(x) dx.$$

If f is a probability density function, then $\langle f, e_k \rangle = E(e_k)$ which can simply be estimated by $n^{-1} \sum_{j=1}^{n} e_k(x_j)$, where x_j, $j = 1, \ldots, n$, is the sample of observations.   ▯

Example 9.26.   General Form of Estimate
In the general context with optimization functional $\mathcal{L}$ we have

$$\mathcal{L}(f) = \mathcal{L}\left(\sum_{e_k \in B_S} c_k e_k \right) = \mathcal{L}(\{c_k\}). \tag{9.30}$$

Since (9.30) is a function of a countable number of variables, $\{c_k\}$, we can find the normal equations and, with the appropriate choice of basis, find a solution. For this we will typically take $\mathcal{L}$ to be twice differentiable with respect to all c_k.   ▯

A traditional choice of basis has been the exponentials. Another choice of basis we suggest is the wavelet basis. A brief description is in order. Let $\psi_{j,k}(x) = 2^{-j/2}\psi(2^{-j}x - k)$ where ψ satisfies the so called admissibility condition, i.e., $\int_{\mathbb{R}} |\hat{\psi}(\lambda)|^2/|\lambda| \, d\lambda < +\infty$. Under certain choices of ψ, the $\psi_{j,k}$ form a doubly indexed orthonormal basis for $L^2(\mathbb{R})$ (actually also for Sobolev spaces of higher order as well as various functional spaces). As we shall see a bit later, a wavelet basis admits an interpretation of a simultaneous time-frequency decomposition of f due to the dilation-translation nature of its basis elements. Moreover using wavelets, fewer basis elements are required for fitting sharp changes or discontinuities. This implies faster convergence in "nonsmooth" situations by the introduction of "localized" basis elements.

Example 9.25 Continued
Notice that

$$c_{j,k} = \langle f, \psi_{j,k} \rangle = \int_{-\infty}^{\infty} \frac{\psi(x/2^j - k)}{2^{j/2}} f(x) \, dx.$$

In the density estimation case

$$c_{j,k} = E\left(\frac{\psi(x/2^j - k)}{2^{j/2}} \right).$$

Thus a natural estimator is

$$\hat{c}_{j,k} = \frac{1}{n2^{j/2}} \sum_{i=1}^{n} \psi\left(\frac{x_i}{2^j} - k \right),$$

where x_i, $i = 1, \ldots, n$, is the set of observations. $\Box$

In the last part of this section, we wish also to mention the relevance of frames. We see two particularly useful settings. First, if we wish to take the union of a finite number of admissible spaces, say S_i, then the union of the bases, $\cup_i B_{S_i}$, is a frame for $\cup_i S_i$. Secondly, if we have an admissible space S with basis B_S, but we wish to add a few additional elements, say $\{g_j, j \in \mathbb{Z}\}$, then $B \cup \{g_j, j \in \mathbb{Z}\}$ is a frame for the enlarged space. This is useful, for example, if we know there are some discontinuities in an otherwise smooth space (e.g., Sobolev space).

We have at this stage alluded to wavelets and frames, but have not really explained why a wavelet decomposition is of any particular interest in nonparametric function estimation. To appreciate this let us look again at the traditional methods of Fourier or harmonic analysis in statistics. Corresponding to a function $f(x)$, there is a function

$$\hat{f}(\lambda) = \int_{-\infty}^{\infty} e^{-i\lambda x} f(x)\, dx,$$

which is the *Fourier transform* of f. The *inverse Fourier transform* can be computed by

$$f(x) = \int_{-\infty}^{\infty} e^{i\lambda x} \hat{f}(\lambda)\, d\lambda. \tag{9.31}$$

We may not always have a complete version of $\hat{f}$ available so it is useful to have a sampled version. In this case,

$$f(x) = \sum_{k=-\infty}^{\infty} c_k e^{i\lambda_k x} \tag{9.32}$$

which is the so-called discrete Fourier transform. Here

$$c_k = \int_{-\infty}^{\infty} e^{-i\lambda_k x} f(x)\, dx$$

are the Fourier coefficients. The fast Fourier transform (FFT) is a fast computational algorithm for computing the discrete Fourier transform. Fourier methods are appropriate for analysis of stationary stochastic processes or time series since stationarity implies that covariance structure is invariant with time. Because the Fourier transform is usually applied to the covariance function to obtain the spectral density, the frequency structure of a stationary process is invariant in time. It is clear that traditional Fourier methods are not suitable for nonstationary or transient stochastic processes. The series in (9.32) is not a parsimonious representation of $f(x)$ and hence will be slow to converge in nonstationary settings.

It is desirable to localize both in time and frequency. One approach in doing so has been to use the *Short Term Fourier Transform* (STFT) or, as it is also known, the *Gabor transform* given by the expression below:

$$Gf(\lambda, \tau) = \int_{-\infty}^{\infty} e^{-i\lambda x} f(x)\overline{w}(x - \tau) \, dx.$$

In other words the STFT is a windowed Fourier transform. There are unfortunately faults with this idea. The STFT is poor at resolving wavelengths longer than the window width, that is, it is poor at resolving low frequencies. Conversely, the STFT is poor at localizing high frequencies because the window average energy over the window width. For fixed window width, the STFT time and frequency resolutions are limited by the Uncertainty Principle. What is needed is a scheme which allows for large window widths at low frequencies and very small window widths at high frequencies.

The basic wavelet idea is to use a transient waveform as in the STFT, but to increase time resolution by keeping a constant relative bandwidth as frequency increases. Practically speaking we choose a prototype wavelet, $\psi(t)$, and consider dilations and translations of this mother wavelet ψ which are the *affine wavelets*

$$\psi_{a,\tau}(t) = \frac{1}{\sqrt{a}} \psi \left(\frac{t - \tau}{a} \right).$$

The *continuous wavelet transform* (CWT) is defined by

$$W_a f(\tau) = \int_{-\infty}^{\infty} f(x)\overline{\psi}_{a,\tau}(x) \, dx$$

and the *inverse wavelet transform* is given by

$$f(x) = C \int_{a>0} \int_{\mathbb{R}} W_a f(\tau)\psi_{a,\tau}(x) \frac{d\tau da}{a^2}. \tag{9.33}$$

This latter equation is sometimes reparametrized with $a = e^{\nu}$. Just as we deal with a Fourier series representation of f, we would like to deal with a wavelet series representation. The parameter a (or its surrogate ν) is the *dilation parameter* and is the analog of the frequency, ω, in the ordinary harmonic case. It is more properly thought of as a scale parameter. The parameter, τ, is a time-location parameter.

As previously indicated, a decrease in scale corresponds to an increase in frequency (large scale for low frequencies, small scale for high frequencies). We may discretize the time and scale parameters by $a = a_0^j$ and $\tau = k a_0^j \tau_0$, where j and k are integers, and the wavelets are

$$\psi_{j,k}(t) = \frac{1}{a_0^{j/2}} \psi \left(\frac{t}{a_0^j} - k\tau_0 \right)$$

with wavelet coefficients given by

$$c_{j,k} = \int_{\mathbb{R}} f(x)\overline{\psi}_{j,k}(x) \, dx.$$

If a_0, τ_0 and $\psi(t)$ have the appropriate properties, we might expect as in the case of Fourier series that

$$f(x) = \sum_{k=-\infty}^{\infty} \sum_{j=-\infty}^{\infty} c_{j,k} \psi_{j,k}(x). \qquad (9.34)$$

This is precisely the form we associate with either an orthonormal basis or a frame. Thus if we can show that the wavelets either form a basis or a frame, then the representation (9.34) will follow. Notice that if a_0 is arbitrarily close to one, then the double sum in equation (9.34) is a Riemann sum approximation to the double integral in (9.33), and hence it is reasonable to believe that such conditions are possible. To make wavelets computationally feasible, however, $a_0 = 2$ is usually taken. Furthermore, when the mother wavelet ψ has compact support, only finitely many terms in the doubly infinite series will contribute to the value of $f(x)$ resulting in great computational advantage.

We end this section by reiterating the connections established: Nonparametric estimation problems can be restated as functional analytic optimization problems. The statistical solutions to these problems can be achieved by finding spanning bases for admissible classes of functions. The localizing property of wavelets bases opens up a range of possibilities not available by conventional methods such as convolution smoothers.

Bibliography

[B1]   J. Benedetto. 1992. "Stationary frames and spectral estimation," in *Probabilistic and Stochastic Methods in Analysis*, NATO-ASI-1991, J.S. Byrnes, ed., Dordrecht: Kluwer Academic Publishers, pp. 117–161.

[B2]   J. Benedetto. 1992. "Irregular sampling and frames," in *Wavelets: A Tutorial in Theory and Applications*, C.K. Chui, ed., New York: Academic Press, pp. 445–507.

[BHe]   J. Benedetto and W. Heller. 1990. "Irregular sampling and the theory of frames I," *Note Math.*, **10**(Suppl. 1): 103–125.

[CHou]   S. Cambanis and C. Houdré. 1992. "On continuous wavelet transforms of second order random processes," University of North Carolina Center for Stochastic Processes Technical Report No. 390.

[Cr]   H. Cramér. 1962. "On some classes of non-stationary stochastic processes," in *Proc. IV Berkeley Symp. on Math. Stat. Prob.*, **2**, J. Neyman, ed., Berkeley: Univ. of California Press, pp. 57–78.

[D]   I. Daubechies. 1988. "Orthonormal bases of compactly supported wavelets," *Commun. Pure Appl. Math.*, **41**: 909–996.

[DLa1]   I. Daubechies and J. Lagarias. 1991. "Two-scale difference equations: I. Existence and global regularity of solutions," *SIAM J. Math. Anal.* **22**: 1388–1410.

[DLa2]   I. Daubechies and J. Lagarias. 1993. "Two-scale difference equations: II. Local regularity, infinite products and fractals," *SIAM J. Math. Anal.*, **23**: 1031–1079.

[DuSc]   R. Duffin and A. Schaeffer. 1952. "A class of nonharmonic Fourier series," *Trans. Amer. Math. Soc.* **72**: 341–366.

[DunSch] N. Dunford and J. Schwartz. 1957. *Linear Operators*, New York: Interscience Publishers.

[F]   P. Flandrin. 1989. "On the spectrum of fractional Brownian Motions," *IEEE Trans. Inform. Theory* **35**(1): 197–199.

[Fu]   H. Furstenberg. 1963. "Noncommuting random products," *Trans. Amer. Math. Soc.* **108**: 377–428.

[FuKe]   H. Furstenberg and H. Kesten. 1960. "Products of random matrices," *Ann. Math. Statist.* **31**: 457–469.

[G]   V.F. Gaposhkin. 1977. "A theorem on the convergence almost everywhere of measurable functions, and its applications to sequences of stochastic integrals," *Math. USSR Sbornik* **33**: 1–17.

[GoGu]   I. Goldsheid and Y. Guivarc'h. 1991. "Dimension de la loi gaussienne pour un produit de matrices aléatoires indépendantes et adhérence de Zariski," *C.R. Acad. Sci.* **313**: 305–308.

[HCo]   C. Heil and D. Colella. 1993. "Dilation equations and the smoothness of compactly supported wavelets," this volume, Chap. 4.

[HWa]   C. Heil and D. Walnut. 1989. "Continuous and discrete wavelet transforms," *SIAM Rev.*, **31**: 628–666.

[Ho]   J. Horowitz. 1977. "Une remarque sur les bimesures," in *Sém. Pro. XI*, Lect. Notes Math. **581**, Berlin: Springer-Verlag, pp. 59–64.

[Hou1]   C. Houdré. 1990. "Harmonizability, V-boundedness, $(2,p)$-boundedness of stochastic processes," *Probab. Theor. Rel. Fields*, **84**: 39–54.

[Hou2]   C. Houdré. 1991. "On the linear prediction of multivariate $(2,p)$-bounded processes," *Ann. Probab.*, **19**: 843–867.

[Hou3]   C. Houdré. 1992. "Path reconstruction of processes from missing and irregular samples," University of North Carolina, Center for Stochastic Processes, Technical Report No. 359, February 1992, *Ann. Probab.*, to appear.

[Hou4]   C. Houdré. 1992. "On the discrete wavelet transform of second order processes," preprint.

[K]   M.I. Kadec. 1964. "The exact value of the Paley–Wiener constant," *Sov. Math. Dokl.* **5**: 559–561.

[Le]   E. Le Page. 1982. *Théorèmes limites pour les produits de matrices aléatoires*, Lect. Notes in Math. **928**, Berlin: Springer-Verlag, pp. 258–303.

[Le]   N. Levinson. 1940. *Gap and Density Theorems*, Colloq. Publ. **26**, New York: Amer. Math. Soc.

[M]   S.G. Mallat. 1989. "Multiresolution approximations and wavelet orthonormal bases of $L^2(\mathbb{R})$," *Trans. Amer. Math. Soc.* **315**: 69–87.

[MaV]   B.B. Mandelbrot and J.W. Van Ness. 1968. "Fractional Brownian motions, fractional noises, and applications," *SIAM Rev.* **10**(4): 422–437.

[Me]   Y. Meyer. 1990. *Ondelettes et Opérateurs*, vols. I, II, Paris: Hermann.

[PWi]   R.E.A.C. Paley and N. Wiener. 1934. *Fourier Transforms in the Complex Domain*, Colloq. Publ. **19**, New York: Amer. Math. Soc.

[Pa]   E. Parzen. 1984. *Time Series Analysis of Irregularly Observed Data*, Lect. Notes in Stat. **25**, New York: Springer-Verlag.

[RZ]   J. Ramanathan and O. Zeitouni. 1991. "On the wavelet transform of fractional Brownian Motion," *IEEE Trans. Inform. Theory.* **37**(4): 1156–1158.

[S]   E. Slud. 1986. "Stability of exponential rate of growth of products of random matrices under local random perturbations," *J. London Math. Soc.* **33**: 180–192.

[St]   G. Strang. 1989. "Wavelets and dilation equations: a brief introduction," *SIAM Rev.* **31**: 614–627.

[W]   E. Wegman. 1984. "Optimal nonparametric function estimation," *J. Stat. Plann. Inference* **9**: 375–387.

[Y]   R. Young. 1980. *An Introduction to Nonharmonic Fourier Series*, New York: Academic Press.

10

Wavelets and adapted waveform analysis

Ronald R. Coifman* and **M. Victor Wickerhauser**[†]

ABSTRACT   Our goal is to describe tools for adapting methods of analysis to various tasks occurring in harmonic and numerical analysis and signal processing. The main point of this presentation is that by choosing an orthonormal basis, in which space and frequency are suitably localized, one can achieve both understanding of structure and efficiency in computation. We describe a fingerprint image segmentation algorithm, an alternative factorization for the FFT, and a wavelet-based denoising algorithm.

CONTENTS

*Research supported in part by DARPA and NSF.
†Research supported in part by AFOSR.

399

Our goal is to describe tools for adapting methods of analysis to various tasks occurring in harmonic and numerical analysis and signal processing. The main point of this presentation is that by choosing an orthonormal basis, in which space and frequency are suitably localized, one can achieve both understanding of structure and efficiency in computation. In fact we claim that the search for computational efficiency is intimately related to efficiency in representation (i.e., compression) and to pattern extraction, or structural understanding.

Traditionally in PDE and harmonic analysis, microlocalization is the main tool for structural understanding of operators. This is done by choosing appropriate partitions of unity in phase space—decomposition of phase space into cells. The waveform analysis approach provides a numerical recipe for microlocalization, in which the partitions are computed to achieve maximum efficiency in describing interactions. Not only does this approach shed light on classical analysis methods, it also suggests new methods of organization and analysis of operators. These include nonstandard forms and discrete phase space approximations. Our goal is to introduce a variety of techniques permitting the mathematician, scientist, or engineer to choose the appropriate analysis method in this catalogue of tools and apply it to practical problems.

In the adapted waveform transform (AWT) method the user is provided with a collection of standard libraries of waveforms—wavelets, wavelet-packets, windowed trigonometric waveforms—which can be chosen to fit specific classes of signals. These libraries come equipped with fast numerical algorithms, enabling real-time implementation of a variety of analysis and signal processing tasks such as data compression, parameter extraction for recognition and diagnostics, and fast transformation and manipulation of digital data.

The process of analysis is usually done by comparing acquired segments of data with stored waveforms. The numerical comparison algorithm itself is fast and perfectly conditioned, always being a factored sparse orthogonal transformation. The most efficient orthonormal basis for compression of the signal is selected and used to extract and manipulate relevant features.

A calculus in compressed variables has been developed enabling the implementation of adapted transform methods for fast numerical algorithms useful for data manipulation and for large scale computation.

10.1 Windowed FFT and adapted window selection

To illustrate our procedures we start with a description of an algorithm to compute the Fourier expansion of a function on $[0, 2]$ from the Fourier expansion of its restrictions to $[0, 1]$ and $[1, 2]$.

Let f be defined on $[0, 2]$ and write $f = f^0 + f^1$, where

$$f^0 = \begin{cases} f & x \in [0, 1] \\ 0 & x \notin [0, 1]. \end{cases}$$

We want to compute

$$\hat{f}_m = \frac{1}{\sqrt{2}} \int_0^2 f(t) e^{-2\pi i m t/2} \, dt$$

in terms of

$$\hat{f}_n^0 = \int_0^1 f(t) e^{-2\pi i n t} \, dt \quad \text{and} \quad \hat{f}_n^1 = \int_1^2 f(t) e^{-2\pi i n t} \, dt.$$

Clearly, when $m = 2n$ we have

$$\hat{f}_{2n} = \frac{1}{\sqrt{2}} \{\hat{f}_n^0 + \hat{f}_n^1\}.$$

For $m = 2n + 1$ we define

$$d_n = \frac{1}{\sqrt{2}} \{\hat{f}_n^0 - \hat{f}_{n+1}^1\}$$

and find

$$\hat{f}_{2n+1} = \frac{1}{\pi i} \sum \frac{d_k}{(n - k + \frac{1}{2})}.$$

In fact,

$$\hat{f}_{2n+1} = \frac{1}{\sqrt{2}} \int_0^1 [f(t) - f(t + 1)] e^{-i\pi t} e^{-2\pi i n t} \, dt.$$

Since d_n are the Fourier coefficients on $[0, 1]$ of $f(t) - f(t + 1)$ and $1/(\pi i(n + 1/2))$ are the coefficients of $e^{-i\pi t}$, we obtain the coefficients of $\hat{f}_{2n+1}$ by convolving these sequences. Yet another way to compute $\hat{f}_{2n+1}$ is to compute the inverse transform on $[0, 1]$ of d_n, multiply by $e^{-it/2}$, and recompute the transform on $[0, 1]$.

We see that in order to compute the Fourier expansion on the large interval, we can start with adjacent pairs of small intervals, combine coefficients to obtain the expansion on their union, and continue until we reach the largest interval at the top level. Along the way we have obtained all dyadic windowed Fourier transforms as intermediate computations. Clearly every disjoint collection of intervals and their orthogonal bases provides us with an orthogonal basis for the union. A natural question that arises in connection with the windowed Fourier transform is how to place the windows. Comparing Figures 10.2 and 3 we see the effect of the window selection on the number of large coefficients in the expansion.

So let us now turn to the question of optimizing the windows to obtain an efficient representation of a function. We can proceed as follows: We start with

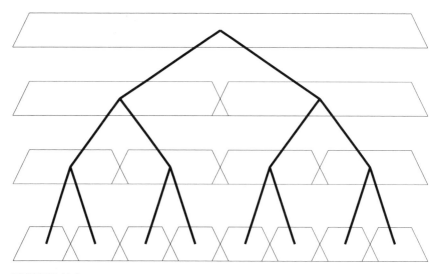

FIGURE 10.1
Schematic description.

the adjacent small intervals and determine the expansion coefficients on each one separately. We then compute the expansion coefficients on the union. Now we can choose that expansion for which the number of coefficients needed to capture 99% of the energy is smallest. Or, we can choose that expansion whose "cost" is smallest: information cost, coding cost, error cost, ... We compare the cost of the chosen expansions on two adjacent unions of pairs to the expansion on their union and again pick the best. We continue until we reach an optimal distribution of windows. For example, see Figure 10.4, where the windows were adapted to digital voice recording. The details of this procedure may be found in [CW].

The procedure described above, although natural, is not very useful if we take the windowed Fourier transform with discontinuous windows. The discontinuity introduces "large" expansion coefficients. A cosine basis on each interval is somewhat better. On the other hand, it is well-known that we cannot find a smooth window function $\omega(x)$ supported on $(-1/2, 3/2)$ such that $\omega(x - k)e^{2i\pi nx}$ are orthogonal. (If we could, it would imply that $\int \omega(x)\omega(x - 1)e^{-2\pi imx} \, dx = 0$ for all m, and thus $\omega(x)\omega(x - 1) \equiv 0$.)

Recently Daubechies, Jaffard, and Journé [DJaJo] and Malvar [M] observed that by taking equal smooth windows and sines or cosines orthogonality can be maintained. Coifman and Meyer [CMe] observed that the windows can be chosen for different sizes. This enables the adaptive constructions described above and in [CMeW], [CW].

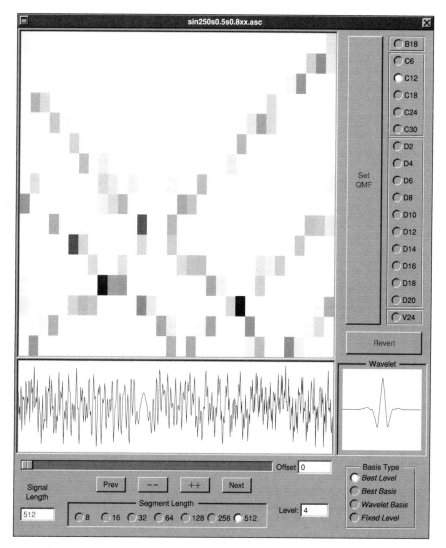

FIGURE 10.2
Optimal window selection.

10.2 Modulated waveform libraries

We start by recalling the concept of a "library of orthonormal bases." For the sake of exposition we restrict our attention to two classes of numerically useful wave-

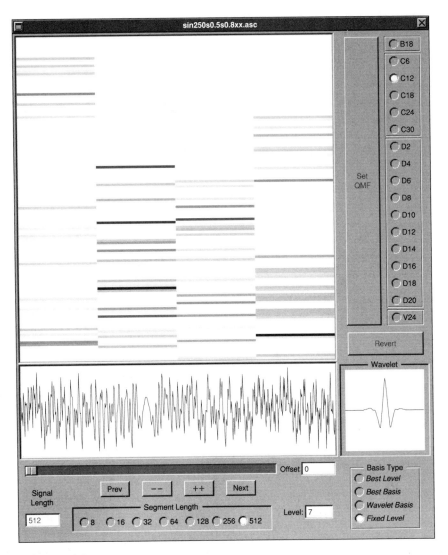

FIGURE 10.3
Large window selection.

forms, introduced recently: *wavelet packets* [CMeW] and *localized trigonometric waveforms* [CMe].

We start with trigonometric waveform libraries. These are localized sine transforms (LST) associated to covering by intervals of **R** or, more generally, of a manifold. We consider a cover $\mathbf{R} = \bigcup_{-\infty}^{\infty} I_i$, where $I = [\alpha_i, \alpha_{i+1})$ and $\alpha_i < \alpha_{i+1}$. Write $\ell_i = \alpha_{i+1} - \alpha_i = |I_i|$ and let $p_i(x)$ be a window function supported in

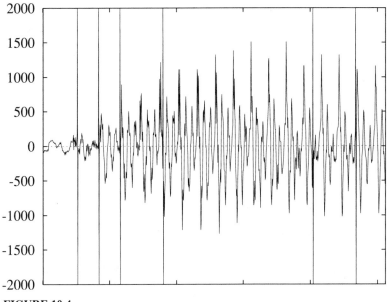

FIGURE 10.4
Adapted window selection for a sound segment.

$[\alpha_i - \ell_{i-1}/2, \alpha_{i+1} + \ell_{i+1}/2]$ such that

$$\sum_{-\infty}^{\infty} p_i^2(x) = 1$$

and

$$p_i^2(x) = 1 - p_i^2(2\alpha_{i+1} - x) \quad \text{for } x \text{ near } \alpha_{i+1}.$$

Then the functions

$$S_{i,k}(x) = \frac{2}{\sqrt{2\ell i}} p_i(x) \sin\left[(2k + 1)\frac{\pi}{2\ell_i}(x - \alpha_i)\right]$$

form an orthonormal basis of $L^2(\mathbf{R})$ subordinate to the partition p_i. The collection of such bases forms a library of orthonormal bases [CMe]. We can likewise form a library of orthonormal local cosine bases:

$$C_{i,k}(x) = \frac{2}{\sqrt{2\ell i}} p_i(x) \cos\left[(2k + 1)\frac{\pi}{2\ell_i}(x - \alpha_i)\right].$$

A few of these basis functions are plotted in Figure 10.5. It is easy to check that if H_i denotes the space of functions spanned by $S_{i,k}$, $k = 0, 1, 2, \ldots$, then $H_i + H_{i+1}$

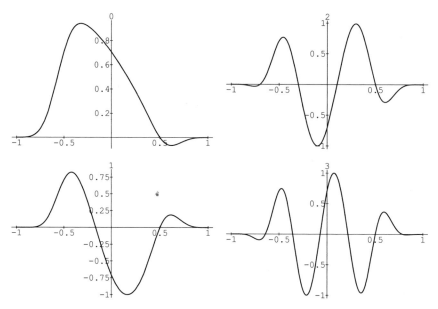

FIGURE 10.5
Local Cosine Transform (LCT) basis functions.

is spanned by the functions

$$P(x) \frac{1}{\sqrt{2(\ell_i + \ell_{i+1})}} \sin \left[(2k+1) \frac{\pi}{2(\ell_i + \ell_{i+1})} (x - \alpha_i) \right],$$

where

$$P^2 = p_i^2(x) + p_{i+1}^2(x)$$

is a "window" function covering the interval $I_i \cup I_{i+1}$.

10.3   Relation to wavelets and wavelet packets

We consider the frequency line **R** split as the union of $\mathbf{R}^+ = [0, \infty)$ and $\mathbf{R}^- = (-\infty, 0)$. On $L^2(\mathbf{R}^+)$ we introduce a window function $p(\xi)$ such that

$$\sum_{k=-\infty}^{\infty} p^2(2^{-k}\xi) = 1 \quad \text{and} \quad \operatorname{supp} p(\xi) \subset \left(\frac{3}{4}, 3 \right).$$

Clearly we can view $p(2^{-k}\xi)$ as window function above the interval $(2^k, 2^{k+1})$ and observe that the functions

$$s_{k,j} = s_{k,j}(\xi) = p(2^{-k}\xi) \sin\left[\left(j + \frac{1}{2}\right)\pi\left(\frac{\xi - 2^k}{2^k}\right)\right]$$

form an orthonormal basis of $L^2(\mathbf{R}^+)$. Similarly

$$c_{k,j} = c_{k,j}(\xi) = p(2^{-k}\xi) \cos\left[\left(j + \frac{1}{2}\right)\pi\left(\frac{\xi - 2^k}{2^k}\right)\right]$$

gives another orthonormal basis, but whose elements are not orthogonal to the functions $s_{k,j}$. However, if we define $S_{k,j}$ as the odd extension to $\mathbf{R}$ of $s_{k,j}$ and $C_{k,j}$ as the even extension of $c_{k,j}$, then we find that $S_{k,j} \perp C_{k',j'}$ for all j,k,j',k'. This permits us to write $C_{k,j} \pm iS_{k,j} = e^{\pm ij\pi\xi/2^k} \hat{\psi}(\xi/2^k)$ where $\hat{\psi}(\xi) = e^{i\pi/2\xi}p(\xi)$ is the Fourier transform of the Meyer wavelet Ψ defined in [Me]. The details of this calculation may be found in [AWeW]. We therefore see that wavelet analysis corresponds to windowing frequency space in "octave" windows $(2^k, 2^{k+1})$.

A natural extension is provided by allowing all dyadic windows in frequency space and adapted window choice. This sort of analysis is equivalent to wavelet packet analysis. The wavelet packet analysis algorithms permit us to perform an adapted Fourier windowing directly in the time domain by successive filtering of a function into different frequency bands. The dual version of the window selection provides an adapted subband coding algorithm. The wavelet packet library is constructed by iterating the wavelet algorithm. This library contains the wavelet basis, Walsh functions, and smooth versions of Walsh functions called wavelet packets [CMeW].

These waveforms are mutually orthogonal; moreover, each of them is orthogonal to all of its integer translates and dyadic rescaled versions. The full collection of these wavelet packets (including translates and rescaled versions) provides us with our library. We may think of these functions as templates or musical notes which we try to match efficiently to signals for analysis and synthesis [CMeW], [D].

We were led to measure the "distance" between a basis and a function in terms of the Shannon entropy of the expansion. This distance measures the efficiency of a particular basis for expanding a given function—in rough terms by judging one expansion to be more efficient than another if its coefficients decrease to zero more rapidly. More generally, let H be a Hilbert space. Let $v \in H$, $\|v\| = 1$, and assume $H = \oplus \sum H_i$ is an orthogonal direct sum. We define

$$\varepsilon^2(v, \{H_i\}) = -\sum \|v_i\|^2 \ln \|v_i\|^2$$

as a measure of distance between v and the orthogonal decomposition. Then ε^2 is characterized by the Shannon equation which is a version of Pythagoras' theorem. If we let $H = \oplus(\sum H^i) \oplus (\sum H_j) = H_+ \oplus H_-$, where H^i and H_j give orthogonal

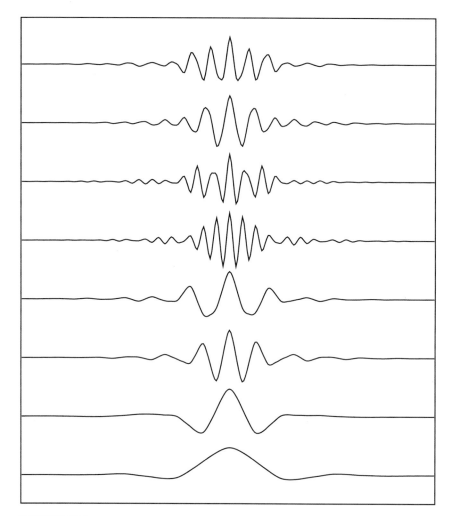

FIGURE 10.6
Wavelet packets.

decompositions $H_+ = \sum H^i, H_- = \sum H_j$, then

$$\varepsilon^2(v; \{H^i, H_j\}) = \varepsilon^2(v, \{H_+, H_-\}) + \|v_+\|^2 \varepsilon^2 \left(\frac{v_+}{\|v_+\|}, \{H^i\} \right)$$

$$+ \|v_-\|^2 \varepsilon^2 \left(\frac{v_-}{\|v_-\|}, \{H_j\} \right).$$

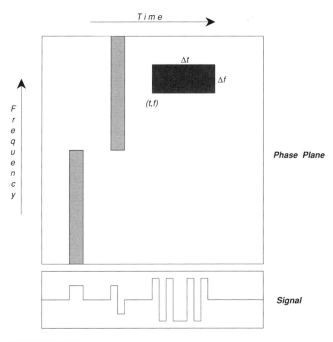

FIGURE 10.7
Cells in the phase plane.

This is Shannon's equation for entropy. If, as in quantum mechanics, we interpret $\|P_{H_+}v\|^2$ as the "probability" that v lies in the subspace H_+, this equation enables us to search for a lowest-entropy expansion of a signal, as described in [CW].

Wavelet packet expansions correspond algorithmically to subband coding schemes and are numerically as fast as the "fast" Fourier transform (FFT). This is true even if we include the complexity of the search for the lowest entropy wavelet packet basis for a given signal or ensemble of signals.

10.4 Time-frequency analysis

To each wavelet packet or local trigonometric function we can associate a time t and a frequency f. These will be uncertain by amounts Δt and Δf, respectively, and Heisenberg's uncertainty principle requires that $\Delta t \cdot \Delta f \geq 1$. The result may be interpreted as a rectangular patch of dimensions Δt by Δf, located around (t, f). We shall call the patch a *Heisenberg cell*; it may be shaded in proportion to the amplitude of the corresponding wavelet packet or local cosine component.

If we choose basis elements which are near-optimally localized in both time and frequency (i.e., for which $\Delta t \cdot \Delta f \approx 1$), then we may depict a basis as a cover of the phase plane by cells of unit area. An orthonormal basis of such *time-frequency atoms* corresponds to a disjoint cover of this idealized phase plane by unit-area cells. Strictly speaking, this correspondence applies to arbitrary disjoint dyadic covers only in the case of the Haar–Walsh or Shannon wavelet packet libraries. The LST and wavelet packet libraries give orthonormal bases only if constrained to certain subsets of the disjoint dyadic covers, since they correspond to a segmentation in space followed by an expansion in frequency, or restriction to a band of frequencies followed by an expansion in space.

Certain bases have characterizations in terms of the shapes of the cells. In Figure 10.8, we see that the Standard basis (in the finite-dimensional case) has optimal time localization and no frequency localization, corresponding to maximally tall and thin cells arranged in one row, while the (discrete) Fourier basis has optimal frequency localization but no time localization, corresponding to maximally short and wide cells stacked into a single column. The wavelet basis is an octave-band decomposition of the phase plane, depicted by the cover in Figure 10.9.

The best-basis of wavelet packets fits a cover to the signal so as to best concentrate the shading into the fewest Heisenberg cells. The compressibility of a sampled signal is easily seen to be the ratio of the total area of the phase plane (N for a signal sampled at N points) divided by the total area of the dark cells (each of area 1). This method allows rectangular Heisenberg cells of all aspect ratios. The best-level or adapted subband basis fits a cover of equal aspect ratio rectangles to the signal, so as to best concentrate the shading. We may automatically analyze

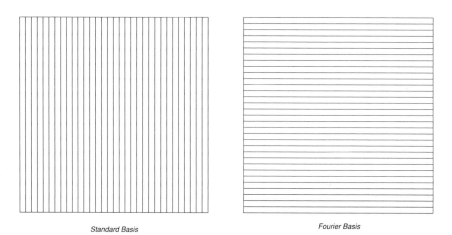

Standard Basis   Fourier Basis

FIGURE 10.8
Phase plane decomposition by the standard and Fourier bases.

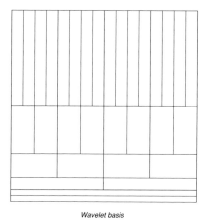

Wavelet basis

FIGURE 10.9
Phase plane decomposition by the wavelet transform.

signals by expanding them in the best basis, then drawing the corresponding phase plane representation. As is clear, the negligible components will not be drawn, as it is not relevant which particular basis is chosen for a subspace containing negligible energy.

Below are certain canonical signals and their automatic analyses by a computer program "WPLab" [W1]. The user selects a quadrature mirror filter from a list of 17 at the right, and the "mother wavelet" determined by that filter is displayed in the small square window at the lower right. The signal is plotted in the rectangular window at bottom, and the phase plane representation is drawn in the large main square window.

We first analyze a relatively smooth transient, spread over 7 samples in a 512 sample signal. Notice that the wavelet analysis at the right correctly localizes the peak in the high-frequency components, but is forced to include poorly localized low-frequency elements as well. The best-basis analysis finds the optimal representation within the library, which in this case is almost a single wavelet packet.

The second signal (in Figure 10.11) is taken from a recording (at 8012 samples per second) of a person whistling. Here the wavelet basis is only able to localize the frequency within an octave, even though the best-basis analysis shows that it falls in a much narrower band. The vertical stripes among the wavelet Heisenberg cells may be used to further localize the frequencies, but the best-basis decomposition performs this analysis automatically.

A chirp is an oscillatory signal with increasing modulation. Some examples are the functions $\sin(250\pi t^2)$ and $\sin(190\pi t^3)$, sampled 512 times on the interval $0 < t < 1$ in Figure 10.12. The modulation increases linearly and quadratically, respectively. The shaded Heisenberg cells in the best bases form a line and a

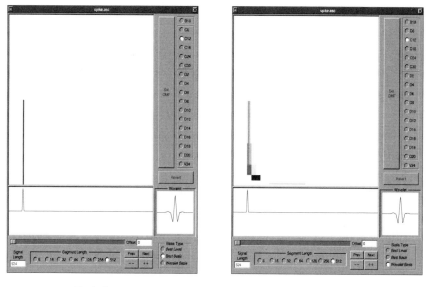

Best basis   Wavelet basis

FIGURE 10.10
Representing a fast transient.

parabolic arc, respectively. In the best-level analyses, all the Heisenberg cells have the same aspect ratio, which is appropriate for a line. In the best-basis analysis, the Heisenberg cells near the zero-slope portion have smaller aspect ratio than those near the large-slope portion.

Such a time-frequency analysis can separate superposed chirps. In Figure 10.13 we see two pairs of linear chirps, differing either by distinct modulation laws or by a time delay. Both are functions on the interval $0 < t < 1$, sampled 512 times. On the left is the function $\sin(250\pi t^2) + \sin(80\pi t^2)$ analyzed in the best wavelet packet basis. Note that the milder-slope chirp is represented by cells of flatter aspect. On the right is $\sin(250\pi t^2) + \sin(250\pi(t - 1/2)^2)$, analyzed by best-level wavelet packets. The downward-sloping line comes from the aliasing of negative frequencies.

10.5   Multidimensional local trigonometric transforms

As seen previously, we can partition the time axis into windows with smooth overlap while maintaining orthogonality of smooth localized cosine or sine waveforms.

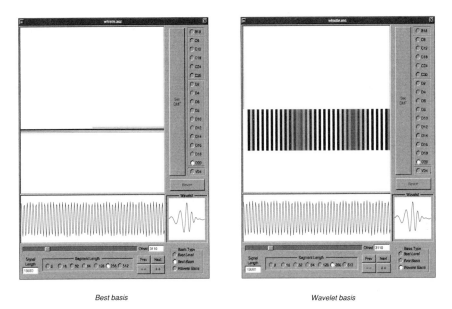

Best basis   Wavelet basis

FIGURE 10.11
Representing a whistle.

A similar procedure is possible in higher dimensions either by simply taking tensor products of the bases in each dimension or, more generally, by segmenting the higher-dimensional phase plane into parallelepipeds of prescribed aspect ratios. The only constraint is that we must assure the compatibility of the various window functions.

For definiteness we can start by considering a basis of the form

$$S_{i,k}(x) = \sqrt{2}p(x-i)\sin\left[(2k+1)\frac{\pi}{2k}(x-i)\right], \quad \text{for } i, k \in \mathbf{Z}^+, i \geq 0,$$

where $\sum p^2(x-i) = 1$ and p is a bell above $[0,1]$ such that $p^2(x) = 1 - p^2(-x)$ near 0, $1 - p^2(2-x) = p^2(x)$ near 1. Then the bivariate functions $S_{i,k}(x)$ and $S_{i',k'}(y)$ form a local trigonometric basis in $\mathbf{R}^2$, based on squares of "size" 1. Assume now that we expand a function on a square of size 2×2 in terms of this basis based on four subsquares of size 1×1 and compare the "cost" of this expansion to the cost on the parent square, choosing the more efficient expansion. Here the bell function chosen for the square $[0,2] \times [0,2]$ would be $P^{(2)}(x,y) = \sqrt{[p^2(x)+p^2(x-1)][p^2(y)+p^2(y-1)]}$. The choice between parent and children can depend on efficiency in coding, or rate of approximation. Such a procedure can continue for several generations leading to a choice of basis based on squares of different sizes.

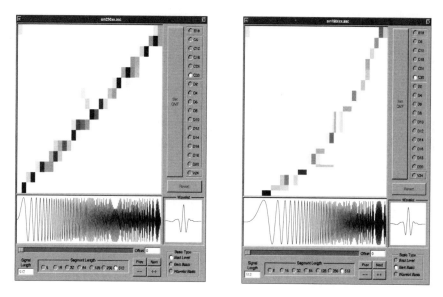

Linear chirp, best level   Quadratic chirp, best basis

FIGURE 10.12
Linear and quadratic chirps.

We apply this procedure to two bivariate analyses. Figure 10.14 shows a fingerprint image to be compressed, and Figure 10.15 shows a matrix to be sparsified. In both of the figures, the selected windows are superposed on the image to indicate how the algorithm chose to segment it. The adapted box expansion for the matrix in Figure 10.15, which is derived from an oscillatory-kernel operator useful in the computation of acoustic scattering in two dimensions, provides an efficient way to describe explicitly the interactions between oscillations and geometry. It shows how regions in space need to be coupled to get an efficient description of interactions. The compressed version leads to a straightforward fast numerical algorithm for applying this image viewed as a matrix to a vector [W2]. It is the analogue in our new language of the traditional decompositions of operators in terms of both space and frequency localizations, which have a long history in the study of singular integrals and Fourier integral operators.

10.6 Multidimensional wavelet packet transforms

The quadrature mirror filter algorithm extends to multidimensional signals by separation of variables. The wavelet packets produced by recursive filtering are products of one-dimensional wavelet packets $W_n(x) = \prod_{i=1}^{d} W_{n_i}(x_i)$, together with

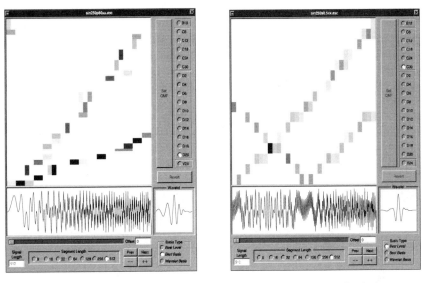

Different slopes, best basis   Different phases, best level

FIGURE 10.13
Superposed chirps.

their isotropic dilations and translations to arbitrary lattice points. Inner products with these multidimensional wavelet packets are computed from averages at the smallest scale, just as in the one-dimensional case. The coefficients may be organized into a stack of d-dimensional intervals, and there is a simple characterization of orthogonal basis subsets. We can restrict our attention to the case $d = 2$. The diagram in Figure 10.16 depicts a decomposition of the unit square into a 3-level wavelet packet tree.

10.7 Denoising and coherent structure extraction

As an application combining these ideas we now describe an algorithm for "denoising" or, more precisely, coherent structure extraction. We chose to view a signal f as being noisy or incoherent relative to a basis $\{\omega_i\}$ if it does not correlate well with the waveforms of the basis, i.e., if its entropy is of the same order of magnitude as $\lg_2 N - \delta$ with small δ, giving a poor compression rate $2^{-\delta}$. From this notion, we are led to the following iterative algorithm based on the several previously defined libraries of orthonormal bases.

We start with a signal f of length N, find the best basis in each library, and select among them the basis minimizing $\varepsilon(f)$. We reorder the coefficients $\alpha_i = \langle f, \omega_i \rangle$

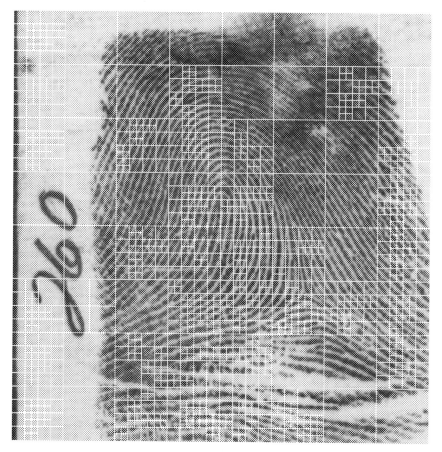

FIGURE 10.14
LCT window selection for a fingerprint compression.

into decreasing order of absolute value $|\alpha_1| \geq |\alpha_2| \geq \cdots \geq |\alpha_{N_0}|$. We may also assume that each α_i for $i \geq N_0$ is below a precision threshold, for example below 0.1% of the total energy. We then decompose $f = c_M + r_M$, where

$$c_M = \sum_1^M \alpha_i \omega_i, \quad \text{and} \quad r_M = \sum_{M+1}^{N_0} \alpha_i \omega_i.$$

We will now say that c_M is coherent and r_M is incoherent if $\varepsilon(r_M) \geq \tau_0$; the threshold τ_0 is chosen to determine if the compression of r_M using $\{\omega_i\}$ is unacceptably bad. We proceed by testing $r_1, r_2, \ldots$ until we reach M for which $\varepsilon(r_M) \geq \tau_0 > 0$,

FIGURE 10.15
LCT window selection for a matrix compression.

$0 < \tau_0 < 1$, or

$$\sum_{i=M_0+1}^{N_0} \frac{\alpha_i^2}{\|r_M\|^2} \lg_2\left(\frac{\|r_M\|^2}{\alpha_i^2}\right) \geq \lg_2(N_0 - M_0 + 1) - \lg_2 \tau_0.$$

Following the "matching pursuit" procedure described by Mallat, we can now consider r_M as a new signal for which we repeat the decomposition, i.e., we pick a new best basis and decompose $r_M = c'_{M_1} + r'_{M_1}$. We can stop after a fixed number of decompositions, or else we can iterate until no new coherent part is obtained. We then superpose the coherent parts to get the coherent part of the signal. What remains is truly "noise" to us, in the sense that it cannot be well-represented by any sequence of our time-frequency atoms.

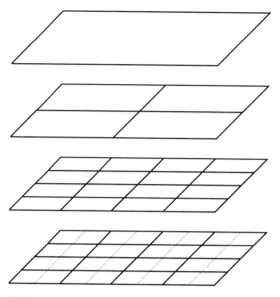

FIGURE 10.16

2-dimensional wavelet decomposition to level 3.

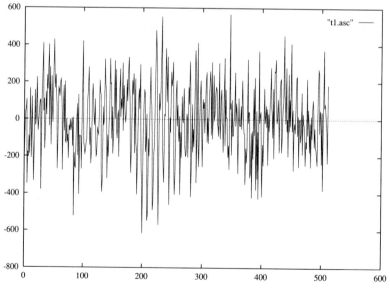

FIGURE 10.17

An underwater sound signal.

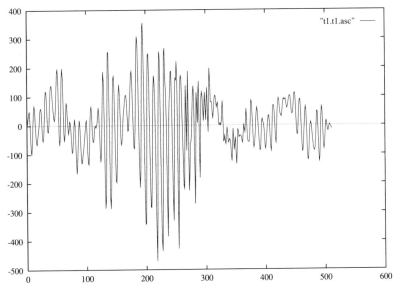

FIGURE 10.18
The first coherent component of the underwater sound signal.

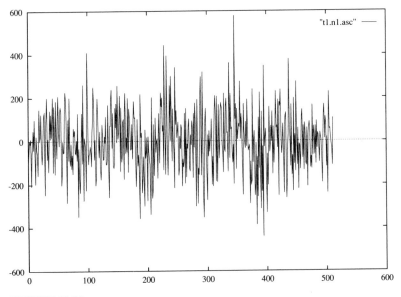

FIGURE 10.19
The first noisy residue of the underwater sound signal.

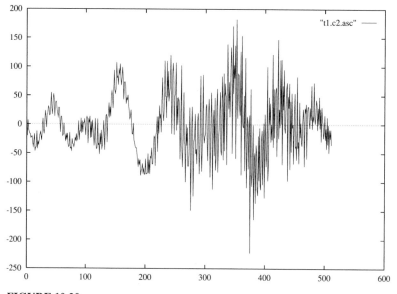

FIGURE 10.20
The coherent part of the preceding noisy part.

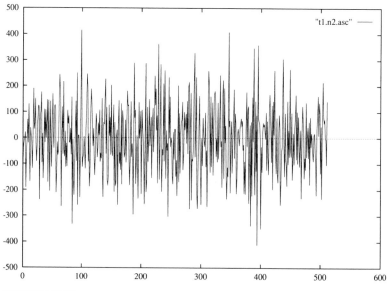

FIGURE 10.21
The second noisy component.

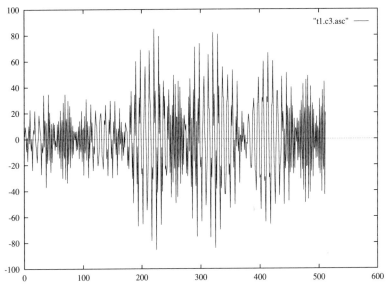

FIGURE 10.22
The third coherent component.

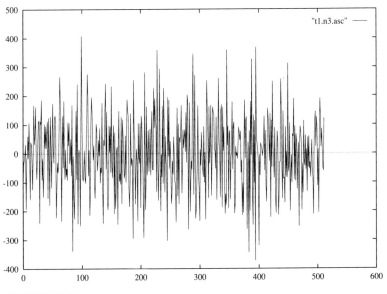

FIGURE 10.23
The third noisy component.

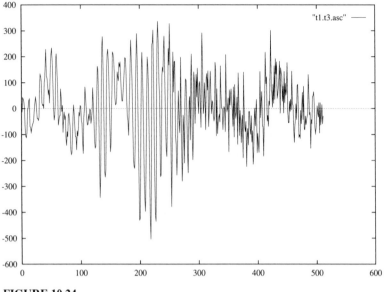

FIGURE 10.24
The sum of the coherent parts.

The denoising algorithm is depicted in Figures 10.17–10.24, which show how a particular signal is peeled into layers as described above. The original signal is a mechanical rumble masked by the noise of aquatic life, recorded through an underwater microphone. The calculations were performed by the program "denoise" [CMaW].

Bibliography

[AWeW] P. Auscher, G. Weiss, and M. V. Wickerhauser. 1992. "Local sine and cosine bases of Coifman and Meyer and the construction of smooth wavelets," in *Wavelets: A Tutorial in Theory and Applications*, C. K. Chui, ed., Boston: Academic Press, pp. 237–256.

[CMaW] R. R. Coifman, F. Majid, and M. V. Wickerhauser. No date. Denoise. Available from FMA&H, 1020 Sherman Ave., Hamden, CT 06514.

[CMe]   R. Coifman and Y. Meyer. 1991. "Remarques sur l'analyse de Fourier à fenêtre," *C. R. Acad. Sci.*, Sér. 1, **312**: 259–261.

[CMeW] R. R. Coifman, Y. Meyer, and M. V. Wickerhauser. 1992. "Wavelet analysis and signal processing," in *Wavelets and Their Applications*, M.

Ruskai, G. Beylkin, R. Coifman, I. Daubechies, S. Mallat, Y. Meyer, L. Raphael, eds., Boston: Jones and Bartlett, pp. 153–178.

[CW]   R. R. Coifman and M. V. Wickerhauser. 1992. "Entropy based methods for best basis selection," *IEEE Trans. Inform. Theory* **32**: 712–718.

[D]   I. Daubechies. 1988. "Orthonormal bases of compactly supported wavelets," *Commun. Pure Appl. Math.* **41**: 909–996.

[DJaJo]   I. Daubechies, S. Jaffard, and J.-L. Journé. 1991. "A simple Wilson orthonormal basis with exponential decay," *SIAM J. Math. Anal.* **22**: 554–572, Erratum **22**: 878.

[M]   H. Malvar. 1990. "Lapped transforms for efficient transform/subband coding," *IEEE Trans. Acoust. Speech Signal Process.* **38**: 969–978.

[Me]   Y. Meyer. 1985–1986. "De la recherche pétrolière à la géometrie des espaces de Banach en passant par les paraproduits," *Seminaire équations aux dérivées partielles*, preprint, École Polytechnique, Palaiseau.

[W1]   M. V. Wickerhauser. 1991. WPLab, version 1.5, *pascal.math.yale.edu*. Available by anonymous ftp in the folder pub/software.

[W2]   M. V. Wickerhauser. 1993. "Computation with adapted time-frequency atoms," in *Proc. Int. Conf. Wavelets Appl.*, Toulouse, 8–13 June 1992, to appear.

11

Near optimal compression of orthonormal wavelet expansions

C.-c. Hsiao, B. Jawerth[*], B.J. Lucier[†], and X. M. Yu

ABSTRACT   We consider the problem of best approximation of a function by finite linear combinations of orthonormal wavelets with the error measured in a standard distribution space. We show that for a large class of spaces an algorithm based only on the size of the coefficients is essentially optimal. This allows us to sharpen some earlier results on nonlinear approximation by wavelets and to establish some new characterizations of functions with a given degree of approximation. We relate these results to the classical problem of comparing the Lebesgue space L_p to the sequence space l_p, and obtain best possible imbeddings in terms of the discrete Lorentz spaces. We also consider compression of operators acting on certain Besov spaces, and characterize the operators that can be compressed well in terms of the size of the (rectangular) wavelet coefficients of the operator kernels.

CONTENTS

[*]This author was partially supported by NSF Grant DMS 8803585, AFOSR Grant 89-0455, and ONR Grant N0001-90-J-1343.
[†]This author was partially supported by NSF Grant DMS-9006219 and ONR Contract N00014-91-J-1152.

11.1   Introduction

The problems we consider in this paper can be motivated and understood in the context of "lossy" image compression, cf. [DeJLu]. For this kind of compression we are given an image and we are interested in removing some of the information in the original image, without degrading the quality too much. In this way we obtain a "compressed" image, which can be stored using less storage and which also can be transmitted more quickly, or using less bandwidth, over a communications channel. The compressed image is thus an approximation of the original. To find a compression algorithm we first decide what kind of approximations we want to consider. To judge the quality of the approximation we next need to decide on how to measure the error between the original and an approximation. Once we have picked the error measure and what kind of approximations we are considering, we are left with the optimization problem how to pick the optimal, or at least near optimal, approximation. To be useful, this near optimal approximation must come with an effective algorithm for constructing it. Finally, once we have chosen an algorithm, it becomes an important problem to understand what kind of images can be compressed well with the algorithm.

We shall consider compression based on the wavelet transform and wavelet decompositions, and address the above mentioned problems when the error is measured in a space which admits a Littlewood–Paley description. In this sense this paper is a continuation of [DeJPo]; in some cases we shall sharpen, and simplify the proofs of, some of the results concerning nonlinear approximation by wavelets obtained there. We shall also establish a number of new results.

Let us start from a fixed function ψ and let

$$\psi_I(x) = 2^{\nu d/2} \psi(2^\nu x - k)$$

for each dyadic cube

$$I = I_{\nu k} = \{x \in \mathbb{R}^d : 2^{-\nu} k_i \leq x_i < 2^{-\nu}(k_i + 1), \, i = 1, \ldots, d\}, \quad \nu \in \mathbb{Z}, \, k \in \mathbb{Z}^d.$$

For a general function f the expansion

$$f = \sum_{I \text{ dyadic}} a_I \psi_I \tag{11.1}$$

is called the *wavelet decomposition* of f. It is well-known that under appropriate conditions on the function ψ it is possible to obtain such decompositions. We shall be particularly interested in *orthonormal wavelets* when the functions ψ_I are orthonormal on $L_2(\mathbb{R})$ (or, more generally, when a finite collection of functions $\psi^{(i)}$, $i = 1, \ldots, N$, together with their dyadic translates and dilates, yield an orthonormal basis $\{\psi_I^{(i)}\}_{i=1, I \text{ dyadic}}^N$ on $L_2(\mathbb{R}^d)$) and form a basis for the space $\mathfrak{F}$ under consideration. Such orthonormal bases have been constructed by [S], [Me], and

[D], as well as others, for most of the classical distribution spaces. The simplest example of a wavelet basis is of course the Haar system on $L_2(\mathbb{R})$.

Note that in the orthonormal case, the coefficients a_I are given by

$$a_I = \langle f, \psi_I \rangle.$$

The transformation

$$f \rightarrow \{\langle f, \psi_I \rangle\}_{I \text{ dyadic}} \tag{11.2}$$

is called the *wavelet transform*.

We shall approximate the original function f by functions with only finitely many nonzero wavelet coefficients. This motivates the following definition. Let $\{\psi_I\}_{I \text{ dyadic}}$ be an orthonormal wavelet basis in some distribution space $\mathfrak{F}$, and let

$$\sigma_N(f)_{\mathfrak{F}} := \inf_{\#\Gamma \leq N} \left\| f - \sum_{I \in \Gamma} b_I \psi_I \right\|_{\mathfrak{F}}, \quad N \geq 1. \tag{11.3}$$

We are thus interested in the following nonlinear approximation problem: how shall we pick N coefficients b_I in order to minimize the error $\sigma_N(f)_{\mathfrak{F}}$ when $\mathfrak{F}$ is one of the standard distribution spaces? We shall solve this, up to an equivalence, for a large class of spaces $\mathfrak{F}$ including the L_p-spaces for $1 < p < \infty$; the Hardy spaces H_p, $0 < p < \infty$; and, more generally, Sobolev spaces based on these spaces. More precisely, in Section 11.3 we shall prove that there are N coefficients $\{\langle f, \psi_I \rangle\}_{I \in \Gamma}$, $\#\Gamma = N$, such that

$$\sigma_N(f)_{\mathfrak{F}} \approx \left\| f - \sum_{I \in \Gamma} \langle f, \psi_I \rangle \psi_I \right\|_{\mathfrak{F}}, \tag{11.4}$$

and the algorithm for choosing $\{\langle f, \psi_I \rangle\}_{I \in \Gamma}$ among all $\langle f, \psi_I \rangle$s is based on the size of the coefficients. It is interesting to note that the proof of this fact requires us to consider the problem, already considered by Banach, about the relation between function spaces, such as L_p, and sequence spaces, such as l_p. Banach [B] proved that L_p and l_p can only be isomorphic when $p = 2$ or $p = \infty$. It is very easy to show, for instance by using the Haar basis, that L_2 is indeed isomorphic to l_2, and in the fifties Pelczynski proved that L_∞ is isomorphic to l_∞. This raises the question what we can say for other ps, $p \neq 2, \infty$. In Section 11.4 we prove that L_p, $1 < p < \infty$, is isomorphic to a space x_p for which we have the continuous inclusions

$$l_{p,\min(p,2)} \subset x_p \subset l_{p,\max(p,2)}.$$

Here $l_{p,q}$ denote the Lorentz spaces based on a infinite countable index set.

As a corollary to the result (11.4) about $\sigma_N(f)_{\mathfrak{F}}$, we immediately obtain, also in Section 11.4, one of the main results in [DeJPo]: under certain conditions on ψ

we have that for each $1 < p < \infty$ and for a certain range of α

$$\sum_{N=1}^{\infty}[N^{\alpha/d}\sigma_N(f)_{L_p(\mathbb{R}^d)}]^{\tau}\frac{1}{N} < \infty \iff f \in B^{\alpha} \tag{11.5}$$

where $B^{\alpha} = B_{\tau}^{\alpha,\tau}$, $\tau = (\alpha/d + 1/p)^{-1}$, are Besov spaces (see Section 11.2 for their definition). In [DeJPo] the limiting case $p = \infty$ was left open. When $p = \infty$ there are in fact two natural candidates to use when measuring the error. The most obvious one is of course L_{∞}. However, it is well-known that the L_p spaces can be renormalized so that BMO, the space of functions of bounded mean oscillation, becomes a limiting space as p tends to infinity. Now, the just-mentioned result suggests that $f \in B^{\alpha}$ is equivalent to $\sum_{N=1}^{\infty}[N^{\alpha/d}\sigma_N(f)_{\text{BMO}(\mathbb{R}^d)}]^{\tau}1/N < \infty$, $\tau = d/\alpha$. In Section 11.4 we shall show that a result of Rochberg–Taibleson immediately yields that this is indeed the case.

Since L_{∞} does not have an unconditional basis, there is no nice description of this space in terms of the size of the wavelet coefficients of its elements. Sometimes BMO is a good substitute. However, for certain applications the Besov space $B_{\infty}^{0,1}$ is better, since a small error in $B_{\infty}^{0,1}$ guarantees a small error in the L_{∞}. In Section 11.5 we shall derive an algorithm which produces a near optimal approximation of a given element f when the error is measured in $B_{\infty}^{0,1}$. In Section 11.6 we exploit the same techniques in the context of compression of operators. More specifically, we show how to obtain an optimal approximation of an operator, acting on the Besov spaces $B_1^{\alpha,1}$, by certain natural, simpler operators, when the error is measured in the operator norm.

In the next section, Section 11.2, we start by collecting some auxiliary results that we need later and by defining the class of spaces $\mathfrak{F}$ we shall consider.

Let us finally note that there are other situations when our techniques and results apply without much change. As we shall see, it is important for us to be able to transfer from function and distribution spaces to corresponding sequence spaces. This is for example possible with biorthogonal wavelets (cf. [CoDFe], [Co]), and, hence, our results carry over to this more general setting with only minor modifications.

A simplification we shall make, for notational simplicity, is to state the results only in the one-dimensional case. When the higher-dimensional result is significantly different we shall point this out explicitly.

11.2   The function spaces

In this section we shall define the spaces that we will consider, and recall some of their properties. We shall be brief and we refer to [FJ] for more details. It is a classical fact that many function spaces share a common underlying structure. It is possible to show that spaces like L_p for $1 < p < \infty$, and more generally the (real) Hardy spaces H_p, $0 < p < \infty$; Sobolev spaces based on these spaces; Besov spaces; and the space BMO of functions of bounded mean oscillation, for instance, all can be defined by using an appropriate Littlewood–Paley expression.

To be more specific, let us look at the case of $L_p = L_p(\mathbb{R})$, $1 < p < \infty$. Classically the norm on this space is of course defined by

$$\|f\|_{L_p} = \left(\int_{\mathbb{R}} |f(x)|^p \, dx \right)^{1/p}.$$

Now let $\{h_I\}_{I \text{ dyadic}}$ be the Haar system on the line. The functions h_I are obtained from the single function

$$h(x) = \begin{cases} 1 & \text{if } 0 \leq x < 1/2, \\ -1 & \text{if } 1/2 < x < 1, \\ 0 & \text{otherwise} \end{cases}$$

by translation and dilation:

$$h_I(x) = 2^{\nu/2} h(2^\nu x - k)$$

when $I = [2^{-\nu}k, 2^{-\nu}(k + 1))$ for some integers ν and k. The Haar system is an orthonormal basis on L_2. As a consequence, a general function $f \in L_2$ has the expansion

$$f(x) = \sum_{I \text{ dyadic}} \langle f, h_I \rangle \, h_I(x),$$

and

$$\|f\|_{L_2} = \left(\sum_{I \text{ dyadic}} |\langle f, h_I \rangle|^2 \right)^{1/2}.$$

The mapping $S_h : f \to \{\langle f, h_I \rangle\}_{I \text{ dyadic}}$ is thus an isometry between L_2 and l_2. For other $p \neq 2$, the Haar system is still an unconditional basis and from Littlewood–Paley theory it follows that

$$\|f\|_{L_p} \approx \left\| \left(\sum_{I \text{ dyadic}} (|\langle f, h_I \rangle| \, |\tilde{\chi}_I(x)|)^2 \right)^{1/2} \right\|_{L_p}. \tag{11.6}$$

Here $\tilde{\chi}_I(x) = \chi(x)/|I|^{1/2}$ is the "L_2-normalized" characteristic function of I. (We use the notation $\|f\| \approx \|g\|$ if there are constants C_1 and C_2 such that $C_1 \|f\| \leq \|g\| \leq C_2 \|f\|$.) Notice that when $p = 2$, by Fubini's theorem the expression on the right reduces to the l_2-norm of $\{\langle f, h_I \rangle\}_{I \text{ dyadic}}$. For other ps the expression still defines a sequence space norm, although much more complicated than the l_p-norm, say, and the sequence space with this more complicated norm is isomorphic to L_p.

The Haar system is the simplest example of an orthonormal wavelet basis. There are other examples $\{\psi_I\}_{I \text{ dyadic}}$ where the ψ_Is are obtained, in an analogous manner, from a function ψ which is arbitrarily smooth and has arbitrarily many vanishing moments, i.e. , $\psi \in C^{(N)}$ and $\int x^\gamma \psi(x)\,dx = 0$ for all $0 \le \gamma \le N$ for any given fixed integer $N \ge 1$. Indeed, using multiresolution analysis (cf. [M], [D]) it is possible to find such ψ which decay very rapidly at infinity, or even has finite support (with the measure of the support being proportional to N). For these more sophisticated bases we still have a characterization of L_p analogous to (11.6). These bases can be used to characterize other spaces as well. The Hardy spaces, for example, coincide with L_p for $p > 1$ and for all p, $0 < p < \infty$, we have, if N is sufficiently large (exactly how large it needs to be depends on p),

$$\|f\|_{H_p} \approx \left\| \left(\sum_{I \text{ dyadic}} \left(|\langle f, \psi_I \rangle| \, \tilde{\chi}_I(x) \right)^2 \right)^{1/2} \right\|_{L_p}. \tag{11.7}$$

The situation is similar for the Sobolev spaces. Let $l \ge 0$ be an integer and let $1 < p < \infty$. The (homogeneous) Sobolev space H_p^l is defined by requiring that the kth derivative is in L_p,

$$\|f\|_{H_p^l} = \|D^k f\|_{L_p} < \infty.$$

For a given l, if N is sufficiently large, we have

$$\|f\|_{H_p^l} \approx \left\| \left(\sum_{I \text{ dyadic}} \left(|I|^{-l} |\langle f, \psi_I \rangle| \, \tilde{\chi}_I(x) \right)^2 \right)^{1/2} \right\|_{L_p}. \tag{11.8}$$

The spaces H_p^l can be generalized in two useful ways by allowing nonintegral values of l and by using the full range $0 < p < \infty$. We let $I^\alpha f$, $\alpha \in \mathbb{R}$, be the Riesz potential of order α of f so that $\widehat{I^\alpha f}(\xi) = |\xi|^{-\alpha} \hat{f}(\xi)$. The space H_p^α, $\alpha \in \mathbb{R}$, $0 < p < \infty$, is the collection of all elements f (or, more precisely, $f \in S'/\mathcal{P}$, tempered distributions modulo polynomials) such that

$$\|f\|_{H_p^\alpha} = \|I^{-\alpha} f\|_{H_p} < \infty.$$

When $\alpha = l$ and $p > 1$, this space coincides, up to a norm equivalence, with H_p^l. The corresponding sequence space is the space h_p^α of all sequences $s = \{s_I\}_{I \text{ dyadic}}$ such that

$$\|s\|_{h_p^\alpha} = \left\| \left(\sum_{I \text{ dyadic}} \left(|I|^{-\alpha} |s_I| \, \tilde{\chi}_I(x) \right)^2 \right)^{1/2} \right\|_{L_p} < \infty. \tag{11.9}$$

(In higher dimensions, the exponent $-\alpha$ is replaced by $-\alpha/d$.) The wavelet transform $f \to \{\langle f, \psi_I \rangle\}_{I \text{ dyadic}}$ is now an isomorphism between H_p^α and h_p^α (if N is sufficiently large, depending on p and α).

The Besov spaces can be defined in a number of equivalent ways, using moduli of smoothness, various Littlewood–Paley functions, etc. We will not need these different descriptions; what we need is the characterization in terms of orthonormal wavelets. Let us first define the appropriate sequence spaces. The space b_p^α, $\alpha \in \mathbb{R}$,

$0 < p \leq \infty$, is the collection of all sequences $s = \{s_I\}_{I \, \mathrm{dyadic}}$ such that

$$\|s\|_{b_p^\alpha} = \left(\sum_{\nu \in \mathbb{Z}} \left\| \sum_{l(I)=2^{-\nu}} |I|^{-\alpha} |s_I| \, \tilde{\chi}_I(x) \right\|_{L_p}^p \right)^{1/p} < \infty. \tag{11.10}$$

(Again, in higher dimensions the exponent $-\alpha$ is really $-\alpha/d$.) There are several facts worth noting about this definition. Since the dyadic intervals of the same side length are pairwise disjoint, we can easily carry out the integration involved in the L_p-norm. In this way we see that

$$\|s\|_{b_p^\alpha} = \left(\sum_{I \, \mathrm{dyadic}} \left(|I|^{-\alpha - 1/2 + 1/p} |s_I| \right)^p \right)^{1/p}.$$

In particular,

$$b_2^0 = l_2,$$

and

$$\|s\|_{b_p^\alpha} = \left(\sum_{I \, \mathrm{dyadic}} |s_I|^p \right)^{1/p} \quad \text{if } \frac{1}{p} = \alpha + \frac{1}{2}. \tag{11.11}$$

We may now define the Besov spaces B_p^α by requiring

$$\|f\|_{B_p^\alpha} = \| \{ \langle f, \psi_I \rangle \}_{I \, \mathrm{dyadic}} \|_{b_p^\alpha} < \infty.$$

The wavelet transform $f \to \{\langle f, \psi_I \rangle\}_{I \, \mathrm{dyadic}}$ is again an isometry and allows us to identify B_p^α and b_p^α.

In this paper, we will mainly be considering the so called homogeneous version of the function spaces. These spaces are traditionally denoted with a dot ($\dot{B}_p^\alpha$, for example); here we shall for notational simplicity omit this dot. In the case of the Besov spaces, for the definition given here to coincide with classical definitions such as those involving modulus of smoothness or Littlewood–Paley descriptions, we need, in general, to put standard restrictions on the parameters N, α, and p (cf. [FJ]). As a further remark, we note that there are obvious analogs of the results we prove for the nonhomogeneous spaces as well.

11.3 A basic compression principle

As we discussed in the previous section, each space $\mathfrak{F}$ we shall consider is isomorphic to a corresponding sequence space $\mathfrak{f}$. For each dyadic cube I we let e_I be the Kronecker sequence

$$(e_I)_J = \begin{cases} 1 & \text{if } J = I, \\ 0 & \text{otherwise.} \end{cases}$$

Given a finite set Γ we also let $1_{\Gamma,\mathfrak{f}}$ be the normalized characteristic sequence

$$1_{\Gamma,\mathfrak{f}} = \sum_{I \in \Gamma} \frac{e_I}{\|e_I\|_{\mathfrak{f}}}.$$

We call the normed space $\mathfrak{F}$ a *p-space*, $1 \leq p \leq \infty$, if there are constants C_1 and C_2 such that

$$C_1 (\#\Gamma)^{1/p} \leq \|1_{\Gamma,\mathfrak{f}}\|_{\mathfrak{f}} \leq C_2 (\#\Gamma)^{1/p}, \tag{11.12}$$

for all finite sets Γ. Since $\mathfrak{F}$ and $\mathfrak{f}$ are isomorphic, this is equivalent to

$$\left\| \sum_{I \in \Gamma} \frac{\psi_I}{\|\psi_I\|_{\mathfrak{F}}} \right\|_{\mathfrak{F}} \approx (\#\Gamma)^{1/p}, \quad N \geq 1. \tag{11.13}$$

For such *p*-spaces we have the following principle.

BASIC PRINCIPLE 11.1
Let $1 \leq p \leq \infty$. If $\mathfrak{F}$ is a p-space, then

$$\frac{1}{c} \|f^N\|_{\mathfrak{F}} \leq \sigma_N(f)_{\mathfrak{F}} \leq \|f^N\|_{\mathfrak{F}}, \quad N \geq 1,$$

where f^N is obtained from the decomposition of f by choosing N terms with $|\langle f, \psi_I \rangle| \, \|\psi_I\|_{\mathfrak{f}}$ as large as possible.

PROOF   Clearly, $\sigma_N(f)_{\mathfrak{F}}$ is dominated by the expression on the right, and we only need to prove the opposite inequality. If we rephrase this on the sequence space level, we see that it is sufficient to show that for each sequence $s = \{s_I\}_{I \text{ dyadic}}$

$$\frac{1}{c} \|s^N\|_{\mathfrak{f}} \leq \Sigma_N(s)_{\mathfrak{f}} := \inf_{\#\{I : s_I^0 \neq 0\} \leq N} \|s - s^0\|_{\mathfrak{f}}, \tag{11.14}$$

where $s^N = \{s_I^N\}_{I \text{ dyadic}}$ is obtained from the sequence s by taking N elements with $s_I \|e_I\|_{\mathfrak{f}}$ as large as possible.

We notice first that if we start using an element from s, then we may just as well use it completely; more precisely, we have that

$$\Sigma_N(s)_{\mathfrak{f}} = \inf_{\#\Gamma \leq N} \|s^\Gamma\|_{\mathfrak{f}}$$

where, for each set Γ, s^Γ is defined by $(s^\Gamma)_I = 0$ if $I \in \Gamma$, $= s_I$ otherwise. For some appropriate set, Γ_0 say, we have $s^N = s^{\Gamma_0}$. Now, if Γ is an arbitrary set with $\#\Gamma \leq N$, then $(s^\Gamma)_I = (s^{\Gamma_0})_I$ when $I \in (\Gamma \cap \Gamma_0) \cup (\Gamma^c \cap \Gamma_0^c)$, and hence, by the triangle inequality,

$$\|s^{\Gamma_0}\|_{\mathfrak{f}} \leq \left\| \{s_I\}_{I \in \Gamma \cap \Gamma_0^c} \right\|_{\mathfrak{f}} + \left\| \{s_I\}_{I \in \Gamma^c \cap \Gamma_0} \right\|_{\mathfrak{f}} \leq \left\| \{s_I\}_{I \in \Gamma \cap \Gamma_0^c} \right\|_{\mathfrak{f}} + \|s^\Gamma\|_{\mathfrak{f}}. \tag{11.15}$$

By the definition of Γ_0, we have

$$\max_{I \in \Gamma \cap \Gamma_0^c} |s_I| \, \|e_I\|_{\mathfrak{f}} \leq \min_{I \in \Gamma_0 \cap \Gamma^c} |s_I| \, \|e_I\|_{\mathfrak{f}}.$$

Since $\mathfrak{F}$ is a p-space, this implies that

$$\left\|\{s_I\}_{I\in\Gamma\cap\Gamma_0^c}\right\|_{\mathfrak{f}} \leq \max_{I\in\Gamma\cap\Gamma_0^c}|s_I|\,\|e_I\|_{\mathfrak{f}}\,\left\|1_{\Gamma\cap\Gamma_0^c,\mathfrak{f}}\right\|_{\mathfrak{f}}$$

$$\leq \min_{I\in\Gamma_0\cap\Gamma^c}|s_I|\,\|e_I\|_{\mathfrak{f}}\,\frac{\#(\Gamma\cap\Gamma_0^c)^{1/p}}{\#(\Gamma_0\cap\Gamma^c)^{1/p}}\,\frac{c_2}{c_1}\,\left\|1_{\Gamma_0\cap\Gamma^c,\mathfrak{f}}\right\|_{\mathfrak{f}} \leq \frac{c_2}{c_1}\,\|s^{\Gamma}\|_{\mathfrak{f}}. \tag{11.16}$$

Inserting this in (11.15) we get $\|s^{\Gamma_0}\|_{\mathfrak{f}} \leq (c_2/c_1 + 1)\,\|s^{\Gamma}\|_{\mathfrak{f}}$, and since Γ is arbitrary, this proves (11.14) with $c = 1 + c_2/c_1$. ∎

THEOREM 11.2
The space H_p^{α}, $\alpha \in \mathbb{R}$, $1 \leq p < \infty$, is a p-space.

PROOF   As we mentioned in the previous section, $H = H_p^{\alpha}$, $\alpha \in \mathbb{R}$, $1 \leq p < \infty$, is isomorphic to $h = h_p^{\alpha}$.

We claim that

$$c_1(\#\Gamma)^{1/p} \leq \left\|\sup_{I\in\Gamma}|I|^{-1/p}\chi_I(x)\right\|_{L_p}. \tag{11.17}$$

Because of the continuous imbedding $l_2 \subset l_{\infty}$, we have

$$\left\|\sup_{I\in\Gamma}|I|^{-1/p}\chi_I(x)\right\|_{L_p} \leq \left\|\left(\sum_{I\in\Gamma}(|I|^{-1/p}\chi_I(x))^2\right)^{1/2}\right\|_{L_p} = \|1_{\Gamma,h}\|_h.$$

Hence, (11.17) would prove one of the inequalities in (11.12).

The inequality (11.17) is easy to establish. We let $F(x) = \sup_{I\in\Gamma}\frac{1}{|I|^{1/p}}\chi_I(x)$ and $\Gamma_k = \{I \in \Gamma : l(I) = 2^{-k}\}$, $k \in \mathbb{Z}$. We have

$$\int F(x)^p\,dx = \sum_{k\in\mathbb{Z}}\int_{2^{k/p}\leq F<2^{(k+1)/p}}F(x)^p\,dx \geq c\sum_{k\in\mathbb{Z}}2^k|\{x : F(x)^p \geq 2^k\}|$$

$$\geq c\sum_{k\in\mathbb{Z}}2^k\sum_{I\in\Gamma_k}|I| = c\sum_{k\in\mathbb{Z}}\#\Gamma_k = c\#\Gamma.$$

This proves (11.17).

Because of the continuous imbedding $l_1 \subset l_2$, to prove the second inequality in (11.12) it is enough to show that

$$\left\|\sum_{I\in\Gamma}|I|^{-1/p}\chi_I(x)\right\|_{L_p} \leq C_2(\#\Gamma)^{1/p} \tag{11.18}$$

Since the dyadic cubes of the same side length are pairwise disjoint, we have, for some appropriate integer k_0 depending on x,

$$\sum_{I\in\Gamma}|I|^{-1/p}\chi_I(x) \leq \sum_{k\leq k_0}2^{k/p} = 2^{k_0/p}\,1/(1 - 2^{-1/p})$$

$$\leq 1/(1 - 2^{-1/p}) \left(\sum_{I \in \Gamma} |I|^{-1} \chi_I(x) \right)^{1/p} .$$

Raising both sides to the pth power and integrating, yields (11.18) with $C_2 = 1/(1 - 2^{-1/p})$ (in higher dimensions $C_2 = 1/(1 - 2^{-d/p})$). ∎

As we shall see later, there are classical spaces which are not p-spaces for any p. An interesting example which is not very classical but relevant in the context of image compression (cf. [DeJLu]) is the space $\mathcal{E} = \mathcal{E}(\{\alpha_\nu\})$ with norm

$$\|f\|_{\mathcal{E}} = \left(\sum_I |\langle f, \psi_I \rangle |I|^{-\alpha_\nu/d}|^2 \right)^{1/2} ,$$

for some fixed sequence $\alpha_\nu \in \mathbb{R}$, $\nu \in Z$. This is Sobolev type space with the smoothness parameter (possibly) changing from scale to scale. When α_ν is not a constant, it is easy to see that this space cannot be associated with one single p. In fact,

$$\sigma_N(f)_{\mathcal{E}} = \|\{\langle f, \psi_I \rangle\}^N\|_{\mathcal{E}}$$

where $\{\langle f, \psi_I \rangle\}^N$ is the sequence obtained from $\{\langle f, \psi_I \rangle\}$ by selecting N coefficients with $|\langle f, \psi_I \rangle |I|^{-\alpha_\nu/d}|$ as large as possible.

11.4 L_p and sequence spaces

Let $l_{p,q}$, $0 < p, q \leq \infty$, be the Lorentz space based on the measure space of all dyadic cubes equipped with counting measure. (For more on the Lorentz spaces, their basic properties, and further references see [BeL].) It is an easy exercise to show that (11.12) is equivalent to the requirement that for any sequence $s = \{s_I\}_I$ dyadic

$$C_1 \|s\|_{l_{p,\infty}} \leq \|\{s_I/\|e_I\|_f\}\|_f \leq C_2 \|s\|_{l_{p,1}} . \tag{11.19}$$

This reformulation is convenient and, for instance, immediately leads to several interpolation results for p-spaces.

Given a pair of "compatible" spaces $\bar{A} = (A_0, A_1)$, we recall (cf. [BeL]) that one way to define the real interpolation spaces uses the best approximation functional E defined by

$$E(\lambda, s; \bar{A}) = \inf_{\|s^0\|_0 \leq \lambda} \|s - s^0\|_1 .$$

The real interpolation space $\bar{A}_{\theta q}$, $0 < \theta < 1$, $1 \leq q \leq \infty$, is the space of all $s \in A_0 + A_1$ with

$$\|s\|_{\bar{A}_{\theta q}} = \left(\int_0^\infty (\lambda^{1-\theta} E(\lambda, s; \bar{A})^\theta)^q \frac{d\lambda}{\lambda} \right)^{1/q} .$$

In the particular case of l_p-spaces, it is well-known (cf. [BeL], ch. 7) that for any fixed $0 < p \leq \infty$ and $0 < r \leq \infty$ we have

$$(l_0, l_{p,r})^\theta_{\theta,q} = l_\tau \tag{11.20}$$

with $1/\tau = 1 - \theta/\theta + 1/p$ and $1/q = 1 - \theta + \theta/p$. Here l_0 is the space of finite sequences equipped with the "norm" $\|s\|_{l_0} = \#\{I : s_I \neq 0\}$.

Combining Theorem 11.2 with (11.20) and using (11.19) immediately yields a characterization of the spaces with a given degree of nonlinear approximation by finite wavelet expansions.

COROLLARY 11.3
For fixed $\alpha \in \mathbb{R}$ and $1 \leq p < \infty$ we let $H = H_p^\alpha$. We have

$$\sum_{N=1}^\infty [N^{\beta/d} \sigma_N(f)_H]^\tau \frac{1}{N} < \infty \iff f \in B_\tau^{\alpha+\beta}$$

with $1/\tau = 1/p + \beta/d$ and $\beta > 0$.

PROOF Note that $\|e_I\|_h = |I|^\gamma$ with $\gamma = 1/p - \alpha/d - 1/2$. Hence, if we let $s = \{\langle f, \psi_I \rangle |I|^\gamma\}_I$ and use (11.19), we obtain that

$$\frac{1}{c} E(N, s; l_0, l_{p,\infty}) \leq \sigma_N(f)_H \leq cE(N, s; l_0, l_{p,1}).$$

By (11.20) and a simple discretization, $\sum_{N=1}^\infty [N^{\beta/d} \sigma_N(f)_H]^\tau 1/N$ is finite if and only if $s \in l_\tau$. However, using (11.11) we readily see that $\{\langle f, \psi_I \rangle |I|^\gamma\}_{I \text{ dyadic}} \in l_\tau$ is equivalent to $f \in B_\tau^{\alpha+\beta}$, and this finishes the proof. ∎

Let us make a brief digression and show how the inclusions (11.19) can be refined, and how this leads to a description of how close we can come to L_p with an l_p-type sequence space. Of course, it is trivial that L_2 is isomorphic to l_2, and Pelczynski [P] proved some thirty years ago that L_∞ is isomorphic to l_∞. As we mentioned in the introduction, it is a classical fact, due to Banach [B], that these are the only possibilities: L_p and l_p can not be isomorphic unless $p = 2$ or $p = \infty$. Next we shall see that by combining some of the inequalities established in the proof of Theorem 11.2 with a simple interpolation result we obtain that L_p, $1 < p < \infty$, is isomorphic to a sequence space which is continuously embedded between the Lorentz spaces $l_{p,\max(p,2)}$ and $l_{p,\min(p,2)}$. More specifically, we shall prove that the space $h_p^{\alpha,2}$, $\alpha/d = 1/p - 1/2$, is continuously embedded between these spaces; of course, $h_p^{\alpha,2} \approx h_p^0 \approx L_p$ so this will be enough. We note that a corresponding result is also true for the Hardy space H_1.

PROPOSITION 11.19
Let $1 \leq p < \infty$ and let $\alpha/d = 1/p - 1/2$. Then we have the continuous inclusions

$$l_{p,\min(p,2)} \subset h_p^\alpha \subset l_{p,\max(p,2)}.$$

PROOF   The spaces h_p^α belong to a more general scale of spaces $h_p^{\alpha,q}$ with norm

$$\|s\|_{h_p^{\alpha,q}} = \left\|\left(\sum_{I \text{ dyadic}} (|I|^{-\alpha/d} s_I \tilde{\chi}_I(x))^q\right)^{1/q}\right\|_{L_p},$$

where $\tilde{\chi}_I = \chi_I/|I|^{1/2}$ denotes the L_2-characteristic function of I. In particular, $h_p^\alpha = h_p^{\alpha,2}$. Expressed in terms of these spaces, the inequality (11.17) readily implies that we have the continuous inclusion $h_p^{\alpha,\infty} \subset l_{p,\infty}$ with $\alpha/d = 1/p - 1/2$; (11.18) similarly shows that $l_{p,\infty} \subset h_p^{\alpha,1}$ with $\alpha/d = 1/p - 1/2$. When $q = p$ we may carry out the integration in the definition of the spaces $h_p^{\alpha,q}$, and this shows that we have $h_p^{\alpha,p} = l_p$. Now the proposition follows by interpolation. For instance, suppose $p \geq 2$. Then

$$l_{p,2} = (l_{p,1}, l_p)_\theta \subset (h_p^{\alpha,1}, h_p^{\alpha,p})_\theta = h_p^{\alpha,2} \tag{11.21}$$

if $1/2 = (1-\theta)/1 + \theta/p$. Here we have used that the Lorentz spaces and the spaces $h_p^{\alpha,q}$ behave in the natural way under interpolation, cf. [R], [Hu], and [FJ]. On the other hand, if $p < 2$, then using the imbedding $l_p \subset l_2$ in the definition of $h_p^{\alpha,q}$ yields

$$l_p = h_p^{\alpha,p} \subset h_p^{\alpha,2}. \tag{11.22}$$

The two inclusions (11.21) and (11.22) together prove the desired left inclusion. The proof of the right one is similar.   ∎

The inclusions in Theorem 11.19 are best possible in the sense that $\min(p,2)$ can not be replaced by a larger q, and $\max(p,2)$ can not be replaced by a smaller. To see this let us fix a sequence $\lambda = \{\lambda_k\}_{k \in \mathbb{Z}}$ of numbers (for simplicity, we shall carry out the argument in one dimension). We first distribute these numbers over the cubes of side length 1: we define $s^{(1)} = \{s_I^{(1)}\}$ by letting

$$s_I^{(1)} = \begin{cases} \lambda_k & \text{if } I = k + [0,1], \\ 0 & \text{otherwise.} \end{cases}$$

Since the dyadic intervals of side length 1 are pairwise disjoint, we see that

$$\|s^{(1)}\|_{h_p^\alpha} = \left(\sum_k |\lambda_k|^p\right)^{1/p}. \tag{11.23}$$

On the other hand, let $\lambda^* = \{\lambda_i^*\}_{i=1}^\infty$ be the sequence λ sorted in decreasing order of magnitude. We now distribute these numbers over the intervals $I \subset [0,1]$ by filling up level by level, starting from $[0,1]$ and λ_1^*. More precisely, we define $s^{(2)} = \{s_I^{(2)}\}$ by letting

$$s_I^{(2)} = \begin{cases} |I|^{1/2-1/p}\lambda_{k+2^\nu}^* & \text{if } I = [k2^{-\nu}, (k+1)2^{-\nu}), \quad \nu \geq 0, \ 0 \leq k < 2^\nu, \\ 0 & \text{otherwise.} \end{cases}$$

A simple calculation shows that

$$\left\| s^{(2)} \right\|_{h_p^\alpha} \approx \left(\sum_{\nu \geq 0} (2^{\nu/p} \lambda_{2^\nu}^*)^2 \right)^{1/2} \approx \|\lambda\|_{l_{p,2}}. \tag{11.24}$$

The two equivalences (11.23) and (11.24) clearly show that the exponents in the Proposition are best possible. In fact, these equivalences show a little bit more. Suppose x is a rearrangement invariant space based on a countably infinite discrete measure space, normalized so that the mass of each point is one. If x is continuously imbedded in h_p^α, then x also imbeds in both l_p and $l_{p,2}$. Conversely, if h_p^α is continuously imbedded in x, then so are both l_p and $l_{p,2}$. In particular, we have the following theorem.

THEOREM 11.5
Let $1 < p < \infty$. Suppose x is a rearrangement invariant space based on a countably infinite discrete measure space, normalized so that the mass of each point is one. If L_p is isomorphic to x, then

$$l_{p,\min(p,2)} \subset x \subset l_{p,\max(p,2)}.$$

An analogous fact is also true for the Hardy space H_1.

PROOF This follows from the discussion before the theorem simply by using that the scale of Lorentz spaces $l_{p,q}$ is increasing in the second parameter: $l_{p,q_1} \subset l_{p,q_2}, q_1 \leq q_2$. ∎

11.5 Compression in BMO

For many applications L_∞ is the natural space in which to measure the error. Unfortunately, L_∞ does not have any unconditional bases and, in particular, it does not have a Littlewood–Paley description in terms of orthonormal wavelets. There are several substitutes for L_∞. One which is often used in harmonic analysis is the space BMO of functions of bounded mean oscillation. This is the space we are going to consider in this section. In the next we shall instead use a Besov space which is also very similar to L_∞.

We shall quickly show how the results of Rochberg–Taibleson [RoT] immediately yields a characterization of the functions which can be approximated well by finite wavelet decompositions, with the error measured in the BMO norm.

We first need the spaces $h_p^{\alpha,q}$ when $p = \infty$; for a finite p they were introduced in the proof of Proposition 11.19. The most obvious extension to $p = \infty$, obtained by simply replacing L_p by L_∞ in the expression for the norm on $h_p^{q,\alpha}$, is unfortunately not the right one. The correct definition is more complicated, cf. [FJ]. We put

$$\|s\|_{h_\infty^{\alpha,q}} = \sup_{I \text{ dyadic}} \left(\frac{1}{|I|} \int_I \sum_{J \subset I} (|J|^{-\alpha/d} |s_J| |\tilde{\chi}_J(x)|)^q \, dx \right)^{1/q}.$$

In the usual way, when $q = \infty$ this means that we replace the sum by the appropriate supremum. In particular, we readily see that $h_\infty^{\alpha,\infty} = b_\infty^\alpha$. When $q < \infty$ we may carry out the integration and obtain that

$$\|s\|_{h_\infty^{\alpha,q}} = \sup_{I \text{ dyadic}} \left(\frac{1}{|I|} \sum_{J \subset I} (|J|^{-\alpha/d-1/2} |s_J|)^q |J| \right)^{1/q}.$$

This makes it clear that in fact $\|s\|_{h_\infty^{\alpha,q}}^q$ is equivalent to the so called Carleson norm of the discrete measure

$$\sum_J (|J|^{-\alpha/d-1/2} |s_J|)^q |J| \delta_{x_J, l(J)},$$

where $x_J = 2^{-\nu} k$ is the "lower left corner" of $J = J_{\nu k}$. One way to express the result in [RoT] is now as follows.

PROPOSITION 11.6
(Rochberg-Taibleson [RoT].) For each fixed $\alpha \in \mathbb{R}$ we have

$$\sum_{N=1}^\infty [N^{\beta/d} \Sigma_N(s)_{h_\infty^{\alpha,q}}]^\tau \frac{1}{N} \approx \sum_{N=1}^\infty [N^{\beta/d} \Sigma_N(s)_{b_\infty^\alpha}]^\tau \frac{1}{N}$$

for each $0 < \tau \leq \infty$ and $\beta > 0$.

   In other words, the spaces $h_\infty^{\alpha,q}$ are so close to each other, for different qs, that all the approximation spaces are the same.
   Let us use this in the case of function spaces. The wavelet transform gives us an isometry between the functions of bounded mean oscillation BMO and $h_\infty^0 = h_\infty^{0,2}$. Hence, Proposition 11.6 immediately yields

PROPOSITION 11.7
We have

$$\sum_{N=1}^\infty [N^{\beta/d} \sigma_N(f)_{\text{BMO}}]^\tau \frac{1}{N} < \infty \iff f \in B_\tau^\beta$$

with $1/\tau = \beta/d$.

PROOF   Going back to the sequence spaces and using Proposition 11.6, this is in fact just a restatement of the fact that $(l_0, l_\infty)_{\theta,q}^\theta = l_p$ with $1/p = (1-\theta)/\theta$ and $1/q = 1 - \theta$, cf. (11.20).   ∎

11.6   Another way to compress near L_∞

The Besov space B_∞^{01} is another space that can be used as a substitute for L_∞. First, it is continuously imbedded in L_∞, and a small error in B_∞^{01} guarantees a small error in L_∞. Furthermore, it has a simple characterization in terms of wavelet expansions. Let b_∞^{01} be the space of all sequences $s = \{s_I\}$ such that

$$\|\{\langle f, \psi \rangle_I\}\|_{b_\infty^{01}} = \sum_{\nu \in \mathbb{Z}} \sup_{l(I)=2^{-\nu}} |s_I|/|I|^{1/2} < \infty.$$

The wavelet–transform is an isomorphism between B_∞^{01} and b_∞^{01}.

We shall thus consider approximation in B_∞^{01} instead of in L_∞. We are interested in finding N coefficients which are almost optimal for $\sigma_N(f)_{B_\infty^{01}}$. As before, we may rephrase this problem on the sequence level instead and it then becomes a special case of the following problem about matrices.

Given two (nonempty) countable index sets I and J, we let $\mathfrak{b} = l_1(l_\infty)$ be the collection of all matrices $a = \{a_{ij}\}_{i \in I, j \in J}$ with

$$\|a\|_{\mathfrak{b}} = \sum_{i \in I} \sup_{j \in J} |a_{ij}| < \infty.$$

We also let

$$\Sigma_N(a)_{\mathfrak{b}} = \inf_{\#\{a_{ij}^0 \neq 0\} \leq N} \|a - a^0\|_{\mathfrak{b}}, \quad N \geq 1. \qquad (11.25)$$

We want to find an almost optimal choice of the matrix a^0. Notice first that for each $\epsilon > 0$ we can find a^0 with $\#\{a^0 \neq 0\} \leq N$ such that $\Sigma_N(a) \leq \|a - a^0\| \leq (1 + \epsilon)\Sigma_N(a)$. Also, we may always assume that a^0 is obtained by choosing N elements from a:

$$(a^0)_{ij} = \begin{cases} a_{ij}, & ij \in \Gamma_N, \\ 0 & \text{otherwise} \end{cases}$$

for some set Γ_N with $\#\Gamma_N = N$. In fact, we can immediately say more. For each $i \in I$ fixed, let N_i be the number of a_{ij}s chosen at the level i. In particular, $0 \leq N_i \leq N$ and $\sum_{i \in I} N_i = N$. Now, we may clearly assume that a^0 is obtained by taking the N_i largest elements (in magnitude) at each level i, $i \in I$. The real issue is thus to decide how many to use at each level.

Our algorithm for finding an almost optimal sequence a^0 to remove from a given a will be based on an auxiliary sequence $\bar{a}$. Let us describe next how $\bar{a}$ is constructed.

Fix the sequence a. For each $i \in I$ let $a_i^* = \{(a_i^*)_k\}_{k=1}^\infty$ be the nonincreasing rearrangement of $\{a_{ij}\}_{j \in J}$. Define the function $a_i^*(t)$ for $t \geq 0$ by

$$a_i^*(t) = (a_i^*)_k \quad \text{for } k - 1 \leq t < k, \qquad (11.26)$$

and let $\bar{a}_i^*(t)$, $t \geq 0$ be the largest convex minorant of $a_i^*(t)$. The sequence $\bar{a} = \{\bar{a}_{ij}\}_{i \in I, j \in J}$ is any fixed sequence with $(\bar{a}_i^*)_k = \bar{a}_i^*(k)$ for all $k \geq 1$.

LEMMA 11.8
Given a sequence a, let $\bar{a}$ be the sequence defined above. We have

$$\Sigma_{2N}(\bar{a})_\flat \leq \Sigma_{2N}(a)_\flat \leq 2\Sigma_N(\bar{a})_\flat, \quad N \geq 1.$$

PROOF   To find $E(2N, a)$ to within an $\epsilon > 0$ we have to select a certain number of the largest elements at each level i. Let us call this number $(2N)_i$, $i \in I$. Now recall that $\bar{a}_i^*$ was a minorant of a_i^*. Hence, if we pick the $(2N)_i$ largest elements of $\bar{a}$ at each level i, then clearly $\Sigma_{2N}(\bar{a})_\flat \leq \sum_i(\bar{a}_i^*)_{(2N)_i} \leq \sum_i(a_i^*)_{(2N)_i} \leq (1+\epsilon)\Sigma_N(a)$. Since $\epsilon > 0$ is arbitrary, this proves the first inequality. The second inequality follows in a similar way from the lemma below (with $\delta = 1/2$). ∎

LEMMA 11.9
Let $f : \bar{\mathbb{R}}_+ \to \bar{\mathbb{R}}_+$, be a nonincreasing function. Then its largest convex minorant $\bar{f}$ satisfies

$$\bar{f}(t) \leq f(t) \leq \frac{1}{1-\delta}\bar{f}(\delta t), \quad t > 0,$$

for each $0 < \delta < 1$.

PROOF   Easy exercise. ∎

Lemma 11.8 allows us to assume that each $a_i^*(t)$, defined by (11.26), is convex. In this situation it is easy to find an optimal decomposition. Given a sequence a with $a_i^*(t)$ convex for each $i \in I$ we define

$$(\Delta a_i^*)_k = (a_i^*)_k - (a_i^*)_{k+1}$$

for $k \geq 1$ and $i \in I$. Since a_i^* is nonincreasing, $(\Delta a_i^*)_k \geq 0$, and, because of the assumed convexity, forms a nonincreasing sequence for each fixed i. Each element a_{ij} of course coincides with a unique $(a_i^*)_k$ for an appropriate $k = k(j)$. With each a_{ij} we associate the quantity $(\Delta a_i^*)_k(j)$, and define the sequence $a_{\text{opt}} = a_{\text{opt}}(N)$ to be the N elements in a with the N largest $(\Delta a_i^*)_k$. Clearly, a_{opt} is a solution to the extremal problem implicit in the definition of $\Sigma_N(a)_\flat$, or, more specifically,

$$\Sigma_N(a)_\flat = \|a - a_{\text{opt}}\|_\flat$$

and

$$\#\{a_{\text{opt}} \neq 0\} \leq N.$$

REMARK 11.10   More generally, using a similar argument we may solve the analogous extremal problem when $\flat = l_q(l_p)$, $0 < p, q \leq \infty$. ∎

Now let us apply this in the particular case of approximation in B_∞^{01}. Fix an integer $N \geq 1$ and the function f with the corresponding coefficients $a = \{\langle f, \psi_I \rangle\}$.

For each $i \in \mathbb{Z}$ collect the coefficients corresponding to intervals of the same length $|I| = 2^{-i}$, divided by the appropriate power of measure of I, in a family a_i:

$$a_i = \{\langle f, \psi_I \rangle / |I|^{1/2}\}_{|I|=2^{-i}}.$$

Now sort the elements in each of these families in nonincreasing order. For each $i \in \mathbb{Z}$ we obtain a nonincreasing sequence $a_i^* = \{(a_i^*)_k\}_{k=1}^{\infty}$. Each $(a_i^*)_k$ then equals $\langle f, \psi_I \rangle / |I|^{1/2}$ for a certain interval I with $|I| = 2^{-i}$. We say that this I corresponds to $(a_i^*)_k$. Now, each sequence a_i^* has a largest convex minorant $\bar{a}_i^*(t)$. Considering all of the $(\Delta a_i^*)_k$, for all is and ks, we remove the N elements with the largest $(\Delta a_i^*)_k$. This corresponds to N specific intervals I. Let us call the family of these intervals Γ_N. We define a_{opt} by

$$(a_{\mathrm{opt}})_I = \begin{cases} \langle f, \psi_I \rangle, & I \in \Gamma_N, \\ 0 & \text{otherwise.} \end{cases}$$

The general discussion above shows that

$$\frac{1}{2} \, \|a - a_{\mathrm{opt}}(2N)\|_{b_\infty^{01}} \le \Sigma_N(a)_{b_\infty^{01}} \le \|a - a_{\mathrm{opt}}(N)\|_{b_\infty^{01}}.$$

Now we may bring this back to the function level. We let

$$\bar{f}^N = \sum_{I \notin \Gamma_N} \langle f, \psi \rangle_I \, \psi_I \tag{11.27}$$

Using that B_∞^{01} and b_∞^{01} are isomorphic via the wavelet–transform, we obtain the following.

THEOREM 11.11
Given a function $f \in B_\infty^{01}$ we let $\bar{f}^N$ be defined by (11.27) above. Then

$$\frac{1}{C} \, \|\bar{f}^{2N}\|_{B_\infty^{01}} \le \sigma_N(f)_{B_\infty^{01}} \le \|\bar{f}^N\|_{B_1^0\infty}, \quad N \ge 1,$$

for some constant C independent of N. ∎

REMARK 11.12* In many situations the third parameter q in $B_q^\alpha p$ is of minor importance. However, this is not the case here. The approximation spaces based on approximation in $B_\infty^0 \infty$, say, are different than the ones based on approximation in B_∞^{01}. This easily follows by considering some simple examples. It becomes most transparent if we consider the analogous fact for matrices. Let $\mathfrak{b}_1 = l_1(l_\infty)$ and $\mathfrak{b}_2 = l_\infty(l_\infty)$ be families of matrices based on the index sets $I = J = \mathbb{Z}$. Define

---

*After this paper was completed, exact descriptions of approximation spaces, with the error measured in a Besov space, have been obtained by B. Jawerth and M. Milman, *Wavelet decompositions and best approximation in Besov spaces*.

the matrix $a = \{a_{ij}\}$ by

$$a_{ij} = \begin{cases} (i)^{-\gamma}, & i \geq 1, j = 0, \\ 0 & \text{otherwise.} \end{cases}$$

for a fixed $\gamma \geq 1$. Then $\Sigma_N(a)_{b_1} = \sum_{i>N} i^{-\gamma} \approx N^{1-\gamma}$ while $\Sigma_N(a)_{b_2} = N^{-\gamma}$. Hence, for each fixed $0 < \theta < 1$ and $0 < \tau \leq \infty$ we may choose γ so that

$$\left(\sum_{N=1}^{\infty} [N^{1-\theta} \Sigma_N(a)_{b_1}^{\theta}]^{\theta} \frac{1}{N} \right)^{1/\tau} = \infty$$

while

$$\left(\sum_{N=1}^{\infty} [N^{1-\theta} \Sigma_N(a)_{b_2}^{\theta}]^{\theta} \frac{1}{N} \right)^{1/\tau} < \infty. \quad \blacksquare$$

11.7 Compression of operators

As we mentioned in Remark 11.10, we readily obtain compression results for certain other (vector-valued) spaces. Some of these are relevant for the compression of classes of bounded operators on certain function spaces.

Let us consider an example. We know that the Besov space $B_1^{\alpha,1}$, $\alpha \in \mathbb{R}$, is isomorphic to a (weighted) l_1 space,

$$\|f\|_{B_1^{\alpha,1}} \approx \|\{\langle f, \psi_J \rangle\}_{J \text{dyadic}}\|_{l_1^{\gamma}}, \quad \gamma = \alpha/d - 1/2,$$

with

$$\|\{s_J\}_J\|_{l_1^{\gamma}} = \sum_J |J|^{-\gamma} |s_J|.$$

Similarly, each bounded operator T on the Besov space corresponds to a bounded operator on the l_1 space with "matrix" $A_T = \{\langle T\psi_J, \psi_I \rangle\}$. This mapping $T \mapsto A_T$ is easily seen to be an isomorphism. More precisely, by the classical result of Schur about bounded operators on l_1, we have (cf. [FJ], section 10)

$$\|T\|_{B_1^{\alpha,1} \to B_1^{\alpha,1}} = \sup_J \sum_I |\langle T\psi_J, \psi_I \rangle| \left(\frac{|J|}{|I|} \right)^{\gamma}.$$

Furthermore, given a matrix $A = \{A_{IJ}\}$ which corresponds to a bounded operator on l_1^{γ}, the operator T defined by

$$Tf = T_A f = \sum_I \sum_J A_{IJ} \langle f, \psi_J \rangle \psi_I \tag{11.28}$$

is bounded on $B_1^{\alpha,1}$. The mappings $T \mapsto A_T$ and $A \mapsto T_A$ are inverse to each other.

Let us define #T to be the number of nonzero entries A_{IJ} in the representation (11.28), and let

$$\sigma_N(T)_{B_1^{\alpha,1} \to B_1^{\alpha,1}} = \inf_{\#S \leq N} \|T - S\|_{B_1^{\alpha,1} \to B_1^{\alpha,1}}.$$

Using the relation with sequence spaces, we have

$$\sigma_N(T)_{B_1^{\alpha,1} \to B_1^{\alpha,1}} = \Sigma_N(A_T)_{l_1^\gamma \to l_1^\gamma} = \inf_{\#A_S \leq N} \|A_T - A_S\|_{l_1^\gamma \to l_1^\gamma}.$$

Hence, we can reduce the problems of finding an optimal S and of characterizing the operators T, which can be approximated to a certain order, to corresponding problems on sequence spaces. Next we shall study these two problems in some detail.

Let us start with two countable index sets I and J. We let $\mathcal{L}$ denote the bounded operators $l_1(J) \to l_1(I)$ with norm

$$\|A\|_{\mathcal{L}} = \sup_j \sum_i |A_{ij}|.$$

Given a matrix A and an integer N, we are first interested in finding a matrix $A^{(N)}$ which is optimal in the sense that

$$\Sigma_N(A)_{\mathcal{L}} = \inf_{\#B \leq N} \|A - B\|_{\mathcal{L}} = \|A - A^{(N)}\|_{\mathcal{L}}.$$

To solve this we shall embed the space $\mathcal{L}$ into the space $l_\infty(\mathbb{N} \times J)$ in the following fashion. Given the matrix $A = \{A_{ij}\}$, we let $A_i^* = \{A_i^*(k)\}_{k \in \mathbb{N}}$ be the nonincreasing rearrangement of the sequence $\{A_{ij}\}_j$ for each $i \in I$. We then define the sequence $a = \{a_{ik}\}_{(i,k) \in I \times \mathbb{N}}$ by

$$a_{ik} = \sum_{l \geq k} A_i^*(l).$$

A moment's reflection shows that we have the following fact.

PROPOSITION 11.13
We have

$$\|A\|_{\mathcal{L}} = \|a\|_{l_\infty(I \times \mathbb{N})},$$

and

$$\Sigma_N(A)_{\mathcal{L}} = E(N, a; l_0, l_\infty) = a^*(N) \tag{11.29}$$

with $l_\infty = l_\infty(I \times \mathbb{N})$ and $l_0 = l_0(I \times \mathbb{N})$. Here a^* denotes the nonincreasing rearrangement of $a = \{a_{ik}\}_{(i,k) \in I \times \mathbb{N}}$.

Of course, we know how to find an optimal approximation of $a \in l_\infty$ by an element in l_0. We simply pick the N largest elements of a. By then going back to the initial matrix A while keeping track of what the N largest elements correspond to, we obtain an optimal $A^{(N)}$.

We can now return to the function spaces. Given an operator T, Proposition 11.13 tells us how to obtain the best approximation S, with the error measured in the operator norm on $B_1^{\alpha,1}$, such that A_S has N nonzero entries. This fact and an application of Hardy's inequality yield the following characterization of the operators that can be compressed well.

THEOREM 11.14
Suppose $\beta > 0$. Let $1/S = \beta$ and $\gamma = \alpha/d - 1/2$. Then

$$\left(\sum_N \left(N^\beta \sigma_N(T)_{B_1^{\alpha,1} \to B_1^{\alpha,1}} \right)^S \frac{1}{N} \right)^{1/S} < \infty$$

if and only if

$$\sum_{I \text{ dyadic}} \sum_{k \geq 1} \left(k^{1+\beta} (A_{I,\cdot}^\gamma)_k^* \right)^{1/\beta} \frac{1}{k} < \infty.$$

Here $A^\gamma = A_T^\gamma = \{ \langle T\psi_J, \psi_I \rangle (|J|/|I|)^\gamma \}_{I,J \text{ dyadic}}$ and $*$ denotes the nonincreasing rearrangement with respect to the counting measure on the set of dyadic squares.

PROOF Note that A is bounded on l_1^γ if and only if A^γ is bounded on l_1. Furthermore,

$$\sigma_N(T)_{B_1^{\alpha,1} \to B_1^{\alpha,1}} = \sigma_N(A_T)_{l_1^\gamma \to l_1^\gamma} = \sigma_N(A_T^\gamma)_{\mathcal{L}},$$

where $\mathcal{L}$ denotes the bounded operators on l_1. By Proposition 11.13,

$$\sigma_N(A_T^\gamma)_{\mathcal{L}} = E(N, a^\gamma; l_0, l_\infty) = (a^\gamma)^*(N),$$

and, hence,

$$\left(\sum_N \left(N^\beta \sigma_N(T)_{B_1^{\alpha,1} \to B_1^{\alpha,1}} \right)^S \frac{1}{N} \right)^{1/S} = \left(\sum_N \left(N^\beta E(N, a^\gamma; l_0, l_\infty) \right)^S \frac{1}{N} \right)^{1/S}$$

$$= \left(\sum_N \left(N^\beta (a^\gamma)^*(N) \right)^S \frac{1}{N} \right)^{1/S} = \| a^\gamma \|_{l_{1/S}}.$$

By Hardy's inequality, this last expression is equivalent to

$$\left(\sum_I \| \{A_{IJ}^\gamma\} \|_{l_{1/(\beta+1),1/\beta}(J)} \right)^{1/\beta}.$$

By using the definition of the discrete Lorentz spaces, we obtain the desired conclusion. ∎

There is also a dual version of this theorem concerning compression of operators T on the Besov spaces $B_\infty^{\alpha,\infty}$; for details we refer to [H].

We note that, at least formally,

$$\langle T\psi_J, \psi_I \rangle = \int K(x,y)\psi_J(y)\,dy\,\psi_I(x)\,dx = \langle K, \psi_I \otimes \psi_J \rangle.$$

Now, $\psi_I \otimes \psi_J$ is a wavelet ψ_R with respect to the rectangle $R = I \times J$, and $\langle K, \psi_I \otimes \psi_J \rangle = \langle K, \psi_R \rangle$ is the wavelet coefficient of the kernel K corresponding to this rectangle. Theorem 11.14 thus characterizes the operators T which can be compressed well in terms of a certain size condition on the rectangular wavelet coefficients of the kernel K of T.

We finally remark that the results of this section can be used to numerically study compression of operators much along the lines of [BeyCoiRok]. This has been carried out in [CH].

Bibliography

[B]   S. Banach. 1932. *Theorie des operations lineaires*. Warsaw.

[BeL]   J. Bergh and J. Löfström. 1976. *Interpolation spaces: an introduction*. New York: Springer Verlag.

[BeyCoiRok] G. Beylkin, R. Coifman, and V. Rokhlin. 1991. "Fast Wavelet Transforms and numerical algorithms I," *Commun. Pure and Appl. Math.* **44**: 141–183.

[CH]   D. Chang and C.-c. Hsiao. 1992. "Compression of operators using rectangular wavelets," Preprint.

[Co]   A. Cohen. 1991. "Biorthogonal wavelets," in *Wavelets: a tutorial in theory and applications*, C. K. Chui, ed., Boston: Academic Press, pp. 123-152.

[CoDFe]   A. Cohen, I. Daubechies, and J. C. Feauveau. 1992. "Biorthogonal bases of compactly supported wavelets," *Commun. Pure Appl. Math.* **45**: 485–560.

[D]   I. Daubechies. 1988. "Orthonormal basis of compactly supported wavelets," *Commun. Pure Appl. Math.* **41**: 909–996.

[DeJLu]   R. DeVore, B. Jawerth, and B. Lucier. 1992. "Image compression through wavelet transform coding," *IEEE Journal on Information Theory* **38**: 719–747.

[DeJPo]   R. DeVore, B. Jawerth, and V. Popov. 1992. "Compression of wavelet decompositions," *Amer. J. Math.* **114**: 737-785.

[FJ]   M. Frazier and B. Jawerth. 1990. "A discrete transform and decompositions of distribution spaces," *J. Funct. Analysis* **93**: 34–170.

[H]   C.-c. Hsiao. 1992.*Rectangular wavelets and compression of operators*, Thesis, Univ. of South Carolina, Columbia.

[Hu]   R. Hunt. 1964. "An extension of the Marcinkiewicz interpolation theorem to Lorentz spaces," *Bull. Amer. Math. Soc.* **70**: 803-807.

[M]   S. Mallat. 1989. "Multiresolution approximation and wavelets," *Trans. Amer. Math. Soc.* **315**: 69–88.

[Me]   Y. Meyer. 1990. *Ondelettes et Opérateurs*, Paris: Hermann.

[P]   A. Pelczynski. 1958. "On the isomorphism of the spaces m and M," *Bull. Acad. Polon. Sci. Sér. Math., Astronom. Phys.* **6**: 695–696.

[R]   N. Riviére. 1966. *Interpolation theory in s-Banach spaces*, Thesis, Chicago.

[RoT]   R. Rochberg and M. Taibleson. 1988. "An averaging operator on a tree," in *Harmonic Analysis and Partial Differential Equations*, J. Garcia-Cuerva, ed., Lect. Notes Math. **1384**, Berlin: Springer-Verlag, pp. 207–213.

[S]   J.-O. Strömberg. 1981. "A modified Franklin system and higher order spline systems on $\mathbb{R}^n$ as unconditional bases for Hardy spaces," in *Conference on harmonic analysis in honor of Antoni Zygmund II*, W. Beckner et al., eds., Chicago: Univ. of Chicago Press, pp. 475–494.

Part III

**Wavelets and Partial
Differential Operators**

12

On wavelet-based algorithms for solving differential equations

G. Beylkin

ABSTRACT   We describe an order N method for computing the Green's function of the two-point boundary value problem for elliptic differential operators in the wavelet "system of coordinates." For simplicity, we consider the ordinary $O(h^2)$ finite-difference scheme, and use wavelets only to perform the "linear algebra." Our main tool is the diagonal preconditioning available for the periodized differential operators in the wavelet bases.

CONTENTS

*The research was partially supported by ONR grant N00014-91-J4037 and a grant from Chevron Oil Field Research Company.

449

12.1   Introduction

The role of the orthonormal wavelet bases in solving integral equations has been studied in [BCoR1], where it was observed that wide classes of operators have sparse representations in the wavelet bases thus permitting a number of fast algorithms for applying these operators to functions, solving integral equations, etc. The operators which can be efficiently treated using representations in the wavelet bases include Calderón–Zygmund and pseudo-differential operators. Let us here summarize several points important for further considerations and refer to [BCoR1], [BCoR2], [Aetal] for the details. If we consider an integral operator (Calderón–Zygmund or pseudo-differential operator),

$$T(f)(x) = \int K(x,y)f(y)dy, \tag{12.1}$$

and construct its matrix representation in a two-dimensional wavelet basis, then we find that the rate of decay of the size of entries as a function of the distance from the diagonal in the sub-blocks of such representation is faster than that of the original kernel. The rate of decay depends on the number of vanishing moments of the basis functions. For example, let the kernel satisfy the conditions

$$|K(x,y)| \le \frac{1}{|x-y|}, \tag{12.2}$$

$$|\partial_x^M K(x,y)| + |\partial_y^M K(x,y)| \le \frac{C_0}{|x-y|^{1+M}} \tag{12.3}$$

for some $M \ge 1$. Then by choosing the wavelet basis with M vanishing moments, the matrices of coefficients $\alpha_{i,l}^j$, $\beta_{i,l}^j$, $\gamma_{i,l}^j$ of the representation of the kernel K in the non-standard form (see [BCoR1]) satisfy the estimate

$$|\alpha_{i,l}^j| + |\beta_{i,l}^j| + |\gamma_{i,l}^j| \le \frac{C_M}{1+|i-l|^{M+1}}, \tag{12.4}$$

for all

$$|i - l| \ge 2M. \tag{12.5}$$

And, if in addition to (12.2) and (12.3),

$$\left| \int\!\!\int_{I \times I} K(x,y)\, dx\, dy \right| \le C|I| \tag{12.6}$$

holds for all dyadic intervals I (the so-called "weak cancellation condition"), then (12.4) is valid for all i,l. Thus, for a given accuracy, the representations of operators which satisfy (12.2) and (12.3) are sparse since we may use banded versions of $\alpha_{i,l}^j$, $\beta_{i,l}^j$, $\gamma_{i,l}^j$ for computing.

We note that considering a banded approximation directly for the kernel satisfying (12.2) does not lead to a satisfactory numerical approximation. The method of [BCoR1] uses the smoothness of the matrix away from the diagonal to increase the rate of decay. Parts of the matrix which can be well approximated by the low degree polynomials are represented by small coefficients in the wavelet system of coordinates since the basis functions have vanishing moments. Once a sparse representation is obtained, fast algorithms are available for a variety of tasks associated with solving integral equations, for example, $O(N)$ algorithm for solving an integral equation or an iterative algorithm for constructing the generalized inverse in $O(N)$ operations ([BCoR1], [BCoR2], [Aetal]).

In this paper we address the question "what are the implications of using wavelet bases for solving differential equations?" The same problem often may be posed both as a problem of solving the boundary value problem for a differential equation and as a problem of solving an integral equation. From the point of view of numerical analysis one would observe a significant difference in these formulations. The discretization of a differential equation leads to a sparse linear algebraic system with a "large" condition number of the corresponding matrix, whereas the discretization of an integral operator leads to a dense matrix with "small" condition number. More precisely, the condition numbers of matrices representing differential operators usually have a polynomial growth with the reduction of the step size (i.e., with the increase of the size of the system). For example, the condition number of the matrix of the second order finite difference operator grows as $1/h^2$ (or N^2), where h is the step size (N is the number of points of discretization). On the other hand, the condition number of the matrix of the linear algebraic system obtained by discretizing the integral equations of the second kind does not grow with the reduction of the step size (usually it actually improves somewhat) but the matrix is dense (full).

We recall that the condition number of a matrix is defined as the ratio of the largest and the smallest singular values. If matrix has a null space (the actual null space or a null space for a given accuracy), then by the condition number we understand the ratio of the largest singular value to the smallest singular value *above* the threshold of accuracy. The condition number controls the rate of convergence of a number of iterative algorithms for solving linear systems; for example the number of iterations of the conjugate gradient method is $O(\sqrt{\kappa})$, where κ is the condition number of the matrix.

Naively, solving the linear system of size $N \times N$ obtained by discretizing the integral equations of the second kind seems to require $O(N^2)$ operations. Solving the sparse linear system obtained by representing differential operators also seems to require $O(N^2)$ operations since the number of iterations is $O(N)$.

As is shown in [BCoR1], the dense matrices obtained by discretizing the integral equations of the second kind may be replaced by sparse matrices in the wavelet system of coordinates (for operators satisfying the conditions (12.2) and (12.3), for example), thus leading to $O(N)$ algorithms for solving such integral equations. As we demonstrate in this paper, if our starting point is a differential equation with boundary conditions then in the wavelet system of coordinates there is a *diagonal* preconditioner which allows us to perform algebraic manipulations only with the sparse matrices whose condition number is $O(1)$, thus also leading to $O(N)$ algorithms for solving the corresponding linear systems.

We describe a method for solving the two point boundary value problem for elliptic differential operators in the wavelet "system of coordinates." To illustrate the difference between our approach and the existing numerical methods for solving the two-point boundary value problems of this kind, such as multigrid (see, e.g., [Bri]) or multilevel (hierarchical) methods or the very simple and elegant algorithm of [GR], we construct the Green's function (the inverse operator) in $O(N)$ operations. We note that the numerical methods mentioned above allow us to find the solution of the problem in $O(N)$ operations. However, since the ordinary matrix representation of the Green's function requires $O(N^2)$ significant entries, fast algorithms for its construction are not readily available. Our method permits solving the problem in $O(N)$ operations as well, but since the representation of the Green's function in the wavelet bases requires (for a given accuracy) only $O(N)$ entries, we concentrate on describing a fast algorithm for its construction.

Once the Green's function is obtained, finding the solution reduces to the matrix-vector multiplication, which in the wavelet system of coordinates is an $O(N)$ procedure. In addition, if the entries of the vector are values of a smooth and nonoscillatory function then the vector is sparse in the wavelet system of coordinates. In this case the number of operations to apply the Green's function to a vector is proportional to the number of significant coefficients of this vector in the wavelet system of coordinates. We illustrate these properties further by considering a modification of the Crank–Nicolson method, which we convert into an explicit and adaptive scheme in the wavelet system of coordinates.

The main tool in our approach is the diagonal preconditioning available for the periodized differential operator in the wavelet bases [B]. The idea of pre-conditioning has long been one of the main ideas in the multilevel and multigrid methods. Among a great number of papers on preconditioning we would like to note [BrPX], where the authors explicitly consider orthonormal chains of sub-spaces (similar to that of the multiresolution analysis) in order to construct the multilevel preconditioners. Apparently unfamiliar with multiresolution analysis and wavelet bases, they remark that "in practice, an orthonormal basis ... is seldom available." In fact, orthonormal wavelet bases provide a very convenient tool for implementing the preconditioners. Moreover, since the inverse operator is sparse in the wavelet bases, it is possible to construct it numerically in $O(N)$ operations.

S. Jaffard in [J] gives a theoretical analysis of solving the elliptic boundary value problem in the wavelet bases and considers the diagonal preconditioning. However, he does not provide a practical method. In this paper our considera-tions are restricted to the two-point boundary value problems, since a practical construction of the wavelet bases in an arbitrary domain is not available at this time. We approach the multidimensional problems using the alternating directions technique, which is modified since we are able to numerically construct the Green's functions of the two-point boundary value problems. We note that our use of the diagonal preconditioning differs from that in [J] since we apply it to the periodized differential operators and solve the boundary value problem by rank-one perturbation.

For simplicity, we consider the ordinary $O(h^2)$ finite-difference scheme for the two-point boundary value problem, and use the periodized wavelets only to perform the "linear algebra." Such an approach enables us to make a clear comparison with other techniques. On the other hand, it also carries some of the limitations of the finite-difference scheme. A more consistent approach which uses the wavelet bases of the interval [Cetal] to achieve an approximation of order h^p, where p is arbitrary, is currently being developed and will be described elsewhere.

12.2   The two-point boundary value problem

Let us consider the two-point boundary value problem

$$\mathcal{L}u \equiv \frac{d}{dx}\left(a(x)\frac{du}{dx}\right) = f(x) \tag{12.7}$$

with the Dirichlet boundary conditions $u(0) = u(1) = 0$. We assume that a is a sufficiently smooth function and $a(x) > 0, x \in (0, 1)$. The method that we describe is applicable to more general elliptic operators, e.g.,

$$\mathcal{L}u = \frac{d}{dx}\left(a(x)\frac{du}{dx}\right) - b(x)u, \tag{12.8}$$

where $b(x) > 0$.

Discretizing this problem on a staggered grid, we obtain the following system of linear algebraic equations

$$a_{i-1/2}u_{i-1} - (a_{i-1/2} + a_{i+1/2})u_i + a_{i+1/2}u_{i+1} = h^2 f_i, \quad i = 1,\dots,N, \tag{12.9}$$

where $u_i = u(x_i)$, $a_{i+1/2} = a(x_{i+1/2})$, $f_i = f(x_i)$, $x_i = ih$ and $x_{i+1/2} = (i + 1/2)h$ and where we explicitly set $u_0 = u_{N+1} = 0$.

We write (12.9) as

$$\mathbf{L}\mathbf{u} = \mathbf{f}, \tag{12.10}$$

where the $N \times N$ matrix $\mathbf{L}$ is as follows

$$\mathbf{L} = \begin{pmatrix} -(a_{1/2} + a_{3/2}) & a_{3/2} & 0 & \cdots & 0 & 0 & 0 \\ a_{3/2} & -(a_{3/2} + a_{5/2}) & a_{5/2} & \cdots & 0 & 0 & 0 \\ \vdots & \vdots & \vdots & \ddots & \vdots & \vdots & \vdots \\ 0 & 0 & 0 & \cdots & a_{N-3/2} & -(a_{N-3/2} + a_{N-1/2}) & a_{N-1/2} \\ 0 & 0 & 0 & \cdots & 0 & a_{N-1/2} & -(a_{N-1/2} + a_{N+1/2}) \end{pmatrix}.$$

(12.11)

There are two reasons for the condition number of the matrix $\mathbf{L}$ to be large. If $a(x) = 1$ in (12.7) and (12.11), then we obtain the central difference matrix representation of the second derivative d^2/dx^2. It is clear that the matrix $\mathbf{L}$ has the condition number $O(N^2)$. On the other hand, noting that the size of the function a might be different in the subintervals of $(0, 1)$, we observe that the condition number of the operator of multiplication by the function $a(x)$ could be arbitrarily large.

Our goal is to construct the matrix $\mathbf{L}^{-1}$ numerically in $O(-N \log \epsilon)$ operations, where ϵ is the desired accuracy. This seemingly hopeless task (it is easy to check for small N that the matrix $\mathbf{L}^{-1}$ is dense in the ordinary representation) has, in fact, a simple solution in the wavelet system of coordinates.

The kernel of the inverse operator for the problem (12.7) (the Green's function for the Dirichlet problem for an elliptic operator) has a sparse representation in the wavelet bases since such a kernel satisfies the estimates of the type in (12.2), (12.3) (see [BCoR1]). Let us show how to construct $\mathbf{L}^{-1}$ numerically starting with the matrix $\mathbf{L}$.

12.3 Reduction to the periodized problem

In the wavelet bases the preconditioner for the periodized differential operator is a diagonal matrix. The condition number of the rescaled operator is $O(1)$ and depends only on the choice of the basis [B]. Moreover, any finite difference matrix representation of periodized differential operators may be rescaled by a diagonal preconditioner. We use this fact to solve the two-point boundary value problem using a fairly standard discretization scheme in (12.10). The wavelets play an auxiliary role in that they provide a system of coordinates in which the condition numbers of the sparse matrices (involved in the computations) are under control. We use such a "mixed" approach for two reasons. First, it provides a simple way to see the advantages of computing in the wavelet bases. Second, it provides a practical way to significantly improve the performance of commonly used finite difference schemes.

In order to use periodized differential operators, we consider the matrix $\mathbf{L}$ as a finite rank perturbation of a periodized matrix. Indeed, we have

$$\mathbf{L} = \mathbf{A} - a_{1/2}\mathbf{e}_1\mathbf{e}_N^T - a_{N+1/2}\mathbf{e}_N\mathbf{e}_1^T, \tag{12.12}$$

where

$$\mathbf{A} = \begin{pmatrix} -(a_{1/2}+a_{3/2}) & a_{3/2} & 0 & \cdots & 0 & 0 & a_{1/2} \\ a_{3/2} & -(a_{3/2}+a_{5/2}) & a_{5/2} & \cdots & 0 & 0 & 0 \\ \vdots & \vdots & \vdots & \ddots & \vdots & \vdots & \vdots \\ 0 & 0 & 0 & \cdots & a_{N-3/2} & -(a_{N-3/2}+a_{N-1/2}) & a_{N-1/2} \\ a_{N+1/2} & 0 & 0 & \cdots & 0 & a_{N-1/2} & -(a_{N-1/2}+a_{N+1/2}) \end{pmatrix}, \tag{12.13}$$

and the unit vectors $\mathbf{e}_1$, $\mathbf{e}_N$ are given by

$$\mathbf{e}_1 = \begin{pmatrix} 1 \\ 0 \\ \vdots \\ 0 \\ 0 \end{pmatrix}, \mathbf{e}_N = \begin{pmatrix} 0 \\ 0 \\ \vdots \\ 0 \\ 1 \end{pmatrix}. \tag{12.14}$$

In this section we consider the case where the size of the function a does not change significantly over the interval $(0,1)$. To illustrate the effect of diagonal preconditioning in the wavelet system of coordinates, let us set $a = 1$ and consider $\mathbf{A} = \mathbf{D}$,

$$\mathbf{D} = \begin{pmatrix} -2 & 1 & 0 & \cdots & 0 & 0 & 1 \\ 1 & -2 & 1 & \cdots & 0 & 0 & 0 \\ \vdots & \vdots & \vdots & \ddots & \vdots & \vdots & \vdots \\ 0 & 0 & 0 & \cdots & 1 & -2 & 1 \\ 1 & 0 & 0 & \cdots & 0 & 1 & -2 \end{pmatrix}. \tag{12.15}$$

In the following two examples we compute the standard form $\mathbf{D}_w$ of the periodized second derivative $\mathbf{D}$ of size $N \times N$, where $N = 2^n$, and rescale it by the diagonal matrix $\mathbf{P}$,

$$\mathbf{D}_w^p = \mathbf{P}\mathbf{D}_w\mathbf{P},$$

where $\mathbf{P}_{il} = \delta_{il}2^j$, $1 \leq j \leq n$, and where j is chosen depending on i,l so that $N - N/2^{j-1} + 1 \leq i,l \leq N - N/2^j$, and $\mathbf{P}_{NN} = 2^n$. The matrix P is illustrated in Figure 12.1.

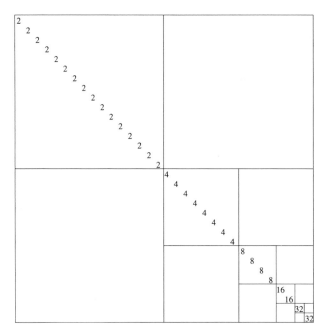

FIGURE 12.1
An example ($n = 5$) of the diagonal matrix **P** used to rescale the matrix of the periodized second derivative $\mathbf{D}_w$ in the wavelet system of coordinates.

Tables 12.1 and 12.2 compare the original condition number κ of **D** (and $\mathbf{D}_w$ since it is obtained by an orthogonal transformation) and κ_p of $\mathbf{D}_w^p$. Since matrices **D**, $\mathbf{D}_w$ and $\mathbf{D}_w^p$ have a null space (of dimension one), the condition numbers are computed on the range of these matrices.

Let us now describe a method for solving the two-point boundary value problem in the wavelet system of coordinates. Denoting the matrix of the discrete wavelet transform by **W** (though the actual transform is applied via a pyramid algorithm, see, e.g., [BCoR1]) and observing that **W** is an orthogonal transformation, we rewrite (12.10) and (12.12) in the wavelet system of coordinates,

$$(\mathbf{A}_w - a_{1/2}\hat{\mathbf{e}}_1\hat{\mathbf{e}}_N^T - a_{N+1/2}\hat{\mathbf{e}}_N\hat{\mathbf{e}}_1^T)\hat{\mathbf{u}} = \hat{\mathbf{f}} \qquad (12.16)$$

where

$$\mathbf{A}_w = \mathbf{WAW}^*, \qquad (12.17)$$

$$\hat{\mathbf{u}} = \mathbf{Wu} \qquad (12.18)$$

$$\hat{\mathbf{f}} = \mathbf{Wf} \qquad (12.19)$$

and $\hat{\mathbf{e}}_l = \mathbf{We}_l$, $l = 1, N$.

Computing the discrete (periodized) wavelet transform of a vector of size $N = 2^n$ and using n scales, we obtain on the most sparse scale a single coefficient

Table 12.1. Condition numbers of the matrix of periodized second derivative (with and without preconditioning) in the system of coordinates associated with Daubechies' wavelets with three vanishing moments $M = 3$.

N	κ	κ_p
32	$0.10409 \cdot 10^3$	8.021
64	$0.41535 \cdot 10^3$	9.086
128	$0.16605 \cdot 10^4$	10.019
256	$0.66405 \cdot 10^4$	10.841
512	$0.26562 \cdot 10^5$	11.562
1024	$0.10625 \cdot 10^6$	12.197

for differences and a single coefficient for averages which we call the total average. We note that the total average of a vector is proportional to the direct sum of the elements of the vector. The sum of the entries in the rows of the matrix $\mathbf{A}$ is identically zero and, therefore, the matrix $\mathbf{A}_w$ has the following structure:

$$\mathbf{A}_w = \begin{pmatrix} \mathbf{B} & \mathbf{0} \\ \mathbf{c}^T & 0 \end{pmatrix}, \tag{12.20}$$

where $\mathbf{B}$ is an $(N - 1) \times (N - 1)$ full rank matrix with the condition number proportional to N^2. Let us now determine the vector $\mathbf{c}^T$. If we compute $\mathbf{A}_w$ by first applying the transformation to the columns of $\mathbf{A}$ we obtain the last row of the

Table 12.2. Condition numbers of the matrix of periodized second derivative (with and without preconditioning) in the system of coordinates associated with Daubechies' wavelets with six vanishing moments $M = 6$.

N	κ	κ_p
32	$0.10409 \cdot 10^3$	5.2002
64	$0.41535 \cdot 10^3$	5.2610
128	$0.16605 \cdot 10^4$	5.2897
256	$0.66405 \cdot 10^4$	5.3035
512	$0.26562 \cdot 10^5$	5.3103
1024	$0.10625 \cdot 10^6$	5.3137

transformed matrix as

$$\rho(a_{N+1/2} - a_{1/2})(e_1^T - e_N^T), \tag{12.21}$$

where ρ is a factor which depends on the size of the matrix **A**. In order to obtain $\mathbf{A}_w$, we have to transform further by applying the wavelet transform to the rows of the intermediate result. Thus, we obtain

$$(\mathbf{c}^T, 0) = \rho(a_{N+1/2} - a_{1/2})(\hat{e}_1^T - \hat{e}_N^T). \tag{12.22}$$

Let us introduce the following notation:

$$\hat{e}_l = \begin{pmatrix} \mathbf{r}_l \\ \rho \end{pmatrix}, \quad l = 1, N, \tag{12.23}$$

where $\mathbf{r}_l$ are vectors of size $N - 1$ and ρ is a scalar factor (common to both vectors),

$$\hat{u} = \begin{pmatrix} \mathbf{d} \\ s \end{pmatrix}, \tag{12.24}$$

and

$$\hat{f} = \begin{pmatrix} \mathbf{f}^d \\ f^s \end{pmatrix}. \tag{12.25}$$

Also, let $2a = a_{1/2} + a_{N+1/2}$, $\alpha = a_{1/2}/(2a)$, $\beta = a_{N+1/2}/(2a)$, so that $\alpha + \beta = 1$. We now rewrite (12.16) as

$$\left[\begin{pmatrix} \mathbf{B} & 0 \\ \mathbf{c}^T & 0 \end{pmatrix} - 2a \begin{pmatrix} \alpha \mathbf{r}_1 \mathbf{r}_N^T + \beta \mathbf{r}_N \mathbf{r}_1^T & \rho(\alpha \mathbf{r}_1 + \beta \mathbf{r}_N) \\ \rho(\alpha \mathbf{r}_N^T + \beta \mathbf{r}_1^T) & \rho^2 \end{pmatrix} \right] \begin{pmatrix} \mathbf{d} \\ s \end{pmatrix} = \begin{pmatrix} \mathbf{f}^d \\ f^s \end{pmatrix}, \tag{12.26}$$

where

$$\mathbf{c}^T = 2a\rho(\beta - \alpha)(\mathbf{r}_1^T - \mathbf{r}_N^T). \tag{12.27}$$

By eliminating s,

$$s = -\frac{(\alpha \mathbf{r}_1^T + \beta \mathbf{r}_N^T)\mathbf{d}}{\rho} - \frac{f^s}{2a\rho^2}, \tag{12.28}$$

we obtain the $(N - 1) \times (N - 1)$ system of linear algebraic equations for **d**,

$$\left[\mathbf{B} + 2a(\alpha^2 \mathbf{r}_1 - \beta^2 \mathbf{r}_N)(\mathbf{r}_1^T - \mathbf{r}_N^T) \right] \mathbf{d} = \mathbf{f}^d - \frac{f^s}{\rho}(\alpha \mathbf{r}_1 + \beta \mathbf{r}_N). \tag{12.29}$$

Our method for solving the two-point boundary value problem (12.7) is based on the fact that the matrix $\mathbf{B}^{-1}$ is sparse and could be computed in $O(N)$ operations. We will construct the matrix $\mathbf{B}^{-1}$ in the next section and let us assume here that it is available. Given the matrix $\mathbf{B}^{-1}$, we solve (12.29) using Sherman–Morrison formula for the rank-one update of the inverse matrix. We obtain

$$\mathbf{d} = \left[\mathbf{B}^{-1} - \sigma \mathbf{B}^{-1}(\alpha^2 \mathbf{r}_1 - \beta^2 \mathbf{r}_N)(\mathbf{r}_1 - \mathbf{r}_N)^T \mathbf{B}^{-1} \right] \left[\mathbf{f}^d - \frac{f^s}{\rho}(\alpha \mathbf{r}_1 + \beta \mathbf{r}_N) \right], \tag{12.30}$$

where

$$\sigma = \frac{2a}{1 + 2a(\mathbf{r}_1 - \mathbf{r}_N)^T \mathbf{B}^{-1}(\alpha^2 \mathbf{r}_1 - \beta^2 \mathbf{r}_N)}. \tag{12.31}$$

REMARK 12.1   The condition number of the sparse matrix **B** after rescaling by **P** is $O(1)$ as is illustrated in Tables 12.1 and 12.2. Thus, the linear system (12.29) may be solved using (12.30) by the standard iterative methods (e.g., conjugate gradient) in $O(N)$ operations since using (12.30) only involves finding the solution of the linear system $\mathbf{Bx} = \mathbf{y}$.   ∎

We look for the inverse operator in the form

$$\mathbf{L}^{-1} = \begin{pmatrix} \mathbf{\Gamma} & \mathbf{p} \\ \mathbf{q}^T & \gamma \end{pmatrix}, \tag{12.32}$$

and obtain

$$\mathbf{\Gamma} = \left[\mathbf{B}^{-1} - \sigma \mathbf{B}^{-1}(\alpha^2 \mathbf{r}_1 - \beta^2 \mathbf{r}_N)(\mathbf{r}_1 - \mathbf{r}_N)^T \mathbf{B}^{-1} \right], \tag{12.33}$$

$$\mathbf{p} = -\frac{1}{\rho} \left[\mathbf{B}^{-1}(\alpha \mathbf{r}_1 + \beta \mathbf{r}_N) - \sigma \kappa_1 \mathbf{B}^{-1}(\alpha^2 \mathbf{r}_1 - \beta^2 \mathbf{r}_N) \right], \tag{12.34}$$

$$\mathbf{q}^T = -\frac{1}{\rho} \left[(\alpha \mathbf{r}_1^T + \beta \mathbf{r}_N^T) \mathbf{B}^{-1} - \sigma \kappa_2 (\mathbf{r}_1 - \mathbf{r}_N)^T \mathbf{B}^{-1} \right], \tag{12.35}$$

$$\gamma = \frac{1}{\rho^2} \left(\kappa_3 - \sigma \kappa_2 \kappa_1 - \frac{1}{2a} \right), \tag{12.36}$$

where

$$\kappa_1 = (\mathbf{r}_1 - \mathbf{r}_N)^T \mathbf{B}^{-1}(\alpha \mathbf{r}_1 + \beta \mathbf{r}_N) \tag{12.37}$$

$$\kappa_2 = (\alpha \mathbf{r}_1^T + \beta \mathbf{r}_N^T) \mathbf{B}^{-1}(\alpha^2 \mathbf{r}_1 - \beta^2 \mathbf{r}_N), \tag{12.38}$$

and

$$\kappa_3 = (\alpha \mathbf{r}_1^T + \beta \mathbf{r}_N^T) \mathbf{B}^{-1}(\alpha \mathbf{r}_1 + \beta \mathbf{r}_N). \tag{12.39}$$

All matrix-vector multiplications in (12.33)–(12.39) involve the sparse matrix $\mathbf{B}^{-1}$ and the sparse vectors $\mathbf{r}_1$ and $\mathbf{r}_N$. Thus, the problem of constructing $\mathbf{L}^{-1}$ is reduced to that of computing the matrix $\mathbf{B}^{-1}$.

12.4   Computing the inverse of the periodized operator

We start by rescaling the $(N-1) \times (N-1)$ matrix **B** by the diagonal matrix **P**, where $\mathbf{P}_{il} = \delta_{il} 2^j$, $1 \le j \le n$, and where j is chosen depending on i, l so that $N - N/2^{j-1} + 1 \le i, l \le N - N/2^j$ (see Figure 12.1). We have

$$\mathbf{B}_p = \mathbf{PBP}, \tag{12.40}$$

FIGURE 12.2
Matrix **B** (in the case **A** = **D**) of size 255×255 in the system of coordinates associated with
the basis of Daubechies' wavelets with 3 vanishing moments. Entries with the absolute
value greater than 10^{-14} are shown black.

and the condition number of the matrix $\mathbf{B}_p$ is $O(1)$ (see Tables 12.1 and 12.2). The
matrices **B** and $\mathbf{B}_p$ are sparse matrices which is illustrated in Figure 12.2. Also,
the matrices **B** and $\mathbf{B}_p$ are full rank.

Our main tool in computing the inverse matrix $\mathbf{B}_p^{-1}$ is the iterative algorithm
[S]

$$\mathbf{X}_{l+1} = 2\mathbf{X}_l - \mathbf{X}_l \mathbf{B}_p \mathbf{X}_l, \tag{12.41}$$

which is initialized by setting

$$\mathbf{X}_0 = \alpha \mathbf{B}_p^*, \tag{12.42}$$

where α is chosen so that $0 < \alpha < 2/\sigma_1$ with σ_1 is the largest singular value of $\mathbf{B}_p$.

For the full-rank matrices the iteration (12.41) converges to $\mathbf{B}_p^{-1}$. The number of
iterations is proportional to the logarithm of the condition number of the matrix $\mathbf{B}_p$
and, thus, is $O(1)$. For the full-rank matrices the iteration (12.41) is self-correcting
and we use this property as described below.

The iteration (12.41) provides an $O(N)$ algorithm to compute the inverse matrix
if $\mathbf{B}_p$, $\mathbf{B}_p^{-1}$ and all the intermediate matrices $\mathbf{X}_l$ have a sparse representation in the
wavelet basis ([BCoR3], [BCoR2], [Aetal]). Since we know in advance that $\mathbf{B}^{-1}$

FIGURE 12.3
Matrix $\mathbf{B}^{-1}$ computed via iterative algorithm of this section with diagonal rescaling. Entries with the absolute value greater than 10^{-9} are shown black and the matrix verifies $\|\mathbf{BB}^{-1}-\mathbf{I}\|$, $\|\mathbf{B}^{-1}\mathbf{B} - \mathbf{I}\| \approx 10^{-9}$.

is sparse in the wavelet basis (for a given accuracy ϵ), we only need to maintain sparsity of the intermediate matrices $\mathbf{X}_l$.

Since the iteration is self-correcting, we first compute the low-accuracy inverse by removing all entries with absolute value less than a given threshold (e.g., 10^{-2}) after each iteration. Once the iteration converges we improve the accuracy of the inverse matrix by continuing the iteration and decreasing the threshold of accuracy. The sparsity of the resulting matrix $\mathbf{B}_p^{-1}$ is illustrated in Figure 12.3.

Finally, to obtain $\mathbf{B}^{-1}$, we have

$$\mathbf{B}^{-1} = \mathbf{P}\mathbf{B}_p^{-1}\mathbf{P}. \tag{12.43}$$

We note that since the matrix $\mathbf{P}$ is a diagonal matrix, there is no loss of accuracy in computing via (12.40) or (12.43), since only the operation of multiplication is involved. In our particular case the multiplication is by the powers of 2 only and, thus, no rounding errors are introduced.

12.5   Various extensions

12.5.1   Preconditioning to compensate for variations in a

In Section 12.3 we assumed that the function a does not change significantly over the interval $(0,1)$. If a is such that the finite difference scheme in (12.9) is appropriate for solving the two-point boundary value problem, then we rescale (12.9) by multiplying the matrix of the system in (12.9) on both sides by the diagonal matrix

$$\mathbf{P}_a = \mathrm{diag}\left(\frac{1}{\sqrt{a_1}}, \frac{1}{\sqrt{a_2}}, \ldots, \frac{1}{\sqrt{a_N}} \right). \tag{12.44}$$

We obtain instead of (12.9),

$$\frac{a_{i-1/2}}{\sqrt{a_{i-1}a_i}} v_{i-1} - \frac{a_{i-1/2} + a_{i+1/2}}{a_i} v_i + \frac{a_{i+1/2}}{\sqrt{a_i a_{i+1}}} v_{i+1} = h^2 \frac{f_i}{\sqrt{a_i}}, \quad i = 1,\ldots,N, \tag{12.45}$$

where

$$v_i = u_i \sqrt{a_i} \quad i = 1,\ldots,N. \tag{12.46}$$

This corresponds to considering the operator

$$\frac{1}{a(x)} \frac{\partial}{\partial x} \left(a(x) \frac{\partial u}{\partial x} \right) \tag{12.47}$$

instead of the operator $\mathcal{L}$ in (12.7).

If a is sufficiently smooth, then we have

$$\frac{a(x - \frac{1}{2}h)}{\sqrt{a(x-h)a(x)}} = 1 + O(h^2), \tag{12.48}$$

$$\frac{a(x - \frac{1}{2}h) + a(x + \frac{1}{2}h)}{a(x)} = 2 + O(h^2), \tag{12.49}$$

and

$$\frac{a(x + \frac{1}{2}h)}{\sqrt{a(x)a(x+h)}} = 1 + O(h^2). \tag{12.50}$$

Thus, the matrix $\mathbf{L}$ corresponding to (12.45) may be written as

$$\mathbf{L} = \mathbf{L}_0 + h^2 \mathbf{R}, \tag{12.51}$$

where

$$
\mathbf{L}_0 = \begin{pmatrix}
-2 & 1 & 0 & \cdots & 0 & 0 & 0 \\
1 & -2 & 1 & \cdots & 0 & 0 & 0 \\
\vdots & \vdots & \vdots & \ddots & \vdots & \vdots & \vdots \\
0 & 0 & 0 & \cdots & 1 & -2 & 1 \\
0 & 0 & 0 & \cdots & 0 & 1 & -2
\end{pmatrix}.
\tag{12.52}
$$

We note that in computing entries of the matrix $h^2\mathbf{R}$ via $h^2\mathbf{R} = \mathbf{L} - \mathbf{L}_0$ one should be careful to obtain a sufficient number of significant digits.

Given the operator $\mathbf{L}_0^{-1}$, we have

$$
\mathbf{L}^{-1} = \mathbf{L}_0^{-1}(\mathbf{I} + h^2\mathbf{L}_0^{-1}\mathbf{R})^{-1}
\tag{12.53}
$$

and, therefore, we need to compute $(\mathbf{I} + h^2\mathbf{L}_0^{-1}\mathbf{R})^{-1}$. Again we use the iteration in Section 12.4 and note that if the largest singular value of the operator $\mathbf{T} = -h^2\mathbf{L}_0^{-1}\mathbf{R}$ is less than one, then the iteration in Section 12.4 takes a particular simple form since we have

$$
(\mathbf{I} - \mathbf{T})^{-1} = \prod_{j=0}^{\infty}(\mathbf{I} + \mathbf{T}^{2^j}).
\tag{12.54}
$$

12.5.2 Additional remarks

REMARK 12.2 It is clear that the generalized inverse $\mathbf{D}^{-1}$ (which on the range of the matrix $\mathbf{D}$ is the matrix $\mathbf{B}^{-1}$ in Figure 12.3) plays a special role in our approach. Therefore, we may compute this matrix in advance. Considering $\mathbf{A}$ in (12.13) as a perturbation of $\mathbf{D}$, we may use the iteration in Section 12.4 to compute the inverse (similar to that in (12.53)). In fact, the matrix $\mathbf{D}^{-1}$ may be stored and used in the nonstandard form [BCoR1], which will result in additional efficiency of computation. ∎

REMARK 12.3 Our approach uses an $O(h^2)$ finite-difference scheme in the wavelet system of coordinates. We may use Richardson extrapolation to improve the accuracy of the solution and, also, of the inverse operator. ∎

12.6 Wavelet-based adaptive Crank–Nicolson scheme

Let us now consider some of the implications of the fact that the Green's functions of the elliptic two-point boundary value problems are available numerically as sparse matrices. As an example, we consider the implicit Crank–Nicolson scheme to solve

$$u_t = \mathcal{L}u \tag{12.55}$$

with the Dirichlet boundary conditions $u(t,0) = u(t,1) = 0$ and the initial condition $u(0,x) = u_0(x)$. Approximating (12.55) by a system of ordinary differential equations, we obtain

$$\frac{d\mathbf{u}}{dt} = \frac{1}{h_x^2}\mathbf{L}\mathbf{u}, \tag{12.56}$$

where the matrix $\mathbf{L}$ is given in (12.11) and h_x is the step size in x-coordinate. Applying trapezoidal rule to (12.56) (in time), we obtain the Crank–Nicolson method

$$\mathbf{u}^{(n+1)} - \mathbf{u}^{(n)} = \frac{h_t}{2h_x^2}(\mathbf{L}\mathbf{u}^{(n+1)} + \mathbf{L}\mathbf{u}^{(n)}), \tag{12.57}$$

where h_t is the step size in time. We have

$$\left(\mathbf{I} - \frac{h_t}{2h_x^2}\mathbf{L}\right)\mathbf{u}^{(n+1)} = \left(\mathbf{I} + \frac{h_t}{2h_x^2}\mathbf{L}\right)\mathbf{u}^{(n)}, \tag{12.58}$$

a well known implicit scheme for solving (12.55).

In the standard Crank–Nicolson method the inverse matrix is never computed since it is a dense matrix. Instead, one solves a tridiagonal linear system at each time step.

We note that the matrix $(\mathbf{I} - (h_t/2h_x^2)\mathbf{L})^{-1}$ is sparse in the wavelet system of coordinates and may be computed explicitly by a procedure similar to that described in the previous sections. Thus, by converting (12.58) into the wavelet basis, we obtain an explicit scheme by computing $(\mathbf{I} - (h_t/2h_x^2)\mathbf{L})^{-1}$. We have

$$\hat{\mathbf{u}}^{(n+1)} = \mathbf{C}\hat{\mathbf{u}}^{(n)}, \tag{12.59}$$

where

$$\mathbf{C} = \left(\mathbf{I} - \frac{h_t}{2h_x^2}\mathbf{WLW}^*\right)^{-1}\left(\mathbf{I} + \frac{h_t}{2h_x^2}\mathbf{WLW}^*\right), \tag{12.60}$$

$$\hat{\mathbf{u}} = \mathbf{W}\mathbf{u}, \tag{12.61}$$

and $\mathbf{W}$ denotes the matrix of the discrete wavelet transform. The matrix $\mathbf{C}$ is a sparse matrix of the structure similar to that of matrices in Figures 12.2 and 12.3.

Such an approach applied to the higher order schemes might have serious advantages. Also, the advantages of the conversion from an implicit to an explicit scheme are apparent if the vector $\hat{\mathbf{u}}^{(n)}$ is sparse in the wavelet system of coordinates. In this case the number of operations per time-step is proportional to the number of the significant entries of the vector $\hat{\mathbf{u}}^{(n)}$.

The vectors $\hat{\mathbf{u}}^{(n)}$ become more and more sparse in the wavelet system of co-ordinates as n (time) increases since the oscillatory modes which correspond to larger eigenvalues of the operator $\mathcal{L}$ decay faster than those corresponding to the smaller eigenvalues. Thus, the speed of the computation via (12.59) is adaptable to the regularity of the solution which is not the case for the standard scheme.

As an example let us consider computing the smallest eigenvalue of the operator $\mathcal{L}$. One of the ways to compute the smallest eigenvalue of the operator $\mathcal{L}$ is to use (12.58) and renormalize the solution after each time-step. Only the modes corresponding to the smallest eigenvalue will remain as a part of the solution after several time-steps. We notice that the eigenvector which corresponds to the smallest eigenvalue is smooth and nonoscillatory. Let us choose the initial condition to be a constant vector. Since such a vector is sparse in the wavelet system of coordinates (in fact, it might be represented by just one number) and since we know that the solution of (12.55) remains smooth and nonoscillatory, we conclude that a time-step requires only $O(1)$ operations.

We note that generalizing this approach to the multidimensional case and using the Crank–Nicolson scheme with alternating directions, obtaining a fast method for computing the smallest eigenvalues for the multidimensional elliptic operators appears feasible.

Bibliography

[Aetal] B. Alpert, G. Beylkin, R.R. Coifman, and V. Rokhlin. 1993. "Wavelet-like bases for the fast solution of second-kind integral equations," *SIAM J. Sci. Statist. Comput.*, **14**(1): 159–184.

[B] G. Beylkin. 1992. "On the representation of operators in bases of compactly supported wavelets," *SIAM J. Numer. Anal.*, **29**(6): 1716–1740.

[BCoR1] G. Beylkin, R.R. Coifman, and V. Rokhlin. 1991. "Fast wavelet transforms and numerical algorithms, I," *Commun. Pure and Appl. Math.*, **44**: 141–183.

[BCoR2] G. Beylkin, R.R. Coifman, and V. Rokhlin. 1992. "Wavelets in numerical analysis," in *Wavelets and Their Applications*, M. Ruskai, G. Beylkin, R. Coifman, I. Daubechies, S. Mallat, Y. Meyer, and L. Raphael, eds., Boston: Jones and Bartlett, pp. 181–210.

[BCoR3] G. Beylkin, R.R. Coifman, and V. Rokhlin. 1993. "Fast wavelet transforms and numerical algorithms, II," in progress.

[BrPX] J. H. Bramble, J. E. Pasciak, and J. Xu. 1990. "Parallel multilevel preconditioners," *Math. Comput.*, **55**(191): 1–22.

[Bri] W. L. Briggs. 1987. *A Multigrid Tutorial*, Philadelphia: Soc. Ind. Appl. Math.

[Cetal]   A. Cohen, I. Daubechies, B. Jawerth, and P. Vial. 1992. "Multiresolution analysis, wavelets and fast algorithms on an interval," *C. R. Acad. Sci.*, sér. 1, to appear.

[GR]   L. Greengard and V. Rokhlin. 1989. "On the numerical solution of two-point boundary value problems," Yale Univ. Technical Report, YALEU/DCS/RR-692.

[J]   S. Jaffard. 1992. "Wavelet methods for fast resolution of elliptic problems," *SIAM J. Numer. Anal.*, **29**(4): 965–986.

[S]   G. Schulz. 1933. "Iterative berechnung der reziproken matrix," *Z. Angew. Math. Mech.*, **13**: 57–59.

13

Wavelets and nonlinear analysis

Stéphane Jaffard

ABSTRACT   We show on a few examples how orthonormal bases of wavelets can be used in the study of some nonlinear problems related to PDEs or the geometry of singularities.

CONTENTS

13.1   Introduction

Our purpose is to show how orthonormal wavelet bases can be used in the study of nonlinear problems. The feasibility of such a program is apparent for those that are acquainted with some basic results in real analysis: indeed, in the thirties, Littlewood and Paley showed that classical Fourier analysis is quite ineffective when dealing with nonlinear quantities. The decomposition that bears their names was constructed to avoid this obstacle. Their basic idea was not to break the function uniformly in the frequency domain (as Fourier series do for instance), but to group together frequencies between 2^j and 2^{j+1}. This has several advantages: It allows a sharp analysis of nonlinear terms, which has proved crucial in the study of nonlinear PDEs (following essentially the work of Bony and his followers, see, for instance, [Bo] or [Me1]), and the poor resolution at high frequencies allows (following the uncertainty principle) the Littlewood–Paley decomposition to be localized in the space domain. This in turn yields a sharp local analysis, which would also be impossible using classical Fourier analysis. This last point has two advantages: for PDEs, the local behavior is often crucial (propagation of singularities, for instance), but also, it facilitates obtaining precise results about the geometry of singularities, their fractal dimensions, etc. Our purpose in this paper is to give some insights and precise results on these various topics.

We shall first recall a few basic properties of orthonormal bases of wavelets, though we refer to [Ma], [Me2], and [D2] for a detailed description. We shall describe more precisely the prerequisites for the study of nonlinear analysis, namely the analysis of differential operators on wavelet bases, and the characterization of some functional spaces (which can also be found in more details in [Me2]). We shall then apply this material to the following problems: the determination of the Hausdorff dimension of the singularities of functions; compensated compactness; and concentration-type problems. We shall see that these subjects are closely related: the proof in the concentration type problem is almost the same as the one used in the determination of the dimension of singularities. This is because the wavelet decomposition of a function gives a natural splitting which is the key to these properties: we group the few large coefficients into a well-localized non-smooth function; what remains is smooth.

13.2   Orthonormal wavelet bases

An orthonormal wavelet basis of $L^2(\mathbb{R})$ is of the form $\psi_{j,k} = 2^{j/2}\psi(2^j x - k)$ where j and k belong to $\mathbb{Z}$, and ψ is a smooth and well-localized function. This last condition needs to be made more precise. For instance, we can suppose that

$$|\psi^{(\alpha)}(x)| \leq C \exp(-\gamma|x|)$$

for $\alpha \leq N$ and a positive γ. The decomposition of a function f in this basis is then local at high frequencies (that is for large js), since

$$f(x) = \sum_{j,k} c_{j,k}\psi_{j,k}(x), \quad \text{where } c_{j,k} = \int f(x)\psi_{j,k}(x)dx$$

and the function $\psi_{j,k}$ has a numerical support essentially of size 2^{-j}, so that the wavelet coefficients for j large yield a local information on f around $k2^{-j}$. An extremely important feature of this decomposition is that it is performed on functions that have cancellation. Actually, for $l < N$,

$$\int x^l \psi(x)dx = 0,$$

which is a straightforward consequence of the orthonormality of the wavelet basis (see [Me1]). This property implies that $\hat{\psi}_{j,k}$ is well-localized near the frequencies of order of magnitude 2^j and -2^j.

The fact that wavelets have vanishing moments allows us to estimate the smoothness of the function f in terms of decay estimates of its wavelet coefficients for large js. Let us give a precise result. But, first, we recall the definition of C^α regularity.

Let $0 < \alpha < 1$. A function f is uniformly C^α if

$$|f(x) - f(y)| \le C|x - y|^\alpha.$$

Then, f is uniformly C^α if and only if

$$|c_{j,k}| \le C2^{-(1/2+\alpha)j}. \tag{13.1}$$

Let us prove this assertion. First, suppose that f is uniformly C^α. Then

$$|c_{j,k}| = \left| \int f(x)\psi_{j,k}(x)dx \right| = \left| \int (f(x) - f(k2^{-j}))\psi_{j,k}(x)dx \right|$$

$$\le C \int |x - k2^{-j}|^\alpha \frac{2^{j/2}}{(1 + |x - k2^{-j}|)^2} dx \le C2^{-(\frac{1}{2}+\alpha)j}.$$

Conversely, suppose that $|c_{j,k}| \le C2^{-(1/2+\alpha)j}$. Let

$$\Delta_j(f) = \sum_k c_{j,k}\psi_{j,k}.$$

Using the localization of the wavelets and of their derivative, we get $|\Delta_j(f)| \le C2^{-\alpha j}$ and $|\Delta_j'(f)| \le C2^j 2^{-\alpha j}$. Let J be such that $2^{-J} \le |x - y| < 2 \cdot 2^{-J}$. Then

$$|f(x) - f(y)| \le \sum_{j \le J} |\Delta_j(f)(x) - \Delta_j(f)(y)| + \sum_{j > J} |\Delta_j(f)(x)| + |\Delta_j(f)(y)|.$$

Using the mean value theorem, the first sum is bounded by

$$C \sum_{j \le J} |x - y|2^j 2^{-\alpha j} \le C|x - y|^\alpha$$

and the second sum is bounded by

$$C \sum_{j > J} 2^{-\alpha j} \le C|x - y|^\alpha.$$

Here, we have a result about global smoothness. Actually, using the localization of the wavelets, we shall prove in Section 13.3.1 a similar result concerning the pointwise smoothness of f (see [J1], [J4]). Notice that in this proof we just use the cancellation of the wavelets and their localization, but not their exact form. We usually won't need it.

We shall now recall a few basic facts about the construction of wavelets, essentially in order to state our notations. A multiresolution analysis is an increasing sequence $(V_j)_{j \in \mathbb{Z}}$ of closed subspaces of L^2 such that

(a) $\cap V_j = \{0\}$

(b) $\cup V_j$ is dense in L^2

(c) $f(x) \in V_j \Leftrightarrow f(2x) \in V_{j+1}$

(d) $f(x) \in V_0 \Leftrightarrow f(x+1) \in V_0$

(e) There is a function g in V_0 such that the $(g(x-k))_{k \in \mathbb{Z}}$ form a basis of V_0.

We also require g to be smooth and well-localized. We shall say that a multiresolution analysis is m-regular if g and its m first derivatives have fast decay. In the following we often won't make precise the regularity needed on the wavelets, supposing they are "smooth enough." However the smoothness actually needed will be clear in the proofs. The sequence of spaces V_j can be interpreted as follows: if $P_j(f)$ is the projection of f on V_j, we "see" on $P_j(f)$ the details of f of size larger than 2^{-j}.

A simple example of multiresolution analysis is obtained using splines (see [L] or [Me2]). Let V_0 be the subspace of L^2 composed of functions which are piecewise polynomials of degree m on each interval $[k, k+1]$ and are C^{m-1}. A possible choice for g is the B-spline σ such that

$$\hat{\sigma}(\xi) = \left(\frac{\sin \xi/2}{\xi/2} \right)^{m+1}.$$

Once a multiresolution analysis is given, it is easy to obtain an orthonormal basis of V_0. Let, for instance

$$\hat{\varphi}(\xi) = \hat{g}(\xi) \left(\sum |\hat{g}(\xi + 2k\pi)|^2 \right)^{-1/2}.$$

Then, the $\varphi(x-k)$ form an orthonormal basis of V_0 and, by dilation, the

$$\varphi_{j,k}(x) = 2^{j/2} \varphi(2^j x - k)$$

form an orthonormal basis of V_j.

Define W_j as the orthogonal complement of V_j in V_{j+1}. The W_j are mutually orthogonal, and their direct sum is equal to L^2. One can then obtain an orthonormal basis of W_0 of the form $\psi(x-k)$. Since the W_j are obtained from each other by dilation, and are mutually orthogonal, the functions $2^{j/2}\psi(2^j x - k)$ form an orthonormal basis of $L^2(\mathbb{R})$.

There are several kinds of wavelets now used: The spline wavelets that we mentioned are used in numerical analysis; the compactly supported wavelets (used in signal analysis, see [D1]), and also the following C^∞ wavelet (see [LMe]), which is not very well-localized (it has "only" fast decay), but is extremely useful for theoretical computations. Let $\hat{\varphi}$ be a C^∞, even function supported by

$[-4\pi/3, 4\pi/3]$ such that

$$\begin{cases} 0 \leq \hat{\varphi} \leq 1 \\ \hat{\varphi}(\xi) = 1 & \text{if } |\xi| \leq 2\pi/3 \\ \hat{\varphi}(\xi)^2 + \hat{\varphi}(2\pi - \xi)^2 = 1 & \text{if } 0 \leq \xi \leq 2\pi. \end{cases}$$

The function φ is in the Schwartz class. The corresponding wavelet ψ is also in the Schwartz class. Its Fourier transform is C^{∞} and supported by the set $[-8\pi/3, -2\pi/3] \cup [2\pi/3, 8\pi/3]$. Yves Meyer in [Me1] calls it the "Littlewood–Paley" wavelet for the following reason. The decomposition $f = \sum_{j\in\mathbb{Z}} \Delta_j(f)$ is similar to the Littlewood–Paley decomposition, which is obtained by splitting f as a sum of functions which have supports in the Fourier domain essentially located on the dyadic intervals $[-2^{j+1}, -2^j] \cup [2^j, 2^{j+1}]$. Many properties of the wavelet decomposition will be a consequence of this dyadic localization in the Fourier domain.

Let us now show how to obtain wavelets in several dimensions. Let $\mathcal{V}_j$ be the subspace of $L^2(\mathbb{R}^2)$ defined by $\mathcal{V}_j = V_j \otimes V_j$. Clearly, we define thus a multiresolution analysis in two dimensions (here $\mathcal{V}_0$ is invariant under translations in $\mathbb{Z} \times \mathbb{Z}$). An orthonormal basis of $\mathcal{V}_j$ is given by the functions

$$\Phi_{j,(k_1,k_2)}(x,y) = \varphi_{j,k_1}(x)\varphi_{j,k_2}(y).$$

Define now $\mathcal{W}_j$ as the orthogonal complement of $\mathcal{V}_j$ in $\mathcal{V}_{j+1}$. A basis of $\mathcal{W}_j$ is given by the functions $\psi_{j,k}^{(i)} = 2^j \psi^{(i)}(2^j x - k)$ where

$$\begin{cases} \psi^{(1)}(x,y) = \psi(x)\varphi(y) \\ \psi^{(2)}(x,y) = \varphi(x)\psi(y) \\ \psi^{(3)}(x,y) = \psi(x)\psi(y). \end{cases}$$

We see that in two dimensions, we have not one but three wavelets. This construction immediately generalizes to the n-dimensional case; in that case, we obtain $2^n - 1$ wavelets.

We remark that another possible basis of $L^2(\mathbb{R}^2)$ is given by the functions $\psi_{j,k}(x)\psi_{j',k'}(y)$. This basis, unlike the one we have just constructed, has unrelated localizations in the x and y directions. These directions play a very privileged role; therefore it seems less natural for decomposing functions. But this drawback can be compensated by the fact that the analyzing functions have cancellation in both directions (integrals in x *and* in y vanish, which is not the case for $\psi^{(1)}$ and $\psi^{(2)}$); therefore, the coefficients of a smooth function may be encoded using less nonvanishing coefficients. This point is discussed in [MPerR] and [JLa].

Let us end this part by defining the notations we shall use. Let λ be the dyadic cube $k2^{-j} + 2^{-j}[0,1]^n$; the wavelet $\psi_{j,k}^{(i)}$ is essentially localized on the cube λ. We shall often write respectively ψ_λ and c_λ instead of $\psi_{j,k}^{(i)}$ and $c_{j,k}^{(i)}$ for the wavelets and the wavelet coefficients.

13.3   Determination of the Hausdorff dimension of singularities

One of the most attractive feature of the wavelets is that they give completely local information on the functions analyzed. We shall study applications of this property.

Let s be a strictly positive real number. Let us recall the usual definition of the Hölder criterion at x_0. A function f belongs to $C_{x_0}^s$ if there exists a polynomial $P(x)$ of degree equal to the integral part of s such that

$$f(x) = P(x - x_0) + O(|x - x_0|^s) \tag{13.2}$$

(we shall sometimes say that f is C^s at x_0).

First, we shall show that we can almost characterize the pointwise smoothness of a function (that is the existence of a Taylor expansion of a given order at a given point) by conditions on the moduli of the coefficients on the wavelets localized near the considered point.

Call E_s the set of points where a function is $C_{x_0}^s$ and not smoother. We shall show how to give a bound of the Hausdorff dimension $\theta(s)$ of the set E_s (the singularity spectrum of f, in the physicist's language) when a global energy estimate on f is known. Finally, we shall show which functions $\theta(s)$ can be singularity spectra.

13.3.1   Analysis of pointwise smoothness

In this section, following [J2], we shall compare the usual definition of pointwise smoothness and a corresponding condition on the wavelet coefficients. The purpose is to show that this last condition provides "a good substitute" for the pointwise Hölder regularity condition: it is easily checkable from the knowledge of the wavelet coefficients, it can be very precisely compared with the Hölder condition, and it has more functional properties. We also give applications of these properties.

THEOREM 13.1
If f belongs to $C_{x_0}^s$,

$$|c_{j,k}| \le C2^{-(\frac{n}{2}+s)j}(1 + |2^j x_0 - k|^s). \tag{13.3}$$

Conversely, if (13.3) holds and f is uniformly C^β for a positive number β, there exists a polynomial P (depending on x_0) of degree less than s such that, if $|x - x_0| \le 1$,

$$|f(x) - P(x - x_0)| \le C|x - x_0|^s \log \frac{2}{|x - x_0|}, \tag{13.4}$$

and this result is optimal.

In this theorem, we are looking for regular points in an irregular background, which is more subtle than the usual approach which consists in finding irregular points in a C^∞ or analytical background (search of the different singular supports). This theorem can be interpreted as a tauberian theorem. We have information on the behavior of averages of f (its Littlewood–Paley decomposition) and a tauberian condition of minimal global regularity, which yields a pointwise result.

PROOF   Suppose that f belongs to $C_{x_0}^s$. Then

$$|c_{j,k}| = \left| \int f(x)\psi_{j,k}(x)dx \right|$$

$$= \left| \int (f(x) - P(x - x_0))\psi_{j,k}(x)dx \right|$$

(because wavelets have vanishing moments)

$$\leq C \int |x - x_0|^s |\psi_{j,k}(x)|dx \quad \text{(using (13.2))}$$

$$\leq C \int |x - x_0|^s \frac{2^{nj/2}}{(1 + 2^j|x - k2^{-j}|)^N} dx$$

(because $\psi(x)$ has fast decay)

$$\leq C 2^{nj/2} \int \frac{|x - k2^{-j}|^s + |k2^{-j} - x_0|^s}{(1 + 2^j|x - k2^{-j}|)^N} dx$$

(because $\forall s > 0, |x + y|^s \leq C(s)(|x|^s + |y|^s)$)

$$\leq C 2^{-(\frac{n}{2} + s)j}(1 + |2^j x_0 - k|^s). \quad \blacksquare$$

Let us now prove the converse result. Define j_0 and j_1 by $2^{-j_0-1} \leq |x-x_0| < 2^{-j_0}$ and $j_1 = s/\beta j_0$. From (13.3), we deduce, for any l,

$$|\Delta_j^{(l)}(f)(x)| \leq C 2^{(l-s)j}(1 + 2^j|x - x_0|)^s.$$

If g is a function, let $T(g)(x_0)$ be the Taylor expansion of g at the order $[s]$ at x_0. Then

$$|f(x) - T(f)(x_0)|$$

$$\leq \sum_{j \leq j_0} |\Delta_j(f)(x) - T(\Delta_j(f))(x_0)| + \sum_{j \geq j_0} |\Delta_j(f)(x)| + \sum_{j \geq j_0} |T(\Delta_j(f))(x_0)|.$$

Let $l = [s] + 1$. The first term is bounded by

$$C|x - x_0|^l \sum_{j \leq j_0} \sup_{[x,x_0]} |\Delta_j^{(l)}(f)(x_0)| \leq C|x - x_0|^l \sum_{j \leq j_0} 2^{(l-s)j} \leq C|x - x_0|^s.$$

For the second term,

$$\sum_{j_0 \leq j < j_1} |\Delta_j(f)(x)| \leq \sum_{j_0 \leq j < j_1} |x - x_0|^s \leq C(j_1 - j_0)|x - x_0|^s$$

and

$$\sum_{j \geq j_1} |\Delta_j(f)(x)| \leq \sum_{j \geq j_1} 2^{-\beta j} \leq C|x - x_0|^s$$

(because f is uniformly C^β). The third term is bounded by

$$\sum_{j \geq j_0} \sum_{|m|} |x - x_0|^m 2^{(m-s)j} \leq C|x - x_0|^s,$$

establishing the converse, since $j_1 - j_0 \leq C \log(2/|x - x_0|)$. The counterexamples that show the optimality of the theorem were discovered by Yves Meyer and can be found in [J2], concluding the proof of the theorem.

Let x_0 given. We have obtained conditions under which the following expansion holds at x_0:

$$f(x) = P(x - x_0) + O(|x - x_0|^s).$$

This can be interpreted as a generalized Taylor expansion : If f were C^n at x_0, Taylor's formula would prove that this expansion holds for $s = n$ and

$$P(x - x_0) = \sum_{|\alpha| \leq n} \frac{f^{(\alpha)}(x_0)}{\alpha!} (x - x_0)^\alpha.$$

This condition with $s > 1$ implies that f is differentiable at x_0, which can be seen by rewriting (13.2) as

$$\frac{|f(x) - P(x - x_0)|}{|x - x_0|} \leq C|x - x_0|^{s-1},$$

where P is a polynomial of degree 1. But, even if s is large, it does not imply more than the differentiability of f at x_0; it may hold while f is not differentiable in a neighborhood of x_0, in which case f is certainly not twice differentiable.

"Good substitutes" for the $C_{x_0}^s$ condition were introduced already in 1961 by A.P. Calderón and A. Zygmund [CZy] as follows. A function f belongs to $T_u^p(x_0)$ if there exists a polynomial P of degree less than u such that

$$\left(\frac{1}{\rho^n} \int_{|x-x_0| \leq \rho} |f(x) - P(x - x_0)|^p dx \right)^{1/p} \leq C\rho^u, \tag{13.5}$$

for ρ small enough. The case $p = \infty$ corresponds to the usual Hölder criterion. If f belongs to $C^u(x_0)$, then f belongs to $T_u^p(x_0)$ for any p. This weak form of pointwise regularity is preserved under fractional integration and singular integral transformations. However, it is not as closely related to the $C_{x_0}^u$ spaces as condition (13.3) is, and a result such as Theorem 13.1 cannot hold for this condition.

Condition (13.3) is a special case of the 2-microlocal spaces. Following Bony [Bo] and [J2], $f \in C_{x_0}^{s,s'}$ if

$$|c_{j,k}| \leq C 2^{-(\frac{n}{2} + s)j} (1 + |2^j x_0 - k|)^{-s'}.$$

If $s' > -s$, then $f \in C_{x_0}^{s,s'}$ implies $f \in C_{x_0}^s$ (the proof is very similar to the proof of Theorem 13.1, see [J2]). Furthermore, if Λ is a partial differential operator of order m, with smooth coefficients and elliptic at x_0, then

$$\Lambda f \in C_{x_0}^{s,s'} \Leftrightarrow f \in C_{x_0}^{s+m,s'}.$$

Using these two last results, we obtain the following result on pointwise regularity of elliptic operators.

COROLLARY 13.2
Let Λ be a partial differential operator of order m, with smooth coefficients and elliptic at x_0. If $\Lambda f = g$ and if g is a function that belongs to $C_{x_0}^s$, then f belongs to $C_{x_0}^{s+m}$.

13.3.2 The singularities of functions in $W^{s,p}$

Let f be a function belonging to the Sobolev space $W^{s,p}(\mathbb{R}^n)$. If $s > n/p$, the Sobolev imbeddings tell us that f is uniformly Hölder of exponent $s - n/p$. This result is optimal since there exists functions in $W^{s,p}(\mathbb{R}^n)$ which do not belong to C^t if $t > s - n/p$. We shall see that this lack of regularity can only happen on a set of points on which the Hausdorff dimension can be sharply bounded (see [J3]).

Let us first recall the definition of Hausdorff dimension. Let A be a bounded set in $\mathbb{R}^n$. Let

$$H_\delta^s(A) = \inf \left\{ \sum \text{diam}(U_i)^s \right\},$$

where the infimum is taken on all possible coverings of A by sets $\{U_i\}$ of diameter at most δ. Then $H^s(A) = \lim_{\delta \to 0} H_\delta^s(A)$ and the Hausdorff dimension of A is

$$d(A) = \inf\{s | H^s(A) = 0\} = \sup\{s | H^s(A) = \infty\}.$$

It suffices to restrict to coverings by balls or dyadic cubes.

THEOREM 13.3
Let $s > 0$ and $1 \le p < \infty$.

    i. If $s - n/p > 0$ and $0 < \alpha < s$, the Hausdorff dimension of the set of points where a function belonging to $W^{s,p}$ is not C^α is bounded by $n - (s - \alpha)p$.

    ii. Conversely, if A is a set of Hausdorff dimension $n - (s - \alpha)p$, there exists f in $W^{s,p}$ such that, for any $\alpha' > \alpha$, the set of points where f is not $C^{\alpha'}$ contains A.

    iii. If $s - n/p < 0$, $\forall \epsilon > 0$ there exists a function in $W^{s,p}$ which is C^ϵ at no point.

    iv. If $s - n/p \le 0$ and $\beta > 0$, the Hausdorff dimension of the set of points where a function belonging to $W^{s,p} \cap C^\beta$ is not C^α is bounded by $(\alpha/\beta)(n - (s - \beta)p)$.

    v. Conversely, if A is a set of Hausdorff dimension $(\alpha/\beta)(n - (s - \beta)p)$, there exists a function f in $W^{s,p} \cap C^\beta$ such that, for any $\alpha' > \alpha$, the set of points where f is not $C^{\alpha'}$ contains A.

It is proved in [Zi] that, if $f \in W^{s,p}$ (and s is an integer), then f coincides with a C^α function except on a set of Hausdorff measure at most $n - (s - \alpha)p$. Furthermore, f belongs to $T^p_\alpha(x_0)$ for all x_0s except on a set of Hausdorff dimension at most $n - (s - \alpha)p$. Nonetheless, from these results one cannot obtain information on the Hölder regularity of f (which can be checked directly or, more simply by remarking that case 3 of the theorem gives the example of a function for which these results hold, but has no Hölder regularity).

PROOF   Let us first check Point 3. Let φ be positive, C^∞, compactly supported, and such that $\varphi(x) = 1$ in a neighborhood of 0. Then $f(x) = (\varphi(x)/|x|^\alpha)$ belongs to $W^{s,p}$ if $\alpha < (n/p) - s$. If (x_k) is a sequence dense in $\mathbb{R}^n$, $\sum_k (1/k^2) f(x - x_k)$ is the expected counterexample, since this series converges in $W^{s,p}$ and is not locally bounded.

Let us sketch the idea of the proofs of Points 1 and 4. Consider the decomposition of f in terms of a wavelet orthonormal basis $\psi_{j,k}$. One shows that, if $f \in W^{s,p}$, for all js, f has "few big" wavelet coefficients $c_{j,k}$. One estimates the Hausdorff dimension of their "influence zone," and one shows that f is smooth outside this zone. Yves Meyer mentioned to us that this technique is similar to the existence proof of slow points of the Brownian motion in [K]. The Brownian motion is decomposed on the Schauder basis (which is a sort of wavelet basis); the coefficients are then independent Gaussians. One isolates the few "big coefficients," and, if x_0 is far away from them, the Brownian will be exactly $C^{1/2}$ at x_0.

We shall use a compactly supported wavelet basis. Suppose first $p > 1$. We shall prove a slightly more general result, namely that Points 2 and 4 still hold if, instead of assuming that f belongs to $W^{s,p}$, we only suppose that f belongs to the Besov space $B_p^{s,\infty}$. Recall the following characterization (we shall prove it when $p = 2$ in Section 13.4.1; for all ps, see [Me2]):

$$f \in W^{s,p} \Leftrightarrow \sum |c_\lambda|^2 2^{2js} 2^{nj} \chi_\lambda(x) \in L^{p/2}$$

where $c_\lambda = \langle f \mid \psi_\lambda \rangle$ and χ_λ is the characteristic function of the cube λ; and

$$f \in B_p^{s,\infty} \Leftrightarrow \forall j \quad \sum_{\lambda \in \Lambda_j} |c_\lambda|^p 2^{(sp - (np/2) - n)j} \leq C. \qquad (13.6)$$

Let $N > 0$ fixed. Let us divide $[\beta, \alpha]$ in N intervals $[\alpha_i, \alpha_{i+1}]$ of equal lengths, with $\alpha_i = \beta + i((\beta - \alpha)/N)$. Since f is uniformly C^β, after eventually multiplying f by a constant, we have (using (13.1)) $|c_\lambda| \leq 2^{-((n/2)+\beta)j}$. Let

$$S^i_j = \bigcup \{\lambda \in \Lambda_j \mid 2^{-((n/2)+\alpha_{i+1})j} \leq |c_\lambda| \leq 2^{-((n/2)+\alpha_i)j}\}$$

and $f^i = \sum_j \sum_{\lambda \in S^i_j} c_\lambda \psi_\lambda$. We have $f = \sum_{i=0}^{N-1} f^i + g$ where the wavelet coefficients of g verify $|\langle g \mid \psi_\lambda \rangle| \leq 2^{-((n/2)+\alpha)j}$, so that g is uniformly C^α. Let

$$D^i_j = \bigcup \{\lambda \in \Lambda_j \mid \text{dist}(\lambda, S^i_j) \leq 2^{-\epsilon_i j}\}$$

where $\epsilon_i < 1$ will be determined later. We shall show that, for a certain choice of ϵ_i, the function f^i is C^α outside $\limsup D^i_j$ (that is, outside a set of points belonging to an infinity of D^i_js). The set D^i_j is the "security zone" around the "big" coefficients localized at S^i_j. We now need the two following lemmas in order to prove Points 1 and 4.

LEMMA 13.4
If $x_0 \notin \limsup D^i_j, f^i$ is C^{α_i/ϵ_i} at x_0.

LEMMA 13.5
The Hausdorff dimension of $\limsup D^i_j$ is bounded by $(n - (s - \alpha_{i+1})p/\epsilon_i)$.

PROOF OF LEMMA 13.4 Let $x_0 \notin \limsup D^i_j$. There exists J such that $\forall j \geq J$, $x_0 \notin D^i_j$. The function $\sum_{j \leq J} \sum_{\lambda \in S^i_j} c_\lambda \psi_\lambda$ is uniformly C^s. We only have to study the regularity at x_0 of $g^i = \sum_{j \geq J} \sum_{\lambda \in S^i_j} c_\lambda \psi_\lambda$. Let

$$Dl(g^i)(x_0) = \sum_{|k| < \alpha} \frac{(x - x_0)^k}{k!} \partial^k g^i(x_0)$$

be the Taylor polynomial of g^i at x_0. Then

$$|g^i(x) - Dl(g^i)(x_0)| \leq \sum_{j \geq J} \sum_{\lambda \in S^i_j} |c_\lambda||\psi_\lambda(x) - Dl(\psi_\lambda)(x_0)|.$$

Since $x_0 \notin D_j$, $|\lambda - x_0| \geq C2^{-\epsilon_i j}$, and, since $\epsilon_i < 1$, x_0 is outside the support of ψ_λ; thus $Dl(\psi_\lambda)(x_0) = 0$. Similarly, $\psi_\lambda(x) = 0$ if $|x - x_0| \leq C2^{-\epsilon_i j}$. Thus, let J' be such that

$$C2^{-\epsilon_i J'} \leq |x - x_0| < C2^{-\epsilon_i(J'-1)};$$

then $J' > J$ and $g^i = \sum_{j \geq J'} \sum_{\lambda \in S^i_j} c_\lambda \psi_\lambda$ so that

$$|g^i(x) - Dl(g^i)(x_0)| \leq \sum_{j \geq J'} \sum_{\lambda \in S^i_j} |c_\lambda||\psi_\lambda(x)|$$

$$\leq \sum_{j \geq J'} 2^{-((n/2)+\alpha_i)j} 2^{(n/2)j} \leq C2^{-\alpha_i J'} \leq C|x - x_0|^{\alpha_i/\epsilon_i}. \quad \blacksquare$$

PROOF OF LEMMA 13.5 If $\lambda \in S^i_j$, $|c_\lambda|^p \geq 2^{-((np/2)+\alpha_{i+1}p)j}$. Thus, because of (13.6), S_j has at most $2^{(n+\alpha_{i+1}p-sp)j}$ terms included in a unit cube so that D^i_j can be covered by $2^{(n+\alpha_{i+1}p-sp)j}$ cubes $Q^{i,j}_k$ of size $2^{-\epsilon_i j}$. Hence,

$$\sum_k (\text{diam } Q^{i,j}_k)^\delta \leq 2^{((\alpha_{i+1}-s)p+n)j} 2^{-\delta \epsilon_i j}$$

and $\sum_{j \geq J} \sum_k \text{diam}(Q^{i,j}_k)^\delta \to 0$ when $J \to \infty$ as soon as $\delta\epsilon_i > (\alpha_{i+1} - s)p + n$. $\blacksquare$

Let us now come back to the proof of the theorem. Let $f \in C^\beta \cap W^{s,p}$; take $\epsilon_i = \alpha_i/\alpha$. We have shown that, if $x_0 \notin \bigcup_{i=0}^{N-1} \lim\sup D_j^i$, then f is C^α at x_0, and the Hausdorff dimension of $\bigcup_{i=0}^{N-1} \lim\sup D_j^i$ is bounded by the supremum of the dimensions of the $\lim\sup D_j^i$, hence by

$$d_N = \sup_{i=0,\ldots,N-1} \alpha\left(\frac{n-(s-\alpha_{i+1})p}{\alpha_i}\right) \leq \sup_{x\in[\beta,\alpha]} \alpha\left(\frac{n-(s-x)p}{x}\right) + O\left(\frac{1}{N}\right).$$

Since N can be chosen arbitrarily large, we get that the Hausdorff dimension of the points where f is not C^α is bounded by $\alpha((n - sp/\beta) + p)$ if $s \leq n/p$, and by $n - sp + p\alpha$ if $s \geq n/p$; hence Point 4. If $s - n/p > 0$, the Sobolev embeddings yield $W^{s,p} = W^{s,p} \cap C^{s-n/p}$ and Point 1 follows by taking $\alpha = s - n/p$.

The theorem still holds if $p = 1$. We use the fact that, if $f \in W^{s,1}$, using the Sobolev embeddings, $f \in W^{s',p'}$ if $1/p' > 1 - (s-s')/n$ so that $p' > 1$ and s' and p' can be chosen arbitrarily close respectively to s and 1. We then pass to the limit.

Before proving Point 2, we remark first that the "devil's staircase" (the primitive of the natural measure on the triadic Cantor set) yields an example where the maximal allowed dimension is reached. This function belongs to $W^{1,1} \cap C^{\log 2/\log 3}$. The theorem immediately implies that the set of points where it is not Hölder of exponent larger than $\log 2/\log 3$ has a Hausdorff dimension at most $\log 2/\log 3$; this is precisely the dimension of the triadic Cantor set, which is the support of these singularities.

Let A be a set of Hausdorff dimension $n - p(s - \alpha) = \delta$. For any $\delta' > \delta$, there exists a sequence $R_i = \{A_j^i\}$ of coverings of A by dyadic cubes such that $\text{diam}(A_j^i) \leq 2^{-i}$ and $\sum_j \text{diam}(A_j^i)^{\delta'} \leq 1$. Let $2^{-l_j^i}$ the size of the cube A_j^i, and $f_i = \sum_j 2^{-((n/2)+\alpha)l_j^i}\psi_{A_j^i}$. The cubes A_j^i being disjoint,

$$\|f_i\|_{W^{s,p}} \leq \int \left(\sum 2^{nl}4^{ls}|c_\lambda|^2\chi_\lambda(x)\right)^{p/2}$$

$$= \sum 2^{npl/2}2^{pls}2^{-((n/2)+\alpha)lp}2^{-nl}$$

$$= \sum_j 2^{-(n+p(\alpha-s))l_j^i}.$$

Since $\sum |A_j^i|^{\delta'} \leq 1$, we have $\sum 2^{-l_j^i\delta'} \leq 1$, so that $\|f_i\|_{W^{s,p}} \leq 1$ when $n+p(\alpha-s) > \delta'$. The function $\sum(1/i^2)f_i(x)$ belongs to $W^{s,p} \cap C^\alpha$ and one immediately checks that it is not Hölder of order larger than δ' at the points where $\lim\sup_i \bigcup_j A_j^i$, which is a set containing A, proving Point 2.

We still have to prove Point 5. Let A be a set of Hausdorff dimension $(\alpha/\beta)(n - (s - \beta)p) = \delta$. For any $\delta' > \delta$, we choose coverings R_i as above. The cube A_j^i being given, we choose a cube B_j^i included in A_j^i of size $2^{-m_j^i}$ where m_j^i is the largest integer smaller than $l_j^i\alpha/\beta$. Let then $f_i = \sum_j 2^{-((n/2)+\beta)m_j^i}\psi_{B_j^i}$. The cubes A_j^i

being disjoint, so are the B_j^is. One checks as above that

$$\|f_i\|_{W^{s,p}} \equiv \sum_j 2^{-(n+p(\beta-s))m_j^i},$$

so that $\|f_i\|_{W^{s,p}} \leq C$ when $\sum_j 2^{-(n+p(\beta-s))(\alpha/\beta)l_j^i} \leq C$, which is the case since $\sum 2^{-l_j^i\delta'} \leq 1$.

Let $f = \sum (1/i^2)f_i$; clearly $f \in W^{s,p} \cap C^\beta$. We still have to check that f is not smoother than C^α at x_0 belonging to $\limsup_i \bigcup_j D_j^i$. Recall that if f is C^α at x_0, $|c_{j,k}| \leq C2^{-((n/2)+\alpha)j}(1 + 2^{\alpha j}|x_0 - k2^{-j}|^\alpha)$, but this upper bound is precisely an equivalence on the cube A_j^i. ∎

We shall end by proving the following result, which can be found in [Zi] when s is an integer.

PROPOSITION 13.6
If $f \in W^{s,p}$ with $s \leq n/p, f$ is bounded except on a set of Hausdorff dimension at most $n - sp$.

PROOF Let $A_j = \{\lambda \mid |c_\lambda| \geq (1/j^2)2^{-nj/2}\}$ and $B_j = \{\lambda/\operatorname{dist}(\lambda, A_j) \leq L2^{-j}\}$. Since $\sum |c_\lambda|^p \leq 2^{-(sp+(np/2)-n)j}$, A_j has at most $j^{2p}2^{(n-sp)j}$ elements, and B_j has at most $(2L)^n j^{2p}2^{(n-sp)j}$ elements. We obtain, as above, that the Hausdorff dimension of $\limsup B_j$ is bounded by $n - sp$.

If $x_0 \notin \limsup B_j$, only the coefficients c_λ such that λ does not belong to A_j bring a contribution at x_0 (using compactly supported wavelets and L large enough). Thus, there exists J_0 such that $|f(x_0)| = |\sum_{j\geq J_0} \sum_k (1/j^2)2^{-nj/2}\psi_\lambda + \sum_{j\leq J_0} c_\lambda \psi_\lambda|$. The second series is a smooth function, and the first is obviously bounded. ∎

We remark that here too, we only need the assumption $f \in B_p^{s,\infty}$.

One cannot determine the Hausdorff dimension of the points where a function f is not C^α by a criterion based on the wavelet coefficients of f that does not take account of their positions. To check that point, we can consider the counterexamples of Points 2 and 5 of the theorem: take the same wavelet coefficients for a given j, but place them as close as possible to the origin. We thus construct a function arbitrarily smooth outside the origin and that has, for each j, the same wavelet coefficients as the counterexample. We see that it is not only the value and the frequencies that matter in the determination of these dimensions, but also the position of the large coefficients.

13.3.3 Construction of multifractal functions with a prescribed spectrum of singularities

Let $s > 0$ be a non-integer. For $s > 0$ we say that $f \in \Gamma^s(x_0)$ if

$$f \in \bigcap_{\epsilon>0} C^{s-\epsilon}(x_0) \quad \text{but } f \notin \bigcup_{\epsilon>0} C^{s+\epsilon}(x_0). \tag{13.7}$$

For each $s > 0$ let $E^{(s)}$ be the set of real numbers x_0 such that f belongs to $\Gamma^s(x_0)$ and let $\theta_f(s)$ be the Hausdorff dimension of $E^{(s)}$. We shall now study the function $\theta_f :]0, +\infty[\rightarrow [0, 1]$. It is important to remark that θ_f is defined point by point.

Multifractal measures have been widely studied (see, for instance, [BrMiPe] and [J4]), and recently, the study of multifractal functions has proved important in several domains of physics. The examples include plots of random walks, interfaces developing in reaction-limited growth processes, or turbulent velocity signals at inertial range (see [ABMu2]). The measured spectra θ have always had the shape of a half-ellipse (see [ABMu1]). It is thus a natural question to wonder if the graph of a singularity spectrum can have essentially any shape or not.

We define $\mathcal{C}$ as the class of functions that can be written as the supremum of a countable set of functions of the form $c1_{[a,b]}(x)$ (where we can have $a = b$). Thus, Riemann integrable functions belong to $\mathcal{C}$, but also, for instance, the indicatrix function of the rationals (but not the indicatrix function of the irrationals).

THEOREM 13.7

Let $\theta :]0, +\infty[\rightarrow [0, 1]$ be a function in $\mathcal{C}$. There exists a continuous function $f : \mathbb{R} \rightarrow \mathbb{R}$ such that $\theta = \theta_f$.

We first construct f when θ is

$$\theta(x) = cx1_{[a,b]}(x) \quad \text{where } 0 < a \leq b < \infty \text{ and } cb \leq 1.$$

We shall actually use three other parameters α, β, γ where $a = \gamma$, $b = \beta\gamma$, and $c = 1/(\alpha\beta\gamma)$. Thus $\gamma > 0$, $\beta \geq 1$, and $\alpha \geq 1$. We thus define $f = f^{(\alpha,\beta,\gamma)}$. The general case will be obtained using a simple "superposition" procedure of the $f^{(\alpha,\beta,\gamma)}$. If θ takes values in $[0, n]$ the construction easily adapts and gives a function f defined on $\mathbb{R}^n$. The parameters α, β, γ being fixed, we denote by f the corresponding function $f^{(\alpha,\beta,\gamma)}$.

Let Λ be the collection of all dyadic intervals $\lambda = [k2^{-j}, (k+1)2^{-j}], j \in \mathbb{Z}, k \in \mathbb{Z}$. We shall construct a subcollection $\Lambda(\alpha, \beta) \subset \Lambda$ and consider the "lacunary" series

$$f(x) = \sum_{\lambda \in \Lambda(\alpha,\beta)} 2^{-(\gamma+1/2)j} \psi_\lambda(x). \tag{13.8}$$

The construction of $\Lambda(\alpha, \beta)$ is performed as follows. Define $\Lambda(\alpha, \beta) = \bigcup_{m \geq 1} \Lambda_m^{(\alpha,\beta)}$ where $\Lambda_m^{(\alpha,\beta)}$ is the set of λs or of couples (j, k) such that $j = [\alpha\beta m]$ and

$$2^{-j}k = \epsilon_1 l_1 + \cdots + \epsilon_m l_m \in F_m$$

where $\epsilon_1 = \pm 1, \ldots, \epsilon_m = \pm 1, l_j = 2^{-[\alpha j]}$; $[x]$ is the integer part of x, and

$$k = 2^{[\alpha\beta m]}(\pm l_1 \pm \cdots \pm l_m)$$

is obviously an integer.

PROPOSITION 13.8
The function f defined by (13.8) belongs to the global Hölder space $C^\gamma(\mathbb{R})$ so that if $s < \gamma$, $E^{(s)}$ is empty. If $\gamma \leq s \leq \beta\gamma$, the Hausdorff dimension of the set of xs such that $f \in \Gamma^s(x)$ is $(s/\alpha\beta\gamma)$. If $s > \beta\gamma$, $E^{(s)}$ is empty.

Because of (13.1), we see that f belongs to $C^\gamma(\mathbb{R})$. Using Theorem 13.1, we see that x belongs to $E^{(s)}$ iff

$$\limsup_{\lambda \in \Lambda(\alpha,\beta)} 2^{-\gamma j}(2^{-j} + \text{dist}(x, \lambda))^{-s-\epsilon} = +\infty, \quad \text{for any } \epsilon > 0 \qquad (13.9)$$

and

$$\limsup_{\lambda \in \Lambda(\alpha,\beta)} 2^{-\gamma j}(2^{-j} + \text{dist}(x, \lambda))^{-s+\epsilon} < \infty, \quad \text{for any } \epsilon > 0. \qquad (13.10)$$

As concerns f defined by (13.8), condition (13.9) becomes $2^{-j} + \text{dist}(x, \lambda) = \eta(\lambda)2^{-(\gamma/s+\epsilon)j}$ where $\liminf \eta(\lambda) = 0$. We suppose $s \geq \gamma$ thus $2^{-j} = o(2^{-(\gamma/s+\epsilon)j})$. The condition $\text{dist}(x, \lambda) = \eta(\lambda)2^{-(\gamma/s+\epsilon)j}$ remains. If $\lambda \in \Lambda_m^{(\lambda,\beta)}$, $j = [\alpha\beta m]$, it becomes

$$\text{dist}(x, F_m) = \eta_m 2^{-(\alpha/\beta\gamma s)m}, \quad (F_m = \{\pm l_1 \pm l_2 \pm \cdots \pm l_m\}) \qquad (13.11)$$

with

$$\liminf \eta_m 2^{-m\epsilon} = 0 \quad \text{for any } \epsilon > 0 \qquad (13.12)$$

and (13.10) becomes

$$\liminf \eta_m 2^{m\epsilon} = +\infty \quad \text{for any } \epsilon > 0. \qquad (13.13)$$

We shall now define the compact set K_α and the sets $E_{\alpha,\delta}$. Let K_α be the compact set of the sums $\sum_1^\infty \epsilon_j l_j$ where $\epsilon_j = \pm 1$. Another equivalent definition is

$$K = \bigcap_1^\infty (F_m + [-\lambda_m, \lambda_m])$$

where $\lambda_m = l_{m+1} + l_{m+2} + \cdots$. The sets $G_m = F_m + [-\lambda_m, \lambda_m]$ are a decreasing sequence.

If $\beta \geq 1$, let $G_m^{(\beta)} \subset G_m$ be defined as $G_m^{(\beta)} = F_m + [-\lambda_m^\beta, \lambda_m^\beta]$ and let $E_{\alpha,\beta}$ belong to an infinity of $G_m^{(\beta)}$. Obviously $E_{\alpha,\beta} \subset K_\alpha$ where $E_{\alpha,\beta} = K_\alpha$ if $\beta = 1$.

LEMMA 13.9
If $\gamma \leq s < \beta\gamma$, then (13.9) is equivalent to $x \in \bigcap_{\delta < (\beta\gamma/s)} E_{\alpha,\delta}$, while if $s \geq \beta\gamma$, (13.9) is equivalent to $x \in K_\alpha$.

Clearly (13.9) is equivalent to (13.11). If (13.11) holds and if $\delta < (\beta\gamma/s)$, let us check that $x \in E_{\alpha,\delta}$. For that, one chooses $\epsilon > 0$ such that $\delta < (\beta\gamma/s) - \epsilon$. Then

$$\text{dist}(x, F_m) = \eta_m 2^{-(\alpha\beta\gamma/s)m} \leq 2^{-((\alpha\beta\gamma/s)-\epsilon)m} \leq l_m^\delta$$

for an infinity of *ms*. Thus $x \in E_{\alpha,\delta}$. If conversely $x \in E_{\alpha,\delta}$ we shall have dist$(x, F_m) \leq l_m^\delta$ so that dist$(x, F_m) \leq 2^{-\alpha\delta m}$ for an infinity of *m*. If $\delta > (\beta\gamma/s) - \epsilon$, we get (13.11). When $s \geq \beta\gamma$, we observe that if $\eta_m > 0$ is an arbitrary sequence such that lim inf $\eta_m = 0$ and if

$$x \in \bigcap_{m \geq 1} F_m + [-\eta_m, \eta_m],$$

then $x \in K_\alpha$. This is because K_α is a compact set, and if $x \notin K_\alpha$, then dist$(x, K_\alpha) = \eta > 0$ so that dist$(x, F_m) \geq \eta$; giving a contradiction. Condition (13.11) is thus equivalent to $x \in K_\alpha$ as soon as $s \geq \beta\gamma$. One proves similarly that

LEMMA 13.10

If $\gamma \leq s \leq \beta\gamma$, Condition (13.10) is equivalent to $x \notin \bigcup_{\delta > (\beta\gamma/s)} E_{\alpha,\delta}$; while if $s > \beta\gamma$, (13.10) is equivalent to $x \notin K_\alpha$.

LEMMA 13.11

The Hausdorff dimension of $E_{\alpha,\beta}$ is $(1/\alpha\beta)$.

LEMMA 13.12

If $\gamma \leq s \leq \beta\gamma$, the Hausdorff dimension of $E^{(s)}$ is $(s/\alpha\beta\gamma)$. If $s > \beta\gamma$, the set $E^{(s)}$ is empty.

In order to prove Lemmas 13.11 and 13.12, we shall construct a probability measure $\mu = \mu_{(\alpha,\beta)}$ supported in a compact set $K_{\alpha,\beta} \in E_{\alpha,\beta}$ such that, for any interval $I \subset \mathbb{R}$, of length $|I| \leq 1/2$, we have

$$\mu(I) \leq C|I|^{1/\alpha\beta} \log\left(\frac{1}{|I|}\right). \tag{13.14}$$

The construction of μ is elementary and left in the next part. Once $\mu_{(\alpha,\beta)}$ is constructed, the proof of Lemma 13.11 is as follows;

$$E_{\alpha,\beta} = \bigcap_{m \geq 1} E_{\alpha,\beta}^{(m)} \quad \text{where } E_{\alpha,\beta}^{(m)} = G_m^{(\beta)} \cup G_{m+1}^{(\beta)} \cup \cdots \text{ and } G_m^{(\beta)} = F_m + [-\lambda_m^\beta, \lambda_m^\beta].$$

For any $\epsilon > 0$, we can cover $E_{\alpha,\beta}$ by the intervals I_q appearing in $G_n^{(\beta)}$, $n \geq m$. One immediately checks that, if $d > (1/\alpha\beta)$, $\sum_0^\infty |I_q|^d \leq C$ where C does not depend on ϵ.

Conversely, if $E_{\alpha,\beta}$ is covered by intervals I_q such that $|I_q| \leq \epsilon$ and $\sum_q |I_q|^d \leq C$ (C doesn't depend on ϵ), then, on one side,

$$\sum_q \mu(I_q) \geq \mu(E_{\alpha,\beta}) = 1$$

and, on the other side,

$$\mu(I_q) \leq C|I_q|^{1/\alpha\beta} \log \frac{1}{|I_q|}.$$

If $0 < d < (1/\alpha\beta)$, we have

$$|I_q|^{1/\alpha\beta} \log \frac{1}{|I_q|} \leq |I_q|^d \epsilon^{(\alpha/\beta-d)} \log \frac{1}{\epsilon}.$$

Thus $1 \leq C\epsilon^{(\alpha/\beta-d)} \log(1/\epsilon)$ which is absurd and concludes the proof of Lemma 13.11.

Let us now prove Lemma 13.12. Suppose first that $\gamma \leq s \leq \beta\gamma$. Then

$$E^{(s)} = \left(\bigcap_{\delta < (\beta\gamma/s)} E_{\alpha,\delta} \right) \setminus \left(\bigcup_{\delta > (\beta\gamma/s)} E_{\alpha,\delta} \right) \quad \text{if } \gamma \leq s < \beta\gamma$$

while if $s = \beta\gamma$, then

$$E^{(s)} = K_\alpha \setminus \left(\bigcup_{\delta > 1} E_{\alpha,\delta} \right).$$

The checking is the same in both cases so that we suppose $\gamma \leq s < \beta\gamma$. Thus $E^{(s)} \subset E_{\alpha,\delta}$ for all $\delta < (\beta\gamma/s)$ so that $\dim(E^{(s)}) \leq (s/\alpha\beta\gamma)$. Let us show by contradiction that $\dim(E^{(s)}) < (s/\alpha\beta\gamma)$ cannot hold. We essentially follow the proof of Lemma 13.9. Let μ be the probability measure $\mu_{(\alpha,\beta\gamma/s)}$. One checks that

$$E^{(s)} \supset E_{\alpha,\beta\gamma/s} \setminus \left(\bigcup_{\delta > \beta\gamma/s} E_{\alpha,\delta} \right) \tag{13.15}$$

and

$$\mu(E_{\alpha,\delta}) = 0 \quad \text{for any } \delta > \beta\gamma/s. \tag{13.16}$$

Property (13.16) is established using the covering of $E_{\alpha,\delta}$ by the intervals appearing in the $G_m^{(\delta)}$. If $(I_q)_{q \geq 0}$ is a covering of $E^{(s)}$ such that $|I_q| \leq \epsilon$ and $\sum_0^\infty |I_q|^d \leq C$ for $d < (s/\alpha\beta\gamma)$, then these intervals form a covering of

$$A = E_{\alpha,\beta\gamma/s} \setminus (\bigcup_{\delta > \beta\gamma/s} E_{\alpha,\delta})$$

and we can use again the proof of Lemma 13.9 since $\mu(A) = 1$. If $s > \beta\gamma$, $E^{(s)}$ is empty since $E^{(s)} = K_\alpha \setminus K_\alpha = \emptyset$.

We shall now construct the measure μ associated to $E_{\alpha,\beta}$. Let $m_1 < m_2 < \cdots$ be an increasing sequence of integers that tends to ∞ quickly enough so that, for any $n \geq 1, m_{n+1} \geq 2^{(\beta-1/\alpha)(m_1+\cdots+m_n)}$, and let

$$K_{(\alpha,\beta)} = \bigcap_{m \geq 1} G_{m_n}^{(\beta)} \tag{13.17}$$

with, now,

$$G_m^{(\beta)} = F_m + [-2^{-[\alpha\beta m]}, 2^{-[\alpha\beta m]}]. \tag{13.18}$$

It means that $G_m^{(\beta)}$ is a finite union of dyadic intervals and the dyadic intervals that form $G_{m+1}^{(\beta)}$ will either be disjoint from the precedents or included in the precedents.

We have $[\alpha\beta m] \geq \beta[\alpha m]$ and the set $G_m^{(\beta)}$ is included in the one used in order to define (and cover) $E_{\alpha,\beta}$.

Let N_n be the number of intervals of length $2.2^{-[\alpha\beta m_n]}$ which can be found in $H_n = G_{m_1}^{(\beta)} \cap \cdots \cap G_{m_n}^{(\beta)}$ and let μ_n be the probability measure $2.2^{-[\alpha\beta m_n]}/N_n dx$ on each of these N_n intervals. One easily checks that $\mu_n \rightharpoonup \mu$ when $n \to \infty$, where $\mu = \mu_{(\alpha,\beta)}$ is carried by $K_{(\alpha,\beta)}$. Obviously $K_{(\alpha,\beta)} \subset E_{\alpha,\beta}$ and we shall check the following

LEMMA 13.13
There exists C such that for every I of length $|I| \leq 1/2$,

$$\mu(I) \leq C|I|^{1/\alpha\beta} \log \frac{1}{|I|} . \tag{13.19}$$

In order to prove Lemma 13.13, one starts by observing that there exists $C > 1$ such that

$$2^{m_n/\alpha} 2^{(1/\alpha - \beta)(m_{n-1} + \cdots + m_1)} \leq N_n \leq 2^{Cn} 2^{m_n/\alpha} 2^{(1/\alpha - \beta)(m_{n-1} + \cdots + m_1)}.$$

Consider then the two cases $2^{-\beta m_n} \leq |I| < 2^{-m_n/\alpha}$ and $2^{-m_n/\alpha} \leq |I| < 2^{-\beta m_{n-1}}$. In the first one, I intersects at most two of the intervals that compose H_n so that

$$\mu(I) \leq CN_n^{-1} \leq C'2^{-m_n/\alpha}m_n,$$

since $m_n \geq 2^{(\beta - 1/\alpha)(m_{n-1} + \cdots + m_1)}$. Thus

$$\mu(I) \leq C|I|^{1/\alpha\beta} \log(1/|I|).$$

In the second case, checking how many intervals of H_n meet I, $\mu(I) \leq (2^{m_n/\alpha}/N_n)|I|$. Thus, since $m_{n-1} \geq 2^{(\beta - 1/\alpha)(m_{n-2} + \cdots + m_1)}$,

$$\frac{2^{m_n/\alpha}}{N_n}|I| \leq |I|2^{(\beta - 1/\alpha)(m_1 + \cdots + m_{n-1})}$$

$$\leq |I|^{1/\alpha\beta} \log(1/|I|) \frac{|I|^{(1 - 1/\alpha\beta)}}{\log(1/|I|)} 2^{(\beta - 1/\alpha)(m_1 + \cdots + m_{n-1})}$$

$$\leq |I|^{1/\alpha\beta} \log(1/|I|) \frac{2^{-(\beta - 1/\alpha)m_{n-1}}}{\beta m_{n-1}} 2^{(\beta - 1/\alpha)(m_1 + \cdots + m_{n-1})}$$

$$\leq C|I|^{1/\alpha\beta} \log(1/|I|), \quad (\text{using } m_{n-1} \geq 2^{(\beta - 1/\alpha)(m_{n-2} + \cdots + m_1)}).$$

This concludes the proof of Lemma 13.13.

Let us now prove the general case in Theorem 13.7. Let $E_1, E_2, \ldots$ be disjoint subsets of $\mathbb{R}$ and suppose that $E_k \subset [a_k, b_k]$ where the $[a_k, b_k]$ are disjoint. Let d_k be the Hausdorff dimension of E_k. Then the Hausdorff dimension of $\bigcup_{k \geq 1} E_k$ is $\sup(d_k)$.

We return to the function $f_{(\alpha,\beta,\gamma)}$. After perhaps multiplying this function by $\phi \in C_0^\infty(\mathbb{R})$, equal to 1 near K_α, we can suppose that $f_{(\alpha,\beta,\gamma)}$ is compactly supported.

After replacing $f(x)$ by $f(px + q)$, we can, without changing θ_f, suppose that the support of f is contained in any given interval $[a, b]$. Let $f_k(x) = f_{(\alpha_k, \beta_k, \gamma_k)}(x)$ be a sequence of functions as in Proposition 13.8 and consider the corresponding functions $\theta_k(s)$. We can suppose that the supports of the $f_k(x)$ are included in $[2^{-k-1}, 2^{-k}]$. We can also replace f_k by $\epsilon_k f_k$ where $\epsilon_k > 0$ tends to 0. Let then $f = \sum_0^\infty \epsilon_k f_k$ and $\theta(s)$ be the supremum of the $\theta_k(s)$. It follows that $\theta = \theta_f$, concluding the proof of Theorem 13.7.

We shall use wavelets to study problems related to PDEs. But we first need some prerequisites concerning the analysis of certain operators and functional spaces on a wavelet basis.

13.4 Analysis of linear operators and functional spaces

13.4.1 An example: Analysis of the Laplacian in $\mathbb{R}^n$

Let $(H, \langle . \mid . \rangle)$ be a Hilbert space and (ϵ_λ) be an orthonormal basis of H. We then consider a linear operator A on H. For x in H, $A(x)$ is computed by

$$A(x) = \sum_\lambda \langle x \mid \epsilon_\lambda \rangle A(\epsilon_\lambda)$$

$$= \sum_\lambda \sum_{\lambda'} \langle A\epsilon_\lambda \mid \epsilon_{\lambda'} \rangle \langle x \mid \epsilon_\lambda \rangle \epsilon_{\lambda'},$$

so that A can be encoded by the '*infinite matrix*' of coefficients

$$(\langle A(\epsilon_\lambda) \mid \epsilon_{\lambda'} \rangle).$$

Depending on the chosen basis, the study of these coefficients may easily reveal certain properties of A. We shall first study the Laplacian on a wavelet basis. As a prerequisite, we first have to define some classes of matrices which we shall use often in the following. Their usefulness will be a consequence of the three following properties:

- (*a*) They will be algebras.

- (*b*) They will be large enough so that a large class of operators we shall consider (for instance differential operators or their inverses) will be products of a matrix in this space times diagonal matrices.

- (*c*) Their properties (such as boundedness on certain functional spaces or estimates on the Green functions) will be straightforward to check using the size of the moduli of the coefficients.

We say that an "infinite matrix" $T(\lambda, \lambda')$ belongs to $\mathcal{M}^\gamma$ if

$$|T(\lambda, \lambda')| \le C\omega_\gamma(\lambda, \lambda')$$

where

$$\omega_\gamma(\lambda, \lambda') = \frac{2^{-|j-j'|(\frac{n}{2}+\gamma)}}{1+(j-j')^2} \left(\frac{2^{-j}+2^{-j'}}{2^{-j}+2^{-j'}+|\lambda-\lambda'|} \right)^{n+\gamma}. \qquad (13.20)$$

An operator U belongs to $\text{Op}(\mathcal{M}^\gamma)$ if its matrix relative to a wavelet basis belongs to $\mathcal{M}^\gamma$, that is if

$$|\langle U(\psi_\lambda) \mid \psi_{\lambda'} \rangle| \le C\omega_\gamma(\lambda, \lambda').$$

The space $\mathcal{M}^\gamma$ is an algebra—that is, if $T \in \mathcal{M}^\gamma$, then $T' \in \mathcal{M}^\gamma \Longrightarrow TT' \in \mathcal{M}^\gamma$. Checking this property is a straightforward computation left to the reader (see [Me2]). Thus, we define

$$\|T\|_\gamma = \frac{1}{\alpha} \sup_{\lambda, \lambda'} \frac{|T(\lambda, \lambda')|}{\omega_\gamma(\lambda, \lambda')}$$

where α is a constant defined by the property that, for all $T, T' \in \mathcal{M}^\gamma$,

$$\|TT'\|_\gamma \le \|T\|_\gamma \|T'\|_\gamma. \qquad (13.21)$$

An operator in $\mathcal{M}^\gamma$ is continuous on l^2, which is a consequence of Schur's Lemma (see [Me1]).

As a first example, we shall now analyze in the Littlewood–Paley wavelet basis the fractional powers of the Laplacian. Recall that, if $s \in \mathbb{R}$, Δ^s is the operator defined in the Fourier transform as follows:

$$\widehat{\Delta^s(f)}(\xi) = |\xi|^{2s}\hat{f}(\xi).$$

Let G be the matrix of this operator in the wavelet basis.

PROPOSITION 13.14
The following decomposition holds: $G = DAD$, where D is the operator diagonal in a wavelet basis such that $D(j,k,j,k) = 2^{sj}$, and A and A^{-1} belong to $\mathcal{M}^\gamma$ for any $\gamma > 0$.

More generally, this result is proved in [J5] for general elliptic operators. It can be interpreted as follows: A diagonal preconditioning makes the condition number for the resolution of $Af = g$ (A elliptic) bounded independently of the discretization.

PROOF If $|j - j'| > 1$, $G(j,k,j',k') = 0$, because the Fourier transforms $\widehat{\psi_{j,k}}$ and $\widehat{\psi_{j',k'}}$ have disjoint supports.
  If $j = j'$,

$$G(j,k,j,k') = 2^{-nj} \int e^{i(k-k')\xi/2^j} |\xi|^{2s} |\hat{\psi}|^2(\xi/2^j)d\xi$$

$$= 2^{2sj} \int e^{i(k-k')\xi} |\xi|^{2s} |\hat{\psi}|^2(\xi)d\xi.$$

Since $|\hat{\psi}|^2$ is in the Schwartz class, its Fourier transform has fast decay. Thus, for any γ,

$$|G(j,k;j',k')| \le \frac{C_\gamma}{(1+|k-k'|)^\gamma}.$$

The proof when $|j-j'|=1$ is similar. Thus $A \in \mathcal{M}^\gamma$. The estimate for A^{-1} is obtained by remarking that it is the estimate on A when changing s into $-s$. $\blacksquare$

As a corollary, let us prove the characterization of Sobolev spaces (see [Me1]).

PROPOSITION 13.15
Let f be in $H^s(\mathbb{R}^n)$. Then the following equivalence holds

$$\|f\|_{H^s}^2 \equiv \sum |c_{j,k}|^2 (1+2^{2j})^s.$$

PROOF   Let $s \ge 0$. Then $\|f\|_{H^s}^2 = \|f\|_{L^2}^2 + \|\Delta^{s/2}f\|_{H^s}^2$.
Let C be the vector of the wavelet coefficients of f. We compute

$$\|\Delta^{s/2}f\|_{H^s}^2 \equiv \langle A(DC) \mid DC \rangle \equiv \|(DC)\|^2 = \sum |c_{j,k}|^2 2^{2js}.$$

This proves the proposition when $s \ge 0$. The case $s < 0$ follows by duality. $\blacksquare$

Let us now introduce the notion of "vaguelettes" (this definition is slightly different from the one that can be found in [Me1]). Functions $(f_{j,k})$ are vaguelettes (of order γ) if their matrix in a wavelet basis (of smoothness larger than γ) belongs to $\mathcal{M}^\gamma$. In this case, for all $l = (l_1, \dots, l_n)$ such that $|l| \le \gamma$,

$$\int (x - k2^{-j})^l f_{j,k}(x)dx = 0$$

and

$$|\partial^l f_{j,k}(x)| \le \frac{c(p)2^{(n/2+|l|)j}}{(1+|2^j x - k|)^\gamma}.$$

We will study the wavelet decomposition of some nonlinear quantities, like the product of two functions, and we draw consequences for the resolution of some nonlinear PDEs. These results are related to the compensated compactness method. The key point of this method will rely on the properties of the real Hardy space $\mathcal{H}^1$, and we shall first define and study this space, following [Me1].

13.4.2   The generalized Hardy space $\mathcal{H}^1$

A sequence (e_n) is a Schauder basis of a Banach space B if any $x \in B$ can be uniquely written

$$x = \alpha_0 e_0 + \cdots + \alpha_k e_k + \cdots$$

where the series converges in B. The basis is unconditional if furthermore, for any sequence ϵ_k such that $|\epsilon_k| \leq 1$, and any sequence β_k,

$$\left\| \sum_0^n \epsilon_k \beta_k e_k \right\| \leq C \left\| \sum_0^n \beta_k e_k \right\|,$$

with C depending only on the space B. For instance, an orthonormal basis of a Hilbert space is unconditional.

Let $\mathcal{H}^1$ be the largest subset in L^1 composed of functions which have an unconditionally convergent wavelet series. If $f \in \mathcal{H}^1$, we have, for any sequence $\epsilon(\lambda)$ taking values ± 1,

$$\left\| \sum_\lambda \epsilon(\lambda) \langle f \mid \psi_\lambda \rangle \psi_\lambda \right\|_1 \leq C. \tag{13.22}$$

We shall denote by $\|f\|_{\mathcal{H}^1}^{(1)}$ the supremum of the left hand-side term for all sequences $\epsilon(\lambda)$. In order to bound (13.22), we shall use the following proposition, called the Khintchine inequality (see [Zy]). Let $A(\epsilon_1, \ldots \epsilon_n) = \sum c_i \epsilon_i$. Let

$$\|A\|_p = \left[\frac{1}{2^n} \sum_{\epsilon_1, \ldots, \epsilon_n} A(\epsilon_1, \ldots, \epsilon_n) \right]^{1/p}$$

where the sum holds on all n-uples taking values ± 1. Then

PROPOSITION 13.16
Let $p > 0$. There exists A_p and B_p such that, for any sequence (c_i),

$$A_p \left(\sum_{i=1}^n c_i^2 \right)^{1/2} \leq \left\| \sum_{i=1}^n c_i \epsilon_i \right\|_p \leq B_p \left(\sum_{i=1}^n c_i^2 \right)^{1/2}. \tag{13.23}$$

PROOF One first proves the right-hand inequality when $p = 2k$, $k \in \mathbb{N}$. Let

$$S_n = \sum_{i=1}^n c_i \epsilon_i.$$

Let E denote the average value on all the n-uples of ± 1 (the mathematical expectation).

$$E(S_n^{2k}) = \sum_{\sum \alpha_i = 2k} \frac{(\alpha_1 + \cdots + \alpha_j)!}{\alpha_1! \ldots \alpha_j!} E(\epsilon_{i_1}^{\alpha_1} \ldots \epsilon_{i_j}^{\alpha_j}) C_{i_1}^{\alpha_1} \ldots C_{i_j}^{\alpha_j};$$

$E(\epsilon_{i_1}^{\alpha_1} \ldots \epsilon_{i_j}^{\alpha_j}) = 1$ if all the α_i are even and $E(\epsilon_{i_1}^{\alpha_1} \ldots \epsilon_{i_j}^{\alpha_j}) = 0$ otherwise. Let

$$A_{\alpha_1 \ldots \alpha_n} = \frac{(\alpha_1 + \cdots + \alpha_j)!}{\alpha_1! \ldots \alpha_j!}.$$

Then

$$E(S_n^{2k}) = \sum_{\sum \beta_i = k} A_{2\beta_1 \ldots 2\beta_j} C_{i_1}^{2\beta_1} \ldots C_{i_j}^{2\beta_j}$$

$$= \sum \frac{A_{2\beta_1 \ldots 2\beta_j}}{A_{\beta_1 \ldots \beta_j}} A_{\beta_1 \ldots \beta_j} C_{i_1}^{2\beta_1} \ldots C_{i_j}^{2\beta_j}$$

$$\leq \sup \left(\frac{A_{2\beta_1 \ldots 2\beta_j}}{A_{\beta_1 \ldots \beta_j}} \right) s_n^{2k}$$

where $s_n^2 = \sum_{i=1}^{n} c_i^2$. We have

$$\frac{A_{2\beta_1 \ldots 2\beta_j}}{A_{\beta_1 \ldots \beta_j}} = \frac{(2k)!}{(2\beta_1)! \ldots (2\beta_j)!} \frac{\beta_1! \ldots \beta_j!}{k!} \leq \frac{2k(2k-1) \ldots (k+1)}{\prod_i 2\beta_i(2\beta_i - 1) \ldots (\beta_i + 1)}$$

$$\leq \frac{2k(2k-1) \ldots (k+1)}{2^{\beta_1 + \ldots \beta_j}} \leq \frac{2k(2k-1) \ldots (k+1)}{2^k} \leq k^k$$

so that $E(S_n^{2k}) \leq k^k s_n^{2k}$, proving the inequality with $B_p = \sqrt{p/2}$. The inequality for all ps holds because $\|S_n\|_p$ is an increasing function of p. The left inequality holds for $0 < p < 2$ because

$$\|S_n\|_p \geq \|S_n\|_2 = s_n.$$

Using the logarithmic convexity of the averages $\|S_n\|_p$, and choosing $r_1, r_2 > 0$ such that $r_1 + r_2 = 1$ and $pr_1 + 4r_2 = 2$, we get

$$s_n^2 = \|S_n\|_2^2 \leq \|S_n\|_p^{pr_1} \|S_n\|_4^{4r_2} \leq \|S_n\|_p^{pr_1} (2^{1/2} s_n)^{4r_2},$$

so that

$$\|S_n\|_p^{pr_1} \geq 4^{-r_2} s_n^{2-4r_2} = 4^{-r_2} s_n^{pr_1}.$$

Hence

$$\|S_n\|_p \geq 4^{-r_2/pr_1} s_n$$

and Proposition 13.16

The left side of (13.23) with $p = 1$ yields

$$\int \left(\sum |\langle f \mid \psi_\lambda \rangle|^2 |\psi_\lambda(x)|^2 \right)^{1/2} dx \leq C.$$

Let

$$\|f\|^{(2)} = \int \left(\sum |\langle f \mid \psi_\lambda \rangle|^2 |\psi_\lambda(x)|^2 \right)^{1/2} dx;$$

thus

$$\|f\|^{(2)} \leq \|f\|^{(1)}. \quad \blacksquare$$

We shall now define three other norms. There exists a dyadic cube $R(\lambda)$ in λ such that $|\psi_\lambda(x)| \geq C > 0$ if $x \in R(\lambda)$. Thus

$$\int \left(\sum |c_\lambda|^2 |\psi_\lambda(x)|^2 \right)^{1/2} dx \geq C \int \left(\sum |c_\lambda|^2 |2^{nj} \chi_{R(\lambda)}(x)|^2 \right)^{1/2} dx$$

and the right hand side defines $\|f\|^{(3)}$. Let

$$\|f\|^{(4)} = \int \left(\sum |c_\lambda|^2 |2^{nj} \chi_\lambda(x)|^2 \right)^{1/2} dx.$$

We have $\|f\|^{(3)} \leq \|f\|^{(2)}$ and $\|f\|^{(3)} \leq \|f\|^{(4)}$.

The last norm we shall use is defined on the atomic decompositions. An atom is a function $a(x) \in L^2(\mathbb{R}^n)$ for which there exists a ball B such that

$$\begin{cases} \text{Supp}\, a \subset B \\ \|a\|_2 \leq |B|^{-1/2} \\ \int_B a(x)dx = 0. \end{cases}$$

The Cauchy–Schwartz inequality implies that $\int |a(x)|dx \leq 1$. A function f belongs to $\mathcal{H}^1_{at}$ if $f = \sum \lambda_j a_j(x)$ with $\sum |\lambda_j| < \infty$ and the a_j are atoms. We define $\|f\|^{(5)}$ as the infimum of $\sum |\lambda_j|$ taken on all possible atomic decompositions of f.

THEOREM 13.17

The five norms defined above are equivalent.

PROOF $\|f\|_1 < \infty$ implies $\|f\|_2 < \infty$ which implies $\|f\|_3 < \infty$; $\|f\|_4 < \infty$ implies $\|f\|_3 < \infty$. Let us prove that $\|f\|_3 < \infty$ implies $\|f\|_5 < \infty$.

The difficulty is that the wavelet decomposition does not yield the required atomic decomposition. We have to group the wavelets. First, we shall derive a condition on the wavelet coefficients of a function which ensures that it is an atom.

If the wavelets are compactly supported, a sequence of coefficients is called an atom if there exists a dyadic cube Q such that $\cup \{\lambda/c_\lambda \neq 0\} \subset Q$ and $\sum |c_\lambda|^2 \leq (1/|Q|)$. One easily checks that then $\sum c_\lambda \psi_\lambda$ is an atom.

Let us now prove that $\|f\|_3 < \infty$ implies $\|f\|_5 < \infty$. Thus we suppose that there exists $\gamma \in (0,1]$ such that, for any dyadic cube λ, there exists $R(\lambda) \subset \lambda$ such that $|R| \geq \gamma |Q|$ and

$$\sigma(x) = \left(\sum |c_\lambda|^2 |\lambda|^{-1} \chi_{R(\lambda)}(x) \right)^{1/2} \in L^1.$$

Let $E \subset \mathbb{R}^n$ such that $|E|$ is finite. We shall first define a larger set E^* containing E as follows. Let $\beta > 0$ be such that $\beta < \gamma$. Let ϕ be the set of all dyadic cubes λ such that $|\lambda \cap E| \geq \beta |\lambda|$ and let E^* be the union of these cubes. Since $|E|$ is finite, the size of the cubes belonging to ϕ is bounded, thus any cube in ϕ is included in a maximal cube $\tilde{Q}$. These maximal cubes form a partition of E^*, so that

$$|E^*| \sum |\tilde{Q}| \leq \frac{1}{\beta} \sum |\tilde{Q} \cap E| \leq \frac{|E|}{\beta}.$$

We have $E \subset E^*$ (up to a set of measure 0), since by Lebesgue's differentiation theorem,

$$\lim_{x \in Q, |Q| \to 0} \frac{|Q \cap E|}{|Q|} = 1 \quad \text{a.e.}$$

Let $E_k = \{x \mid \sigma(x) > 2^k\}$. Then

$$
\begin{aligned}
\int \sigma(x)dx &= \sum_{-\infty}^{+\infty} \int_{x \in E_k \backslash E_{k+1}} \sigma(x)dx \\
&\geq \sum 2^k(|E_k| - |E_{k+1}|) \\
&= \sum 2^k|E_k| - \frac{1}{2}\sum 2^k|E_k|,
\end{aligned}
$$

so that

$$\sum 2^k|E_k| \leq 2 \int \sigma(x)dx.$$

Let ϕ_k be the set of dyadic cubes λ such that $|\lambda \cap E_k| \geq \beta|\lambda|$. We define E_k^* as above, so that $|E_k^*| \leq (1/\beta)|E_k|$. Hence

$$\sum 2^k|E_k *| \leq \frac{2}{\beta} \int \sigma(x)dx. \tag{13.24}$$

Let $Q(k,l)$ be the maximal dyadic cubes of ϕ_k, and let $\Delta_k = \phi_k \backslash \phi_{k+1}$. We shall write $\lambda \in \Delta(k,l)$ if $\lambda \in \Delta_k$ and $\lambda \subset Q(k,l)$. Let us first prove

$$\sum_{\lambda \in \Delta(k,l)} |c_\lambda| \leq \frac{\gamma^2}{\gamma - \beta} \int_{Q(k,l) \backslash E_{k+1}} \sigma^2(x)dx \leq \frac{\gamma^2}{\gamma - \beta} 4^{k+1}|q(k,l)|. \tag{13.25}$$

If $\lambda \in \Delta(k,l)$, then $\lambda \subset Q(k,l)$ and $\lambda \notin \phi_{k+1}$ so that $|\lambda \cap E_{k+1}| < \beta|\lambda|$. Since $R(\lambda) \subset \lambda$ and $|R(\lambda)| \geq \gamma|\lambda|$, we have

$$|R(\lambda) \cap E_{k+1}| \leq \beta|\lambda| \leq \frac{\beta}{\gamma}|R(\lambda)|$$

so that $|R(\lambda)/E_{k+1}| \geq (1 - (\beta/\gamma))|R(\lambda)| \geq (\gamma - \beta/\gamma^2)|\lambda|$.

From the definition of σ, we get

$$\sigma^2(x) \geq \sum_{\lambda \in \Delta(k,l)} |c_\lambda|^2 |\lambda|^{-1} \chi_{R(\lambda)}(x),$$

so that

$$\int_{Q(k,l)\setminus E_{k+1}} \sigma^2(x) \geq \sum_{\lambda \in \Delta(k,l)} |c_\lambda|^2 |\lambda|^{-1} |R(\lambda)\setminus E_{k+1}| \geq \frac{\gamma - \beta}{\gamma^2} \sum_{\lambda \in \Delta(k,l)} |c_\lambda|^2.$$

This proves the first inequality of (13.25).

The second inequality is just a consequence of $\sigma(x) \leq 2^{k+1}$ if $x \notin E_{k+1}$, hence (13.25). Let $\lambda(k,l) = |Q(k,l)|^{1/2} (\sum_{\lambda \in \Delta(k,l)} |c_\lambda|^2)^{1/2}$. We define the atoms $\alpha_{k,l}(\lambda)$ by

$$\alpha_{k,l}(\lambda) = \frac{c_\lambda}{\lambda(k,l)} \quad \text{if } \lambda \in \Delta(k,l)$$

$$= 0 \quad \text{otherwise.}$$

We still have to check that $\sum \sum \lambda(k,l) \leq C \|\sigma\|_{L^1}$. We have

$$\sum \sum \lambda(k,l) \leq \frac{2\gamma}{\sqrt{\gamma - \beta}} \sum \sum 2^k |Q(k,l)|$$

$$= \frac{2\gamma}{\sqrt{\gamma - \beta}} \sum 2^k |E_k^*| \quad \text{(because the } Q(k,l) \text{ are a partition of } E_k^*\text{)}$$

$$\leq \frac{4\gamma}{\beta\sqrt{\gamma - \beta}} \|\sigma\|_{L^1} \quad \text{(using (13.24)).}$$

We have thus realized, on the wavelet coefficients, the atomic decomposition of f.

More precisely, we have shown that

$$\left\| \sum c_\lambda \psi_\lambda \right\|_{L^1} \leq C \left\| \left(\sum |c_\lambda|^2 |\lambda|^{-1} \chi_{R(\lambda)} \right)^{1/2} \right\|_{L^1}.$$

Since the right hand side is only a function of $|c_\lambda|$, we obtain

$$\left\| \sum c_\lambda \epsilon_\lambda \psi_\lambda \right\|_{L^1} \leq C \left\| \left(\sum |c_\lambda|^2 |\lambda|^{-1} \chi_{R(\lambda)} \right)^{1/2} \right\|_{L^1},$$

so that $\|f\|_3 < \infty$ implies $\|f\|_1 < \infty$. Let us check that $\|f\|_5 < \infty$ implies $\|f\|_4 < \infty$. Suppose that f is an atom associated to the ball B centered at x_0 and of radius $R > 0$. Let $\tilde{B} = B(x_0, CR)$ and $R_k = \{x / 2^k CR \leq |x - x_0| < 2^{k+1} CR\}$, where $C > 1$ will be fixed later.

Let $S(f)(x) = (\sum |c_\lambda|^2 |\psi_\lambda|(x)^2)^{1/2}$. We have

$$\int_{\tilde{B}} S(f) \leq |\tilde{B}|^{1/2} \|S(f)\|_{L^2} \leq C |B|^{1/2} \|f\|_{L^2} \leq C.$$

In order to estimate the integral on R_k, we notice that $c_\lambda = 0$ if $m\lambda$ does not meet B, and, if $m\lambda$ does not meet R_k, we do not have to consider the cube λ in the evaluation of the integral of $S(f)$ over R_k. But, if $m\lambda$ meets B and R_k, then

$2^{-j} \geq CR2^k$. Hence

$$
|c_\lambda| = \left| \int_B f\psi_\lambda \right| = \left| \int_b f \left(\psi_\lambda(\lambda) + \int_{x_0}^x \nabla \psi'_\lambda(\theta) d\theta \right) dx \right|
$$
$$
= \left| \int_b f \left(\int_{x_0}^x \nabla \psi'_\lambda(\theta) d\theta \right) dx \right| \leq \|f\|_{L^1} \sup_{x \in B} \left| \int_{x_0}^x \nabla \psi'_\lambda(\theta) d\theta \right| \leq CR2^{nj/2}2^j.
$$

Thus, if $x \in R_k$, $S(f)(x) \leq C2^{-k(n+1)}r^{-n}$; adding up, we obtain $\int S(f)(x)dx \leq C$. The last implication being checked for the atoms, it also holds for any function in $\mathcal{H}^1$, which ends the proof of the theorem. ∎

Let us finish with some remarks. It is sometimes a problem to suppose that atoms are compactly supported, hence the following definition. A function f is called a molecule of width d if it has a vanishing integral and if for $s > n$,

$$
\int |f(x)|^2 \left(1 + \frac{|x - x_0|}{d} \right)^s dx \leq d^{-n}
$$

This can be useful in many situations; for instance, in order to prove that an operator is continuous on $\mathcal{H}^1$, one can to prove that it maps atoms to molecules.

The dual of $\mathcal{H}^1$ is called **BMO** (for bounded mean oscillations); it can be characterized by the following condition. For any ball B,

$$
\int_B |f - \bar{f}|^2 \leq C|B|
$$

where $\bar{f}$ is the average of f on B. BMO is also characterized by the following condition on the wavelet coefficients (see [Me2]). For any dyadic cube Q,

$$
\sum_{\lambda \subset Q} |c_\lambda|^2 \leq |Q|.
$$

13.5 Wavelets and PDEs

13.5.1 Renormalization of the product

We shall now use the wavelet decomposition of a product of functions to prove that one can pass to the weak limit in certain nonlinear (quadratic) quantities. The idea of using harmonic analysis and wavelet methods to obtain and improve results previously obtained through compensated compactness methods appeared in [Cetal] and [Do], and we follow the presentation given in [Do] (see also [Me2] and [Me3]).

We use the following compactly supported wavelets. Let P_j be the orthogonal projection on V_j and Q_j the orthogonal projection on W_j. Let $f, g \in L^2(\mathbb{R}^n)$. Since

$P_j f \to f$ when $j \to +\infty$ and $P_j f \to 0$ when $j \to -\infty$, in L^2,

$$fg = \sum_j (P_{j+1}fP_{j+1}g) - (P_jfP_jg) = \sum_j P_jfP_jg + \sum_j Q_jfP_jg + \sum_j Q_jfQ_jg$$

where the convergence is in L^1. Let $P_1(f,g)$ be the first sum, $P_2(f,g)$ be the second sum, and we split the third one into

$$P_3(f,g) = \sum_j \sum_k \sum_{l \neq k} \langle f \mid \psi_{j,k} \rangle \langle g \mid \psi_{j,l} \rangle \psi_{j,k} \psi_{j,l}$$

and

$$R(f,g) = \sum_j \sum_k \langle f \mid \psi_{j,k} \rangle \langle g \mid \psi_{j,k} \rangle |\psi_{j,k}|^2.$$

We have

$$P_1(f,g) = \sum_j \sum_k \sum_l \langle f \mid \phi_{j,k} \rangle \langle g \mid \psi_{j,l} \rangle \phi_{j,k} \psi_{j,l}$$

and

$$P_2(f,g) = \sum_j \sum_k \sum_l \langle f \mid \psi_{j,k} \rangle \langle g \mid \phi_{j,l} \rangle \psi_{j,k} \phi_{j,l}$$

Let then $P^\#(f,g) = fg - R(f,g)$ (we have subtracted from the product all the terms that do not have a vanishing integral); $P^\#(f,g)$ is called the renormalized product of f and g. This definition is borrowed from the quantum mechanics notion of renormalization (see [F]).

We remark that, for instance,

$$\phi_{j,k}(x)\psi_{j,l}(x) = 2^{nj/2}\omega^l_{j,k}(x),$$

where the $(\omega^l_{j,k})$ are vaguelettes and $\omega^l_{j,k} = 0$ if $|k - l| \geq M$. Thus, the sums on l have at most $(2M)^n$ terms.

The operators P_1, P_2 and P_3 are clearly continuous from $L^2 \times L^2$ to L^1. We shall prove

PROPOSITION 13.18
The operators P_1, P_2 and P_3 are continuous from $L^2 \times L^2$ to $\mathcal{H}^1$.

PROOF The result for P_3 is straightforward since P_3 is already written as an atomic decomposition, because

$$\sum_{j,k,l} |\langle f \mid \psi_{j,k} \rangle||\langle g \mid \psi_{j,l} \rangle| \leq (2M)^n \|f\|_{L^2} \|g\|_{L^2}.$$

In order to prove the continuity of P_1, we first use the fact that the $\omega^l_{j,k}$ are renormalized molecules. Thus, the operator $\psi_{j,k} \to \omega^l_{j,k}$ is continuous on $\mathcal{H}^1$, so that we only have to prove that

$$\sum \langle f \mid \phi_{j,k} \rangle \langle g \mid \psi_{j,k+l} \rangle 2^{nj/2} \psi_{j,k} \in \mathcal{H}^1,$$

equivalently that

$$(|\langle f \mid \phi_{j,k}\rangle|^2 |\langle g \mid \psi_{j,k+l}\rangle|^2) 2^{2nj} \chi_{j,k})^{1/2} \in L^1.$$

Let $f^*(x)$ be the maximal Hardy–Littlewood function defined by

$$f^*(x) = \sup_{\{Q|x\in Q\}} \frac{1}{|Q|} \int_Q |f(y)|dy.$$

Then (see [St]), if $1 \le p < \infty$,

$$\|f^*\|_{L^p} \le C\|f\|_{L^p}.$$

It is sufficient to prove that

$$\sup_{\{j,k|x\in Q_{j,k}\}} \langle f \mid 2^{nj/2}\phi_{j,k}\rangle \left(\sum_{j,k} |\langle g \mid \psi_{j,k+m}\rangle|^2 2^{nj} \chi_{j,k}(x)\right)^{1/2} \in L^1.$$

But

$$\sup_{\{j,k|x\in Q_{j,k}\}} |\langle f \mid 2^{nj/2}\phi_{j,k}\rangle| \le Cf^*(x)$$

and

$$\left(\sum_{j,k} |\langle g \mid \psi_{j,k+m}\rangle|^2 2^{nj} \chi_{j,k}(x)\right)^{1/2} \in L^2$$

because

$$\int \sum_{j,k} |\langle g \mid \psi_{j,k+m}\rangle|^2 2^{nj} \chi_{j,k}(x)dx = \sum\sum |\langle g \mid \psi_{j,k+m}\rangle|^2 \le \|g\|_2^2.$$

Hence P_1 is continuous. The continuity of P_2 is obtained by permuting the roles of f and g. ∎

We shall now show that renormalization is "compatible" with weak convergence. Define VMO as the closure of $\mathcal{D}(\mathbb{R}^n)$ in BMO. Wavelets are clearly total in VMO.

PROPOSITION 13.19
If $f_q \rightharpoonup f$ in L^2 and $g_q \rightharpoonup f$ in L^2, then $P^\#(f_q, g_q) \rightharpoonup P^\#(f, g)$ for the weak $(\mathcal{H}^1, VMO)$ topology.

PROOF First note that if $f_q \rightharpoonup f$ in L^2 and $g_q \rightharpoonup f$ in L_2, the sequence $P_1(f_q, g_q)$ is bounded in $\mathcal{H}^1$ (because f_q and g_q are bounded in L^2). The same holds for P_2 and P_3. Thus, we only have to check that, for each m,

$$\sum \langle f_q \mid \phi_{j,k}\rangle \langle g_q \mid \psi_{j,k+l}\rangle 2^{nj/2}\psi_{j,k} \rightharpoonup \sum \langle f \mid \phi_{j,k}\rangle \langle g \mid \psi_{j,k+l}\rangle 2^{nj/2}\psi_{j,k}.$$

This result holds because, if h_q is bounded in $\mathcal{H}^1$, then $h_q \rightharpoonup h$ for the weak $(\mathcal{H}^1, VMO)$ topology iff $\forall j, k \ \langle h_q \mid \psi_{j,k} \rangle \to \langle h \mid \psi_{j,k} \rangle$. Thus, Proposition 13.19 is proved. ∎

We are now in a position to prove the so-called theorem of the Jacobian.

THEOREM 13.20
If f and g belong to the Sobolev space $H^1(\mathbb{R}^2)$, then the Jacobian

$$J(f,g) = \frac{\partial f}{\partial x} \frac{\partial g}{\partial y} - \frac{\partial f}{\partial y} \frac{\partial g}{\partial x}$$

belongs to $\mathcal{H}^1$. If f_q and g_q are bounded in $H^1(\mathbb{R}^2)$, and if ∇f_q and ∇g_q converge weakly to ∇f and ∇g in L^2 then $J(f_q, g_q)$ converges weakly to $J(f, g)$ for the $(\mathcal{H}^1, VMO)$ topology.

PROOF We start by renormalizing the products $(\partial f/\partial x)(\partial g/\partial y)$ and $(\partial f/\partial y)(\partial g/\partial x)$. The renormalized products belong to $\mathcal{H}^1$ and we have compatibility with weak convergence. We only have to consider the remaining part

$$J_B(f,g) = \sum \left(\left\langle \frac{\partial f}{\partial y} \mid \psi_\lambda \right\rangle \left\langle \frac{\partial g}{\partial x} \mid \psi_\lambda \right\rangle - \left\langle \frac{\partial f}{\partial x} \mid \psi_\lambda \right\rangle \left\langle \frac{\partial g}{\partial y} \mid \psi_\lambda \right\rangle \right) |\psi_\lambda(x)|^2.$$

Let $f = \sum \alpha_\lambda \psi_\lambda$ and $g = \sum \beta_\lambda \psi_\lambda$. Since f and g belong to the Sobolev space $H^1(\mathbb{R}^2)$, $(2^j \alpha_\lambda) \in l^2$, and $(2^j \beta_\lambda) \in l^2$. Let

$$\mu_x(\lambda', \lambda) = \left\langle \frac{\partial \psi_{\lambda'}}{\partial x} \mid \psi_\lambda \right\rangle,$$

$$\mu_y(\lambda', \lambda) = \left\langle \frac{\partial \psi_{\lambda'}}{\partial y} \mid \psi_\lambda \right\rangle,$$

and

$$J_B(f,g)(x) = \sum_\lambda \left(\sum_{\lambda'} \sum_{\lambda''} \alpha_{\lambda'} \beta_{\lambda''} \mu_y(\lambda', \lambda) \mu_x(\lambda'', \lambda) \right.$$
$$\left. - \mu_x(\lambda', \lambda) \mu_y(\lambda'', \lambda) \right) |\psi_\lambda(x)|^2.$$

We have $2^{-j'} \mu_x(\lambda', \lambda) \in \mathcal{M}^\gamma \ \forall \gamma$ and $2^{-j'} \mu_y(\lambda', \lambda) \in \mathcal{M}^\gamma \ \forall \gamma$ because $2^{-j}(\partial \psi_\lambda / \partial x)$ and $2^{-j}(\partial \psi_{\lambda'} / \partial y)$ are vaguelettes. Let

$$m_{\lambda', \lambda''}(x) = \frac{2^{-j'} 2^{-j''}}{\omega_\gamma(\lambda, \lambda'')} \sum_\lambda (\mu_y(\lambda', \lambda) \mu_x(\lambda'', \lambda) - \mu_x(\lambda', \lambda) \mu_y(\lambda'', \lambda)) |\psi_\lambda(x)|^2.$$

We have

$$J_B(f,g)(x) = \sum_{\lambda'} \sum_{\lambda''} 2^{j'} \alpha_{\lambda'} 2^{j''} \beta_{\lambda''} \omega_\gamma(\lambda', \lambda'') m_{\lambda', \lambda''}(x).$$

We shall see that we have an atomic decomposition of J_B. Let us prove that $m_{\lambda',\lambda''}(x)$ belongs to L^1 and has a vanishing integral:

$$\sum_\lambda \int |\mu_y(\lambda',\lambda)\mu_x(\lambda'',\lambda) - \mu_x(\lambda',\lambda)\mu_y(\lambda'',\lambda)||\psi_\lambda(x)|^2$$

$$\leq C2^{j'+j''}\sum_\lambda \omega_\gamma(\lambda',\lambda)\omega_\gamma(\lambda'',\lambda) \leq C2^{j'+j''}\omega_\gamma(\lambda',\lambda'').$$

Thus the functions $m_{\lambda',\lambda''}$ are uniformly in L^1. Let us integrate term by term the series defining $m_{\lambda',\lambda''}$; we have

$$\int \omega_\gamma(\lambda',\lambda'')m_{\lambda',\lambda''} = \sum_\lambda \mu_y(\lambda',\lambda)\mu_x(\lambda'',\lambda) - \mu_x(\lambda',\lambda)\mu_y(\lambda'',\lambda)$$

$$= \sum \left(\left\langle \frac{\partial \psi_{\lambda'}}{\partial y} \,\Big|\, \psi_\lambda \right\rangle \left\langle \frac{\partial \psi_{\lambda''}}{\partial x} \,\Big|\, \psi_\lambda \right\rangle \right.$$

$$\left. - \left\langle \frac{\partial \psi_{\lambda'}}{\partial x} \,\Big|\, \psi_\lambda \right\rangle \left\langle \frac{\partial \psi_{\lambda''}}{\partial y} \,\Big|\, \psi_\lambda \right\rangle \right)$$

$$= \left\langle \frac{\partial \psi_{\lambda'}}{\partial y} \,\Big|\, \frac{\partial \psi_{\lambda''}}{\partial x} \right\rangle - \left\langle \frac{\partial \psi_{\lambda'}}{\partial x} \,\Big|\, \frac{\partial \psi_{\lambda''}}{\partial y} \right\rangle$$

$$= \int J(\psi_{\lambda'},\psi_{\lambda''}) = 0$$

(the integral of a Jacobian vanishes). One checks then that $(m_{\lambda',\lambda''})_{\lambda'}$ are vaguelettes and that

$$\sum 2^{j'+j''}|\alpha_{\lambda'}||\beta_{\lambda''}|\omega_\gamma(\lambda',\lambda'') < \infty.$$

One concludes by remarking that the condition for belonging to $\mathcal{H}^1$ is verified (since vaguelettes are L^2-normalized molecules). ∎

Let us now give a few applications of Theorem 13.20 that are useful in PDEs (see [Cetal]).

PROPOSITION 13.21
If $u \in (H^1(\mathbb{R}^2))^2$, $\det(\nabla u) \in H^{-1}(\mathbb{R}^2)$.

PROOF We first check that the Sobolev space $H^1(\mathbb{R}^2)$ is included in VMO. Since $\mathcal{D}(\mathbb{R}^2)$ is dense in $H^1(\mathbb{R}^2)$, we only have to check that $H^1(\mathbb{R}^2)$ is included in BMO, which we do using the wavelet characterization of these spaces. If $f \in H^1(\mathbb{R}^2)$, $\sum |c_\lambda|^2(1 + 2^{2j}) \leq \|f\|_{H^1}^2$. Hence

$$\sum_{\lambda \subset Q} |c_\lambda|^2 \leq \frac{\|f\|^2}{1 + 2^{2j_0}} \quad \text{(if Q is of size 2^{-j_0})}$$

$$\leq |Q|\|f\|^2.$$

By duality, we obtain that $\mathcal{H}^1 \hookrightarrow H^{-1}(\mathbb{R}^2)$ (using the fact that $\mathcal{H}^1$ is the dual of VMO, see [Cetal]), proving Proposition 13.21. ∎

PROPOSITION 13.22
Let ϕ be the solution of

$$\Delta\phi = \det(\nabla u), \quad u \in (H^1(\mathbb{R}^2))^2.$$

Then $\hat{\phi} \in L^1$.

To prove this, we first remark that $(\partial^2\phi/\partial x_i\partial x_j) \in \mathcal{H}^1$, because

$$\frac{\partial^2\phi}{\partial x_i\partial x_j} = R_iR_j(\det\nabla u),$$

where R_i is the Riesz transform, defined by $\widehat{R_if} = (\xi_i/\xi)\hat{f}$. One checks that $R_i, R_j \in \mathcal{M}^\gamma$ for all γ; thus, R_i and R_j are continuous on $\mathcal{H}^1$ and the result holds because, if $f \in \mathcal{H}^1(\mathbb{R}^2)$, $(1/|\xi|^2)\hat{f}(\xi) \in L^1(\mathbb{R}^2)$ (see [Cetal]).

PROPOSITION 13.23
If $u \in H^2_{loc}(\mathbb{R}^2)$ and $\det(\nabla u) \geq 0$ a.e., then

$$\det(\nabla u)\log(\det(\nabla u)) \in L^1_{loc}$$

This result holds because, if ϕ is a positive function locally in $\mathcal{H}^1$, $\phi\log\phi \in L^1_{loc}$ (see [Cetal]).

13.5.2 Wavelets and concentration

In several types of nonlinear PDEs (for instance Boltzmann [DL], Euler [DM], [PnMaj], [PLi], or Klein–Gordon [Z]), one wishes to prove the existence of solutions starting from a minimizing sequence for which an energy estimate holds, like

$$\|f_n\|_{W^{s,p}} \leq C.$$

The goal is to deduce that f_n is bounded in L^q for a q larger than the one given by the Sobolev imbeddings (namely $((1/p) - (s/n))^{-1}$). We shall see that it is possible outside a "small" set. The proof will be very similar to the one given in Section 13.3.2.

THEOREM 13.24
Let f_n be a bounded sequence in $W^{s,p}$. Let r and q such that $r < s$ and $q > p$. Let η_1 and $\eta_2 > 0$ be fixed and $d > n - (pq(s-r)/q-p)$. There exists a subsequence f_{θ_n} of f_n such that $f_{\theta_n} = g_{\theta_n} + h_{\theta_n}$, with g_{θ_n} bounded in $W^{r,q}$, and for any ball B of radius 1, $(\cup_n \text{Supp}\, h_{\theta_n}) \cap B$ can be covered by dyadic cubes Q_k such that

diam $Q_k \leq \eta_1$ *and*

$$\sum (\text{diam } Q_k)^d \leq \eta_2.$$

We shall prove this theorem in two steps. In the first one we show how to split a function f into two parts as above, using its wavelet decomposition. In the second, we show how to make this decomposition on a sequence of functions. We shall use compactly supported wavelets.

Coming back to the wavelet characterization of $W^{s,p}$ given in Section 13.3.2, we have

$$f \in W^{s,p} \Rightarrow \sum_{\lambda \in \Lambda_j} |c_\lambda|^p 2^{(sp+(np/2)-n)j} \leq C \qquad (13.26)$$

Suppose that there exists $\epsilon > 0$ such that $\sum_{\lambda \in \Lambda_j} |c_\lambda|^q 2^{(rq+(nq/2)-n)j} 2^{\epsilon j} \leq C$. Then

$$\left\| \sum_{\lambda \in \Lambda_j} |c_\lambda|^2 2^{(n+2r)j} \chi_\lambda(x) \right\|_{L^{q/2}}^{q/2} \leq C 2^{-\epsilon j},$$

so that

$$\sum_{\lambda \in \Lambda} |c_\lambda|^2 2^{(n+2r)j} \chi_\lambda \in L^{q/2}.$$

Thus

$$\sum_{\lambda \in \Lambda_j} |c_\lambda|^q 2^{(rq+(nq/2)-n)j} 2^{\epsilon j} \leq C \Rightarrow f \in W^{r,q}. \qquad (13.27)$$

Let us now prove the announced splitting for one function. Let $\epsilon > 0$. We split Λ into Λ' and Λ'' as follows: $\lambda \in \Lambda'$ if $|c_\lambda| \leq 2^{((ps-rq-\epsilon/q-p)-(n/2))j}$, and otherwise $\lambda \in \Lambda''$. Then $\sum_{\lambda \in \Lambda'} c_\lambda \psi_\lambda \in W^{r,q}$ because

$$\sum_{\lambda \in \Lambda' \cap \Lambda_j} |c_\lambda|^q 2^{(rq+(nq/2)-n)j} 2^{\epsilon j} \leq \sum_{\lambda \in \Lambda_j} |c_\lambda|^p 2^{(q-p)((ps-rq-\epsilon/q-p)-(n/2))j} 2^{(rq+(nq/2)-n)j} 2^{\epsilon j}$$

$$\leq \sum_{\lambda \in \Lambda_j} |c_\lambda|^p 2^{(ps+(np/2)-n)j} \leq C \|f\|_W^{s,p}.$$

Let J be such that $2^{-J} \leq \eta_1$ and $\theta = \sum_{\lambda \in \Lambda'', j \leq J} c_\lambda \psi_\lambda$. Then θ is arbitrarily smooth. There remains $h = \sum_{\lambda \in \Lambda'', j \geq J} c_\lambda \psi_\lambda$. Let N be the cardinality of $\{\lambda \in \Lambda'' \cap \Lambda_j\}$. Since

$$\sum_{\lambda \in \Lambda_j} |c_\lambda|^q 2^{(ps+(np/2)-n)j} \leq C,$$

we have

$$N 2^{p(((ps-rq-\epsilon)/(q-p))-(n/2))j} 2^{(ps+(np/2)-n)j} \leq C,$$

so that

$$N \leq C2^{n-(((pq(s-r)-p\epsilon)/(q-p))j}.$$

If the wavelet is supported by the cube centered at the origin and of size A, we cover the support of h by cubes $\tilde{\lambda} = A\lambda$. Thus

$$\sum (\text{diam } \tilde{\lambda})^d \leq \sum_{j \geq J} CA^d 2^{n-(((pq(s-r)-p\epsilon)/(q-p))j} 2^{-dj}.$$

If J is large enough, the sum can be arbitrarily small if $d > n-(pq(s-r)-p\epsilon/q-p)$. Since ϵ can be chosen arbitrarily small, the condition is

$$d > n - \frac{pq(s-r)}{q-p}.$$

This proves the theorem for one function. Note that one obtains $d = 0$ if $(1/q) = (1/p) - ((s-r)/n)$, which is the case when $W^{s,p} \subset W^{r,q}$.

Let us now prove the theorem. First, we can suppose that f_n is supported in a given ball, because, if not, we localize using a decomposition of the identity. The result holds for each localized sequence, hence for f_n after extracting subsequences. Let η_1 and η_2 be fixed, and J be such that

$$2^{-J} \leq \eta_1 < 2.2^{-J}.$$

We now decompose f_n as above in three functions

$$f_n = g_n + h_n + \theta_n,$$

where $\theta_n = \sum_{j \leq J} c_\lambda \psi_\lambda$. Thus, using the Bernstein inequality, for s' arbitrarily large we have

$$\|\theta_n\|_{W^{s',p}} \leq C(s,s',p,J)\|f_n\|_{W^{s,p}}.$$

Let $\epsilon = (1/2)(d - n + (pq(s-r)/q-p))$ and $d' = d - \epsilon$. One defines Λ'_n and Λ''_n as above. Thus

$$\left\| \sum_{\lambda \in \Lambda'_n, j \geq J} c_\lambda \psi_\lambda \right\|_{W^{r,q}} = \|g_n\|_{W^{r,q}} \leq \|f_n\|_{W^{s,p}}.$$

The support of h_n is covered by cubes $\lambda \in \Lambda''_n$ of size at most η_1 such that

$$\sum_{\lambda \in \Lambda''_n} (\text{diam } \lambda)^d \leq \eta_2.$$

We shall now describe the procedure to extract a subsequence of f_n. For each $j \geq J$ we shall define an extraction "of order j," and we shall then apply the diagonal procedure. Suppose the extraction has been performed up to the order $j - 1$. There exists a finite number of cubes of size 2^{-j} that belong to $\cup_n \Lambda''_n$. Consider each of these cubes successively. If λ belongs to a finite number of Λ''_n, we discard the corresponding ns, and if λ belongs to an infinity of Λ''_n, we keep those ns. Once this extraction has been performed for all cubes of size 2^{-j}, we have obtained a subsequence $(f_{\theta_j(n)})$. Let $\omega(n) = \theta_{J+n}(n)$. The support of $h_{\omega(n)}$ is included in $\cup_{\lambda \in \Lambda''_{\omega(n)}} \lambda$ and

$$\sum_{\lambda \in \Lambda''_{\omega(n)}} (\text{diam } \lambda)^{d'} \leq \eta_2.$$

Furthermore, if $n \leq p$, the cubes $\lambda \in \Lambda''_{\omega(n)}$ of size smaller than 2^{-n} are the same as those in $\Lambda''_{\omega(p)}$. We say that λ of size 2^{-j} belongs to Λ'' if it belongs to $\Lambda''_{\omega(j)}$.

Let $A(n) = \sum_{\lambda \in \Lambda''_{\omega(n)}} (\text{diam } \lambda)^{d'}$ and $A = \sum_{\lambda \in \Lambda''} (\text{diam } \lambda)^{d'}$. We have $A(n) \leq \eta_2$. Let $J_0 > J$; then

$$\sum_{\lambda \in \Lambda'', j \leq J_0} (\text{diam } \lambda)^{d'} \leq \sum_{\lambda \in \Lambda''_{\omega(j)}} (\text{diam } \lambda)^{d'} \leq A(j),$$

so that

$$A \leq \liminf A(j) \leq \eta_2.$$

Let $B(n) = \sum_{\lambda \in \Lambda''_{\omega(n)}} (\text{diam } \lambda)^d$ and $B = \sum_{\lambda \in \Lambda''} (\text{diam } \lambda)^d$. Then

$$B(n) \leq \sum_{\lambda \in \Lambda''_{\omega(n)}, j \leq n} (\text{diam } \lambda)^d + \sum_{\lambda \in \Lambda''_{\omega(n)}, j > n} (\text{diam } \lambda)^d.$$

The value of the first sum is $\sum_{\lambda \in \Lambda'', j \leq n} (\text{diam } \lambda)^d \leq B$ and the second sum is bounded by $2^{-n\epsilon} \sum_{\lambda \in \Lambda'', j > n} (\text{diam } \lambda)^{d'} \leq \eta_2 2^{-n\epsilon}$. The union of the E_n is covered by

$$\Lambda''' = \Lambda'' \cup (\cup_{n=J}^{\infty} \{\lambda \in \Lambda''_{\omega(n)} \mid \text{diam } \lambda \leq 2^{-n}\})$$

and

$$\sum_{\lambda \in \Lambda'''} (\text{diam } \lambda)^d \leq \eta_2 + \sum_{n=J}^{\infty} \eta_2 2^{-n\epsilon} \leq \eta_2 \left(1 + \frac{2^{-\epsilon J}}{1 - 2^{-\epsilon J}} \right).$$

This proves Theorem 13.24.

Bibliography

[ABMu1] A. Arnéodo, E. Bacry, and J.F. Muzy. 1991. "Direct determination of the singularity spectrum of fully developed turbulence data," preprint.

[ABMu2] A. Arnéodo, E. Bacry, and J.F. Muzy. 1993. "Singularity spectrum of fractal signals from wavelet analysis: exact results," *J. Stat Phys.* **70**: 635.

[BMaPa]   E. Bacry, S. Mallat, and G. Papanicolaou. 1991. "Time and space wave-
          let adaptive scheme for partial differential equations," paper presented
          at *Wavelets and Turbulence*, Princeton, June.

[Bo]   J.M. Bony. 1984. "Second microlocalization and propagation of sin-
          gularities for semilinear hyperbolic equations," in *Tanaguchi Symp.
          HERT*, Katata, pp. 11–49.

[BrMiPe] G. Brown, G. Michon, and J. Peyrière. 1991. "On the multifractal
          analysis of measures," preprint.

[CZy]   A.P. Calderón and A. Zygmund. 1957. "Singular integral operators and
          differential equations," *Amer. J. Math.* **79**: 901–921.

[Cetal]   R.R. Coifman, P.L. Lions, Y. Meyer, and S. Semmes. 1993. "Com-
          pensated compactness and Hardy spaces," *J. Math. Pures Appl.*, to
          appear.

[D1]   I. Daubechies. 1988. "Orthonormal bases of compactly supported
          wavelets," *Commun. Pure Appl. Math.* **41**: 909–996.

[D2]   I. Daubechies. 1992. *Ten Lectures on Wavelets*, CBMS-NSF Reg. Conf.
          Ser. Appl. Math. **61**, Philadelphia: Soc. Ind. Appl. Math,

[Do]   S. Dobyinski. 1992. Thèse, Université Paris Dauphine, CEREMADE.

[F]   P. Federbush. 1987. "Quantum field theory in ninety minutes," *Bull.
          Amer. Math. Soc.* **107**: 319.

[J1]   S. Jaffard. 1989. "Exposants de Holder en des points donnés et
          coefficients d'ondelettes," *C. R. Acad. Sci.*, Sér. 1 **308**: 79–81.

[J2]   S. Jaffard. 1991. "Pointwise smoothness, two-microlocalization and
          wavelet coefficients," *Publ. Mat.* **35**: 155–168.

[J3]   S. Jaffard. 1991. "Sur la dimension de Hausdorff des points singuliers
          d'une fonction," *C. R. Acad. Sci.*, Sér. 1 **314**: 31–36.

[J4]   S. Jaffard. 1992. "Construction de fonctions multifractales ayant un
          spectre de singularités prescrit," *C. R. Acad. Sci.*, Sér. 1 **315**: 19–24.

[J5]   S. Jaffard. 1992. "Wavelet methods for fast resolution of elliptic
          problems," *SIAM J. Numer. Anal.* **29**(4): 965–986.

[JLa]   S. Jaffard and Ph. Laurençot. 1992. "Orthonormal wavelets, analysis
          of operators, and applications to numerical analysis," in *Wavelets: A
          Tutorial in Theory and Applications*, C.K. Chui, ed., Boston: Academic
          Press, pp. 543–601.

[K]   J.P. Kahane. 1968. *Some Random Series of Functions*. Cambridge:
          Cambridge Univ. Press.

[L]   P.G. Lemarié. 1988. "Ondelettes à localisation exponentielle," *J. Math.
          Pures Appl.* **67**: 227–236.

[LMe]   P.G. Lemarié and Y. Meyer. 1986. "Ondelettes et bases hilbertiennes,"
          Rev. Mat. Iberoamericana **2**: 1–18.

[MPerR] Y. Maday, V. Perrier, and J.C. Ravel. 1991. "Adaptativité dynamique sur bases d'ondelettes pour l'approximation d'équations aux dérivées partielles," *C. R. Acad. Sci.*, sér. 1 **312**: 405–410.

[Ma] Mallat S. 1989. "Multiresolution approximations and wavelet orthonormal bases of $L^2(\mathbb{R})$," *Trans. Amer. Math. Soc.* **315**: 69–88.

[Me1] Y. Meyer. 1990. *Ondelettes et Opérateurs*, Paris: Hermann.

[Me2] Y. Meyer. 1981. "Remarques sur un théorème de J.M. Bony," *Suppl. Rendiconti Circ. Mat. Palermo* **1**: 1–20.

[Me3] Y. Meyer. 1990. "Wavelets and applications," in *Proceedings of the International Congress of Mathematics*, Kyoto.

[PLi] R.J. di Perna and P.L. Lions. 1989. "On the Cauchy problem for Boltzmann equation. Global existence and weak stability," *Ann. Math.* **130**: 321–366.

[PnMaj] R.J. di Perna and A. Majda. 1988. "Reduced Hausdorff dimension and concentration-cancellation for two-dimensional incompressible flows," *J. Amer. Math. Soc.* **1**: 59–98.

[St] E. Stein. 1970. *Singular Integrals and Differentiability Properties of Functions*. Princeton: Princeton Univ. Press.

[Z] Y. Zheng. 1991. "Concentration in sequences of solutions to the nonlinear Klein–Gordon equation," *Ind. Univ. Math. J.* **40**(1): 86–147.

[Zi] W.P. Ziemer. 1989. *Weakly Differentiable Functions*, Berlin: Springer-Verlag.

[Zy] A. Zygmund. 1959. *Trigonometric Series*, 2nd ed., Cambridge: Cambridge Univ. Press.

14

Scale decomposition in Burgers' equation

Frédéric Heurtaux,[*] **Fabrice Planchon,**[*] **and Mladen Victor Wickerhauser**[†]

ABSTRACT   The wavelet representation of a time-dependent signal can be used to study the propagation of energy between the different scales in the signal. Burgers' evolution operator (in 1 and 2 dimensions) can itself be described from this scaling point of view. Using wavelet-based algorithms we can depict the transfer of energy between scales. We can write the instantaneous evolution operator in the wavelet basis; then large off-diagonal terms will correspond to energy transfers between different scales. We can project the solution onto each fixed-scale wavelet subspace and compute the energy; then the rate of change of this energy by scale can detect and quantify any cascades that may be present. These methods improve the classical Fourier-transform-based scale decomposition which uses the notion that wavenumber equals scale. The wavelet basis functions underlying our scale decompositions have finite, well-defined position uncertainty (i.e., scale) whereas Fourier basis functions have formally unbounded position uncertainty.

CONTENTS

[*]Supported in part by the Internship Office of the École Polytechnique.
[†]Supported in part by the Air Force Office of Scientific Research.

14.1   Introduction

The difficult problem of turbulence in fluids has spawned a rich variety of ideas and notions of measurement. A fluid flow contains a very large number of degrees of freedom, and yet there are certain identifiable features which can be described at least approximately with a small number of parameters. For example, in viscous fluids the large regions of high vorticity seem to evolve in a coherent manner. The number of such regions is necessarily limited and their sizes and shapes change quite slowly. These have been quite successfully modelled by point vortices and particle mechanics [BeL], point vortices with a few extra moments [Beetal], and contour dynamics [Z], all of which replace the grid point approximation by a much lower rank approximation and rewrite the equations of motion in terms of the new parameters. It is also possible to compute various averages and show from first principles that they are conserved, or change monotonically, or have specified behavior when a particular term of the differential equation is dominant. Such quantities include total energy, total enstrophy and average velocity, which can be calculated using Newton's laws of motion or other conservation principles.

The famous 1941 paper by Kolmogorov [K] predicted that the energy spectrum of the velocity field of fully developed turbulence should be modelled by a $k^{-5/3}$ power law (in the wavenumber k). One intuitively compelling explanation for this relationship is that energy flows from large-scale coherent velocity fields to small-scale dissipative eddies. However, it has been difficult to quantify this notion, partly because it is difficult to define scale in a satisfactory manner. Traditionally, scale has been set equal to wavenumber k in the Fourier transform of the velocity field. The energy at a given scale was then defined to be the sum of the squares of the Fourier coefficients in a certain range of wavenumbers. Unfortunately, the operation of restriction to a range of wavenumbers is ill-conditioned in any norm but L^2; it is equivalent to projection onto basis functions which have no well-defined scale, and can mislead the aforementioned intuition simply by its mathematical ill-behavior. Our goal in this paper is to replace the Fourier–wavenumber "scale" projection with a mathematically better-behaved alternative of wavelets. We can then exploit this new wavelet scale projection to measure energy transfers between scales in a particular solution to Burgers' equation, a simplified fluid flow equation [Bu].

For turbulent flows, Kolmogorov, Kraichnan [Kr] and Batchelor [B] predicted that the energy repartition depends on the scale of the structures that appear in the flow. Wavelets are an especially suitable tool for problems involving scale, since every wavelet has a well-defined scale of its own. We can define the *energy of a function at a scale* σ simply by summing the squares of the amplitudes of all wavelets of scale σ in the function. We can then ask how energy is transferred between scales in an evolving turbulent flow. To approach this question, we will consider Burgers' equation in one dimension, where exact analytic expressions of the solution are known for some initial condition. We will first see how the small scale wavelets in the decomposition of the solution collect more and more energy as the evolution proceeds and approaches singularity. We will also describe the time-varying Burgers evolution operator corresponding to a known evolution. We will conjugate it into a matrix with respect to the wavelet basis, where scales can be easily identified. We will observe those matrix coefficients which couple the different scales and transfer energy between them. We can thus observe the creation and destruction of small-scale phenomena such as rapid fluctuations or "shocks." We will also repeat a part of this analysis for the two-dimensional Burgers equation, to see if our surmises hold in that more complicated situation.

In related work [Fetal], an initial parameter reduction by projection onto just the largest wavelet packet coefficients was compared to parameter reduction by projection onto low wavenumbers. This provides another test of the notion that energy cascades from large wavelengths to small. The main purpose of that paper, however, was to test how well the reduced-parameter flow predicts the original, reference flow; it demonstrated that projection onto large wavelet packets is substantially better than projection onto low wavenumbers in both a deterministic and statistical sense. In other related work [LT] an adaptive algorithm used wavelets to perform the numerical integration of Burgers' equation. When the energy in some range of small scales exceeds a threshold in a region, the solution is resampled more finely in that region. While the methods discussed in this present paper are intended only to suggest how to quantify certain intuitive notions, they are oriented toward improving the numerical solution of fluid dynamics problems by similar adapted algorithms.

14.2 Burgers' evolution equation

Burgers' equation is the first part of the following initial value problem:

$$\frac{\partial F}{\partial t}(x,t) = -\frac{1}{2}\frac{\partial}{\partial x}F^2(x,t) + \nu\Delta F(x,t); \qquad F(x,0) = F_0(x). \qquad (14.1)$$

The constant ν is the viscosity of the fluid and the function $F_0 = F_0(x)$ is the initial state at time $t = 0$.

Let us consider one example: $F_0(x) = \sin(2\pi x)$. The graph in Figure 14.1 shows the evolution of this function at times $0, 0.08, 0.16, 0.32, 0.5, 0.75$, and 1.00. The two arcs of the sine, one positive the other negative, are propagating in opposite directions to produce a steep slope at $x = 32/64$. The dissipative term ΔF produces the vanishing effect: the total energy in the solution tends to 0 as time increases. Without dissipation the slope at $x = 32/64$ would become infinite and a discontinuity would appear; the viscosity term controls how close the solution gets to singularity before dissipating. The apparition of a near-discontinuity means that the amplitudes of small-scale wavelets in the solution are increasing, since they contribute the large derivatives. We can effectively see this phenomenon in

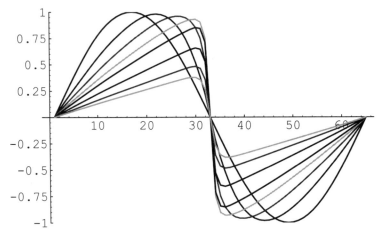

FIGURE 14.1
Evolution of $\sin(2\pi x)$ over $[0; 1]$.

Figures 14.2 and Figure 14.3. The graduations between 0 and 100 represent time; the others show the index of the wavelet coefficients. The first wavelet coefficient is the mean of the signal (actually 0), the second is the biggest-scale difference coefficient, and so on. The last 32 are the smallest-scale difference coefficients, since we took 64 samples of the signal. We use the "Coiflet" wavelets based on a quadrature mirror filter with 30 taps [D] because they have a large number of vanishing moments and are nearly symmetric.

We computed the evolution with a Godounov scheme applied to the 1-periodic signal, using a space-step of 1/64 and a time-step of 1/100. Figures 14.2 and 14.3 show the evolution of wavelet coefficients. They indicate that the energy in the biggest-scale wavelets is decreasing while the energy in the smallest-scale ones is increasing. In the view from below (Figure 14.3), we observe that one of the big-scale amplitudes already begins to decrease at time zero. Figure 14.2 shows that the maxima of the smaller-scale amplitudes are reached later and later with decreasing scale. This last aspect can be better seen on Figure 14.4 which shows the absolute value of the wavelet coefficients in gray scale: white is zero, black is the maximum.

Ultimately, all the wavelet coefficients decrease to 0, because through dissipation the energy in the signal decreases to 0.

14.3   Burgers' evolution operator

We will now try to extract inter-scale energy transfer information from the evolution operator of Burgers' equation. In fact, we will use the infinitesimal

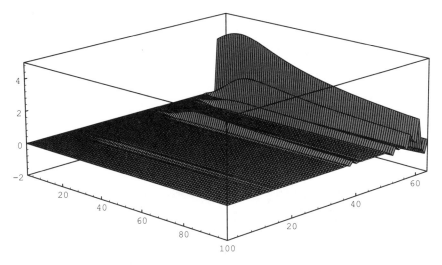

FIGURE 14.2
Evolution of the wavelet coefficients.

generator of the evolution, which is the Jacobian of the right-hand side of Equation 14.1.

Let G be the matrix of our transformation using the Godounov scheme in the normal space, and W be the matrix of our wavelet transform. Let $u(x, t)$ (or simply $u(t)$ but actually $u(i\Delta x, n\Delta t)$) be our signal, and $U(t)$ be its wavelet form. We

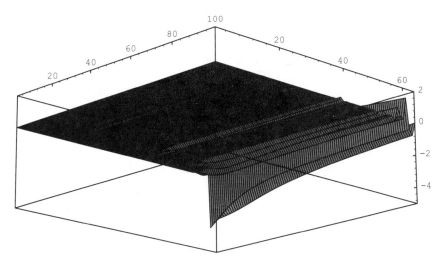

FIGURE 14.3
Same 3-dimensional representation, seen from below.

Abs[wavelet coef]

initial condition:sin(2*Pi*x)

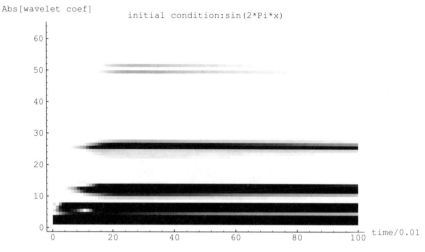

FIGURE 14.4
Time evolution of the wavelet coefficients in gray scale.

have:

$$u(t + \Delta t) = Gu(t); \qquad U(t + \Delta t) = (WGW^{-1})U(t). \qquad (14.2)$$

Figure 14.5 shows the coefficients of $K = WGW^{-1}$ at times $t = 0.00$ on the left and $t = 0.33$ on the right. In the density plots, black means the minimum and white means the maximum, with gray being zero. We are using the classical

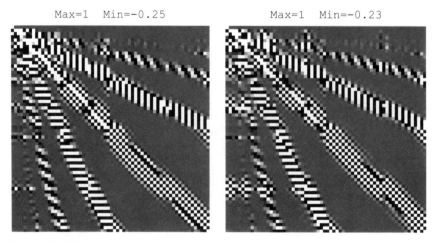

FIGURE 14.5
Burgers' evolution operating on wavelet components.

form of simple operators in the wavelet basis, the so-called "standard form" of
G. Beylkin, R. Coifman, and V. Rokhlin (see M. Ruskai, G. Beylkin, R. Coifman,
I. Daubechies, S. Mallat, Y. Meyer, L. Raphael, [BeyCR]). The "fingers" which
lay across the matrix density plots represent strong coupling between wavelet
scales. The main diagonal is the coupling between wavelets at equal scales, the
upper triangle fingers are couplings from large scales to small scales, and the
lower triangle fingers are couplings from small scales to large scales.

The isolated dark points in the last row and last column of each matrix are not
artifacts of the periodization, but are due to the periodicity of each block itself in
the matrix, and also to the discontinuity introduced by the Godounov scheme in
the derivative at this point.

Unfortunately, it is not easy to read Figure 14.5. We see in the right-hand
picture that the scale interactions mainly occur in the high-gradient zone of the
signal near the "midpoint," where the large amplitudes in K show up as fingers.
We can observe that the strongest interactions take place between nearby scales,
since the variation from gray is strongest along the middle fingers and decreases
away from the main diagonal. We can also say that the interactions between
the largest and smallest scales become stronger as the solution evolves, since the
outside fingers become thicker as t increases. But it is impossible to say what
kind of interactions these are, since the coefficients are alternately positive and
negative.

To see where the energy is concentrated, we now study the effects of Burgers'
evolution operator on the absolute value of the wavelet coefficients in our signal.
The sum of these absolute values is what we shall call "energy." Let $|U|$ denote
the absolute value of these coefficients, let S be the diagonal "sign matrix" made
of the signs of these coefficients, and let I be the identity matrix. We have:

$$\frac{\partial |U|(t)}{\partial t} \approx \frac{|U|(t + \Delta t) - |U|(t)}{\Delta t} \tag{14.3}$$

$$= \frac{1}{\Delta t}(|WGu| - |Wu|)(t) \tag{14.4}$$

$$= \frac{1}{\Delta t}(|WGW^{-1}Wu| - SWu)(t) \tag{14.5}$$

$$= \frac{1}{\Delta t}(|WGW^{-1}SSWu| - SWu)(t) \tag{14.6}$$

$$\approx \frac{1}{\Delta t}(SWGW^{-1}S - I)|U|(t) \tag{14.7}$$

The last approximation assumes that $WU(t + \Delta t)$ and $WU(t)$ have the same sign
matrix. Since the wavelet coefficients are continuous, slowly varying functions of
time, this assumption will only be violated by coefficients close to 0 whose error
contribution is therefore small. The new matrix $R = SWGW^{-1}S - I$ is not really
simple to understand either, even if we only keep the sign of its most significant
coefficients–the ones greater than 0.5% of the maximum coefficient, as in the right

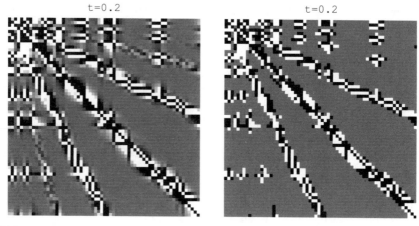

FIGURE 14.6
Evolution matrix of $|U|$.

picture of Figure 14.6. The matrices shown in that figure have the following block structure:

$-1/-1$	$0/-1$	$1/-1$	$2/-1$	$\cdots$	$5/-1$
$-1/0$	$0/0$	$1/0$	$\cdots$	$\cdots$	$5/0$
$-1/1$	$0/1$	$1/1$	$\cdots$	$\cdots$	$5/1$
$\cdot$	$\cdot$	$\cdots$	$\cdots$	$\ddots$	$\cdots$
$-1/5$	$0/5$	$1/5$	$2/5$	$\cdots$	$5/5$

Each block i/j represents the part of the matrix corresponding to the influence of scale i on what the scale j will be at the next time step. Scale -1 is the mean of the signal, a single value. Scale 5 is the smallest scale and contains 32 components.

In order to get more readable results, we now forget about the space localization of the signal and only look at how the different scales act upon each other. We sum the amplitudes of each block of the evolution operator to one characteristic value, the contribution of all wavelets at one scale i to the sum of the absolute values of the wavelet coefficients for another scale j. To do this we first set to zero all coefficients of the evolution matrix outside the block (i/j), then multiply it by $|U|$ and sum all the coefficients of the result. We thereby obtain a "matrix of energy transfers between scales." It displays the expected phenomenon: the energy of the biggest scales flows to the smallest ones. Figure 14.7 shows an example of this matrix at one particular time. It is possible to compute this matrix at each time step and to prepare an animation of the result; this has been done, and the data is available by anonymous ftp [W].

The contributions of the different scales can determined from the sign of the histogram in Figure 14.7. The coefficients above the diagonal are negative, while those below the diagonal are positive. The negative superdiagonal elements show that the small scale components draw energy from larger scales. Likewise, the

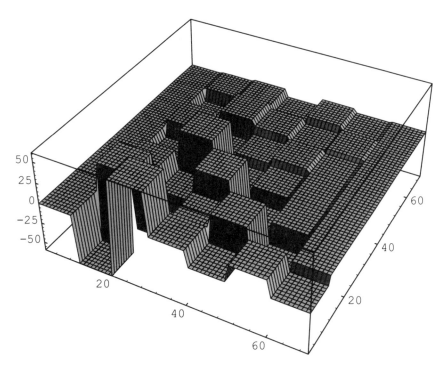

FIGURE 14.7
Matrix of interactions between scales at time equals 0.2.

positive subdiagonal coefficients indicate that big-scale components contribute energy to the creation of small-scale components. The structure of this energy transfer matrix changes rapidly between $t = 0.00$ and $t = 0.15$, and then remains stable in the form shown above.

If we now want to know how the energy concentrates itself in certain scales, we only have to multiply this matrix by the vector $(1, 1, \ldots, 1, 1)$ to sum over all contributions. We then obtain the results in Figure 14.8. From this collection of snapshots, taken at different times, we can see the derivative evolution of the sum of the absolute values of the wavelet coefficients for each scale: the scale whose energy increases fastest changes with time, from scale 3 (where there are 4 coefficients) at time $t = 0.00$ to scale 5 (where there are 16 coefficients) at time $t = 0.70$. Note that the scale graduations are shifted by one. The energy comes from the biggest scales, where it always decreases. The last three pictures show the final phase of the evolution, when the signal collapses. All appears as if Burgers' evolution operator is propagating the energy down through the scales from the biggest ones to the smallest ones.

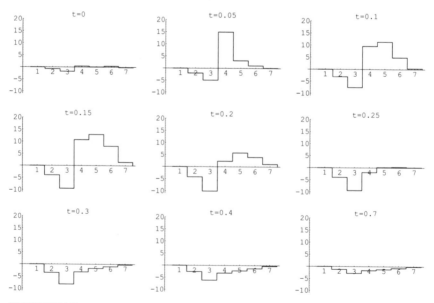

FIGURE 14.8
Time derivative of the "energy" contained in each scale.

14.4 Other examples

We now study a less particular example, one in which there is propagation of the region where the small scales develop. This example is obtained simply by subtracting the constant $1/\sqrt{2}$ from our sine function.

Figure 14.9 shows the propagation we get with this signal. The graphs show the signal at times 0.00, 0.08, 0.16, 0.24, 0.32, 0.50, 0.75 and 0.99. The two inflection points of the signal translate with a slowly varying speed which we may call the *speed of propagation*. Because of this propagation, we observe dyadic artifacts due to the relative motion between the signal and the wavelet centers. For instance, when the function $\sin(2\pi x)$ is translated along the x-axis by amounts which are less than the nominal support of the smallest wavelet, its periodic wavelet transform on $[0, 1]$ changes rather dramatically with each translation step. Such translation produces not only phase shifts in the wavelet coefficients, as is the case with Fourier coefficients, but also rather complicated amplitude variations. The amplitudes at each wavelet scale oscillate several times with a mean period roughly equal to the scale length divided by the propagation speed.

Figure 14.10 shows the evolution of the wavelet coefficients of the signal in Figure 14.9 in gray scale. Notice that the large-amplitude, small-scale wavelets cluster around the signal's region of largest derivatives, which propagates. Note too that the smaller-scale wavelets develop significant amplitudes later and later in the evolution. Compare this picture with Figure 14.4 and observe that the lack of symmetry in the solution does not interfere with the appearance of small-scale wavelets.

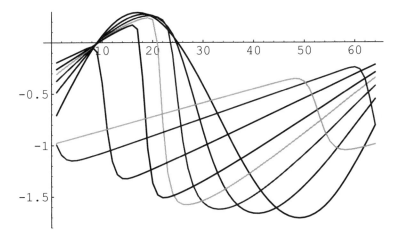

FIGURE 14.9
Evolution of $\sin(2\pi x) - 0.707$, $x \in [0, 1]$.

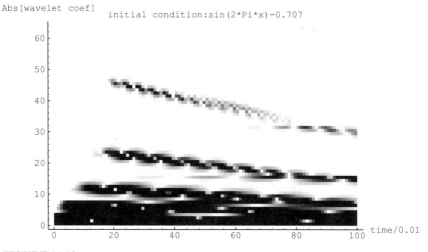

FIGURE 14.10
Time evolution of the absolute value of the wavelet coefficients in the previous figure, in gray scale.

The spatial localization of wavelets is used only to define a mathematically acceptable notion of scale. Since we are interested only in the scale of the energetic components, and not in their spatial location, we can average the absolute value of the wavelet amplitudes at a given scale over all shifts to get a shift-invariant measure of energy in that scale. That is to say, we define the energy in a scale i to be the average over all translates by 1 of the sum of the absolute values of the amplitudes of all wavelets at scale i. We can compute these sums simply by making an average over all the possible shifts of the signal. In fact, it is possible to do this average over two shifts for the smallest scale, four shifts for the next bigger, and so on. We need all the shifts only for the biggest scale; the sum at scale s (with $s = 1$ being the smallest scale) is a 2^s-periodic function of the shift. Such averaging will smooth out most but not all of the oscillations; there will still be some oscillations which are caused by translations on a scale smaller than that of the finest spacing of our dyadic grid. This unavoidable portion of the oscillation phenomenon can be seen on Figure 14.11, showing the same thing as Figure 14.8, but for the shifted sine propagation.

Oscillations make reading the energy difficult. We must determine the amplitude of the oscillations despite their irregularity in order to do as in Figure 14.8. After averaging over shifts we obtain a mean period for the remaining oscillations. The results of this averaging are shown in Figure 14.10. The six first pictures can be interpreted exactly the same way as for $\sin(2\pi x)$, but the three last show that this interpretation breaks down with time. For example, between $t = 0.20$ and

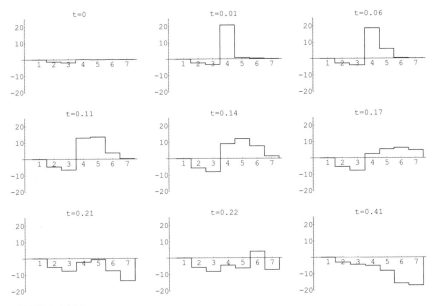

FIGURE 14.11
Time derivative of the "energy" contained in each scale, with averaging over shifts.

$t = 0.21$ there is an important variation of the derivative of the energy located in each scale.

The interaction matrices in Figures 14.12–14.15 are quite interesting, too. They show that as soon as each scale develops enough energy, it positively affects the immediately finer scale. At the beginning, we can also see that some big scales transfer their energy to almost all finer scales. We can also see that after a certain time, the smallest scales contribute negatively to the greater scales.

Although the experiments are not included in this article, we remark that the same type of energy transfer occurs if we start with a Gaussian instead of a sine or a shifted sine function.

14.5 Two dimensions

We have also computed an analogous wavelet decomposition of the energy transfer phenomenon in the two-dimensional case. We started with two bumps, one negative and the other positive, and crashed them into one another much as in the unidimensional case. The evolving signal, sampled on a 64×64 grid over 100 time steps, is displayed in the left-hand parts of Figures 14.16–14.19. The associated 4096×4096 instantaneous evolution operator is far too complicated

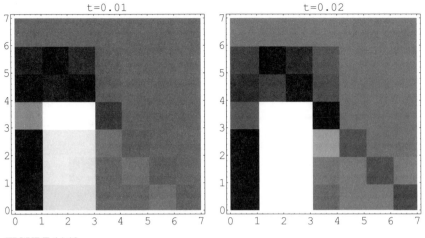

FIGURE 14.12
Matrix of interaction between scales: time steps 1, 2.

to analyze in detail, so we limited ourselves to studying transfers of total energy among the 6 (isotropic) scales in that picture. This gave the much simpler 6×6 energy transfer matrix visible in the right-hand parts of Figures 14.16, 14.17, 14.18, and 14.19. The bidimensional signal exhibits similar variations of its summed absolute wavelet amplitudes as the unidimensional signal.

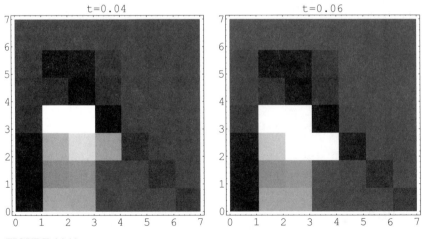

FIGURE 14.13
Matrix of interaction between scales: time steps 4, 6.

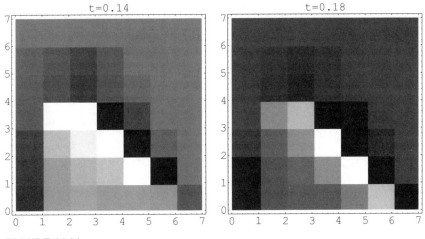

FIGURE 14.14
Matrix of interaction between scales: time steps 14, 18.

14.6   Future directions

These computations make rigorous the notion of energy transfer between scales within a solution of the 1-dimensional Burgers equation. To go further in this direction, we could refine the analysis to determine how energy is transferred between scales at each spatial location. This might introduce some additional

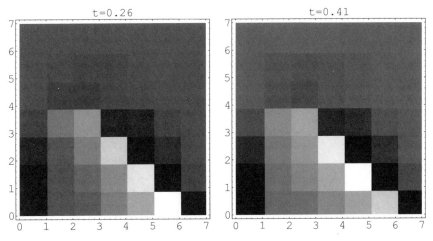

FIGURE 14.15
Matrix of interaction between scales: time steps 26, 41.

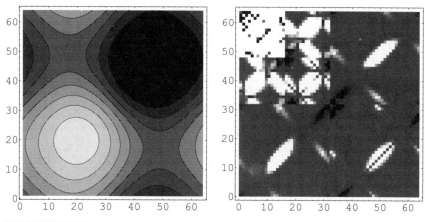

FIGURE 14.16
Signal and time derivative of wavelet coefficient absolute values at $t = 0.10$.

dyadic grid artifacts, such as were already present in the propagating solution analysis. It is thus difficult to describe the instantaneous evolution operator at a fixed time step.

Even the scale interaction matrices are only able to show accurately the average contribution of one scale to another, namely the sum over all wavelets of a given scale.

The few tools we built and used for this brief analysis show how it is possible to use wavelets to measure energy transfers between scales in a sampled solution to a model equation. Even our very simple examples contain some obstacles

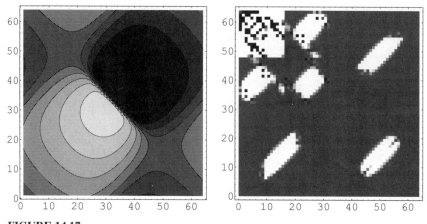

FIGURE 14.17
Signal and time derivative of wavelet coefficient absolute values at $t = 0.50$.

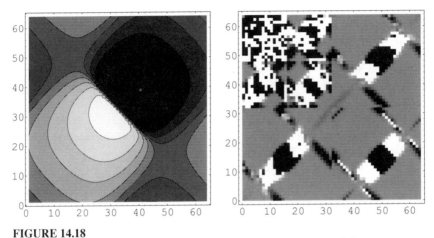

FIGURE 14.18
Signal and time derivative of wavelet coefficient absolute values at $t = 0.60$.

to a straightforward analysis, such as shift artifacts. Nevertheless these first results could be considered rather encouraging. Moreover, considering our very preliminary results, the same wavelet description of the propagation phenomenon seems to be useful even in the two-dimensional case.

Acknowledgments. All computations were performed on an unloaded NeXT workstation, using a combination of *Mathematica* [Woetal] and the *Adapted Waveform Analysis* library of C functions [A].

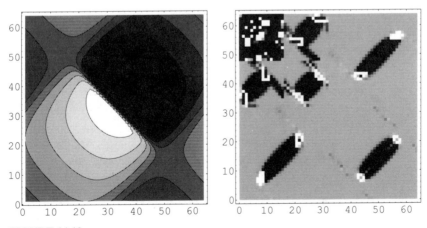

FIGURE 14.19
Signal and time derivative of wavelet coefficient absolute values at $t = 1.00$.

Bibliography

[A]   *Adapted Waveform Analysis Library, Version 2.0.* 1992. Software Documentation. Hamden, Connecticut: Fast Mathematical Algorithms and Hardware Corporation.

[B]   C.K. Batchelor. 1946. *Nature,* **158**: 883. (Contribution to the Turbulence Symposium at the Sixth International Congress for Applied Mathematics, Paris, September, 1946.)

[Beetal] R. Benzi, M. Briscolini, M. Colella, and P. Santangelo. 1991. "A simple point vortex model for two-dimensional decaying turbulence," preprint, Universitá di Roma.

[BeL]   R. Benzi and B. Legras. 1987. "Wave-vortex dynamics," *J. Phys. A* **20**(15): 5125–5144.

[BeyCR]   G. Beylkin, R.R. Coifman, and V. Rokhlin. 1991. "Fast wavelet transforms and numerical algorithms I," *Commun. Pure Appl. Math.* **44**: 141–183.

[Bu]   J. M. Burgers. 1923. *Proc. R. Acad. Sci., Amsterdam,* **26**: 582.

[D]   I. Daubechies. 1988. "Orthonormal bases of compactly supported wavelets," *Commun. Pure Appl. Math.* **41**: 909–996.

[Fetal]   M. Farge, E. Goirand, Y. Meyer, F. Pascal, and M.V. Wickerhauser. 1992. "Improved predictability of two-dimensional turbulent flows using wavelet packet compression," *Fluid Dyn. Res.,* **10**: 229–250.

[K]   A. N. Kolmogorov. 1941. "The local structure of turbulence in incompressible viscous fluids for very large Reynolds numbers," *Dokl. Akad. Nauk SSSR,* **30**: 301–305. In Russian.

[Kr]   R.H. Kraichnan. 1975. "Statistical dynamics of two-dimensional flow," *J. Fluid Mech.,* **67**(part 1): 155–175.

[LT]   J. Liandrat and Ph. Tchamitchian. 1990. "Resolution of the 1-D regularized Burgers equation using a spatial wavelet approximation—algorithm and numerical results," *ICASE report.* (Was also published with V. Perrier in a shorter form in [Retal].)

[Retal]   M. Ruskai, G. Beylkin, R. Coifman, I. Daubechies, S. Mallat, Y. Meyer, and L. Raphael, eds. 1992. *Wavelets and Their Applications,* Boston: Jones and Bartlett.

[W]   Washington University in St. Louis. 1991–present. wuarchive. InterNet Anonymous File Transfer (ftp) Site wuarchive.wustl.edu [128.252.135.4].

[Woetal] T.W. Gray, D. Stein, and S. Wolfram. 1992. *Mathematica: A System for Doing Mathematics by Computer* (Version 2.0 for NeXT computers), Champaign, IL: Wolfram Research Incorporated.

[Z]   N.J. Zabusky. 1984. "Contour dynamics: a method for inviscid and nearly inviscid two-dimensional flows," in *Proceedings of the IUTAM Symposium on Turbulence and Chaotic Phenomena*, T. Tatsumi, ed., Amsterdam: North-Holland, pp. 251–257.

15

The Cauchy singular integral operator and Clifford wavelets

Lars Andersson,[*] **Björn Jawerth,**[†] **and Marius Mitrea**

ABSTRACT We give an elementary, self-contained real-variable proof of the L^2-boundedness of the Cauchy singular integral operator on a Lipschitz surface. The main new feature is the role played by a system of Clifford algebra valued wavelets adapted to the geometry of the surface.

CONTENTS

15.1 Introduction

The Cauchy singular integral operator (CSIO for short) plays a fundamental role in many aspects of harmonic analysis and partial differential equations. For example, the boundedness of the singular double layer potential corresponding

[*]This author was partially supported by NFR and ONR Grant N00014-90-J-1343.
[†]This author was partially supported by ONR Grant N00014-90-J-1343 and DARPA grant AFOSR 89-0455.

to a Lipschitz surface follows from the boundedness of the CSIO on the surface. In fact, by using the rotation method, the boundedness in the general case of a Lipschitz surface can be reduced to the special case of a Lipschitz curve. That the CSIO on a Lipschitz curve is bounded is the celebrated result due to Calderón [C], for small Lipschitz constants, and Coifman–McIntosh–Meyer [CoMMe], in the general case. In [CoJS] Coifman–Jones–Semmes present two elementary proofs of this result, one based on complex-variable arguments and one based on wavelet methods.

It is natural to try to consider the CSIO on a surface directly, without relying on the rotation method. The continuity properties of this operator were apparently first studied by Iftimie [I] (cf. [LMS]) who proved in 1965 that the CSIO on a Lyapunov surface is a bounded operator on Hölder continuous functions. By considering the framework of Clifford algebras, Coifman, Murray, and McIntosh [Mu], [M] established directly the L^2-boundedness of the CSIO on a Lipschitz surface. Very recently, Li, McIntosh, and Semmes [LMS] have extended and given a new proof of this result, following the outline of the complex-variable approach presented in [CoJS].

In this paper we shall follow the second approach in [CoJS] (cf. also [T]) and prove the L^2-boundedness of the CSIO on a Lipschitz surface Σ. The idea is to construct a set of Clifford-valued wavelets by adapting the standard inner product and the standard Haar system for $L^2(\mathbb{R}^n)$ to the geometry of Σ. Then, by relying on Littlewood–Paley theory and well-known arguments, the boundedness will be a consequence of Schur's lemma.

In the final part of the paper we shall briefly indicate how the method we exploit can also be used to give a proof of the Clifford T(b) theorem (see [DJoS] and [MMe]).*

15.2   Clifford algebra rudiments

Here we shall set up the general Clifford algebra framework needed throughout the paper (a much more detailed account can be found in [BDeSo] and in [GMu]).

Let us start by recalling the definition of Clifford algebras over $\mathbb{R}$ (or $\mathbb{C}$). Fix a nonnegative integer n and let $e_0, e_1, \ldots, e_n$ be the standard basis in $\mathbb{R}^{n+1}$ (or $\mathbb{C}^{n+1}$).

DEFINITION 15.1   *The 2^n-dimensional Clifford algebra $\mathbb{R}_{(n)}$ (or $\mathbb{C}_{(n)}$) is the algebra over $\mathbb{R}$ (or $\mathbb{C}$ respectively) freely generated by $e_0, e_1, \ldots, e_n$ subject to the relations:*

   1. e_0 is the multiplicative identity;

---

*Added in proof. After this paper was completed, a paper by Ausher and Tchamitchian [AT] was called to our attention. Their paper discusses wavelets in the context of the Clifford algebra and presents results similar to ours, but our explicit construction differs considerably from their approach.

2. for $1 \leq j, k \leq n$, $e_j e_k + e_k e_j = -2\delta_{jk} e_0 = \begin{cases} 0, & \text{if } j \neq k, \\ -2e_0, & \text{if } j = k. \end{cases}$

In particular, the Clifford algebras $\mathbb{R}_{(0)}$, $\mathbb{R}_{(1)}$, and $\mathbb{R}_{(2)}$ are the real numbers, complex numbers, and quaternions, respectively.

We embed $\mathbb{R}^{n+1}$ in $\mathbb{R}_{(n)}$ (or $\mathbb{C}_{(n)}$) by

$$\mathbb{R}^{n+1} \ni x = (x_0, x_1, \ldots, x_n) \mapsto \sum_{j=0}^{n} x_j e_j \in \mathbb{R}_{(n)} \subset \mathbb{C}_{(n)}.$$

The image of $\mathbb{R}^{n+1}$ under this embedding is called *the set of Clifford numbers* in $\mathbb{R}_{(n)}$ (or $\mathbb{C}_{(n)}$).

For $A \subseteq \{1, 2, \ldots, n\}$, $A = \{i_1 < i_2 < \cdots < i_k\}$, we set $e_A := e_{i_1} e_{i_2} \cdots e_{i_k}$. We also write $e_\emptyset = e_0 = 1$. Clearly $\{e_A\}_{A \subseteq \{1,2,\ldots,n\}}$ is a basis for $\mathbb{R}_{(n)}$ and $\mathbb{C}_{(n)}$, therefore each element x has a unique representation $x = \sum_A x_A e_A$. We shall often write the term $x_\emptyset e_\emptyset$ as $x_0 e_0$ or simply x_0, and refer to x_0 as the scalar part of x, i.e., $x_0 =: \operatorname{Re} x$. Note that the product of two elements $x = \sum_A x_A e_A$ and $y = \sum_B y_B e_B$ is $xy = \sum_{A,B} x_A y_B e_A e_B$.

If we want the above mentioned basis in $\mathbb{R}_{(n)}$ (or $\mathbb{C}_{(n)}$) to be orthonormal, then we are lead to the inner product

$$\langle x, y \rangle := \sum_A x_A \overline{y_A}$$

and the Euclidean norm $|x| := (\sum_A |x_A|^2)^{1/2}$. With this norm, $\mathbb{R}_{(n)}$ and $\mathbb{C}_{(n)}$ become normed algebras since, e.g., $|xy| \leq 2^{n-1} |x||y|$, for all $x, y \in \mathbb{R}_{(n)}$.

The conjugation $\bar{}$ is an involution of $\mathbb{R}_{(n)}$ (or $\mathbb{C}_{(n)}$) defined by

$$\bar{x} := \sum_A \overline{x_A} \, \overline{e_A},$$

where $\overline{e_A} := \pm e_A$ with the sign uniquely determined by imposing $e_A \overline{e_A} = \overline{e_A} e_A = 1$. More precisely $\overline{e_A} = (-1)^{|A|(|A|+1)/2} e_A$, with $|A|$ standing for the cardinality of $A \subseteq \{1, 2, \ldots, n\}$.

Note that $\overline{x} \overline{y} = \overline{y} \overline{x}$ and that $(x, y) = \operatorname{Re} \overline{y} x$. Also, for $x \in \mathbb{R}^{n+1} \subseteq \mathbb{R}_{(n)}$, we have $x\bar{x} = |x|^2$ and, as a consequence, any nonzero element x in $\mathbb{R}^{n+1}$ has a multiplicative inverse given by $x^{-1} = |x|^{-2} \bar{x}$ (in general this is false in $\mathbb{C}^{n+1}$; for example, $(e_0 + ie_1)(e_0 - ie_1) = 0$). The multiplicative group generated by the set of all nonzero Clifford numbers in $\mathbb{R}^n$ is called the *Clifford group*. It is also worth noting that $|xy| = |x| |y|$ if x is an element in the Clifford group and y an element in $\mathbb{C}_{(n)}$.

Now, consider an open set Ω in $\mathbb{R}^{n+1}$ and a Clifford-valued function $f : \Omega \to \mathbb{R}_{(n)}$ (or $\mathbb{C}_{(n)}$), $f = \sum_A f_A e_A$, of class $\mathcal{C}^1$. We define the first-order differential operator

$$D := \sum_{j=0}^{n} \partial_j e_j$$

acting on f from the left by

$$Df := \sum_{j,A} \partial_j f_A e_j e_A.$$

Similarly, the action from the right is given by

$$fD := \sum_{j,A} \partial_j f_A e_A e_j.$$

We say that f is *left monogenic* (*right monogenic*) if $Df = 0$ ($fD = 0$). Note that for $n = 2$, in which case $\mathbb{R}_{(1)} \equiv \mathbb{C}$, D becomes the Cauchy–Riemann operator $\partial_0 + i\partial_1$. In other words, f is left monogenic (or right monogenic) if and only if f is holomorphic.

Next we shall briefly discuss the natural extension of the usual Cauchy's theorem and Cauchy's integral formula to the Clifford algebra setting. We consider a Lipschitz domain Ω in $\mathbb{R}^{n+1}$ (i.e., a domain whose boundary is locally the graph of a Lipschitz function after an appropriate rotation of the coordinates). We denote by $n(X)$ the exterior unit normal, which is defined a.e. for $X \in \partial\Omega$ with respect to the surface measure dS.

PROPOSITION 15.2
With the above notations, if Ω is a bounded Lipschitz domain, then any left monogenic function f in a neighborhood of $\overline{\Omega}$ has the reproducing property

$$\frac{1}{\sigma_n} \int_{\partial\Omega} \frac{\overline{Y - X}}{|Y - X|^{n+1}} n(Y) f(Y) \, dS(Y) = \begin{cases} f(X) & \text{for } X \in \Omega, \\ 0 & \text{for } X \notin \overline{\Omega}, \end{cases}$$

and f has the cancellation property[*]

$$\int_{\partial\Omega} n(X) f(X) \, dS(X) = 0.$$

PROOF   In the case of a smooth domain Ω the argument is classical, see, e.g., [BDeSo]. The general case follows by a limiting argument as in [N] or [V].   ∎

15.3   The higher-dimensional Cauchy integral

Consider Σ the graph of a Lipschitz function $g : \mathbb{R}^n \to \mathbb{R}$. We embed $\mathbb{R}^{n+1}$ in the 2^n-dimensional Clifford algebra $\mathbb{R}_{(n)}$ (or even in $\mathbb{C}_{(n)}$) in the standard way. In

---

[*]Note that the integrands have to be interpreted in the sense of multiplication in the complex Clifford algebra $\mathbb{C}_{(n)}$. We also remark that there is a version of the theorem for right monogenic functions as well.

particular, by making the identification $\Sigma \ni X = (g(x), x) \mapsto g(x)e_0 + \sum_i x_i e_i \in \mathbb{R}_{(n)}$, we have

$$\Sigma := \{X = z(x); x \in \mathbb{R}^n\} \subset \mathbb{R}^{n+1} \hookrightarrow \mathbb{R}_{(n)},$$

where $z(x) := (g(x), x)$, $x \in \mathbb{R}^n$, is the graph map of g. Note that $z : \mathbb{R}^n \to \Sigma$ is a bi-Lipschitz homeomorphism.

The exterior unit normal $n(X) := N(z^{-1}(X))/|N(z^{-1}(X))|$, with $N(x) := (1, -\nabla g(x))$ is then a well-defined Clifford number for almost all $X \in \Sigma$.

The higher-dimensional Cauchy singular integral operator (CSIO) corresponding to Σ is defined in the principal value sense as

$$\mathcal{C}f(X) := \text{p.v.} \frac{2}{\sigma_n} \int_\Sigma f(Y)n(Y)\frac{\overline{Y - X}}{|Y - X|^{n+1}} \, dS(Y), \quad X \in \Sigma,$$

where f is a Clifford algebra valued function on Σ. Here σ_n is the area of the unit sphere in $\mathbb{R}^{n+1}$ and dS is the canonical surface measure on Σ. Note that the integrand has to be interpreted in the sense of multiplication in the complex Clifford algebra $\mathbb{C}_{(n)}$.

Also, in local coordinates the operator becomes

$$\mathcal{C}f(x) := \text{p.v.} \frac{2}{\sigma_n} \int_{\mathbb{R}^n} f(y)N(y)\frac{\overline{z(y) - z(x)}}{|z(y) - z(x)|^{n+1}} \, dy, \quad x \in \mathbb{R}^n,$$

where f is a $\mathbb{C}_{(n)}$-valued function on $\mathbb{R}^n$.

Our goal is to show that $\mathcal{C}$ is a bounded operator on $L^2(\mathbb{R}^n)_{(n)} := L^2(\mathbb{R}^n) \otimes \mathbb{C}_{(n)}$, the Clifford algebra version of $L^2(\mathbb{R}^n)$.

For complex-valued functions f on Σ we have $\text{Re}\,\mathcal{C}f = \mathcal{D}f$, where

$$\mathcal{D}f(X) := \text{p.v.} \frac{2}{\sigma_n} \int_\Sigma \frac{\langle n(Y), Y - X \rangle}{|Y - X|^{n+1}} f(Y)\,dS(Y), \quad X \in \Sigma,$$

is the (singular) double layer potential operator on Σ. Hence, the boundedness of $\mathcal{C}$ immediately yields that $\mathcal{D}$ is bounded as well. This is important for example when solving the Dirichlet problem on a Lipschitz domain using the method of layer potentials. Let us make a brief digression to indicate why.

Fix an unbounded Lipschitz domain Ω in $\mathbb{R}^{n+1}$ and consider the Dirichlet problem for Δ, the Laplacean in $n + 1$ coordinates, in Ω:

$$(D) \begin{cases} \Delta u = 0 \text{ in } \Omega, \\ u^* \in L^2(\partial\Omega, dS), \\ u|_{\partial\Omega} = f \in L^2(\partial\Omega, dS), \end{cases}$$

where * is the usual nontangential maximal operator; see e.g., [V]. The basic idea behind the method of layer potentials is to look for the solution expressed as layer potential extensions of a certain function defined on the boundary. More precisely, the goal is to show that the solution is given by the formula

$$u(X) := \frac{2}{\sigma_n} \int_\Sigma \frac{\langle X - Y, n(Y) \rangle}{|X - Y|^{n+1}} [(I + \mathcal{D})^{-1} f](Y) dS(Y), \quad X \in \Omega.$$

For this approach to work we clearly need to establish two facts, namely that $\mathcal{D}$ is bounded on L^2 and that $I + \mathcal{D}$ is invertible. In this paper we are thus discussing a simple way of establishing the first of these facts. The second is due to Verchota [V]. We note that using double layer potentials in this fashion is not the only way to solve the Dirichlet problem on a Lipschitz domain. In fact, historically another method was first used. The solvability of the Dirichlet problem had been a longstanding open problem when in 1977, Dahlberg used very precise estimates for the Green's function to establish this fact. Subsequently, the theory developed rapidly through the works of Dahlberg, Kenig, Fabes, Jerison, Verchota, and others; we refer to [V] for a more complete historical account and references.

Let us return to our main theme, and next explain some of the notational conventions we will follow. For positive functions $F(s)$ and $G(s)$, $s \in S$, we write $F \ll G$ if there exists a constant $c > 0$, depending only on n and $\|\nabla g\|_\infty$, such that $F(s) \leq cG(s)$ for all $s \in S$. Also, $F \sim G$ stands for $F \ll G$ and $G \ll F$.

We let $\mathcal{F}$ denote the collection of all dyadic cubes of $\mathbb{R}^n$:

$$Q = Q_{k,v} = \{x \in \mathbb{R}^n \; 2^{-k} v_i \leq x_i \leq 2^{-k}(v_i + 1), \; i = 1, 2, \ldots, n\},$$

for $k \in \mathbb{Z}$ and $v \in \mathbb{Z}^n$. Each dyadic cube Q has 2^n "children":

$$\{Q^i\}_{j=1}^{2^n} := \{Q' \in \mathcal{F} \; l(Q') = \frac{1}{2} l(Q), \; Q' \subset Q\},$$

where $l(Q)$ is the side length of Q. We also set $\mathcal{F}_k := \{Q \in \mathcal{F}; \; l(Q) = 2^{-k}\}$.

Next, consider $m(Q) := |Q|^{-1} \int_Q N(x) dx$, $|Q|$ being the volume of Q. Since $N(x)$ is a Clifford number with $\operatorname{Re} \check{N}(x) = 1$ and $|N(x)| \sim 1$, we have that $m(Q)$ is a Clifford number as well with $\operatorname{Re} m(Q) = 1$ and $|m(Q)| \sim 1$. We let $\langle \cdot, \cdot \rangle_\Sigma$ be the "bilinear" form

$$\langle f_1, f_2 \rangle_\Sigma := \int_{\mathbb{R}^n} f_1(x) N(x) f_2(x) dx \tag{15.1}$$

defined for $\mathbb{C}_{(n)}$-valued functions f_1 and f_2.

15.4   Square function estimates

We first construct a modified Haar system which behaves like an (bi)orthogonal set of functions with respect to the form (15.1). (This construction might be carried out in greater generality, see [Mi1], but we shall confine ourselves to the particular context relevant here.) For $Q \in \mathcal{F}$ and $i = 1, 2, \ldots, 2^n - 1$ we define the

$\mathbb{C}_{(n)}$-valued functions $\{\beta^L_{Q,i}\}_{Q,i}$ and $\{\beta^R_{Q,i}\}_{Q,i}$ by

$$\beta^L_{Q,i} := 2^{n/2}|Q|^{-1/2}M(Q,i)\left\{i^{-1}m\left(\bigcup_{\nu=1}^{i}Q^\nu\right)^{-1}\left(\sum_{\nu=1}^{i}\chi_{Q^\nu}\right) - m(Q^{i+1})^{-1}\chi_{Q^{i+1}}\right\},$$

and

$$\beta^R_{Q,i} := 2^{n/2}|Q|^{-1/2}\left\{i^{-1}m\left(\bigcup_{\nu=1}^{i}Q^\nu\right)^{-1}\left(\sum_{\nu=1}^{i}\chi_{Q^\nu}\right) - m(Q^{i+1})^{-1}\chi_{Q^{i+1}}\right\}M(Q,i),$$

where $M(Q,i) := (\{i^{-1}m(\bigcup_{\nu=1}^{i}Q^\nu)^{-1} + m(Q^{i+1})^{-1}\}^{-1})^{1/2}$. That the square root exists is guaranteed by the properties of the integral mean of $N(x)$ and the following.

LEMMA 15.3
For any Clifford number λ there exists a Clifford number μ with $\mu^2 = \lambda$.

See [Mi1] for a proof of a stronger result. The present case is a simple consequence of the observation that if μ is a Clifford number, then μ^2 is also a Clifford number.

The basic properties of the functions $\{\beta^L_{Q,i}\}_{Q,i}$ and $\{\beta^R_{Q,j}\}_{Q,j}$ are collected in the next proposition.

PROPOSITION 15.4
The two families of $\mathbb{C}_{(n)}$-valued functions $\{\beta^L_{Q,i}\}_{Q,i}$, $\{\beta^R_{Q,j}\}_{Q,j}$ for $Q \in \mathcal{F}$ and $i,j = 1,\ldots,2^n - 1$, satisfy the following.

1. $\operatorname{supp}\beta^L_{Q,i}$, $\operatorname{supp}\beta^R_{Q,j} \subseteq Q$ and $\beta^L_{Q,i}$, $\beta^R_{Q,j}$ are constant on each "child" of Q.

2. $|\beta^L_{Q,i}|$, $|\beta^R_{Q,j}| \ll |Q|^{-1/2}$.

3. $\langle \beta^L_{Q,i}, 1 \rangle_\Sigma = 0$ and $\langle 1, \beta^R_{Q,j} \rangle_\Sigma = 0$.

4. $\langle \beta^L_{Q,i}, \beta^R_{Q',j} \rangle_\Sigma = \delta_{QQ'}\delta_{ij}e_0$.

5. $\|\sum_{i=1}^{2^n-1}\alpha_i\beta^L_{Q,i}\|_2^2 \sim \sum_{i=1}^{2^n-1}|\alpha_i|^2 \sim \|\sum_{j=1}^{2^n-1}\beta^R_{Q,j}\alpha_j\|_2^2$, $\alpha_i \in \mathbb{C}_{(n)}$, $1 \leq i \leq 2^n - 1$.

6. *Consider $f = \sum_{i=1}^{2^n}\alpha_i\chi_{Q^i}$ with $\alpha_i \in \mathbb{C}_{(n)}$, $i = 1, 2, \ldots, 2^n$. If $\langle f, 1 \rangle_\Sigma = 0$, then*

$$f = \sum_{i=1}^{2^n-1}\langle f, \beta^R_{Q,i}\rangle_\Sigma \beta^L_{Q,i};$$

and, if $\langle 1, f \rangle_\Sigma = 0$, then

$$f = \sum_{j=1}^{2^n-1}\beta^R_{Q,j}\langle \beta^L_{Q,j}, f\rangle_\Sigma.$$

PROOF   The properties (1)-(3) are direct consequences of the way the βs were constructed. Next, note that unless $\beta_{Q,i}^L$ or $\beta_{Q',j}^R$ have the same pair of subscripts, one of them is constant on the support of the other one. Hence, (4) follows from the vanishing moment property in (3).

For (5), we have that

$$\sum_j |\alpha_j| = \sum_j \left| \left\langle \sum_i \alpha_i \beta_{Q,i}^L, \beta_{Q,j}^R \right\rangle_\Sigma \right| \ll \left\| \sum_i \alpha_i \beta_{Q,i}^L \right\|_2 \sum_j \|\beta_{Q,j}^R\|_2$$

$$\ll \left\| \sum_i \alpha_i \beta_{Q,i}^L \right\|_2 \ll \sum_i |\alpha_i| \|\beta_{Q,i}^L\|_2 \ll \sum_i |\alpha_i|,$$

since $\|\beta_{Q,i}^L\|_2, \|\beta_{Q,j}^R\|_2 \ll 1$. Since there are only finitely many α_is, the ℓ^2-sum is comparable with the ℓ^1-sum, so this proves (5).

Using the explicit expressions of the β^Ls, we readily see that χ_{Q^2} is spanned by χ_{Q^1} and $\beta_{Q,1}^L$ in the set of $\mathbb{C}_{(n)}$-valued functions defined on $\mathbb{R}^n$ with its natural structure as a left Clifford module. Continuing this inductively yields that any characteristic function χ_{Q^i} is spanned by χ_{Q^1} and $\beta_{Q,1}^L, \ldots, \beta_{Q,2^n-1}^L$. Now, if $f = \lambda_1 \chi_{Q^1} + \sum_i \gamma_i \beta_{Q,i}^L$, then the fact that $\langle f, 1 \rangle_\Sigma = 0$ implies that $\lambda_1 = 0$. Hence, by (3), it follows that

$$\gamma_j = \langle f, \beta_{Q,j}^R \rangle_\Sigma, \qquad j = 1, 2, \ldots, 2^n - 1$$

and we are done. ∎

We define the "conditional expectation" operators E_k^L and E_k^R by

$$E_k^L f(x) := |Q|^{-1} \left(\int_Q f(y) N(y)\, dy \right) m(Q)^{-1}, \quad \text{if } x \in Q \in \mathcal{F}_k,$$

and

$$E_k^R f(x) := |Q|^{-1} m(Q)^{-1} \left(\int_Q N(y) f(y)\, dy \right), \quad \text{if } x \in Q \in \mathcal{F}_k,$$

respectively. The usual argument shows that if $f \in L^2(\mathbb{R}^n)_{(n)}$, then

$$E_k^L f \to f \text{ as } k \to +\infty, \quad \text{and} \quad E_k^L f \to 0 \text{ as } k \to -\infty.$$

A similar result is true for E_k^R as well.

We also set

$$\Delta_k^L f := \sum_{Q \in \mathcal{F}_k} \sum_{i=1}^{2^n-1} \langle f, \beta_{Q,i}^R \rangle_\Sigma \beta_{Q,i}^L,$$

and

$$\Delta_k^R f := \sum_{Q \in \mathcal{F}_k} \sum_{i=1}^{2^n-1} \beta_{Q,i}^R \langle \beta_{Q,i}^L, f \rangle_\Sigma.$$

The relation between these operators and E_k^L and E_k^R is the following.

PROPOSITION 15.5
We have that $\Delta_k^L = E_{k+1}^L - E_k^L$ *and* $\Delta_k^R = E_{k+1}^R - E_k^R$.

PROOF By restricting attention to one dyadic cube $Q \in \mathcal{F}_k$ at a time, we easily see that $\langle E_{k+1}^L f - E_k^L f, 1 \rangle_\Sigma = 0$. Also, note that $E_{k+1}^L f - E_k^L f$ is constant on each dyadic subcube of Q and that $E_k^L f$ is constant on Q. Thus, by Proposition 15.4 (6) and (3), we need to prove that $\langle E_{k+1}^L f, \beta_{Q,j}^R \rangle_\Sigma = \langle f, \beta_{Q,j}^R \rangle_\Sigma$.

However, if $\beta_{Q,j}^R = \sum_i \lambda_i^j \chi_{Q^i}$ and $E_{k+1}^L f = \sum_i f_{Q^i} m(Q^i)^{-1} \chi_{Q^i}$ where we set $f_{Q^i} := |Q^i|^{-1} \int_{Q^i} f N$, then

$$\langle E_{k+1}^L f, \beta_{Q,j}^R \rangle_\Sigma = \sum_i |Q^i| f_{Q^i} \lambda_i^j = \sum_i \int_{Q^i} f N \lambda_i^j \, dx = \langle f, \beta_{Q,j}^R \rangle_\Sigma. \quad \blacksquare$$

We immediately get the following corollary.

COROLLARY 15.6
For each $f \in L^2(\mathbb{R}^n)_{(n)}$,

$$f = \sum_k \Delta_k^L f = \sum_{Q \in \mathcal{F}} \sum_{i=1}^{2^n - 1} \langle f, \beta_{Q,i}^R \rangle_\Sigma \beta_{Q,i}^L,$$

and

$$f = \sum_k \Delta_k^R f = \sum_{Q \in \mathcal{F}} \sum_{i=1}^{2^n - 1} \beta_{Q,i}^R \langle f, \beta_{Q,i}^L \rangle_\Sigma.$$

The next result shows that in fact the βs behave much like an orthonormal basis for $L^2(\mathbb{R}^n)_{(n)}$.

THEOREM 15.7
If $f \in L^2(\mathbb{R}^n)_{(n)}$ *then*

$$\|f\|_2^2 \sim \sum_{Q \in \mathcal{F}} \sum_{i=1}^{2^n - 1} |\langle f, \beta_{Q,i}^R \rangle_\Sigma|^2 \sim \sum_{Q \in \mathcal{F}} \sum_{i=1}^{2^n - 1} |\langle \beta_{Q,i}^L, f \rangle_\Sigma|^2.$$

Before we prove this, we consider the case $\Sigma \equiv \mathbb{R}^n$, i.e., $g(x) = 0$, and, consequently, $N(x) = (1, 0, \ldots, 0)$. We set

$$h_Q^i := 2^{n/2} |Q|^{-1/2} \left(\frac{i}{i+1} \right)^{1/2} \left\{ \frac{1}{i} \sum_{\nu=1}^i \chi_{Q^\nu} - \chi_{Q^{i+1}} \right\}, \quad i = 1, \ldots, 2^n - 1, \ Q \in \mathcal{F}.$$

The family $\{h_Q^i\}_{Q,i}$ is easily checked to be an orthonormal basis for $L^2(\mathbb{R}^n)$ with the standard inner product $(\cdot, \cdot)$. In this special case, we denote Δ_k^L (or Δ_k^R) and

E_k^L (or E_k^R) by Δ_k and E_k, respectively. As in the general case, $\Delta_k = E_{k+1} - E_k$. Since

$$\Delta_k f = \sum_{Q \in \mathcal{F}_k} \sum_{i=1}^{2^n - 1} \langle f, h_Q^i \rangle h_Q^i,$$

we have that

$$\sum_{k=-\infty}^{+\infty} \int_{\mathbb{R}^n} |\Delta_k f|^2 \, dx = \|f\|_2^2, \quad f \in L^2(\mathbb{R}^n)_{(n)}. \tag{15.2}$$

PROOF OF THEOREM 15.7   Note that $E_k^L f = E_k(fN)E_k(N)^{-1}$. Hence,

$$
\begin{aligned}
|\Delta_k^L f| = |E_{k+1}^L f - E_k^L f| &= |E_{k+1}(fN)E_{k+1}(N)^{-1} - E_k(fN)E_k(N)^{-1}| \\
&\le |E_{k+1}(fN)E_{k+1}(N)^{-1} - E_k(fN)E_{k+1}(N)^{-1}| \\
&\quad + |E_k(fN)E_{k+1}(N)^{-1} - E_k(fN)E_k(N)^{-1}| \\
&\ll |\Delta_k(fN)| + |E_k(fN)||E_{k+1}(N)^{-1} - E_k(N)^{-1}| \\
&\ll |\Delta_k(fN)| + |E_k(fN)||\Delta_k(N)|,
\end{aligned}
\tag{15.3}
$$

since $|E_k(N)| \sim 1$. Now, by definition,

$$\Delta_k^L f = \sum_{Q \in \mathcal{F}_k} \sum_{i=1}^{2^n - 1} \langle f, \beta_{Q,i}^R \rangle_\Sigma \beta_{Q,i}^L.$$

Using that dyadic cubes of a fixed size are pairwise disjoint and that $\operatorname{supp} \beta_{Q,i}^L \subset Q$ we see that

$$\int_{\mathbb{R}^n} |\Delta_k^L f(x)|^2 \, dx = \sum_{Q \in \mathcal{F}_k} \| \sum_{i=1}^{2^n-1} \langle f, \beta_{Q,i}^R \rangle_\Sigma \beta_{Q,i}^L \|_2^2 \sim \sum_{Q \in \mathcal{F}_k} \sum_{i=1}^{2^n-1} |\langle f, \beta_{Q,i}^R \rangle_\Sigma|^2 \tag{15.4}$$

by (5) of Proposition 15.4. In order to prove that

$$\sum_{k=-\infty}^{+\infty} \int_{\mathbb{R}^n} |\Delta_k^L f|^2 \, dx \ll \|f\|_2^2,$$

(15.2), (15.3), and (15.4) show that it is enough to prove that

$$\sum_{k=-\infty}^{+\infty} \int_{\mathbb{R}^n} |E_k(fN)|^2 |\Delta_k(N)|^2 \, dx \ll \|f\|_2^2. \tag{15.5}$$

The proof of this is more or less standard; we shall follow Christ [Ch] with minor alterations. We make the following definition.

DEFINITION 15.8 *We call a sequence of positive, measurable functions on $\mathbb{R}^n$, $\omega = \{\omega_k\}_k$, Carleson if*

$$\|\omega\|_C := \sup_{Q \in \mathcal{F}} |Q|^{-1} \int_Q \left\{ \sum_{\{k \in \mathbb{Z}; 2^{-k} \leq l(Q)\}} \omega_k(x)\, dx \right\} < +\infty.$$

To complete the proof of Theorem 15.7 we need the following two (essentially) well known lemmas.

LEMMA 15.9
For each $b \in BMO(\mathbb{R}^n)_{(n)}$, $\{|\Delta_k(b)|^2\}_k$ is Carleson with norm $\ll \|b\|_{BMO}^2$.

LEMMA 15.10
Suppose $\omega = \{\omega_k\}_k$ is Carleson. Then, for $p \in (1, +\infty)$,

$$\int_{\mathbb{R}^n} \sum_k |E_k f(x)|^p \omega_k(x)\, dx \ll \|\omega\|_C \|f\|_p^p.$$

Accepting these results for a moment and using the fact that $N \in L^\infty(\mathbb{R}^n)_{(n)} \subset BMO(\mathbb{R}^n)_{(n)}$, the inequality (15.5) immediately follows. Hence, we have shown that

$$\sum_{Q \in \mathcal{F}} \sum_{i=1}^{2^n-1} |\langle f, \beta_{Q,i}^R \rangle_\Sigma|^2 \ll \|f\|_2^2, \tag{15.6}$$

and a similar argument also yields

$$\sum_{Q \in \mathcal{F}} \sum_{i=1}^{2^n-1} |\langle \beta_{Q,i}^L, f \rangle_\Sigma|^2 \ll \|f\|_2^2. \tag{15.7}$$

Having established (15.6) and (15.7), the converse inequalities are proved by a kind of polarization argument. Let $f \in L^2(\mathbb{R}^n)_{(n)}$ with $\|f\|_2 = 1$. If $\{f_A\}_A$ are the components of f with respect to the standard basis $\{e_A\}_{A \subseteq \{1,\ldots,n\}}$ in $\mathbb{C}_{(n)}$ and we set $g_A := N^{-1}\overline{f_A} e_A$ for each $A \subseteq \{1, 2, \ldots, n\}$, then we have that $\|g_A\|_2 \ll 1$ for each A. Furthermore,

$$1 = \|f\|_2^2 = \sum_A \int |f_A|^2 = \sum_A |\text{Re}\langle f, N^{-1}\overline{f_A} e_A \rangle_\Sigma| \ll \sum_A |\langle f, g_A \rangle_\Sigma|$$

$$\ll \sum_A \left| \sum_{Q \in \mathcal{F}} \sum_{i=1}^{2^n-1} \langle f, \beta_{Q,i}^R \rangle_\Sigma \langle \beta_{Q,i}^L, g_A \rangle_\Sigma \right|$$

$$\ll \left(\sum_{Q \in \mathcal{F}} \sum_{i=1}^{2^n-1} |\langle f, \beta_{Q,i}^R \rangle_\Sigma|^2 \right)^{1/2} \sum_A \left(\sum_{Q \in \mathcal{F}} \sum_{i=1}^{2^n-1} |\langle \beta_{Q,i}^L, g_A \rangle_\Sigma|^2 \right)^{1/2}$$

$$\ll \left(\sum_{Q\in\mathcal{F}}\sum_{i=1}^{2^n-1}|\langle f,\beta_{Q,i}^R\rangle_\Sigma|^2\right)^{1/2}\left(\sum_A\|g_A\|_2\right)$$

$$\ll \left(\sum_{Q\in\mathcal{F}}\sum_{i=1}^{2^n-1}|\langle f,\beta_{Q,i}^R\rangle_\Sigma|^2\right)^{1/2}.$$

The proof of the remaining inequality is essentially the same, and this completes the proof of Theorem 15.7.   ∎

For the reader's convenience, we shall include the proofs of Lemmas 15.9 and 15.10.

PROOF OF LEMMA 15.9   We need to estimate

$$|Q|^{-1}\int_Q\left(\sum_{\{k\in\mathbb{Z};\,2^{-k}\le l(Q)\}}|\Delta_k(b)|^2\,dx\right)$$

for an arbitrary, fixed dyadic cube Q. Since Δ_k annihilates constants, we may suppose that b_Q, the integral mean of b over Q, is in fact zero. Write $b \in BMO(\mathbb{R}^n)_{(n)}$, as $b = b_0 + b_\infty$ with $b_0 := \chi_Q b$ and $b_\infty := \chi_{\mathbb{R}^n\setminus Q}b$. Then, by (15.2) and the John–Nirenberg inequality,

$$|Q|^{-1}\int_Q\sum_{2^{-k}\le l(Q)}|\Delta_k(b_0)(x)|^2\,dx \le |Q|^{-1}\int_{\mathbb{R}^n}\sum_{k\in\mathbb{Z}}|\Delta_k(b_0)(x)|^2\,dx$$

$$= |Q|^{-1}\|b_0\|_2^2 = |Q|^{-1}\int_Q|b|^2\,dx = |Q|^{-1}\int_Q|b-b_Q|^2\,dx \ll \|b\|_{\mathrm{BMO}}^2.$$

Next we want to estimate $|\Delta_k(b_\infty)(x)|$ for a fixed $x \in Q$ and $l(Q) \le 2^{-k}$. We note that there are unique dyadic cubes $Q' \in \mathcal{F}_k$ and $Q'' \in \mathcal{F}_{k+1}$ which contain x, and these cubes must be contained in Q. Since $b_\infty \equiv 0$ on Q we have

$$\Delta_k b_\infty(x) = |Q''|^{-1}\int_{Q''}b_\infty\,dx - |Q'|^{-1}\int_{Q'}b_\infty\,dx = 0,$$

and the proof is complete.   ∎

PROOF OF LEMMA 15.10   We have

$$\int_{\mathbb{R}^n}\sum_k|E_kf(x)|^p\omega_k(x)\,dx = p\int_0^{+\infty}\sum_k\omega_k(\{x;\,|E_kf(x)|>\lambda\})\lambda^{p-1}\,d\lambda. \quad (15.8)$$

For each fixed λ we choose a maximal collection $\{Q_j\}$ of pairwise disjoint dyadic cubes among the dyadic cubes Q such that

$$\left||Q|^{-1}\int_Q f\,dx\right| > \lambda.$$

If we let $Mf(x) := \sup_{k \in \mathbb{Z}} |E_k f(x)|$, then clearly $|\{x; |Mf(x)| > \lambda\}| = \sum_j |Q_j|$. Furthermore, if $\|Q\|^{-1} \int_Q f \, dx > \lambda$ for some $Q \in \mathcal{F}_k$, then, by the maximality of the Q_js, Q is contained in exactly one of the Q_js. As a consequence,

$$\sum_k \omega_k(\{x; |E_k f(x)| > \lambda\}) \ll \sum_j \sum_{\{k; 2^{-k} \le l(Q_j)\}} \omega_k(Q_j) \ll \|\omega\|_c \sum_j |Q_j|$$

$$\ll \|\omega\|_c |\{x; |Mf(x)| > \lambda\}|.$$

Inserting this in (15.8), the maximal theorem shows that

$$\int_{\mathbb{R}^n} \sum_k |E_k f(x)|^p \omega_k(x) \, dx \ll \|\omega\|_c \|Mf\|_p^p \ll \|\omega\|_c \|f\|_p^p. \quad \blacksquare$$

15.5 Schur's lemma

For a fixed $\delta > 0$ and for $f \in L^2(\mathbb{R}^n)_{(n)}$ we define the operators T_δ^L and T_δ^R by

$$T_\delta^L f(x) := \frac{1}{\sigma_n} \int_{\mathbb{R}^n} f(y) N(y) \frac{\overline{z(y) - z(x) - \delta e_0}}{|z(y) - z(x) - \delta e_0|^{n+1}} \, dy,$$

and

$$T_\delta^R f(x) := \frac{1}{\sigma_n} \int_{\mathbb{R}^n} \frac{\overline{z(y) - z(x) + \delta e_0}}{|z(y) - z(x) + \delta e_0|^{n+1}} N(y) f(y) \, dy,$$

respectively. Obviously, T_δ^L and T_δ^R are both bounded operators from L^2 to L^∞.

The main estimate in the proof of the L^2-boundedness of the CSIO is contained in the next theorem.

THEOREM 15.11
For each fixed $\epsilon > 0$ we let $w(Q) := |Q|^{1/2 - \epsilon}$ where Q runs over the set of all dyadic cubes in $\mathbb{R}^n$. Then for $1/n > \epsilon > 0$ we have

$$\sup_{Q,i} w(Q)^{-1} \sum_{Q'} \sum_{j=1}^{2^n - 1} w(Q') |\langle T_\delta^L \beta_{Q,i}^L, \beta_{Q',j}^R \rangle_\Sigma| \le C < +\infty, \quad (15.9)$$

$$\sup_{Q',j} w(Q')^{-1} \sum_{Q} \sum_{i=1}^{2^n - 1} w(Q) |\langle T_\delta^L \beta_{Q,i}^L, \beta_{Q',j}^R \rangle_\Sigma| \le C < +\infty, \quad (15.10)$$

with C independent of δ. Similar inequalities are also true for T_δ^R.

We shall postpone the proof of this for a moment and first derive some of its consequences.

An application of (a version) of Schur's test immediately yields the following.

COROLLARY 15.12

The operators T_δ^L and T_δ^R are bounded on $L^2(\mathbb{R}^n)_{(n)}$ with bounds independent of δ.

The operators T_δ^L and T_δ^R are closely related to $\mathcal{C}$, the CSIO. For a fixed $\delta > 0$ let $\mathcal{C}_\delta^L$ and $\mathcal{C}_\delta^R$ denote truncated versions of the (left and right) CSIO:

$$\mathcal{C}_\delta^L f(X) := \frac{2}{\sigma_n} \int_{\substack{|X-Y|\geq\delta \\ Y\in\Sigma}} f(Y)n(Y) \frac{\overline{Y-X}}{|Y-X|^{n+1}}\, dS(Y), \quad X \in \Sigma,$$

$$\mathcal{C}_\delta^R f(X) := \frac{2}{\sigma_n} \int_{\substack{|X-Y|\geq\delta \\ Y\in\Sigma}} \frac{\overline{Y-X}}{|Y-X|^{n+1}} n(Y)f(Y)\, dS(Y), \quad X \in \Sigma.$$

Simple, direct computations involving the Cauchy kernel $\bar{X}|X|^{-n-1}$ show that

$$|T_\delta^L f(x) - \mathcal{C}_\delta^L(f \circ z^{-1})(z(x))| \ll \mathcal{M}f(x), \quad x \in \mathbb{R}^n,$$

uniformly in δ. Here $\mathcal{M}$ is the standard Hardy–Littlewood maximal operator. Similarly, the difference between T_δ^R and $\mathcal{C}_\delta^R$ is majorized by $\mathcal{M}$. In particular, $\{\mathcal{C}_\delta^L\}_\delta$ and $\{\mathcal{C}_\delta^R\}_\delta$ are bounded in $L^2(\mathbb{R}^n)_{(n)}$ with bounds independent of δ. As a consequence, the Cauchy kernel is a Calderón–Zygmund kernel; see the appendix of [CoMMe].

With this and some standard arguments from the theory of the Calderón–Zygmund operators, as presented in, e.g., [Me], we get the following.

THEOREM 15.13

The Cauchy singular integral operators $\mathcal{C}^L$ and $\mathcal{C}^R$ defined for all $f \in L^p(\Sigma, dS)_{(n)} := L^p(\Sigma, dS) \otimes \mathbb{C}_{(n)}$ and almost all $X \in \Sigma$ by

$$\mathcal{C}^L f(X) := \lim_{\delta\to+0} \mathcal{C}_\delta^L f(X) \quad and \quad \mathcal{C}^R f(X) := \lim_{\delta\to+0} \mathcal{C}_\delta^R f(X)$$

are well-defined and bounded on $L^p(\Sigma, dS)_{(n)}$ for each $1 < p < \infty$.

If we now recall that the (singular) double-layer potential operator is the scalar part of these CSIOs, we obtain the following.

COROLLARY 15.14

The (singular) double-layer potential operator $\mathcal{D}$ on Σ is bounded on $L^p(\Sigma, dS)$, $1 < p < \infty$.

Using simple, classical arguments and the operators T_δ and $\mathcal{C}_\delta$, as in [LMS] for instance, we obtain the next theorem.

THEOREM 15.15

For $f \in L^p(\mathbb{R}^n)_{(n)}$, $1 < p < +\infty$, and almost all $x \in \mathbb{R}^n$, we have

$$\lim_{\delta\to\pm0} \frac{1}{\sigma_n} \int_{\mathbb{R}^n} f(y)N(y) \frac{\overline{z(y) - z(x) - \delta e_0}}{|z(y) - z(x) - \delta e_0|^{n+1}}\, dy = \frac{1}{2}\left(F(X) \pm \mathcal{C}^L F(X)\right)$$

and

$$\lim_{\delta \to \pm 0} \int_{\mathbb{R}^n} \frac{\overline{z(y) - z(x) - \delta e_0}}{|z(y) - z(x) - \delta e_0|^{n+1}} N(y)f(y)\, dy = \frac{1}{2}\left(F(X) \pm C^R F(X)\right)$$

where $F := f \circ z^{-1}$, $X = z(x)$, $Y = z(y)$.

We remark that it is not much harder to work in the more general setting of weighted measures with a weight ω belonging to the Muckenhoupt class A_p. This shows that the analogues of Theorem 15.13, Theorem 15.15, and Corollary 15.14 with $L^p(\mathbb{R}^n)_{(n)}$ replaced by $L^p(\mathbb{R}^n, \omega\, dx)_{(n)}$ are still true; cf. [Me] and [Mi2].

The rest of this section is devoted to the proof of Theorem 15.11. First we shall establish some basic estimates. We emphasize that all the constants involved are independent of δ.

LEMMA 15.16
For any dyadic cube Q and any $i = 1, 2, \ldots, 2^n - 1$, we have:

$$|T_\delta^L \beta_{Q,i}^L(x)| \ll l(Q)|Q|^{1/2}|x - x_Q|^{-(n+1)}, \quad \text{if } x \notin 2Q, \tag{15.11}$$

and

$$|T_\delta^L \beta_{Q,i}^L(x)| \ll |Q|^{-1/2} \log \frac{c_n l(Q)}{\text{dist}(x, \partial_0 Q)}, \quad \text{if } x \in 2Q, \tag{15.12}$$

where $\partial_0 Q := \bigcup_{j=1}^{2^n} \partial Q^j$, *with* $\{Q^j\}_j$ *the "children" of Q, and x_Q is the "leftmost corner" of Q, i.e., for $Q = Q_{k,v}$, $x_Q := 2^{-k}v$. Here c_n denotes a constant which depends only on n. Analogous estimates hold for $T_\delta^R \beta_{Q,j}^R$ as well.*

PROOF Inequality (15.11) is a consequence of the fact that β's have vanishing "moment" so that

$$|T_\delta^L \beta_{Q,i}^L(x)| \approx \left| \int_{\mathbb{R}^n} \beta_{Q,i}^L(y) N(y) \left(\frac{\overline{z(y) - z(x) - \delta e_0}}{|z(y) - z(x) - \delta e_0|^{n+1}} - \frac{\overline{z(x_Q) - z(x) - \delta e_0}}{|z(x_Q) - z(x) - \delta e_0|^{n+1}} \right) dy \right|.$$

Hence, the inequality follows by using the mean-value theorem in estimating the expression inside the innermost parentheses and noting that for $x \notin 2Q$, $|x - x_Q| \ll \text{dist}(x, Q)$.

Next we consider (15.12). If $x \in 2Q$, we have

$$|T_\delta^L \beta_{Q,i}^L(x)| \approx \left| \int_{\mathbb{R}^n} \beta_{Q,i}^L(y) N(y) \frac{\overline{z(y) - z(x) - \delta e_0}}{|z(y) - z(x) - \delta e_0|^{n+1}}\, dy \right|$$

$$\ll |Q|^{-1/2} \sum_{j=1}^{2^n} \left| \int_{Q^j} N(y) \frac{\overline{z(y) - z(x) - \delta e_0}}{|z(y) - z(x) - \delta e_0|^{n+1}}\, dy \right|.$$

Let $d := \text{dist}(x, \partial_0 Q) \leq \text{dist}(x, \partial Q^j), \forall j$. For $x \in 2Q \setminus Q$ majorize each integral in the above sum by

$$\int_{d < |x-y| \leq d + \text{diam}(Q)} |x - y|^{-n} \, dy \ll \log(2n^{1/2} l(Q)/d),$$

which is of the right order.

Assume now that $x \in Q^j$ for some j. This time we split each integral as

$$\int_{Q^j} = \int_{|X-Y| > d} + \int_{|X-Y| < d} =: I + II,$$

where $X := z(x)$, $Y := z(y)$. Since $|x - y| \sim |X - Y|$, we get that I has the right size in the same way as before.

As for II, notice that if Ω_- is the domain in $\mathbb{R}^{n+1}$ below the surface Σ, then, by the monogenicity of the Cauchy kernel (see Proposition 15.2),

$$II = -\int_{\substack{|X-Y|=d \\ Y \in \Omega_-}} \frac{Y - X}{|Y - X|} \frac{\overline{Y - X - \delta e_0}}{|Y - X - \delta e_0|^{n+1}} \, d\sigma_d(Y),$$

where $d\sigma_d$ is the standard surface measure on the sphere of radius d centered at the origin of $\mathbb{R}^{n+1}$. Hence,

$$|II| \ll \int_{\substack{|X-Y|=d \\ Y \in \mathbb{R}^{n+1}}} \frac{1}{|X - Y|^n} \, d\sigma_d(Y) = \sigma_n,$$

which completes the proof. ∎

LEMMA 15.17
For $f \in L^\infty_{\text{comp}}(\mathbb{R}^n)_{(n)}$ we have $\langle T^L_\delta f, 1 \rangle_\Sigma = 0$ and $\langle 1, T^R_\delta f \rangle_\Sigma = 0$.

PROOF These are both simple consequences of Cauchy's theorem (cf. Proposition 15.2) applied to the domains Ω_-, Ω_+ located below and above Σ, respectively, and to the functions $T^L_\delta f$, $T^R_\delta f$ which are right monogenic in a neighborhood of Ω_- and left monogenic in a neighborhood of Ω_+, respectively. That this works is guaranteed by the good decay of the functions at ∞. ∎

LEMMA 15.18
We have $\langle T^L_\delta f, f' \rangle_\Sigma = -\langle f, T^R_\delta f' \rangle_\Sigma$, for functions $f, f' \in L^\infty_{\text{comp}}(\mathbb{R}^n)_{(n)}$. In particular, $\langle T^L_\delta \beta^L_{Q,i}, \beta^R_{Q',j} \rangle_\Sigma = -\langle \beta^L_{Q,i}, T^R_\delta \beta^R_{Q',j} \rangle_\Sigma$.

PROOF This is immediate by Fubini's theorem since the double integral

$$\frac{1}{\sigma_n} \int\int f(y) N(y) \frac{\overline{z(y) - z(x) - \delta e_0}}{|z(y) - z(x) - \delta e_0|^{n+1}} N(x) g(x) \, dx dy$$

is absolutely convergent. ∎

Let us now consider a dyadic cube $Q_{k,v}$. We let Φ be the linear rescaling, mapping $Q_{k,v}$ onto the standard unit cube in $\mathbb{R}^n$, i.e.,

$$\Phi(y) := w := 2^k y - v.$$

For an arbitrary dyadic cube Q we set $Q^* := \Phi(Q)$. We have that Q^* is still dyadic as long as $l(Q) \leq 2^{-k}$. Furthermore, if $z^*(w) := 2^k z(2^{-k}(w+v))$ for $w \in \mathbb{R}^n$, it is easy to show that z^* is bi-Lipschitz with constants comparable to those of the initial z.

Let Σ^* be the graph of z^*; as a general rule, we agree that the superscript $*$ is used to label objects, related to Σ^*, which are analogous to those we have constructed in connection with Σ.

Direct calculation shows that

$$\beta_{Q^*,i}^{L^*}(w) = 2^{-kn/2}\beta_{Q,i}^{L}(y), \qquad T_\delta^{L^*}\beta_{Q^*,i}^{L^*}(w) = 2^{-kn/2}T_{2^{-k}\delta}^{L}\beta_{Q,i}^{L}(y),$$

and

$$\langle T_{2^k\delta}^{L^*}\beta_{Q^*,i}^{L^*}, \beta_{Q',j}^{R^*}\rangle_{\Sigma^*} = \langle T_\delta^{L}\beta_{Q,i}^{L}, \beta_{Q',j}^{R}\rangle_{\Sigma}.$$

In addition, $w^*(Q^*) = 2^{nk(1/2-\epsilon)}w(Q)$, so that

$$w^*(Q'^*)^{-1} \sum_{\{Q\in\mathcal{F};l(Q^*)\leq l(Q'^*)\}} \sum_{i=1}^{2^n-1} w^*(Q^*)|\langle T_\delta^{L^*}\beta_{Q^*,i}^{L^*}, \beta_{Q',j}^{R^*}\rangle_{\Sigma^*}|$$

$$= w(Q')^{-1} \sum_{\{Q\in\mathcal{F};l(Q)\leq l(Q')\}} \sum_{i=1}^{2^n-1} w(Q)|\langle T_{l(Q')\delta}^{L}\beta_{Q,i}^{L}, \beta_{Q',j}^{R}\rangle_{\Sigma}|,$$

for each $Q' \in \mathcal{F}$ and $j = 1, 2, \ldots, 2^n - 1$.

This shows that in order to finish the proof of (15.9), it is enough to prove that

$$\forall j, \quad \sum_{\{Q\in\mathcal{F};1\leq l(Q)\}} \sum_{i=1}^{2^n-1} w(Q)|\langle T_\delta^{L}\beta_{Q,i}^{L}, \beta_{[0,1]^n,j}^{R}\rangle_{\Sigma}| \leq C < +\infty,$$

and

$$\forall i, \quad \sum_{\{Q\in\mathcal{F};l(Q)\leq 1\}} \sum_{j=1}^{2^n-1} w(Q)|\langle \beta_{[0,1]^n,i}^{L}, T_\delta^{R}\beta_{Q,j}^{R}\rangle_{\Sigma}| \leq C < +\infty,$$

where C may depend on $\|\nabla g\|_\infty$ but is independent of δ. If we combine these estimates with the analogous ones needed to complete the proof of (15.10), we find that we must prove that

$$\forall i,j, \quad \sum_{Q\in\mathcal{F}} w(Q)|\langle T_\delta^{L}\beta_{Q,i}^{L}, \beta_{[0,1]^n,j}^{R}\rangle_{\Sigma}| \leq C < +\infty \tag{15.13}$$

and

$$\forall i,j, \quad \sum_{Q\in\mathcal{F}} w(Q)|\langle \beta_{[0,1]^n,i}^{L}, T_\delta^{R}\beta_{Q,j}^{R}\rangle_{\Sigma}| \leq C < +\infty, \tag{15.14}$$

with C possibly depending on $\|\nabla g\|_\infty$ but not on δ.

The proofs of (15.13) and (15.14) are virtually the same, and we confine ourselves to showing (15.13).

PROOF OF (15.13)   We need to discuss several cases.

Case I

$l(Q)$ "large" and Q "clearly" disjoint from the standard unit cube: $l(Q) \geq 1$ and $2Q \cap [0,1]^n = \emptyset$.   ∎

Using (15.11) and that $|x_Q| \ll |x - x_Q|$, we get

$$w(Q)|\langle T_\delta^L \beta_{Q,i}^L, \beta_{[0,1]^n,j}^R \rangle_\Sigma| \ll |Q|^{1/2 - \epsilon} l(Q) |Q|^{1/2} |x_Q|^{-(n+1)}. \tag{15.15}$$

Now if $Q = Q_{k,v}$, our hypotheses imply that $v \neq 0$, so that for $\epsilon > 0$ the right-hand side of (15.15) is majorized by

$$\sum_{k=0}^{-\infty} \sum_{\substack{v \neq 0 \\ v \in \mathbb{Z}^n}} 2^{-kn(1/2 - \epsilon)} 2^{-k} 2^{-kn/2} 2^{k(n+1)} |v|^{-(n+1)}$$

$$= \left(\sum_{k=0}^{-\infty} 2^{kn\epsilon} \right) \left(\sum_{\substack{v \neq 0 \\ v \in \mathbb{Z}^n}} |v|^{-(n+1)} \right) < +\infty,$$

which proves that the corresponding piece in (15.13) satisfies the right estimate.

Case II

$l(Q)$ is "large," i.e., $l(Q) \geq 1$, but $[0,1]^n \cap 2Q \neq \emptyset$.   ∎

Note that $[0,1]^n \cap 2Q \neq \emptyset$ implies that there exists a fixed nonnegative integer M_0 (3^n will do), such that for any k, $\mathcal{F}_k$ contributes with at most M_0 dyadic cubes to this case. Now, by Lemma 15.18, (15.11), and (15.12),

$$w(Q)|\langle T_\delta^L \beta_{Q,i}^L, \beta_{[0,1]^n,j}^R \rangle_\Sigma| = w(Q)|\langle \beta_{Q,i}^L, T_\delta^R \beta_{[0,1]^n,j}^R \rangle_\Sigma|$$

$$\ll |Q|^{1/2 - \epsilon} |Q|^{-1/2} \left(\int_{Q \cap 2[0,1]^n} |T_\delta^R \beta_{[0,1]^n,j}^R(x)| \, dx + \int_{Q \setminus 2[0,1]^n} |T_\delta^R \beta_{[0,1]^n,j}^R(x)| \, dx \right)$$

$$\ll |Q|^{-\epsilon} \left(\int_{2[0,1]^n} \log \frac{c_n}{\text{dist}(x, \partial_0 [0,1]^n)} \, dx + \int_{\mathbb{R}^n \setminus 2[0,1]^n} |x|^{-(n+1)} \, dx \right) \sim |Q|^{-\epsilon}.$$

Hence,

$$w(Q)|\langle T_\delta^L \beta_{Q,i}^L, \beta_{[0,1]^n,j}^R \rangle_\Sigma| \ll |Q|^{-\epsilon}. \tag{15.16}$$

Using this, the part of (15.13) corresponding to this case can be estimated by

$$\sum_{k=0}^{-\infty} 2^{kn\epsilon} \sum_{v \in \text{finite set}} 1 \sim \sum_{k=0}^{-\infty} 2^{kn\epsilon} < +\infty,$$

as desired.

Case III
$l(Q)$ "small" and Q "clearly separated" from the standard unit cube: $l(Q) < 1$ and $2Q \cap [0,1]^n = \emptyset$. ☐

This time we have, by using (15.11) again,

$$w(Q)|\langle T_\delta^L \beta_{Q,i}^L, \beta_{[0,1]^n,j}^R \rangle_\Sigma| \ll |Q|^{1/2-\epsilon} \int_{[0,1]^n} |T_\delta^L \beta_{Q,i}^L(x)| \, dx$$

$$\ll l(Q)|Q|^{1-\epsilon}|x_Q|^{-(n+1)}.$$

Suppose $Q = Q_{k,v}$. Since $2Q \cap [0,1]^n = \emptyset$, we must have $2^k \ll |v|$. The appropriate part of the sum in (15.13) is now majorized by

$$\sum_{k=0}^{+\infty} 2^{-k} 2^{-kn(1-\epsilon)} 2^{k(n+1)} \sum_{2^k \ll |v|} |v|^{-(n+1)} \sim \sum_{k=0}^{+\infty} 2^{k(n\epsilon-1)} < +\infty,$$

provided $0 < \epsilon < 1/n$.
There remains

Case IV
$l(Q) < 1$ and $Q \subset [-3,3]^n$. ☐

Let us first analyze the situation when $Q \subset [-3,0] \times [-3,3]^{n-1}$.
To estimate $|\langle T_\delta^L \beta_{Q,i}^L, \beta_{[0,1]^n,j}^R \rangle_\Sigma|$, notice that since $\beta_{[0,1]^n,j}^R$ is a linear combination of the characteristic functions of the "children" of the standard unit cube, it is enough to control

$$|\langle T_\delta^L \beta_{Q,i}^L, \chi_{Q'} \rangle_\Sigma| \ll \int_{Q'} |T_\delta^L \beta_{Q,i}^L(x)| \, dx, \tag{15.17}$$

for an arbitrary fixed "child" Q' of $[0,1]^n$.
The first possibility is that the boundary of Q has no common points with the hyperplane $\{x^1 = 0\}$. Then we may use (15.11) in (15.17) combined with the fact that $|x_Q^1| \ll |x - x_Q|$ for $x \in [0,1]^n$ to dominate the integral by a multiple of

$$l(Q)|Q|^{1/2} \int_{|x_Q^1|}^{+\infty} r^{-(n+1)} r^{n-1} \, dr = l(Q)|Q|^{1/2}|x_Q^1|^{-1}.$$

Hence, the contribution to the sum in (15.13) from this part has the upper bound

$$\sum_{k=0}^{+\infty} 2^{-kn(1/2-\epsilon)} 2^{-k} 2^{-kn/2} 2^k \sum_{\substack{i=2,\dots,n \\ -3 \cdot 2^k \le v_i \le 3 \cdot 2^k - 1}} \sum_{v_1=1}^{-2^{k+1}} |v_1|^{-1} \sim \sum_{k=0}^{+\infty} k 2^{-k(1-n\epsilon)} < +\infty,$$

for $1/n > \epsilon > 0$.
The second possibility is that Q is adjacent to the hyperplane $\{x^1 = 0\}$. Let us write $Q' = Q_1' \cup Q_2'$ with $Q_1' = 2Q \cap Q'$ and $Q_2' = Q' \setminus Q_1'$. On the Q_2' part

we still use (15.11) to majorize the integrand in (15.17). This and the fact that $|x - x_Q| \geq l(Q)$ show that

$$\int_{Q'_2} |T_\delta^L \beta_{Q,i}^L(x)| \, dx \ll l(Q)|Q|^{1/2} \int_{Q'_2} |x - x_Q|^{-(n+1)} \, dx$$

$$\ll l(Q)|Q|^{1/2} \int_{l(Q)}^{+\infty} r^{-n-1} r^{n-1} \, dr \sim |Q|^{1/2}.$$

For each fixed $k \geq 0$ there are $\mathcal{O}(2^{k(n-1)})$ dyadic cubes which fit into this case, and

$$\sum_{k=0}^{+\infty} 2^{-kn(1/2-\epsilon)} 2^{-kn/2} 2^{k(n-1)} = \sum_{k=0}^{+\infty} 2^{-k(1-n\epsilon)} < +\infty,$$

for $1/n > \epsilon > 0$. Hence, we conclude that this part of the sum in (15.13) satisfies the right estimate as well.

As for the Q'_1 part in (15.17), using (15.12) and $\text{dist}(x, \partial_0 Q) \geq x^1 \geq 0$, we find

$$\int_{Q'_1} |T_\delta^L \beta_{Q,i}^L(x)| \, dx \ll |Q|^{-1/2} \int_{Q'_1} \log \frac{c_n}{\text{dist}(x, \partial_0 Q)} \, dx$$

$$\ll |Q|^{-1/2} \int_{Q'_1} \log \frac{c_n}{x^1} \, dx \ll |Q|^{-1/2} l(Q)^{n-1} \int_0^{l(Q)} \log \frac{c_n}{x^1} \, dx^1$$

$$= |Q|^{-1/2} \mathcal{O}(l(Q)^{n-1+\alpha}),$$

since, for any $\alpha \in (0, 1)$, $|t(\log(1/t)+1)| \ll t^\alpha$ uniformly for $t \in (0, 1)$. We choose $n\epsilon < \alpha < 1$. Then

$$\sum_{k=0}^{+\infty} 2^{-kn(1/2-\epsilon)} 2^{kn/2} 2^{-k(n-1+\alpha)} 2^{k(n-1)} = \sum_{k=0}^{+\infty} 2^{-k(\alpha-n\epsilon)} < +\infty.$$

We have now finished the proof when $Q \subset [-3, 0] \times [-3, 3]^{n-1}$. Next we use the invariance of the boundedness of

$$\sum_Q w(Q) |\langle T_\delta^L \beta_{Q,i}^L, \chi_{Q'} \rangle_\Sigma|$$

under translations, permutations of coordinates, and symmetries with respect to coordinate axis. This allows us to reduce the problem to the one we just finished whenever Q and Q' have disjoint interiors. In fact, the only remaining possibility we need to check in Case IV is that $Q \subseteq Q' = [0, 1/2]^n$. We have

$$|\langle T_\delta^L \beta_{Q,i}^L, \chi_{[0,1/2]^n} \rangle_\Sigma| = |\langle T_\delta^L \beta_{Q,i}^L, \chi_{\mathbb{R}^n \setminus [0,1/2]^n} \rangle_\Sigma|$$

$$\ll |\langle T_\delta^L \beta_{Q,i}^L, \chi_{\mathbb{R}^n \setminus 2[0,1/2]^n} \rangle_\Sigma| \tag{15.18}$$

$$+ |\langle T_\delta^L \beta_{Q,i}^L, \chi_{2[0,1/2]^n \setminus [0,1/2]^n} \rangle_\Sigma|.$$

The last term can be estimated by decomposing $2[0, 1/2]^n \setminus [0, 1/2]^n$ as a finite union of cubes which are disjoint from Q and for which we can apply the argument described in the preceding paragraph. To estimate the first term we let B_R be the ball of radius R and centered at the origin of $\mathbb{R}^n$. We see that

$$|\langle T_\delta^L \beta_{Q,i}^L, \chi_{\mathbb{R}^n \setminus 2[0,1/2]^n} \rangle_\Sigma| = \lim_{R \to +\infty} |\langle T_\delta^L \beta_{Q,i}^L, \chi_{B_R \setminus 2[0,1/2]^n} \rangle_\Sigma|$$

$$= \lim_{R \to +\infty} |\langle \beta_{Q,i}^L, T_\delta^R \chi_{B_R \setminus 2[0,1/2]^n} \rangle_\Sigma|$$

$$= \lim_{R \to +\infty} |\langle \beta_{Q,i}^L, T_\delta^R \chi_{B_R \setminus 2[0,1/2]^n} - T_\delta^R \chi_{B_R \setminus 2[0,1/2]^n}(x_Q) \rangle_\Sigma|.$$

Now, since $|(\partial/\partial x_j) T_\delta^R \chi_{B_R \setminus 2[0,1/2]^n}(x)| \ll 1$ uniformly in R for $x \in [0, 1/2]^n$, $j = 1, 2, \ldots, n$, we may bound the limit by $|Q|^{1/2} l(Q)$. Each $\mathcal{F}_k$ contains $\mathcal{O}(2^{kn})$ dyadic cubes inside $[0, 1/2]^n$, and

$$\sum_{k=0}^{+\infty} 2^{-kn(1/2-\epsilon)} 2^{-kn/2} 2^{-k} 2^{kn} = \sum_{k=0}^{+\infty} 2^{-k(1-n\epsilon)} < +\infty$$

for $1/n > \epsilon > 0$. Hence, this finishes Case IV, and we have thus completed the proof of (15.13) and the proof of Theorem 15.11. ∎

Let us finally note that the techniques exploited here enable us to give a new proof of the T(b) theorem, see, e.g., [DJoS] and [MMe], for Clifford algebra–valued functions.

Somewhat more specifically, suppose b takes on values which are Clifford numbers and that it is a para-accretive function on $\mathbb{R}^n$. We then work with the "bilinear" form

$$\langle f_1, f_2 \rangle_b := \int_{\mathbb{R}^n} f_1(x) b(x) f_2(x) \, dx.$$

As a substitute for the accretivity of our old $N(x)$, following G. David, we should modify the sequence of σ-algebras with respect to which the conditional expectation operators were defined (for more detail see [CoJS]).

Next, assuming $T(b) =^t T(b) = 0$, we have to estimate $|\langle T\beta, \beta \rangle_b|$, rather than $|T(\beta)(x)|$ as in Lemma 15.16, and then use Schur's test.

The general case reduces to this one subtracting off paraproduct-like operators. In fact, these operators will not have standard kernels, but the logarithmic-rate blow up is good enough to control them.

Bibliography

[AT]   P. Auscher and Ph. Tchamitchian. 1989. "Bases d'ondelettes sur les courbes corde-arc, noyau de Cauchy et espaces de Hardy associes," *Rev. Mat. Iberoamericana* **5**: 139–170.

[BDeSo]   F. Bracks, R. Delanghe, and F. Sommen. 1982. *Clifford Analysis, Res. Notes Math.* **76**, Pitman Advanced Publishing Program.

[C]   A.P. Calderón. 1977. "Cauchy integrals on Lipschitz curves and related operators," *Proc. Natl. Acad. Sci. U.S.A.* **74**.

[Ch]   M. Christ. 1990. *Lecture on Singular Integral Operators*, CBMS-NSF Reg. Conf. Ser. Appl. Math. **77**.

[CoJS]   R.R. Coifman, P. Jones, and S. Semmes. 1989. "Two elementary proofs of the L^2 boundedness of the Cauchy integrals on Lipschitz curves," *J. Amer. Math. Soc.* **2**: 553–564.

[CoMMe]   R.R. Coifman, A. McIntosh, and Y. Meyer. 1982. "L'intégrale de Cauchy définit un opérateur borné sur L^2 pour les courbes lipschitziennes," *Ann. Math.* **116**: 361–387.

[DJoS]   G. David, J.-L. Journé, and S. Semmes. 1985. "Operateurs de Calderón–Zygmund, fonctions para-accretives et interpolation," *Rev. Mat. Iberoam.* **1**: 1–56.

[GMu]   J.E. Gilbert and M.A.M. Murray. 1991. *Clifford Algebras and Dirac Operators in Harmonic Analysis*, Cambridge Studies Adv. Math. **26**, Cambridge: Cambridge University Press.

[I]   V. Iftimie. 1965. "Fonctions hypercomplexes," *Bull. Math. Soc. Sci. Math. R. Soc. Roum.* **4**: 279–332.

[LMS]   C. Li, A. McIntosh, and S. Semmes. 1992. "Convolution singular integrals on Lipschitz surfaces," *J. Amer. Math. Soc.* **5**: 455–481.

[M]   A. McIntosh. 1989. "Clifford algebras and the higher dimensional Cauchy integral," in *Approximation Theory and Function Spaces*, Banach Center Publications **22**, Warsaw: Polish Scientific Publishers, pp. 253–267.

[MMe]   A. McIntosh and Y. Meyer. 1985. "Algebres d'operateurs definis par des integrales singuliers," *C. R. Acad. Sci., Ser. 1* **301**: 395–397.

[Me]   Y. Meyer. 1990. *Ondelettes et Operateurs*, Paris: Hermann.

[Mi1]   M. Mitrea. 1991. Ph.D. Thesis, University of South Carolina.

[Mi2]   M. Mitrea. 1991. "Weighted Hardy spaces and Clifford algebras." Preprint.

[Mu]   M.A.M. Murray. 1985. "The Cauchy integral, Calderón commutators and conjugations of singular integrals in $\mathbb{R}^m$," *Trans. Amer. Math. Soc.* **289**: 497–518.

[N]   J. Nečas. 1967. *Les Methodes Directes en Théorie des Equations Elliptiques*, Prague: Academia.

[T]   Ph. Tchamitchian. 1989. "Ondelettes et integrale de Cauchy sur les courbes Lipschitziennes," *Ann. Math.* **129**: 641–649.

[V]   G.C. Verchota. 1984. "Layer potentials and regularity for the Dirichlet problem for Laplace's equations in Lipschitz domains," *J. Funct. Anal.* **59**: 572–611.

16

The use of decomposition theorems in the study of operators

Richard Rochberg

ABSTRACT   We describe several ways in which wavelet decompositions and related decomposition techniques can be used to study operators which are mixtures of potential operators and multiplication operators.

CONTENTS

16.1   Introduction and summary

The representation of functions using wavelets and other closely related representations gives a powerful tool in the study of individual functions and in the study of classical function spaces. We can also use these same tools to study operators between function spaces and that is what I will discuss here. Of course this is not a new idea. All that is relatively new here is the focus on certain tools that are part of this general technique and some of the particular details.

*Supported in part by NSF grant DSM 9007491.

There are many ways of using decomposition theorems to study operators. Although I will mention a number of these, the main attention will be on specific approaches with which I have been involved. I hope to give the general flavor of the ideas in the context of particular examples and show some of the details. The presentations will be informal and sketchy with references to the detailed proofs which are presented elsewhere. We will concentrate on operators which are a mixture of multiplication operators and differential operators. That is exactly the context in which we expect wavelets and related ideas to be useful. The basic theme is that although the two classes of operators can't be simultaneously diagonalized, the decomposition theorems do about as good a job as can be done of simultaneously approximately diagonalizing both classes.

One way to use wavelets or other decomposition theorems to study an operator T acting on a function f in a function space X is to use the decomposition theorem to split f into pieces, study how T acts on each piece, and then look at the question of how to combine the resulting terms to find out about all of Tf. We will see an example of that type of approach worked out in detail in Section 16.3.2. One can take that approach further still. Suppose $\{\psi_\alpha\}_{\alpha \in A}$ is a basis of X and $\{\tilde{\psi}_\beta\}_{\beta \in B}$ is a basis for the dual space. We could study the operator by looking at the matrix $\{\langle T\psi_\alpha, \tilde{\psi}_\beta \rangle\}$. This approach, using either wavelets or more general sets of functions, is very effective for studying Calderón–Zygmund operators and other closely related classes. We won't give any details about it and just refer to the books and papers [Me1], [Me2], [Da], [FrJa], [FrJaWe], [P], [JL], and [H]. The operators we will look at are a bit rougher than those often considered in the Calderón–Zygmund theory and in the references just mentioned. In particular the matrices of our operators don't quite satisfy the "almost diagonal" criteria of [Me1], [Me2], and [FrJa].

Another way to use decomposition methods to study T would be to break the operator into pieces. If the operator T depends linearly on a symbol function b then the decomposition of b into simple pieces may do the same for the operator; see, for example, [RSe1]. More generally, if T is an integral operator given by integral kernel $k(x, y)$ then various decomposition techniques can be used to split $k(x, y)$ into simple pieces which generate relatively simple operators. Work of this sort has been done in [BCoRo], [P], [RSe2], and [H]. In Section 16.3.1 we will give an example of that sort of analysis.

The wavelet representation of a function can be regarded as a discrete version of the Calderón reproducing formula. The Calderón reproducing formula gives a representation of the identity operator as a (continuous) sum of one dimensional projections; $I = \int P_\zeta d\mu(\zeta)$. This representation of the identity can be used in a number of ways to study operators. One way is to follow the pattern used in the definition of Fourier multiplier operators, that is, to insert a multiplication operator before doing the reconstruction implicit in the integration. Formally, for a multiplier $b(\zeta)$, we define an operator $T_b = \int b(\zeta) P_\zeta d\mu(\zeta)$. This type of construction is familiar in physics as the "coherent state representation" of the operator [KlSk], [D]. The theory is a bit more complicated than the theory of Fourier multipliers because the P_ζ aren't a commuting family of projections and hence we don't get a commuting family of operators. This class of operators—called Calderón–Toeplitz operators for reasons which will be explained later—are closely related to Calderón–Zygmund operators. In Section 16.3.3 we will look at some operators that arise through this construction.

Thus, in Section 16.3.1 and 16.3.2 we will use decomposition theorems to study naturally occurring operators. In Section 16.3.3 we will try to see what happens when we use decomposition ideas to build simple models of operators of interest.

In Section 16.2 we set up notation, collect some background results, and introduce the operators we will study. In the three parts of Section 16.3 we give some details of three different versions using decomposition techniques to get boundedness results, eigenvalue, and singular value estimates for operators. In Section 16.4 we discuss briefly how the ideas of Section 16.3 can be used in a more refined analysis of operators. In particular we will discuss construction of quasi-particles for the Lipman–Schwinger equation and size estimates for eigenvectors. The final section contains some closing thoughts and speculation.

16.2 Background

16.2.1 The operators

One of the operators we will look at is the Schrödinger operator acting on $L^2(\mathbb{R}^3)$. That is, let Δ be the Laplacian and let V be the operator of multiplication by the nonnegative potential function $v(x)$. We will look at the operator $S = -\Delta - V$. In some circumstances S will have a set of negative eigenvalues with no accumulation points except for (possibly) zero. The relation between the properties of v and the distribution of the eigenvalues is of great interest. For a positive λ let $N(\lambda) = N(\lambda, V)$ be the number of eigenvalues of S which are less than $-\lambda$. A nice introduction to the study of $N(\lambda)$ is given in [F] where, among other things, we see a fundamental relation between estimating $N(\lambda)$ and weighted norm estimates for potential operators. Here, we follow the lead of Birman and Schwinger and move from consideration of the unbounded operator S to an associated positive compact operator. For fixed positive λ suppose that f is an eigenvector of S with eigenvalue $-\lambda$; $Sf = -\lambda f$. Thus,

$$(-\Delta - V)f = -\lambda f.$$

Algebraic rearrangement of this equation and the introduction of a new function $g = v^{1/2}f$ yields

$$V^{1/2}(\lambda - \Delta)^{-1}V^{1/2}g = g.$$

Thus, at least on a formal level, eigenvectors of S with eigenvalue λ are related to eigenvectors with eigenvalue one of the related operator

$$T = T_\lambda = V^{1/2}(\lambda - \Delta)^{-1}V^{1/2}.$$

In fact more is true. Let $M(\lambda) = M(\lambda, V)$ be the number of eigenvalues of T_λ which are at least 1.

PROPOSITION 16.0 *(Birman, Schwinger)*

$$N(\lambda) \leq M(\lambda).$$

PROOF   See page 87 of [Si] or Theorem 5.3 of [S1] for the proof and for more background.   ∎

In Section 16.3.1 we will study the operator T_λ by decomposing its integral kernel.

T_λ is a product of three operators. The first and third are multiplication operators and the one in the middle is a function (via the functional calculus) of $D = (-\Delta)^{1/2}$. Such combinations are often denoted symbolically by $f(X)g(D)h(X)$; here $f(X)$ is the operator of multiplication by $f(x)$ and similarly for $h(X)$; $g(D)$ is the function of Δ defined by the functional calculus. Such operators, as well as the slightly simpler operators $f(X)g(D)$ and more complicated operators $f(X)g(-i\nabla)$, arise in a number of places. They are discussed systematically in Chapter 4 of [Si] and in the survey [BiKaSo]. One of the basic results in the area is the theorem of Cwikel which gives sharp conditions on f and g to ensure that $f(X)g(D)$ and related operators are in a weak Schatten ideal. In Section 16.3.2 we investigate what happens when we try to recapture Cwikel's result using wavelets. In Section 16.3.3 we look at the analog of Cwikel's result for Calderón–Toeplitz operators which are simplified models for $f(X)g(D)$.

16.2.2   Singular values

In addition to looking at boundedness criteria for operators, we will also be looking at their eigenvalues, and, more generally, their singular values. We now recall the relevant definitions. For a bounded operator T acting on a Hilbert space and for $n = 0, 1, \ldots$, we define $s_n(T)$, the nth singular number of T, by

$$s_n(T) = \inf\{\|T - R_n\| : R_n \text{ an operator of rank at most } n\}.$$

If T is compact then the sequence $\{s_n(T)\}$ will be the eigenvalues of $|T| = (T^*T)^{1/2}$ arranged in decreasing order. In particular, if T is a positive compact operator then the $s_n(T)$ are the eigenvalues of T in decreasing order. The Schatten–von Neumann ideal S_p consists of those compact operators for which the sequence $\{s_n(T)\}$ is in the Lebesgue sequence space l^p. We let $S_{p,q}$ be the space of operators for which $\{s_n(T)\}$ is in the Lebesgue–Lorentz space $l^{p,q}$. In particular $S_{p,\infty}$ is the space of operators for which $\{s_n(T)\}$ is in the weak-type space $l^{p,\infty}$.

16.2.3   Wavelets

Let $\mathcal{Q}$ be the collection of dyadic cubes in $\mathbb{R}^n$. For $Q \in \mathcal{Q}$ let ζ_Q be the lower left corner of Q, η_Q the side length of Q and $|Q|$ the measure of Q. Let ψ be a mother wavelet. Thus, the set of functions $\psi_Q(x) = |Q|^{-1/2}\psi((\cdot - \zeta_Q)\eta_Q^{-1}), Q \in \mathcal{Q}$, are an orthonormal basis. (We are, of course, suppressing an index in the case $n > 1$.) We have the reconstruction formula

$$f = \sum \langle f, \psi_Q \rangle \psi_Q. \tag{16.1}$$

16.2.4   The Calderón reproducing formula

The Calderón reproducing formula is the continuous analog of (16.1). Let G be the translation and dilation group which we realize as $G = \{\zeta = (v,t) : v \in \mathbb{R}^n, t > 0\} = \mathbb{R}^{n+1}_+$. On G we work with the measure $d\mu = t^{-n-1}dvdt$ (=left invariant Haar measure on G=hyperbolic measure on $\mathbb{R}^{n+1}_+$). Suppose that ψ is a smooth function with mean zero and good decay at infinity. For $\zeta \in G$ define $\psi_\zeta(x) = t^{-n/2}\psi((x - v)/t)$. If ψ satisfies the appropriate normalization condition,

$$\int_0^\infty |\hat{\psi}(t\lambda)|^2 t^{-1} dt = 1$$

for all unit vectors λ, then we have the following continuous analog of (16.1):

$$f = \int \langle f, \psi_\zeta \rangle \psi_\zeta d\mu(\zeta). \tag{16.2}$$

16.2.5   NWO sequences, weights

We say that a sequence of functions $\{e_Q\}_{Q \in \mathcal{Q}}$ is nearly weakly orthonormal (NWO) if the maximal function which maps $f(x)$ to

$$f^*(x) = \sup_{|\zeta_Q - x| < \eta_Q} \{|Q|^{-1/2}|\langle f, e_Q \rangle|\}$$

is bounded on $L^2(\mathbb{R}^n)$.

   Here is a basic class of examples of such sequences; in fact, it is the only class of examples we will use. Suppose K is a fixed large number; for Q in $\mathcal{Q}$ let KQ be the cube with the same center and orientation as Q but K times the side length. Suppose we are given a sequence of functions $\{e_Q\}_{Q \in \mathcal{Q}}$ with $\text{supp}(e_Q) \subset KQ$ and that there is an $r, 2 < r \le \infty$, and $c > 0$ so that for all Q

$$\|e_Q\|_r \le c|Q|^{1/r-1/2}. \tag{16.3}$$

From Section 7 of [RSe2] we have

PROPOSITION 16.1
If (16.3) holds then $\{e_Q\}_{Q \in \mathcal{Q}}$ is NWO.

Given a sequence of numbers $\{a_Q\}_{Q \in \mathcal{Q}}$ we define a new sequence $\{a_Q^*\}_{Q \in \mathcal{Q}}$ by

$$a_Q^* = \sum_{\substack{R \in \mathcal{Q} \\ R \subset Q}} \frac{|R|}{|Q|} |a_R|.$$

If $\{a_Q^*\}$ is bounded we write

$$\|\{a_Q\}\|_{CM_d} = \sup\{a_Q^*\}.$$

(The subscript is for "discrete Carleson Measures.")
We will be interested in operators acting on L^2 which can be written as

$$T = \sum_{Q \in \mathcal{Q}} a_Q \, e_Q \otimes f_Q$$

where $\{e_Q\}_{Q \in \mathcal{Q}}$ and $\{f_Q\}_{Q \in \mathcal{Q}}$ are NWO. By that notation we mean

$$Tf = \sum_{Q \in \mathcal{Q}} a_Q \langle f, f_Q \rangle e_Q.$$

From [RSe2] we have

PROPOSITION 16.2
$\|T\| \leq c \|\{a_Q\}\|_{CM_d}$.

If $\{a_Q^*\}$ tends to zero then T is compact and we have information about approximation of T by finite rank operators. If the sequence $\{a_Q^*\}$ tends to zero we denote its nonincreasing rearrangement by $\{a^*(n)\}$. Suppose N is a positive integer and $a_{Q_j}^* = a^*(j), j = 1, 2, \ldots, N$. Set $F_N = \sum_{j=1}^{N} \alpha_{Q_j} e_{Q_j} \otimes f_{Q_j}$. F_N is a good rank N approximation to T.

PROPOSITION 16.3
$\|T - F_N\|_{op} \leq ca^*(N+1)$.

Hence,

$$s_N(T) \leq ca^*(N+1), \quad N = 0, 1, \ldots.$$

Using this one shows

PROPOSITION 16.4
If $1 \leq p < \infty$, and $1 \leq q \leq \infty$, then $\{a_Q\} \in l^{p,q}$ implies $\{a_Q^*\} \in l^{p,q}$ which in turn implies $T \in S_{p,q}$.

A weight is a nonnegative function on $\mathbb{R}^n$. We will be interested in weights which belong to the Muckenhoupt class A_∞ which consists of weights for which there are constants C and δ so that for any cube Q and any measurable $E \subset Q$

$$\frac{\int_E w}{\int_Q w} \leq C \left[\frac{|E|}{|Q|} \right]^{\delta}. \tag{16.4}$$

We will need some facts about these weights. The proofs can be found in [GFra].

LEMMA 16.5

(a) *Suppose Q is a cube in $\mathbb{R}^d$, Q' is a cube of the same size and with a common face, and Q'' is a subcube of Q with edge half that of Q. For w in A_∞ set $m = \int_Q w, m' = \int_{Q'} w$, and $m'' = \int_{Q''} w$. The ratios m/m' and m/m'' are bounded above and below by positive constants which can be chosen to depend only on the constants appearing in the (16.4).*

(b) *(Reverse Holder Inequality) Suppose w is in A_∞. There are positive constants C and δ which can be taken to depend only on the constants in (16.4) so that for all cubes Q in $\mathbb{R}^d$*

$$\left[\frac{1}{|Q|} \int_Q w(t)^{1+\delta} dt \right]^{1/(1+\delta)} \leq C \left[\frac{1}{|Q|} \int_Q w(t) dt \right].$$

One of the relationships between A_∞ weights and NWO sequences is

LEMMA 16.6
Suppose w^2 is in A_∞. For $Q \in \mathcal{Q}$ set

$$e_Q(x) = \|w\chi_Q\|_2^{-1} w\chi_Q(x).$$

The set $\{e_Q\}$ is NWO.

PROOF It suffices to verify (16.3). (16.3) follows from the Reverse Holder Inequality in the previous lemma. ∎

16.3   Analysis of operators

16.3.1   NWO sequences and eigenvalue estimates for Schrödinger operators

We will give an example of an analysis of an operator, in this case the Birman–Schwinger operator associated to the Schrödinger equation, based on decomposition of the integral kernel. This will lead to a representation of (a slightly simplified model of) the operator as a sum of rank one operators. Thus, we will be looking at an operator of the form

$$Af = \sum c_\alpha \langle f, h_\alpha \rangle k_\alpha.$$

If $\{h_\alpha\}$ and $\{k_\alpha\}$ were orthonormal then the boundedness of A on L^2 would follow from the boundedness of the sequence $\{c_\alpha\}$. In fact, that would still be true if $\{h_\alpha\}$ and $\{k_\alpha\}$ were the images of orthonormal sets under a bounded map; for instance, if they were the functions used in the decomposition theorems of [FrJa]. Just as is the case with the decomposition theorems using wavelets our sequences will be indexed by the dyadic cubes and the hs and ks will be, roughly, localized on their index cube. However, our functions lack smoothness and cancellation typical of wavelets and of the functions used in [FrJa]. Hence, a Carleson-measure condition (namely, Proposition 16.2), slightly stronger than the boundedness of $\{c_\alpha\}$, is needed to ensure boundedness of the operator A. This approach was developed in [RSe1], [RSe2] and is surveyed in [R1]. The results we present here summarize some of the work in [R5]. The results we obtain, in particular the eigenvalue estimate Theorem 16.10, are already known. Our main contribution here is to show that this type of result follows in a relatively straightforward manner once we decide to apply decomposition methods to the kernel of the integral operator.

Our goal now is to relate the asymptotics of the eigenvalues to the properties of the potential function. As we noted before we can instead look at the compact operator $T = T_\lambda = -V^{1/2}(\lambda - \Delta)^{-1}V^{1/2}$. For notational convenience we will write $v = w^2$. First consider the case in which $\lambda = 0$. In that case $(\lambda - \Delta)^{-1}$ is a Riesz potential operator and is given by convolution with a constant multiple of $|x|^{-1}$. Hence, (ignoring multiplicative constants) we are looking at an operator with integral kernel $K(x, y)$—i.e., $Tf(x) = \int K(x, y)f(y)dy$ with

$$K(x, y) = \frac{w(x)w(y)}{|x - y|}.$$

To analyze T we break the kernel into a sum of terms. Let $\{Q_i \times Q_i'\}$ be a Whitney covering of $\{\mathbb{R}^3 \times \mathbb{R}^3 \setminus \text{diag}\} = \{(x, y) \in \mathbb{R}^3 \times \mathbb{R}^3 : x \neq y\}$ by dyadic cubes (p. 169 of [St]). The diameter of each covering cube is comparable to its distance to the diagonal. Hence, for each $Q_i \times Q_i'$ we have

$$|x - y|^{-1} \sim (\text{side length of } Q_i)^{-1} = \eta_{Q_i}^{-1}.$$

Our goal for now is not actually to study T, rather it is to replace T with a simpler model operator. The first step is based on the previous estimates. With them in mind we modify K by replacing $|x - y|^{-1}$ on $Q_i \times Q_i'$ by $\eta_{Q_i}^{-1}$. Noting that

$$\sum_i \chi_{Q_i \times Q_i'}(x, y) = 1 \text{ a.e. } dxdy \tag{16.5}$$

and using the fact that $\chi_{Q_i \times Q_i'}(x, y) = \chi_{Q_i}(x)\chi_{Q_i'}(y)$, we obtain the modified kernel

$$\sum w(x)w(y)\eta_{Q_i}^{-1}\chi_{Q_i}(x)\chi_{Q_i'}(y). \tag{16.6}$$

The geometry of the Whitney cover ensures that the distance between Q_i and Q_i' is of the same order as their side length. Hence, in our move toward a simpler model we drop the distinction and consider the analog of (16.6) with $Q_i = Q_i'$.

Further, we make the operator larger and sum over all Q in $\mathcal{Q}$. Thus, the operator we actually study is the one with kernel

$$\sum_{Q \in \mathcal{Q}} w(x)w(y)\eta_Q^{-1}\chi_Q(x)\chi_Q(y).$$

Set

$$e_Q(y) = \|w\chi_Q\|^{-1}w(y)\chi_Q(y)$$

and

$$f_Q(x) = \|w\chi_Q\|^{-1}w(x)\chi_Q(x).$$

Thus, we are looking at the operator R with kernel

$$\sum_{Q \in \mathcal{Q}} \eta_Q^{-1}\|w\chi_Q\|^2 f_Q(x)e_Q(y);$$

that is,

$$R = \sum_{Q \in \mathcal{Q}} \eta_Q^{-1}\|w\chi_Q\|^2 f_Q \otimes e_Q. \tag{16.7}$$

We want to be able to use Lemma 16.6. To do that we need to assume

$$v = w^2 \text{ is in } A_\infty. \tag{16.8}$$

THEOREM 16.7
Suppose (16.8) holds. In order for R, and also T, to be bounded from L^2 to L^2 it is sufficient that the coefficients in (16.7) be bounded. Explicitly, recalling that $w^2 = v$, it is sufficient that

$$A = \sup_{Q \in \mathcal{Q}} \eta_Q^{-1}\left[\int_Q v(x)dx\right] < \infty.$$

In this case $\|R\| \leq cA, \|T\| \leq cA$.

PROOF First we look at R. Because we have (16.8) we can use Lemma 16.6 to see that $\{e_Q\}$ and $\{f_Q\}$ are NWO. Hence, Proposition 16.2 applies. We have

$$a_Q = \eta_Q^{-1}\left[\int_Q w(x)^2 dx\right].$$

Also,

$$a_Q^* = \sum_{\substack{R \in \mathcal{Q} \\ R \subset Q}} \frac{|R|}{|Q|}|a_R| = \sum_{n=0}^{\infty}\sum_{\eta_R=\eta_Q/2^n} 2^{-3n}|a_R|.$$

For fixed n the inner sum is

$$2^{-3n}2^n\eta_Q^{-1}\sum_{\eta_R=\eta_Q/2^n}\left[\int_R w^2\right].$$

But these small cubes, R, form a partition of Q; hence, the previous expression is

$$2^{-2n}\eta_Q^{-1}\left[\int_Q w^2\right] = 2^{-2n}a_Q.$$

Thus, we sum over $n \geq 0$ and find $a_Q^* \leq ca_Q$. Proposition 16.2 now gives the required estimate and the proof for R is complete.

Because the integral kernel for T is equal to a sum of kernels each of which is subject to the same analysis that we just gave for R, the same result holds for T. For details of such an argument (i.e., the passage from the model operator R to the operator of interest T) are a bit technical. They can be done as in [RSe2] or by majorizing T by an operator that looks like R but is built using characteristic functions of dilated cubes. ∎

Hence, by the theorem of Birman and Schwinger in Section 16.2.1, if the constant A in Theorem 16.7 is small enough then the number of negative eigenvalues must be less that 1 and hence must be 0. There is great interest in giving sharp numerical values on constants such as A which would force such a conclusion. However, such "sharp constant" results don't seem accessible by these methods.

We now return to the general case $\lambda \neq 0$. The form of the analysis is the same. The difference is in the estimates for the integral kernel of the operator. $(\lambda - \Delta)^{-1}$ is a Bessel potential operator and the integral kernel, $G(x,y)$, for that operator satisfies the estimates, setting $\mu = \lambda^{1/2}$,

$$G(x,y) \sim \begin{cases} c_1|x-y|^{-1} & \text{for } \mu|x-y| \leq 1 \\ c_2|x-y|^{-1}\exp(-\mu|x-y|) & \text{for } \mu|x-y| \geq 1. \end{cases}$$

Here c_1 and c_2 are constants that are independent of μ. (See, for instance, formula (v) on p. 131 of [St].) The lack of homogeneity leads to a slightly more intricate pattern of analysis.

Fix μ. Exactly as before we reduce to a model operator. We are led to consider the boundedness of the operator

$$S = \sum_{Q \in \mathcal{Q}} b_Q \|w\chi_Q\|^2 f_Q \otimes e_Q$$

where

$$b_Q = \begin{cases} \eta_Q^{-1} & \text{for } \mu\eta_Q \leq 1 \\ \eta_Q^{-1}\exp(-\mu\eta_Q) & \text{for } \mu\eta_Q > 1. \end{cases}$$

THEOREM 16.8
Suppose (16.8) holds. Then

$$\|S\| \leq c \sup_{\substack{Q \in \mathcal{Q} \\ \mu\eta_Q \leq 1}} \eta_Q^{-1}\left[\int_Q v\right]. \tag{16.9}$$

The same estimates hold for $T = T_\lambda$.

PROOF Set $I_Q = [\int_Q w^2]$. The line of argument of the previous proof gives $\|S\| \le c \sup\{(b_Q I_Q)^*\}$ and we need to study $(b_Q I_Q)^*$. As before the sum defining $(b_Q I_Q)^*$ can be organized by cube size and the subsum involving cubes of the same size can be explicitly summed. This leads to the needed estimate for small cubes, those with $\mu\eta_Q \le 1$. For the large cubes we have the estimate

$$(b_Q I_Q)^* \le \left[\sum_{\mu\eta_Q \le 2^n} 2^{-3n}(2^{-n}\eta_Q)^{-1} + \sum_{\mu\eta_Q > 2^n} 2^{-3n}(2^{-n}\eta_Q)^{-1} \exp(-\mu 2^{-n}\eta_Q) \right] I_Q.$$

(In both cases the summation index is n.) The first sum is a geometric series with sum dominated by $c\mu^{-2}\eta_Q^{-3}$. To estimate the second sum we need a lemma.

LEMMA 16.9
Suppose $A \ge 1$ and β is positive. There is a constant C which depends on β but not A so that

$$\sum_{A > 2^n} e^{-A2^{-n}} 2^{-n\beta} \le CA^{-\beta}.$$

Applying the lemma with $\beta = 2$ lets us estimate the second sum by $c\mu^{-2}\eta_Q^{-3}$. Thus, by Proposition 16.2

$$\|S\| \le c \sup_{Q \in \mathcal{Q}} \min(\mu^{-1}, \eta_Q)^2 \eta_Q^{-3} \int_Q w^2.$$

Now we need to see why we can ignore the large cubes in computing the supremum. The reason is that the term involving a cube Q with $\mu\eta_Q > 1$ is the average of the terms involving cubes of side length approximately μ^{-1} which are contained in Q. (We omit the straightforward computation.) Hence, the large cube can't give the supremum. This finishes the proof sketch for R. Again the passage to an analogous result for T is mechanical but technical and will be omitted. Again [RSe2] is a reference. ∎

Actually, we have been digressing. We don't want boundedness estimates, we need to estimate the number of eigenvalues larger that 1.

Suppose λ is the nth smallest eigenvalue of the Schrödinger equation with which we started. Then, as we noted, $s_n(T_\lambda) \ge 1$. Hence, by Proposition 16.3 there is a constant c such that there must be a set K of at least n cubes Q for which $a_Q^* \ge c$. During the previous proof we developed the estimate

$$a_Q^* \le c \min(\mu^{-1}, \eta_Q)^2 \eta_Q^{-3} \left[\int_Q w^2 \right].$$

Hence, on the n cubes of K,

$$c \leq \min(\mu^{-1}, \eta_Q)^2 \eta_Q^{-3} \left[\int_Q w^2 \right].$$

Using the geometry of the set of dyadic cubes we can find a big subset K' of K so that each Q in K' contains a subcube of half its side length which is disjoint from all the other subcubes obtained this way. We first pass to K' and then pass to those subcubes which we denote K''. By Lemma 16.6 A,

$$c' \leq \min(\mu^{-1}, \eta_Q)^2 \eta_Q^{-3} \left[\int_Q w^2 \right]$$

for those subcubes. Finally, as we noted in the proof of the previous theorem, if we have a cube of side length larger that μ^{-1} on which the inequality holds then it contains a cube of length μ^{-1} on which the estimate holds.

Collecting all this information we have the following conclusion. There is a fixed constant d_1, independent of n, λ, v, so that if we set

$$N'(\lambda) = \sup \operatorname{card} \left\{ Q \in \mathcal{Q} : \eta_Q \leq \lambda^{-1/2}, \eta_Q^{-1} \int_Q v > d_1, \text{ the } Q\text{s are disjoint} \right\}$$

then we have the following:

THEOREM 16.10
Suppose v is in A_∞. There is a constant c so that $N(\lambda) \leq cN'(\lambda)$.

Again we have just outlined things for the model operator but the result extends to the operator T. The details are in [R5]. Much of the recent interest in results of this type derives from the results by C. Fefferman and Phong presented in [F]. The results there are of the same form as Theorem 16.10 but involve pth power means for p slightly above 1. However, since we assume our v is in A_∞, our results (which appear slightly stronger than those in [F]) are actually equivalent. Extensions of and alternatives to the results in [F] include [ChWiWo], [KeS], [S2].

If we wanted to apply our techniques to v which are not assumed to be in A_∞ we would have to start by replacing the potential by an A_∞ majorant and then going forward. (Actually, that is done at one point in [F].)

Converse estimates are also known. The methods we have been using can be used to give converse estimates (several instances are in [RSe2]), but the proofs in this context would end up being quite similar to earlier proofs such as those in [F]. On the other hand our direct proofs seem rather different in flavor from the other proofs of the direct results.

Finally we note that our analysis in this section is in $\mathbb{R}^3$ and is for the operator $-\Delta - V$. However, there is no problem with working in n dimensions nor is there any reason not to do an analogous analysis if $-\Delta$ is replaced by a different Hamiltonian H. All that is needed is good estimates for the integral kernel of $(H + \lambda)^{-1}$. For example, it would be interesting to develop analogous results for the relativistic Schrödinger operator in [CMaSi]. This pattern of analysis could be used if appropriate estimates were obtained for the resolvent kernel associated with the relativistic Hamiltonian.

16.3.2 Wavelets and Cwikel's estimates for $f(X)g(i\nabla)$

We are interested in operators on $L^2(\mathbb{R}^n)$ which are denoted in [Si] by $T = f(X)g(-i\nabla)$. As before $f(X)$ denotes the operator of multiplication by $f(x)$ and now $g(-i\nabla)$ is the operator $g(-i\nabla)h = (g\hat{h})^\vee$. Thus, if $g(\xi) = g(|\xi|)$, then $g(-i\nabla)$ is the operator $g(D)$ discussed earlier. The result we want to look at is the following fundamental theorem of Cwikel.

THEOREM 16.11
Suppose $2 < p < \infty$. If f is in L^p and g is in $L^{p,\infty}$, then $T = f(X)g(-i\nabla)$ is in $S_{p,\infty}$ and

$$\|T\|_{p,\infty} \le c\|f\|_p\|g\|_{p,\infty}.$$

We refer to Chapter 4 of [Si] and to [BiSo] for background and motivation.

This result is sharp in the sense that f in $L^{p,\infty}$ is not enough to even ensure T is compact.

We want to see what happens when we try naively to prove this using wavelets. We will get a partial result but not the complete theorem. To help focus on the main issues we suppose $n = 1$ and that $g(\xi)$ only depends on $|\xi|$. Those simplifications don't help with the real problem which apparently is dealing with the lack of smoothness of g. Without loss of generality we may suppose f is positive. The reason is that $f(X)$ is $|f|(X)$ followed by a unitary operator; hence, it is enough to estimate the $S_{p,\infty}$ norm of the operator built using $|f|$. In fact, it would be enough to prove the theorem for $F(X)g(D)$ where $F(X) \ge f(x)$ a.e. and we return to this point later. Finally by composing on the right with a unitary operator we may also suppose $g \ge 0$.

We start with the wavelet decomposition (16.1). Thus, for h in L^2 we have

$$h = \sum \langle h, \psi_Q \rangle \psi_Q.$$

Hence,

$$Th = \sum \langle h, \psi_Q \rangle f(X)g(D)\psi_Q. \tag{16.10}$$

We think of $\hat{\psi}_Q$ as concentrated near $\{\xi : |\xi| \sim l_Q^{-1}\}$. (We are now using l_Q for the side length of Q.) Hence, if g is smooth, then at the level of metaphor

$$g(D)\psi_Q = g(l_Q^{-1})\psi_Q. \tag{16.11}$$

If we take this as fact instead of metaphor we are led to the operator:

$$Rh = \sum g(l_Q^{-1})\langle h, \psi_Q \rangle f\psi_Q. \tag{16.12}$$

We haven't yet specified what wavelets we are using. However, if we select them to be continuous and with compact support and if we assume

$$f^2 \in A_\infty; \tag{16.13}$$

then, by (an elementary extension of) Lemma 16.6 the set of vectors

$$\varphi_Q = \|f\psi_Q\|_2^{-1} f\psi_Q, \quad Q \in \mathcal{Q}$$

is NWO. Set $f_Q = \|f\psi_Q\|_2$. We are looking at the operator

$$R = \sum g(l_Q^{-1}) f_Q \, \psi_Q \otimes \varphi_Q.$$

LEMMA 16.12
In addition to $f \in L^p$, $g \in L^{p,\infty}$, we also suppose g is smooth in the following sense. Given $\lambda > 0$, let $\Lambda = \{n : g(2^{-n}) > \lambda\}$. Then

$$|\{x : g(x) > \lambda\}| \sim \sum_{n \in \Lambda} 2^{-n}. \tag{16.14}$$

In this case the set of numbers $\{g(l_Q^{-1}) f_Q\}$ is in $l^{p,\infty}$.

Once we have this we will have:

THEOREM 16.13
Suppose f is in L^p and satisfies (16.13); g is in $L^{p,\infty}$, depends only on $|\zeta|$, and is smooth in the sense of the previous lemma. Then R is in $S_{p,\infty}$ with the natural norm estimates.

PROOF By Lemma 16.6 $\{\varphi_Q\}$ is NWO. $\{\psi_Q\}$ are orthonormal and hence are NWO. By the previous lemma the coefficients are in $l^{p,\infty}$. Hence, by Proposition 16.4, we are done. ∎

First we'll prove the Lemma then discuss how close we have come to proving Cwikel's theorem.

PROOF First note that the right hand side of (16.14) is comparable to 2^{-N} where $N = \min \Lambda$. By the assumption that g is in $L^{p,\infty}$ the left-hand side is dominated by $C\lambda^{-p}$. Thus, $\lambda^{-p} \geq c2^{-\min \Lambda}$. Hence, if $g(2^{-n}) > \lambda$ then $2^{-n} < 2^{-N} < c\lambda^{-p}$. That is, for $s = 2^{-n}$, $g(s) \leq cs^{-1/p}$. We want to show that the numbers $\{g(l_Q^{-1}) f_Q\}$ are in $l^{p,\infty}$ or, equivalently, $\{g(l_Q^{-1})^p f_Q^p\}$ are in $l^{1,\infty}$. We only make things harder by replacing g with the majorant. Thus, we are trying to show $\{l_Q f_Q^p\}$ is in $l^{1,\infty}$. Now note, for some large K,

$$f_Q^p = \|f\psi_Q\|_2^p \leq c \left(\frac{1}{|Q|} \int_{KQ} f^2 \right)^{p/2} \leq c(m_{KQ}(f^2)^{1/2})^p.$$

Here m_{KQ} denotes the operation of taking the mean over KQ. Because f is in L^p, f^2 is in $L^{p/2}$. Now $p/2 > 1$ and hence

$$f^+(x) = \sup_{x \in Q} m_{KQ}(f^2)$$

is in $L^{p/2}$ (by the Hardy–Littlewood maximal theorem). Consider the function $H(x,t)$ defined on $\mathbb{R} \times \mathbb{R}^+$ which takes the value $l_Q f_Q^p$ on the set $\hat{Q} = Q \times [l_Q/2, l_Q]$. Showing $\{l_Q f_Q^p\}$ is in $l^{1,\infty}$ is equivalent to showing H is in $L^{1,\infty}(\mathbb{R}_+^2, t^{-2} dx dt)$. Thus, we want to estimate

$$\iint_{H(x,t) > \lambda} t^{-2} dx dt.$$

We replace the region of integration by the region $\{(x,t); tf^+(x)^{p/2} > \lambda\}$, which has comparable or greater measure, and estimate

$$\int \left(\int_{tf^{+p/2} > \lambda} t^{-2} dt \right) dx = \int \frac{1}{\lambda} f^{+p/2} dx = \frac{1}{\lambda} \int (f^+)^{p/2} dx.$$

But we know f^+ is in $L^{p/2}$ so the lemma is proved. ∎

How close have we come to proving Cwikel's theorem? Although at the level of metaphor we have done it, there is some distance between what we have and what we would like. The restriction (16.13) isn't a problem. If we denote the Hardy–Littlewood maximal operator by M then the function $F = M(f^{p/2})^{2/p}$ is a pointwise majorant of f (and we noted earlier that it suffices to prove the theorem for such a majorant) and is in A_∞ ([CoR]). The restrictions to $n = 1$ and g which only depend on $|\xi|$ were for bookkeeping convenience. The assumption that g is smooth is more troublesome. Even if g is smooth (16.11) isn't true; however, if g is smooth we could try to build a useful approximation or majorant for our operator starting with (16.11) as a guide. It is not clear, however, how to even start the analysis if, for example, g is unbounded in every interval. It seems an interesting problem to try to find a proof of Cwikel's theorem which fits comfortably in the language of wavelets and NWO sequences.

16.3.3 Calderón–Toeplitz models for $f(X)g(D)$

Calderón–Toeplitz operators are operators on $L^2(\mathbb{R}^n)$ constructed by putting a multiplier in (16.2). That is, given a function $b(x,t)$ defined on the upper half space, $\mathbb{R}_+^{n+1} = \mathbb{R}^n \times \mathbb{R}_+$, the associated Calderón–Toeplitz operator is the operator T_b defined by

$$T_b f = \int_{\mathbb{R}_+^{n+1}} b(\zeta) \langle f, \psi_\zeta \rangle \psi_\zeta d\mu(\zeta). \tag{16.15}$$

The idea of using this reproducing formula to build operators is in [KlSk] and is discussed in [Da], Ch. 2. The name is based on the fact that there is a strong formal resemblance to Toeplitz operators on the Bergman space—the main difference being the use of the Calderón reproducing formula in place of the Bergman reproducing formula. Steps toward study of these operators in the context of harmonic analysis are in [No1], [No2], [R1], [R2], [R3]. In this section we try to construct operators T_b which are models for the operators $f(X)g(D)$ which we discussed in the previous section. Most of what we say here is from [No1].

As in the previous section, rather than consider a general $f(X)g(-i\nabla)$, we will suppose f and g are nonnegative, $g(\xi)$ depends only on $|\xi|$, and we are working on $L^2(\mathbb{R})$.

In the previous sections we used decomposition ideas to break operators of interest into simpler pieces. Here we start with the simple pieces and try to build models of the operators of interest. Hence, although the tools are similar the viewpoints differ.

Now let's look at (16.15) when $b(x,t) = d(t)$. Roughly we are multiplying the part of the transform that contains information about scale t (and hence frequency $1/t$) by $d(t)$. T_b commutes with translation and hence must be a multiplier. A Fourier transform computation shows that T_b is a multiplier:

$$T_b f = (m\hat{f})^\vee$$

with

$$m(\xi) = \int_0^\infty d(s)|\hat{\psi}(s\xi)|^2 \frac{ds}{s}.$$

This is in line with the informal analysis. The normalization of ψ ensures that $|\hat{\psi}(s\xi)|^2$ is a probability density with respect to the measure $s^{-1}ds$. Also, if we, as is common, select ψ with $\hat{\psi}$ concentrated near the unit sphere then $|\hat{\psi}(s\xi)|^2$ will be concentrated near the sphere of radius s^{-1}.

Suppose now that we look at T_b when $b(x,t) = h(x)$. Roughly, T_b resembles M_h, multiplication by h. For example, $\|T_b\| \sim \|h\|_\infty$ and such a T_b will never be compact. Also, if b is smooth and decays at infinity then $T_b - M_h$ will be compact and, depending on the smoothness of b, in various Schatten ideals. However, the match between T_b and M_h, while good, is not excellent. (Actually $T_b = M_h +$ a paraproduct with h.) However, it is interesting to think of T_b as a type of generalized multiplication operator.

Before we go on, let's try to understand T_b a bit better for general b. We decompose $\mathbb{R}^2_+$ as $\cup_{Q\in\mathcal{Q}}\hat{Q}$. Thus,

$$T_b = \sum_{Q\in\mathcal{Q}} \int_{\hat{Q}} (\cdots)d\mu.$$

All the $\hat{Q}$ have the same size $(d\mu)$. On $\hat{Q}$, ψ_ζ changes little; hence, we can think of it as represented by the value at ζ_Q, the center of $\hat{Q}$. Writing $m_{\hat{Q}}(b)$ for the mean of b on $\hat{Q}$, this leads to,

$$T_b \sim \sum m_{\hat{Q}}(b)\, \psi_{\zeta_Q} \otimes \psi_{\zeta_Q}. \tag{16.16}$$

The functions $\{\psi_{\zeta_Q}\}$ are not wavelets—they need not be orthonormal. However, they are almost orthonormal in various technical senses. For instance, the operators $\psi_{\zeta_Q} \otimes \psi_{\zeta_Q}$, which are rank one projections, satisfy the hypotheses of the Cotlar–Stein lemma. Hence, we expect from (16.16) that the size of the numbers $m_{\hat{Q}}(b)$ will give insight into the boundedness and singular valves of T_b.

On the basis of what has been said we can think of $T_{g(1/t)}$ as a model for $g(D)$ and can think of $T_{f(x)}$ as closely related to $f(X)$. Hence, even though the Calderón–Toeplitz operators don't have a perfect symbol calculus, i.e., $T_a T_b \neq T_{ab}$ in general, it is still interesting to compare $T_{f(x)g(1/t)}$ to $f(X)g(D)$.

Let $b(x,t) = f(x)g(1/t)$ with f and g positive, f in L^p, and g in $L^{p,\infty}$. Suppose for the moment that g is smooth. In that case, as in the previous section, $g(s) \leq cs^{-1/p}$. We make our operator larger by replacing g with $s^{-1/p}$. Hence, the quantities $m_{\hat{Q}}(b)$ of (16.16) can be estimated by

$$m_{\hat{Q}}(b) \leq cl_Q^{1/p} m_Q(f),$$

where $m_Q(f)$ is the mean of f on Q. By the Cauchy–Schwartz inequality, $m_Q(f) \leq m_Q(f^2)^{1/2}$. Thus,

$$m_{\hat{Q}}(b) \leq cl_Q^{1/p} m_Q(f^2)^{1/2}.$$

This is the estimate we had in the previous section. Using the fact that f is in L^p for some $p > 2$ the previous estimate is enough to ensure $\{m_{\hat{Q}}(b)\} \in l^{p,\infty}$. We could now use this fact in (16.15).

Rather than try to fill in details to turn this discussion into a proof we summarize the results of Nowak in [No1], [No2].

Suppose $b(x,t) \geq 0$. Fix $\eta > 0$. For $\zeta = (x,t)$ let $m_\zeta(b)$ be the integral of b over the hyperbolic ball in $\mathbb{R}_+^2$ centered at ζ. If ψ is reasonable (a bit of smoothness is needed to use a version of the Cotlar–Stein Lemma) then:

THEOREM 16.14
For $1 < p < \infty$ and $1 \leq q \leq \infty$, $T_b \in S_p$ if and only if $m_\zeta(b) \in L^p(\mathbb{R}_+^2, t^{-2}dxdt)$.

COROLLARY 16.15
If $b(x,t) = f(x)g(1/t)$ then

$$\|T_b\|_{p,\infty} \leq c\|f(x)g(1/t)\|_{p,\infty} \leq c\|f\|_p\|g\|_{p,\infty}.$$

The fact that the conclusion of the theorem involves averages of g removes the need we had in the previous section to assume that g is smooth.

It is not clear how much this actually tells us about $f(X)g(D)$. Alternatively it isn't clear if this T_b could be used as a substitute for $f(X)g(D)$ in some contexts. There is some indication some progress in this direction may be possible. See, for instance, the discussion of the Schrödinger operator in [D].

16.4   Beyond size estimates

In the previous section we looked at variations on the idea of studying the size of operators. We now look at an example and we study the operator in more detail. These ideas are from [R4], [R5]. We will be quite informal and refer to those papers for details.

Again we look at the Schrödinger operator $S = -\Delta - V$ on $L^2(\mathbb{R}^3)$.

In some circumstances S has generalized eigenvectors. That is, for positive k there may be a function ψ which is not in L^2 but such that $S\psi = k^2\psi$. One way to study such ψ is as a perturbation of the exponential function $\exp(ikx)$ which has the required properties if $V = 0$. This leads to the modified Lipman–Schwinger equation for ψ which we now describe. (For more background see Ch. 5 of [Si].) Let $R = R_k$ be the operator with integral kernel

$$K(x,y) = v^{1/2}(x)v^{1/2}(y)\exp(ik|x-y|)|x-y|^{-1}.$$

It is straightforward to check that, formally, $\psi = (I+R)^{-1}(v^{1/2}\exp(ikx))$. In favorable circumstances $\|R\| < 1$ and the series development $(I+R)^{-1} = \sum(-R)^n$, called the "Born series" in this context, converges and is used to take the analysis further. Another common situation is that R may have large norm but is compact. In such a case one can write $R = F + M$, a finite rank operator F plus an operator M with small norm, $\|M\| < 1$. One then writes $(I+R)^{-1} = (I+M)^{-1}[I+F(I+M)^{-1}]^{-1}$. The second factor is a finite rank perturbation of the identity and is studied algebraically. The simplest way to obtain the required splitting is to use the eigenvectors of R. However, if the eigenvectors are not known any splitting $R = F + M$ can be used. The finite number of rank one operators making up F are sometimes called "quasiparticles" in the physics literature. See [W] and [N] for further discussion.

Let's write $W^2 = V$. We again suppose (16.8) holds. Following the pattern that led to (16.6) gives

$$K(x,y) \sim \sum_Q \eta_Q^{-1}e^{ik|x-y|}w(x)w(y)\chi_Q(x)\chi_Q(y)$$

where "$\sim$" means that the kernel on the right is a model for the one on the left. Let $\widetilde{M}$ be the operator with the kernel on the right. Pick a small ϵ and let $W = \{Q_1,\ldots,Q_n\}$ be a listing of the Q in $\mathcal{Q}$ for which $\eta_Q^{-1}\int_Q v > \epsilon$. (If the eigenvalues of S tend to zero then the estimates converse to Theorem 16.10 ensure that W is finite.) This induces a splitting of $\widetilde{M}$:

$$\widetilde{M} = M_1 + M_2,$$

where M_1 has kernel obtained by summing over Q in W. We now claim that M_2 is small. This doesn't quite follow from the methods described in Section 16.2.5 because of the factors $\exp(ik|x-y|)$. However, if we denote by M_2^+ the operator with kernel

$$\sum_{Q \notin R} \eta_Q^{-1} w(x) w(y) \chi_Q(x) \chi_Q(y),$$

then, for any f, g in L^2,

$$|\langle M_2 f, g \rangle| \le |\langle M_2^+ |f|, |g| \rangle|.$$

The methods of Section 16.2.5 apply to M_2^+ and this gives the desired estimates for M_2.

We now need to study M_1 a bit more. For each Q_i in W we pick a smooth bump φ_i supported on $10Q_i$ which is identically 1 on Q_i. We then expand $\varphi_i e^{ik|x-y|}$ as a multiple Fourier series on $10Q_i$. This lets us split the operator $M_{1,i}$ corresponding to that term of the kernel as an infinite sum of rank one operators with rapidly decreasing coefficients. We split that sum as F_i, a finite rank operator plus a small remainder. We do this for each i. Collecting the finite sum of rank one operators for each i and lumping the small remainder with M_2, we finally get

$$\widetilde{M} = F + M$$

where F is a finite sum of rank one operators and M has small norm.

The reason we didn't just invoke the ideas of [RSe2] is that this situation is a bit different from the one there. The reason is that $e^{ik|x-y|}$ is highly oscillatory for large k or for fixed k and large $|x - y|$. When we expand $\varphi_i e^{ik|x-y|}$ in a Fourier series we get estimates which involve derivatives of $\varphi e^{ik|x-y|}$ on $10Q$. This involves factors of $k|x - y|$. The only reason this factor doesn't undermine the analysis is that we only have a finite set of Qs in W; hence, we can bound $|x - y|$.

A similar type of analysis can be used to study eigenvectors. Recall that if $(-\Delta - V)f = \lambda f$, then $g = v^{1/2} f$ satisfies

$$V^{1/2} (\lambda - \Delta)^{-1} V^{1/2} g = g$$

which we summarize as $Tg = g$.

At the model level T is described by a kernel

$$\sum \exp(-\sqrt{\lambda} \eta_Q) \eta_Q^{-1} w(x) w(y) \chi_Q(x) \chi_Q(y).$$

As before we split

$$R = F + M,$$

where F is obtained by selecting finitely many terms of this sum in such a way that the remainder is small. We have

$$g = Fg + Mg.$$

Hence,

$$g = (I - M)^{-1} Fg = \sum M^n Fg = Fg + MFg + \cdots. \qquad (16.17)$$

The fact that $\|M\|$ is small ensures convergence. The point now is that $(I - M)^{-1}$ is applied to a function Fg which is in a finite dimensional space spanned by functions of the form $w\chi_Q$. Because of the explicit form we can make some progress. Look at the special case in which $Fg = w\chi_{Q_0}$ for a fixed Q_0. When we apply M to $w\chi_{Q_0}$ there are three types of terms produced. There are the terms involving $Q \subseteq Q_0$; we collect those as $\sum_1$. There are terms with $Q \supset Q_0$; we call those $\sum_2$. The remaining case gives no contribution. Thus,

$$M(w\chi_{Q_0}) = \sum\nolimits_1 \exp(-\sqrt{\lambda}\eta_Q)\eta_Q^{-1}\left(\int_Q w^2\right)w\chi_Q$$
$$+ \sum\nolimits_2 \exp(-\sqrt{\lambda}\eta_Q)\eta_Q^{-1}\left(\int_{Q_0} w^2\right)w\chi_Q.$$

When we multiply by w^{-1} to get back to f, the actual eigenvector, we find

$$w^{-1}(Mw\chi_{Q_0}) = \sum\nolimits_1 \exp(-\sqrt{\lambda}\eta_Q)\eta_Q^{-1}\left(\int_Q w^2\right)\chi_Q$$
$$+ \sum\nolimits_2 \exp(-\sqrt{\lambda}\eta_Q)\eta_Q^{-1}\left(\int_{Q_0} w^2\right)\chi_Q.$$

Thus, we have two terms: $\sum_1$, which is relatively complicated and is supported on Q_0, a cube in phase space associated with a large average value of the potential; and $\sum_2$, which is constant on Q_0, is relatively tame, and decays exponentially as we go to ∞. To see this a bit more clearly look at the case $\eta_{Q_0} = \lambda = 1$. For the terms in $\sum_1$ we drop the exponential factor. For each x in Q_0 we sum $\eta_Q^{-1}\int_Q w^2$ over Q in Q_0 containing x. That sum is majorized by the potential operator $(-\Delta)^{-1} = I_2$ applied to $w^2\chi_{Q_0}$. To estimate $\sum_2$ we note that for fixed x far from Q_0 most terms are small. We end up with

$$w^{-1}(Mw\chi_{Q_0})(x) \leq cI_2(w^2\chi_Q)(x) + c\left(\int_{Q_0} w^2\right)\exp(-c\,\mathrm{dist}(x, Q_0)).$$

It may be possible to carry this type of analysis through to the end. The conclusion we would hope to find is that the nth eigenfunction is estimated by $O(n)$ islands on which it is controlled by $I_2(v)$ plus a global estimate that decays exponentially away from the islands. Although the analysis would be complicated, it has been possible to carry such analysis through for slightly simpler operators of this sort [RSe2].

16.5 Afterthoughts and speculation

16.5.1 Other operators

Wavelet and related decomposition are now known to be very useful for analyzing Calderón–Zygmund operators as well as other singular integral operators of roughly the same sort. There are other types of operators where these ideas haven't been tried but might be useful. In particular, these ideas might be useful in studying composition operators. A typical question is the following. Suppose X and Y are function spaces on $\mathbb{R}$ for which wavelets give unconditional bases. Suppose φ is a homeomorphism of $\mathbb{R}$. Let C_φ be the composition operator defined by $C_\varphi(f)(x) = f(\varphi(x))$.

Question

For which φ does C_φ map X to Y? (Or, more generally, is $C_\varphi^\alpha(f) = f(\varphi(x))\varphi'(x)^\alpha$ bounded from X to Y?) $\Box$

A relation to wavelets is suggested by the following incorrect statement: Suppose $\{\psi_Q\}$ are the wavelet basis for X. Then $C_\varphi(\psi_Q) \sim \psi_{\varphi^{-1}(Q)}$. Here $\varphi^{-1}(Q)$ is being used to represent a dyadic cube which is approximated by the set $\varphi^{-1}(Q)$.

Although this isn't really so, what it suggests is that the matrix $\langle C_\varphi(\psi_Q), \psi_R \rangle$ will, approximately, have only a small number of nonsmall elements in each row and each column. (A related fact is that composition operators acting on Hilbert spaces with reproducing kernels have the property that their adjoints permute the kernel functions. The analogy between wavelets and reproducing kernels is discussed systematically in [R1].) Here is a related question that is a bit more explicit and perhaps easier. The homeomorphism φ can be used to induce a mapping $\tilde{\varphi}$ on $\mathcal{Q}$, the dyadic cubes. One can then build a map $\tilde{C}_\varphi$ or $\tilde{C}_\varphi^{(\alpha)}$ which maps sequence spaces defined on $\mathcal{Q}$ to other such sequence spaces. The question is when $\tilde{C}_\varphi^{(\alpha)}$ maps the space $\tilde{X}$ of wavelet coefficients of functions in X to $\tilde{Y}$, the analogous space for Y.

16.5.2   Eigenvectors

What can be said about eigenvectors of singular integral operators? The Hilbert transform acting on $L^2(\mathbb{R})$ is a singular integral operator and, according to some definitions, so is the identity. Hence, the question is probably too general.

However, consider the following operator acting on $H^2(\mathbb{R})$, the Hardy space. A function f in H^2 can be regarded an analytic function $f(z)$ on $\mathbb{R}^2_+$. Let $b(z)$ be holomorphic function on $\mathbb{R}^2_+$ with boundary values in $BMOA(\mathbb{R})$. Define $\prod(b, \cdot)$ by

$$\prod(b,f)(z) = \int_{i\infty}^{z} b'(\zeta)f(\zeta)d\zeta.$$

This is an example of a paraproduct—a type of singular integral operator (although this formulation disguises that fact). It is known that $\prod(b, \cdot)$ is bounded if b is in $BMOA$ and is compact and in the Schatten ideal S_p if b is in the diagonal Besov space B_p. A nice thing about this example is that we can solve the eigenvalue equation. If

$$\prod(b,f) = \lambda f,$$

then

$$f = \exp(b/\lambda).$$

This isn't actually in H^2 because $b(x) \to 0$ at ∞. However, this example exhibits an operator in S_p with the property that the singularities of the generalized eigenvectors are no worse than

$$O\left(\exp\left(\left(\log \frac{1}{|x|}\right)^{1/p'}\right)\right) \tag{16.18}$$

at the origin (here $1/p + 1/p' = 1$).

I speculate that this example is part of a pattern. The fact that a singular integral operator T is compact and in S_p is a statement about the rate of approximation of T by finite rank operators. If that finite rank approximation can be done in the context of the decomposition theorems then an analysis of the type described in the previous section for the Schrödinger operator could perhaps be used to estimate the size of the eigenvectors. In [R4] such an analysis is carried out for the particularly simple operator given on $L^2(\mathbb{R})$ by

$$Tf = \sum_{Q \in \mathcal{Q}} \lambda_Q \left(\frac{1}{|Q|} \int_Q f\right) \chi_Q$$

where $\{\lambda_Q\} \in l^p$. The result is that the eigenvectors can be estimated by expressions such as (16.18).

Bibliography

[BCoRo] G. Beylkin, R. Coifman, and V. Rokhlin. 1992. "Wavelets in numerical analysis: compression of operators," in *Wavelets and Their Application*, M. Ruskai, G. Beylkin, R. Coifman, I. Daubechies, S. Mallat, Y. Meyer, and L. Raphael, eds., Boston: Jones and Bartlett, pp. 181–210.

[BiSo] M. Birman and M. Solomyak. 1991. "Estimates for the number of negative eigenvalues of the Schrödinger operator and its generalizations," *Adv. Sov. Math.* **7**: 1–55.

[BiKaSo] M. Birman, G. Karadzhov, and M. Solomyak. 1991. "Boundedness Conditions and Spectrum Estimates for the operators $b(X)a(D)$ and their Analogs," *Adv. Sov. Math.* **7**: 85–106.

[CMaSi] R. Carmona, W.C. Masters, and B. Simon. 1990. "Relativistic Schrödinger operators: Asymptotic behavior of the eigenfunctions," *J. Funct. Anal.* **91**: 117–142.

[ChWiWo]   S.Y.A. Chang, J.M. Wilson, and T.H. Wolff. 1985. "Some weighted norm inequalities concerning the Schrödinger operator," *Commun. Math. Helv.* **60**: 217–246.

[CoR]   R. Coifman and R. Rochberg. 1980. "Another characterization of BMO," *Proc. Amer. Math. Soc.* **79**: 249–254.

[D]   I. Daubechies. 1992. *Ten Lectures on Wavelets*, CBMS-NSF Reg. Conf. Ser. Appl. Math. **61**, Philadelphia: Soc. Ind. Appl. Math,

[Da]   G. David. 1991. *Wavelets and Singular Integrals on Curves and Surfaces*, *Lect. Notes Math.* **1465**, Berlin: Springer-Verlag.

[F]   C. Fefferman. 1983. "The uncertainty principle," *Bull. Amer. Math Soc.* **9**: 129–206.

[FrJa]   M. Frazier and B. Jawerth. 1990. "A discrete transform and decompositions of distribution spaces," *J. Funct. Anal.* **93**: 34–170.

[FrJaWe]   M. Frazier, B. Jawerth, and G. Weiss. 1991. *Littlewood-Paley Theory and the Study of Function Spaces*, CBMS Conf. Lect. Notes **79**, Providence: Amer. Math. Soc.

[GFra]   J. Garcia-Cuerva and J.L. Rubio de Francia. 1985. *Weighted Norm Inequalities and Related Topics*, North-Holland Math. Studies **116**, Amsterdam: North-Holland.

[H]   Y.S. Han. 1992. "Calderón type reproducing formula and the Tb theorem," preprint.

[He]   L.I. Hedberg. 1972. "On certain convolution inequalities," *Proc. Amer. Math. Soc.* **36**: 505–510.

[JL]   S. Jaffard and Ph. Laurençot. 1992. "Orthonormal wavelets, analysis of operators and applications to numerical analysis," in *Wavelets: A Tutorial in Theory and Applications*, **2**, C.K. Chui, ed., Boston: Academic Press, pp. 543–601.

[KeS]   R. Kerman and E. Sawyer. 1986. "The trace inequality and eigenvalue estimates for Schrödinger operators," *Ann. Inst. Fourier, Grenoble* **36**: 207–228.

[KlSk]   J. Klauder and B.-S. Skagerstam. 1985. *Coherent States: Applications in Physics and Mathematical Physics*, Singapore: World Scientific.

[Me1]   Y. Meyer. 1989. "Wavelets and operators," in *Analysis at Urbana*, E. Berkson et al., eds., London Math. Soc. Lect. Note **137**, Cambridge: Cambridge Univ. Press, pp. 256–364.

[Me2]   Y. Meyer. 1990. *Ondelettes et Operateurs*, 3 vols., Paris: Hermann.

[N]   R. Newton. 1966. *Scattering Theory of Waves and Particles*, New York: McGraw-Hill.

[No1]   K. Nowak. 1991. "Toeplitz operators and commutators based on the Calderón reproducing formula," Ph.D. Thesis, Washington University, St. Louis.

[No2]   K. Nowak. 1992. "On Calderón–Toeplitz operators," *Monat. fur Math.*, to appear.

[P]   L. Peng. 1993. "Wavelets and paracommutators," to appear.

[Pe]   C. Perez. 1990. "Two weighted norm inequalities for Riesz potentials and uniform L^p-weighted Sobolev inequalities," *Ind. Univ. Math. J.* **39**: 31–44.

[R1]   R. Rochberg. 1990. "Toeplitz and Hankel operators, wavelets, NWO sequences, and almost diagonalization of operators," *Proc. Symp. Pure ath.* **51**(I): 425–444.

[R2]   R. Rochberg. 1992. "A correspondence principle for Toeplitz and Calderón–Toeplitz operators," in *Interpolation Spaces and Related Topics*, Isr. Math Conf. Proc., **5**, M. Cwikel, M. Milman, and R. Rochberg, eds., Providence: Amer. Math. Soc., pp. 229–243.

[R3]   R. Rochberg. 1992. "Eigenvalue estimates for Calderón–Toeplitz operators, function spaces," *Lect. Notes Pure Appl. Math.* **136**, K. Jarosz, ed., New York: Marcel Dekker, pp. 345–357.

[R4]   R. Rochberg. 1992. "Size estimates for eigenvectors of singular integral operators," preprint.

[R5]   R. Rochberg. 1993. "NWO sequences, weighted potential operators, and Schrödinger eigenvalues," *Duke Math. J.*, to appear.

[RSe1]   R. Rochberg and S. Semmes. 1986. "A decomposition theorem for functions in BMO and applications," *J. Funct. Anal.* **67**: 228–263.

[RSe2]   R. Rochberg and S. Semmes. 1989. "Nearly weakly orthonormal sequences, singular value estimates, and Calderón–Zygmund operators," *J. Funct. Anal.* **86**: 237–306.

[S1]   M. Schechter. 1986. *Spectra of Partial Differential Operators*, 2nd ed., Amsterdam: North-Holland.

[S2]   M. Schechter. 1989. "The spectrum of the Schrödinger operator," *Trans. Amer. Math. Soc.* **312**: 115–128.

[Si]   B. Simon. 1979. *Trace Ideals and Their Applications*, Cambridge: Cambridge Univ. Press.

[St]   E. Stein. 1970. *Singular Integrals and Differentiability Properties of Functions*, Princeton: Princeton Univ. Press.

[V]   I.E. Verbitsky. 1992. "Weighted norm inequalities for maximal operators and Pisier's theorem on factorization through $L^{p\infty}$," *Integr. Equat. Oper. Theory* **15**: 124–152.

[W]   S. Weinberg. 1963. "Quasiparticles and Born series," *Phys. Rev.* **131**: 440–460.

Index